Designing Unmanned Aircraft Systems:
A Comprehensive Approach

Designing Unmanned Aircraft Systems: A Comprehensive Approach

Jay Gundlach
Aurora Flight Sciences
Manassas, Virginia

AIAA EDUCATION SERIES

Joseph A. Schetz, Editor-in-Chief
Virginia Polytechnic Institute and State University
Blacksburg, Virginia

Published by the
American Institute of Aeronautics and Astronautics, Inc.
1801 Alexander Bell Drive, Reston, Virginia 20191-4344

American Institute of Aeronautics and Astronautics, Inc., Reston, Virginia

1 2 3 4 5

Library of Congress Cataloging-in-Publication Data

Gundlach, Jay, 1975-
Designing unmanned aircraft systems : a comprehensive approach/Jay
Gundlach. – 1st ed.
p. cm. – (AIAA education series)
Includes bibliographical references and index.
ISBN 978-1-60086-843-6
1. Drone aircraft. 2. Micro air vehicles–Control systems. I. American Institute of Aeronautics
and Astronautics. II. Title.
UG1242.D7G86 2012
623.74'69–dc23
2011033783

ISBN 978-1-60086-843-6

AIAA EDUCATION SERIES

Editor-in-Chief
Joseph A. Schetz
Virginia Polytechnic Institute and State University

We are exceedingly pleased to present *Designing Unmanned Aircraft Systems: A Comprehensive Approach* by Dr. Jay Gundlach. This textbook is a broad and deep treatment of a very important subject area and contains 20 chapters with homework problems. In addition to the usual coverage of aerodynamics, structures, propulsion, and vehicle control, this book features a broader treatment—including pulse jets, electric propulsion, fuel cells, solar power, avionics, communications, sensing, and so forth that are essential to UAV design.

Dr. Gundlach is particularly well qualified to write on this subject given his extensive experience in the field. It is with great enthusiasm that we present this new book to our readers.

Editorial Board

To my devoted wife,
whose encouragement and support made
this book possible.

CONTENTS

PREFACE

The field of unmanned aircraft systems (UASs) is an exciting segment of the aerospace industry. Recent technology advances and customer acceptance have led to the widespread employment of these amazing systems. Many engineering professionals entering the aerospace industry today will work on UASs. In fact, most new military aircraft development programs are UASs.

Designing Unmanned Aircraft Systems: A Comprehensive Approach provides a comprehensive coverage of all elements of unmanned aircraft systems, architectural options, and design drivers across diverse system classes. The end-to-end unmanned aircraft system is described, rather than just the aircraft. The chapters highlight the system element interactions that impact top-level system performance. For example, the interactions between sensor resolution, acoustic detection, and unmanned-aircraft (UA) sizing are detailed. The reader will gain a deep appreciation for the multidisciplinary nature of unmanned aircraft system design. She/he will be able to conduct cross-discipline trade studies to yield robust, well-balanced systems that provide superior operational utility.

This text provides detailed analysis of system elements unique to unmanned aircraft. By combining all of these disciplines in one work, this should serve as a single resource for unmanned systems analysis. Many important system element analysis methods are either not published, not covered in sufficient depth elsewhere, or exist in single-discipline books. Topics include the following:

- Approach for developing competitive, balanced unmanned aircraft systems through a multidisciplinary systems philosophy
- Data-driven analysis of system components, technology trends, unmanned aircraft configurations, and unmanned aircraft capabilities
- Extensive survey and analysis of unmanned aircraft launch and recovery techniques, along with selection and design processes
- Derivations of electric aircraft performance equations
- Show an approach for creating custom weight estimating relationships for unusual design problems
- Method for defining UAS products through market analysis, gap identification, shaping customer requirements, and development of system design solutions

- Method for integrating diverse payload types into unmanned aircraft at the conceptual design phase and for legacy system upgrades
- Design of the aircraft geometry and payload placement ensures mission-driven field-of-regard requirements are satisfied
- Airborne remote-sensing physics, sensor assembly design, and operational techniques
- Attributes of the electromagnetic spectrum are explored, such as phenomenology that affects payload products at various bands
- Exploration of system-level mission systems' effectiveness across multiple payload classes
- Exploration of the full motion video imagery chain for image quality, data management, metadata insertion points, and round-trip latency
- Overview of ground control station types and functionality provided

Designing Unmanned Aircraft Systems: A Comprehensive Approach can support a capstone course on aircraft design with an emphasis on UAS projects. Rather than just focus on UA design, this text brings in numerous other engineering disciplines that are required to produce a successful system. A UAS educational project can bring in students from a variety of engineering departments.

This text is also suitable for professional continuing education for those interested in UASs. Professionals with the need to learn about UASs may come from diverse backgrounds. Aerospace engineers entering the UAS field may need to learn more about payloads and communications systems, for example. Electrical engineers responsible for payload design may require more knowledge of UA design and systems integration. The broad treatment of all aspects of UAS design should be a good single resource for professionals of different backgrounds.

The methods presented in this book are suitable for academic study and conceptual design, but are not intended to be sufficient for engineering design that produces flight hardware. Although the techniques may have use in a hardware development effort, this coverage does not provide preliminary and detail design methods. The objective is to provide insights into UAS technology and understanding of the associated design drivers across multiple disciplines.

This work attempts to comprehensively cover all major design disciplines that are needed for a successful unmanned aircraft *system*. The emphasis is on a systems approach, where interdependencies are detailed. Beyond traditional aircraft design topics, this book describes system architectures, launch and recovery methods, physics of remote sensing, payloads, missions, communication systems, ground control, reliability, supportability, cost analysis, system synthesis, and product development, among other topics. This work combines many of the relevant topics as they relate to UAS into a single source, if only at the introductory level.

To cover such a broad scope in a single text, it is necessary to sacrifice some depth. Topics such as communications engineering and remote sensing are treated as introductions. This should provide an appreciation of the physics and system impacts. This should serve as a simple starting point for readers who wish to learn about these topics.

Although this text covers numerous design disciplines beyond the UA, it is also intended to be useful for aircraft design. Several notable dedicated aircraft design texts provide deeper treatment of traditional aircraft design. This book is meant to supplement other aircraft design texts by providing UAS-specific design information.

UASs share much in common with manned aircraft. This book does not seek to comprehensively cover all aspects of aircraft design, especially where it overlaps with manned aircraft. Rather, system design choices and attributes that mostly apply to unmanned systems are emphasized. The approaches and features are identified to illuminate design rationale.

Designers may call the UA design or system architecture an art. This is an expression of the complexity involved in bringing so many competing disciplines together effectively into a winning solution. This book attempts to demystify the art of unmanned aircraft system design so that an appreciation of the design drivers from the various system elements can be readily understood.

Some methods covered deviate from traditional aircraft design texts in order to take advantage of advances in personal computers and software. These algorithms can be readily implemented in spreadsheets, programming languages, or other computation environments. Manned aircraft conceptual design makes liberal use of empirical equations based on over a century of design knowledge. Quite frequently, these established methods are not applicable to unmanned aircraft. It is therefore important to employ more analytical methods that build upon first principles even at the conceptual design phase.

The most general forms of equations are used when practical. This allows users flexibility in use of units, such as English or SI. This flexibility comes at the cost of requiring the reader to perform unit conversions.

The book contains numerous examples to illustrate the key analysis methods presented. Each chapter concludes with problems. Instructors using this book are aided by a solutions manual.

I have collected UAS books and other materials for many years. The data presented here are based on this open-source information. Some of these data are incorporated into a database that is used to substantiate trends in UA characteristics, subsystems, and payloads. This book was reviewed and approved by the Office of Security Review to ensure that there is no sensitive or export-controlled information in this work.

As an aerospace engineer with a specialization in aircraft design, I have a particular affinity for UAs. This is tempered by an appreciation for all of the aspects of UAS that define a successful system. That is why the unmanned aircraft only consumes less than half of the material.

Jay Gundlach
Director of Conceptual Design
Aurora Flight Sciences
December 2011

ACKNOWLEDGMENTS

This book was made possible through the support and contributions of many people. I am especially grateful to my beautiful wife who encouraged me to start this project and made enormous sacrifices during the year that I wrote this text. This book is possible thanks to the years of patient mentorship from Rick Foch, the world's most prolific UAS developer and a leading aircraft designer. My employer, Aurora Flight Sciences, supported this long endeavor with enthusiasm. Insitu was generous with images and reviews. Many of my colleagues provided content recommendations, subject matter expert inputs, and thorough review. I would especially like to thank Greg Keogh, Rip Ripley, Steve Gardner, Rick Foch, Tom Koonce, and my mother for their time and energy to make this book accurate and readable.

NOMENCLATURE

Item	Definition
A	area; availability; ground area covered in a mission; radar antenna area, m^2; conversion between radians and minutes of arc
A_a	achieved availability
A_{bound}	bounded area for a closed section
A_d	IR detector sensitive area, m^2
A_{eff}	effective antenna area, length^2
A_i	inherent availability
A_O	operational availability; UA availability
A_p	propeller disk area, length^2
A_{Rate}	area coverage rate
A_r	effective collection area of optical receiver
A_{Surf}	surface area
AR	aspect ratio
AR_{Wet}	wetted aspect ratio
AR'	aspect ratio along spanwise path
a	UA acceleration; maximum fuselage cross-section width; speed of sound; detector characteristic dimension
a_{wa}	radar mainlobe width metric
a_{wr}	radar mainlobe width metric
a_x	acceleration along the x direction (acceleration)
B	acuity gain due to binoculars; boom area; effective noise bandwidth of receiving process, Hz
$B_{Doppler}$	Doppler bandwidth (time^{-1})
B_N	effective noise bandwidth of the receiving process
B_T	radar signal bandwidth (time^{-1})
$BSFC_{SL}$	brake specific fuel consumption at sea level
b	web length; wing span; maximum fuselage cross-section height
b_w	wing span
b'	span without dihedral
C	cost of contractor services and products, \$
C_D	drag coefficient
$C_{D,Chute}$	parachute drag coefficient

$C_{D,Fuse,Tail}$	drag coefficient of the fuselage and tail
$C_{D,gear}$	landing-gear drag
C_{Di}	induced drag coefficient
$C_{Di,H}$	horizontal tail induced drag coefficient
C_{Dprof}	three-dimensional profile drag coefficient
$C_{D,0}$	zero-lift drag coefficient
C_{D0}	zero-lift drag coefficient
$C_{D0,Wing}$	wing zero-lift drag coefficient
$C_{d,frict}$	two-dimensional friction drag coefficient
C_{di}	local induced drag coefficient
$C_{d,press}$	two-dimensional pressure drag coefficient
$C_{d,prof}$	two-dimensional profile drag coefficient
$C_{D,sphere}$	sphere-drag coefficient
$C_{D,wave}$	wave-drag coefficient
C_f	skin-friction coefficient
$C_{f,lam}$	laminar skin-friction coefficient
$C_{f,seg}$	segment flat-plate skin-friction drag coefficient
$C_{f,turb}$	turbulent skin-friction coefficient
C_L	three-dimensional lift coefficient
C_l	two-dimensional lift coefficient
$C_{L,H}$	horizontal stabilizer lift coefficient
$C_{L,Land}$	landing lift coeffcient
$C_{L,Loiter}$	loiter lift coefficient
$C_{L,max}$	maximum lift coefficient
$C_{L,w}$	wing lift coefficient
$C_{L\delta e}$	lift-curve derivative with respect to elevator deflection (angle^{-1})
C_M	pitching-moment coefficient
$C_{M,CL}$	pitching moment due to lift
$C_{M,flap}$	pitching moment due to flap deflection
$C_{M,fuse,nac}$	pitching moment due to fuselages and nacelles
$C_{M,0L}$	zero-lift pitching moment
C_m	pitching-moment coefficient
$C_{m\alpha}$	pitching-moment derivative with respect to angle of attack (angle^{-1})
C_P	coefficient of power
C_T	coefficient of thrust
C_0	target inherent contrast
C_θ	required visual contrast
Capacity	battery capacity, current-time
CT	total cost of contractor and government services and products, \$

c	chord; chord length; wing chord; speed of light, 3.00×10^8 m/s
c_{Radome}	airfoil-shaped radome chord; lift-curve slope (angle^{-1})
c_r	root chord
c_t	tip chord
D	dish antenna diameter; UA drag; propeller diameter; ground swath; IR aperture diameter, m; lens diameter
D_{Ball}	EO/IR ball diameter
D_{Chute}	parachute diameter
D_{Cool}	drag due to cooling drag
$D_{Diffraction}$	diameter or blur spot
D_i	induced drag
D_{Pow}	equivalent drag for electrical power generation
D_{Ram}	ram drag, force
D^*	S/N per unit S for detector, $(\text{Hz}^*\text{m}^2)^{1/2}/W\alpha$
$Data_{Collected}$	data that are output from the camera
$Data_{Discarded}$	data that are deleted
$Data_{Stored}$	data that are stored on a data storage device
$Data_{Trans}$	data that are transmitted to the ground
Deg	degree portion of longitude or latitude
Deg_{Dec}	decimal degrees of longitude or latitude
Dir	sign of longitude or latitude
$Dist$	total flight distance, nm
d	antenna cylinder diameter; mac y distance from the wing root; length of the focal plane array
d_c	target characteristic dimension
d_{GS}	great circle distance
dA	differential area
dP/dx	pressure gradient
ds	differential perimeter length
E	endurance; modulus of elasticity
E_{Max}	maximum endurance, hrs
E_{max}	maximum endurance
E_{Net}	energy absorbed by the net
$E_{Recovery}$	energy change of the UA at recovery
E_{seg}	endurance of mission segment
E_{Spec}	battery specific energy
E_{Tot}	total endurance from launch to recovery, hrs
E_{tot}	total endurance
Eb/No	energy per bit to noise power spectral density ratio, dB

$EIRP$	effective isotropic radiated power, dBm
$Energy_{Batt}$	energy stored in the battery
e	span efficiency factor
F	force; noise factor
F_{Act}	actuator linear force
F_{Cable}	total cable force
F_{Comp}	composites factor
$F_{Compression}$	image compression factor
$F_{Compress,Stored}$	compression factor for stored data
$F_{Compress,Trans}$	compression factor for transmitted data; shear modulus
F_M	vertical force on main landing gear
F_N	vertical force on nose landing gear
$\mathbf{F}_{Net}$	net force
F_{Piston}	force acting on the pneumatic piston
$F_{Shuttle}$	force acting upon the shuttle
Fee	contractor fee rate
FF_{fuse}	fuselage form factor
FF_{seg}	segment form factor
FH_{Sched}	scheduled flight hours
f	electromagnetic frequency, Hz; focal length; frequency
f_{Bleed}	bleed air factor
$f_{Install}$	installation factor
f_P	pulse repetition frequency (time^{-1})
f_T	thrust multiplication factor
f_{Usable}	battery usable energy factor
$f(\ )$	function
G	gain; gear ratio; multiples of the acceleration of gravity; acoustic antenna gain IR detection system processing gain; image postprocessing noise gain
G_A	transmitter antenna gain
G_a	signal-to-noise gain due to azimuth processing
G_R	receiver antenna gain, no units or dBi
G_r	signal-to-noise gain due to range processing
G_T	transmitter antenna gain, no units or dBi
GFX	government furnished (equipment, information, services, personnel), \$
GR	ground distance from the UA to the target
GSD_H	horizontal ground sample distance
GSD_V	vertical ground sample distance
g	acceleration of gravity

H	geometric mean system postprocessing edge overshoot factor; fuel heating value; ground antenna height above ground level, ft
H_k	measurement matrix
H_{Pix}	number of horizontal pixels in the focal plane array
H_{Pixels}	number of horizontal pixels in a row
H_{Tgt}	height of target as viewed by the imager
Hr	person-hours
h	UA height above ground level; altitude above the target; mean sea-level altitude; height of the web
h_{Loss}	height loss after net recovery
$h_{\max}$	maximum altitude
h_{Sat}	satellite orbit height above Earth's surface
h_T	altitude of the tropopause
h/D_{Ball}	height-to-diameter ratio for EO/IR ball
l	characteristic length; light intensity
$I_{\text{Max,Cell}}$	battery cell maximum current, A
$I_{\text{Max,Pack}}$	battery pack maximum current, A
I_{Mot}	motor current, A
I_T	hearing sound intensity threshold, W/cm^2
I_{Tot}	rocket total impulse, time
I_x, I_y, I_z, I_{xz}	moments of inertia about component center of gravity, mass-length^2
I_{xx}	moment of inertia about x axis (length^4)
I_{xy}	moment of inertia in the x-y plane (length^4)
$I_{x0}, I_{y0}, I_{z0}, I_{xz0}$	moments of inertia about reference center of gravity, mass-length^2
I_{yy}	moment of inertia about y axis (length^4)
I_0	light intensity at center of the Airy disk
I_{0L}	zero-load motor current, A
$IFOV$	instantaneous field of view of one detector, rad
i	incidence angle
J	polar moment of inertia (length^4); advance ratio
J_0	radiant intensity of target above the background, W/sr
K_k	Kalman gain
K_V	motor voltage constant, rpm/V
k	spring constant; Boltzmann's constant, 1.38054 E-23 J/K
L	fuselage length; tail moment arm; length; lift force; temperature lapse rate; characteristic length; lift; synthetic aperture length; target dimension, m

L_{Accel}	length of the acceleration phase of dolly
L_{Arm}	servo arm length
L_{atm}	two-way atmospheric loss factor for atmospheric propagation
L_{Booms}	boom length
L_{Decel}	length of dolly deceleration
L_{Extend}	linear extension of net cable
L_{Horn}	control horn length
L_{Launch}	launch stroke
L_P	absorptive propagation loss, no units or dB
$L_{P,Atm}$	atmospheric absorption losses, dB
L_{Piston}	piston stroke
$L_{P,Precip}$	precipitation absorption losses, dB
$L_{P,Tot}$	total absorptive propagation loss, dB
L_R	receiver losses, no units or dB
$L_{R,Line}$	receiver line losses, dB
$L_{R,Point}$	receiver pointing losses, dB
L_{Rail}	launcher total rail length
$L_{Recovery}$	stroke length of recovery deceleration
$L_{R,Polar}$	receiver polarization losses, dB
$L_{R,Radome}$	receiver radome losses, dB
$L_{R,Spread}$	receiver spreading implemention losses, dB
L_r	range processing losses
L_{radar}	radar transmission loss factor due to sources other than the atmosphere
L_{Stroke}	stroke length
L_{Struct}	fuselage structural length
L_T	transmitter losses, no units or dB
$L_{T,Line}$	transmitter line losses, dB
$L_{T,Point}$	transmitter pointing losses, dB
$L_{T,Radome}$	transmitter radome losses, dB
L_{Tot}	overall UA length
L/D	lift-to-drag ratio
L/D_{Max}	maximum lift-to-drag ratio
LR	loss rate—crashes per 100,000 flight hours
l	actual antenna length
M	mass; bending moments (force $\times$ length); Mach number; pitching moment; thermal emittance, $W/cm^2\text{-}\mu m$
$\bar{M}$	mean active maintenance time
M_{Batt}	battery mass
M_{crit}	critical Mach number

$\bar{M}_{ct}$	corrective maintenance time
M_{DD}	drag-divergence Mach number
M_{max}	maximum Mach number
M_{tip}	propeller tip Mach number
MF_{Empty}	empty weight mass fraction
MF_{Energy}	energy source mass fraction
MF_{Fuel}	fuel mass fraction
$MF_{Fuel,Tot}$	mass fraction of the total fuel weight
MF_{H2}	hydrogen mass fraction
MF_{Prop}	propulsion mass fraction
$MF_{Storage}$	mass fraction of the fuel and fuel storage
MF_{Struct}	structures mass fraction
MF_{Subs}	subsystems mass fraction
Min	minute portion of longitude or latitude
m	UA mass; mac leading edge x distance from the root chord leading edge; exponent in superellipse equation
$\dot{m}_a$	air mass flow rate through propulsion system
$\dot{m}_{Bleed}$	bleed air mass flow rate
$\dot{m}_c$	core mass flow rate
$\dot{m}_{Fuel}$	fuel mass flow rate
$\dot{m}_f$	fan mass flow rate
$\dot{m}_{Ram}$	cooling mass flow
mac	mean aerodynamic chord
N	number of cycles across the target; number of items; system noise, W
$N_{Ab,Mx}$	number of aborts due to maintenance issues
N_{AV}	number of UA
N_{Blades}	number of propeller blades
N_{ClassA}	number of Class A mishaps
N_{Cells}	number of battery cells
N_{Cx}	number of cancellations
N_{Hops}	number of hops per cargo aircraft per deployment flight
N_{Props}	number of propellers
N_{Rep}	number of repair activities
N_r	noise power at radar receiver
N_{Segs}	number of mission segments
N_{Series}	number of battery cells in series
$N_{Sorties}$	number of sorties
$N_{Strings}$	number of battery strings
N_{Ult}	ultimate load factor (multiples of gravity)

N_z	vertical acceleration (multiples of gravity)
NF	noise figure, dB
Ni	noise, W
N_{50}	number of cycles across the target for 50% discrimination task success
N_z	ultimate load factor
n	exponent in superellipse equation; propeller rotational rate, cycles/s
P	axial force; tank pressure differential; atmospheric pressure; propulsion power; detector pitch; radar power, W; pixel pitch; cross-section perimeter
$P_{abs,prop}$	power absorbed by propeller
P_{Batt}	battery power
$P_{Carrier}$	effective carrier power, dBm
P_{cr}	critical buckling load
P_{Eng}	engine power
P_{ESC}	motor controller input power
$P_{Extract}$	shaft power extracted by other than propeller
P_e	pressure at propulsion system exit
P_{Gen}	generator power output
P_k	covariance matrix
P_l	laser pulse power
P_{Max}	maximum engine power
P_{Mot}	motor output power
$P_{Mot,Max}$	maximum motor power
P_{Noise}	effective noise power, dBm
P_{Other}	other power consumers
P_{Out}	actuator shaft power
$P_{PL,Max}$	maximum payload power
$P_{Propulsion}$	propulsion power
P_r	received signal power for a single pulse
P_S	static pressure, pressure
P_{Shaft}	shaft power
$P_{Shaft,Max}$	maximum shaft power
$P_{Shaft,Max,SL}$	maximum shaft power at sea level
$P_{Shaft,Prop}$	shaft power absorbed by propeller
$P_{Shaft,SL}$	shaft power at sea level
P_{SL}	sea-level atmospheric pressure
P_T	atmospheric pressure at the tropopause; transmitted power, W or dBm; total pressure
P_{Thrust}	thrust power
P_t	transmitted signal power for a single pulse

P_0	radiated acoustical power in narrowband, W
P_∞	ambient pressure
$Pavg$	average transmitting power
$Prof$	profit, $
$P()$	probability of achieving target discrimination task
P/W_{Avion}	avionics power-to-weight ratio
$P/W_{Controller}$	power-to-weight ratio of the motor controller
$P/W_{FuelCell}$	power-to-weight ratio of the fuel-cell stack
P/W_{Motor}	power-to-weight ratio of the motor
$P/W_{Powerplant}$	power-to-weight ratio of the powerplant
$P/W_{Propeller}$	power-to-weight ratio of the propeller
P/W_{Ref}	reference power-to-weight ratio
P/W_{TO}	power-to-weight ratio at takeoff
Q	first moment from neutral axis (length^3)
Q_{cool}	cooling heat load
Q_k	disturbance covariance
Q_{seg}	segment interference factor
q	shear flow; dynamic pressure
q_b	basic shear flow
q_{Dive}	maximum dynamic pressure in a dive
$q_{s,0}$	shear flow at the cut section
R	rate, $/hr or $/day; range; radius; gas constant (287.05 N-m/kg-K); communications line-of-sight range; slant range; mission radius; acoustic detection range, km; visual detection range, km; reliability
R_{Data}	data rate, bits/s
R_E	Earth's radius
R_{Earth}	Earth's radius
R_k	sequence covariance
$R_{Mission}$	mission reliability
R_{Mot}	motor resistance, Ω
R_{seg}	range of segment
R_{tot}	total range
R_0	radar range where $S/N = 1$, m
$Range$	flight range, n miles
$Rate_{Data}$	rate of imagery data production, bits/time
$RateFrame$	number of frames per unit time, time^{-1}
Re	Reynolds number
Re_{tr}	transition Reynolds number
ROC_{seg}	rate of climb during the mission segment
ROC_{Max}	maximum rate of climb
RPM_{Mot}	motor rotational rate, rpm

RPM_{Prop}	propeller rotational rate, rpm
S	planform; radar signal strength, W
S_A	takeoff air distance
S_{CL}	climb-out distance
S_{CS}	control surface planform area
S_{Emp}	empennage planform area
$S_{Exposed}$	exposed area
$S_{Fuse,Tail}$	reference area of the fuselage and tail
S_G	takeoff ground roll distance
S_{NGR}	takeoff distance with nose gear on ground
S_{Paint}	surface area that is painted
S_R	rotation distance
S_{ref}	reference area, area
S_{TO}	takeoff distance
S_{TR}	transition distance
S_{Wet}	UA wetted area, area
S_w	wing planform area; shear force on the web
$S_{wet,fuse}$	fuselage wetted area
$SChute$	parachute reference area
Sec	second portion of longitude or latitude
Si	signal strength, W
SNR	signal-to-noise ratio, dB
SNR_{Avail}	available signal-to-noise ratio, dB
SNR_{image}	signal-to-noise ratio of the SAR image
SNR_{Req}	required signal-to-noise ratio, dB
SNR_r	signal-to-noise ratio at the SAR receiver
s	semispan
s'	panel semispan before dihedral correction
sm	static margin
T	thrust; air temperature, absolute temperature; atmospheric temperature torque (force $\times$ length); time, hrs
T_{Act}	actuator torque
T_{Cable}	net corner cable tension
T_{Cable}	cable tension
$T_{Collect}$	collection time
T_{comp}	component temperature
T_{eff}	pulse effective duration
T_{Hinge}	torque applied to hinge line
T_{Line}	line tension
T_{Max}	maximum engine thrust
T_{max}	maximum thrust

$T_{\min}$	minimum thrust
T_{Nozzle}	turboprop nozzle thrust
$\boldsymbol{T}_{\text{Propulsion}}$	propulsion thrust
T_{prop}	propeller thrust
$T_{\text{Rep,Tot}}$	sum of the repair time (time)
T_{req}	required thrust
$\boldsymbol{T}_{\text{Rocket}}$	rocket thrust
T_S	atmospheric temperature in the stratosphere
T_{SL}	sea-level atmospheric temperature
TR	time ratio
TVC	tail volume coefficient
TVC_H	horizontal tail volume coefficient
$T/W_{\text{Powerplant}}$	thrust-to-weight ratio of the powerplant
T/W_{Ref}	reference thrust-to-weight ratio
T/W_{TO}	thrust-to-weight ratio of the aircraft at takeoff
$\overline{TVC}$	modified tail volume coefficient
t	time required to bring the UA to rest; skin thickness; time interval
t_{Min}	minimum gauge thickness
t_{RT}	roundtrip time between transmit and receiving radar pulse
t_X	time required to arrest horizontal motion
t/c	airfoil-shaped radome thickness-to-chord ratio
t/c_{Root}	wing root thickness-to-chord ratio
V	true airspeed; shear force; flight velocity
V_{AV}	flight velocity magnitude
$\boldsymbol{V}_{\text{AV}}$	flight velocity
V_{Cell}	battery cell voltage, V
V_{Comp}	component volume
V_{Eq}	equivalent airspeed
V_e	mass-averaged exit velocity; cooling exit velocity
$V_{\text{Eq,Max}}$	maximum equivalent airspeed
$V_{\text{eq,max}}$	maximum equivalent airspeed
V_{Fuel}	fuel volume
V_g	ground velocity
V_{Land}	landing flight velocity
V_{Launch}	UA velocity at the end of launch
V_{Max}	maximum voltage, V
$V_{\min}$	minimum airspeed
V_{Mot}	voltage across the motor leads, V
V_{Pack}	battery pack voltage, V
V_{Pix}	number of vertical pixels in the focal plane array

V_{Pixels}	number of vertical pixels in a column
$V_{Platform}$	platform velocity
$V_{Release}$	UA velocity upon release from launch dolly
V_r	rotation speed
$V_{Structure}$	volume enclosed by structure
V_T	true airspeed; steady-state descent velocity under parachute
V_{Tank}	volume of pressure vessel tank
V_{Wind}	wind velocity
V_{WOD}	wind over deck velocity
V_w	wind velocity
V_x	flight velocity in the direction of the synthetic aperture path
Vol	volume of fuselage
$Voltage$	battery voltage, V
val	parameter value
val_{Max}	maximum parameter value
val_{Min}	minimum parameter value
W	weight, lb
W_{APU}	auxiliary power system weight
W_{Arm}	armament weight
W_{Aux}	auxiliary gear weight
W_{AV}	UA weight
W_{Avion}	avionics weight
$W_{Avionics}$	uninstalled weight of avionics
W_{ai}	air induction system weight
W_{avg}	average UA weight
W_{Ball}	EO/IR ball weight
$W_{Ballast}$	ballast weight
W_{Booms}	weight of booms connecting tails to wings
W_{Cant}	cantilevered weight
$W_{Carried}$	weight carried within the fuselage
W_{Comms}	communication system weight
W_{Comp}	component weight
$W_{Comp,Install}$	installed component weight
W_{crew}	crew weight
W_{DG}	design gross weight
W_E	empty weight
W_{ECS}	environmental control system weight
W_{Elec}	electrical system weight
W_{Emp}	empennage weight, force
W_{Empty}	empty weight
W_{Energy}	energy source weight

W_{Engine}	engine weight
$W_{Eng,Installed}$	weight of installed engine
$W_{EO/IR}$	EO/IR ball weight
W_{EPS}	weight of electric propulsion system
W_{ESC}	motor controller weight
W_{FCS}	flight control system weight
W_{Fixed}	weights that do not vary with takeoff gross weight; fixed weights
W_{FTI}	flight-test instrumentation weight
W_{Fuel}	fuel weight
$W_{FuelSys}$	fuel system weight
W_{Furn}	furnishings weight
W_{Fuse}	fuselage weight
$W_{Fuse,Tail}$	weight of the fuselage and tail structure
W_f	fuel weight at takeoff; final weight at end of segment
$\dot{W}_f$	fuel flow rate
W_{feq}	fixed equipment weight
W_{fuel}	fuel weight at takeoff
W_{fuelb}	fuel burned during the mission
$W_{fuel,max}$	maximum fuel weight
$W_{fuel,seg}$	fuel burned during mission segment
W_{Gear}	gearbox weight
W_{Gen}	generator weight
$W_{Handling}$	baggage and cargo handling equipment weight
W_{HT}	horizontal tail weight
W_{Hyd}	hydraulic system weight
W_{H2}	weight of the hydrogen
W_{Inst}	instrumentation weight
W_i	initial weight at beginning of segment
W_{Jet}	jet-engine weight
$W_{Jet,Installed}$	installed jet-engine weight
W_{Land}	landing weight
W_{LG}	landing-gear weight
W_{Motor}	motor weight
W_{Nac}	nacelle weight
W_{Ops}	operational items weight
W_{Other}	other weights
W_{Oxy}	oxygen system weight
W_{Paint}	paint weight
W_{PL}	payload weight
$W_{PL,max}$	maximum payload weight
$W_{PowSource}$	power source weight
W_{Prop}	propulsion system weight

W_{Props}	propeller weight
W_{PropSys}	balance of propulsion system weight
W_{pwr}	propulsion system weight
W_{Recovery}	weight of the UA at recovery
W_{Rocket}	weight of rocket booster
W_{Shuttle}	shuttle weight
W_{Skins}	skin weight
W_{Storage}	weight of the fuel storage
W_{Struct}	structures weight
W_{struct}	structures weight
W_{Subs}	subsystems weight
W_{Tgt}	width of the target as viewed by the imager
W_{TO}	takeoff gross weight
$W_{\text{TO,Calc}}$	calculated takeoff gross weight, force
W_{Tot}	total weight
W_{tfo}	trapped fuel and oil weight
W_{Useful}	useful load
W_{VT}	vertical tail weight
W_{Weaps}	weapons provisions, launchers, and guns
W_{Wing}	wing weight
W_{Wiring}	wiring harness weight
W_{0F}	zero-fuel weight
W_1	weight at the beginning of the mission segment
W_2	weight at the conclusion of the mission segment
WEF	weight escalation factor
WM	weight maturity
X_{ac}	x location of the wing aerodynamic center in body coordinates
$X_{\text{CG}}, Y_{\text{CG}}, Z_{\text{CG}}$	x,y,z coordinates of the center of gravity
$X_{\text{LE,b}}$	wing root leading-edge x location in body coordinates
X_{NP}	x location of the neutral point is body coordinates
X_P, Y_P, Z_P	coordinates of a point on the airframe
X_S, Y_S, Z_S	coordinates of the sensor
X_{Tgt}	downrange distance to the target
x	sombrero function parameter; fuel-to-air ratio
x_b, y_b, z_b	body coordinates
$x_{\text{CG}}, y_{\text{CG}}, z_{\text{CG}}$	center-of-gravity position is body coordinates
x', y', z'	spanwise coordinates
$\hat{\mathbf{x}}_k$	state matrix as estimated by Kalman filter
x/c	chord ratio
x/c_{tr}	transition location relative to chord length
Y_{Tgt}	crossrange distance to the target

y	distance along the span
y_t	y location of the wing tip in body coordinates
z_k	measurement matrix
α	atmospheric attenuation at the acoustic frequency, 1/km required visual contrast curve fit parameter; angle of attack; bypass ratio; detector angular subtense
α_{twist}	wing twist
α_{0L}	zero-lift angle of attack
β	sideslip angle; atmospheric attenuation, km^{-1}
Γ	dihedral angle
Γ_k	process noise distribution matrix
Γ_{Max}	maximum SATCOM coverage latitude
$\Gamma_{\text{V-Tail}}$	V-tail dihedral angle
γ	shear strain; ratio of specific heats; flight-path angle; aircraft still-air climb angle
ΔE_{AV}	change in UA energy
$\Delta E_{\text{Braking}}$	energy lost to braking
ΔE_{Launch}	change in energy from launch
$\Delta E_{\text{Launcher}}$	launcher useful work
ΔE_{Line}	line energy change
ΔE_{Losses}	energy losses
$\Delta E_{\text{Propulsion}}$	propulsion useful work
ΔE_{Stored}	changed in stored energy
ΔEnergy	energy change
Δh	height change
$\Delta h_{\text{Recovery}}$	height change at recovery
$\Delta h_{\text{Release}}$	height change upon UA release from dolly
ΔKE	change in kinetic energy
ΔPE	change in potential energy
ΔR_a	resolvable azimuth resolution
ΔR_r	resolvable range resolution
ΔR_y	resolvable range resolution projected on the ground plane
ΔT_{Climb}	thrust increment required for climb
Δt	time interval
$\Delta \mathbf{V}$	velocity of the UA relative to the platform
$\Delta W_{\text{Fuel,Climb}}$	change in fuel weight due to climb
ΔX	line length change
$\Delta X, \Delta Y, \Delta Z$	distance from the sensor to the point of the airframe
$\Delta \lambda$	difference in longitude
$\Delta \sigma$	spherical angular distance between two waypoints

δ	air pressure ratio
δ_L	roll control
δ_M	pitch control
δ_N	yaw control
δ_T	thrust control
δe	elevator deflection
ε	emissivity of the thermal source; strain; downwash angle
ε_{Min}	minimum SATCOM elevation angle above horizon
ε_{Sat}	SATCOM elevation angle above horizon
ε_0	downwash angle at zero angle of attack
η	dynamic pressure ratio
η_a	atmospheric transmission efficiency
$\eta_{antenna}$	antenna efficiency
η_{ap}	aperture efficiency
η_{Batt}	battery efficiency
η_{Cycle}	cycle efficiency
η_{Dist}	power distribution efficiency
η_{ESC}	motor controller efficiency
η_{Gen}	generator efficiency
η_{gear}	gearbox efficiency
η_H	ratio of dynamic pressures at horizontal stabilizer
η_{motor}	motor efficiency
η_O	transmission efficiency of the laser system optics
η_p	propeller efficiency
$\eta_{p,ideal}$	ideal propeller efficiency
$\eta_{p,nonideal}$	nonideal propeller efficiency factor
$\eta_{propulsive}$	propulsion efficiency
η_{th}	thermal efficiency
η_v	engine volumetric efficiency
η'	semispan distance ratio in spanwise coordinates
θ	temperature ratio; semispan ratio parameter; beamwidth angle; angle from the Airy disk center; sensor elevation angle; IR detection system scan rate, rad/s
θ_{Align}	polarization misalignment angle
θ_{Az}	antenna beam azimuth angle
θ_{Cable}	cable angle
θ_{Diff}	Airy disk blur spot angle
θ_h	horizontal beamwidth angle
$\theta_{Installed}$	antenna blocking elevation angle due to installation
$\theta_{Launcher}$	launcher elevation angle

θ_{Look}	look angle; sensor look angle
θ_{Maneuver}	antenna blocking elevation angle due to maneuvering
θ_{Nom}	propeller pitch angle at 75% radius
θ_{Rocket}	rocket thrust elevation angle
θ_v	vertical beamwidth angle
θ_{4dB}	mainlobe beam width for 4 dB
κ_a	airfoil technology factor
Λ	sweep angle
$\Lambda_{c/2}$	sweep angle at half-chord
$\Lambda_{c/4}$	wing quarter-chord sweep
λ	wing taper ratio; incident rate (time^{-1}); wavelength; radar wavelength, m; sound wavelength; electromagnetic wavelength
λ_{eq}	equivalent taper ratio for a multisegment wing
μ	kinematic viscosity; viscosity
μ_{g}	rolling-friction coefficient
ρ	air density; material density; radial distance from centroid
ρ_{Foam}	foam density
ρ_{Fuel}	fuel density, lb/gal
ρ_{Matl}	material density
ρ_{SL}	sea-level air density, density
ρ_t	effective Lambertian reflectivity of the target
σ	radar cross section; attenuation factor; air density ratio; ratio of density to sea-level density; stress; target radar cross section, m^2
σ_{hoop}	hoop stress
σ_{long}	longitudinal stress
σ_{Ult}	ultimate stress
σ_0	distributed target reflectivity
τ	shear stress; pulsewidth time of the radar
τ_A	transmission of the atmosphere
τ_O	transmission of the IR optics
τ_U	ultimate shear stress
Φ_k	state transition matrix
ϕ	UA bank angle; local dihedral; twist angle
ϕ_{bank}	bank angle
ϕ_f	final waypoint latitude
ϕ_s	starting waypoint latitude
ψ	sensor azimuth angle
$\psi_{\text{AV-Platform}}$	angle between V_{AV} horizontal plane and X axis
ψ_{gr}	grazing angle
ψ_{Sq}	squint angle

Ω	rotational rate
Ω_{Shaft}	servo angular rate
$\Omega_{Surface}$	control surface angular rate

Subscripts

Accident	accidents
AFee	award fee
ALRE	aircraft launch and recovery equipment
AMPR	aeronautical manufacturer's planning report (AMPR)
AV	air vehicle
Avion	avionics
BFee	Base fee
BLOS	beyond-line-of-sight communications
Bot	bottom
BSFC	brake specific fuel consumption
bend	bending material
C	canard
Cargo	cargo aircraft
CG	center of gravity
Code	software coding
Comp	component
ConMatl	consumable material
Crew	crew
c	centroid
$c/4$	quarter-chord
Defl	deflection
Deploy	deployment
Depot	depot
Des	design
DetDes	detailed design (software)
Dir	direct
Disp	disposal
Eng	engines
Engr	engineering
Fac	facilities
Fin	finance
Flt	flight
FM	forces and moments
FT	flight test
FTO	flight-test operations
Fuel	Fuel
f	fuselage

Gear	landing gear
GS	ground station
H	horizontal stabilizer
HW	hardware
hp	hardpoint
Ind	indirect
IT	integration and test
Kill	shoot-down event
LC	life cycle
LE	leading edge
LossYr	losses per year
Low	lower surface
L&R	launch and recovery
Maint	maintenance
Manf	manufacturing
Matl	materials
Mb	megabit
Misc	miscellaneous
m	midpoint
mp	mass properties
Ops	operations
Payl	payload
PerfCon	contractor performance
Pers	personnel
Pitch	pitch
POL	petroleum, oil, and lubricants
Prod	production
ProdDes	product design (software)
ProdStd	production standard
QC	quality control
RDTE	research, development, test and evaluation
Ref	reference
Replace	replacement
r	root
Side	side
Subs	subsystems
Supt	support equipment
SW	software
Sys	systems
shear	shear material
TE	trailing edge
Tool	tooling

Top	top
t	tip; divided by thickness
UASsys	UAS systems
Up	upper surface
V	vertical stabilizer
W	weight
WL	winglet
w	wing
wb	wing and body combined
web	web
x	X direction
y	Y direction
z	Z direction

ACRONYMS

BHP	brake horsepower
BMEP	brake mean effective pressure
BSFC	brake specific fuel consumption
CEF	cost escalation factor
CP	contract performance
CY	cost year
DR	data rate
ETOS	effective time on station
FOM	figure of merit
FOS	factor of safety
FH	flight hours
FOV	field of view
GSD	ground sample distance
MDT	maintenance downtime
MMH/FH	maintenance man hours per flight hour
MR	mishap rate (time^{-1})
MTBF	mean time between failures
MTBM	mean time between maintenance
MTBO	mean time between overhauls
MTTR	mean time to repair
NIIRS	National Imagery Interpretability Rating Scale metric value
RER	geometric mean normalized edge response
ROD	rate of descent
SNR	signal-to-noise ratio of unprocessed imagery
ROC	rate of climb, velocity
TSFC	thrust specific fuel consumption, time^{-1}

Chapter 1
Overview of Unmanned Aircraft Systems

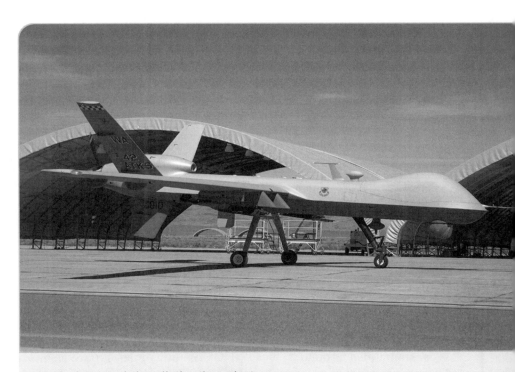

- Understand the distinctions between unmanned aircraft and other systems
- See a brief history of UAS technology and operations
- Understand UAS system architectures

General Atomics MQ-9 Reaper multimission UAS. (U.S. Air Force photo by Senior Airman Larry E. Reid Jr./released.)

1.1 Introduction

The field of unmanned aircraft systems (UASs) is very broad, covering myriad missions and system types. Those who are new to UASs may think of one particular system that embodies the essence of unmanned systems. It might be the Predator medium-altitude, long-endurance UAS that achieved fame in the recent conflicts in Afghanistan or Iraq. The system might be a hand-launched, man-portable system such as the Raven. Others may think of UASs as sophisticated versions of model aircraft or as simply aircraft without pilots. All of these perceptions are true to some extent, but these do not paint the full picture of UASs' diversity, complexity, and capabilities.

In this chapter we will seek to define UASs. We will examine their major features and attributes. This forms a basic understanding for the more detailed treatment in the remainder of the book. This chapter also explores the history of UASs, which reveals how far these systems have come.

1.2 Defining an Unmanned Aircraft

The U.S. Department of Defense (DOD) defines an unmanned aircraft as follows:

> A powered vehicle that does not carry a human operator, can be operated autonomously or remotely, can be expendable or recoverable, and can carry a lethal or nonlethal payload. Ballistic or semi-ballistic vehicle, cruise missiles, artillery projectiles, torpedoes, mines, satellites, and unattended sensors (with no form of propulsion) are not considered unmanned vehicles. Unmanned vehicles are the primary component of unmanned systems [1].

This definition covers multiple forms of unmanned systems including UASs, unmanned ground vehicles (UGVs), unmanned surface vehicles (USVs), and unmanned underwater vehicles (UUVs). To narrow this definition to a UAS, the unmanned system must include a flying vehicle.

The definition excludes sensors with no form of propulsion. This exclusion is likely intended to eliminate sonobuoys and unattended ground sensors (UGSs). Nonpowered aircraft, known as gliders, can perform many useful missions and are generally considered to be UASs. They have flight controls, are capable of navigation, and carry payloads. One could argue that a glider's propulsion is the conversion of potential to kinetic energy.

UASs have taken many names and forms over their long history. Unmanned aircraft have been dubbed drone, remotely piloted vehicle (RPV), unmanned aerial vehicle (UAV), uninhabited combat aerial vehicle (UCAV), organic aerial vehicle (OAV), uninhabited combat aircraft system (UCAS), remotely piloted aircraft (RPA), remotely piloted helicopter (RPH), aerial robotics, and micro aerial vehicle (MAV), to list a few. The aircraft portion of the system is known as the air vehicle (AV), remotely piloted aircraft (RPA), or simply the unmanned aircraft (UA). Here we adopt the term unmanned aircraft or UA. Whatever called, unmanned aircraft systems have an airborne component that performs at least one mission role without a pilot onboard. Additionally, unmanned aircraft must be capable of controlled and sustained flight; otherwise, the first cast stone is also the first unmanned aircraft.

Let us explore the distinction between a cruise missile and an unmanned aircraft further. A UAS can drop bombs or launch weapons and then return to base, while a cruise missile flies the integral warhead into the target. High-speed, jet-powered targets often have similar characteristics as cruise missiles, except that the payload and mission differ. The launch methods overlap, where both classes can be launched conventionally, ground rocket launched, ship or sub launched via rockets, or air launched. One might present a nebulous argument that it must look like a cruise missile, but propeller-driven aircraft with integral warheads certainly perform a cruise missile mission. Antiradar aircraft, once referred to as "harassment drones" or "lethal UAVs," are operationally indistinguishable from cruise missiles. Arguably (and there certainly has been much argument), a cruise missile is a form of UAS, and when a platform crosses the line to become a cruise missile, it depends on the judgment of the organization making the determination.

General attributes of unmanned aircraft are the following:

* Smaller size potential
* High versatility
* Greater performance than manned aircraft—design flexibility

1.3 Motivation for Employment of Unmanned Aircraft

It is often said that unmanned aircraft save lives. The lives in question are flight crew in the equivalent manned aircraft that is replaced and those on the ground who get operationally critical support from the UAS. There is no crew onboard the UAS, and therefore the loss of the aircraft does not result in the loss of the flight crew. The operators in the ground control station (GCS) are quite unscathed when the unmanned aircraft crashes, with the possible exception of postincident paperwork.

One might assume that reducing the pilot's risk of death or capture would be warmly received by the aviator community. However, this is not generally the case for one simple reason: UASs put traditional pilots out of a job. Most military aviators understand and accept the risks of the job and are willing to fight in combat or perform dangerous civil missions. To many, it is preferable to have the dangerous job of a combat aviator or a search and rescue pilot than to be safe behind a desk or in a ground control station. If UASs keep pilots safe by flying manned aircraft missions, then there is diminished need for manned aviation. So naturally UASs are a threat to flight crew professions, though pilots are unlikely to be replaced entirely.

Flight crews are a liability. A captured flight crew can create a major political, tactical, or diplomatic setback, and can change the course of public support for a war or endanger delicate treaty negotiations. Downed flight crew members behind enemy lines are frequently followed by dangerous rescue operations where lives might be lost. A downed BlackHawk helicopter in Mogadishu led to a deadly rescue attempt where many American soldiers were killed. The public was horrified to see bodies dragged through the streets by angry mobs. The campaign was terminated shortly after this incident due to faltering public support. Francis Gary Powers survived the shoot-down of his U-2 over the USSR in 1960 and was subsequently taken prisoner, producing a major diplomatic embarrassment for the United States and further escalating Cold War tensions. Major General Kenneth Israel, the former director of the Defense Airborne Reconnaissance Office (DARO) perhaps says it best:

> The key premise is this: UAVs saved lives. They not only supply intelligence, but take the pilot out of high risk situations. When F-16 pilot Scott O'Grady was shot down, it was a crisis, but when a $2 million Predator UAV was shot down, it was a curiosity. Who is going to tell a parent that their child is not worth $2 million? [2]

Nobody mourns a lost UAS, nor is there public outrage if one is captured. There is no notification of next of kin and no funerals. Enemy forces will not get news coverage if they get an autopilot to express a message "10110010." Search and rescue teams will not be lost because there are no people to save. The motto "no man left behind" does not apply to autopilots. A captured autopilot is not a POW, and the largest risk is that the technology will be exploited.

Because there is no flight crew onboard UASs, these systems can be used for roles that would be simply unacceptable or less effective with manned aircraft. These mission classes are often characterized as "dull, dirty, and dangerous" roles. Let us examine examples of each role:

- *Dull*—These missions have little stimulation, where human flight crew performance can suffer due to low workload. Such missions include very long-duration target coverage, communications relay, or air sampling.
- *Dirty*—These missions involve flight through contaminated air that would be harmful to humans. The contaminants can include radiation,

biological agents, or chemical agents. The UAS might be required to sample or measure the contents of the air or perform other missions in this environment.

- *Dangerous*—These missions would put a human life at risk. Aerial targets, cruise missiles, and suppression of enemy air defenses (SEAD) are among these risky missions. Other dangerous operations might involve other missions such as surveillance in enemy-controlled airspace. Dangerous civil missions might include flight at very low altitudes or inside an active volcano crater.

1.4 Distinction Between Manned and Unmanned Aircraft

The most obvious difference between an unmanned aircraft and a manned aircraft is the lack of a pilot or other flight crew. The systems required to accommodate and support the flight crew can be much greater than just the weight of the humans. Let us list some systems that are required for manned aircraft that can be left on the runway (that is, excluded from the unmanned aircraft) for a dedicated UAS type. The reference mission of interest is a tactical intelligence, surveillance, and reconnaissance (ISR) mission performed by a tactical UAS today. The manned aircraft that would be required to perform this mission would likely be a two-seat aircraft and resemble forward air controller aircraft such as a Cessna O-2.

The total comes to 600 lb in this example, as shown in Table 1.1.

Table 1.1 Human-Related Manned Aircraft Weight Components

Component	Weight, lb	Notes
Pilot	160	Power for flight controls
Second crew member	160	Likely radio or ISR systems operator
Windows	20	Visibility for pilot and crew
Furnishings	60	Not ejection seats
Doors	30	
Instrumentation and avionics	100	Includes guidance and navigation equipment
Control interface	20	
Cabin environmental controls	30	
Survival kit	10	
Portion of electrical system	10	For powering avionics and cabin environmental controls

Table 1.2 Component Weights Associated with Unmanned
Tactical UASs

Component	Weight, lb	Notes
Autopilot and other avionics	15	
Flight control actuation	15	Electromechanical actuators
Line-of-sight communications	10	C2 and payload links
Portion of electrical system	5	Generator and power system

The total for the unmanned aircraft components is 45 lb, as shown in Table 1.2. Note that the same systems' functions can weigh a small fraction of a pound for micro unmanned aircraft. The difference between the manned and unmanned system weights is 555 lb. As we will see in Chapter 3, this weight difference will have a larger overall weight impact on the unmanned aircraft.

The manned aircraft window geometry is driven by field-of-regard constraints of the pilot and other crew. The window area depends on the distance between the observer's head and the window surface and the required viewing angles. The weight is determined by the window area, material type, and loads. A bubble canopy on a supersonic fighter will have greater aerial weight than a small observation aircraft due to the higher dynamic pressure and higher accelerations. Windows might require heating to prevent fogging or ice buildup.

The furnishings in the tactical ISR example are relatively simple seats with mechanical adjustment features and safety harnesses. An ejection seat on a fighter or attack aircraft can weigh several hundred pounds per occupant. Flight suits and parachutes add additional weight for some applications.

The cabin containing the humans consumes volume. The density of the cabin, or the weight of the components contained within the cabin divided by the cabin volume, is low relative to avionics bays in UASs. The volume required for any manned aircraft fuselage is high relative to that of an equivalent UAS. This volume results in higher surface area, which generates more drag.

The cabin requires environmental conditioning to accommodate humans. The environmental control system (ECS) maintains the air at the necessary temperature and pressure to enable effective crew operations. Manned aircraft might additionally require backup oxygen systems.

The weight associated with humans on board an aircraft could approach 2000–3000 lb per person on a tactical military aircraft. This is weight that could be used for fuel or payload if the platform is unmanned, which would provide enhanced mission utility.

Because humans make program decisions, there is still an unavoidable bias towards manned aircraft in new system acquisitions. After all, it is

Table 1.3 Performance Comparison of the Human Pilot vs UAS Avionics

Attribute	Human Pilot	UAS Avionics
Acceleration limits	5–10 g peak	Design criteria. Some are designed to over 10,000-g shock.
Temperature limits	Body temperature must be maintained at 98.6 deg. Cabin temperatures typically 60 to 100°F.	Design criteria. Many avionics are designed to operate at −20 to 120°F.
Pressure limits	Typically 0.75–1 atm	Design criteria. Space electronics can operate in vacuum.
Orientation	Pilots are generally upright (sitting upright). Rare aircraft have prone pilot (lying down on belly with face forward)	Avionics and sensors can be designed for a wide range of orientations. Unmanned aircrafts can fly inverted for extended duration.
Mathematical operations	Humans have a limited ability to perform calculations in flight.	Avionics can perform millions of calculations per second, with an ever-increasing capability trend.
Upgradability	The Mark I human pilot capability experiences almost no change over relevant timescales. Variation exists in a population, but the distribution is nearly constant over time. Without controversial future biomechanical technology (i.e., cyborgs), pilots can't be upgraded.	Major revolutions in avionics and software occur about once a decade.
Reasoning	Pilots can flexibly react to unexpected situations. Pilot situational awareness is high due to visual cues, sounds, smell, and physical feel. Pilots have an intuitive feel for the aircraft sometime referred to as "seat of the pants."	Avionics can only react as programmed. Human operators in the ground stations can handle situations within the available interfaces if communications exist. The situational awareness of humans is limited to data received from the unmanned aircraft.
Visual sensor performance	The neck can enable a field of regard beyond the forward hemisphere. The eye can see in the visual band (400–700 nm), but has no	The visual sensor performance is a design criterion. Payload optics can often resolve features of a few inches at ranges of 1–20 miles. The field

(Continued)

Table 1.3   *Continued*

Attribute	Human Pilot	UAS Avionics
	capability in short-wave IR, mid-wave IR, and long-wave IR.	of regard is typically less than a full hemisphere, but greater than what a human neck permits. Optics can be made to operate in any band of the entire atmospheric transmission window.
Acoustic sensor performance	The human ear can detect sound in the 0.02–20-kHz range. The frequency range and threshold sound pressure levels vary among individuals.	Most UASs have no acoustic sensors. Microphones could be added if necessary with performance superior to the human ear.
Radio-frequency detection	Humans have no ability to detect RF signals.	UAS can transmit or receive line-of-sight and satellite RF signals with the necessary sensitivity to permit receipt of data. Typical RF communications frequencies range from 72 MHz (VHF) through 30 GHz (Ka band).
Errors	Pilot error is a major cause of manned aircraft losses. Human decision making is intuitive and imperfect.	Pilot errors are a major cause of UAS losses when pilots are in control of the unmanned aircraft. Autonomous UAS losses due to system logic errors are considered design errors.
Work duration	8–12-hr shifts, rest required between shifts.	Only limited by unmanned aircraft flight performance or reliability. 5-year duration solar-powered aircraft are feasible.
Operational service	20–30-yr career	Technology obsolete in 5–10 yrs, component service life of 100–10,000 hr
Bravery	Pilots are generally courageous and can be motivated to perform highly dangerous or suicidal (i.e., Kamikaze) missions, but pilots also have an inherent will to live.	An autopilot is indifferent to risk or self-sacrifice.
Value	Immeasurable (except by actuaries)	Hundreds to millions of U.S. dollars

unlikely that people would flock to the theaters if the movie *Top Gun* was about UCAVs. The leading character would be an autopilot whose key line is "10011010. . ." rather than "great balls of fire." The class rankings for the Top Gun competition would be remarkably close if all of the autopilots were built to spec and had the same operational flight program. Well-designed flight controls and mission management systems more closely represent Ice Man's cold precision rather than Maverick's intuitive tactical genius (unless fuzzy logic is applied). If the mission management computer ("Goose" equivalent) had a critical failure, it would simply be replaced, and the autopilot would have no emotional performance degradation in the next flight. Unmanned aircraft will likely have more appeal to strategists and engineers than to aviators and the public for many years to come.

Now let us compare the UAS avionics to the human pilot. Here the avionics are considered elements that are required to replicate the functions performed by the human pilot. Using this information, ask yourself if it is a good system trade to use an autopilot that weighs 180 lb (with large manufacturer variation) with the human pilot characteristics. In some cases the UAS functionality might partially reside on the ground. The capabilities and limitations of human pilots are well documented [3–5]. Some of the key attributes of human pilots and UAS avionics are shown in Table 1.3.

Some additional favorable attributes of unmanned aircraft include the following:

* *Better transportability due to smaller size*—The smallest UASs are capable of being carried by a single person. There is no practical manned aircraft that can be carried in a similar manner.
* *Lower cost expectations*—UASs usually have a lower acquisition cost than manned aircraft, with the possible exception of very large UASs such as high-altitude long endurance (HALE) UASs and UCAVs. This is largely due to smaller size for a given mission and less stringent design requirements. The operations costs for UASs are not necessarily lower than manned aircraft due to the often high manpower requirements.
* *Speed of implementation*—UASs are often faster to prototype and to bring into operations.

However, there are some additional unfavorable attributes. These can include the following:

* The satellite communications gimbaled antenna consumes a large field of regard that requires a large internal volume. This generates drag in a manner similar to a manned aircraft cabin.
* The avionics, communication system, and payload power requirements are high for the UA size. This requires power generation that increases

fuel burn from the engine. The cooling drag can degrade the aerodynamic efficiency.
* UASs have no limited ability to handle unexpected situations. What capability exists comes in the form of avionics robustness and human operator actions in the ground stations.

UASs offer aircraft designers a means of designing multiple aircraft within a career. The author has had the good fortune of working on over 30 distinct UASs that have achieved at least first flight. This opportunity is generally not possible with manned aircraft, where a defense procurement system might take 20 years to go through full development. A designer in the 1940s or 1950s could work on dozens of manned aircraft in a career, but today's manned aircraft are more complex, and there are fewer acquisition programs. The developmental time frame is generally shorter with smaller and simpler systems, which often contributes to the fast development of UASs. This difference in development timelines is reflected in acquisition planning, where manned aircraft are planned 10–20 years in advance and UASs' developments are expected to take 2–4 years. A designer has no way of knowing whether design assumptions are valid unless she/he sees the outcome of those decisions. The UAS designer is in a unique position within the aerospace industry of applying lessons learned from past programs to new programs.

In principle, the same level of analysis could be applied to a MAV and a sophisticated UCAV. The main difference in design effort can be attributed to the MAVs relative simplicity, smaller budget, and lower consequences of crashing than the UCAV.

Smaller UASs such as MAVs or hand launch systems are often flown with relatively little aerodynamic analysis and no wind-tunnel testing. Perhaps too frequently, the design tools can consist of nothing more than a spreadsheet. Although at first this might appear cavalier, the atmosphere is the most accurate wind tunnel for assessing aircraft flight performance and vehicle stability and control. If shortfalls are discovered, new design iterations of small prototype aircraft can cost substantially less than a reconfigurable wind-tunnel model and tunnel entry. New Federal Aviation Administration (FAA) regulations are making this approach much more difficult because airspace access is now more difficult to obtain.

UASs typically lag manned aircraft in sophistication. To date, no jet engines have been developed specifically for unmanned aircraft, while new manned aircraft programs often have parallel engine development efforts. Technology generally migrates from manned aircraft to unmanned aircraft. Much of this phenomenon is the result of lower cost expectations of unmanned aircraft. An exception to the jet engine example is cruise missile design, where low-cost single-use engines have no manned counterpart.

Despite the myriad design benefits of unmanned aircraft, few systems leverage these advantages. If we take endurance off the table, most unmanned aircraft have flight performance characteristics similar to pre-World War II manned aircraft. The speed, maximum altitude, and payload capacity are comparable or worse. Where is the low-cost, 20-*g*, unmanned air superiority fighter with advanced air-to-air missiles that can outmaneuver any manned aircraft? Where is the long-range, high-efficiency, heavy-lift unmanned cargo aircraft? Where is the low-flying, highly maneuverable, close-air-support unmanned aircraft that has dozens of high-precision munitions under its wings? Current-generation unmanned aircraft tend to be subsonic, have low maneuverability, and have limited payload capacity relative to their manned counterparts. The good news is that unmanned aircraft have growth opportunities and much potential for fulfilling future missions. Perhaps the readers of this book will bring about these advances.

1.5 Moral, Ethical, and Legal Issues for UASs

As we will see throughout this book, UASs offer numerous advantages over manned aircraft and can enable new missions. Depending on perspective, it can be argued that UASs are potentially capable of performing missions that they should not perform. Most issues in question are related to the conditions under which UASs can obtain intelligence and take human lives or wound humans in combat.

There is increasing interest in robot ethics. In his book entitled *Wired for War* [6], P. W. Singer explores this domain. A more humorous perspective can be found in *How to Survive a Robot Uprising* [7], where the mere title highlights the dangers of autonomous systems. Some potential issues include the following:

- The use of unmanned systems in combat gives the enemy a perception that the forces employing the systems are too cowardly to fight directly.
- The use of unmanned systems can negatively affect the goodwill of the civilian population.
- By not placing human flight crew at risk, leadership can become more inclined to start hostilities.
- Unmanned systems technology can become uncontrollable.
- Autonomous targeting and weapons delivery (no human in the loop) might be unethical.
- Use of UAS for civil surveillance and law enforcement might violate civil liberties.

1.6 Brief History of Unmanned Aircraft

It is difficult to appreciate the amazing UAS capabilities fully without understanding what came before. There have been numerous bursts of innovation and eras of widespread employment. There have also been long periods where the technologies were all but forgotten. Brilliant achievements have been followed by decades of dormancy. Systems that we now take for granted usually have predecessors that achieved some success or were utterly unsuccessful. Only in the past decade have we achieved a true golden era where UASs have established a large presence in military operations.

UAS development has paralleled manned aviation from the beginning of flight and cruise missiles since World War I. The technologies migrate between these domains from one program to the next.

The long history of unmanned aviation has seen thousands of programs, though fewer than a hundred have seen operational service. The author has identified over 1,300 systems in a database, though hundreds of additional systems are not yet entered. Here, we will focus on the systems that made it to operations and those that represent the largest investments of the time.

Those interested in learning more about the history of unmanned aviation have many resources [2, 8–10]. Mueller et al. [11] describe key technology innovations in chronological order with an emphasis on small UAS applications. Werrell [12] covers the history of cruise missile development, which overlaps that of UASs.

1.6.1 Early Years (1896–1939)

In 1896 Samuel Pierpont Langley launched an unmanned, steam-powered aircraft over the Potomac River. The catapult launcher was positioned atop a houseboat, much as modern pneumatic launchers are used for ship-based UASs today. The unmanned aircraft, dubbed the *Aerodrome*, flew for almost a mile. The Aerodrome had no guidance system, but it was a UAS in all other regards. This historic flight predates the Wright Brothers' first flight by seven years.

The missing element of Langley's flight was a guidance, navigation, and control (GN&C) system. Elmer Sperry used the gyroscope as the core of a simple autopilot for manned and then unmanned aircraft in 1918.

The first serious attempt to widely employ unmanned aircraft in combat was the Kettering Bug program of World War I. This reciprocating-engine-powered, unmanned biplane took off from a horizontal rail. Upon autonomously navigating up to 40 miles to the target, the wings were to be shed, and the fuselage would deliver a 180-lb warhead to the ground. World War I ended before the system could be fielded.

The interwar period saw limited new development. Autopilots and actuators underwent incremental improvements. Most UASs were aerial targets used for gunnery practice. The British produced a number of unmanned target types. Another important technology introduced at this time was remote control via radio-frequency (RF) transmissions.

1.6.2 World War II (1939–1945)

The largest unmanned aircraft strides of World War II came from Germany. Germany developed air-to-ground guided munitions and cruise missiles that greatly advanced UAS technology. Perhaps the greatest and most well-known system was the Fiesler V-1 "Buzz Bomb." The Luftwaffe used radio-controlled glide bombs to destroy ships, including the Italian battleship Roma that was surrendering to the Allied forces. The German technology formed the foundation for U.S. cruise missile development and the space program after the war.

The U.S. Navy has been working on carrier-based attack UASs for the past decade in the form of the UCAV programs, yet at the time of publication these systems have not yet operated from a carrier. Many find it surprising that U.S. carrier-based unmanned attack aircraft were used in combat during World War II. The TDR-1 Assault Drone was a modified Interstate Aircraft twin-engine manned aircraft. It took off under its own power from the carrier deck. A TBF Avenger control aircraft guided the TDR-1 to the target area using radio control and a television camera. The Assault Drone would then drop bombs or a torpedo on the target.

Project Aphrodite was an ambitious U.S. program to convert bombers such as the B-17 and PB-4Y to unmanned flying bombs. The bomber would take off under pilot control. Once aloft, a crew member would arm the warhead. Control would be relinquished to a chase plane that communicated via a RF link and used a camera inside the UAS cockpit to read instruments and provide the view of the flight path. The crew aboard the flying bomb would parachute to safety while the UAS would continue to be remotely guided to the target. The purpose was to engage hardened targets in German-occupied territory, such as a giant cannon intended to fire on London. Later these unmanned bombers were fitted with instrumentation and used to fly through mushroom clouds after nuclear tests, which is a classic "dirty" mission.

The most prolific small UASs of World War II, and of history, were Radioplane's gunnery targets. More than 12,000 of these small targets were produced.

1.6.3 Korean War, Vietnam War, and the Early Cold War (1945–1972)

After World War II, unmanned aircraft were mostly relegated to target missions (then called "target drones") until the Vietnam War. The Korean War saw limited use of unmanned aircraft in the form of unmanned conversions of manned aircraft that were loaded with explosives for attacks against ground targets.

Cruise missile development took center stage during much of this period. The systems started as incremental improvements, or even copies of, the Fiesler V-1. Very large jet-powered cruise missiles were developed, such as the Navaho, Matador, and Snark. These large missiles owed much of their size to the bulky avionics that were carried. Few programs achieved operational status, as developers struggled with technologies that were stubbornly immature. These cruise missiles paved the way for the smaller BGM-109 Tomahawk and AGM-86 Air Launched Cruise Missile (ALCM) cruise missiles that were developed in the 1970s and are still in use today.

The bright spot of this era is the Teledyne Ryan Firebee/Firefly family. The contract for the Firebee, then called the Q-2, was awarded to the Ryan Aeronautical Company in 1948. Over 7000 of these systems have been produced in myriad variants. These systems were usually air launched from a modified C-130 and recovered by a combination of parachute and airborne helicopter snag. All of the multiple variants were jet powered. The payloads employed in Vietnam included wet film cameras and ELINT systems. Over 3,500 sorties were flown during the war, arguably creating the first golden era of unmanned aviation.

The first operational unmanned helicopter was developed and deployed in this period. The Gyrodyne QH-50 was a coaxial rotor design initially powered by a reciprocating engine and then a turboshaft. Its mission was to carry torpedoes, sonobuoys, cameras, flares, cargo, and smoke generators from destroyers. Although widely deployed, it had limited effectiveness due to still immature technologies. Approximately 800 QH-50s were produced.

The Lockheed D-21 was the most impressive system of the era. This Mach 3+ ramjet-powered UAS was air launched first from a modified A-12 (predecessor to the SR-71 Blackbird) and later from a B-52. This long-range platform was designed to fly preprogrammed missions autonomously. Only the camera payload and avionics were recovered. The module housing these systems was jettisoned from the aircraft, descended under a parachute, and was snagged by a fixed-wing aircraft using the Mid-Air Recovery System (MARS). Decades after the system was retired, we still have no equivalent to this remarkable UAS.

1.6.4 U.S. Hiatus, Israel Carries the Torch (1972–1988)

If we use a civilization analogy, the Vietnam War was like the days of the Roman Empire. The years 1972–1988 were more like the Dark Ages. The impressive technology gains and operational systems were shelved. Defense budgets shrank after the Vietnam War, and new UASs were either not developed to completion or were visible programmatic failures.

The forerunners to today's medium-altitude long-endurance (MALE) and high-altitude long-endurance (HALE) UAS classes were first attempted in this period. The MALE Compass Dwell program saw Martin Marietta and E-Systems develop unmanned variants of Scweitzer gliders and motor-gliders. Boeing and Teledyne Ryan developed jet-powered HALE UA for the Compass Cope program. Although both companies were successful, the Teledyne Ryan version achieved an endurance record of over 28 h. The Compass Dwell and Compass Cope programs involved direct government funding of two contractor teams for initial demonstrators. Most recent UAS competitions have required that contractors self-fund their own systems.

The Teledyne Ryan Compass Arrow was another major accomplishment. This system built upon the Firebee/Firefly heritage, but with a clean sheet design. This 54-ft span, jet-powered, high-altitude UAS was designed for long-range performance. The program was terminated after President Nixon made inroads with China, thus eliminating its primary mission.

The most sustained UAS development during this period was the Lockheed Aquila (YMQM-105A) tactical UAS. The size and performance capabilities were roughly equivalent to the currently fielded Shadow 200. The program was cancelled in 1987, representing the nadir of the period. Up until this point unmanned aircraft were called remotely piloted vehicles (RPVs), but the name changed to unmanned aerial vehicles (UAVs) after the Aquila program. The negative program outcome has been widely studied, and a major culprit was that system requirements were not compatible with available technologies.

While the United States was struggling with diminished budgets in the 1970s and floundering programs in the 1980s, Israel was honing highly effective UAS designs and exploring new concepts of operations. Israel developed the Mastiff, Scout, and Searcher tactical UAS systems. These were capable of runway operations, had line-of-sight communications, and had gimbaled video payloads. Prior to these systems, operational UASs employed optical payloads for non-real-time reconnaissance, where still images were processed on the ground after the flight. The Israeli systems were capable of real-time surveillance. Descendants of these systems were later procured by the United States in the form of the Pioneer and

Hunter systems, which filled a critical gap resulting from unsuccessful indigenous efforts. Other nations such as the United Kingdom, Italy, France, Germany, China, and Russia also developed UAS systems in this period, but Israel carried the torch.

1.6.5 Modern UAS Ascendancy (1988–2000)

The late 1980s and early 1990s saw the maturation of enabling technical innovations such as digital electronics, global positioning systems (GPS), digital data links, and satellite communications. Starting in the late 1990s, these revolutionary systems started operations. This period can best be likened to the Renaissance.

U.S. military leadership realized the potential of UASs in Operation Desert Storm. Systems used in that conflict included the tactical Pioneer, low-cost Exdrone, high-speed BQM-74E Chukar, and hand-launched Pointer. The Chukar served as a decoy, supporting the air campaign. The star of the war was the Pioneer, which operated off of battleships for artillery spotting and surveillance. The total number of UASs, the sortie rate, and total flight hours were very small relative to recent campaigns, but the operational benefits were clear. The systems employed were the product of 1980s development programs or acquisition of modified Israeli systems.

DARPA sponsored the Boeing Condor program in the late 1980s. This 200-ft span HALE UAS attained a maximum altitude of 67,000 ft and a maximum endurance of 59 hrs. It had two reciprocating engines with triple turbocharger stages. The Condor was the first UAS to use GPS navigation and auto-land technology. Although Condor did not reach operational status, the technologies set the bar for later systems.

The Joint Program Office (JPO) and DARPA oversaw the development of three remarkable systems starting in 1994: General Atomics Predator (Tier II) MALE, Teledyne Ryan Global Hawk (Tier II+) HALE, and Lockheed DarkStar (Tier III-) low observable UASs. All programs were funded under the Advanced Concept Technology Demonstration (ACTD) contract vehicle, which gave the contractors great flexibility in developing the systems. The Predator later became the MQ-1 and the Global Hawk the RQ-4A/B, which provided remarkable operational service. The DarkStar was cancelled after initial flight tests. The Predator, Global Hawk, and DarkStar all had high-bandwidth digital satellite communications links, enabling beyond line-of-sight operations for the first time. With these remarkable systems, the United States regained the world's leadership position for UAS technology and systems. The JPO also oversaw the acquisition of the TRW (now Northrop Grumman) Hunter and General Atomics Gnat, which also saw operational service. The Hunter and Predator systems provided valuable intelligence during the Balkans campaign.

While the Air Force filled its stable with high-performance Predator and Global Hawk systems, the Army and Navy added few new systems in this period. In the aftermath of the Aquila experience, the Army saw three successive competitive procurements for tactical systems. First, the Hunter system was selected but was cancelled after problems in testing. (The system saw extensive operational service after it was terminated.) In the next competitive procurement the Alliant Techsystems Outrider was selected, but it too was cancelled after it experienced major developmental issues. Finally in 2000, the AAI Shadow 200 system was competitively selected. The Shadow 200 was successfully developed and has seen extensive operations.

The Navy considered numerous program starts during this period, including an air-launched system called the medium-range UAV (MRUAV) that was cancelled after encountering development problems. In the late 1990s, the FireScout system was selected for the Navy's VTUAV competition. The FireScout is now seeing limited operational service after over a decade of development.

The U.S. Naval Research Laboratories (NRL) pioneered many critical technologies in this period under the leadership of Rick Foch, the world's most prolific UAS designer. These technologies included miniature digital avionics, GPS-based autopilots, low-Reynolds-number unmanned aircraft design, miniature payloads, in-flight reconfiguration technologies, electric propulsion, and micro unmanned aircraft.

NASA sponsored the Environmental Research and Sensor Technology (ERAST) program in the 1990s. The purpose of the program was to develop high-altitude UASs and payloads for scientific observations of the upper atmosphere. The urgent need at the time was to measure the ozone hole phenomenon, which was not well understood. Several UASs were developed under this program including the Aurora Flight Sciences Perseus A/B, Aerovironment Pathfinder solar-powered aircraft, General Atomics Altus, and Scaled Composites Raptor. Ground-breaking airframes and propulsion systems emerged from the program. Related NASA efforts yielded the Aurora Flight Sciences Theseus, Aerovironment Helios, and General Atomics Altair. Many of these efforts were funded with shoestring budgets and often required contractor cost sharing. Although these high-altitude platforms were not widely used for their intended scientific missions, the technologies developed advanced the state of the art substantially.

In 1999, a diminutive UAS surprised the world by flying over 2000 miles across the Atlantic Ocean on 1.5 gal of gas. The Insitu Aerosonde has a 10-ft wing span and is powered by a four-stroke gasoline reciprocating engine. The lightweight composite structure and high-efficiency low-Reynolds-number aerodynamic design enabled this historic flight. The flight was monitored via a low-bandwidth Iridium satellite communication link.

By the conclusion of this period, the United States had the following operational military UASs:

* Aerovironment Pointer
* BAI Exdrone
* AAI Shadow 200 (RQ-7A)
* Northrop Grumman Hunter (MQ-5A)
* General Atomics Predator A (RQ-1A)
* Northrop Grumman Global Hawk (RQ-4A)
* Northrop Grumman Chukar (BQM-74E)

1.6.6 UAS Golden Era (2001 – Present)

Now we are in the UAS golden era. Over the past decade unmanned systems have flourished. UAS saw prolific use in Operation Enduring Freedom (OEF) and Operation Iraqi Freedom (OIF). The number of operational systems and the cumulative flight hours expanded dramatically in this period. By 2002, the United States and Israel each surpassed 100,000 cumulative flight hours. By 2011, the U.S. Army and Air Force each passed 1,000,000 cumulative flight hours. UASs are now essential military systems that have proven their value.

The full impact of the technologies developed in the 1990s was not realized until the days and years following the 9/11 attacks. Like the 1941 attack on Pearl Harbor (or hitting a hornet's nest with a stick), 9/11 awakened the innovative spirit and industrial might of the United States. The widespread use of UASs today is directly attributable to the response to terrorism.

The General Atomics Predator and Reaper systems did much of the heavy lifting. In addition to ISR, these systems were also utilized in an offensive role. The unmanned aircraft can carry Hellfire missiles and other armaments to directly engage ground targets. The satellite communications link permitted mission operators to be located outside of the theater of operations. Predators and Reapers have high-performance EO/IR and SAR payloads that can find and track targets. After engaging the target, these payloads can perform battle damage assessment. These capabilities combined with long endurance provided unprecedented deadly persistence.

This period also saw the development of more survivable armed UASs. The DARPA-led Joint Unmanned Combat Aircraft System (J-UCAS) built upon earlier efforts to develop high-performance, low-observable (stealthy) armed UASs for the Navy and Air Force. The two systems that emerged were the Northrop Grumman X-47B and Boeing X-45C (now called the Phantom Ray). The Navy picked up the program in the form of the Naval UCAS Demonstrator program, which was won by the Northrop Grumman

X-47B. The future Navy program of record system using UCAS technology is called UCLASS, which has not yet gone through source selection.

The Navy caught up with UAS development with multiple programs. The Northrop Grumman FireScout unmanned helicopter development continues. Northrop Grumman won the Navy's Broad Area Maritime Surveillance (BAMS) program with a Global Hawk derivative. The Navy also employed small, fixed-wing Insitu ScanEagle UAS from ships under a services contract. Later, the larger Insitu Integrator won the Navy/Marine Corps Small Tactical Unmanned Aircraft System (STUAS) competition, which provides long-duration, runway-independent ISR from land and ships.

The Army filled out its stable of UASs with multiple systems. Aerovironment's hand-launched Raven A/B is a smaller version of the earlier Pointer. Ravens are widely used at the unit level. The AAI Shadow 200 development was completed, and the system saw substantial operational service. General Atomics Gnat long-endurance systems were employed until General Atomics won the Extended Range Multi-Purpose (ERMP) program with the Gray Eagle, an enhanced variant of the Predator. A comprehensive system of systems program called Future Combat Systems (FCS) sought to develop multiple UASs. FCS was ultimately cancelled, but the Textron MAV ducted-fan vertical takeoff and landing small UAS developed for FCS ultimately went into service.

In this period, the following systems became operational or are on the path to becoming operational. Note that systems fielded before 2001 are not repeated:

* Aerovironment Wasp
* Textron MAV
* Aerovironment Raven A/B
* Aerovironment/NRL Dragon Eye
* Aerovironment Puma
* Insitu ScanEagle (services contracts)
* Insitu Integrator (RQ-21A)
* Northrop Grumman FireScout (RQ-8A)
* General Atomics Gray Eagle (MQ-1C)
* General Atomics Reaper (MQ-9A)
* Northrop Grumman Global Hawk Block 20-40 (RQ-4B)
* Northrop Grumman BAMS

Other nations have not stood still. Israel developed a broad range of UASs that spans most of the classes employed by the United States, from MAVs to large MALE systems. These systems are employed by the Israeli Defense Forces, but are also widely sold internationally. China has recently begun ambitious development of a bevy of new UASs across multiple system classes. The United Kingdom, Germany, France, Italy, and Sweden are developing a number of MALE and UCAV systems, often with international cooperation. The array of European UCAV and UCAV demonstrator programs is quite impressive: Dassault Filur, Saab Sharc, BAE Corax, BAE Taranis, Alenia Sky-X, and EADS Barracuda. The twin-engine BAE Mantis MALE rivals the General Atomics Reaper in payload capacity and performance. Other key UAS developments have occurred in Russia, South Africa, and South Korea.

Despite the remarkable advances in UAS technology, there are few civilian UASs in service today. The regulatory and airspace access issues are the primary culprits. Although UAS developers might be frustrated with regulators blocking their business models, most airline passengers would not want to encounter a stray UAS. The utility and market for civilian UASs is quite strong, but critical safety concerns must first be addressed. Solving UAS airspace access is a major initiative today.

1.6.7 The Future

It is always difficult to predict the future of unmanned systems. Revolutionary systems can rapidly emerge. What is clear is that the vast potential of these remarkable systems is not yet realized. Will we see a decline in UAS capabilities following the conflicts in Iraq and Afghanistan, just as we did after the Vietnam War? Will these systems experience advances in artificial intelligence and flight performance capabilities, relegating manned aircraft to few nice roles? Will commercial use of UASs overtake military operations, darkening the civil airspace with revolutionary systems? Will air freight and then airlines eliminate pilots in favor of autonomous systems? Will future air battles be fought exclusively by opposing unmanned fighters with supermaneuverability, intelligent swarming, and networked situational awareness? Will hydrogen-powered UASs fly for weeks without landing or aerial refueling? Will solar-powered aircraft fly for years?

It is possible that the United States might not retain its global leadership position in unmanned systems. Various European nations and China are developing remarkable UAS capabilities that rival existing U.S. capabilities. The rate of new programs in the United States has decreased because most major system categories are in the field and performing satisfactorily. Although the overall U.S. UAS budget is strong, much of this is for additional procurement of mature systems and operations support.

This is not without precedent. World militaries follow a rearmament cycle, where systems are developed sporadically and then operated until the next burst of development replaces the fielded systems. Generally, a new cycle is driven by external events. Consider the fighter analogy of World War II. The United States and Britain had completed a rearmament cycle in the mid-1930s while Germany and Japan developed more modern aircraft in the late 1930s. The U.S.-developed Brewster Buffalo, P-40 Warhawk, and Grumman Wildcats were outclassed by the Japanese Mitsubishi Zero at the commencement of hostilities. Similarly, the British Fairey Battle and early Hawker Hurricanes were decimated over mainland Europe by the then-modern Messerschmitt BF-109. Both the United States and Britain rapidly underwent a rearmament cycle early in the war and produced excellent fighters such as the P-51 Mustang, P-47 Thunderbolt, P-38 Lightning, Grumman Hellcat, Vought Corsair, and later variants of the Supermarine Spitfire. These newer fighters were generally superior to their adversaries and were built in larger quantities. Germany underwent a rearmament cycle towards the end of the war that included the revolutionary Messerschmitt 262 jet fighter, but too few were produced to affect the war's outcome. Israel began their UAS rearmament cycle in the 1970s, leapfrogging ahead of the rest of the world. The United States saw a remarkable UAS rearmament cycle from 1994–2011. It appears that many European nations and China are at the beginning stages of a cycle today. What new UAS systems are developed in the future will be governed by this phenomenon.

Perhaps the future will be brightest for civil and scientific applications of unmanned systems. The possibilities include climate science, telecommunications, air freight, forest fire protection, law enforcement, news coverage, agricultural missions, and disaster relief. Tentative demonstrations have revealed the potential of UASs for such uses, but airspace access is still a limiting factor. Practical new technologies must be developed to enable safe airspace integration. Lawmakers will need to carefully consider the appropriate limits of civil UAS usage to prevent Orwellian scenarios. UASs must also be made more affordable to economically compete with manned aircraft.

1.7 Introduction to System Elements and Architectures

A UAS is much more than the unmanned aircraft. Key elements include the unmanned aircraft, payload, communications systems, control stations, launch and recovery equipment, and support equipment. We will look at each of these elements later in this book.

The communication system is a peculiar element because it physically resides in multiple places. Nodes are present on the unmanned aircraft and on various ground elements. Communications can connect the system in numerous ways. A major discriminator among UASs is the ability to use satellite communications, effectively dividing unmanned aircraft into two categories. The unmanned aircraft might have the ability to fly autonomously without communications, but most applications require communications with one or more control stations and other payload data consumers. The ground stations will usually communicate to outside systems via networks to receive mission tasking and disseminate data.

The unmanned aircraft is the most visible segment of system, but the system is highly dependent on all segments. One rarely sees images of ground control stations or payloads when a system is described, even though these elements are just as important. With the full importance of the overall system, one would expect to see more desk models of GCSs or payloads.

Professionals who are responsible for system elements other than the unmanned aircraft have expressed the role of the unmanned aircraft in the context of the overall system in humorous but enlightening ways. Some notable descriptions of the unmanned aircraft include the following:

Dust cover [for other critical systems]	Communications and payloads perspective
Antenna farm	Communications perspective
Aperture	Payload perspective
Pickup truck	Payloads perspective
Egg carton	Payload perspective
Airborne node	Communications perspective
When we [payloads] need to go somewhere, you [unmanned aircraft] give us a ride.	Payloads perspective
It's the payloads, stupid	Payloads perspective
The camera	GCS mission payload operator perspective
It's not about the glory of flight.	Business perspective

References

[1] Dept. of Defense, *Unmanned Systems Roadmap 2007-2032*, Office of the Secretary of Defense, Washington, D.C., 10 Dec. 2007.

[2] McDaid, H., and Oliver, D., *Smart Weapons, Top Secret History of Remote Controlled Airborne Weapons*, Barnes and Noble, Inc., New York, 1997.

[3] Hurst, R., *Pilot Error, A Professional Study of Contributory Factors*, Granada Publishing Limited, London, 1976.

[4] O'Hare, D., and Roscoe, S., *Flightdeck Performance, The Human Factor*, Iowa State Univ. Press, Ames, IA, 1990.

[5] Roscoe, S. N., *Aviation Psychology*, Iowa State Univ. Press, Ames, IA, 1980.

[6] Singer, P. W., *Wired for War, The Robotics Revolution and Conflict in the 21st Century*, Penguin Press, New York, 2009.

[7] Wilson, D. H., *How to Survive a Robot Uprising, Tips on Defending Yourself Against the Coming Rebellion*, Bloomsbury Publishing Plc., U.K., 2005.

[8] Gordon, Y., *Soviet/Russian Unmanned Aerial Vehicles*, Midland Publishing, London, England, U.K., 2005.

[9] Yenne, B., *Attack of the Drones, A History of Unmanned Aerial Combat*, Zenith Press, St. Paul, MN, 2004.

[10] Ehrhard, T. P., *Air Force UAVs, The Secret History*, Mitchell Inst. Press, 2010.

[11] Mueller, T. J., Kellogg, J. C., Ifju, P. G., and Shkarayev, S. V., *Introduction to the Design of Fixed-Wing Micro Air Vehicles, Including Three Case Studies*, AIAA, Reston, VA, 2007.

[12] Werrell, K. P., *The Evolution of the Cruise Missile*, Air Univ. Press, Maxwell Air Force Base, AL, 1985.

Problems

1.1   Compare the Developmental Sciences SkyEye family (R4D and R4E) to the Insitu ScanEagle and Integrator. What similarities exist in technologies, roles, and product families? What technologies changed in the decades separating the systems?

1.2   Identify at least four non-U.S. MALE UASs currently in development or operational service.

1.3   Identify at least five non-U.S. tactical UASs with capabilities similar to the RQ-7B.

1.4   Trace the lineage of the Northrop Grumman BAMS back to 1960. Identify all of the unmanned aircraft that most influenced the succeeding aircraft.

Chapter 2

Unmanned Aircraft Categories

- Understand the most common UAS categories
- Understand applications for the system categories
- See example systems within each category

Fig. 2.1 An array of small, tactical, and target UASs. (U.S. Navy photo by Photographer's Mate 2nd Class Daniel J. McLain.)

2.1 Introduction

This chapter describes a small subset of the diverse unmanned aircraft categories to demonstrate the broad scope of the field, like those shown in Fig. 2.1. The category names are not exclusively tied to official designations, as these frequently change and vary among nations and branches of the armed services. Although the organization presented here is UA centric, there are broader implications for the overall system. These system categories will be referenced in upcoming chapters. System characteristics shown here are based on manufacturer's marketing materials and the media.

UAS design space is a continuum. Defining categories is challenging and yields imperfect groupings of systems. The attributes that distinguish categories from one another are weight, size, mission altitude, survivability, vehicle design approach (i.e., helicopter vs fixed wing), and mission (i.e., strike vs ISR).

So what is the best UAS category? The answer, as you might suspect, depends upon the application. A high-altitude long-endurance UAS might provide better imagery than a small UAS for unit level support, but the larger UA is too expensive and cumbersome for that use. A good rule of thumb is that the best UAS class is the smallest and most affordable that can satisfactorily perform the desired role.

2.2 Micro Air Vehicles

A micro UA (MAV) is an unmanned aircraft with a maximum physical length of 6 in., according to the definition in the Defense Advanced Research Projects Agency (DARPA) MAV program. However, manufacturers have claimed UA having wing spans up to 2.5 ft as falling within this category. New categories of progressively smaller unmanned aircraft have been dubbed "pico air vehicles" and "nano air vehicles." For the purposes of this discussion, MAVs will be considered to capture a range of the smallest unmanned aircraft. Mueller [1, 2] provides detailed description of MAV technologies and capabilities.

MAVs tend to have several common characteristics. The challenging low-Reynolds-number flight regime yields poor aerodynamic efficiency compared with larger, unmanned aircraft. Small size, weight, and power (SWAP) budgets necessitate miniature avionics, communications systems, and payloads with significant performance limitations. Finally, MAVs hold promise of low unit costs in production.

The small physical scale and low flight speeds result in fixed-wing MAV chord Reynolds numbers generally less than 10^5 and sometimes less than 5×10^4. Chapter 5 will detail the boundary-layer phenomena, but the typical lift-to-drag ratio of these UA is 3-7. Increasing the chord increases the Reynolds number to reduce the profile drag and also reduces the UA wing loading (weight per wing area)—both of which reduce the propulsion power. These favorable contributors are partially offset by the higher induced drag of the lower wing aspect ratio.

MAVs generally weigh less than 0.5 lb, which leaves little provision for avionics or mission systems. Perhaps 25 – 35% of the gross weight can be allocated to these components, and capability must be maximized within the stringent SWAP constraints. Although initial flight experiments might use radio control, operational systems require autopilots and mission systems. These systems might need to be tightly integrated into custom electronics, driving up the development costs. Low-RF radiated power generally limits the communications range to 1 – 15 km.

Chemical battery electric propulsion is the most prevalent type used. Rechargeable batteries are common for reusable systems. Small glow fuel reciprocating engines are occasionally used for quick demonstrations, but starting difficulties and this nontraditional fuel type limit their use in operational systems. Air-launched MAV gliders, such as the NRL Cicada, have been successfully demonstrated.

MAVs are not steady sensor platforms. Their low mass and slow flight speeds make them particularly sensitive to gusts and turbulence, even with well-designed flight controls. High-drift-rate microelectromechanical-systems (MEMS) inertial sensors exacerbate the problem. Payload mechanical stabilization is often not possible because of mass constraints and hysteresis. The resulting sensor quality can be inferior to small UASs unless sophisticated image processing is performed on the ground.

Many high-value MAV missions exist, despite the platform's inherent limitations. MAVs can provide short-range tactical imagery. Because of their maneuverability and small size, these unmanned aircraft might be capable of flying in urban canyons or even inside buildings. MAVs can make up for their small size by operating in swarms for area searches and other purposes. The low cost and mass production potential provides the opportunity to field multiple UAs, and quantity is a valuable quality.

Aircraft Profile: AeroVironment Wasp III

AeroVironment has been at the forefront of MAV development starting with the DARPA MAV program. The Wasp III (Fig. 2.2, Table 2.1) spans 28.5 in., which is larger than the original MAV definition. However, miniature avionics, payloads, and communication systems of MAV pedigree are utilized in this useful UA. The system uses a portable ground station.

Fig. 2.2   Aerovironment Wasp III MAV. (U.S. Navy photo by Mass
Communication Specialist 3rd Class Daisy Abonza.)

Table 2.1   AeroVironment Wasp III Characteristics

Characteristic	Value
Span	2.375 ft
Takeoff gross weight	0.95 lb
Maximum payload capacity	—
Endurance	0.75 hr
Maximum altitude	1000 ft above ground level
Maximum airspeed	38 kt
Launch method	Hand-held catapult and hand launch
Recovery method	Belly landing
Propulsion	Battery-electric
Communications	Line of sight

2.3 Small Unmanned Aircraft

Small unmanned aircraft systems (SUAS), as defined here, range from 1–55 lb gross weight. These are larger than MAVs and smaller than small tactical UASs. Sometimes this general category is also called *Mini UAS*. The USMC defines a Tier I, which is now filled by the AeroVironment Dragon Eye and AeroVironment RQ-11B Raven B. The boundaries of this designation might appear arbitrary, but it represents a host of technologies and mission capabilities. The 55-lb upper takeoff gross weight limit corresponds to the Academy of Model Aeronautics (AMA) upper weight limit for hobbyist aircraft, though commercial and military UAS operations fall outside the scope of AMA activities. The lowest weight systems are generally electric powered, 0.5–2 hr endurance aircraft that are launched by hand or with the aid of elastic lines. The middle- and upper-weight UA are medium- to long-endurance, low-altitude aircraft that are taking over the missions of older tactical UAS.

SUAS are the most prolific category in terms of the most distinct systems flown. The simplest explanation is that the cost and technical barriers to entry are low. Many systems referred to as unmanned aircraft never progress beyond radio control. Many autonomous small UASs are simply model aircraft with low-cost autopilots. Such systems are the focus of the do-it-yourself (DIY) drone movement for hobbyists and numerous university research projects. Despite the hundreds of known SUAS platforms, very few see operational use.

Systems miniaturization now make SUAS operationally useful, effectively performing missions relegated to much larger systems as late as the 1990s. Lightweight, GPS-based autopilot technologies originally demonstrated on MAVs offer SUAS the ability to carry high useful loads. Very small gimbaled camera payloads weighing 1–5 lb were developed specifically for this class of UA using new design approaches. A new generation of line-of-sight communications systems offers high-bandwidth analog or digital video at ranges of 10–100 km, while weighing less than 5 lb. These and other technologies enable operationally useful SUAS systems. The Aurora Flight Sciences Skate (Fig 2.3) is an example of a SUAS.

Aircraft Profile: AeroVironment RQ-11B Raven

AeroVironment's Raven (Fig 2.4, Table 2.2) family can trace its lineage to the larger Pointer UAS that was developed in the 1980s and used in the first Gulf War. The Pointer had a span of 9 ft, weighed 10 lb, and used NiCd batteries. It could be hand launched, though the large size made this task challenging for some.

Fig. 2.3 Aurora Flight Sciences Skate SUAS. (Photo courtesy of Aurora Flight Sciences.)

Raven uses lithium batteries with much better specific energy, complemented by a higher-efficiency brushless motor. The newer propulsion technologies, combined with miniature GPS-based autopilot and small cameras,

Fig. 2.4 AeroVironment Raven RQ-11B hand launch. (Photo courtesy of Department of Defense, photographer unknown.)

Table 2.2 AeroVironment Raven B Characteristics

Characteristic	Value
Span	4.5 ft
Takeoff gross weight	4.2 lb
Maximum payload capacity	0.36 lb
Endurance	1–1.5 hr (with rechargeable batteries)
Maximum altitude	14,000 ft (maximum launch altitude)
Maximum airspeed	50 kt
Launch method	Hand launch
Recovery method	Deep-stall belly landing
Propulsion	Battery-electric
Communications	Line of sight

enabled the Raven to achieve a much smaller size than its predecessor. The Raven system has been widely employed with the U.S. Army.

2.4 Small Tactical Unmanned Aircraft Systems

Small tactical unmanned aircraft systems (STUASs) can be considered UASs that are larger than SUASs and smaller than tactical UASs. Weights typically range from 55–200 lb. The United States Marine Corps (USMC) presently has a Tier II category that includes STUASs, as well as SUASs like the Insitu ScanEagle. Note that in the 1990s Tier II represented medium-altitude long-endurance UASs like the General Atomics Predator A, though that definition is no longer in use.

The joint U.S. Navy and Marine Corps STUAS program highlighted the high expectations placed on these systems as well as the robust competitive landscape. The STUAS competition was won by the Insitu Integrator (now the RQ-21A). The contenders weighed between 55–200 lb. The author created the initial conceptual design of the Boeing-Insitu Integrator and led its development during his tenure as Insitu's Vice President of Advanced Development. Other impressive competitors included the Raytheon Killer Bee, an enhanced variant of the AAI Aerosonde, and the UAS Dynamics Storm.

Aircraft Profile: Insitu Integrator (RQ-21A)

The Integrator system (Fig 2.5, Table 2.3) was developed as a larger stable mate to the venerable ScanEagle SUAS. Both systems use a pneumatic launcher and the SkyHook^{TM} vertical cable recovery system to achieve runway independence, as described in Chapter 11. Integrator derives its

Fig. 2.5 Insitu Integrator small tactical UAS.

name from the goal of easily integrating new payloads. This is achieved by modular nose and center-of-gravity payload bays and two wing hardpoints. The UA comes apart into several modules for compact storage and transportability. The Integrator's 24-hr endurance is made possible by lightweight composite structures, high-aspect-ratio wings, aerodynamic refinement, light wing loading, and propulsion system efficiency features. Integrator has over three times longer endurance than larger operational tactical UAS.

2.5 Tactical Unmanned Aircraft Systems

The U.S. Department of Defense (DOD) [3] defines the tactical class with takeoff gross weight between 55 and 1320 lb. However, we will consider tactical unmanned aircraft systems (TUASs) as covering systems weighing

Table 2.3 Insitu Integrator Characteristics

Characteristic	Value
Span	16 ft
Takeoff gross weight	135 lb
Maximum payload capacity	50 lb
Endurance	24 hrs
Maximum altitude	20,000 ft
Maximum airspeed	90 kt
Launch method	Pneumatic launcher
Recovery method	SkyHook™ vertical line recovery system
Propulsion	8-hp reciprocating engine
Communications	Line of sight

between 200–1000 lb, given the distinct STUAS class. The USMC defines a Tier III class, which is now filled by the AAI RQ-11B Shadow. The U.S. Army has a Tier III class, which is also filled by Shadow. The TUAS category has been a major focus of system development and operational employment since the 1970s. The flight performance of most systems in this category has generally held constant at 5–12 hrs at altitudes less than 20,000 ft despite the impressive progression of enabling technology state of the art across multiple disciplines. The experience with these systems in the 1980s and 1990s varies from numerous U.S. acquisition failures to successful widespread operational employment by Israel in the same period. The United States fielded the Israeli-developed Pioneer UAS starting in 1988. The U.S. Army and Marine Corps now operate the AAI Shadow 200 (RQ-11B), which has performed well in the field.

Aircraft Profile: AAI Shadow 200 (RQ-7B)

The AAI Shadow 200 (Fig 2.6, Table 2.4) won the Army's TUAV competition in 2000. Supported by years of system development leading to the contract award, the system was quickly fielded and put into operational service. The ground station is housed within a vehicle-mounted shelter, which includes UA operator (AVO) and mission payload operator (MPO)

Fig. 2.6 AAI Shadow 200 Tactical UAS. (U.S. Army photo by Spc. Kimberly K. Menzies.)

Table 2.4 AAI Shadow 200 Characteristics

Characteristic	Value
Span	14 ft
Takeoff gross weight	375 lb
Maximum payload capacity	55.7 lb
Endurance	6–7 hrs
Maximum altitude	19,000 ft
Maximum airspeed	123 kt
Launch method	Conventional runway or pneumatic rail launcher
Recovery method	Conventional runway with arresting cables or a net
Propulsion type	Rotary engine using gasoline
Communications	Line of sight

workstations. The ground stations were upgraded to utilize the Army One System, which is compliant with the STANAG 4586 interoperability standard. The primary payload is an IAI POP 300 EO/IR ball.

2.6 Medium-Altitude Long Endurance

Medium-altitude long-endurance (MALE) UASs generally have substantially greater mission capabilities than TUAS and other smaller systems. Gross weights may vary from 1,000–10,000 lb, with payload capacities of 200–1,000 lb, and an endurance range of 12–40 hrs. This category of aircraft closely matches the DOD [3] definition of Theater UAS, which has a takeoff gross weight greater than 1,320 lb. In the 1990s the DOD defined Tier II as MALE. Today the U.S. Army puts MALE UASs in a Tier III category. The operational altitude varies by propulsion type with 15,000–30,000 ft service ceiling for reciprocating engines and 30,000–50,000 ft with turboprops. The line between high-altitude long-endurance and MALE UASs is blurred for heavy turboprop systems such as the MQ-9 Reaper or the BAE Mantis. Many MALE systems are capable of carrying high-bandwidth satellite communications systems and external weapons loads. Using external payloads tends to increase drag significantly, effectively turning long-endurance platforms to medium-endurance platforms. MALE UAS characteristics and operations are described by Yenne [4] and Martin [5].

Today's MALE systems typically perform ISR, SIGINT, and strike missions, though they are capable of expanded roles. Relative to smaller UASs, there is greater operational utility and payload flexibility. New payload types generally start off large and are then miniaturized over time. Therefore, MALE systems can be the first adopters of new mission capabilities.

The large size gives a larger footprint. Hangars are usually required to shelter these UA and enable maintenance activities. Support equipment is needed to assemble and disassemble the airframe. Most MALE systems must be transported in large containers that are manipulated by heavy ground equipment. These systems usually employ a conventional runway launch and recovery technique. So far, the large launch and capture energy for these large systems has precluded runway independent techniques for non-vertical-takeoff-and-landing (VTOL) systems.

There have been relatively few MALE programs compared to smaller UAS categories. The development investment is high, and there are more technical barriers to entry. Despite these hurdles, many of today's operational MALE systems were developed largely on company funding. MALE system development can take several years.

Aircraft Profile: General Atomics MQ-1C Gray Eagle

The Army held the Extended Range/Multi-Purpose (ER/MP) competition for a MALE UAS. The two final contenders were the General Atomics Warrior and the Northrop Grumman Hunter II. The Northrop Grumman Hunter II was a modified version of the Israeli IAI Heron.

The Warrior, later renamed the MQ-1C Gray Eagle (Fig 2.7, Table 2.5), is a major upgrade to the Predator A platform. A 135-hp heavy fuel engine

Fig. 2.7 General Atomics MQ-1C Gray Eagle. (U.S. Army photo by Spc. Roland Hale.)

Table 2.5 General Atomics MQ-1C Gray Eagle Characteristics

Characteristic	Value
Span	56 ft
Takeoff gross weight	3200 lb
Maximum payload capacity	575 lb internal, 500 lb external
Endurance	30+ hrs
Maximum altitude	29,000 ft
Maximum airspeed	135 kt
Launch method	Conventional runway
Recovery method	Conventional runway
Propulsion	135-hp general-aviation reciprocating engine using heavy fuel
Communications	High-bandwidth SATCOM and line of sight

was added, which enabled an increase in takeoff gross weight and use of logistics fuels. The onboard systems including avionics and communications were extensively upgraded for this program of record. Early versions of the system are in the field for initial operations. The Gray Eagle doubles the Hellfire load of its predecessor to four missiles.

2.7 High-Altitude Long Endurance

High-altitude long-endurance (HALE) UASs are sophisticated, high-performance systems. HALE systems typically fly above 50,000 or 60,000 ft with a total endurance greater than 24 hrs. The system weight is above 5,000 lb in most cases. The Defense Airborne Reconnaissance Office (DARO) defined a Tier II+ class in the mid-1990s that was filled by the Northrop Grumman RQ-4A Global Hawk. Today Global Hawk is the sole HALE system in the U.S. arsenal if we categorize the General Atomics Reaper as a MALE.

Aircraft Profile: Northrop Grumman Global Hawk RQ-4

No discussion of HALE would be complete without the Northrop Grumman Global Hawk (Fig 2.8, Table 2.6). This UA was developed by Teledyne Ryan starting with a 1994 DARPA ACTD contract, building upon its vast heritage of high-performance jet-powered UASs such as the Firebees and Compass Cope-R. Teledyne Ryan was later acquired by Northrop Grumman. The first Global Hawk flight occurred in 1998. The ACTD systems were essentially prototypes intended to prove the system

Fig. 2.8 Northrop Grumman RQ-4B Global Hawk. (U.S. Air Force photo by Jim Shryne.)

capabilities, though the 9/11 attacks spurred their rapid deployment in support of Operation Enduring Freedom and Operation Iraqi Freedom. The systems became strategic assets, proving their worth with broad area coverage and target location capabilities. Global Hawk generated 55% of

Table 2.6 Northrop Grumman Global Hawk (RQ-4B) Characteristics

Characteristic	Value
Span	130.9 ft
Takeoff gross weight	32,250 lb
Maximum payload capacity	3,000 lb
Endurance	33 hrs
Maximum altitude	65,000 ft
Average airspeed	310 kt at 60,000 ft
Launch method	Conventional runway
Recovery method	Conventional runway
Propulsion	Rolls-Royce AE3007H turbofan engine
Communications	High-bandwidth SATCOM and line of sight

the time-sensitive targeting used to attack the Iraqi air defenses. Drezner and Leonard [6] detail the development of the Global Hawk ACTD system.

The Global Hawk underwent a more robust development in what became the RQ-4B. The wing span increased from 116.2 to 130.9 ft, and the payload capacity grew, among many changes. The RQ-4B was more operationally supportable than its ACTD predecessor. A variant of the Global Hawk won the Navy's Broad Area Maritime Surveillance (BAMS) program, which will provide shore-based ISR missions. Numerous nations have expressed interest in procuring the Global Hawk.

2.8 Ultra Long Endurance

Ultra long endurance (ULE) is a new system category that is enabled by a convergence of propulsion, structures, and aerodynamics technologies. The endurance is greater than 5 days, and the altitude is 25,000 ft or higher, covering the altitude domains of MALE and HALE systems. Payloads are typically greater than 500 lb. No ULE systems are operational today, though several are in development including the Aurora Flight Sciences Orion, Boeing Phantom Eye, and AeroVironment Global Observer. The Orion uses heavy fuel whereas the Phantom Eye and Global Observer use hydrogen fuel.

The primary discriminators among the systems are the propulsion type. If we exclude solar-powered aircraft (covered separately), these designs use heavy fuel or hydrogen as energy sources. Heavy fuel designs use high-efficiency reciprocating engines. Hydrogen-fueled UA can use fuel cells or reciprocating engines to generate power.

Aircraft Profile: Aurora Flight Sciences Orion

Aurora Flight Sciences adapted the high-altitude, hydrogen-powered Orion design to operate at medium altitude with conventional heavy fuel (Fig 2.9, Table 2.7). A combination of high-efficiency reciprocating engines, light wing loading, large fuel mass fraction, and high aerodynamic efficiency permit this conventional aircraft to attain persistence many times higher than other MALE systems. The system was rolled out in 2010 in preparation for flight tests.

2.9 Uninhabited Combat Aerial Vehicles

The DOD defines a class of Combat UAS [3] that is designed from inception as a strike platform with internal bomb bays or external weapon pylons, a high level of survivability, and a takeoff gross weight of greater than

Fig. 2.9 Aurora Flight Sciences Orion Ultra Long Endurance UAS. (Photo courtesy of Aurora Flight Sciences.)

1,320 lb. This class of UA has been called uninhabited combat aerial vehicle (UCAV) and later an uninhabited combat aircraft system (UCAS). In this book we call these systems UCAVs. The primary discriminators between UCAVs and weaponized MALE UASs are increased speed

Table 2.7 Aurora Flight Sciences Orion Characteristics

Characteristic	Value
Span	132 ft
Takeoff gross weight	—
Maximum payload capacity	2500 lb
Endurance	120 hrs (5 days)
Maximum altitude	20,000 ft
Maximum airspeed	—
Launch method	Conventional runway
Recovery method	Conventional runway
Propulsion type	Twin Austro AE300 168-hp general-aviation heavy fuel reciprocating engines
Communications	High-bandwidth SATCOM and LOS

enabled by jet engines and improved survivability through low observable design features.

United States' examples include the Boeing X-45A and X-45C (Phantom Ray), Northrop Grumman X-47A and X-47B, and General Atomics Avenger. The United Kingdom has flown the BAE Corax and is developing the BAE Taranis. Sweden developed the SAAB Sharc. Germany developed the EADS Barracuda demonstrator.

Aircraft Profile: Boeing Phantom Ray

Boeing and Northrop Grumman both participated in the DARPA J-UCAS program that sought to develop UCAV systems for both the Navy and Air Force. After the program was cancelled, the Navy funded the Northrop Grumman X-47B demonstrator for carrier UCAV demonstrations. Boeing continued to develop the X-45C (Fig 2.10, Table 2.8) under company funding, renaming it the Phantom Ray. The Phantom Ray system is nearing flight-test status at the time of publication.

2.10 Manned Aircraft Conversions

From the earliest days of unmanned flight to today, manned aircraft are frequently converted to unmanned aircraft. Although the motivations vary,

Fig. 2.10   Boeing Phantom Ray. (U.S. Navy photo by Photographer's Mate 2nd Class Daniel J. McLain.)

Table 2.8 Boeing Phantom Ray Characteristics

Characteristic	Value
Span	50 ft
Takeoff gross weight	36,500 lb
Maximum payload capacity	—
Range	1,500 miles
Maximum altitude	40,000 ft
Maximum airspeed	Mach 0.85
Launch method	Conventional runway
Recovery method	Conventional runway
Propulsion	One General Electric F404-GE-102D jet engine
Communications	Line of sight, SATCOM unknown

cost savings and short time to operations relative to custom unmanned aircraft development are considerations. A stringent regulatory environment for unmanned flight in civilian airspace now makes optionally piloted aircraft (OPA) based on certified manned aircraft attractive. Most mission needs can find a suitable airframe because manned aircraft range from ultralights to strategic bombers. Platform categories include not just fixed-wing aircraft, but also helicopters, autogyros, lighter than air, and parafoils, to name a few. Candidate aircraft might be in production, out of production, or available as kits.

The roles performed by manned aircraft conversions span a wide range of the robotic mission spectrum. The most prevalent use is target drone missions. Other notable roles include radiation sampling, ground-launched cruise missiles, cargo delivery, and ISR.

The economics are often favorable compared with dedicated unmanned aircraft. Leveraging a mature manned aircraft eliminates most development and testing costs associated with the airframe, propulsion, and many subsystems. New engineering documentation and manuals' efforts can be greatly reduced for unchanged features. Finally, the operations and maintenance costs can be lowered if the source aircraft is certified by leveraging existing support infrastructure and training resources.

Organizations with a mature avionics/software base suitable for manned aircraft conversion and appropriate hardware-in-the-loop simulation environments can realize very rapid time to flight. The primary difference between conversion programs relates to the aircraft model and development of the flight control laws. Much of the hardware can remain unchanged between platforms, much like humans can fit in any aircraft.

Development can be accelerated by including onboard pilots during initial flight tests. This can also be true for dedicated unmanned aircraft capable of carrying a pilot, as was shown on the Teledyne Ryan Model 410 and the Scaled Composites Raptor. The pilot can switch the autopilot on and off, acting as an onboard safety pilot. A flight software failure that would result in a crash or activation of the flight termination system for a dedicated UAS can be remedied by a switch to safety pilot control for the equivalent OPA. Landing and, to a lesser extent, takeoff are far more challenging for autopilots than other normal flight modes. Pilots can handle these challenging modes until the control laws are validated in other flight modes.

Piloted aircraft can greatly accelerate mission systems development. Mission systems such as payloads, payload data communications links, and mission management systems on the air and ground frequently compete with the flight controls and UA management systems for development critical path. Use of piloted aircraft fitted with developmental hardware—either as the final product airframe or as a dissimilar surrogate platform—allows parallel development outside of the integration laboratory. Otherwise, a serial approach is generally adopted where the flight critical systems are first validated on the unmanned aircraft and are then followed by mission systems.

OPAs, as the name suggests, are aircraft that are capable of flight with or without an onboard pilot. Such aircraft must include both pilot interfaces and full, unmanned flight control systems. A major advantage of the OPA is the ability to fly in controlled or otherwise congested airspace as a manned platform, thus reducing regulatory factors and risk to other air traffic. Additionally, these UA can transit long distances without dependence on ground control stations or beyond line-of-sight communications links. OPAs pay an additional systems size, weight, power, and complexity cost over pure unmanned conversions as they must retain the human interface, which typically weighs 150–400 lb for general-aviation aircraft. This penalty is relatively small for larger aircraft. Two notable examples of recent OPAs are the Boeing-Gulfstream G-550 business jet (Fig. 2.11) OPA entry into the Navy's Broad Area Maritime Surveillance (BAMS) competition and the Aurora Flight Sciences Centaur OPA, which is a converted Diamond Aircraft DA-42.

Manned aircraft major airframe components or system elements can be partially applied to otherwise custom unmanned aircraft. It is rare for new, unmanned aircraft within the weight class of manned aircraft to not use manned aircraft elements. For example, unmanned aircraft above 1,000 lb gross weight generally use manned aircraft propulsion systems and subsystems such as landing gear. Less frequently, major structural components such as wings or fuselage components are taken directly from manned aircraft, as was seen on the 1970s Compass Dwell program where two contractors converted Schweitzer aircraft. Large unmanned rotorcraft usually reuse

Fig. 2.11 This Gulfsteam G550 was a BAMS program contender as an optionally piloted vehicle.

power systems and rotors from manned helicopters, like the FireScout use of the Schweitzer 333 helicopter.

Thanks to the bounded range of human physical characteristics used to design cockpits, the process can be remarkably similar across diverse UA classes. The most direct means of conversion is for the autopilot to mechanically drive human control interfaces (yoke or column, rudders, and throttle) via actuators. Another approach is for the actuators to be coupled with the control cables as is common for general-aviation autopilots. Finally, the flight control surfaces and other effectors such as engine controls can be directly driven by autopilot-commanded actuators, yielding the greatest control effectiveness.

Despite the speed of development and affordability benefits of unmanned conversions, there are often flight-performance penalties relative to dedicated unmanned aircraft. Manned aircraft configurations are largely shaped by accommodation of humans. Human-carrying cabins are generally less dense than is required for avionics and mission systems, leading to larger, heavier fuselages that generate more drag. Canopies or windows are heavier than structural skins. Large doors create inefficient structural load paths. Certification requirements such as FAR Part 23 or 25 might be more stringent than is necessary for many unmanned applications, leading to more complex and heavy airframes than a custom unmanned aircraft. These and other considerations can result in less flight performance, payload capacity, and payload effectiveness utility. However, careful selection of the manned platform or nonstressing system requirements might make the utility penalties insignificant.

The elimination of the pilot and any other human-related systems can be added to the useful load, augmenting the payload capacity or fuel load. The gross weight can be increased beyond the manned limit when operating in the unmanned mode if permitted. The total useful load is then limited by the airframe's payload and fuel adaptability, structural load limits, and performance constraints. The payload can be limited by structural attachment weight provisions and center-of-gravity travel constraints. The fuel load can be limited by the fuel tank capacity or the ability to add auxiliary fuel tanks. The UA can be flown at a higher gross weight with reduced structural factor of safety (i.e., reduction from 1.5 to 1.25), or further limiting the flight envelope. Performance constraints such as field performance, climb gradients, service ceiling, or controllability can also limit the acceptable gross weight.

Aircraft Profile: Aurora Flight Sciences Centaur

Aurora Flight Sciences developed an OPA conversion of the Diamond DA42 Multi Purpose Platform (MPP) (Fig 2.12, Table 2.9). The endurance and payload capacity are significantly enhanced when the pilot is not onboard, providing MALE UAS capabilities. Unlike a dedicated UAS, this OPA can be ferried to mission location with a pilot onboard. The ground control station and ground-based communications equipment can fit within the aircraft for self-deployment of the complete system. By utilizing a certified manned aircraft, the development cost and operations cost are minimized. The MPP platform is designed with multiple payload mounting options, facilitating mission systems integration.

Fig. 2.12 Aurora Flight Sciences Centaur OPA. (Photo courtesy of Aurora Flight Sciences.)

Table 2.9   Aurora Flight Sciences Centaur Characteristics

Characteristic	Value
Span	44 ft
Takeoff gross weight	—
Maximum payload capacity	800 lb
Endurance	24 hrs
Maximum altitude	27,500 ft
Maximum airspeed	175 kt
Launch method	Conventional runway
Recovery method	Conventional runway
Propulsion	Twin general-aviation heavy fuel reciprocating engines
Communications	Line-of-sight and low-bandwidth SATCOM

2.11   Air-Launched Unmanned Aircraft

Air-launched UASs include all unmanned aircraft that are released from a host platform. The payload capacity, flight envelope, and propulsion types vary greatly. Small electric and gas-powered air-launched systems weighing 1–20 lb have become prevalent, though few are fielded today. The Teledyne Ryan Firebee/Firefly jet-powered reconnaissance UASs were widely employed during the Vietnam War. The pinnacle of all air-launched UAS performance is the Lockheed's D-21 ramjet-powered supersonic UAS.

Aircraft Profile: L-3 Cutlass

The L-3 Cutlass (Fig 2.13, Table 2.10) system is air launched from a 120–150 mm tube dispenser system. The wings are stowed along the top of the fuselage and pivot for deployment. The tails pivot out of the fuselage. The electric propulsion system utilizes folding propeller blades.

2.12   Targets

Aerial targets come in three primary forms: low-speed surface gunnery targets, high-speed targets, and manned aircraft conversions. The low-speed targets have been developed in the highest quantities, with the demand spiking in World War II to train anti-aircraft gunners on land and at sea. As the threats that surface-based air defenses faced evolved from propeller aircraft to jet-powered aircraft and high-speed cruise

Fig. 2.13   L-3 Cutlass air-launched UAS.

missiles, the targets also changed to faster systems as well. Fighter aircraft also use high-speed targets to simulate threat aircraft for missile testing and training. Jet-powered targets can match the flight envelope of the threats that they simulate. Finally, unmanned conversions of manned aircraft are frequently used in this role.

Converted manned aircraft are well-suited as targets because they closely represent enemy aircraft. The subject aircraft are generally obsolete types or have exceeded their useful life. High-performance types such as fighters and attack aircraft are most appropriate. Widely fielded manned aircraft have the potential to yield significant quantities of targets, thus justifying the development investment. Example U.S. target drone conversions

Table 2.10   L-3 Cutlass Characteristics

Characteristic	Value
Span	4.6 ft
Takeoff gross weight	12–15 lb
Maximum payload capacity	3 lb
Endurance	1 hr
Maximum altitude	—
Maximum airspeed	85 KEAS
Launch method	Air launch or surface launch
Recovery method	Not provided
Propulsion	Battery electric
Communications	Line of sight

include the QF-86, QF-100, QF-4, and, most recently, the QF-16 program. The *Q* preceding the *F* is a letter designation for unmanned aircraft. These target drone conversions occur in most developed nations, including the United Kingdom, Australia, Russia, and China. More information on the target drone mission systems is found in Chapter 14.

Aircraft Profile: CEi BQM-167

The CEi BQM-167A (Fig 2.14, Table 2.11) was selected by the U.S. Air Force (USAF) as the Air Force Subscale Aerial Target (AFSAT) in 2002. Unlike the legacy system that it replaces, the BQM-167A is extensively constructed of carbon-fiber composites.

2.13 Rotorcraft

Rotorcraft, also called helicopters, span the size range from micro air vehicles to classes that overlap manned helicopters. Their roles include ISR, antisubmarine warfare, strike, cargo delivery, and sensor emplacement. Important attributes of rotorcraft include the ability to take off and land vertically and hover. The technology is well-established and can be considered the conventional approach for VTOL. Frequently airframes can

Fig. 2.14 CEi BQM-167 during rocket launch. (U.S. Air Force Photo by Bruce Hoffman, CIV.)

Table 2.11 CEi BQM-167 Characteristics

Characteristic	Value
Span	11 ft
Takeoff gross weight	—
Maximum payload capacity	350 lb internal, >250 lb on each wing
Endurance	3+ hr
Maximum altitude	50,000 ft
Maximum airspeed	Mach 0.91
Launch method	Rocket launch
Recovery method	Parachute recovery
Propulsion	Turbojet engine
Communications	Line-of-sight and low-bandwidth SATCOM

be selected from radio-controlled model helicopters or manned helicopters. Viable propulsion types include battery-electric, reciprocating engines, and turboshaft engines. A helicopter's open rotors can pose safety hazards that often take this class of platform out of consideration in favor of other forms of VTOL or runway-independent fixed-wing UAs. Another limitation faced by rotorcraft is a short-endurance potential relative to fixed-wing alternatives. That being said, helicopters tend to enjoy an endurance advantage over other forms of VTOL. Helicopters are limited in top speed by the retreating blades.

Aircraft Profile: Northrop Grumman Firescout (MQ-8B)

The Firescout system (Fig 2.15, Table 2.12) won the Navy's Vertical Takeoff and Landing Tactical Unmanned Aerial Vehicle (VTUAV) competition in 2000, and flight testing began in 2002. FireScout performed autonomous shipboard landings starting in 2006. The UAS is based on a Schweizer Model 333 manned helicopter. The VTUAV program experienced a number of budgetary and other challenges that slowed system fielding. The Firescout was selected as the Class IV UAS for the Army's Future Combat System before the overall program was cancelled. At the time of publication, the Firescout is beginning initial ship-based deployments.

2.14 Other Vertical Takeoff and Landing

The realm of VTOL is by no means limited to helicopters. A short list of methods used to generate lift at low speeds includes jet-engine thrust vectoring, tilted propellers, flapped wing-propeller interactions, shrouded lift fans,

Fig. 2.15   Northrop Grumman Firescout rotorcraft UAS. (U.S. Navy photo by Kurt Lengfield.)

and rockets. Often a VTOL platform will incorporate multiple techniques. The lifting thrust mechanism can be the same one used for forward flight thrust, eliminating the need for separate dedicated propulsion systems.

A VTOL platform must generate a thrust-to-weight ratio of more than one at the takeoff condition in order to leave the ground. If the propulsion

Table 2.12   Northrop Grumman RQ-8B Firescout Characteristics

Characteristic	Value
Rotor diameter	27.5 ft
Takeoff gross weight	3,150 lb
Maximum payload capacity	500+ lb
Endurance	8+ hr
Maximum altitude	20,000 ft
Maximum airspeed	125 kt
Launch method	Vertical takeoff
Recovery method	Vertical landing
Propulsion	One Rolls-Royce 250-C20W turboshaft engine
Communications	Line of sight

system is the same for forward flight and vertical flight, this high thrust can be used to obtain high flight speeds. The Aurora Flight Sciences Excalibur uses a pivoting jet engine to augment electric lift fans in hover and then for high-speed forward flight. Ducted fans can improve thrust at hover by utilizing a diffuser, but the fan exit must change geometry to a converging nozzle in order to sustain high thrust at high speeds.

Weight is especially critical for VTOL platforms. The high thrust dictates a high engine weight within a given propulsion class. The heavy engine reduces the useful load that is composed of the payload and fuel weight. Once in forward flight, the thrust requirements diminish, and the large propulsion system often operates at lower efficiency that corresponds to a high fuel flow rate. High fuel consumption and low fuel capacity reduce flight endurance relative to a fixed-wing platform. If we view UASs as systems rather than platforms, there is a trade between UA capability and ground footprint. VTOL platforms, including helicopters, can be thought of as carrying their launch and recovery equipment on the platform rather than placing it on the ground. This VTOL penalty is unavoidable, though many inventors and designers have worked on ways to conquer this enduring challenge.

Aircraft Profile: Aurora Flight Sciences GoldenEye 80

The GoldenEye 80 (Fig 2.16, Table 2.13) was first developed under the DARPA Organic Air Vehicle II (OAV II) program. It is part of a family of GoldenEye VTOL systems scaled to meet different missions. GoldenEye 80 carries a high-resolution mission equipment package (MEP) EO/IR ball as well as other payloads. The system was developed to minimize acoustic detection by the targets. Wings can be added to the GoldenEye UA for extended duration.

2.15 Solar-Powered Aircraft

Solar-powered aircraft can be considered a distinct class because of the unique vehicle attributes relative to non-solar-powered UA. The ultimate objective of solar flight is to provide continuous flight through multiple day–night (diurnal) cycles. So far, eternal flight has remained elusive but tantalizingly close with emerging technologies. Indeed, the aerospace industry has been a few years away from this goal for almost three decades.

Solar-powered aircraft have numerous technical challenges to achieve long-duration flight at high latitudes and in strong winds. These aircraft tend to have a low wing loading to reduce the power required to fly while increasing the wing-mounted solar-array collection area. Solar-cell

Fig. 2.16 Aurora Flight Sciences GoldenEye 80 VTOL UAS. (Photo courtesy of Aurora Flight Sciences.)

Table 2.13 Aurora Flight Sciences GoldenEye 80 Characteristics

Characteristic	Value
Height	5.4 ft
Takeoff gross weight	230 lb
Maximum payload capacity	25 lb
Endurance	3 hrs
Maximum altitude	—
Maximum airspeed	80 kt
Launch method	Vertical takeoff
Recovery method	Vertical landing
Propulsion	Rotary engine
Communications	Line of sight

efficiency, energy-storage roundtrip efficiency, and energy-storage specific energy are three propulsion drivers. Lightweight, gossamer structures are required, using advanced composite materials and thin skin material. If the UA can be built to support extreme endurance, then system reliability is the ultimate endurance limiter.

Solar-powered UASs are as old as solar-powered flight. The Astro Flight Sunrise I and Sunrise II were the first practical demonstrators in the late 1970s. AeroVironment built a series of increasingly capable solar-powered aircraft from the 1980s through the early 2000s including the Pathfinder, Centurion, and Helios. QinetiQ developed the Zephyr platforms, which are smaller solar-powered UASs intended for battlefield communications relay and other missions. The most recent sizable solar-powered UAS effort is the DARPA Vulture program. Boeing, Aurora Flight Sciences (Fig. 2.17), and Lockheed Martin developed system designs in Phase I, and Boeing was selected to develop the Phase II system.

2.16 Planetary Aircraft

Unmanned aircraft can be designed to fly within the atmospheres of other planets or moons. Most emphasis is placed on Mars, Venus, and Saturn's moon Titan (Fig. 2.18). The gas giants—Jupiter, Saturn, Neptune, and Uranus—have dense atmospheres that could also support some form of flight. These vehicles are often called *flyers* instead of aircraft because the very name aircraft implies that the vehicle operates in air, which only Earth possesses in our solar system. The flyer designs are as different as the atmospheres that they fly within. One thing that all planetary flyers have in common is that none use traditional airbreathing propulsion systems.

Fig. 2.17 Odysseus Z-Wing configuration maximizes solar collection during the day and aerodynamic efficiency at night. (Photos courtesy of Aurora Flight Sciences.)

Fig. 2.18 NRL/Aurora Flight Sciences Titan Flyer. (Photo courtesy of Aurora Flight Sciences.)

2.17 Lighter Than Air

Lighter-than-air (LTA) vehicles and semibuoyant vehicles use buoyant lift. A volume of lifting gases with lower density than the surrounding air generates a lifting force. The most common lifting gas is helium, followed by hydrogen. Helium is an inert gas, making handling safer than combustible hydrogen. However, helium is twice as dense as hydrogen (H_2) making it less potent for lift generation. Helium is also a nonrenewable gas whose supply will ultimately be depleted. The advantage of LTA over fixed-wing and powered-lift vehicles is that no power is required for lift generation. The disadvantage is that the large buoyant gas volume produces large drag that must be overcome for forward flight or stationkeeping in winds. Major types of buoyant vehicles include the following:

- *Tethered aerostats*—These tethered platforms are not UASs by most definitions. They are not powered or guided, though they generally carry payloads and communications systems. The purpose of the tether is to permit stationkeeping by overcoming drag by the line tension forward component. For long-duration missions over a fixed area, aerostats might be a more affordable option than UAS for equivalent coverage.
- *Balloons*—The LTA vehicles are also not UASs because they are not guided in a traditional sense. It is possible for a balloon to adjust its

buoyancy by release of ballast, thereby controlling altitude. Balloons are mostly at the mercy of the winds, with little or no flight-path control.

* *Blimps*—These LTA vehicles have a pressurized envelope that forms the body shape. Blimps are controlled, satisfying the definition of UAS. Propulsion systems give these vehicles the ability to penetrate winds and navigate.
* *Airships*—Airships have a rigid hull with internal gas envelopes. This arrangement is rarely used for UAS applications.
* *Hybrid airships*—These semibuoyant vehicles generate a portion of the lift from buoyant forces and the remainder from aerodynamic forces. The motivations for this configuration over blimp or airship are 1) the hull can be made smaller due to the reduced gas volume and 2) the heavier-than-air arrangement enables conventional takeoff and landing on runways, greatly simplifying ground handling operations.

The propulsion types vary depending on the mission and operational environment. High-altitude airships generally use solar propulsion, as the large envelope has ample solar collection area. Low-altitude airships frequently use reciprocating engines that burn hydrocarbon fuels.

2.18 Research Unmanned Aircraft

Academic UAS projects often fall within the small and small tactical categories. The relatively small size limitation is dictated by short program durations and limited budgets for development and operations. The objectives of the programs are student competitions, payload demonstrations, or flight research. A student competition or capstone design project can take place over one to two semesters. Graduate research typically lasts one to four years.

Fig. 2.19 Virginia Tech's 1996 entry into the AIAA Student Design, Build, Fly Competition.

Fig. 2.20 Aurora Flight Sciences Perseus B high-altitude research UAS. (NASA Photo by Jim Ross.)

Two significant student UAS competitions include the AIAA Student Design, Build, Fly (DBF) (Fig. 2.19) and the SAE Heavy Lift competitions. This competition is invaluable for learning how to design UA, rapidly building the designs, and then determining how close the flight hardware came to the design predictions.

Beyond academics, research UASs are used for a variety of technology research and scientific purposes. The UA itself can demonstrate aircraft features or capabilities in the areas of advanced configurations, flight controls, aerodynamics, or propulsion technologies. Scientific research depends more on the payloads than the UA technologies, resulting in more conventional designs. These missions span many scientific interest areas including climate science and species monitoring. Figure 2.20 shows the Aurora Flight Sciences Perseus B UAS (Fig. 2.20) developed under the NASA ERAST program to study the upper atmosphere.

References

[1] Mueller, T. J., *Fixed and Flapping Wing Aerodynamics for Micro Air Vehicle Applications*, AIAA, Reston, VA, 2001.
[2] Mueller, T. J., Kellogg, J. C., Ifju, P. G., and Shkarayev, S. V., *Introduction to the Design of Fixed-Wing Micro Air Vehicles, Including Three Case Studies*, AIAA, Reston, VA, 2007.
[3] Dept. of Defense, Unmanned Systems Roadmap 2007–2032, Office of the Secretary of Defense, Washington, D.C., 10 Dec. 2007.

[4] Yenne, B., *Birds of Prey, Predators, Reapers and America's Newest UAVs in Combat*, Specialty Press, North Branch, MN, 2010.

[5] Martin, M. J., and Sasser, C. W., *Predator, The Remote-Control Air War over Iraq and Afghanistan: A Pilot's Story*, Zenith Press, Minneapolis, MN, 2010.

[6] Drezner, J. A., and Leonard, R. S., *Global Hawk and Darkstar*, Executive Summary and Vols. 1–3, RAND, Arlington, VA, 2002.

Problems

2.1   A special operations force requires rapid response ISR. What class of UAS should be considered and why?

2.2   A general responsible for a multination zone wants to gain an understanding of the threat radar capabilities in her area of responsibility. What class of UAS should be considered?

2.3   A Navy wishes to cover the economic exclusion zone over a large coast line continuously. What is the appropriate UAS class?

2.4   A military security force operating at an austere desert forward base wishes to quickly locate the source of mortar attacks. What class of UAS should be used?

2.5   A telecommunications company wishes to provide cell phone coverage over a city for multiple years continuously. What class of UAS should be considered?

2.6   A military commander wishes to provide an initial strike and then support strike missions after achieving air superiority. What classes of UASs should be considered?

2.7   A police force is attempting to end a hostage crisis. They need to know how many hostages are present and the types of arms employed by the hostage taker. They do not want to alert the hostage taker to the UAS presence. What class of UAS should be employed?

Chapter 3

Initial Unmanned-Aircraft Sizing

- Understand UA sizing
- Understand methods for assessment of aerodynamic parameters
- Use performance equations for various propulsion classes

The Aurora Flight Sciences Orion sizing was driven by its 120-hour endurance and large payload capacity. (Photo courtesy of Aurora Flight Sciences.)

Introduction

U A sizing is the synthesis of multiple design disciplines to define the major attributes of the design. A rigorous analysis of the UA using detailed methods is necessary to ensure that the solution is compliant with the system requirements. However, simple methods can be used to gain visibility into the design space and better understand characteristics of the UA solution. Flexibility is important for early sizing of unmanned aircraft that can span several orders of magnitude in weight and size.

Simple Weight Relationships

UA takeoff gross weight W_{TO} is a major parameter that largely defines the vehicle class and from which much can be inferred. Later chapters will demonstrate how W_{TO} has implications for life-cycle cost, competitive environment, and suitability for various operations. Therefore, the sizing discussion begins with intuitive methods for determining this important parameter.

The first objective is to derive rapid, closed-form methods for estimating W_{TO} that are suitable for all classes of unmanned aircraft, from MAVs to very large vehicles. While enabling broad applicability and expediency might compromise accuracy, these methods are suitable for many applications. Some example uses include initially assessing new

Navmar Tigershark		**Navmar Mako**		**Swift Killer Bee 3**		**Swift Killer Bee 2**	
W_{TO}	285 lbs	W_{TO}	140 lbs	W_{TO}	95 lbs	W_{TO}	43 lbs
W_{PL}	30 lbs	W_{PL}	30 lbs	W_{PL}	15-30 lbs	W_{PL}	7-15 lbs
Span	17 ft	Span	12.7 ft	Span	9.2 ft	Span	6.5 ft

BAI Viking 400		**BAI Viking 100**		**Griffon Broadsword**		**Griffon Outlaw**	
W_{TO}	493 lbs	W_{TO}	150 lbs	W_{TO}	550 lbs	W_{TO}	120 lbs
W_{PL}	60 lbs	W_{PL}	20 lbs	W_{PL}	120 lbs	W_{PL}	40 lbs
Span	20 ft	Span	12 ft	Span	22.5 ft	Span	13.5 ft

Fig. 3.1 Many manufacturers have families of unmanned aircraft, each with similar configurations and design features.

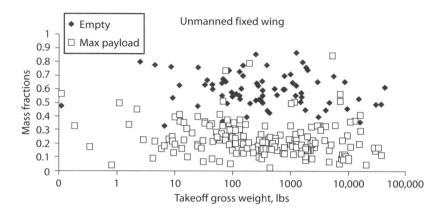

Fig. 3.2 Empty and maximum payload mass fractions for fixed-wing unmanned aircraft.

requirements, determining technology impacts, evaluating the veracity of competitor's marketing claims, or providing a "sanity check" for the outputs of more detailed methods. The simple weight sizing relationships are well suited to scaling aircraft within a family of UA, like those shown in Fig. 3.1.

Mass fraction scaling of major UA system weight groupings is a convenient and intuitive way to perform weight estimation. Mass fractions are simply the ratio of weights. For example, the empty weight mass fraction is the ratio of the empty weight W_{Empty} to W_{TO}.

$$MF_{\text{Empty}} = \frac{W_{\text{Empty}}}{W_{\text{TO}}} \tag{3.1}$$

Similarly, the maximum payload mass fraction $MF_{\text{PL,Max}}$ is the ratio of the maximum payload weight $W_{\text{PL,Max}}$ to W_{TO}. Figures 3.2 and 3.3 show these two mass fraction parameters for fixed-wing and helicopter unmanned aircraft across several orders of magnitude in W_{TO}. The data points shown do not distinguish between powerplant type, flight regime, or launch and recovery methods. Narrowing selection criteria produce less scatter for the comparative systems. However, it can be seen that these mass fractions do not have a strong correlation with W_{TO} for the widely varying unmanned systems shown.

To begin the development of the simple weight sizing methods, all of the major weights are identified. W_{TO} is simply the summation of all weights on the UA at the launch condition.

$$W_{\text{TO}} = \sum_{i=0}^{\text{Max}} W_i \tag{3.2}$$

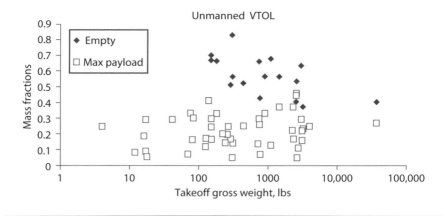

Fig. 3.3 Empty and maximum payload mass fractions for VTOL unmanned aircraft.

The weight elements will now be expanded to more useful weight groupings.

$$W_{TO} = W_{Struct} + W_{Subs} + W_{Prop} + W_{Avion} + W_{Other} + W_{PL} + W_{Energy} \quad (3.3)$$

where W_{Struct} is the structures weight, W_{Subs} is the subsystems weights, W_{Prop} is the propulsion system weight, W_{Avion} is the avionics weight, W_{Other} is other weights not captured by other categories, W_{PL} is the payload weight, and W_{Energy} is the weight of the stored energy. So far, only a definition of weights is presented with no assumptions. The only requirement is that all weights are contained within one of the groups. Note the weight groups shown here do not necessarily need to comply with the Society of Allied Weights Engineers (SAWE) recommended practices. This potential deviation from established weights accounting standards is acceptable to best capture the weight scaling behavior described later. However, the weights can be appropriately regrouped after the calculation for reporting purposes.

Now the weights groups are allocated to one of two categories: weights that scale linearly with W_{TO} and fixed weights that do not vary. This is a major assumption about the the weight groups' sizing relationship behavior. Obviously, important factors such as UA geometry and operating environment appear to be neglected. Further elaboration will demonstrate that many factors are appropriately captured or have secondary impacts.

The structures group includes major airframe elements such as the wing, fuselage, and nacelles. Landing gear or alternative launch and recovery provisions can be included here or in the subsystems. It is assumed here that the structural weight is proportional to W_{TO} and the scaling factor is the structural mass fraction MF_{Struct}.

$$W_{Struct} = \frac{W_{Struct}}{W_{TO}} \cdot W_{TO} = MF_{Struct} \cdot W_{TO} \qquad (3.4)$$

Unmanned aircraft can utilize various propulsion types, each with different sizing characteristics. This derivation includes both the power generator and propulsor in the propulsion system. Energy storage is included in the fuel weights. As with the structures, the propulsion weight is assumed to be proportional to W_{TO}.

$$W_{Prop} = \frac{W_{Prop}}{W_{TO}} \cdot W_{TO} = MF_{Prop} \cdot W_{TO} \qquad (3.5)$$

The linear form of the group weight relationships permits linear expansion of the mass fraction terms, providing an opportunity to add physically meaningful relationships. Here the propulsion mass fraction is expanded to incorporate UA-propulsion sizing and propulsion system characteristics for various propulsion types.

Power is the major sizing parameter for propulsion systems using propellers, such as reciprocating engines, electric motor-driven propellers, and turboprops. A linear relationship between propulsion systems power and weight is an example of what is called "rubber engine" sizing, which assumes that a perfectly matched engine will be available for the aircraft. The aircraft power-to-weight ratio is also applied as a parameter. The power referenced for sizing is generally the maximum sea-level static uninstalled shaft power.

A propulsion system consisting of a reciprocating engine or turboshaft engine directly driving a propeller includes the engine mechanical installation and auxiliary systems. Here it is assumed that all items beyond the powerplant scale directly with the engine power, which is captured by the $f_{Install}$ factor that is greater than unity. Items such as digital engine controls can be captured either as a contributor to the $f_{Install}$ factor or as separate fixed weights if they are known. The propulsion mass fraction can now be written as

$$MF_{Prop} = f_{Install} \cdot \frac{(P/W_{TO})_{Aircraft}}{(P/W_{Powerplant} + P/W_{Propeller})} \qquad (3.6)$$

The power-to-weight ratio of select propeller-driven unmanned aircraft is shown in Fig. 3.5. The power-to-weight ratio of various fueled powerplants is shown in Fig. 3.4.

The relationship for a chemical battery-driven aircraft is similar, but the powerplant is replaced by the motor and motor controller. The battery will

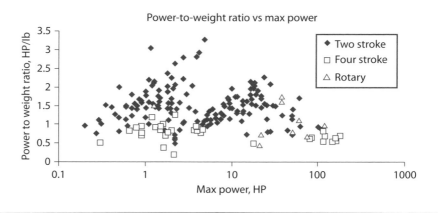

Fig. 3.4 Power-to-weight ratios of two-stroke, four-stroke, and rotary engines.

be included in separate energy storage terms. The referenced power is generally the maximum motor shaft power. The propulsion mass fraction becomes

$$MF_{\text{Prop}} = f_{\text{Install}} \cdot \frac{(P/W_{\text{TO}})_{\text{Aircraft}}}{(P/W_{\text{Motor}} + P/W_{\text{Controller}} + P/W_{\text{Propeller}})} \qquad (3.7)$$

A simple nonregenerative fuel-cell system where a fuel-cell stack provides power directly to a motor-propeller propulsor adds an additional term to the preceding equation to account for the fuel cell. The hydrogen or other fuel sources are once again covered separately in energy weight. The fuel-cell weight includes the fuel-cell stack plus any additional hardware required for the fuel cell to function. The propulsion mass fraction of a nonregenerative fuel cell system is

$$MF_{\text{Prop}} = f_{\text{Install}} \cdot \frac{(P/W_{\text{TO}})_{\text{Aircraft}}}{(P/W_{\text{FuelCell}} + P/W_{\text{Motor}} + P/W_{\text{Controller}} + P/W_{\text{Propeller}})}$$

$$(3.8)$$

Moving on to thrust-dominated systems, it becomes necessary to change the form of the propulsion mass fraction equations. Turbojet, turbofan, and pulsejet systems scale with thrust rather than power. Once again, an installation factor f_{Install} covers mechanical installation and miscellaneous propulsion-related systems. This factor can include inlets, nozzles, and perhaps nacelles for jet engines. Here the referenced thrust is generally the maximum sea-level static uninstalled thrust. The propulsion mass fraction for thrust dominated systems is

$$MF_{\text{Prop}} = f_{\text{Install}} \cdot \frac{(T/W_{\text{TO}})_{\text{Aircraft}}}{(T/W_{\text{Powerplant}})} \tag{3.9}$$

Performance-based selection of aircraft power-to-weight ratios for propeller-driven aircraft or thrust-to-weight ratio for thrust-dominated aircraft can be performed by analysis. More detailed conceptual level analysis involves a trade between the appropriate propulsion parameter (T/W or P/W) and wing loading (W/S). Several constraints must be satisfied, including takeoff distance, climb gradients, service ceiling, dash speed, stall speed, sustained turning flight, and landing distance. These methods are described in more detail by Lan and Roskam [1], Raymer [2], and Mattingly [3], and are not repeated here. A more generalized and robust method for determining flight performance constraints that help size propulsion will be presented in Chapter 8.

Now a similarity method is explored in which the aircraft power-to-weight ratio or thrust-to-weight ratio is selected via comparison with similar aircraft. Existing aircraft—especially those that are operational—provide great insight into design attributes. These aircraft have progressed far beyond the early conceptual design stage and have worked through challenges of the design, build, and test phases. Put another way, flying aircraft represent more design fidelity than can be captured by simplified conceptual design tools. Therefore, these mature systems are valuable starting points for selection of relative propulsion sizing. Conversely, comparison of a conceptual design with other conceptual designs does not necessarily yield high design point selection realism.

Determining a "similar aircraft" involves many considerations. Selections should have similar size, physical characteristics, technologies, cost approach, operating environments, propulsion type, launch and recovery methods, and mission sets. For example, two tactical, conventional takeoff and landing, reciprocating engine-powered, conventional wing-tail configuration UASs would be similar. Two dissimilar aircraft, for example, would be the aforementioned tactical UAS and a large HALE jet-powered UAS. Figure 3.5 shows comparisons of power-to-weight ratios for the tactical UAS example, where all UA have achieved flight status but not necessarily operational status. If the anticipated takeoff gross weight range is 300–600 lb, then an aircraft power-to-weight ratio of 0.08 would be reasonable.

The subsystems weight group includes the electrical power system, environmental control system, flight control system, pneumatic and hydraulic system, and potentially the landing gear. High-fidelity sizing

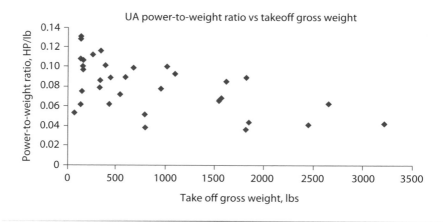

Fig. 3.5 Power-to-weight ratios for fixed-wing, conventional wing-tail configuration, and reciprocating-engine unmanned aircraft.

relationships incorporate many factors such as operating environment, flight performance, duty cycles, propulsion system characteristics, system complexity, electrical load requirements, and other key drivers that are not proportional to W_{TO}. These considerations will be evaluated further in Chapter 10. Within a bounded class of UA, subsystem configuration, and technology level, it is appropriate to assume a linear subsystem weight relationship with W_{TO} for the purposes of initial sizing. This assumption yields the following subsystem sizing relationship:

$$W_{Subs} = \frac{W_{Subs}}{W_{TO}} \cdot W_{TO} = MF_{Subs} \cdot W_{TO} \qquad (3.10)$$

The energy mass fraction is linked with UA performance. For a design space where the aerodynamic and propulsion efficiencies are constant, the energy mass fraction is constant regardless of unmanned aircraft size. The methods presented here assume that secondary power consumption drivers such as bleed air or power generation must have weight-proportional or negligible impacts as well. The energy mass fraction is derived from performance models that we will discuss later in this chapter. In the case of hydrocarbon fuels, the stored energy medium is simply fuel. Alternative forms are chemical batteries, hydrogen or hydrazine, among others. Storage devices such as fuel tanks are typically included in the subsystems weight, though an alternative formulation for hydrogen storage will be shown. The energy weight is proportional to W_{TO} using an energy mass fraction MF_{Energy} for scaling.

$$W_{Energy} = MF_{Energy} \cdot W_{TO} \qquad (3.11)$$

The energy mass fraction for hydrocarbon-fueled aircraft such as traditional reciprocating engines, turboprops, or jets is simply the fuel mass fraction:

$$MF_{Energy} = MF_{Fuel} \qquad (3.12)$$

Similarly, the energy mass fraction for chemical batteries is the battery mass fraction.

$$MF_{Energy} = MF_{Batt} \qquad (3.13)$$

Two forms of the energy mass fraction equation will now be presented for hydrogen propulsion systems. The first takes the same form as the preceding two examples, but it substitutes hydrogen as the energy source.

$$MF_{Energy} = MF_{H2} \qquad (3.14)$$

Hydrogen-storage device weight is a significant first-order aircraft sizing driver, and it can be proportional to the amount of hydrogen stored as an approximation. Hydrogen-storage device effectiveness is often provided as a storage mass fraction:

$$MF_{Storage} = \frac{W_{H2}}{W_{H2} + W_{Storage}} \qquad (3.15)$$

Now the energy mass fraction that includes both the hydrogen-storage device and the hydrogen becomes

$$MF_{Energy} = \frac{MF_{H2}}{MF_{Storage}} \qquad (3.16)$$

The avionics and payload weights are generally fixed. Avionics weight is the summation of the avionics components that are driven by hardware options and system requirements. Payload weight capacity is often defined directly in a customer requirement. Alternatively, the payload performance is specified, and the system developer selects the hardware. The payload is a fixed weight in either case. Another parameter W_{Other} is intended to catch any other fixed weights that are not easily included in other categories.

Now Eq. (3.2) can be rewritten in terms of weights that are fixed weights and proportional to W_{TO}.

$$W_{TO} = W_{PL} + W_{Avion} + W_{Other}$$
$$+ \left(MF_{Struct} + MF_{Subs} + MF_{Prop} + MF_{Energy} \right) \cdot W_{TO} \qquad (3.17)$$

Rearranging the terms to solve for W_{TO} results in the following equation:

$$W_{\mathrm{TO}} = \frac{W_{\mathrm{PL}} + W_{\mathrm{Avion}} + W_{\mathrm{Other}}}{1 - \left(MF_{\mathrm{Struct}} + MF_{\mathrm{Subs}} + MF_{\mathrm{Prop}} + MF_{\mathrm{Energy}}\right)} \quad (3.18)$$

This seemingly simple equation has many powerful implications. Despite the bold assumption that structures, propulsion, and energy weights are proportional to W_{TO}, the relationship can be surprisingly accurate and robust across diverse UA classes. This equation is of great importance to aircraft designers, and it merits careful study.

Note that the equation takes the form of a hyperbola. An asymptote exists where the summation of the mass fractions in the denominator equals one, causing the denominator to equal zero. Feasible design space exists only where the denominator value is less than one and greater than zero.

It can also be shown that the takeoff gross weight is proportional to the summation of the fixed weights. Therefore, doubling the fixed weights doubles the takeoff gross weight. Later chapters will demonstrate that specifications for payload capacities rather than payload integrated performance can generate unnecessarily heavy and expensive aircraft.

The weight escalation factor (WEF) is the rate of W_{TO} growth with respect to additional fixed weight. The WEF for this derivation is

$$WEF = \frac{1}{1 - \left(MF_{\mathrm{Struct}} + MF_{\mathrm{Subs}} + MF_{\mathrm{Prop}} + MF_{\mathrm{Energy}}\right)} \quad (3.19)$$

To be applicable to more complex models using nonlinear weight-estimating relationships, the WEF can be described more generally as

$$WEF = \partial W_{\mathrm{TO}} / \partial W_{\mathrm{Fixed}} \quad (3.20)$$

A WEF value of 5 means that every pound of fixed weight added will result in 5 lb of W_{TO}. Requirements satisfaction and system affordability become increasingly challenging as the WEF grows. Unmanned aircraft weight growth and resulting size simply become unmanageable beyond a point. This parameter should be less than $8-10$ for a reasonable chance of program success. Modern design practices employed by typical design teams simply cannot cope with greater weight sensitivity.

Unmanned aircraft design problems rarely have the luxury of developing a custom engine that exactly matches the system requirements. Rather, the designer must select from one or more distinct candidate engines. The propulsion system and its installation impact are then known fixed weights for the initial sizing. The sizing equation becomes

$$W_{\text{TO}} = \frac{W_{\text{PL}} + W_{\text{Avion}} + W_{\text{Propulsion}} + W_{\text{Other}}}{1 - \left(MF_{\text{Struct}} + MF_{\text{Subs}} + MF_{\text{Energy}}\right)} \qquad (3.21)$$

This formulation change is illustrated to highlight the design implications. Consider a jet-powered aircraft that requires a minimum thrust-to-weight ratio of 0.33 to meet the performance constraints. The fixed weights are known to be 500 lb, the structural mass fraction is 0.35, and the subsystem's mass fraction is 0.1. Both rubber engines and fixed engines will be considered, and all have an installed thrust-to-weight ratio of 6 ($f_{\text{Install}} = 1$). The energy mass fraction is the fuel mass fraction, which is allowed to vary for this example. The fixed engines are 1, 2, and 4 unit combinations of a 1,000-lb thrust engine, totaling 1,000; 2,000; and 4,000 lb, respectively. The results are plotted in Fig. 3.6.

An interesting series of relationships emerge. The rubber-engine curve represents a perfectly matched jet and UA combination, resulting in the minimum takeoff gross weight and fully satisfied T/W constraints. For each of the fixed-engine cases, the T/W is greater than the rubber engine at low fuel mass fractions. At fuel mass fractions greater than the fixed- and rubber-engine curve intersection value, the fixed-engine UA T/W is too low. A larger thrust is required, and a larger fixed-engine thrust combination must be selected. Figure 3.6 shows the overall fixed-engine curve follows the minimal sized engine that satisfies the T/W requirements, which requires discrete switching among the three candidates as the fuel mass fraction increases. The greatest engine thrust margin exists immediately after the engine thrust combination jumps higher. A fixed engine that has little thrust margin, which is close to the fixed- and rubber-engine curve intersection, is at risk of missing performance requirements or requiring an engine selection change should the design experience difficulties.

These weight methods apply best to well-understood design spaces or for variations from a reference design. The accuracy depends greatly on quality of the component weights and mass fractions.

Despite these caveats, this method is often better than inappropriately extrapolating from more detailed methods. For example, the structural mass fraction of a MAV and HALE UAS will generally be within 10–20% of each other despite the size and weight differences of three and five orders of magnitude, respectively. However, applying empirically based general-aviation structures weight-estimating relationships with unspecified parameter bounds will likely produce poor quality results for the aforementioned UAS classes.

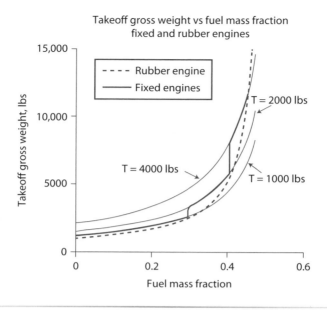

Fig. 3.6 Fixed- and rubber-engine sizing of an example jet-powered UA.

Although the simple weight method is robust across UA classes and provides great insight into design relationships, it is inappropriate for engineering analysis as the design process progresses. More appropriate mass properties processes are described in Chapter 6. There it will be shown that an iterative weights convergence is required for nonlinear relationships.

3.3 Flight Performance

The interfaces between the mass properties method just described and aircraft performance are the energy mass fraction and either aircraft P/W or T/W. This section provides simple methods to generate the energy mass fraction. For now, the propulsion requirements are attained using similar P/W or T/W as representative aircraft. Later chapters provide more detailed performance, propulsion, and aerodynamics methods.

Starting with the case of hydrocarbon-fueled propulsion systems, the fuel mass fraction used by the sizing analysis is the ratio of the fuel weight consumed in the sizing mission to the takeoff gross weight of the UA. Most mission profiles consist of several segments with different performance objectives such as range, endurance, dash speed, climb, and descent. Each segment consumes fuel that contributes to the overall fuel usage.

This section considers nondimensional mission segment fuel mass fractions rather than total fuel weight consumed. The total fuel mass fraction for a mission MF_{Fuel} is found by

$$(1 - MF_{\text{Fuel}}) = \prod_{i=1}^{NSegs} (1 - MF_{\text{Fuel,i}}) \tag{3.22}$$

where N_{Segs} is the number of mission segments and $MF_{\text{Fuel,i}}$ is the fuel mass fraction for each segment. Solving for MF_{Fuel}, the equation becomes

$$MF_{\text{Fuel}} = 1 - \prod_{i=1}^{NSegs} (1 - MF_{\text{Fuel,i}}) \tag{3.23}$$

The segment fuel mass fractions for various types of mission are provided next for multiple propulsion classes. The range and endurance equations are applied once to the applicable segment, though more detailed methods that evaluate performance characteristics and constraints multiple times per segment will be presented in Chapter 9.

3.3.1 Reciprocating Engines and Turboprop Propulsion

Range and endurance equations for turboprops or reciprocating engines driving propellers are applicable to any liquid-fueled propulsion system where the power generator converts the fuel's chemical energy into propeller shaft power. The UA loses weight as it expends the fuel mass after combustion as exhaust. For example, these equations can be applied to a hydrazine-fueled engine that drives a propeller. The range and endurance equations presented are rearranged to solve for fuel mass fraction.

Range for a propeller aircraft at constant angle of attack is

$$R = \frac{L/D \cdot \eta_p}{BSFC} \cdot \ell n \left(\frac{1}{1 - MF_{\text{Fuel}}} \right) \tag{3.24}$$

where R is the range, L/D is the aircraft lift-to-drag ratio, η_p is the propeller efficiency, and $BSFC$ is the power generator brake specific fuel consumption. Solving this equation for fuel mass fraction results in

$$MF_{\text{Fuel}} = 1 - \exp \left(\frac{-R \cdot BSFC}{L/D \cdot \eta_P} \right) \tag{3.25}$$

Endurance for a propeller aircraft with a constant angle of attack and constant velocity is

$$E = \frac{L/D \cdot \eta_p}{V \cdot BSFC} \cdot \ell n \left(\frac{1}{1 - MF_{\text{Fuel}}} \right) \tag{3.26}$$

where E is the endurance and V is the true airspeed. Solving the preceding equation for fuel mass fraction yields

$$MF_{\text{Fuel}} = 1 - \exp\left(\frac{-E \cdot V \cdot BSFC}{L/D \cdot \eta_p}\right) \qquad (3.27)$$

Alternatively, the fuel mass fraction equation can be rewritten in terms of wing loading, endurance parameter ($C_L^{3/2}/C_D$), and air density ρ.

$$MF_{\text{Fuel}} = 1 - \exp\left(\frac{-E \cdot BSFC}{C_L^{3/2}/C_D \cdot \eta_p}\sqrt{\frac{2 \cdot W_{\text{TO}}/S_w}{\rho}}\right) \qquad (3.28)$$

Endurance for a propeller aircraft with constant angle of attack flying at a constant altitude is

$$E = \frac{\eta_p}{BSFC} \cdot \sqrt{2 \cdot \rho \cdot S_w} \cdot \frac{C_L^{3/2}}{C_D} \cdot \left(\frac{1}{\sqrt{W_2}} - \frac{1}{\sqrt{W_1}}\right) \qquad (3.29)$$

where S_w is the wing planform area, W_1 is the UA weight at the beginning of the mission, and W_2 is the UA weight at the conclusion of the segment. This can be arranged as

$$E = \frac{\eta_p}{BSFC} \cdot \sqrt{\frac{2 \cdot \rho}{W_{\text{TO}}/S_w}} \cdot \frac{C_L^{3/2}}{C_D} \cdot \left(\frac{1}{\sqrt{1 - MF_{\text{Fuel}}}} - 1\right) \qquad (3.30)$$

Solving this equation for fuel mass fraction gives

$$MF_{\text{Fuel}} = 1 - 1 \left/ \left(\frac{E \cdot BSFC}{\eta_P \cdot C_L^{3/2}/C_D}\right)\sqrt{\frac{W_{\text{TO}}/S_w}{2 \cdot \rho}} + 1 \right. \qquad (3.31)$$

Note that the $BSFC$ is often provided in units of pounds/horsepower-hour, which must be divided by 325.66 to get units of nautical miles^{-1}.

The range and endurance equations presented here do not account for engine generator power. Directly calculating the power consumed by the engines greatly complicates these simple equations and changes their basic form. The best way to account for the generator power is to multiply the brake specific fuel consumption by a load factor f_{Load}. The load factor is equal to unity when there is no generator load applied. Other parasitic load impacts such as shaft-driven engine cooling fans can be included in f_{Load} as well. Nonpropulsion engine loads can be 0–15% for many UAS applications.

3.3.2 Jet Propulsion

Jet propulsion performance methods are applicable to thrust-dominated systems that consume fuel, such as turbojets, turbofans, and pulsejets. No assumptions are made about the engine cycle; only the net performance in terms of the thrust specific fuel consumption $TSFC$ is required.

Range for a jet aircraft at constant velocity and constant angle of attack is

$$R = \frac{V \cdot L/D}{TSFC} \cdot \ell n \left(\frac{1}{1 - MF_{\text{Fuel}}} \right) \tag{3.32}$$

This is known as the Breguet range equation. Solving this equation for fuel mass fraction gives

$$MF_{\text{Fuel}} = 1 - \exp \left(\frac{-R \cdot TSFC}{V \cdot L/D} \right) \tag{3.33}$$

Aircraft endurance is the range divided by the true airspeed. For a jet aircraft the endurance is

$$E = \frac{L/D}{TSFC} \cdot \ell n \left(\frac{1}{1 - MF_{\text{Fuel}}} \right) \tag{3.34}$$

The associated fuel mass fraction for jet endurance is

$$MF_{\text{Fuel}} = 1 - \exp \left(\frac{-E \cdot TSFC}{L/D} \right) \tag{3.35}$$

Again, the jet range and endurance equations do not include nonpropulsion load penalties. The thrust specific fuel consumption can be multiplied by f_{Load} to account for generator power, mechanical power takeoff, and bleed air extraction.

3.3.3 Battery-Powered Electric Propulsion

The preceding sections on fueled propulsion types cover well-established performance equations. The derivations can be found in most good aircraft performance texts. In contrast, performance methods for electric-powered aircraft are less commonly covered in the literature. Using this gap as motivation, this section derives various forms of simple electric aircraft performance equations.

The endurance for a battery-powered electric aircraft is

$$E = \frac{Energy_{\text{Batt}}}{P_{\text{Batt}}} \tag{3.36}$$

where $Energy_{Batt}$ is the battery usable energy and P_{Batt} is the average battery power output. The usable energy of a battery is expanded to

$$Energy_{Batt} = Capacity \cdot Voltage \cdot \eta_{Batt} \cdot f_{Usable} \tag{3.37}$$

Capacity is the rated current-time integral for the battery pack and is often provided in units of Amp-hours. *Voltage* is the nominal voltage across the battery leads. The battery efficiency η_{Batt} accounts for heating losses and is a function of output current and internal resistance. The f_{Usable} factor accounts for the permissible battery pack depth of discharge, which can be a very significant endurance driver. The depth of discharge is the ratio of battery usable energy to the total stored energy. The battery efficiency and permissible depth of discharge depend on battery chemistry and design attributes.

Using battery-specific energy, E_{spec} is a more general approach that facilitates more intuitive battery applications across broad problem spaces. This parameter is the battery energy-to-mass ratio. The common units are Watt-hours per kilogram. The usable battery energy can now be shown as

$$Energy_{Batt} = E_{spec} \cdot M_{Batt} \cdot \eta_{Batt} \cdot f_{Usable} \tag{3.38}$$

where M_{Batt} is the battery mass. In most cases, the battery provides power to the propulsion system and other loads such as avionics and payloads.

$$P_{Batt} = P_{Propulsion} + P_{Other} \tag{3.39}$$

For now the other power loads will be considered negligible, and the resulting battery power loads come only from the propulsion system.

The aircraft thrust power generated by the propeller is

$$P_{Thrust} = T \cdot V \tag{3.40}$$

For level flight the thrust is equal to drag. The drag can be found by dividing the weight by lift-to-drag ratio. Electric aircraft do not change mass in flight like fueled aircraft, and so the flight weight remains constant at the takeoff gross weight. Therefore, the thrust power generated by the propeller becomes

$$P_{Thrust} = D \cdot V = \frac{W_{TO}}{L/D} \cdot V \tag{3.41}$$

Many system elements produce, manage, or convert forms of the power between the battery terminals and the propeller, each with losses. The propeller also has inefficiencies in converting the shaft power to thrust power.

These elements can include power distribution, motor controllers, motors, gearboxes, and the propeller. The relationship between battery power and thrust power is

$$P_{\text{Batt}} = P_{\text{Propulsion}} = \frac{P_{\text{Thrust}}}{\prod \eta} \tag{3.42}$$

The product of the propulsion system efficiencies is

$$\prod \eta = \eta_p \cdot \eta_{\text{gear}} \cdot \eta_{\text{motor}} \cdot \eta_{\text{ESC}} \cdot \eta_{\text{Dist}} \tag{3.43}$$

where η_p is the propeller efficiency, η_{gear} is the motor gearbox efficiency, η_{motor} is the electric motor efficiency, η_{ESC} is the electronic speed control efficiency, and η_{Dist} is the efficiency of the power distribution system.

Combining these relationships gives

$$P_{\text{Batt}} = \frac{W_{\text{TO}} \cdot V}{L/D \cdot \prod \eta} \tag{3.44}$$

Using the relationships $L/D = C_L/C_D$ and

$$V = \sqrt{\frac{2 \cdot W_{\text{TO}}/S_w}{\rho \cdot C_L}} \tag{3.45}$$

the battery power equation can be rewritten as

$$P_{\text{Batt}} = W_{\text{TO}} \sqrt{\frac{2 \cdot W_{\text{TO}}/S_w}{\rho \cdot C_L}} \Big/ \left(\prod \eta \cdot \frac{C_L}{C_D} \right) = W_{\text{TO}} \sqrt{\frac{2 \cdot W_{\text{TO}}/S_w}{\rho}} \Big/ \left(\prod \eta \cdot \frac{C_L^{3/2}}{C_D} \right) \tag{3.46}$$

Here the familiar wing loading and endurance parameter terms emerge. This form of the battery power equation is now used to build the electric aircraft endurance equation.

$$E = \frac{Energy_{\text{Batt}}}{P_{\text{Batt}}} = \frac{Energy_{\text{Batt}}}{W_{\text{TO}}} \cdot \prod \eta \cdot \frac{C_L^{3/2}}{C_D} \sqrt{\frac{1/2 \cdot \rho}{W_{\text{TO}}/S_w}} \tag{3.47}$$

The battery-specific energy uses mass rather than weight. The battery mass is the weight divided by the acceleration of gravity. One must be careful not to directly convert from units of mass to weight here (i.e., 1 kg is equivalent to 2.2 lb on the Earth's surface), as this will result in an error when dividing by the acceleration of gravity. To arrive at an endurance equation that uses battery mass fraction terms, the battery energy equation is rewritten as

$$Energy_{\text{Batt}} = E_{\text{spec}} \cdot \frac{W_{\text{Batt}}}{g} \cdot \eta_{\text{Batt}} \cdot f_{\text{Usable}} \tag{3.48}$$

Substituting this weight-based battery energy equation back into the endurance equation

$$E = \frac{E_{\text{Spec}}}{g} \cdot \frac{W_{\text{Batt}}}{W_{\text{TO}}} \cdot \prod \eta \cdot \eta_{\text{Batt}} \cdot f_{\text{Usable}} \cdot \frac{C_L^{3/2}}{C_D} \cdot \sqrt{\frac{1/2 \cdot \rho}{W_{\text{TO}}/S_w}} \qquad (3.49)$$

The ratio of the battery weight to the takeoff gross weight is the battery mass fraction MF_{Batt}. This substitution gives the final form of the endurance equation.

$$E = \frac{E_{\text{Spec}}}{g} \cdot MF_{\text{Batt}} \cdot \prod \eta \cdot \eta_{\text{Batt}} \cdot f_{\text{Usable}} \cdot \frac{C_L^{3/2}}{C_D} \cdot \sqrt{\frac{1/2 \cdot \rho}{W_{\text{TO}}/S_w}} \qquad (3.50)$$

The battery mass fraction is

$$MF_{\text{Batt}} = \frac{E \cdot g}{E_{\text{Spec}} \cdot \prod \eta \cdot \eta_{\text{Batt}} \cdot f_{\text{Usable}} \cdot C_L^{3/2}/C_D} \cdot \sqrt{\frac{W_{\text{TO}}/S_w}{1/2 \cdot \rho}} \qquad (3.51)$$

Range can be found by multiplying the endurance by the airspeed. Rewriting the final endurance equation in terms of velocity gives

$$E = \frac{E_{\text{Spec}}}{g} \cdot MF_{\text{Batt}} \cdot \prod \eta \cdot \eta_{\text{Batt}} \cdot f_{\text{Usable}} \cdot \frac{L/D}{V} \qquad (3.52)$$

Multiplying this endurance equation by airspeed gives us the range equation.

$$R = \frac{E_{\text{Spec}}}{g} \cdot MF_{\text{Batt}} \cdot \prod \eta \cdot \eta_{\text{Batt}} \cdot f_{\text{Usable}} \cdot L/D \qquad (3.53)$$

The range equation can now be solved for the battery mass fraction.

$$MF_{\text{Batt}} = \frac{R \cdot g}{E_{\text{Spec}} \cdot \prod \eta \cdot \eta_{\text{Batt}} \cdot f_{\text{Usable}} \cdot L/D} \qquad (3.54)$$

3.4 Simple Aerodynamics Methods

The performance equations require aerodynamic performance inputs in the form of the lift-to-drag ratio or endurance parameter. Here simple methods are provided for rapid assessment of these aerodynamic parameters. To eliminate the complexities of compressibility, only the subsonic flight regime is considered now.

First, it is assumed that the UA's drag coefficient has a parabolic relationship with the lift coefficient. This parabolic drag polar assumption is common for early conceptual design, though the limitations and alternatives will be explored in Chapter 5. For now, the UA drag coefficient is

$$C_D = C_{D0} + \frac{C_L^2}{\pi \cdot AR \cdot e} \qquad (3.55)$$

The parameter C_{D0} is the zero-lift drag coefficient, which assumes that the minimum drag occurs when the aircraft is not lifting. Drag polars of aircraft with highly cambered wings, such as conventional configuration long-endurance UA, do not follow this behavior. Other aircraft such as flying wings, delta wings, and low-camber conventional designs more closely follow the model. At very early stages of design or analysis, this approximation is appropriate.

The aircraft aspect ratio AR is defined as

$$AR = \frac{b^2}{S_w} \qquad (3.56)$$

where b is the wing span and S_w is the wing reference area. The wing reference area must be the same value used to define the lift and drag coefficients. The wing reference area is conventionally defined as the projected area of the main wing. This is also known as the *planform area*. Any wing area covered by bodies such as fuselages and nacelles is included in the total wing reference area.

The parameter e is commonly referred to as the "span efficiency factor," with the implication that it only captures induced drag deviation from an ideal elliptical wing. However, the lift dependency of the drag polar is due to much more than just induced drag. In breaking with long-established tradition, the author considers this parameter to be merely a calibration factor. Other contributions to the lift-varying drag include the wing-profile drag behavior, trim drag, fuselage drag, and interference drag. Rick Foch, the DOD Senior Scientist for Expendable Systems, recommends a range of 0.65–0.72 for initial sizing purposes. Lan and Roskam [1] show an e correlation of 0.67–0.93 for various manned aircraft drag polars. By comparison, true span efficiency factor that accounts only for induced drag is typically 0.95–1 for planar wings. Figure 3.7 shows a comparison

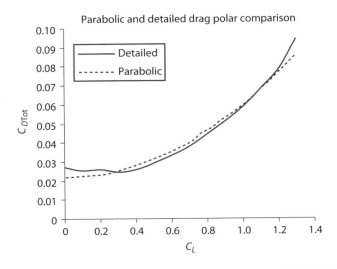

Fig. 3.7 Comparison of a detailed and parabolic drag polar.

between a detailed drag polar using airfoil data and an idealized parabolic drag polar.

The zero-lift drag coefficient can be estimated from a variety of methods ranging from similarity to other aircraft to a detailed drag buildup. The similarity methods are suitable because measured aerodynamic performance of flying vehicles includes all miscellaneous drag factors that are frequently neglected in conceptual drag buildup methods. Similarity methods are described here, and more detailed aerodynamic methods are explored in Chapter 5.

The maximum lift-to-drag ratio vs wetted aspect ratio for several aircraft is presented in Fig. 3.8. Wetted aspect ratio is defined as

$$AR_{\text{Wet}} = \frac{b^2}{S_{\text{Wet}}} \qquad (3.57)$$

The wetted area S_{Wet} is the total exposed surface area of the UA. Note that the wetted aspect ratio is different from the aspect ratio, which uses the planform area as the reference area. The graph shows a band of maximum lift-to-drag ratios with a clear relationship to wetted aspect ratio. Interestingly, no distinction is made between configurations in the data points, suggesting that configuration selection is a secondary aerodynamic driver compared to wetted aspect ratio. The difference between the minimum and maximum values of maximum lift-to-drag ratio for a given wetted aspect ratio can be driven by flight regime, level of design refinement, propulsion integration impacts, and emphasis on aerodynamic efficiency relative to competing design drivers. The designer selects the most representative value for the design approach for the selected or calculated wetted aspect ratio.

The curves plotted in Fig. 3.8 show the predicted maximum lift-to-drag ratio as a function of wetted aspect ratio. Each curve has a value of C_f/e, and e and the mean skin-friction coefficient $\bar{C}_f$ are based on the wetted area.

$$L/D_{\text{Max}} = \sqrt{\frac{\pi \cdot AR_{\text{Wet}}}{4 \cdot (\bar{C}_f/e)}} \qquad (3.58)$$

Torenbeek [4] provides additional background for this method as well as insights into the groupings of manned aircraft classes. Maximum lift-to-drag ratio data of aircraft and other flying vehicles can be found in Thomas [5], Raymer [2], Tennekes [6], Lan and Roskam [1], and Johnstone and Arntz [7], to name a few resources.

The maximum lift-to-drag ratio occurs when the lift-dependent drag term is equal to the zero-lift drag.

$$C_{D0} = \frac{C_{L,\text{Best}L/D}^2}{\pi \cdot AR \cdot e} \qquad (3.59)$$

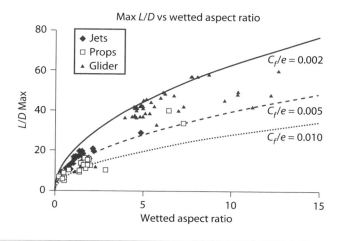

Fig. 3.8 Maximum lift-to-drag ratio vs wetted aspect ratio.

where $C_{L,\text{Best}L/D}$ is the lift coefficient at which the maximum L/D is achieved. Solving for this lift coefficient gives

$$C_{L,\text{Beast}L/D} = \sqrt{C_{D0} \cdot \pi \cdot AR \cdot e} \tag{3.60}$$

The aircraft total drag coefficient at the maximum lift-to-drag ratio condition is equal to twice the zero-lift drag coefficient. The lift-to-drag ratio is simply the ratio of the aircraft lift coefficient to the drag coefficient.

$$L/D_{\text{Max}} = \frac{C_{L,\text{Best}L/D}}{2 \cdot C_{D0}} = \frac{\sqrt{C_{D0} \cdot \pi \cdot AR \cdot e}}{2 \cdot C_{D0}} = \sqrt{\frac{\pi \cdot AR \cdot e}{4 \cdot C_{D0}}} \tag{3.61}$$

Solving for the zero-lift drag coefficient yields

$$C_{D0} = \frac{\pi \cdot AR \cdot e}{4 \cdot (L/D_{\text{Max}})^2} \tag{3.62}$$

Most forms of the propeller aircraft endurance equations utilize the endurance parameter $C_L^{3/2}/C_D$. The maximum endurance parameter value for a parabolic drag polar is

$$\left(\frac{C_L^{3/2}}{C_D} \right)_{\text{Max}} = \frac{3^{3/4}}{4} \cdot \frac{(\pi \cdot AR \cdot e)^{3/4}}{C_{D0}^{1/4}} \tag{3.63}$$

The values of maximum lift-to-drag ratio and endurance parameter can be used directly in the applicable performance equations only if the aircraft operates at these points. Note that the propulsion efficiency terms must also apply to these conditions. UAs frequently operate off the optimal aerodynamic points due to constraints or a mismatch between the propulsion system and aerodynamic best design conditions.

3.5 Initial UA Sizing Process

The process for the initial UA sizing varies greatly among design problems. The mass properties, performance, and aerodynamics methods described are tools that can facilitate the analysis, though often with modifications required by the problem. The general approach for conducing the initial UA sizing is shown in Fig. 3.9.

Note that geometry is not directly tied to mass properties methods. The aerodynamic performance is estimated from wetted aspect ratio, but this geometric parameter can be achieved by many alternative configurations. Some of the performance equations require that the wing loading is specified, so that the calculated takeoff gross weight can be used afterwards to determine the required wing area. Geometry constraints such as transportability or storage are also not handled by these methods.

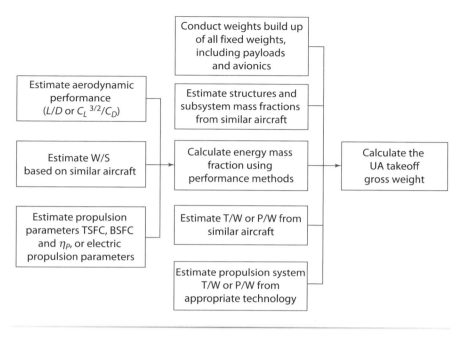

Fig. 3.9 Initial UA sizing process.

3.6 Examples

Example 3.1 Selection of Reciprocating Engine Technology

Problem:

An unmanned aircraft company is exploring a potential new tactical UAS commercial development intended to penetrate the military ISR market by replacing obsolete operational systems produced by their competitors. The company has access to advanced, scalable two-stroke and four-stroke heavy fuel engine technologies, but is unsure which is most desirable. The new aircraft will leverage the design approach, configuration, geometric proportions, and many components of their successful smaller UAS product. The existing product has the following characteristics:

$MF_{\text{Struct}} = 0.25$
$MF_{\text{Subs}} = 0.05$
$P/W_{\text{Aircraft}} = 0.05 \text{ hp/lb}$
Loiter $L/D = 15$

Loiter true airspeed at constant altitude = 70 KTAS
Propeller efficiency at loiter condition = 0.7

The lightweight two-stroke engine class has an installed power-to-weight ratio of 1 hp/lb and a brake specific fuel consumption of 1 lb/hp-h at the loiter condition. The more efficient four-stroke engine class has an installed power-to-weight ratio of 0.5 hp/lb and a brake-specific fuel consumption of 0.5 lb/hp-h at the loiter condition. The installed weight includes engine support systems, mechanical installation, and the propeller.

The total fixed weight for the new design is 100 lb, which includes all avionics, payloads, and other nonvarying weights.

Which engine technology yields the lightest aircraft for a mission consisting of a single 24-hr loiter segment? What power ratings are required for each engine class?

Solution:
Here the desired output is the takeoff gross weight for the two engine technologies.

First, consider the takeoff gross weight sizing equation.

$$W_{TO} = \frac{W_{PL} + W_{Avion} + W_{Other}}{1 - (MF_{Struct} + MF_{Subs} + MF_{Prop} + MF_{Energy})}$$

All weight components in the numerator are provided as a single value of 200 lb. The structural and subsystems mass fractions from the reference aircraft are applied directly here.

The propulsion mass fraction can be determined from the reciprocating engine equation.

$$MF_{Prop} = f_{Install} \frac{(P/W_{TO})_{Aircraft}}{(P/W_{Powerplant} + P/W_{Propeller})}$$

Here the $P/W_{Powerplant}$ includes all installation items and the propeller, and so $f_{Install}$ is unity, and $P/W_{Propeller}$ is set to zero. Rearranging with the new terms, we get

$$MF_{Prop} = \frac{(P/W_{TO})_{Aircraft}}{P/W_{Powerplant}}$$

Using the predecessor aircraft as a reference for P/W_{TO} gives

$$MF_{Prop} = \frac{0.05 \, hp/lb}{1 \, hp/lb} = 0.05$$

for the two-stroke engine. The four-stroke engine propulsion mass fraction is 0.1 by the same method.

Now the known quantities are substituted into the takeoff-gross-weight sizing equation. The resulting two-stroke engine equation is

$$W_{TO} = \frac{100 \, lb}{1 - (0.25 + 0.05 + 0.05 + MF_{Energy})} = \frac{100 \, lb}{0.65 - MF_{Energy}}$$

And the four-stroke engine equation is

$$W_{TO} = \frac{100 \, lb}{1 - (0.25 + 0.05 + 0.1 + MF_{Energy})} = \frac{100 \, lb}{0.6 - MF_{Energy}}$$

The only unknown is the energy mass fraction, which is the same as the fuel mass fraction in this case. Clearly the four-stroke engine must have a much lower fuel mass fraction to result in a lighter aircraft.

The fuel mass fraction equation for an aircraft flying at constant angle of attack and airspeed with specified endurance is

$$MF_{Fuel} = 1 - \exp\left(\frac{-E \times V \times BSFC}{L/D \times \eta_p}\right)$$

The flight velocity, lift-to-drag ratio, and propeller efficiency will be the same as the reference UA. The endurance is specified at 24 hrs. If we use units of knots for velocity and hours for endurance, the brake specific fuel consumption must be converted to units of inverse nautical miles by dividing by 326. The fuel mass fraction for the two-stroke engine is

$$MF_{Fuel} = 1 - \exp\left[\frac{-24 \, hr \times 90 \, kt \times (1 \, lb/hp - hr)/326}{15 \times 0.6}\right] = 0.52$$

The four-stroke engine calculation by the same method results in a fuel mass fraction of 0.31.

Now these fuel mass fractions can be put back into the takeoff gross weight equation's energy mass fraction term. Solving these equations shows that the takeoff gross weight is 769 lb for the two-stroke engine and 345 lb for the four-stroke engine. Despite the four-stroke engine's heavy characteristics, the low fuel consumption more than compensates in the UA sizing.

For the second question about power rating, the following equation must be used to calculate the sea-level uninstalled peak power of each engine

$$P = W_{TO} \times \left(P/W_{TO}\right)_{Aircraft}$$

Solving for the two-stroke engine

$$P = 769 \, lb \times 0.05 \, lb/hp = 38.5 \, hp$$

Using the same equation on the four-stroke engine shows that a 17.3-hp engine is required.

Example 3.2 MAV Requirements Analysis

Problem:

An engineer at a military program office is developing system requirements for a new expendable, container-launched micro-UA program. Because of currently certified physical constraints of the container, the MAV must weigh less than 0.5 lb. A key stakeholder organization responsible for flight safety and airspace integration demands that navigation and anticollision lights be placed on every unmanned aircraft regardless of size or role. The engineer performs an exhaustive industry search and government lab investigation, which showed that the smallest compliant lighting system using emerging technology will weigh 0.2 lb.

Other aspects of the design are known from separate feasibility studies. The smallest avionics suite, which also incorporates all subsystems functions, weighs 0.05 lb. The structural mass fraction is 0.2, the battery energy storage mass fraction is 0.2, and the electric motor and propeller mass fraction is 0.1. Except for the payload, all weights are fully accounted for within these parameters.

What is the maximum payload capacity of the MAV?

Solution:

The takeoff gross weight sizing equation is rearranged to adapt to the new information. The subsystems mass fraction is eliminated because this weight is captured in the fixed avionics weights. The fixed lighting system weight W_{Lighting} replaces the W_{Other} parameter because it is the only other fixed weight not covered by another parameter. The weight sizing equation becomes

$$W_{\text{TO}} = \frac{W_{\text{PL}} + W_{\text{Avion}} + W_{\text{Lighting}}}{1 - \left(MF_{\text{Struct}} + MF_{\text{Prop}} + MF_{\text{Energy}}\right)}$$

Solving the equation for payload weight gives

$$W_{\text{PL}} = W_{\text{TO}}\left[1 - \left(MF_{\text{Struct}} + MF_{\text{Prop}} + MF_{\text{Energy}}\right)\right] - W_{\text{Avion}} - W_{\text{Lighting}}$$

Substituting the parameters in

$$W_{\text{PL}} = 0.5\,\text{lb} \times \left[1 - (0.2 + 0.1 + 0.2)\right] - 0.05\,\text{lb} - 0.2\,\text{lb} = 0\,\text{lb}$$

the 0.2 lb of lighting consumed all of the available payload capacity. Unless the lighting requirement is dropped or the design changes in some way such as lighter structures or shorter endurance, this system will yield no military utility.

References

[1] Lan, C.-T. E., and Roskam, J., *Airplane Aerodynamics and Performance*, Design, Analysis and Research Corp., Lawrence, KS, 2003, p. 193.
[2] Raymer, D. P., "Thrust-To-Weight Ratio and Wing Loading," *Aircraft Design: A Conceptual Approach*, 2nd ed., AIAA, Reston, VA, 1992, p. 22, 77–100.
[3] Mattingly, J., *Aircraft Engine Design, Second Edition*, AIAA, Reston, VA, 2002.
[4] Torenbeek, E., *Synthesis of Subsonic Airplane Design*, Delft Univ. Press, Delft, The Netherlands, 1984, pp. 56, 148–155.
[5] Thomas, F., *Fundamentals of Sailplane Design*, College Park Press, College Park, MD, 1989, p. 168.
[6] Tennekes, H., *The Simple Science of Flight, From Insects to Jets*, MIT Press, Cambridge, MA, 1992, pp. 65–89.
[7] Johnstone, R., and Arntz, N., "CONDOR – High Altitude Long Endurance (HALE) Automatically Piloted Vehicle (APV)," AIAA Paper 90-3279, 1990.

Problems

3.1   Using the UA parameters described in Example 3.1, at what endurance will the two-stroke and four-stroke engine options provide equivalent takeoff-gross-weight values?

3.2   A design team is conducting a trade study of recovery options for a rocket-launched high-speed battlefield reconnaissance system. The mission radius is 400 n miles with 1-h time on station. The jet propulsion system used after initial launch has a T/W of 5, and a TSFC of 1.2 lb/lb-hr. The UA T/W without the rocket is 0.4. The UA has a structural mass fraction of 0.3, 100-lb payload, and 40 lb of all other fixed weights combined. The recurring production cost of the UA is $2,000 per pound of takeoff gross weight. The L/D for loiter is 10, and the L/D for cruise is 8. The UA cruises at 300 kt. The two landing-gear options are 1) a destructive recovery without landing gear that has no weight penalty and 2) a retractable landing gear with a mass fraction of 4%. What is the most cost-effective recovery option for five sorties?

3.3   An UA has $MF_{Struct} = 0.3$, $MF_{Subs} = 0.05$, and MF_{Prop} of 0.1. At what fuel mass fraction will the weight escalation factor equal 5?

3.4   A tailless aircraft has a wing span of 10 m, aspect ratio of 15, a fuselage length of 5 m, and a fuselage length-to-diameter ratio of 5. What is the wetted aspect ratio? What is the likely L/D_{max} if the UA is a) sailplane, b) jet, and c) piston-prop?

3.5   Using the results of problem 3.4, generate a drag polar assuming that the span efficiency factor is 0.7.

3.6   A jet-powered UA has $MF_{\text{Struct}} = 0.3$ and $MF_{\text{Subs}} = 0.05$. The total avionics, subsystems, and payload weight is 1,000 lb. The minimum UA thrust-to-weight is 0.33. The selected engine has 1,000 lb of thrust and weighs 200 lb installed. The number of engines is allowed to vary. Plot W_{TO} as a function of fuel mass fraction such that the thrust-to-weight constraint is not violated.

3.7   Demonstration project. Follow all applicable laws and safety precautions. Build a radio-controlled (R/C) electric-powered model airplane kit. Equip the aircraft with battery eliminator circuitry that allows the aircraft to remain controllable after a motor cutoff voltage is reached. Weigh all components, and calculate the major geometric properties. Have a properly qualified R/C pilot fly the aircraft. Fully charge the battery within manufacturer recommendations. Use the minimum power required to maintain level flight, and measure the flight endurance. Estimate the system aircraft endurance parameter and overall electric system efficiencies.

Chapter 4

Unmanned-Aircraft Geometry and Configurations

- Understand unmanned-aircraft geometric relationships
- Graphically depict unmanned-aircraft geometry from parametric definition
- Understand wing system configurations
- Fuselage configurations

Fig. 4.1 NASA's HiMat unmanned research aircraft in flight. (Photo courtesy of NASA.)

4.1 Introduction

Thre is no single best UA configuration. (Figure 4.1 shows one example of a workable UA configuration.) Well-written aircraft design requirements have many suitable answers, as is seen in varied industry responses to competitive procurements. Configuration selection is about much more than the best aerodynamic efficiency or other quantitative strengths. In this chapter we will see some of the pressures and influences that help guide the configuration selection.

Here, we will define UA geometry relationships and parameters. These relationships are important for configuration design, UA sizing, aerodynamics analysis, and mass properties' estimation.

4.2 Aircraft Geometry Relationships

4.2.1 Coordinate Systems

Coordinate systems are used to describe the UA geometry. Many systems exist for various purposes. In this chapter we will explore drafting and body coordinate systems. The reader is encouraged to consult Kirschbaum and Mason [1] for further details.

Drafting and mass properties use a common coordinate system to define geometry. This system, which we will call the drafting coordinate

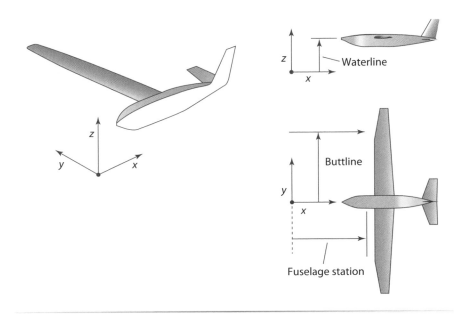

Fig. 4.2 Drafting coordinate system.

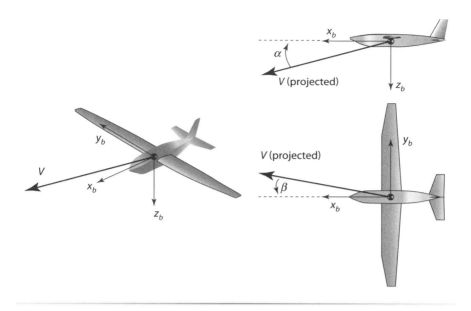

Fig. 4.3 Body-axis coordinate system.

system, is shown in Fig. 4.2. We will use the drafting coordinate system throughout the book unless noted otherwise. The x-z plane cuts vertically through the middle of the aircraft. For symmetric configurations, the x-z plane is the plane of symmetry. The x axis is positive looking aft. The y axis is positive out the right wing. The z axis is positive up. The origin is located ahead and below the nose of the aircraft. The reasons for locating the origin at this spot are so that the x and z coordinates of the UA are always positive. The x distance between the origin and the UA nose should be large enough so that the nose will be positive even if it grows. The same is true for the y-axis origin position. Typically, the nose is located at 100 length units in the x direction and y direction. The units are whatever length units are used, such as inches or centimeters.

Now we will introduce some geometric terminology specific to aircraft. The distance from the origin along the x axis is the *fuselage station*, which is abbreviated as FS. The vertical distance from the origin is the *waterline*, which is abbreviated as WL. The y distance from the origin is the *buttline*, which is abbreviated as BL. The y distance is also known as the *wing station*, with the abbreviation WS.

A common mistake is to place the coordinate system x origin to be located at the aircraft neutral point. The problem with this is that the neutral-point location estimate will be refined over time and the point may bear no apparent relationship to observable UA features.

The body axis system is used for stability and control analysis (see Fig. 4.3). The origin sits at the center of gravity. If the center of gravity changes, the body axis origin moves with it. The x axis points towards the nose, which has the opposite sign as drafting coordinates. The y axis points out the right wing (also known as the starboard wing for those who use a nautical analogy). The z axis points down, which again is opposite of the convention used for drafting.

The body axis is generally not aligned with the flight velocity vector. The misalignment in pitch is called angle of attack, which is denoted by α. The angle defined by a side component of the velocity vector on the body axis is called sideslip, which is denoted by β.

The UA geometry is created in the drafting coordinate system, but stability and control analysis is performed with the body axis. It is therefore necessary to convert from the drafting to body coordinate systems. The center-of-gravity location in the drafting coordinate system is x_{CG}, y_{CG}, and z_{CG}. The conversions are given by

$$x_b = x_{CG} - x \tag{4.1}$$

$$y_b = y \tag{4.2}$$

$$z_b = z_{CG} - z \tag{4.3}$$

This conversion assumes that the horizontal angular orientation is identical in both coordinate systems.

4.2.2 Wing Planform Geometry

Wing planform geometry for a simple single taper wing is shown in Fig. 4.4. This wing geometry discussion applies equally well to right-left

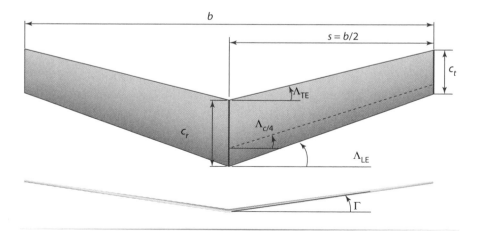

Fig. 4.4 Simple wing geometric parameters.

symmetric wings, horizontal stabilizers, and canards. So, it will therefore be treated generically at first.

The taper ratio λ is defined as

$$\lambda = \frac{c_t}{c_r} \tag{4.4}$$

The planform area is the projection of the wing at zero angle of attack onto a horizontal surface. This can be thought of at the area of the wing's shadow. The planform area S is given by

$$S = \frac{b}{2} \cdot (c_r + c_t) \tag{4.5}$$

The planform area in terms of taper ratio is

$$S = \frac{b}{2} \cdot c_r \cdot (1 + \lambda) \tag{4.6}$$

The wing aspect ratio AR is a measure of slenderness relationship between the span and chord. It is defined by

$$AR = \frac{b^2}{S} \tag{4.7}$$

The aspect ratio in terms of taper ratio and root chord is

$$AR = \frac{b}{c_r} \cdot \frac{2}{1 + \lambda} \tag{4.8}$$

The wing sweep for a trapezoidal wing is referenced to the normalized distance along the chord. The wing sweeps that are most commonly needed are referenced to the leading edge (LE), quarter-chord ($c/4$), and trailing edge (TE). The leading edge, quarter-chord, and trailing-edge relative chord values (x/c) are 0, 0.25, and 1, respectively. The way to convert sweeps from one relative chord value to another (x/c_1 to x/c_2) is

$$\Lambda_2 = \Lambda_1 + \tan^{-1}\left[\frac{c_r - c_t}{s} \cdot \left(\frac{x}{c_1} - \frac{x}{c_2}\right)\right] \tag{4.9}$$

This can also be expressed in terms of taper ratio and aspect ratio:

$$\Lambda_2 = \Lambda_1 + \tan^{-1}\left[\frac{4}{AR} \cdot \left(\frac{x}{c_1} - \frac{x}{c_2}\right) \cdot \frac{1 - \lambda}{1 + \lambda}\right] \tag{4.10}$$

Example 4.1   Wing Sweep Conversion

Problem:

A wing has the following characteristics:

$b = 20$ ft
$c_r = 2$ ft
$\lambda = 0.5$
$\Lambda_{c/4} = 15$ deg
What is the trailing-edge sweep?

Solution:

We will use Eq. (4.9) to find the leading-edge sweep. First, we must find the unknown parameters. The semispan is

$$s = \frac{b}{2} = \frac{20\,\text{ft}}{2} = 10\,\text{ft}$$

The tip chord is

$$c_t = c_r \cdot \lambda = 2\,\text{ft} \cdot 0.5 = 1\,\text{ft}$$

At the trailing edge x/c is 1. At the quarter-chord x/c is 0.25. Now we can solve for the trailing-edge sweep.

$$\Lambda_2 = \Lambda_1 + \tan^{-1}\left[\frac{c_r - c_t}{s} \cdot \left(\frac{x}{c_1} - \frac{x}{c_2}\right)\right]$$

$$= 15\,\text{deg} + \tan^{-1}\left[\frac{2\,\text{ft} - 1\,\text{ft}}{10\,\text{ft}} \cdot (0.25 - 1)\right] = 10.71\,\text{deg}$$

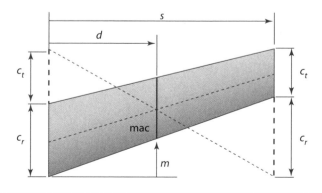

Fig. 4.5   Mean aerodynamic chord for a single panel.

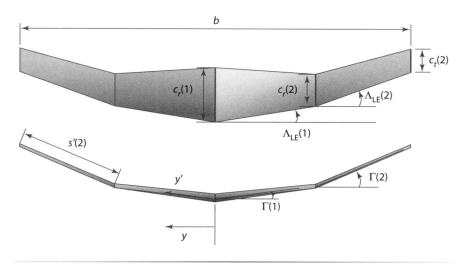

Fig. 4.6 Two-panel multipanel wing geometry definition.

The mean aerodynamic chord is an area weighted average chord that is used for stability and control analysis. The geometry used for finding the mean aerodynamic chord for a single wing panel is shown in Fig. 4.5. The mean aerodynamic chord length *mac* is found by

$$mac = \frac{2}{3} \cdot \left(c_r + c_t - \frac{c_r \cdot c_t}{c_r + c_t} \right) \qquad (4.11)$$

The *x* distance from the wing panel root to the leading edge of the mean aerodynamic chord *m* is found by

$$m = \frac{s}{3} \cdot \frac{c_r + 2 \cdot c_t}{c_r + c_t} \qquad (4.12)$$

The *y* distance from the wing root to the mean aerodynamic chord *d* is found by

$$d = \frac{s}{3} \cdot \frac{c_r + 2 \cdot c_t}{c_r + c_t} \qquad (4.13)$$

Wings often have more than one panel. A multipanel wing example with two panels is shown in Fig. 4.6. A multipanel wing has N panels, and an index j where $1 \leq j \leq N$. The wing is defined in both drafting coordinates (x, y, z) and a new span coordinate system (x', y', z'). The y' direction follows the wing from the root to the tip in the x-y plane. The z' direction is normal to the local y' direction in the x-y plane. The x and x'

directions are parallel. Each panel j has a root chord $c_r(j)$, tip chord $c_t(j)$, panel length $s'(j)$, y location of the root $y_r(j)$, y location of the tip $y_t(j)$, and normal planform area $S'(j)$, and sweep about x/c $\Lambda_{x/c}(j)$, and dihedral angle $\Gamma(j)$.

The span is the distance between the wing tips. This is expressed by

$$b = 2 \cdot y_t(N) \tag{4.14}$$

The span can also be found by

$$b = \sum_{j=1}^{N} s'(j) \cdot \cos\left[\Gamma(j)\right] \tag{4.15}$$

The span as measured along the span path b' is found by

$$b' = \sum_{j=1}^{N} s'(j) \tag{4.16}$$

When the dihedral angle is zero for all panels, b and b' are equal. The equivalent taper ratio λ_{eq} for the wing is found by

$$\lambda_{\text{eq}} = \frac{s'(1)}{b'/2} \cdot \left[1 + \frac{c_t(1)}{c_r}\right] + \left[\sum_{j=2}^{N} \frac{s'(j)}{b'/2} \cdot \frac{c_r(j) + c_t(j)}{c_r}\right] - 1 \tag{4.17}$$

The summation term is not evaluated when there is only one panel ($N = 1$). Note that c_r is the same as $c_r(1)$. Also, $c_r(j)$ is equal to $c_t(j\text{-}1)$ for a wing without spanwise chord distribution discontinuities. It is often convenient to parameterize the multipanel wing with panel taper ratios with respect to the root chord $\lambda(j)$, which is defined by

$$\lambda(j) = \frac{c_t(j)}{c_r} \tag{4.18}$$

The ratio of the distance y' along the span to the wing spanwise coordinate system semispan $b'/2$ is defined by

$$\eta' = \frac{y'}{b'/2} \tag{4.19}$$

This ratio for panel j root and tip locations is found by

$$\eta'_r(1) = 0 \quad \text{and} \quad \eta'_r(j) = \sum_{i=1}^{j-1} s'(j) \quad \text{for } j > 1 \tag{4.20}$$

$$\eta'_t(j) = \sum_{i=1}^{j} s'(j) \tag{4.21}$$

Using panel taper, the wing equivalent taper λ_{eq} can be found by

$$\lambda_{eq} = \eta'_t(1) \cdot [1 + \lambda(1)] + \left\{ \sum_{j=2}^{N} [\eta'_t(j) - \eta'_r(j)] \cdot [\lambda(j-1) - \lambda(j)] \right\} - 1$$

(4.22)

Recall that the planform area S is the area projected onto a horizontal surface. For a multisegment wing, this is found by

$$S = \sum_{j=1}^{N} S'(j) \cdot \cos[\Gamma(j)]$$

(4.23)

The planform area component normal to the spanwise path S' is found by

$$S' = \sum_{j=1}^{N} S'(j)$$

(4.24)

The aspect ratio along the spanwise path AR' is defined by

$$AR' = \frac{b'^2}{S'}$$

(4.25)

The major defining parameters for the wing in spanwise coordinates are b', c_r, AR', and S'. Compound taper wings can use λ_{eq} as a shape function. Given any two defining parameters, the other two can be calculated. Table 4.1 shows the conversions for a wing given b'. Other forms of the table (given c_r, AR', or S') are left as an exercise. Note for a wing with no dihedral that b and b', S and S', and AR and AR' are equivalent.

The process for calculating the mean aerodynamic chord and its location involves evaluation and averaging of multiple panels (see Fig. 4.7).

Table 4.1 Wing Parameter Calculations Given b'

| Find this parameter via equation | Given b' and this parameter | | |
	AR'	c_r	S'
AR'	—	$\dfrac{2}{(1+\lambda_{eq})} \cdot \dfrac{b'}{c_r}$	$\dfrac{b'^2}{S'}$
c_r	$\dfrac{2}{(1+\lambda_{eq})} \cdot \dfrac{b'}{AR'}$	—	$\dfrac{2}{(1+\lambda_{eq})} \cdot \dfrac{S'}{b'}$
S'	$\dfrac{b'^2}{AR'}$	$\dfrac{(1+\lambda_{eq})}{2} \cdot c_r \cdot b'$	—

The mean aerodynamic chord for panel j, $mac(\ )$, is found by

$$mac(j) = \frac{2}{3} \cdot \left[c_r(j) + c_t(j) - \frac{c_r(j) \cdot c_t(j)}{c_r(j) + c_t(j)} \right] \tag{4.26}$$

The distance from the wing root to the leading edge of $mac(j)$, $m(j)$, is found by

$$m(j) = \frac{s(j)}{3} \cdot \frac{c_r(j) + 2 \cdot c_t(j)}{c_r(j) + c_t(j)} + \sum_{i=1}^{j-1} s(i) \cdot \tan[\Gamma_{LE}(i)] \tag{4.27}$$

where $\Gamma_{LE}(i)$ is the leading-edge sweep of panel i. Note that the summation term is set to zero when $j = 1$. The area of a panel is given by

$$S(j) = \frac{c_r(j) + c_t(j)}{2} \cdot s(j) \tag{4.28}$$

The wing mean aerodynamic chord can now be found by

$$mac = \sum_{j=1}^{N} mac(j) \cdot S(j) \Big/ \sum_{j=1}^{N} S(j) \tag{4.29}$$

The distance from the wing root to the wing mean aerodynamic center is found by a similar method:

$$m = \sum_{j=1}^{N} m(j) \cdot S(j) \Big/ \sum_{j=1}^{N} S(j) \tag{4.30}$$

The parameter d for the wing becomes

$$d = \frac{s(j)}{3} \cdot \frac{c_r(j) + 2 \cdot c_t(j)}{c_r(j) + c_t(j)} + \sum_{i=1}^{j-1} s(i) \tag{4.31}$$

Once again, the summation term is not evaluated when $j = 1$.

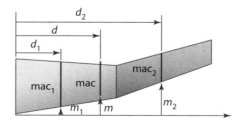

Fig. 4.7 Mean aerodynamic chord for a two-panel wing.

The aerodynamic center of the wing is located at the quarter-chord of the wing mean aerodynamic chord. This point is important for stability and control analysis. In the body coordinate system, the x location of the wing aerodynamic surface X_{ac} is

$$X_{ac} = X_{LE,b} - m - \frac{1}{4} \cdot mac \tag{4.32}$$

where $X_{LE,b}$ is the x location of the wing leading edge in the body coordinate system.

4.2.3 Tail Parametric Sizing

Tail volume method of tail-area sizing relates tail area to the wing area, tail moment arm, a wing characteristic length, and a scaling coefficient. Tail volume methods are often used in early conceptual design to produce tails with geometries that are reasonably likely to support acceptable handling qualities if historical norms are followed. Detailed stability and control analysis that comes after conceptual design can lead to modifications of the tail sizing and proportions.

The horizontal tail volume coefficient TVC_H is defined by

$$TVC_H = \frac{S_H \cdot L_H}{S_w \cdot mac_w} \tag{4.33}$$

where S_H is the horizontal stabilizer planform area, S_w is the wing area, and mac_w is the wing mean aerodynamic chord. The horizontal tail moment arm L_H is defined by

$$L_H = X_{CG} - X_{ac,H} \tag{4.34}$$

where $X_{ac,H}$ is the x location of the horizontal stabilizer aerodynamic center in body coordinates. The horizontal stabilizer area can be calculated by rearranging the tail volume coefficient equation.

$$S_H = \frac{TVC_H \cdot S_w \cdot mac_w}{L_H} \tag{4.35}$$

From this equation, we can see that increases in wing area or wing mean aerodynamic chord will increase the horizontal stabilizer area, and a longer tail moment arm reduces the stabilizer area. Higher-aspect-ratio wings have lower horizontal stabilizer areas because the wing mean aerodynamic chord is smaller for a given wing area. The inverse proportional relationship to tail moment arm reveals that close coupled tails, that is, tails with a short moment arm, will have larger areas than those with longer moment arms.

The canard tail volume coefficient TVC_C is

$$TVC_C = \frac{S_C \cdot L_C}{S_w \cdot mac_w} \tag{4.36}$$

where S_C is the canard planform area. The canard moment arm LC is defined by

$$L_C = X_{CG} - X_{ac,C} \tag{4.37}$$

where $X_{ac,C}$ is the x location of the canard aerodynamic center in body coordinates. Note that the moment arm is negative if the canard is located ahead of the center of gravity. Therefore, TVC_C is expected to be a negative number.

The vertical tail volume coefficient TVC_V is scaled by the wing span instead of the wing mean aerodynamic chord. It is given by

$$TVC_V = \frac{S_V \cdot L_V}{S_w \cdot b} \tag{4.38}$$

where S_V is the vertical tail planform area and L_V is the longitudinal distance between the aircraft center of gravity and the vertical tail aerodynamic center. The vertical tail moment arm L_V is defined by

$$L_V = X_{CG} - X_{ac,V} \tag{4.39}$$

where $X_{ac,V}$ is the x location of the vertical stabilizer aerodynamic center in body coordinates.

4.2.4 Airfoil Geometry

Airfoils are two-dimensional cross sections of a wing. The aerodynamic performance of the UA is strongly dependent upon this geometry. Airfoils define the aerodynamic efficiency (lift-to-drag ratio or endurance parameter) and landing speeds, among other performance characteristics.

Two major parameters that characterize an airfoil are thickness and camber. The thickness and camber distribution is the art and science of airfoil design, as we will see in Chapter 5. For now we will merely explore how the geometry is defined.

Airfoil data files provide the upper and lower surface coordinates normalized to the chord length. The chord station parameter is x/c, where x is measured from the leading edge. The leading edge and trailing edge of the airfoil have x/c values of 0 and 1, respectively. The upper and lower surface height is parameterized as y/c. Note that airfoil y coordinates have no relationship to drafting or body coordinates. A common convention for defining the y/c coordinates is to start at the trailing edge ($x/c = 1$), move along the upper surface to the leading edge ($x/c = 0$), and then go back to the trailing edge ($x/c = 1$) on the lower surface.

UA geometry software tools read in the airfoil geometry files. The coordinates are scaled to the local chord of the aerodynamic surface. The incidence might need to be adjusted to account for wing incidence angle and washout distribution. Geometric interpolation is sometimes necessary when different airfoils are applied across the aerodynamic surface.

The National Advisory Committee for Aeronautics (NACA), the predecessor to NASA, produced several families of airfoils whose geometry can be produced with parametric mathematical formulas. Although these sections are often inferior to more specialized airfoil sections, NACA airfoils do sometimes become part of UAS designs even today. The author admits to using symmetrical NACA four digit airfoils on tail sections frequently.

The numbers in the name of a four-series NACA airfoil provide the parameters that are required to generate the airfoil coordinates. The first digit is the percentage of maximum camber. M is value of the maximum camber. The second digit represents the location of the position of the maximum camber, measured in 10% increments. P is the x/c value of the maximum camber. A NACA 2410 airfoil has 2% camber, maximum camber located 40% of the chord length from the leading edge, and has a 10% thickness-to-chord ratio. Here $M = 0.02$, and $P = 0.4$.

Here we will explore the NACA four-digit airfoil family to better understand the major defining airfoil geometric parameters. The upper surface x and y coordinates, x_u and y_u, respectively, are given by

$$x_u = x - y_t(x) \cdot \sin(\theta), \quad y_u = y_c(x) + y_t(x) \cdot \cos(\theta) \tag{4.40}$$

The lower surface coordinates are given by

$$x_l = x + y_t(x) \cdot \sin(\theta), \quad y_l = y_c(x) - y_t(x) \cdot \cos(\theta) \tag{4.41}$$

The angle θ is given by

$$\tan(\theta) = \frac{dy_c}{dx} \tag{4.42}$$

The parameter y_c is the height of the camber line, and the derivative dy_c/dx is the slope of the camber line. For $x/c < P$,

$$y_c = c \cdot \frac{M}{P^2} \cdot \left[2 \cdot P \cdot (x/c) - (x/c)^2\right] \quad \text{and} \quad \frac{dy_c}{dx} = \frac{2 \cdot M}{P^2} \cdot [P - (x/c)] \tag{4.43}$$

For $x/c \geq P$,

$$y_c = c \cdot \frac{M}{(1 - P)^2} \cdot \left[1 - 2 \cdot P + 2 \cdot P \cdot (x/c) - (x/c)^2\right] \tag{4.44}$$

and

$$\frac{dy_c}{dx} = \frac{2 \cdot M}{(1 - P)^2} \cdot [P - (x/c)] \tag{4.45}$$

The thickness parameter y_t is found by

$$y_t = c \cdot (t/c) \cdot \left[a_0 \cdot \sqrt{x/c} - a_1 \cdot (x/c) - a_2 \cdot (x/c)^2 + a_3 \cdot (x/c)^3 - a_4 \cdot (x/c)^4 \right]$$

(4.46)

The coefficients are as follows: $a_0 = 1.4845$, $a_1 = 0.6300$, $a_2 = 1.7580$, $a_3 = 1.4215$, and $a_4 = 0.5075$.

NACA also provided five-series airfoils that can be parametrically generated. The NACA families have been modified for a number of applications, including flying wing airfoils with reflexed (s-shaped) camber lines.

4.2.5 Fuselage Geometry

The fuselage geometry is defined by a few characteristic dimensions, the distribution of width and height, and the cross-section shapes. From these geometric dimensions and shape parameters other useful information can be derived, such as the surface area and internal volume. The geometric definition must be graphically represented in a CAD system or aircraft design tool user interface for proper design visualization.

Much like an airfoil, the contours of the upper and lower fuselage can be described by the local ratio of vertical dimension to the overall fuselage length. Because fuselages are three dimensional, the local width to fuselage length is also an important parameter. These three parameters are z_{Top}/L, z_{Bot}/L, and y_{Side}/L, respectively. Note that x, y, and z are parallel to the drafting axes.

The fuselage is generally broken into multiple discrete segments for analysis. The breakpoints for the segmentation may correspond to the points at which the shape parameters are provided. Cross section i has a forward cross section given the index i and a rear cross section with the index $i + 1$. The volume of the segment $Vol(i)$ can be approximated by

$$Vol(i) \approx \frac{A_{\text{cross}}(i) + A_{\text{cross}}(i + 1)}{2} \cdot [x(i + 1) - x(i)]$$

(4.47)

where A_{cross} is the cross-section area. The surface area of the segment $A_{\text{Surf}}(i)$ is approximated by

$$A_{\text{Surf}}(i) \approx \frac{P(i) + P(i + 1)}{2} \cdot [x(i + 1) - x(i)]$$

(4.48)

where P is the perimeter of the cross section. The segment volume and surface areas are approximations because the relationships vary with cross-section shape family. The total volume and surface area are found by summing all of the contributing segments.

Designers can use a variety of different cross-section families to define the fuselage geometry. The most simple are the elliptical or rectangular cross sections. Other cross-section families are useful for more complex blending and aerodynamics considerations.

Conic sections are commonly used for aircraft constructed of sheet metal. The conic family of curves permits wrapping sheets of flat material across sections of an aircraft. Most modern UASs have outer skins made of molded composite materials rather than sheet metal. Molded structures permit compound curvature, and the constraints of conic sections are no longer required, yielding much greater design flexibility. However, this shape family is still useful for fuselage design. Raymer [2] provides a methodology for generating these curves.

The superellipse family can generate cross sections similar to conics and other shapes as well. The general form of the superellipse cross-section equation is

$$\left(\frac{z}{a}\right)^{2+n} + \left(\frac{y}{b}\right)^{2+m} = 1 \tag{4.49}$$

Solving for z/b yields

$$\frac{z}{b} = \left[1 - \left(\frac{y}{a}\right)^{2+m}\right]^{1/(2+n)} \tag{4.50}$$

This equation is an ellipse when n and m are both equal to 0. Several superellipse shapes are plotted in Fig. 4.8.

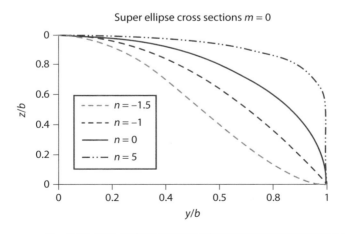

Fig. 4.8 Superellipse cross sections.

The cross-sectional area of the superellipse can be found through a numerical integration.

$$A_{\text{cross}} = \frac{b}{N} \cdot \left\{ \sum_{i=2}^{N} \left| z[y(i)] + z\left[y(i) - \frac{b}{N}\right] \right|_{\text{Top}} + \sum_{i=2}^{N} \left| z[y(i)] + z\left[y(i) - \frac{b}{N}\right] \right|_{\text{Bot}} \right\}$$

(4.51)

Rectangular cross sections are quite common in UASs. Here the cross section is simply a rectangle. Fuselage drag can be reduced substantially if the corners are rounded, even with a radius of 10% of the maximum width or height. Sharp corners are easier to manufacture if sheets of material are used for the side. Otherwise, the rounded corners are easier to build in molds. Often the rectangular cross section progresses to a rounded cross section at the nose. The cross-sectional area of a rectangular cross section with rounded corners is

$$A_{\text{cross}} = W \cdot H + (\pi - 4) \cdot r^2 \tag{4.52}$$

Very detailed geometry can be defined with few parameters. Figure 4.9 shows a fuselage that is created by generating multiple cross sections for each body. This aircraft has a main fuselage, fuselage landing-gear fairings, and four rotor nacelles. Conic sections were used, where shape parameters ρ defined the fullness of the cross section. Each fuselage is represented by an array of data in the following format:

x/L	z/Lm	z/Lt	z/Lb	y/Ls	rhot	rhob
0.000	0.000	0.000	0.000	0.000	0.414	0.414
0.005	0.000	0.003	−0.003	0.003	0.414	0.414
....	......	....	....	......	....	......
$x/L_f(j)$	$z/L_{m,f}(j)$	$z/L_{t,f}(j)$	$z/L_{b,f}(j)$	$y/L_f(j)$	$\rho_t(j)$	$\rho_b(j)$

Fig. 4.9 Fuselage geometry defined by multiple cross sections.

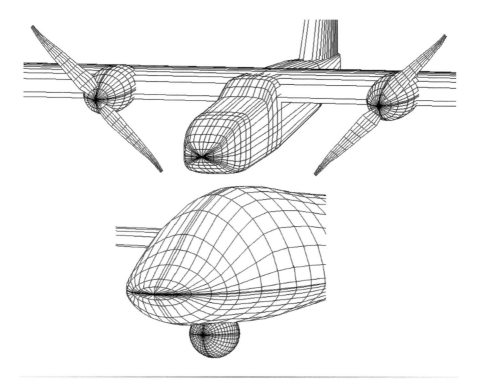

Fig. 4.10   Fuselage geometry approximations of the NRL Dragon Eye and General Atomics Predator.

The x coordinate moves from the nose to the tail, the y coordinate moves out the right wing, and the z coordinate is positive in the up direction. All x, y, and z data are normalized to the fuselage length. Three z-coordinate data points are defined for each x coordinate. The first is a midline, which defines the z location of the widest portion of the fuselage for the section. The second is the upper surface of the fuselage, and the third is the lower surface of the fuselage. The single y-coordinate data point defines the maximum width of the fuselage. The ρ parameters define the fullness of the fuselage cross section such that the surface between adjacent fuselage segments is a conic section. A ρ value of 0 corresponds to a straight line (triangle), 1 is a rectangle, and 0.414 is an ellipse. The ρ parameter is defined separately for the upper and lower surfaces.

The top section of Fig. 4.10 shows a conic representation of the NRL Dragon Eye, and the bottom section shows the Predator A nose region. The Dragon Eye's main fuselage has a conic parameter of nearly 1. As drawn, the Predator's upper conic parameter ρ for the main fuselage is 0.35, and the EO/IR ball has a conic parameter of 0.414. The EO/IR ball is the sphere under the Predator fuselage.

4.3 Configuration Drivers

Nearly every subject covered in this book influences the UA configuration, with the possible exception of the ground control station. Indeed, it is difficult to separate the mission from the UA layout. Unmanned aircraft have design freedom relative to their manned aircraft counterparts, but also many additional constraints. We will explore configuration drivers that are common to both domains, as well as considerations that are unique to UASs.

Literature on the benefits and drawbacks of various wing system configurations abound [2–4]. Conventional wing-tail, canard, tandem wings, flying wings, and perhaps more exotic configurations are compared for aerodynamic efficiency, high lift, or other metrics. Some studies include structural sizing and stability and control as well. You can conclude from reviewing numerous configuration studies that the conventional wing-tail configuration is hard to beat. From a purely aerodynamics perspective this might be true. However, we will see that there are myriad systems considerations that can sway the configuration selection towards an unexpected alternative, even if there is an aerodynamic penalty for doing so. For example, which configuration has the best packaging efficiency within a tube? The wing system configuration can impact the UA takeoff gross weight by perhaps 5–10%, but other design choices are on the same level of impact.

Aesthetics are important for generating customer enthusiasm, but many UAS competition winners look rather unremarkable. There can be no evaluation criteria for aesthetics in competently managed competitive source selections. Source selections do involve a degree of subjectivity because humans are still required to evaluate proposals. After all, there are many exceptions to the cliché "if it looks good, it will fly good." Form generally follows function, and refined geometry implies a higher degree of design maturity. Customers would prefer to not be embarrassed by homely desk models because they will invest a portion of their careers with the system. Another cliché is "beauty is in the eye of the beholder," so it is wise to recognize the preferences of the customer. Graceful, sailplane-like lines and menacing, angular contours bristling with weapons might each appeal to different customers.

Companies might be forced to use a particular configuration because of intellectual property, the desire to recoup previous investments, a competition requirement for high system maturity, or preferences of senior company leadership. If possible, companies should perform unbiased configuration trade studies to find the best match of configuration and mission requirements to ensure that the best offering is provided.

Companies can have preferred configurations. For example, most of General Atomics' designs have employed close-coupled V-tails (with some inverted) and a single fuselage. AAI's Shadow 200, 400, and 600 also have remarkably similar appearances. Signature configurations can support branding and product recognition. Designs might also appear to be scaled versions of an earlier platform, despite having different mission capabilities. This similarity of appearance could justify a sole-source system upgrade procurement rather than a new system competitive procurement, which is highly desirable for the incumbent. Chapter 3 shows several UA families that share a common configuration.

Companies can reuse configurations to prove system maturity or recoup prior investments in the products. New contract opportunities can arise suddenly, and a company might not have time to start from a clean sheet of paper on a new design. Finding opportunities for reuse can make the difference between bidding or not. A company could make the case that a reuse of a proven configuration reduces risk for a new product, even if there are differences in scale or flight envelope. A direct scaling of UA geometry can permit use of a common aerodynamic database and support reuse of flight control logic. For example, Swift Engineering developed the Killer Bee 2 and 3, which appear to be direct scaling of the outer mold line.

Some configurations have worked well on other successful UASs, and so they are adopted for new designs. Such arrangements are unlikely to generate unwanted controversy with customers. These archetypal configurations come to be what is expected. The conventional wing-tail twin-boom pusher engine configuration is the most notable example of this phenomenon. If you ever sense that there sure are a lot of conventional, twin-boom pusher UASs you are not wrong, as can be seen in Fig. 4.11. Successful 1980s system such as the IAI Mastiff, IAI Scout, and then the Pioneer UAS all had this configuration. This seems to have inspired a generation of UAS designers. Customers usually feel comfortable with familiar-looking designs and will perceive that the familiar is likely to succeed. The risk of looking too similar to the norm is that your system will not stand out. It is a tradeoff between the influences of "nobody ever got fired for buying a conventional twin boom pusher" and a "me too" design.

Occasionally an aircraft designer might find that it becomes necessary to evaluate an unusual configuration. The author recommends early use of very simple sheet balsa wood prototypes to evaluate feasibility—a practice evangelized by Rick Foch of Naval Research Labs. The models can use balsa sheets of 1/16th–3/32nd-in. thickness cut to the shape of the

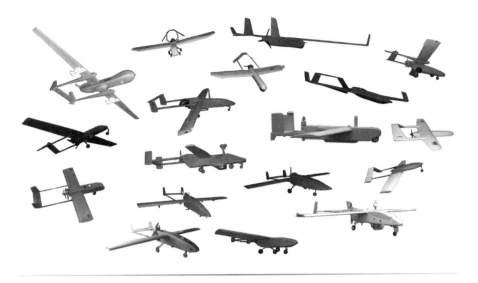

Fig. 4.11   Plethora of conventional, twin-boom pushers.

wings, tails, and bodies. Clay or metal washers can be applied to adjust the longitudinal center of gravity. The horizontal stabilizers or elevator surfaces can be adjustable. The models typically span less than 10 in. and can be hand-launched indoors. Much can be learned from these simple models that can take less than 2 hrs to construct and test. The required center-of-gravity location for positive static margin can be quickly established. The lateral-directional stability characteristics are immediately evident, such as spiral stability and Dutch roll. Some designs cannot be trimmed, can consistently fly inverted, or swap ends in yaw. Such behavior is generally an indication that the design must be modified or that development will be challenging. Quick model modifications can solve these problems with minimal investment.

Unconventional configurations often come with great promises by their advocates. However, great claims should be supported by great proof. So far, no configuration has solved all of aviations' problems in all circumstances. Unusual configurations might possess undesirable characteristics that are not immediately apparent. Any new configuration should be considered as high risk until proven on multiple flight vehicles.

Unmanned aircraft often offer more configuration freedom than manned aircraft. However, there are also many additional drivers such as sensor field of regard, payload flexibility, satellite communications, transportability, maintainability, and launch and recovery provisions. The configuration should be influenced by the design requirements to enable a compliant system.

4.4 Wing System Configurations

The wing system configuration is the combination of lifting surface and pitch trim surfaces. These functions are performed by separate surfaces for most configurations, but are combined in the case of a flying wing. Most UA are conventional configurations with the tail behind the wing, though flying-wing designs rank second in prevalence. Conventional, canard, three-surface, and flying-wing configurations are described in this section, though numerous alternatives exist. Figure 4.12 shows the primary wing system configuraitons.

For this discussion of the pitch stability and control, the geometry is defined in the body coordinate system. The positive x direction points towards the nose from the center of gravity. A point aft of the center of gravity has a negative x-coordinate value. Many stability and control texts put the x direction towards the tail from the leading edge of the mean aerodynamic chord for the purposes of longitudinal stability exploration. Hopefully by using the body coordinate system here, some confusion will be avoided when considering forces, moments, and rates for stability and control analysis, which uses the body coordinate system.

Pitch static stability is the tendency of the UA to return to its pitch state after being disturbed. The neutral point is the longitudinal location about which the aircraft is neutrally stable in pitch to a disturbance. If the center of gravity is located ahead of this point, then the UA has positive pitch stability. A center of gravity aft of the neutral point has negative pitch stability. The neutral point x location in the body coordinate system X_{NP} for an aircraft with a main wing plus N additional horizontal surfaces is

$$\frac{X_{NP}}{mac_w} = \underbrace{\frac{X_{ac,w}}{mac_w}}_{\text{Wing}} + \underbrace{\frac{C_{ma,f}}{C_{L\alpha,w}}}_{\text{Fuselage}} - \underbrace{\sum_{j=1}^{N} \eta(j) \cdot \frac{S(j)}{S_w} \cdot \left[\frac{X_{CG} - X_{ac}(j)}{mac_w}\right] \cdot \frac{C_{L\alpha}(j)}{C_{L\alpha,w}} \cdot \left[1 - \frac{d\varepsilon}{d\alpha}(j)\right]}_{\text{Stabilizers}}$$

(4.53)

where

$C_{L\alpha}(j)$ = lift-curve slope of surface j (derivative of surface j C_L with respect to α)

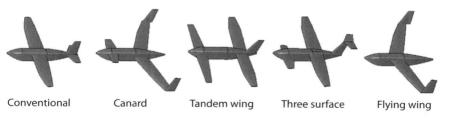

| Conventional | Canard | Tandem wing | Three surface | Flying wing |

Fig. 4.12 Primary wing system configurations.

$C_{L\alpha,w}$ = wing lift-curve slope (derivative of wing C_L with respect to α)

$C_{m\alpha,f}$ = fuselage pitching-moment coefficient derivative with respect to angle of attack

$d\varepsilon/d\alpha(j)$ = derivative of downwash angle with respect to angle of attack for surface j

$X_{ac}(j)$ = x location of surface j's aerodynamic center in body coordinates

X_{CG} = x location of the center of gravity in body coordinates

$S(j)$ = planform area of surface j

$\eta(j)$ = ratio of dynamic pressure at surface j to the freestream dynamic pressure.

For a flying wing with no fuselage, the neutral point and the wing aerodynamic center the same point. Fuselages generally have positive $C_{m,\alpha,f}$, and the wing lift-curve slope is positive, so that fuselages tend to move the neutral point forward. Horizontal stabilizers located behind the center of gravity move the neutral point aft, whereas canards move the neutral point forward. Note that the term

$$\frac{S(j)}{S_w} \cdot \frac{[X_{CG} - X_{ac}(j)]}{mac_w} \tag{4.54}$$

is the tail volume coefficient for horizontal stabilizers where the horizontal surface is located behind the center of gravity, and it is also the tail volume coefficient for canards.

The static margin is a measure of the pitch stability. The stick-fixed static margin *sm* is defined as

$$sm = \frac{X_{CG} - X_{NP}}{mac_w} \tag{4.55}$$

A center of gravity located ahead of the neutral point results in a positive static margin and positive longitudinal static stability. The static margin for many UASs can span 5–15%. Highly maneuverable aircraft, such as air-to-air combat UCAVs, can have a neutral or negative static margin.

Now, consider a generic case where there is one main wing and N additional horizontal surfaces. The index for each wing is j, where $1 \le j \le N$. The pitching moment about the center of gravity C_m is

$$C_m = C_{m,w} + C_{m,fuse} + \sum_{j=1}^{N} C_m(j) \tag{4.56}$$

The pitching moment of the horizontal surfaces can be expanded to

$$C_m(j) = \left\{ C_{m0}(j) - \eta(j) \cdot \left[\frac{X_{CG} - X_{ac}(j)}{mac_w} \right] \cdot C_L(j) \right\} \cdot \frac{S(j)}{S_w} \tag{4.57}$$

where

$C_L(j)$ = lift coefficient of surface j reference to area $S(j)$
$C_{m,w}$ = wing pitching-moment coefficient
$C_{m0}(j)$ = zero lift pitching moment of surface j referenced to area $S(j)$

Recombining terms gives

$$C_m = C_{m,w} + C_{m,f} + \sum_{i=1}^{N} \left\{ C_{m0}(j) - \eta(j) \cdot \left[\frac{X_{CG} - X_{ac}(j)}{mac_w} \right] \cdot C_L(j) \right\} \cdot \frac{S(j)}{S_w}$$

(4.58)

Here, it can be seen that positive lift on a surface located behind the center of gravity will yield a nose-down pitching moment (negative pitching moment). The opposite is also true: positive lift on a surface located ahead of the center of gravity will produce a nose-up pitching moment (positive pitching moment).

The lift coefficient of a horizontal surface can be found by

$$C_L(j) = C_{L\alpha}(j) \cdot \left\{ i(j) - i_w - \varepsilon_0(j) + \left[1 - \frac{d\varepsilon}{d\alpha}(j) \right] \cdot \alpha \right\} + C_{L\delta e}(j) \cdot \delta e(j)$$

(4.59)

where

$C_{L\delta e}(j)$ = derivative of the tail lift coefficient with respect to elevator deflection
$i(j)$ = incidence angle of the horizontal stabilizer
i_w = wing angle of the wing
$\delta_e(j)$ = elevator deflection
$\varepsilon_0(j)$ = downwash angle at surface j for zero angle of attack

For a given angle of attack, the pitching-moment contribution of surface j can be affected by adjusting the incidence angle or the elevator deflection. Horizontal surfaces that can adjust incidence for control are called *full flying stabilizers*. Usually a horizontal surface is either a full flying stabilizer or has an elevator, but not both.

The wing pitching moment about the center of gravity is

$$C_{m,w} = C_{m0,w} - C_{L,w} \cdot \left(\frac{X_{CG} - X_{ac,w}}{mac_w} \right)$$

(4.60)

where $C_{m0,w}$ is the wing zero-lift pitching-moment coefficient about the wing aerodynamic center, $C_{L,w}$ is the wing lift coefficient, and $X_{ac,w}$ is the wing aerodynamic center x location in body coordinates. The wing zero-lift pitching-moment coefficient is caused by the airfoil pitching-moment coefficient for wings with zero quarter-chord sweep. Swept wings have an additional zero-lift pitching-moment contribution because of the lift distribution because the wing might have portions of the wing with positive and negative lift that cancels.

This can also be expressed in terms of angle of attack α

$$C_{m,w} = C_{m0,w} - C_{L\alpha,w} \cdot \left(\frac{X_{CG} - X_{ac,w}}{mac_w} \right) \cdot \alpha \qquad (4.61)$$

The fuselage contribution to the pitching moment is

$$C_{m,f} = C_{m0,f} + C_{m\alpha,f} \cdot \alpha \qquad (4.62)$$

where $C_{m0,f}$ is the fuselage pitching-moment coefficient at zero angle of attack. This assumes that the fuselage generates negligible lift. Methods of estimating the fuselage moments and forces are covered in Chapter 5. The relationship between α and $C_{L,w}$ is

$$\alpha = \frac{C_{L,w}}{C_{L\alpha,w}} + \alpha_{0L} \qquad (4.63)$$

where α_{0L} is the wing zero-lift angle of attack. The fuselage pitching moment can now be put in terms of $C_{L,w}$:

$$C_{m,f} = C_{m0,f} + C_{m\alpha,f} \cdot \left(\frac{C_{L,w}}{C_{L\alpha,w}} + \alpha_{0L} \right) \qquad (4.64)$$

By combining multiple equations to put the $C_{m,CG}$ equation in terms of the wing lift coefficient, we get

$$C_m = C_{m0,w} - C_{L,w} \cdot \left(\frac{X_{CG} - X_{ac,w}}{mac_w} \right) + C_{m0,f} + C_{m\alpha,f} \cdot \left(\frac{C_{L,w}}{C_{L\alpha,w}} + \alpha_{0L} \right)$$
$$+ \sum_{i=1}^{N} \left\{ C_{m0}(j) - \eta(j) \cdot \left[\frac{X_{CG} - X_{ac}(j)}{mac_w} \right] \cdot C_L(j) \right\} \cdot \frac{S(j)}{S_w} \qquad (4.65)$$

The UA is trimmed when the pitching moment about the center of gravity is zero.

The preceding relationships evaluate pitching moment about the center of gravity where the lift forces act over the distance between the center of gravity and surface aerodynamic centers. Although this is somewhat intuitive, the static margin is a major design parameter that should be applied to the design. To do so, it is necessary to define a wing-body neutral point $X_{NP,wb}$. This is found by

$$X_{NP,wb} = X_{ac,w} + \frac{C_{m\alpha,f}}{C_{L\alpha,w}} \cdot mac_w \qquad (4.66)$$

The pitching moment about the center of gravity becomes

$$C_m = C_{m0,w} - C_{L,w} \cdot sm + C_{m0,f} + C_{m\alpha,f} \cdot \left(\frac{C_{L,w}}{C_{L\alpha,w}} + \alpha_{0L} \right)$$

$$+ \sum_{i=1}^{N} \left\{ C_{m0}(j) - \eta(j) \cdot \left[\frac{X_{NP,wb} - X_{ac}(j)}{mac_w} \right] \cdot C_L(j) \right\} \cdot \frac{S(j)}{S_w} \qquad (4.67)$$

For aft horizontal stabilizers, a modified tail volume coefficient is defined as

$$\overline{TVC}(j) = \left[\frac{X_{NP,wb} - X_{ac}(j)}{mac_w} \right] \cdot \frac{S(j)}{S_w} \qquad (4.68)$$

The relationship between the tail volume coefficient and this new version is

$$TVC(j) = \overline{TVC}(j) - \left(\frac{X_{CG} - X_{NP,wb}}{mac_w} \right) \frac{S(j)}{S_w} \qquad (4.69)$$

The UA lift coefficient C_L is

$$C_L = C_{L,w} + \sum_{j=1}^{N} C_L(j) \cdot \frac{S(j)}{S_w} \qquad (4.70)$$

Conventional configurations can have a negative lift coefficient on the tail (lifting downwards), which will result in C_L being less than $C_{L,w}$. A canard configuration will generally have a positive lift coefficient on the canard surface, and so the UA C_L will be greater than $C_{L,w}$. For more detailed treatment of longitudinal stability and control, the reader is encouraged to review texts dedicated to the subject [3–6].

4.4.1 Conventional Configuration

The conventional wing configuration is characterized by a main wing and a smaller horizontal stabilizer located aft of the main wing. Perhaps the reason why this arrangement is considered conventional today is that it is well understood, simple to analyze, and offers good performance. A vast experience base exists for the conventional wing configuration, making it a relatively low risk configuration option. Most UASs adopt the configuration, some of which can be seen in Fig. 4.13.

We begin with the neutral point and trim calculation methods. The x location of the neutral point for a conventional configuration can be found by

$$X_{NP} = \underbrace{X_{ac,w}}_{\text{Wing}} + \underbrace{\frac{C_{m\alpha,fuse}}{C_{L\alpha,wing}} \cdot mac_w}_{\text{Fuselage}} - \underbrace{\eta_H \cdot TVC_H \cdot \frac{C_{L\alpha,H}}{C_{L\alpha,wing}} \cdot \left(1 - \frac{d\varepsilon}{d\alpha} \right) \cdot mac_w}_{\text{Horizontal\_Tail}}$$

$$(4.71)$$

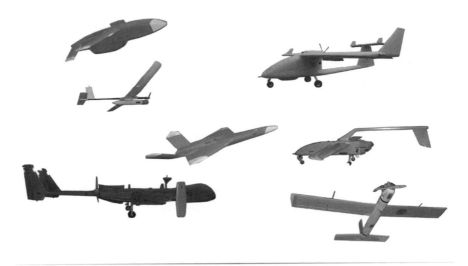

Fig. 4.13 Conventional configuration examples.

where $C_{L\alpha,H}$ is the lift-curve slope of the horizontal tail and η_H is the dynamic pressure ratio at the horizontal stabilizer.

The pitching moment about the center of gravity is expressed by

$$C_m = C_{m,w} + C_{m,f} + C_{m,H} \tag{4.72}$$

The horizontal tail pitching moment about the center of gravity is

$$
\begin{aligned}
C_{m,H} &= C_{m0,H} \cdot \frac{S_H}{S_w} - \eta_H \cdot \left(\frac{X_{CG} - X_{ac,H}}{mac_w} \right) \cdot \frac{S_H}{S_w} \cdot C_{L,H} \\
&= C_{m0,H} \cdot \frac{S_H}{S_w} - \eta_H \cdot TVC_H \cdot C_{L,H} \tag{4.73}
\end{aligned}
$$

The zero-lift pitching moment of the horizontal tail is generally considered to be negligible. With this simplification the pitching moment of the horizontal stabilizer becomes

$$C_{m,H} \approx -\eta_H \cdot TVC_H \cdot C_{L,H} \tag{4.74}$$

Combining terms and setting the pitching moment about the center of gravity give

$$C_m = 0 = C_{m0,w} - C_{L,w} \cdot \left(\frac{X_{CG} - X_{ac,w}}{mac_w} \right) + C_{m,f} - \eta_H \cdot TVC_H \cdot C_{L,H} \tag{4.75}$$

The tail lift coefficient required to trim the UA at a given $C_{L,w}$ and c.g. position is

$$C_{L,H} = \frac{C_{m0,w} + C_{m,f} - C_{L,w} \cdot (X_{CG} - X_{ac,w}/mac_w)}{\eta_H \cdot TVC_H} \tag{4.76}$$

This can also be put in terms of static margin.

$$C_{L,H} = \frac{C_{m0,w} + C_{m,f} - C_{L,w} \cdot sm}{\eta_H \cdot \overline{TVC}_H} \tag{4.77}$$

Example 4.2 Conventional Configuration Tail Sizing

Problem:

The stressing horizontal stabilizer design condition for the horizontal tail configuration is a peak lift coefficient of the main wing. The following parameters are known about the UA:

$C_{Lmax,w} = 1.5$
Maximum negative horizontal lift coefficient, $C_{Lmin,H} = -0.7$
$C_{m0,w} = -0.01$
$C_{m,f} = 0$
$sm = 10\%$
$mac_w = 6$ ft
$\lambda_w = 1$
$AR_w = 10$
$\eta_H = 0.95$
Distance between the center of gravity and wing-body neutral point $(X_{CG} - X_{NP,wb}) = 2$ ft
Tail moment arm $(X_{CG} - X_{ac,H}) = 16$ ft

1. What is the minimum horizontal tail area?
2. What is the horizontal tail volume coefficient?
3. What is the UA maximum lift coefficient?

Solution:

The pitching moment about the center of gravity is zero

$$C_m = 0 = C_{m0,w} - C_{L,w} \cdot sm + C_{m,f} - \eta_H \cdot \overline{TVC}_H \cdot C_{L,H}$$

The tail produces its greatest negative lift when the wing produces its maximum lift. Solving for $\overline{TVC_H}$ gives

$$\overline{TVC_H} = \frac{C_{m0,w} - C_{L,w} \cdot sm + C_{m,f}}{\eta_H \cdot C_{L,H}} = \frac{-0.01 - 1.5 \cdot 0.1 + 0}{0.95 \cdot (-0.7)} = 0.457$$

$\overline{TVC_H}$ is defined by

$$\overline{TVC_H} = \left[\frac{X_{NP,wb} - X_{ac}(j)}{mac_w}\right] \cdot \frac{S_H}{S_w}$$

Solving for S_H gives

$$S_H = \overline{TVC_H} \cdot S_w \cdot \left(\frac{mac_w}{X_{NP,wb} - X_{ac,H}}\right)$$

The wing area must be found. The wing is straight ($\lambda = 1$), and so the average chord is the same as the mean aerodynamic chord. The wing area can be found by

$$AR = \frac{b_w^2}{S_w} = \left(\frac{Sw}{c_{avg}}\right)^2 \cdot \frac{1}{S_w} = \frac{S_w}{c_{avg}^2}$$

$$Sw = AR \cdot c_{avg}^2 = 10 \cdot (6\text{ ft})^2 = 360\text{ ft}^2$$

The other unknown is the distance $(X_{NP,wb} - X_{ac,H})$. This can be found by

$$(X_{NP,wb} - X_{ac,H}) = -(X_{CG} - X_{NP,wb}) + (X_{CG} - X_{ac,H})$$
$$= -(2\text{ ft}) + (16\text{ ft}) = 14\text{ ft}$$

Now we can find the horizontal stabilizer area.

$$S_H = 0.457 \cdot 360\text{ ft}^2 \cdot \left(\frac{6\text{ ft}}{14\text{ ft}}\right) = 70.5\text{ ft}^2$$

The horizontal tail volume coefficient is

$$TVC_H = \overline{TVC_H} - \left(\frac{X_{CG} - X_{NP,wb}}{mac_w}\right)\frac{S_H}{S_w}$$

$$= 0.457 - \left(\frac{2\text{ ft}}{6\text{ ft}}\right)\frac{70.5\text{ ft}^2}{360\text{ ft}^2} = 0.392$$

The UA lift coefficient is

$$C_L = C_{L,w} + C_{L,H} \cdot \frac{S_H}{S_w}$$

$$= 1.5 + (-0.7) \cdot \frac{70.5\text{ ft}^2}{360\text{ ft}^2} = 1.36$$

Conventional wing configurations are conceptually simple and well understood by design experience. The wing is designed to produce lift, and a horizontal surface located behind the wing trims the UA. The pitch control surface is usually solely on the horizontal stabilizer, taking the form of an elevator or full flying horizontal stabilizer (adjustable incidence angle). The wing can be designed for lifting capability or aerodynamic efficiency without imposing pitch trim requirements on the wing. High-lift devices such as flaps or slats are conceptually simple to incorporate from a trim perspective. Also, the influence of the tail on the wing is generally negligible.

Conventional configurations can yield very high aerodynamic efficiency in terms of lift-to-drag ratio (L/D) and endurance parameter. The highest L/D sailplanes all utilize this configuration. With the tail providing trim, the main wing is able to generate high lift coefficients. No surfaces ahead of the wing generate undesirable downwash on the wing or a turbulent wake.

Conventional configurations can be made adaptable to wingspan extensions because they do not double as a pitch trim device. The most common approach is to have a strong inboard section with a continuous spar and two outboard panels. When short wings are applied, the inboard wing spar can be sized for gust loads at higher speed. The extended span comes with more wing area, reducing the top speed. The inboard wing must accommodate the increased bending stresses from the extended span. Increased horizontal and vertical tail area might be required to maintain the appropriate stability and control characteristics. Examples of conventional wing configuration wing extensions include the Northrop Grumman Hunter ER vs the standard Hunter, and the Lockheed Martin SkySpirit and SkySpirit ER.

4.4.2 Canards

The canard configuration has a single horizontal surface located ahead of the wing. This horizontal surface is called a *canard*. The canard surface provides a positive lift to trim out the UA, and therefore all horizontal surfaces provide lift that counteracts weight. The first successful manned aircraft, the Wright Flyer, had a canard configuration. However, the greater preponderance of wing-aft tail configurations has given that later configuration the title "conventional." The canard configuration is relatively rare for UAS applications because there are no aerodynamic advantages over the conventional configuration and it has more parts than a flying wing (Fig. 4.14). Secondary design drivers such as packaging or special payload integration are usually needed to justify adopting this approach. Generally

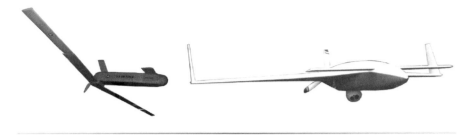

Fig. 4.14   Canard configuration UAS.

considered to look "futuristic," the canard's aesthetic appeal might be sufficient motivation.

The x location of the neutral point for a canard configuration can be found by

$$X_{\text{NP}} = \underbrace{X_{\text{ac,w}}}_{\text{Wing}} + \underbrace{\frac{C_{m\alpha,\text{fuse}}}{C_{L\alpha,\text{wing}}} \cdot mac_w}_{\text{Fuselage}} - \underbrace{\eta_C \cdot TVC_C \cdot \frac{C_{L\alpha,C}}{C_{L\alpha,\text{wing}}} \cdot \left(1 - \frac{d\varepsilon}{d\alpha}\right) \cdot mac_w}_{\text{Canard}}$$

(4.78)

where $C_{L\alpha,C}$ is the lift-curve slope of the canard surface. Because TVC_C is negative, the canard surface pushes the neutral point forward. Here the efficiency of the canard surface η_C can be set to one because the canard is the front wing exposed to undisturbed air.

The pitching-moment coefficient about the center of gravity for a canard configuration is

$$C_m = C_{m,w} + C_{m,f} + C_{m,C}$$

(4.79)

Expanding the terms and setting the pitching moment about the center of gravity to zero give

$$C_m = 0 = C_{m0,w} - C_{L,w} \cdot \left(\frac{X_{\text{CG}} - X_{\text{ac,w}}}{mac_w}\right) + C_{m,f} - TVC_C \cdot C_{L,C}$$

(4.80)

The canard lift coefficient required to trim the UA at a given $C_{L,w}$ and c.g. position is

$$C_{L,C} = C_{m0,w} + C_{m,f} - C_{L,w} \cdot \left(\frac{X_{\text{CG}} - X_{\text{ac,w}}}{mac_w}\right) \Big/ TVC_C$$

(4.81)

Recall that the canard tail volume coefficient is negative because the canard is ahead of the center of gravity. The canard lift coefficient can also be put in terms of static margin.

$$C_{L,C} = \frac{C_{m0,w} + C_{m,f} - C_{L,w} \cdot sm}{\overline{TV}C_C} \tag{4.82}$$

The canard lift coefficient is positive when the wing lift coefficient and static margin are both positive.

The total lift coefficient for the UA is

$$C_L = C_{L,w} + C_{L,C} \cdot \frac{S_C}{S_w} \tag{4.83}$$

The canard must stall before the wing so that the nose drops and airspeed can be regained. If the wing were to stall first, the loss of lift behind the center of gravity would cause the nose to rise further. This self-reinforcing increase in angle of attack is known as an *unstable pitch break*. The main wing must have an angle-of-attack margin when the canard—and therefore the UA—stalls. The wing is not permitted to reach its stall C_L. The canard lift is not adequate to compensate for the lost wing lift potential, and so a canard configuration will have a lower maximum lift coefficient than an equivalent conventional design. The lower C_{Lmax} requires a larger wing area for a given stall speed. Conversely, for a given wing area a canard configuration will stall at a higher speed than a conventional configuration.

UA seek to maximize the endurance parameter ($C_L^{3/2}/C_D$) for long-duration flight, which benefits from a high C_L. The canard configuration stall margin requirement for the main wing drives the $C_{L,w}$ lower than what can be expected with a conventional configuration, and therefore the endurance parameter might not be as high. So while both the canard and main wing both provide positive lift, the canard is usually more heavily loaded than the main wing in terms of lift coefficient. Put another way, the main wing of a canard configuration underperforms.

Another negative attribute of the canard configuration is that the canard produces downwash on the main wing behind it. The downwash orients the wing local lift force vector aft, which increases the induced drag.

The net effect of the underloaded main wing and canard downwash is that the aerodynamic performance of a canard is generally inferior to a conventional configuration. Although remarkable canard designs exist, such as the Rutan Voyager that flew around the world nonstop and unrefueled, attempts to better conventional designs with the canard are usually unsuccessful.

The center of gravity required for a positive static margin is located ahead of the wing aerodynamic center and behind the canard surface.

A payload located at the center of gravity can be largely unobstructed by the wing for some canard designs, providing a favorable side field of regard.

The main wing is frequently swept aft, even when the UA operates below the transonic flight regime. The motivations for this sweep might be to do the following:

* Improve aesthetics
* Provide a vertical tail arm for wing-tip-mounted vertical stabilizers
* Enable a pusher-mounted engine located near the desired center-of-gravity location
* Allow a compact fuselage, where the distance between the canard and main wing carry-through structures is reduced while maintaining an appropriate canard moment arm
* Enable inboard wing fuel tanks without excessive aft moment arm from the center of gravity

Canard configurations are well suited to short fuselages. Most canards have a shorter canard moment arm than conventional configuration tail moment arm. Therefore most canards are close-coupled in pitch. The aft fuselage in a conventional configuration is often sparsely populated with systems so that the center of gravity can be maintained. This largely empty volume contributes to the wetted area (and drag) while providing no other utility than to extend the tail arm. The canard configuration can make more effective use of the fuselage volume because the moment arm between the main wing and canard crosses the center of gravity.

Variable weight items such as fuel or payloads can introduce challenges on canard designs. Wing hardpoints and wing internal fuel tanks are usually located aft of the UA center of gravity. Changes in these weights in flight will result in a shift in the UA center of gravity. For example, as wing tank fuel is burned off, the center of gravity moves towards the nose. One approach to maintain the center of gravity with fuel burn is to have a fuel tank located ahead of the center of gravity, though this requires controlled depletion of the fore and aft tanks. The fuselage volume can accommodate a fuel tank near the center of gravity, though it takes up valuable fuselage volume while leaving wing volume empty of fuel. Another common approach, which was pioneered on Burt Rutan's canard designs such as the Long EZ, is to sweep the wings aft and provide a thick strake with substantial fuel volume located near the center of gravity.

Many of the canard UAS designs are conversions of manned canard aircraft. Some examples include the L-3 Mobius and Proxy SkyRaider. The original aircraft is usually a home-built or general-aviation aircraft made of composite materials. The space near the center of gravity formerly occupied by the pilot can be filled with payloads or additional fuel tanks.

Control canards were used in some early cruise missile designs. The guidance system is located in the nose, and so this arrangement offered the minimal distance to the pitch control surfaces. Today's use of lighter-weight wiring harnesses reduces the benefit of this approach.

4.4.3 Tandem Wing

The tandem-wing configuration has two wings of similar area arranged one in front of the other. The distinction between a tandem wing, conventional configuration, and canard is somewhat arbitrary. A tandem wing could be characterized as a conventional configuration with a very large horizontal stabilizer. Alternatively, it is also a canard configuration with an oversized canard surface.

Like a canard, the forward wing produces a downwash field on the rear wing. This generates higher induced drag on the aft wing. The rear wing can operate in the wake of the forward wing. As we will see momentarily, the rear wing can be lightly loaded, resulting in relatively little lift contribution for the profile drag produced. These characteristics usually translate to a lower aerodynamic efficiency relative to a conventional wing configuration for an equivalent wetted aspect ratio.

The tandem wing configuration might experience poor stall behavior. The rear wing will stall first if the two wings have equivalent lift loads, aspect ratios, airfoils, and incidence angles. Rear-wing stall results in a nose-up pitching moment and unstable pitch break, which can be difficult for an autopilot to control. It is therefore desirable to adapt the configuration so that the front wing stalls first. Most tandem-wing UAS applications require a positive static margin, which dictates that the forward wing has a higher lift load (if areas are equivalent). For a given wing system geometry, the static margin is controlled by center-of-gravity placement. The higher front wing lift load can be accomplished by increasing the front-wing incidence angle or using more highly cambered airfoils, where the former brings the front wing closer to the stall lift coefficient. Another method to help the front-wing stall first is to increase the front-wing aspect ratio relative to the rear wing so that the rate of lift coefficient increase with angle of attack is greater. The autopilot can avoid the unstable pitch break by adopting an *alpha limiter*, where the control laws place an upper limit on the commanded angle of attack to stay away from the problematic condition.

The tandem-wing configuration does have advantages that make it a suitable choice for some UAS applications. The center of gravity is located between the two wings, allowing good side field of regard without wing obstruction. The length of the fuselage section can be adjusted on

Fig. 4.15 Tandem-wing configurations.

modular aircraft, where the distance between the wings increases with longer central payload sections.

Tandem wings have favorable characteristics for UA that deploy from a stowed state in flight. These UA can fold the forward wings aft and the rear wings forward along the fuselage. Each wing semispan can be equal to the fuselage length because the wing pivots are located at near the front and rear of the fuselage. This generally yields slightly greater span and more than twice the wing area than for conventional configurations. The ACR Coyote and AeroVironment Switchblade designs take this approach as shown in Fig. 4.15.

4.4.4 Three-Surface Configuration

The three-surface configuration has a main wing with a canard ahead of the wing and a horizontal stabilizer aft of the wing (Fig. 4.16). This configuration can be considered a combination of the conventional and canard configurations. This design can offer benefits in aerodynamic efficiency and center-of-gravity range. The major drawback relative to other alternatives is that there are more parts to manufacture.

The x location of the neutral point for a three-surface configuration can be found by

$$X_{\mathrm{NP}} = \underbrace{X_{\mathrm{ac,w}}}_{\text{Wing}} + \underbrace{\frac{C_{m\alpha,\text{fuse}}}{C_{L\alpha,\text{wing}}} \cdot mac_w}_{\text{Fuselage}} - \underbrace{\eta_C \cdot TVC_C \cdot \frac{C_{L\alpha,C}}{C_{L\alpha,\text{wing}}} \cdot \left(1 - \frac{\mathrm{d}\varepsilon}{\mathrm{d}\alpha}\right) \cdot mac_w}_{\text{Canard}}$$

$$\underbrace{- \eta_H \cdot TVC_H \cdot \frac{C_{L\alpha,H}}{C_{L\alpha,\text{wing}}} \cdot \left(1 - \frac{\mathrm{d}\varepsilon}{\mathrm{d}\alpha}\right) \cdot mac_w}_{\text{Horizontal\_Tail}} \tag{4.84}$$

The three-surface pitching moment about the center of gravity is

$$C_m = C_{m0,w} - C_{L,w} \cdot \left(\frac{X_{CG} - X_{ac,w}}{mac_w} \right) + C_{m,f} - TVC_C \cdot C_{L,C} - \eta_H \cdot TVC_H \cdot C_{L,H}$$

(4.85)

The alternative form is

$$C_m = C_{m0,w} - C_{L,w} \cdot sm + C_{m,f} - \overline{TVC}_C \cdot C_{L,C} - \eta_H \cdot \overline{TVC}_H \cdot C_{L,H}$$

(4.86)

Setting the pitching moment about the center of gravity to zero yields

$$\overline{TVC}_C \cdot C_{L,C} + \eta_H \cdot \overline{TVC}_H \cdot C_{L,H} = C_{m0,w} - C_{L,w} \cdot sm + C_{m,f} \qquad (4.87)$$

Unlike the conventional configuration or canard, there is no single combination of $C_{L,C}$ and $C_{L,H}$ that uniquely trims the UA. The canard and horizontal stabilizer lift coefficients can be optimized to minimize drag or increase C_{Lmax}. It is possible for a three-surface design to have better aerodynamic performance than an equivalent conventional design.

The reasons for the lack of widespread adoption are the increased complexity caused by the additional surface and the difficulty in analyzing the stability and control. The extra surface adds manufacturing cost and has negative supportability impacts. The potential aerodynamic benefits are offset by the complexity of analysis, where the time required to optimize the configuration for best performance is considered excessive. Another consideration is that most designers are not as familiar with this approach.

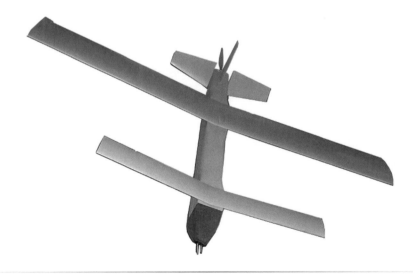

Fig. 4.16 Three-surface configuration UAS.

4.4.5 Flying Wings

The flying-wing configuration uses a single wing as the only horizontal lifting surface. The wing satisfies both the lift and trim functions without a need for additional horizontal surfaces. Flying wings can have vertical surfaces, and the wing and fuselage are often blended to the wing. A key motivation for flying wings is simplicity; there are fewer parts to manufacture, transport, and maintain. Also, this configuration offers advantages for low RCS when suitably designed. Flying wings (Fig. 4.17) are the second most prevalent UAS configuration, just behind the conventional configuration (Fig. 4.17).

The x location of the neutral point for a flying-wing configuration can be found by

$$X_{\mathrm{NP}} = \underbrace{X_{\mathrm{ac,w}}}_{\text{Wing}} + \underbrace{\frac{C_{m\alpha,\mathrm{fuse}}}{C_{L\alpha,\mathrm{wing}}} \cdot mac_w}_{\text{Fuselage}} \tag{4.88}$$

The pitching-moment coefficient about the center of gravity is

$$C_m = C_{m0,w} - C_{L,w}(\delta e) \cdot \left(\frac{X_{\mathrm{CG}} - X_{\mathrm{ac,w}}}{mac_w}\right) + C_{m\delta e} \cdot \delta e + C_{m,f} \tag{4.89}$$

Note that $C_{L,w}$ is the same as C_L because there are no other horizontal control surfaces.

$$C_m = C_{m0,w} - C_L(\delta e) \cdot sm + C_{m\delta e} \cdot \delta e + C_{\mathrm{mNP,f}} \tag{4.90}$$

The elevators are wing trailing-edge control surfaces. The elevators are generally placed towards the wing tips on swept flying wings. The elevator

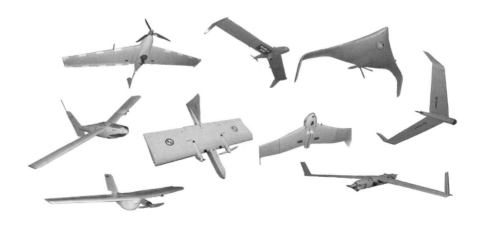

Fig. 4.17 Flying-wing UA.

control function is often shared with the aileron control function on the same surface, which is called an *elevon*. Because the elevator is located on the wing, deflections will create changes in wing lift coefficient.

Flying wings can be unswept, forward swept, or aft swept. Unswept flying wings are sometimes referred to as *flying planks*. Examples of unswept flying wings are the NRL Dragon Eye and the Lockheed Martin/Boeing DarkStar. The vast majority of flying wing UASs have aft swept wings. The author is unaware of any forward-swept flying wing UASs, and this arrangement is very rare with manned aircraft.

Let us first consider an unswept flying wing. For simplicity, assume that each elevon runs across the entire wing semispan. Here the zero-lift pitching moment of the wing is caused only by the airfoil pitching moment. The lift distribution acts directly at the wing aerodynamic center because there is no sweep. A suitable airfoil will have a zero-lift pitching moment of zero, which requires either a symmetrical airfoil or one with reflex camber (more on reflex camber airfoils in Chapter 5). With a positive static margin, positive lift generates a nose-down pitching moment that must be countered by upward deflection of the elevons. This upward elevon deflection reduces the wing lift coefficient at a given angle of attack, which is why C_L is expressed as a function of δ_e.

The spanloader is a special type of unswept flying wing that distributes the weight across the wing. The weight is lifted by the wing locally, keeping the shear force on the spar from building up and thereby greatly reducing the bending moment. With reduced static loading, the spar weight can be greatly reduced. The structural sizing is generally driven by aeroelastic effects. The AeroVironment Pathfinder, Centurion, and Helios solar-powered UASs are all spanloader designs (see Fig. 4.18).

The lift distribution on a swept flying wing affects the pitching moment. A good comparison between the arrangements can be seen in the unswept flying-wing Lockheed Martin/Boeing Darkstar and the swept flying-wing Lockheed Martin Polecat. The lift distribution is affected by sweep, taper, washout (local incidence relative to the root), and airfoil camber. A positive static margin increases the lift at the wing root relative to the tip for trim. Reduced or negative static margin can improve the wing load distribution to reduce induced drag, but this places more demands on the flight control system.

A pure flying wing has no distinct fuselage or vertical surfaces. The internal volume of the wing is used to house all components normally placed in the fuselage. Appropriate lateral-directional handling qualities are possible without vertical surfaces for swept flying wings. The asymmetric drag distribution along the wing in sideslip provides a favorable yawing moment. The asymmetric lift distribution caused by sideslip also provides the appropriate roll response.

Fig. 4.18   AeroVironment Helios is a spanloader flying-wing configuration. (NASA photo by Nick Galante.)

Unswept flying wings or designs with large fuselages can require vertical surfaces for appropriate lateral-directional stability and control. Swept flying wings frequently use vertical surfaces at the wing tips, called winglets, to serve this function. The aft-sweep creates a yaw moment arm relative to the center of gravity for the vertical surface to act upon. Unswept flying wings have a small moment arm at the wing tips, so that winglets are not effective for lateral-directional purposes. Instead, unswept flying wings use the fuselage or booms to affix vertical surfaces.

The combined effects of the lift reduction caused by upward elevon deflection and required wing lift distribution for trim limits the achievable C_{Lmax}. A flying wing generally produces substantially less lift per area than a conventional configuration. The reduced effective lifting capacity results in either a higher takeoff and landing speed or increased wing area.

The reduced lifting ability of the flying wing relative to a conventional configuration of equivalent wing area tends to produce a lower endurance parameter. This makes flying wings less desirable for endurance missions. Partially offsetting this impact is the potential for the flying wing to have a lower wing loading as a result of lower structural weight. On balance, the flying wing is inferior to the conventional configuration for endurance missions.

Best range performance occurs at a lower lift coefficient than for best endurance. Flying wings do not have the wetted area of canards or stabilizers, and these drag contributors are eliminated. Flying wings become competitive for range missions, particularly for jet-powered UAS.

Perhaps the greatest advantage for this configuration is that there are fewer major assemblies. There are no horizontal stabilizers to manufacture, store, or maintain. The relative simplicity can make packaging for air-launch and storage easier. The UA can be assembled more quickly, as fewer parts need to be attached. The eliminated components have no weight.

Often the fuselage is blended with the wing, and there is not a distinct fuselage. With high degrees of blending and an elongated center section, this is called a blended wing body (BWB). The unmanned X-48B research aircraft explored the flight dynamics of the BWB configuration (Fig. 4.19). When the center section is less pronounced, the configuration is still considered a flying wing. Many UCAVs are flying wings with blended fuselages. The inboard sections of flying wings often have high leading-edge sweep to support forward placement of components for balance.

Fuselages—whether distinct or blended—introduce challenges for flying wings. The fuselage can create an aeroelastic mode called pitch-plunge, where the fuselage pitches relative to the wing. Long fuselages can introduce inertial coupling, where roll rate induces a pitch up or pitch down due to centripetal forces acting on the fuselage. Fuselages are also destabilizing in pitch, which increases the demands on the pitch effectors. Vortices produced by the fuselage at high angles of attack, particularly for flat or straked bodies, can contribute to an unstable pitch break.

Very low-aspect-ratio flying wings (AR ~ 1) are commonly used for MAVs. The planforms take many shapes, including rectangles, tapered wings, delta wings, ellipses, and trochoids. These low-aspect-ratio flying wings have highly three-dimensional flow structures across the wing span, where vortex structures dominate. By comparison, high-aspect-ratio flying wings can assume that the local airflow is two-dimensional at a given spanwise station. Low-aspect-ratio flying wings operate at relatively high angles of attack, and the interaction of the tip vortex on the wing can generate high lift coefficients. This phenomena was first explored in research that led to the Chance Vought V-173 "Flying Pancake." Low-aspect-ratio wings are generally more gust resistant than higher-aspect-ratio wings, which is important for imagery quality.

Delta wings are highly swept low-aspect-ratio flying wings. They became common in target drones because of their simplicity. Later ISR variants emerged as evolutions of the targets, as seen on the BAI Exdrone. Some manufacturers of subsonic delta-wing UASs have attributed high maximum speed to this configuration, but this is simply not the case. Some ISR variants apparently attempt to decrease the stall speed and improve aerodynamic efficiency by adding tails to delta wings (i.e., DRS Sentry and Northrop Grumman Huntair). Delta wings are similar to

Fig. 4.19   X-48B unmanned BWB demonstrator. (NASA photo by Carla Thomas.)

blended wing bodies in that the high root thickness provides volume for fuel and systems integration.

4.4.6 Other Wing System Configurations

Although the wing-tail configuration is considered to be conventional, some alternatives are more unconventional than others. The canard, tandem, three-surface, and flying-wing configurations are unconventional but still prevalent. Other configurations or aspects of wing system configurations are presented here.

When one or more wings operate outside of a single horizontal plane, the wing system is known as *nonplanar wings*. Moderate wing dihedral does not qualify as nonplanar because this can be closely approximated by planar-wing analysis. Vertical surfaces that do not produce normal aerodynamic forces in the absence of sideslip, such as centerline vertical stabilizers, do not qualify either. Biplanes, triplanes, many joined wings, and even winglets can qualify as nonplanar wing systems.

To illustrate the potential of nonplanar wings, we will begin with an analysis of the venerable biplane. Consider two identical aircraft flying along parallel paths but with a large vertical separation. The induced drag from the two UAs combined is twice that of a single UA.

$$D_{i,\text{Tot}} = 2 \cdot D_{i,1} = 2 \cdot C_{Di,1} \cdot q \cdot S_{w1} = 2 \cdot \frac{C_{L,1}^2}{\pi \cdot AR_1 \cdot e_1} \cdot q \cdot S_{w,1} \qquad (4.91)$$

Here the parameters with subscript 1 are with respect to a single UA. Now let us rigidly attach these two UAs with a zero-mass, zero-drag structural member such that the two UAs become one. We now have a biplane where the interference effects are negligible as a result of the large vertical separation. The wing area is the area of the upper and lower wings combined.

$$S_w = S_{w,1} + S_{w,2} = 2 \cdot S_{w,1} \qquad (4.92)$$

Because our aircraft are identical, the biplane wing area is twice the single UA wing area. The span b is the maximum span of either wing. Because the wings are identical, $b = b_1$. The weight is now twice the weight of the single UA. It can be shown that when lift equals weight the lift coefficient is

$$C_L = C_{L,1} \cdot \frac{S_{w,1}}{S_w} \cdot \frac{W}{W_1} = C_{L,1} \cdot \frac{1}{2} \cdot \frac{2}{1} = C_{L,1} \qquad (4.93)$$

The new aspect ratio is defined by

$$AR = \frac{b^2}{S_w} = \frac{b_1^2}{2 \cdot S_{w,1}} = 1/2 \cdot AR_1 \qquad (4.94)$$

The induced drag for this new combined UA is

$$D_i = \frac{C_L^2}{\pi \cdot AR \cdot e} \cdot q \cdot S_{w,1} = D_{i,\text{Tot}} \qquad (4.95)$$

The total induced drag in terms of the combined UA is

$$D_{i,\text{Tot}} = 2 \cdot \frac{C_{L,1}^2}{\pi \cdot AR_1 \cdot e_1} \cdot q \cdot S_{w,1} = 2 \cdot \frac{C_L^2}{\pi \cdot (2 \cdot AR) \cdot e_1} \cdot q \cdot \left(\frac{1}{2} \cdot S_w \right)$$

$$= \frac{C_L^2}{\pi \cdot AR \cdot 2 \cdot e_1} \cdot q \cdot S_w \qquad (4.96)$$

By setting D_i equal to $D_{i,\text{Tot}}$ and solving for e, we find that

$$e = 2 \cdot e_1 \qquad (4.97)$$

This shows that when wing interference effects are neglected the induced drag coefficient is half that of a monoplane, even though the total induced drag is the same. Aircraft work on forces rather than coefficients, so there is no net gain in this example. Interference effects reduce e as the vertical separation between the two wings is reduced. Therefore, the biplane in this example will have greater induced drag as a system.

Now let us approach the problem from the perspective of comparing a biplane to a monoplane where both UA have the same span, wing area, and weight. Assume that both biplane wings have equivalent area, span, and lift load. Note that each wing has half the total area, so that $AR_1 = AR_2 = 2 \times AR$. Let us also assume that e_1 and e_2 are the same and equal to half of e. The total induced drag is the combination of the upper and lower wing induced drag.

$$D_{i,\text{Biplane}} = D_{i,1} + D_{i,2} = 2 \cdot D_{i,1}$$

$$= 2 \cdot \frac{\left(\frac{1}{2} \cdot C_L\right)^2}{\pi \cdot (2 \cdot AR) \cdot e_1} \cdot q \cdot \left(\frac{1}{2} \cdot S_w\right) = \frac{1}{2} \cdot \frac{C_L^2}{\pi \cdot AR \cdot e_1} \cdot q \cdot S_w$$

(4.98)

The induced drag of the equivalent monoplane is

$$D_{i,\text{Monoplane}} = \frac{C_L^2}{\pi \cdot AR \cdot e_{\text{Monoplane}}} \cdot q \cdot S_w$$
(4.99)

This shows that if the e_1 is equivalent to $e_{\text{Monoplane}}$, then this biplane will have half the induced drag. As mentioned earlier, the biplane wing interference effects diminish this induced drag benefit.

Each biplane wing in this example has twice the aspect ratio of the equivalent monoplane wing. The chord is half as long, which reduces the Reynolds number thereby increasing the profile drag and reducing the maximum lifting capability. Without external bracing, the higher-aspect-ratio individual biplane wings will have higher structural weight. External bracing adds drag, but potentially reduces combined wing weight relative to the monoplane.

Other nonplanar wing concepts that act much like a biplane include the box wing and joined wing. These configurations feature two wings that are offset horizontally and vertically. The wing tip of the aft wing connects to the main wing by a vertical surface for the box wing or through direct intersection with a joined wing. The joining of the wings offers potential benefits for structural weight reduction. These configurations are much like tandem wings in their longitudinal stability and control if no additional horizontal surfaces are present. Boeing explored joined-wing SensorCraft UAS designs. The Alliant Techsystems Outrider is an example of a box wing augmented with an aft horizontal stabilizer.

Nonplanar wing arrangements might be desirable when the span is constrained. This is particularly true if the span is limited and the wings are not removable from the fuselage. The problems with these configurations are that they can be more difficult to analyze than conventional monoplanes.

Wings that are free to rotate about the aerodynamic center but are aerodynamically and mass balanced are known as *freewings*. The wings remain at the commanded angle of attack relative to the airflow independent of aircraft orientation. The wing airfoils are typically reflexed airfoils with trailing-edge control surfaces. The airfoil pitching moment about the aerodynamic center is always zero, where the trailing-edge control surface adjusts the wing angle of attack. The wings align themselves with the airflow. One of the motivations for this approach is improved turbulence response that provides better sensor ride quality. One drawback of this approach is that, like an unswept flying wing, the C_{Lmax} is limited compared to a rigidly mounted wing. It is possible to lock the wing in place for landing so that the landing speed is reduced. Another approach is to combine the freewing with thrust vectoring, as was done on the Freewing Scorpion.

UA can adopt asymmetric configurations, where the left and right sides of the UA are not symmetric with each other. The configuration must be able to trim in pitch, yaw, and roll. Developing the asymmetric UA proportions and controls can be especially challenging. However, elimination of the symmetry constraint opens up substantial design freedom.

The oblique wing is an example of an asymmetric configuration. Here the wing is swept forward on one side and swept aft on the other. The wing sweep can be fixed or used with a wing pivot mechanism that adjusts the sweep angle in flight. The left and right wing drag moment arm must be the same to avoid a yawing moment. The left and right lift moment arms must also be the same to prevent a rolling moment. The lateral-directional control is difficult to design for the entire angle-of-attack range. Also, the stall behavior should prevent an unstable pitch break, as well as uncorrectable roll or yaw moments. The motivation for an oblique wing might be to improve packaging for air-launched UASs or to reduce wave drag in transonic or supersonic flight.

Guided parafoils are commonly used as recovery devices for UASs. However, they are also increasingly used in the place of wing systems for air-launched gliders such as JPADS or ground-based special-purpose UASs such as the MMIST SnowGoose. A key advantage is that the airframe can package quite effectively for transportation and storage. The large effective wing area permits slow flight and recovery speeds. A major drawback is that the aerodynamic efficiency can be low relative to fixed-wing aircraft (L/D is around 3–5). The power required for flight is a strong function of wing loading, so that endurance and engine size are often acceptable for many applications.

Launch and recovery of parafoil UASs have limitations. The wing system is the parafoil, and it cannot maintain the flight geometry while static. The transition from the static to flight configurations is generally performed in an air launch where a drogue chute assists with the parafoil deployment, or with the support of a moving ground vehicle that can inflate the canopy. The MMIST SnowGoose can be air-launched from cargo aircraft or ground deployed via a HMWWV.

4.5 Tail Configurations

Numerous tail configurations exist, and the options are generally identical to those of manned aircraft. This book will not attempt to exhaustively detail all of the alternatives, though some observations are relevant for unmanned aircraft.

Tails are the artistic pallet of the designer. The UA drag is relatively insensitive to the tail styling—particularly for the vertical tail, provided that they are appropriately sized. Often the chief designer of an aircraft can be readily identified by the tail configuration. At other times certain tail styles are widely popular across the industry for a given period— many twin-boom UASs from the 1970s–1990s have vertical tails that closely resemble the Cessna 337 for example.

Tails can have single or multiple attachment points to the rest of the UA. The single tail attachment points are generally to a single fuselage or single boom. With multiple attachment points the tails generally attach to two tail booms that extend from the wing. The single attachment point options include conventional, T-tail, cruciform, V-tail, inverted V-, Y-, and H-tail, as shown in Fig. 4.20. Twin attach options include conventional, inverted U-, inverted V-, and H-tail arrangements, as shown in Fig. 4.21.

Single-attachment inverted V-tails have the obvious disadvantage of requiring long landing gear to prevent tail strike. This is mitigated for high-aspect-ratio wings that have limited rotation angles. The most

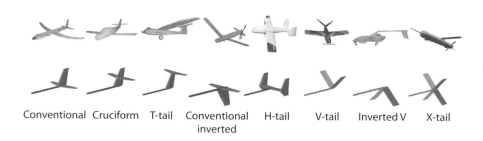

Conventional Cruciform T-tail Conventional inverted H-tail V-tail Inverted V X-tail

Fig. 4.20 Single-attach point tail configurations.

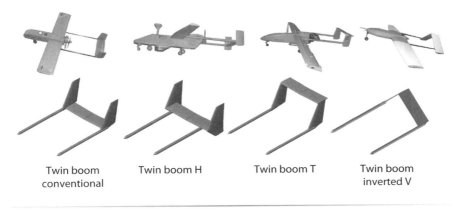

| Twin boom conventional | Twin boom H | Twin boom T | Twin boom inverted V |

Fig. 4.21 Twin-boom tail configurations.

famous example of this arrangement is the General Atomics Predator A. Inverted V-tails are well suited to in-flight reconfigurable UAs with high wings; the tails can deploy before the wing without interference.

Twin booms have the option of attached or disconnected tails. Disconnected tails can have the same arrangements as for the single boom, but also self-asymmetric arrangements. The original Insitu Integrator design had a vertical tail extending upwards and a single horizontal surface extending outboard from each boom. A twin-boom T-tail was adopted as the design matured

Sweeping tails aft from the fuselage or boom attachment point increase the moment arm between the center of gravity and the tail aerodynamic center and therefore reduce the tail size required for a fixed tail volume coefficient. Swept tails can produce more drag at high angles relative to the airflow, which has a stabilizing influence.

Aft tails should not stall before the wing in order to prevent an unstable pitch break. This generally necessitates a lower aspect ratio than the wing to permit broader range of angle of attack before stall.

Zero-loaded tails are designed such that the lift coefficient on the tail for the design condition is zero. Thus, there is no induced drag on the tail, and all of the drag is from profile drag, which can be comparatively low relative to a lifting tail (positive or negative lift). The center of gravity is farther aft in a zero-loaded tail configuration, permitting use of more fuselage volume for payload. Zero-loaded tail UASs were pioneered by NRL, but are rarely used on operational systems today.

Tails can provide substantial engine noise and IR signature blocking. To accomplish this, the tails should provide line-of-sight blockage between the engine components (i.e., propellers or jet nozzles) and the threat. The Global Hawk V-tails provide blockage of the jet nozzle from the ground.

V-tails are particularly attractive because of the reduction in parts count. The tail height is reduced relative to a single vertical tail. The dihedral angle of the V-tail $\Gamma_{V\text{-Tail}}$ can be found by considering the equivalent ratios of vertical and horizontal tail areas.

$$\Gamma_{V-Tail} = \tan^{-1}\left(\frac{S_V}{S_H}\right) \tag{4.100}$$

If we use the tail volume coefficient method, this can be put in the following form:

$$\Gamma_{V-Tail} = \tan^{-1}\left(\frac{TVC_V \cdot b_w}{TVC_H \cdot mac}\right) \tag{4.101}$$

Note that the moment arm between the center of gravity and the tail aerodynamic center is identical for both the horizontal and vertical equivalent surfaces, so that it cancels and is not included in this equation.

The vertical tail configurations can be independent of the other horizontal tail surfaces. Several vertical tail arrangements are shown in Fig. 4.22.

Winglets can serve the function of vertical stabilizer as well as a drag-reduction device. The winglet aerodynamic center must be behind the center of gravity in order to generate a moment arm. Canard configurations and flying wings with aft swept wings commonly use this technique.

UAs typically have numerous RF transmitters and receivers for flight and payload operations. These can be for communications or as part of a payload, such as a SIGINT receiver. Vertically polarized dipole antennas are quite frequently used. Vertical surfaces such as winglets or vertical tails can house these antennas with no drag impact. Having multiple vertical tails and winglets offers great flexibility, provided that they are constructed of dielectric (RF transparent) materials.

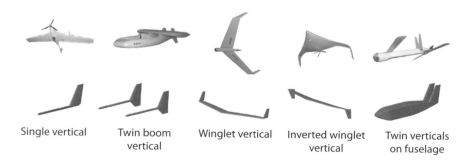

| Single vertical | Twin boom vertical | Winglet vertical | Inverted winglet vertical | Twin verticals on fuselage |

Fig. 4.22   Vertical stabilizer configurations.

4.6 Fuselage System Configurations

Fuselage design is an art. The aesthetics are critical, as most images of UAs feature the fuselage and vertical tail more than the wings. The fuselage must perform many functions and house numerous systems. Yet all the fuselage can do is detract from the performance due to weight and drag impacts. Therefore, the negative impacts must be minimized by making the fuselage compact and given low-drag shaping.

Items housed within the fuselage can include avionics, subsystems, propulsion system, landing gear, payloads, and fuel. The placement of these or other components must support aircraft balance. Payload bays that can accommodate variable weights and fuel tanks clamor for longitudinal locations near the center of gravity during conceptual design because the center-of-gravity travel must be kept within a narrow range. A long fuselage that extends behind the wing to act as a tail attachment will generally need to avoid placing heavy items aft of the center of gravity, leaving the aft portion of the fuselage largely empty.

Fuselages serve to connect UA major elements—such as wings, tails, engines, and landing gear. The loads from these components are transferred through the fuselage(s). In other words, the fuselage holds the UA together.

A fuselage system can consist of one or multiple fuselages. The most common forms include the single fuselage, single fuselage with twin booms, twin fuselage, and single fuselage with twin pods.

4.7 Propulsion Integration

4.7.1 Jet Engines

UA have numerous options for integration of jet engines. These configurations largely overlap with those available to manned aircraft. However, UAS jet engines are usually mounted on the fuselage. Key installation drivers are propulsion performance, maintainability, survivability, structural efficiency, packaging, and compatibility with other configuration features.

The vast majority of jet-powered UASs have a single engine, requiring centerline thrust. The best mounting option for a single engine is on a central fuselage. Wing externally mounted jet engines are rarely used on UASs, though this would be suitable for multi-engine designs.

Let us first consider external fuselage-mounted engines first. Several external fuselage-mounted jet engine configurations are shown in Fig. 4.23. Single external engines can be located in pods above the wings (Boeing and Teledyne Ryan Compass cope designs), on the upper aft fuselage

| Lower front | Lower mid | Lower aft | Upper mid | Upper aft | Twin side |

Fig. 4.23 Fuselage externally mounted jet engine configurations.

(Northrop Grumman Global Hawk and Teledyne Ryan Compass Arrow), under the front of the fuselage (Northrop Grumman Firebee), or under the aft fuselage. Twin external fuselage-mounted engines are generally pod engines installed on either side of the aft fuselage, much like the common business jet arrangement. Podded engines are easy to integrate, maintain, permit a clean wing design, and provide minimal inlet losses to the engine. The drawbacks to this arrangement are increased wetted area due to the nacelles and poor RCS.

Jet engines are often buried inside the fuselage. These engine installations are characterized by the inlet type and nozzle arrangements. Major inlet types include lower inlets, upper inlets, nose inlets, side inlets, and bifurcated side inlets. Side inlets make the upper and lower surfaces of the fuselage available for payload or communications field of regard. Fuselage internally mounted engines have more complex inlet and exhaust ducting than external engine pods. The engine and its ducting can consume substantial fuselage volume. Dedicated engine bays are sometimes necessary to house high-temperature jet engines.

S-shape ducts permit vertical separation between the engine centerline and the inlet face. The motivation is either to support engine integration where the fuselage ahead of the engine is occupied, or to reduce the RCS. For the latter consideration, line-of-sight blockage precludes a direct RF path to the high RCS engine face. A problem with S-ducts is that the curvature introduces distortion, often limiting maximum jet bypass ratio.

Other inlet types include the following:

- *Nose inlets*—The inlet is mounted in the nose with a relatively straight path to the engine. The Regulus cruise missile and Teledyne Ryan Q-2 used this approach. The drawback of the nose inlet approach is that substantial fuselage volume is consumed.
- *Bifurcated inlets*—Bifurcated inlets are two inlets that join together to provide airflow to a single engine. These inlets can be difficult to design due to the combination of two flow sources.

- *NACA inlets*—NACA style inlets are not as efficient, but do not penetrate the OML. The Matador missile (Fig. 4.24) used this inlet type. No configuration changes are required if the fuselage must fit within a container.
- *Pop-out inlets*—These inlets are contained within the fuselage while stowed and then projected from the fuselage in the flight configuration. The inlet face is off the centerline of the engine, and so this is in effect an S-duct. The motivation for pop-out inlets is to reduce the fuselage cross section when stored, as might be the case for a tube-launched UAS.
- *Holes in the fuselage*—Occasionally, poor inlet designs consist of nothing more than holes in the fuselage without proper ducting.

Generally, it is desirable for military UASs to shield the exhaust to reduce noise and the IR signature. Both are detection mechanisms, and the latter can be used to track and target the platform, and so this is an important survivability consideration. Engines mounted on top of the fuselage help provide this blockage as seen on the Northrop Grumman Global Hawk and the Teledyne Ryan Compass Arrow.

UCAV designs with reduced RCS tend to hide the inlets. For UCAV designs with blended wing and fuselage, it is difficult to distinguish between fuselage and wing integration, though bulges in the wing surface are usually necessary to integrate the jet engines. Other UCAVs with distinct wings and fuselages place the inlet on the upper fuselage, as demonstrated on the General Atomics Avenger.

The inlet and nozzle can be on opposite sides of the fuselage. In particular, sometimes the inlet is on the lower surface, and the exhaust is on the upper surface as can be seen on the Northrop Grumman BQM-74 E/F. The tail cone aft of the inlet can house systems such as target system payloads or parachutes. This aft body shields the hot jet exhaust and can be used to mount the stabilizers with an increased moment arm.

Fig. 4.24 Matador cruise missile with NACA inlets on the lower fuselage.

4.7.2 Reciprocating-Props and Turboprops

Reciprocating-props and turboprops combine the engine and the propeller, thereby colocating the powerplant and propulsor into a single system. The propeller and engine can be coupled directly or with a gearbox or shaft extension, but they are mechanically a single unit. The propeller can be arranged as a tractor or a pusher. Numerous installation options exist for single and multiple engines integrated as tractors or pushers.

Tractor propellers have clean airflow to the propeller. This leads to higher propeller efficiency than the pusher configuration. Tractors can also be quieter because there is no wake impingement upon the propeller. Tractors enable a large tail moment arm due to the forward engine location. One drawback is that the propeller can obstruct payload forward field of regard and preclude missile installation behind the propeller disk. Another potential problem is that the forward engine exhaust and leaked fluids can contaminate payloads such as optical systems and airborne samplers.

The simplest method of integrating a single tractor engine is at the front of a single fuselage. The single propeller provides centerline thrust with minimal fuselage system frontal area. This arrangement has the worst payload impacts.

A pod tractor installation enables favorable payload field of regard for a single engine design. The pod can be easily accessed for maintenance and can be removable. Common tractor pod locations include above the forward fuselage and in over-wing pods. The pods separate propulsion from the rest of the fuselage to provide more room for the payload. Pods are good for seaplanes where the upper installation protects the engine and propeller from water spray. The propulsion provides a nose-down moment that must be overcome by the horizontal tail, resulting in larger horizontal stabilizers and higher trim drag. Another disadvantage is higher total fuselage frontal area, producing higher drag than a nose tractor installation.

Wing installation is common when two or more engines are used. The engines are installed in nacelles that aerodynamically fair the engine and structurally interface the engine assembly to the wing. Twin engines are generally tractor configurations, though pusher arrangements are possible as well. The tails must be sized for engine-out conditions, where a strong yawing moment is caused by the thrust on the operating engine and drag on the inoperable engine nacelle. The flight controls must accommodate single-engine failure, which can be a challenge for manned aircraft. If flight on one engine results in a crash, then the UA has half the reliability of a single-engine aircraft. Twin wing-mounted engines—and twin engines in general—are relatively rare on UASs. Some suitable manned aircraft candidates for UAS conversion can be twins, as can be seen in the Diamond Aircraft DA-42 that was converted to an unmanned aircraft as the Aeronautics Dominator and Aurora Flight Sciences Centaur. A dedicated UAS twin example is the IAI Hermes 1500.

Most UASs with propellers use pusher installations, which is largely due to beneficial payload integration options. The forward field of regard is unobstructed by the propeller, missile launch paths are clear, and there is no payload contamination in flight. The drag of the body ahead of the propeller can be reduced from the flow entrainment, which is a favorable aerodynamics-propulsion interaction. The wake impingement from the body ahead of the propeller reduces the propeller efficiency and increases the noise. Pusher propellers provide pitch and yaw stabilizing effects. Cooling can be a problem for pushers because the aft cowling generally has an adverse pressure gradient as the shape curves in to meet the spinner. Parachute risers or tethers can interfere with pusher propellers without proper design provisions. An aft propeller can create challenges for balance, where the powerplant aft location tends to make the UA tail heavy.

Let us consider the balance considerations of a pusher propeller installation. The challenge is to reduce the aft moment arm of the propulsion system so that ballast in the front of the aircraft is not required to maintain the center-of-gravity limits. Several common pusher engine installation options are shown in Fig. 4.25.

The pusher engine-prop can be fitted to the rear of a single fuselage aft of the tails. This yields large, close-coupled tails and elongated noses. A minimum weight in the nose is needed to counter the engine moment, which is generally satisfied by avionics, communications systems, or payload. The General Atomics Predator A, Warrior, and Reaper systems take this approach.

Another pusher installation approach for a conventional configuration is to bring the engine closer to the center of gravity through the use of one or multiple booms extending from outside the propeller disk to the tails. The conventional twin-boom pusher has become almost a default standard for reciprocating engine UAS. Single booms can extend above or below the prop disk as well.

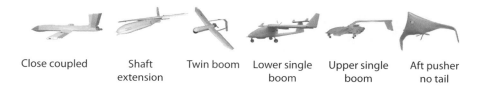

Close coupled   Shaft   Twin boom   Lower single   Upper single   Aft pusher
                  extension   boom   boom   no tail

Fig. 4.25   Pusher engine installation options.

Flying-wing and canard UA can place the engine close to the fuselage without the complexity of booms. No horizontal surfaces are located aft of the engine. The vertical surfaces can take the form of winglet vertical stabilizers.

A shaft extension between the engine and propeller can allow the propeller to be located far aft on the fuselage while the heavier engine is located closer to the center of gravity. The shaft extension approach permits a large tail moment arm for a single fuselage and less aggressive closure angles of the fuselage aft of the engine. The nose length does not need to be as long as a close-coupled pusher. Mechanical design of the shaft extension system can be challenging, though not insurmountable. The Aurora Flight Sciences Perseus A/B (Fig. 4.26) and NRL Finder UA use shaft extensions in this manner.

Parasol wings with pusher propeller and pods located behind the wing permit a continuous fuselage below the wings. The vertical offset between the propeller centerline and the lower fuselage is driven by the propeller radius plus a clearance. This configuration is practical when the propeller diameter is small relative to the fuselage length such as in direct-drive electric motors for low-power or high-speed applications. The propeller must pass through the wake of the wing, which can increase the noise. Frequently the aft fuselage is a tube that can be removed for transportation. The Aero-Vironment Raven is an example of this configuration.

Fig. 4.26   Aurora Flight Sciences Perseus B UAS uses a shaft extension from a central engine. (NASA photo by Tom Tschida.)

Twin pusher engines can be mounted off the side of the fuselage, much like the common business jet configuration. The advantage relative to twin-wing mounted engines is that the wing design is not complicated by the engine fairings, permitting a lower drag wing. A faired pylon structurally connects the propulsion pods to the fuselage. The offset is defined by the propeller diameter and minimum tip clearance. The BAE Mantis employs this configuration.

A push–pull configuration uses a tractor and a pusher engine both mounted along the centerline. The motivation is to eliminate the engine-out yawing moment that is experienced by off-centerline installations. This configuration has many of the pusher and tractor engine disadvantages combined plus the complexity of two separate engine installations. The Northrop Grumman Hunter (RQ-5A) has a push–pull configuration with a twin boom for the pusher engine, as shown in Fig. 4.27.

Ducted propellers have a ring shroud that can serve the function of both horizontal and vertical stabilizer. If the ring must perform pitch and yaw control as well, then vanes downstream of the propeller can serve this purpose. Shrouded pusher propellers can improve ground handling safety, prevent prop interference with nets, serve the function of horizontal and vertical stabilizers, and modify the acoustic characteristics. The acoustic signature can be reduced from the side but augmented in the front and rear. Efficiency is largely a function of tip clearance and duct geometry. The Lockheed Aquila and Yakovlev Shmel use ducted propeller tails. Ducted propellers are common for VTOL applications, where the duct can also serve as the fuselage when the diameter is large relative to the UA.

Engine installations require substantial volume beyond the core engine. The engine bay is separated from the rest of the UA with a firewall. Some propulsion systems that are housed within the engine bay include engine mounts, ducting, engine controls, fuel delivery systems, and header tanks. Reciprocating engines will generally also have mufflers and cylinder head cooling ducts.

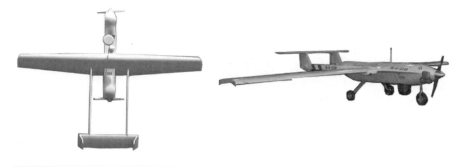

Fig. 4.27 Push–pull engine designs.

4.7.3 Electric-Props

Electric-props include at least one electric motor driving a propeller and a power source. The power source can be batteries, fuel cells, or a generator driven by an engine. In each case the motor-propeller propulsor is separated from the power source. The battery is a fixed weight, so that it can be placed away from the center of gravity, leaving more room for payloads at this prime location. The propulsor is lightweight compared to reciprocating engines or turboprops and can therefore be integrated on the aircraft with greater flexibility. The propulsor can be divided into many units and distributed across the airframe. Some propulsor-mounting location options include the tip of V-tails (Fig. 4.28), tip of horizontal stabilizers, tip of a vertical stabilizer, at the horizontal-vertical intersection of T-tails (Fig. 4.28), distributed along the wings, and in pods that can be mounted in many locations. The lightweight propulsor also permits mounting at the aft end of a fuselage without requiring that the tail becomes close coupled.

Batteries do not change weight as they are depleted, unlike fuel. This important property enables batteries to be placed far away from the center of gravity. This opens up design freedom compared to fueled aircraft that must place the fuel tank center of gravity close to the UA center of gravity to prevent exceeding balance limitations as the fuel is burned.

4.8 Launch and Recovery System Integration

4.8.1 Conventional Landing Gear

Landing gear enables the UA to launch and recover conventionally and supports ground taxi. Numerous landing-gear configurations exist. The landing gear can be fixed, retractable, or partially retractable.

Landing-gear integration impacts the UA design substantially, much in the way it does for manned aircraft. The main landing-gear location is close to the center of gravity, which is where interchangeable payloads

Fig. 4.28 T-tail (left) and V-tail tip (right) electric propulsor-mounting options.

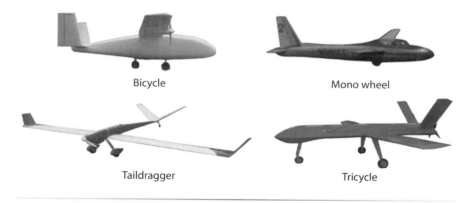

Fig. 4.29 Common landing-gear configurations.

are generally placed. Care must be taken to prevent sensor field-of-regard obstruction. Retractable landing gear consumes valuable internal real estate that would otherwise be used for variable weight items such as fuel or droppable internally carried payloads.

The landing-gear configurations available to unmanned aircraft are no different than for manned aircraft. The primary types include tricycle, tail-dragger, bicycle, and monowheel configurations as shown in Fig. 4.29.

The tricycle configuration is by far the most common for unmanned aircraft because of the simplicity for takeoff, landing, and ground handling. This arrangement consists of two main wheels behind the center of gravity and a nose wheel ahead of the center of gravity. The nosewheel is usually steerable, and braking is often applied to the main wheels. Curry [7] and other landing-gear system literature focus mostly on this configuration, which can be regarded as conventional.

Taildragger configurations, by contrast, might have complex takeoff logic and risk ground-loops. This arrangement consists of two main wheels located ahead of the center of gravity and a smaller trailing tail wheel. Takeoff autopilot logic can be more complicated because the aircraft usually lifts the tail wheel up after sufficient speed is reached and the aircraft continues the ground acceleration until takeoff, whereas the tricycle configuration can fully accelerate without this second step. Taildraggers are notorious for ground loops, where the rear of the aircraft spins around the main gear horizontally. Manned aircraft require pilot training to overcome this phenomenon, but unmanned aircraft must use autopilot logic or remote human-controlled taxi.

Bicycle landing gear has two main landing gears located along the UA symmetry plane on either side of the center of gravity. These can be accompanied by outrigger gear for ground stability or wing-tip skids.

Any combination of the front and rear wheels can be steerable, permitting satisfactory ground handling and runway tracking.

Monowheel landing gear consists of a single wheel located below the center of gravity. The advantage of this arrangement is simplicity and a lower drag installation. However, monowheels are not capable of taxi, requiring external intervention. Also, the wings must be kept level prior to the ground roll.

A droppable launch dolly can provide the wheels but with reduced integration challenges and weight impacts compared to permanent landing gear. The German Messerschmitt 163 Komet manned rocket fighter used a wheeled dolly for the ground roll and released this device after takeoff. Droppable dolly systems require an alternative form of recovery. The propulsion system must be sized to provide required takeoff acceleration with the dolly attached. The air-dropped dolly can present a safety hazard to personnel on the ground.

Fixed landing gear is relatively light and simple, but comes at the expense of higher drag and high RCS. The drag can be partially reduced by faired gear legs and wheel pants.

Retractable landing gear stows the landing gear to reduce drag. Retracted gear also supports radar signature reduction for improved survivability. The drawbacks of retractable landing gear are increased weight, more complexity, higher cost, and lower reliability relative to fixed landing gear. The main gear can retract into the wing, fuselage, or propulsion nacelles. The nose gear usually retracts aft into a centerline fuselage.

Landing gear for large UASs are generally taken from manned aircraft, for example, the BAE Mantis main gear from Piaggio P.180. Leveraging mature landing gear cuts down on development cost and helps compress the development schedule. The nose gear and main gear need not be from the same donor aircraft, as the retract geometry, UA proportions, and other major parameters can vary from the UAS and the aircraft source of the landing gear.

4.8.2 Rail Launchers

UA that use rail launchers must have designs that are compatible with these launch systems. In addition to launch, many prelaunch ground operations take place while the UA sits on the launcher. The UAS system must support placing and removing the UA on the launcher.

The UA interfaces to the launcher via an adapter. The launcher adapter is affixed to the shuttle, which travels along the rail. This adapter imparts the launch loads to the aircraft. The UA must clear the launcher interface at the conclusion of the acceleration phase. There may be a trade between other UA configuration drivers and the launcher interface complexity. A simple launcher interface has no moving parts, which is enabled by no airframe obstructions as it leaves the launcher. Consider an UA with two wing-mounted hardpoints that the interface pushes against while the tail structure and propeller arc are entirely inboard of this point. This configuration does not have any interference with the launcher interface postrelease. Now consider an UA with a pusher propeller and a lower fuselage hardpoint for the launcher adapter, where the hardpoint is within the propeller disk. Here the launcher adapter will undergo a geometry change to permit propeller clearance after release.

Many static operations must occur prior to launch. These activities can include preflight activities, ground support, and engine start. The configuration must support access to the UA while maintaining personnel safety. The launcher is prepared for launch with pressurized accumulators, and there are many potential hazards associated with releasing this stored energy. A pressurized launcher with an UA on it has the potential to injure or kill a human in the launch path. The configuration should allow normal ground operations without requiring work in a hazard zone. For example, an access panel on a winglet or tail is much better than one on the forward fuselage ahead of the wing.

The UA must be mounted on the launcher interface. Small UASs can be simply placed on the interface by hand. This operation becomes more complex for larger UASs that are too heavy to lift. Support equipment can lower the UA on the launcher, requiring support equipment interfaces such as hoist points. Another approach is for the UA to roll onto the launcher, pushed or pulled by human power. This operation requires a place on the structure that the human operator can apply the necessary force.

Most UA are not equipped with self-starting engines, and so personnel must use external, hand-held starters. These starters lock in torque with the propeller shaft and need positive contact pressure. Care must be taken to keep personnel from hitting starter components into the propeller. Tractor propellers are especially difficult to start on launchers because the ground personnel might need to stand in front of the aircraft. A pusher engine located between tail booms offers a similar risk because ground crew could be trapped between the empennage and the engine, or the UA could be accidentally launched. Inverted twin-boom V-tails or high horizontal tails partially eliminate this risk. If ground starting is required while on the launcher, engines located at the aft end of the aircraft are easiest to make safe.

Preflight activities can occur while the UA sits on the launcher. Ground power can be applied to the UA using an umbilical power cable.

Cooling air can be provided to the engine, avionics, or payload with a ground air supply. Ground computers can be physically connected as well. It is common to provide an interface panel to enable simple access and the ability to rapidly disconnect cables and hoses. Many of the electrical and physical support interfaces can come directly from the launcher.

4.9 Survivability Impacts on Configuration

Basic strategies for improving the UA survivability are well established. The primary signatures of interest include radar, acoustic, visual, and infrared (IR). In this section we will consider the configuration design influences. More in-depth treatment of survivability features can be found in Raymer [2], Jenn [8], and Ball [9].

Configuration features that can help reduce RCS include the following:

* Eliminate vertical surfaces
* Use a flying wing or BWB configuration
* Use planform alignment of leading and trailing edges, such that most sweep angles align
* Blended wing and fuselage with gradual slope changes
* Buried jet engines with line-of-sight blockage
* Retractable landing gear
* Internal payload bays
* Eliminate external features. (Antennas, hardpoints, protruding air data sensors and any other external protuberances should be eliminated.)

Many survivable UCAV designs eliminate the tail. Examples include the X-45A, X-45C, X-47A, and X-47B for the United States. Foreign systems such as the Dassault Filur and BAE Taranis follow this approach as well. Differential induced drag from asymmetric elevon deflection or spoilers provide directional control. Swept flying wings with positive static margin can have stable lateral-directional characteristics without vertical surfaces, provided that there is not much side area in the front of the aircraft. Several LO UCAV configurations are shown in Fig. 4.30.

The primary driver of the UA's acoustic signature is the propulsion system. Configuration features that reduce the acoustics include the following:

* Tractor propellers
* Shaping upstream of pusher propellers to reduce wake impingement
* Line-of-sight blockage of noise sources

The primary configuration feature that can help reduce the IR emissions is physical blockage of the heat source.

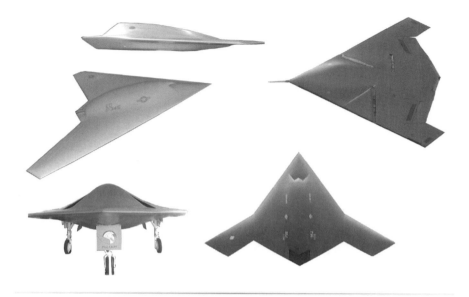

Fig. 4.30 UCAV configurations with blended fuselages and no vertical tail surfaces.

4.10 Transportability and Storage Impacts on Configuration

Because most UASs do not self-deploy, they rely upon other transportation systems to get to the point of operations. The UASs are generally stored when not in use. These two considerations drive the need for UASs to fit within some form of storage or transportation container. These two functions are usually combined such that the storage boxes are also used for transportation. Chapter 17 provides more detail on the system supportability. Here we will look at how fitting an UA in a box drives the configuration.

Once the box dimensions and associated offsets are established, the UA must be capable of fitting within this geometry when disassembled. A good objective is to minimize the number of parts that must be detached. The transportation joints should be the same as the manufacturing breaks to minimize complexity. The box should be kept as small as possible to reduce the transportation footprint. Therefore the UA packaging should be volumetrically efficient.

Some options for transportation breaks include the following:

* Payload modules might be separated from the fuselage to allow for greater shock and vibration isolation.

- The fuselages can be broken into multiple segments, and the central segment can remain attached to the wing.
- Wing halves and tails can come off at the fuselage intersection.
- Wings can be made of flexible materials that can be rolled into tubes, as is done on the ARA NightHawk.

The designer should consider how humans will assemble the UA. Force must be applied to the parts in order to put them in place. The parts must have suitable handling points that are easy to lift and can withstand the human grasp. Hoist points and other mechanical interfaces are needed for components too heavy for a person to manipulate.

4.11 In-Flight Reconfiguration

In-flight reconfiguration is quite common for unmanned systems because mission requirements can demand launch from a container, air-launch, or major unmanned aircraft performance attribute changes during the mission. Changing the geometry in flight is much more constraining than transportability considerations. Here all geometry transformations must be undertaken in a dynamic environment and without human or external assistance.

Joints are used to effect the reconfiguration. Joint mechanisms enable rotation, translation, or a combination of rotation and translation between two parts. Some of the most common joints are the following:

- *Pivots*—This rotating joint generally refers to a rotation in the plane of the aerodynamic surface. For example, a wing has a vertical axis of rotation such that the wing can align with the fuselage when stowed and then be normal to the flight direction when deployed. Pivots are common for wings, vertical stabilizers, horizontal stabilizers, or other more exotic tails. Wings generally pivot about the wing root where the right and left halves are continuous. When the right and left wings pivot separately, this is known as a *swing wing*. Stabilizer surfaces pivot about a point at the root as a semispan unit.
- *Hinged Surfaces*—The aerodynamic surface, such as a wing, rotates about an axis parallel to the chord line. The surface can fold upon itself with one segment resting on the other segment when stowed. Alternatively, the surface can be hinged such that it rests along the side of a fuselage. With wing hinges the outboard wing segment generally folds under the inboard section when stowed. Upon activation, the outboard panel initially rotates downward with a gravity assist. The aerodynamic lifting force of the outboard surface helps it to fly into the horizontal position. A latching

mechanism secures the wing in place. Outboard panels that rest on top of the inboard panel must overcome both gravity and aerodynamic forces to rotate into the flight position.

- *Telescoping wings*—One wing fits within another in a manner that allows one wing to slide (translate) out to increase span, much like a collapsible fishing rod. Usually the exterior wing is attached to the fuselage while the smaller interior wing translates outward, as is the case with the Boeing ScanEagle Compressed Carriage (SECC) UA. A guide mechanism aligns the wings and reduces friction for deployment. Care must be taken to ensure that the wings do not bind. The exterior wing is hollow, devoid of vertical structural members. The shear loads must therefore be transferred through the leading and trailing edges, which is a structurally inefficient approach.
- *Translating mechanisms*—These mechanisms allow for one body to slide relative to the other. A translating wing root position along the fuselage permits a wing span or semispan nearly equal to the fuselage length while positioning the wing aerodynamic center at the appropriate position along the fuselage. The L-3 Cutlass uses this method.
- *Chord extension*—The wing trailing edge can extend aft from the stowed to deployed configuration along guides to increase the chord length. This approach is common for slotted flaps, though slots or large camber changes are not necessary. The effect of increasing the chord length is to increase the wing area, which permits lower speed flight.

Other nontraditional reconfiguration options include the following:

- *Inflatable wings*—A membrane is densely folded when stowed and then attains the wing flight geometry when pressurized. The pressurized gas makes the structure rigid. A drawback of inflatable wings is that smooth contours are difficult to attain, increasing the drag and reducing the achievable C_{Lmax}. The SAIC LEWK used inflatable wings.
- *Parafoils*—These specialized devices combine the packaging of a parachute with many attributes of a wing. The canopy is a wing with a downward arc. The fabric is passively inflated by the airflow such that airfoil geometry is attained. The body containing the payload, systems, and propulsion is suspended beneath the canopy.
- *Rigallo wings*—This configuration uses a simple structure consisting of leading-edge members, a centerline member, and bracing structure to support a fabric delta wing. The leading-edge members can be stowed parallel to the fuselage. The Rigallo wing has fallen out of use as a result of the use of parafoils that offer equivalent or better performance but with reduced complexity.

Joints are potential failure mechanisms. When high reliability is sought, the UA should have as few and least complex joints as practical. Passive deployment approaches might be more reliable than those that require controllers and intricate sequencing. Proven joint technologies should be employed when possible.

The UA must not lose control during the in-flight reconfiguration. One approach is to make the UA fully controllable during the deployment sequence in all geometric permutations. This is often not possible. Another approach is to make problematic deployment steps occur so rapidly that the flight controls can compensate for the destabilizing transients once a fully controllable state is attained. Tails should generally be deployed first to stabilize the UA, whereby the nose of the aircraft is aligned with the flow. The UA follows a ballistic trajectory until the wings deploy. Another approach is to use a drogue parachute or ballute (inflatable envelope that acts much like a parachute) to stabilize the UA during reconfiguration.

UA must generally fold lengthwise to fit within the storage container or provide minimal drag when externally carried on a host aircraft. Loitering UA generally need to increase the span and wing area to reduce required power and fuel flow. It is therefore desirable to make the wing span longer than the fuselage to get optimum performance. This must be balanced against the weight penalty for the joint mechanisms and system reliability. High-speed UA generally do not require as large of wings and are therefore easier to package.

The flight-configuration wing position is quite often near the center of the fuselage, which yields a tail moment arm equal to approximately half of the fuselage length. This is a close-coupled arrangement, which drives the need for a large tail area to achieve desired tail volume coefficients. Unfortunately, the packaging constraints often limit the effective tail area. The small tail volume can limit the speed range and introduce nontrivial controls challenges. The horizontal stabilizer is often sized to enable a pull-out maneuver.

The wing-fuselage and tail-fuselage intersections can be large drag contributors for in-flight reconfigurable aircraft. Pop-out fairings can be employed, though with increased complexity. Creative configuration design might yield low drag installations. Otherwise, this drag contributor must be accounted for in the drag buildup.

The propulsion system might require configuration changes between the stowed and flight states. Propellers can fold forward or aft near the hub. The centripetal acceleration passively unfolds the blades, though care must be taken to ensure that the propellers do not impact other structural members. Jet engine inlets can be made to extend from the fuselage after launch.

4.12 Modularity and Airframe Growth

Successful UA will undergo many design modifications to improve performance or add new mission capabilities. The configuration should be designed with sufficient modularity to enable this design evolution. Modularity is also important for maintainability.

Parts that are interchangeable are desirable for support. Interchangeable means that the part can be fitted to the UA without modification to the part or the attachment point on the UA. Replaceable means that some adjustments are permissible. The difference between interchangeable and replaceable comes primarily from part manufacturing tolerances. The supply chain is simplified by interchangeable parts, and the maintainers spend less time swapping components, thereby improving mission availability.

A truly modular UA has interchangeable modules that can be replaced over the life of the UA. It is entirely possible for one UA to remain in operational service indefinitely but have all of the parts replaced. The concept of tracking a tail number has little applicability in such a system.

4.13 Manufacturing Configuration Drivers

An UA should be as inexpensive to manufacture as the design can permit. To help achieve this goal, the UA should minimize the number of major assemblies and simplify the construction. For example, a flying wing will likely have fewer major assemblies than a conventional design, which in turn is simpler than a three-surface UA. A blended-wing body design could combine the molds for the wing and fuselage, thus decreasing the parts count.

Investment in tooling is often a major cost driver. Hand-built prototypes are common for early development where capital or external nonrecurring engineering funding is in short supply. Available tooling can therefore drive new unmanned aircraft. An example is the Shadow 200, which reused the Pioneer wing tooling for the new design [10]. This generally leads to a nonoptimized design, but the penalties might be acceptable.

Packaging density drives complexity and cost. Large UASs are less volume constrained and can use COTS boxes (ATR sizes) with good access for maintainability. When significant performance and electronic capability is placed on smaller UAS classes, tight integration is often required. Avionics are often custom developed to fit peculiar geometries, with a substantial initial development. This is much the same for missiles.

Adopting a constant fuselage cross section for a portion of the fuselage length can allow a simplified internal structure and more comprehensible mechanical interface for avionics and new payloads. The constant cross section allows straight longerons and common bulkhead tooling.

Structural members should be combined when possible. For example, a main fuselage bulkhead can be combined with a wing carry-through bulkhead. Ribs in the wings can be combined with hardpoints. Eliminating parts makes the structure lighter and less expensive to manufacture. The manufacturing breakpoints can be the same feature as field breaks to reduce the overall number of interfaces. Combining internal structural features often requires external changes in the configuration.

4.14 Rotorcraft Configurations

The range of vertical takeoff and landing (VTOL) configurations is quite broad, with dozens of notable arrangements. Here we will only mention rotorcraft configurations rather than attempt to cover the waterfront of VTOL arrangements.

Helicopters have unducted rotors to generate vertical lift and to provide a forward-thrust component. Each individual rotor generates a torque that must be counteracted. Additionally, the helicopter must be controllable. Several notable rotor configurations include the following:

* *Conventional*—The conventional helicopter configuration uses a main lift rotor whose torque is countered by a side-mounted tail rotor.
* *Tandem rotor*—A tandem rotor has two lifting rotors, one behind the other. Each rotor spins in the opposite direction to cancel the torque.
* *Coaxial rotor*—One rotor is mounted on top of the other along the same axis of rotation. The rotors spin in the opposite direction to cancel the torque.
* *Meshing rotor*—Here a rotor is mounted on either side of the helicopter tilted to the side by an angle. The rotor disks intermesh, but gearing prevents rotor collisions. The rotors spin in opposite directions. This configuration is used by Kaman helicopters, such as the K-max.

References

[1] Kirschbaum, N., and Mason, W. H., *Aircraft Design Handbook, Aircraft Design Aid and Layout Guide*, 1993-1994 ed., Dept. of Aerospace and Ocean Engineering, Virginia Polytechnic Inst. and State Univ., Blacksburg, VA, 1994, pp. 1-1–1-20.

[2] Raymer, D. P., *Aircraft Design: A Conceptual Approach*, 2nd ed., AIAA, Reston, VA, 1992, pp. 123–139.

[3] Roskam, J., *Airplane Design, Part II: Preliminary Configuration Design and Integration of the Propulsion System*, Roskam Aviation and Engineering Corp., KS, 1989.

[4] Roskam, J., *Airplane Flight Dynamics and Automatic Flight Controls, Part I*, DARcorp., Lawrence, KS, 1995.

[5] Etkin, B., and Reid, L. D., *Dynamics of Flight, Stability and Control*, 3rd ed., Wiley, New York, 1996.

[6] Nelson, R. C., *Flight Stability and Automatic Control*, McGraw-Hill, New York, 1989.

[7] Curry, N. S., *Aircraft Landing Gear Design: Principles and Practices*, AIAA, Reston, VA, 1988.

[8] Jenn, D. C., *Radar and Laser Cross Section Engineering*, AIAA, Reston, VA, 1995.

[9] Ball, R. E., *The Fundamentals of Aircraft Combat Survivability Analysis and Design*, AIAA, Reston, VA, 1985.

[10] Palumbo, D. J., "Design Evolution of the Shadow 200 Tactical Unmanned Air Vehicle," *Proceedings of the Association of Unmanned Vehicle Systems International 2000 Conference*, Orlando, FL, July 2000.

Problems

4.1 Construct a table of the same format as Table 4.1 for a single panel wing with no dihedral. Given aspect ratio (AR) and taper ratio λ, generate the matrix to solve for b, S, and c_r.

4.2 A two-segment wing has a span of 20 ft and a root chord of 2 ft. There is a taper break at midspan. The midspan taper is 0.5, and the tip taper is 0.3. The leading-edge sweep is 30 deg. What are the wing area and aspect ratio? What is the mean aerodynamic chord length? Calculate the location parameters d and m.

4.3 Use the wing from problem 4.2. A horizontal tail has a root leading edge located 5 ft behind the wing root trailing edge. The tail has a root chord of 1 ft, tip chord of 0.5 ft, b' of 5 ft, quarter-chord sweep of 0 deg. What is the horizontal moment arm? What is the horizontal tail volume coefficient?

4.4 Calculate the coordinates for a NACA 4412 airfoil.

4.5 A fuselage has a constant cross section. The length is 10 ft, the height is 1 ft, and the width is 0.5 ft. It is symmetric top-bottom and left-right. The cross section is defined by a superellipse with $m = 0$ and $n = 1$. Calculate the cross-section perimeter and cross-section area. Calculate the fuselage surface area and volume.

4.6 Identify the UA in Fig. 4.11.

4.7 Find three tail configurations not covered by this chapter.

4.8 Sketch five solar-powered aircraft configurations that facilitate effective solar collection. These can be based on manned and unmanned solar-powered aircraft that have flown or are concepts.

4.9   A conventional UA has a $C_{m0,w} = -0.02$, $sm = 0.08$, $C_{m,f} = 0.01$, and $\eta_H = 0.95$. Find the modified tail volume coefficient required to achieve a tail lift coefficient of -0.5 the wing C_L is 1.2.

4.10 A canard UA has a canard surface stall lift coefficient of 1.5 and a wing stall lift coefficient of 1.8. The UA has a $C_{m0,w} = -0.02$, $C_{m,f} = 0.005$, and $\overline{TVC}_C = -0.2$. The wing area is 10 ft^2, the mean aerodynamic chord is 1 ft, and the canard's aerodynamic center is located 3 ft ahead of the wing-body neutral point. Plot the UA's maximum lift coefficient as a function of the static margin, where the static margin ranges from 0 to 0.2.

4.11 A three-surface UA has a $C_{m0,w} = -0.01$, $sm = 0.1$, $C_{m,f} = 0.005$, $\overline{TVC}_C = 0.2$, $\overline{TVC}_C = -0.3$, and $C_{L,w} = 1$. Assume that the tail efficiency is 1. Plot the required horizontal lift coefficient as a function of the canard lift coefficient such that the UA is trimmed.

4.12 Demonstration project. Follow all applicable laws and safety precautions. Build a tandem-wing balsa model with a span of 10 in. The wings and vertical tail should be made of 1/16th-in. balsa sheet. The fuselage should use 3/32nd-in. balsa sheet. Use modeling clay or small metal washers to adjust the nose weight until the model trims in pitch so that a steady glide can be attained. Adjust the forward-wing incidence as necessary. Note: a downward arc flight path indicates that the center of gravity is too far forward, and a sudden pitch up indicates that the center of gravity is too far aft.

Chapter 5

Aerodynamics

- Define flight regimes for various UAS classes
- Provide a simplified method for estimating lift-and-drag performance
- Generate aerodynamics tables for performance analysis
- Identify more detailed methods for preliminary and detail design phases

Fig. 5.1 NASA's Drones for Aerodynamic and Structural Testing (DAST) Firebee in flight. (Photo courtesy of NASA.)

5.1 Introduction

I n Chapter 3 we apply a simple method of calculating the maximum lift-to-drag ratio as a function of the wetted aspect ratio. Although this approach provides a rapid way of estimating the potential performance and is based on historical aircraft, many important design details are obscured by this general treatment. Relevant aerodynamics drivers include aircraft configuration, wing planform geometry, airfoil selection, fuselage shaping, propulsion type, propulsion integration, internal vs external landing gear, payload integration geometry, and a host of other considerations. This chapter identifies more detailed methods that can be applied in conceptual design, as well as an understanding of higher-fidelity methods that can be used in later design phases.

5.2 Flight Regime

When one thinks of flight regime, Mach number comes to mind. Mach number is the ratio of the flight velocity V to the speed of sound a, which is defined as

$$M = \frac{V}{a} \qquad (5.1)$$

The speed of sound for a given gas or mixture of gasses is a function of temperature T alone.

$$a = \sqrt{\gamma \cdot R \cdot T} \qquad (5.2)$$

The ratio of specific heats γ is a property of the gas, and R is the gas constant. For air γ is 1.4, and R is 287.05 N-m/kg-K. Flight regimes are segregated by Mach number (see Table 5.1). The vast majority of UAS applications are subsonic. Those with Mach numbers less than around 0.3 can be considered to operate in incompressible flow. HALE UAS, cruise missiles, and high-speed targets can be transonic. Transonic flight must contend with compressibility effects. Few reconnaissance UASs and targets are supersonic. Today only research UASs travel in the hypersonic regime.

Table 5.1 Flight Regimes Defined by Mach Number

Mach Range	Flight Regime
Less than 1	Subsonic
Approximately 0.7 to approximately 1.3	Transonic
Greater than 1	Supersonic
Greater than approximately 4	Hypersonic

The Reynolds number is a key governing parameter that relates the relative influences of inertial to viscous forces. Much like the Mach number, it characterizes a flight regime and is unitless. The Reynolds number Re is defined as

$$Re = \frac{\rho \cdot V \cdot l}{\mu} \tag{5.3}$$

where ρ is the air density, V is the flight velocity, l is a characteristic length, and μ is the kinematic viscosity of the fluid. The air density and kinematic viscosity are usually found through atmospheric table look-ups as a function of altitude. Alternatively, the Sutherland formula is a simple way to estimate the kinematic viscosity as a function of only absolute temperature T (in degrees Kelvin).

$$\mu = 0.00001716 \cdot \left(\frac{T}{273.1}\right)^{1.5} \cdot \frac{383.7}{T + 110.6} \qquad N - s/m \tag{5.4}$$

Unmanned aircraft can operate at Reynolds numbers that overlap with manned aircraft. Small UASs and MAVs operate in a Reynolds regime substantially lower than manned aircraft, which is known as low Reynolds number (LRN). Mueller [1] describes the aerodynamic phenomena of the smallest UAS classes.

Example 5.1   Flight in Extraterrestrial Atmospheres

Problem:

Consider a rocket-propelled aircraft (sometimes called a "flyer" if operating in a non-Earth atmosphere) that flies in various planetary atmospheres. The mass is 40 kg, the wing area is 1 m², the average chord length is 0.2 m, and the cruise lift coefficient is 0.5. The notional planetary and moon atmospheric properties at the flight conditions of interest are shown in Table 5.2. These atmospheric properties are provided for the purposes of this exercise only, and any planetary flight missions should consult other sources. Planetary and moon atmospheres vary with altitude, location, time of day (called a sol on other planets or moons), season, and other influences. Calculate the flight velocity, Mach number, and Reynolds number.

Solution:

The velocity is found by

$$V = \sqrt{\frac{2 \cdot m \cdot g}{S_w \cdot \rho \cdot C_L}}$$

Table 5.2   Notional Planetary and Moon Atmospheric Properties for Example

Planet or Moon	Gravity, m/s^2	Density, kg/m^3	Speed of Sound, m/s	Kinematic Viscosity, kg/m-s
Earth (sea level)	9.81	1.225	340.3	1.7894×10^{-5}
Earth (30 km)	9.81	0.0184	301.7	1.4753×10^{-5}
Mars (surface)	3.57	0.0155	~244	1.2×10^{-5}
Titan (surface)	1.35	5.4	~200	$~5.9 \times 10^{-6}$
Venus (surface)	8.87	65	~426	$~3 \times 10^{-5}$

The results of all required parameters are shown in the following table.

Planet or Moon	V, m/s	M	Re
Earth (sea level)	35.8	0.105	4.9×10^5
Earth (30 km)	292.1	0.968	7.3×10^4
Mars (surface)	192.0	0.768	5×10^4
Titan (surface)	6.3	0.032	1.158×10^6
Venus (surface)	4.7	0.011	2.025×10^6

5.3   Boundary Layers

The unmanned aircraft geometry generates a flowfield on the surrounding air. At a much smaller scale, the molecules that comprise the air at the boundary of the surface have zero relative velocity. Outside of the boundary layer the flow has the velocity U_e. A thin layer separates the flow at the edge of the boundary layer, $u = U_e$, and zero relative velocity at the surface $u = 0$. This is known as a boundary layer. The boundary-layer thickness δ is the height above the surface at which $u = 0.99\, U_e$.

Boundary layers are either laminar or turbulent. Laminar boundary layers have substantially lower drag, as can be seen in Fig. 5.2. The skin-friction coefficient for laminar flow on a flat plate is given by

$$C_{f,\text{lam}} = \frac{0.664}{\sqrt{Re}} \tag{5.5}$$

The incompressible turbulent skin-friction coefficient on a flat plate can be approximated by

$$C_{f,\text{turb}} = 0.0583 \cdot Re^{-0.2} \tag{5.6}$$

A flat plate is often used for friction drag calculations because it is a relatively simple case. There are many theoretical analyses and experimental

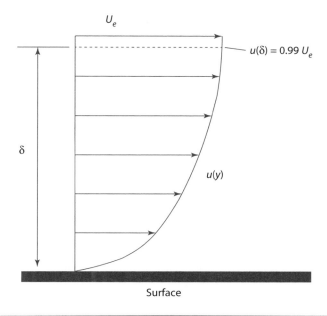

Fig. 5.2 Boundary layer.

data that can be readily leveraged. A flat plate has no pressure gradient, that is, $dP/dx = 0$. As we will see, the flat-plate case can be related to more complex geometries such as airfoils or bodies through shape influence approximations.

A laminar boundary can change to turbulent flow before reaching the total length. The change between these states is called *transition*. The transition Reynolds number Re_{tr} for a flat plate is around 5×10^5. This value is higher for a favorable pressure gradient ($dP/dx < 0$) and lower for an adverse pressure gradient ($dP/dx > 0$). The skin-friction coefficient for a flat plate with transition is approximated by

$$C_f = C_{f,\text{lam}} \cdot x/c_{tr} + C_{f,\text{turb}} \cdot (1 - x/c_{tr}) \qquad (5.7)$$

Note that the laminar and turbulent skin-friction coefficients are evaluated at the overall Reynolds number. Here x/c_{tr} is the ratio

$$\frac{x}{c_{tr}} = \frac{Re_{tr}}{Re} \qquad (5.8)$$

when $Re_{tr} < Re$, and $x/c_{tr} = 1$ otherwise.

Given the significant drag benefits of laminar flow, aerodynamicists often seek to increase the laminar extent. The benefits of laminar flow for skin-friction reduction can be seen in Fig. 5.3. In other words, x/c_{tr} should

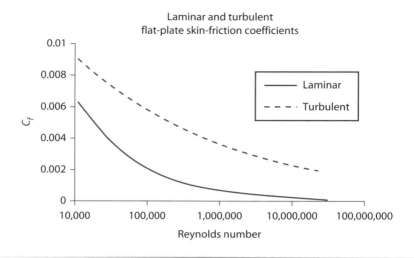

Fig. 5.3   Laminar and turbulent flat-plate skin-friction drag coefficients.

be made as large as practical to minimize drag. Some of the many contributors to transition include the following:

* Wing sweep (Tollman–Schlicting waves and crossflow instabilities cause transition.)
* Bug debris
* Ice buildup
* Ice crystals in stratospheric clouds
* Water accumulation due to rain
* Steps, gaps, and cracks due to access panels, doors, skin panel intersections, and control surfaces
* Surface waviness
* Surface roughness
* Intersections of two bodies (i.e., wing-fuselage juncture)
* Adverse pressure gradients
* Compressible shocks
* Propeller wash
* Vibration
* Acoustic energy (noise)

Separation is a phenomenon where the flow detaches from surface, leaving a region of low-energy air with large-scale eddies. Separated regions have substantially greater drag than nonseparated regions. Separation occurs where the boundary layer does not have sufficient energy to contend with an adverse pressure gradient. An adverse pressure gradient has increasing pressure with length along the surface ($dP/dx > 0$). Laminar flow is less able to adapt to adverse pressure gradients because of low energy within the boundary layer. Turbulent boundary layers are thicker and more energetic and are therefore more able to withstand adverse pressure gradients without separation.

The boundary layer should transition from laminar to turbulent flow before reaching an adverse pressure gradient that can cause separation. Forcing transition is not a problem at high Reynolds numbers, but low-Reynolds-number flows might experience difficulties transitioning. In other words, low-Reynolds-number flows can be too laminar, resulting in excessive separation regions. Methods of intentionally tripping the boundary layer include adding surface roughness elements or trip strips. Schetz [2] can be consulted for further reading about boundary layers.

5.4 Coordinate Systems and Aerodynamic Coefficients

The body-axis system used for stability and control is defined in Chapter 4. Table 5.3 defines aerodynamic forces, moments, and rates within this reference frame.

Relatively few aerodynamic parameters are required for flight performance analysis. As we saw in Chapter 3, the lift and drag coefficients are

$$C_L = \frac{L}{1/2 \cdot \rho \cdot V^2 \cdot S_w} \quad \text{and} \quad C_D = \frac{D}{1/2 \cdot \rho \cdot V^2 \cdot S_w} \quad (5.9)$$

Note that the L here refers to lift force and not rolling moment. In flight performance it is assumed that the pitching moment is zero. The definition of pitching-moment coefficient is

$$C_M = \frac{M}{1/2 \cdot \rho \cdot V^2 \cdot S_w \cdot mac} \quad (5.10)$$

Many more aerodynamic parameters are required for stability and control analysis. Many of these are derivatives of coefficients with respect

Table 5.3 Body-Axis Aerodynamic Parameters

	Roll Axis, x_b	Pitch Axis, y_b	Yaw Axis, z_b
Angular rates	p	q	r
Velocity	u	v	w
Aerodynamic forces	X	Y	Z
Aerodynamic moments	L	M	N
Moment of inertia	I_x	I_y	I_z
Products of inertia	I_{yz}	I_{xz}	I_{xy}

Table 5.4   Longitudinal Stability and Control Parameters

Parameter	Derivative of	With respect to
$C_{D,M}$	Drag coefficient, C_D	Mach, M
$C_{D,\text{Ref}}$	Drag coefficient, C_D	
$C_{D,\alpha}$	Drag coefficient, C_D	Angle of attack, α
$C_{L,M}$	Lift coefficient, C_L	Mach, M
$C_{L,q}$	Lift coefficient, C_L	Pitch rate, q
$C_{L,\alpha}$	Lift coefficient, C_L	Angle of attack, α
$C_{L,\dot{\alpha}}$	Lift coefficient, C_L	Angle of attack rate, $\dot{\alpha}$
$C_{M,\dot{\alpha}}$	Pitching-moment coefficient, C_M	Angle of attack rate, $\dot{\alpha}$
$C_{T,M}$	Thrust coefficient, C_T	Mach, M
$C_{T,V}$	Thrust coefficient, C_T	Velocity, V
$C_{T,\delta M}$	Thrust coefficient, C_T	Pitch control input, δM
$C_{T,\delta T}$	Thrust coefficient, C_T	—

to a parameter. The longitudinal parameters are shown in Table 5.4, and the lateral-directional parameters are shown in Table 5.5. C_L refers to lift coefficient in the longitudinal context, and C_l is rolling moment coefficient in the lateral-directional context.

Estimating these stability derivatives requires techniques that are beyond the scope of this book. A vortex lattice method (VLM) is commonly used to predict many of these parameters. Empirical curve fits are another

Table 5.5   Lateral-Directional Aerodynamic Parameters

Parameter	Derivative of	With respect to
$C_{l,p}$	Rolling-moment coefficient, C_l	Roll rate, p
$C_{l,r}$	Rolling-moment coefficient, C_l	Yaw rate, r
$C_{l,\beta}$	Rolling-moment coefficient, C_l	Sideslip angle, β
$C_{l,\delta L}$	Rolling-moment coefficient, C_l	Roll control input, δL
$C_{l,\delta N}$	Rolling-moment coefficient, C_l	Yaw control input, δN
$C_{n,r}$	Yawing-moment coefficient, C_n	Yaw rate, r
$C_{n,\beta}$	Yawing-moment coefficient, C_n	Sideslip angle, β
$C_{n,\delta L}$	Yawing-moment coefficient, C_n	Roll control input, δL
$C_{n,\delta N}$	Yawing-moment coefficient, C_n	Yaw control input, δN
$C_{Y,p}$	Side-force coefficient, C_Y	Roll rate, p
$C_{Y,r}$	Side-force coefficient, C_Y	Yaw rate, r
$C_{Y,\beta}$	Side-force coefficient, C_Y	Sideslip angle, β
$C_{Y,\delta N}$	Side-force coefficient, C_Y	Yaw control input, δN

common approach, where the stability and control DATCOM [3] is regarded as a standard, though not very applicable to low-speed or high-aspect-ratio aircraft. Roskam [4] provides empirical methods for generating the stability derivatives based on DATCOM. CFD and panel methods can also be employed. Large UAS programs will often use wind tunnels to generate these parameters.

5.5 Airfoils

In Chapter 4 we covered airfoil geometry definition. The main parameters such as thickness to chord, camber, and their distributions were described. Now we will take a look at airfoil aerodynamic performance and selection.

Airfoils are often provided in families. The NACA 4-series, whose geometry is described in Chapter 4, is one example of a family. Other NACA series such as the 6-series are also available. More modern attempts at airfoil families stress design regimes such as low Reynolds numbers or transonic flight. These series of airfoils have a similar geometric philosophy and are usually developed by a single airfoil designer. A single designer can have numerous airfoil families, but the designer's initials generally identify the source. As examples, several designer designations include Eppler (E), Selig (S), Selig-Donovan (SD), Martin Hepperle (MH), Somers-Maughmer (SM), and Wortmann (FX). Some aircraft designers will stick with airfoils developed by their favorite airfoil designers.

Airfoil design is a complex endeavor. Generating a good airfoil involves much more than simply minimizing drag coefficient at a given operating point consisting of lift coefficient and Reynolds number. The airfoil must have acceptable drag at multiple operating points. Additional considerations include the following:

- C_d shape of the usable C_l region
- Behavior of the drag bucket for laminar airfoils (high Reynolds numbers)
- Stall behavior—sharp vs gradual
- Pitching-moment coefficient—can drive trim drag
- Compatibility with flaps
- Effectiveness with control surfaces—should not have dead bands
- Thickness to chord—for structures and internal volume
- Ability to operate over entire flight envelope (*Re*, Mach)
- Ease of construction (Thin, cusped trailing edges are hard to build.)
- Tolerance to manufacturing imperfections
- Robustness against icing, rain, and bugs (Laminar flow airfoils can have dramatic loss of maximum lift coefficient when the boundary layer is tripped.)

Table 5.6 Typical Chord Reynolds Numbers for Aviation Applications

Typical Chord Reynolds-Number Range	Applications
$3 \times 10^4 - 10^5$	MAVs and small model aircraft
$10^5 - 3 \times 10^5$	Small UASs and sport model aircraft
$2 \times 10^5 - 6 \times 10^5$	Small tactical UASs and giant scale model aircraft
$6 \times 10^5 - 3 \times 10^6$	MALE, tactical UASs, and manned sailplanes
$2 \times 10^6 - 8 \times 10^6$	Large MALE, HALE, and general aviation
$5 \times 10^6 - 2 \times 10^7$	UCAVs and manned tactical aircraft

The airfoil designer must start with a two-dimensional airfoil code such as XFOIL or MSES. These are two-dimensional boundary-layer codes that relate the boundary-layer thickness to the circulation distribution and can predict transition, laminar separation bubbles, the onset of separation, and some postseparation effects. Some codes can analyze multi-element airfoils or compressibility effects. A useful design tool is an inverse design method that defines the geometry based on a desired pressure distribution. Often a similar airfoil is used as a starting point and incrementally modified until the desired characteristics are attained.

The Reynolds number is especially important for airfoil performance. This parameter determines the achievable section maximum lift coefficient and lift-to-drag ratio. It also bounds the maximum thickness-to-chord ratio, beyond which point the airfoil will have unacceptable performance. Table 5.6 shows typical Reynolds-number ranges for various UAS applications.

There are many more UAS projects than there are airfoil designers. Most programs simply select an airfoil from available sources. These sources include public domain resources such as the University of Illinois at Urbana Champaign (UIUC) online airfoil database [5]. Companies often have proprietary airfoil designs that are reused on new programs. When a program can afford the cost and time of having a custom airfoil designed by an expert, the unmanned aircraft will usually have improved performance, lower weight, and other favorable attributes.

Example 5.2 Airfoil Selection

Problem:

An aerodynamicist is selecting an airfoil for a wing. The average section lift coefficient in cruise is 0.5 and 1.0 in loiter. Approximately 25% of the mission is spent in cruise and 75% in loiter. The maximum nonflapped

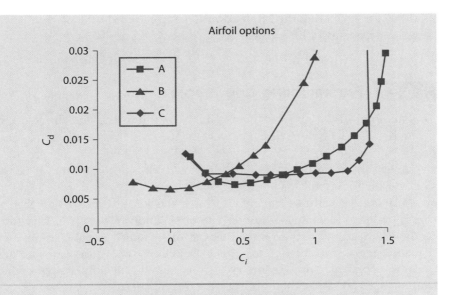

Fig. 5.4 Airfoil drag polars.

section lift coefficient must be at least 1.2. Using the airfoils shown in Fig. 5.4, select the best airfoil. Note that we are neglecting many considerations.

Solution:

Airfoil B is not capable of achieving the section lift coefficient of 1.2, and so it is eliminated from further consideration. Next, we will create an objective function to evaluate the two remaining sections. We seek to minimize the objective function.

$$Obj = 25 \cdot C_{d,\text{cruise}} + 0.75 \cdot C_{d,\text{loiter}}$$

The drag coefficients and objective function for the two airfoils are shown in the following table:

Airfoil	$C_{d,\text{cruise}}$	$C_{d,\text{loiter}}$	Obj
A	0.0075	0.0111	0.0102
C	0.0091	0.0092	0.0092

Airfoil C is the best choice based on these criteria. It performs best in loiter, which is most heavily weighted.

5.6 Three-Dimensional Lift Distribution Methods

Airfoils provide two-dimensional lift, which is equivalent to the characteristics of a section of an infinite span wing. Real wings, which is to say finite span wings, behave quite differently. This section describes methods for predicting the lift distribution and induced drag of finite span wings and systems of lifting surfaces.

5.6.1 Prandtl Lifting-Line Theory

The Prandtl lifting-line theory (LLT) predicts the lift distribution through a Fourier sine series. The general theory is treated in Bertin and Smith [6], Anderson [7], and Katz and Plotkin [8]. Anderson [7] provides a numerical lifting-line theory that can address nonlinear lift-curve slopes, which is useful for analyzing wings near and beyond stall angle of attack. Here we will use a matrix form of the Prandtl lifting-line theory that is practical for many subsonic UAS applications. This lift distribution method is relatively easy to implement, it offers speed of calculation, and the accuracy is satisfactory for many problems. The disadvantages are that the applicable wing geometry is constrained and only a single wing can be analyzed rather than multiple horizontal surfaces.

The wing must meet the following criteria:

- The wing must have no sweep about the quarter-chord line or negligible sweep ($<$10 deg).
- The wing must have no dihedral or negligible dihedral.
- The wing must have at least moderate aspect ratio ($>$5).
- The flow is incompressible.
- The airfoils have linear lift-curve slopes and are not stalled.

The lifting-line method presented (Fig. 5.5) allows varied chord, camber, and twist distributions along the span. Literature on this method frequently assumes a single taper, and so this approach is more general. The general formulation is

$$\frac{\pi \cdot c(\theta)}{2 \cdot b_w}[\alpha + \alpha_{\text{twist}}(\theta) - \alpha_{0L}(\theta)]\sin\theta = \sum_{n=1,\text{odd}}^{\infty} A_n \sin(n\theta)\left[\frac{\pi \cdot c(\theta) \cdot n}{2 \cdot b_w} + \sin\theta\right]$$

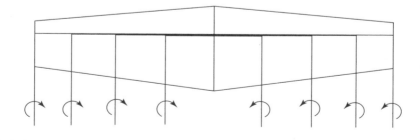

Fig. 5.5 Lifting-line horseshoe vortex representation.

where

A_n = influence coefficient
b_w = wing span
c = chord
α = angle of attack
α_{twist} = washout angle
α_{0L} = zero-lift angle of attack of the airfoil

The angle θ is a parameterization of the semispan ratio at a distance y from the wing root.

$$\theta = \cos^{-1}\left(\frac{-2 \cdot y}{b_w}\right)$$

Then, v is divided up into N segments between 0 and $\pi/2$ radians, and n assumes odd integer values from 1 to $2N - 1$. Note that in effect the integer $2N-1$ replaces ∞: in the preceding general equation. The chord at the given θ, $c(\theta)$, is found through linear interpolation across the semispan.

The following matrix formula is solved for x:

$$Ax = b$$

where

$$A(i, j) = \sin[n(j) \cdot \theta(i)] \cdot \left\{\frac{\pi \cdot c(i) \cdot n(j)}{2 \cdot b_w} + \sin[\theta(i)]\right\} \tag{5.11}$$

$$b(i) = \frac{\pi \cdot c(i)}{2 \cdot b_w} \cdot [\alpha + \alpha_{twist}(i) - \alpha_{0L}(i)] \cdot \sin[\theta(i)] \tag{5.12}$$

The parameter $n(j)$ is equal to

$$n(j) = 2 \cdot j - 1 \tag{5.13}$$

The indices i and j go from 1 to N. Notice that the parameter n is always odd numbered (1, 3, 5, ...).

The coefficients $x(i)$ are determined by the matrix operation:

$$x = A^{-1}b \tag{5.14}$$

The total lift coefficient for the wing is

$$C_L = AR \cdot \pi \cdot x(1) \tag{5.15}$$

and the induced drag is

$$C_{Di} = \frac{C_L^2}{\pi \cdot AR \cdot e} \tag{5.16}$$

where

$$e = 1 \bigg/ 1 + \sum_{j=2}^{N} n(j) \cdot [x(j)/x(1)]^2 \qquad (5.17)$$

The section lift coefficient at station i is

$$C_l(i) = \frac{4 \cdot b_w}{c(i)} \cdot \sum_{j=1}^{N} x(j) \cdot \sin[\theta(i)] \qquad (5.18)$$

The LLT can only analyze a single wing at a time, and so downstream downwash effects on tails cannot be evaluated directly. The tail can be modeled as a new wing without an upstream wing, which creates inaccuracies. The downstream surface is always more affected by the upstream surface than vice versa because the upstream wing has trailing vortices and a trailing downwash. A wing downwash model that creates a new tail effective local angle-of-attack distribution would improve the tail lift predictions. The inherent limitations of the LLT are that the tail effectiveness is overpredicted, and the tail contribution to induced drag is underpredicted.

Example 5.3   Lifting-Line Theory

Problem:
Consider a wing with no sweep about the quarter-chord line (see Fig. 5.6). The span is 10 m, the aspect ratio is 10, and the taper ratio is 0.6. The zero-lift angle of attack is -1 deg for all airfoils along the span. The wing has a washout angle of 2 deg at the tip (positive washout is a nose-down twist, or negative angle of incidence relative to the root), and the washout distribution is linear from the root to the tip. Use five evenly distributed panels along the semispan. What is the lift coefficient and induced drag coefficient when the wing is at 5-deg angle of attack relative to the root chord?

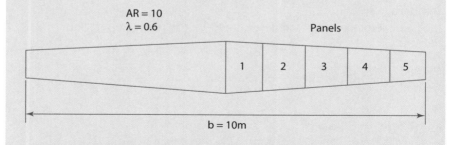

Fig. 5.6 Wing geometry and paneling.

Solution:

The semispan is 5 m, which is half of the overall span. Recall from Chapter 4 the relationship between root chord, span, aspect ratio, and taper ratio:

$$AR = \frac{b}{c_r} \cdot \frac{2}{1+\lambda}$$

This can be rearranged to solve for the root chord.

$$c_r = \frac{b}{AR} \cdot \frac{2}{1+\lambda} = \frac{10\,\text{m}}{10} \cdot \frac{2}{1+0.6} = 1.25\,\text{m}$$

The tip chord is

$$c_t = c_r \cdot \lambda = 1.25\,\text{m} \cdot 0.6 = 0.75\,\text{m}$$

First, let us find several of the key parameters for the panels:

Panel Index, i	y, m	c, m	α_{twist}, deg	θ, deg
1	0.5	1.2	−0.2	84.26
2	1.5	1.1	−0.6	72.54
3	2.5	1	−1	60
4	3.5	0.9	−1.4	45.57
5	4.5	0.8	−1.8	25.84

The elements of the A and b matrix are

$$A = \begin{bmatrix} 1.178 & 1.075 & 0.886 & 0.611 & 0.245 \\ -1.491 & -0.899 & 0.000 & 0.780 & 0.794 \\ 1.700 & 0.086 & -1.430 & -1.054 & 0.825 \\ -1.768 & 1.153 & 1.702 & -1.118 & -0.021 \\ 1.670 & -2.312 & 0.000 & 1.525 & -1.244 \end{bmatrix} \quad \text{and}$$

$$b = \begin{bmatrix} 0.01899 \\ 0.01553 \\ 0.01187 \\ 0.00811 \\ 0.00402 \end{bmatrix}$$

Solving for the x matrix gives

$$x = \begin{bmatrix} 0.01463 \\ -0.00007 \\ 0.00090 \\ 0.00012 \\ 0.00002 \end{bmatrix}$$

The two-dimensional lift coefficient and planform area for each panel are as follows:

Panel Index, i	$C_l(i)$	Area(i), m^2
1	0.5232	1.2
2	0.5472	1.1
3	0.5464	1.0
4	0.5006	0.9
5	0.3438	0.8

From a simple numerical integration, the wing three-dimensional lift coefficient is $C_L = 0.50$. This compares well with the formula

$$C_L = AR \cdot \pi \cdot x(1) = 10 \cdot \pi \cdot 0.01463 = 0.4595$$

The span efficiency factor e is found by

$$e = 1 \bigg/ 1 + \sum_{j=2}^{N} n(j) \cdot [x(j)/x(1)]^2 = 0.979$$

The induced drag is

$$C_{Di} = \frac{C_L^2}{\pi \cdot AR \cdot e} = \frac{(C_L)^2}{\pi \cdot AR \cdot e} = 0.00686$$

5.6.2 Vortex Lattice Methods

The vortex lattice method (VLM) offers the ability to evaluate more complex configurations than the LLT. Wing sweep and dihedral are modeled. Configurations with multiple surfaces can be analyzed simultaneously, where the interactions are fully captured. For example, the downwash of the wing on the tail is modeled. Also, fuselage lift and side-force contributions can be analyzed. VLM is useful for more than just lift distributions; it can also estimate aerodynamic derivatives required for stability and control analysis.

The vortex lattice methods of Bertin and Smith [6] are suitable for conceptual design software implementation. The three-dimensional formulation determines the streamwise, spanwise, and vertical velocities (u_m, v_m, and w_m) induced by the circulation by ensuring that there is no flow passing through each control point. The general paneling is shown in Figs 5.7 and 5.8 for two different VLM codes.

The VLM method takes the matrix form

$$\mathbf{Ax} = \mathbf{b} \tag{5.19}$$

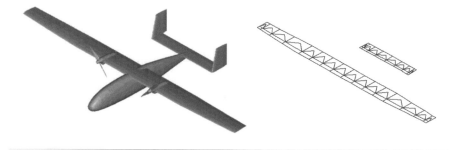

Fig. 5.7 Example UA VLM paneling. Note that the tip of the triangles at the $\frac{3}{4}$-chord location is the control point.

Here A is a set of geometric influence coefficients that describes how the panel vortices affect one another, x is the matrix of circulation strengths for the panels ($\Gamma_1, \ldots, \Gamma_N$), and b is the forcing function that includes angle of attack. The elements matrices are presented in Bertin and Smith [6]. The matrix equations are solved such that the boundary conditions are satisfied.

VLM is likely to yield unrealistic results if the trailing vortex from an upstream panel approaches a downstream panel control point, and so it

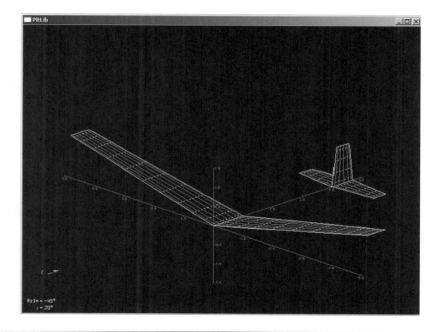

Fig. 5.8 AVL geometry plot showing panels for a wing-tail combination.

is important to ensure sufficient vertical separation between tandem aerodynamic surfaces. The user must control this through inspection or separation checking algorithms for arbitrary configurations. If the trailing vortex is too close to a control point, then the paneling must be modified.

If a single panel is used for the entire chord, then the local lift coefficient is found by

$$C_l(i) = \frac{2 \cdot x[i]}{V \cdot c(i)} \qquad (5.20)$$

The total unmanned aircraft lift coefficient is found by

$$C_L = \sum_{i=1}^{N} \frac{2 \cdot C_l(i) \cdot c(i) \cdot \Delta y(i) \cdot \cos\left[\phi(i)\right]}{S_w} \qquad (5.21)$$

where the factor 2 accounts for Δy being summed over only one semispan by assuming symmetry in the lift distribution. Usually multiple panels are distributed along the chord, as shown in the AVL code image in Fig. 5.8.

The induced drag is found by evaluating the downwash in the Trefftz plane and then relating it back to the downwash at the wing surface. This is similar to the approach taken in the Trefftz plane analysis, which is described later. The Trefftz plane can be thought of as the conditions associated with the ideal flow at a location infinitely downstream of the lifting surfaces. The downwash on the bound vortex is half that of the associated downwash infinitely downstream. The induced downwash influence coefficient matrix A_i can be determined for a position 100 main span lengths downstream of the origin as a practical approximation for infinity. The equivalent nondimensional velocity in the Trefftz plane is determined by the matrix b.

$$A_i x = b \qquad (5.22)$$

Note that the A_i matrix inverse does not need to be evaluated in this operation because x for this system are the values known from the VLM lift distribution analysis. This system of equations is solved directly for b, from which the downwash may angle $\varepsilon(i)$ for each panel can be determined. The local induced drag $C_{di}(i)$ is found by

$$C_{di}(i) = C_l(i) \cdot \varepsilon(i) \qquad (5.23)$$

The total induced drag is found by

$$C_{Di} = \sum_{i=1}^{N} \frac{2 \cdot C_{di}(i) \cdot c(i) \cdot \Delta y(i)}{S_w} \qquad (5.24)$$

5.6.3 Other Three-Dimensional Aerodynamic Methods

5.6.3.1 Trefftz Plane Analysis

The Trefftz plane analysis evaluates the lift distribution in the Trefftz plane, which is a plane perpendicular to the direction of flight located infinite distance behind the unmanned aircraft. The general theory and algorithms are described by Grasmeyer [9] and NASA SP 405 [10]. This analysis method offers the advantage of automatically determining the optimal trimmed lift distribution for minimum induced drag at a given C_L. LLT and VLM must find the appropriate elevator deflections to ensure trim. The Trefftz plane analysis does not require knowledge of the elevator deflections. There are, however, several disadvantages. First, the method does not determine how this optimal lift distribution is attained or if it is even possible to achieve. Second, the twist and camber distribution that can achieve the optimal characteristics for one lift condition might not provide the optimal lift distribution at a second condition. In other words, the method is optimistic by assuming that a single unmanned aircraft can generate optimal lift distributions at more than one lifting condition without variable geometry. Third, no angle-of-attack information is provided to or generated by this method directly.

Despite its limitations, the Trefftz plane analysis is suitable for quickly analyzing and assessing the optimized results of a lifting system with little definition. This makes the Trefftz plane approach ideal for conceptual design. Methods like LLT and VLM become more beneficial when further unmanned aircraft definition becomes available. The algorithms provided by Grasmeyer [9] are well documented and relatively easy to implement in a new software code that is part of a conceptual design analysis.

5.6.3.2 Panel Methods

Panel methods represent the entire outer geometry of the unmanned aircraft with rectangular panels. Typically, the fuselage outer mold line is divided into numerous panels around the perimeter and multiple segments along the length. The upper and lower wing and tail surfaces are paneled individually, rather than simply modeling the camber line with the VLM method. The geometry analyzed by panel methods more closely resembles the actual geometry than VLM, providing greater insight into the flow conditions. Notable panel method codes include PAN AIR, VSAERO, QUADPAN, and PMARC. These methods model the elemental doublets and sometimes sources to ensure that no flow penetrates the panel control points. Some of these codes can be linked to three-dimensional boundary-layer analysis to predict the drag and separation. Katz and Plotkin [8] describe the panel method theory and provide algorithms that can be implemented for aircraft analysis.

Despite the increased level of detail that panel methods provide relative to VLM, panel methods are rarely used today. Although VLM geometry deviates from the actual geometry in many practical cases, the number of panels required to analyze an unmanned aircraft are fewer in number and therefore have a lower computational burden. VLM is used as a work-horse for conceptual design and stability and control derivative generation. Today computational fluid dynamics (CFD) has largely supplanted panel methods for detailed analysis.

5.6.3.3 CFD

CFD models the volume of air surrounding the unmanned aircraft. This volume is divided into grid elements, where the greatest grid density occurs near the surface and where the greatest flow condition variation is expected. The flow properties are found at every grid location using powerful matrix solvers. The flow is subject to boundary conditions at the unmanned aircraft surface. CFD can be solved using inviscid Euler equations or viscous Navier–Stokes equations. The Euler method has a much faster run time, but is not capable of predicting drag directly.

CFD can be used to predict many aerodynamic phenomena. These include the following:

- Total lift and drag forces and pitching moments acting on the unmanned aircraft
- Lift distribution
- Forces and moments acting on a component
- Aerodynamic derivatives for stability and control analysis, including effects of control surface deflections
- Aerodynamic-propulsion interactions (i.e., prop wash, jet intakes)
- Flow visualization
- Prediction of separation, though often with difficulty

Today CFD is used for detailed unmanned aircraft analysis after the geometry is established. These methods are brought in for preliminary and detail design phases, but less often for conceptual design. Depending on the computers, CFD code, and complexity of the problem, the CFD solutions can take hours or even days to run. With current computer technology, it is impractical to link CFD to an aircraft MDO code where thousands of aircraft sizing function calls are made.

CFD tools are becoming more user-friendly and practical for everyday use. Newer codes such as STAR CCM+ can be run on inexpensive desktop computers with parallel processors rather than depending on super-computers. Modern CFD tools can import geometry files from CAD programs and perform automatic grid generation. In the past the process of grid generation might require separate tools and great individual expertise.

The absolute drag values produced by CFD must be carefully scrutinized. Most CFD codes do not predict transition from laminar to turbulent flow, and so the transition location must be input directly, or the boundary layer is defined as all laminar or all turbulent. Without a reliable method of determining the transition, the drag predictions are often unreliable. However, the relative drag between two configurations is more useful. Say a configuration change occurs, such as aft fuselage shaping. Here the change in drag between configurations is useful design information. The lift distribution of the wing and tails can be passed to a parasite drag code that numerically integrates two-dimensional airfoil polars along the span. Often the CFD fuselage drag is simply the best predictive method available, and most UAS fuselages are nearly fully turbulent.

5.7 Pitching Moment

For trimmed flight, the sum of the pitching moments must be equal to zero. This is accomplished through a pitch control surface such as an elevator or elevons. The LLT and VLM methods can provide lift distributions and induced drag estimations for a given configuration at a given angle of attack and elevator deflection. However, there is no guarantee that the configuration is trimmed at the given condition, and so the required pitch control surface deflection is found. Recall that the Trefftz plane analysis automatically finds the trimmed state. The trimmed flight condition must be characterized for performance estimation. The primary purpose of the aerodynamics model is to determine the trimmed drag polar for use in the fuel burn calculations in the performance model.

The trimmed condition occurs only when the summation of the pitching moments about the unmanned aircraft center of gravity is equal to zero. The pitching-moment contributors are the lifting moments, airfoil zero-lift pitching moments, flap pitching moments, and pitching moments due to fuselages and nacelles. Note that the empennage lifting moments for a conventional configuration, or elevator-affected lift distribution on a flying wing, counteract all other contributions to attain the trimmed condition. The means of determining the trim requirements are handled by the lift distribution methods.

$$C_M = C_{M,CL} + C_{M,0L} + C_{M,\text{Flap}} + C_{M,\text{Fuse,Nac}} \tag{5.25}$$

The pitching moment due to lift for each panel is determined by

$$C_{M,CL} = \sum_{i=1}^{N} 2 \cdot C_l(i) \cdot \frac{c(i) \cdot \Delta y(i)}{S_w} \cdot \frac{x_{CG} - x_{c/4}(i)}{mac_W} \cdot \cos[\phi(i)] \tag{5.26}$$

There are N panels for a semispan. The zero-lift pitching moment of the airfoil sections is determined by

$$C_{M,0L} = \sum_{i=1}^{N} 2 \cdot C_{M,0L}(i) \cdot \frac{c(i) \cdot \Delta y(i)}{S_w} \cdot \cos[\phi(i)] \qquad (5.27)$$

The profile pitching-moment contribution due to flap deflection is

$$C_{M,\text{flap}} = \sum_{i=1}^{N} 2 \cdot \frac{\partial C_{M,\text{flap}}}{\partial C_l}(i) \cdot \Delta C_{l,\text{flap}}(i) \cdot \frac{c(i) \cdot \Delta y(i)}{S_w} \cdot \cos[\varphi(i)] \qquad (5.28)$$

The flap pitching moments and lift increments come from experimental data or empirical equations. Alternatively, VLM, panel methods, or CFD can provide the flap pitching-moment estimates.

The pitching moment due to fuselages and nacelles is determined by semi-empirical methods found in Nelson [11], DATCOM [3], VLM, panel methods, or CFD. Note that fuselage and nacelle pitching-moment contribution methods are treated in the same manner, so that the typical fuselage and nacelle subscript is covered by the fuselage subscript alone. The pitching moment is sometimes found by

$$C_{M,\text{fuse}} = \sum_{\text{Fuselages}} \left(C_{M0,\text{fuse}} + \frac{\partial C_{M,\text{fuse}}}{\partial \alpha} \cdot \alpha \right) \qquad (5.29)$$

where $\partial C_{M,\text{fuse}}/\partial \alpha$ is the derivative of the fuselage pitching moment with respect to angle of attack. The reference line used to determine fuselage angle of attack is the zero-incidence fuselage reference.

For a given angle of attack, the only means of trimming the UA is through flap/elevator deflections. Though deflections create profile pitching moments, their greatest contribution comes from redistributing the lift on the surfaces multiplied by the moment arm about the center of gravity. The elevator deflection that creates a zero pitching moment about the c.g. is found via a single variable search algorithm such as the secant method. If the required elevator deflection falls outside the permissible elevator deflection limitations, the solution is infeasible. Similarly, if the required surface local C_l for trim is greater than $C_{l,\text{max}}$, this is an infeasible trim case.

The Trefftz plane analysis requires a pitching moment input prior to calculating the trimmed lift distribution. The zero-lift profile pitching moment and fuselage pitching-moment coefficients are input into the method, and the trimmed lift distribution solution ensures that the total pitching moment about the center of gravity is zero.

5.8 Drag

Drag is force that acts in opposite direction of the flight velocity. For steady level flight, the thrust must equal the drag to maintain equilibrium. The thrust generation consumes fuel or other energy and requires weight-bearing propulsion systems, which in turn impact the unmanned aircraft weight, size, and cost. With few exceptions, such as increasing the rate of descent for landing, drag should be minimized.

There are numerous drag contributors that comprise the total drag. We explored techniques for estimating the induced drag in the lifting-line theory, VLM, and Trefftz plane analysis sections. Here we will look into other major drag contributors.

5.8.1 Friction and Parasite Drag

Friction drag is caused by shear forces acting in the boundary layer normal to the surface. The boundary-layer section showed the laminar and turbulent skin friction for a flat plate. A flat plate is rarely encountered on practical unmanned aircrafts. Even thin flat sheets are not quite flat plates due to the finite thickness and blunt leading and trailing edges.

Common bodies such as wings and fuselages are three dimensional. The shaping of these bodies plays an important role in the drag. Pressure drag becomes a significant contributor. The pressure drag is the resultant drag from an integration of the pressure and area normal to the flight path. An inviscid calculation, such as VLM or a panel method, will show no pressure drag. However, separation caused by adverse pressure gradients and other viscous effects will produce pressure drag.

The combination of pressure and friction drag is called profile drag. Airfoil drag polars show the profile drag vs lift coefficient. This is also known as parasite drag.

$$C_{d,prof} = C_{d,frict} + C_{d,press} \tag{5.30}$$

A parabolic drag polar is a simple way to approximate the variation of drag with lift coefficient. We used this approach in Chapter 3 for initial unmanned aircraft sizing. The drag is expressed as

$$C_D = C_{D0} + \frac{C_L^2}{\pi \cdot AR \cdot e} \tag{5.31}$$

where C_{D0} is the zero-lift drag coefficient. This parameter can be approximated by

$$C_{D0} = \frac{1}{S_{ref}} \sum C_{f,seg} \cdot FF_{seg} \cdot Q_{seg} \cdot S_{wet} \tag{5.32}$$

where $C_{f,seg}$ is the flat-plate skin-friction coefficient, FF_{seg} is the shape form factor, Q_{seg} is the interference factor, and S_{wet} is the wetted area of the body or aerodynamic surface of interest. The form factor is a multiplication factor that is applied to the flat-plate skin-friction drag to give profile drag, which accounts for the pressure drag contribution. The Q factor accounts for adverse interference between bodies, such as a wing and fuselage.

Note that wave drag contributions (drag caused by compressibility) are not under consideration for the purposes of this discussion. This formulation requires that all lift-dependent drag sources are included in the second term on the right-hand side of the first equation. The span efficiency factor e is usually considered to be independent of lift, but it can vary with geometry. C_{D0} accounts for the profile drag at a Reynolds number. C_{D0} is also considered to be independent of lift.

The preceding formulation works reasonably well for aircraft with all-turbulent or fixed transition boundary layers, but it is not suitable for many practical applications. A more suitable model is as follows:

$$C_D = C_{Dprof} + \frac{C_L^2}{\pi \cdot AR \cdot e} \tag{5.33}$$

The profile drag is the summation of the profile drag contributors.

$$C_{Dprof} = \frac{1}{S_{ref}} \sum C_{d,prof,seg} \cdot Q_{seg} \cdot S_{seg} \tag{5.34}$$

where S_{seg} is the planform area of the segment. The profile drag is readily obtainable from airfoil wind-tunnel test data libraries and two-dimensional boundary-layer codes such as XFOIL.

Significant attention is given to profile drag, because it is a major drag contributor, especially at low Reynolds numbers. Using a realistic profile drag will yield an aircraft drag polar that generally does not have a parabolic form at all Reynolds numbers. The upper and lower transition locations are a strong function of lift coefficient and Reynolds number. The transition location strongly affects the friction drag. The pressure drag coefficient is also a function of lift coefficient and Reynolds number.

Figure 5.9 shows data for the symmetrical NACA 0009 airfoil at very low (6×10^4), low ($10^4-3 \times 10^5$), and high ($3 \times 10^6-6 \times 10^6$) Reynolds numbers. Note that the parabolic features of the drag polar become much more pronounced as the Reynolds number decreases. At the very low Reynolds number of 6×10^4, the parabolic behavior breaks down. The NACA 0009 is a very predictable airfoil compared to those with camber. The data are based on Abbot and Von Doenhoff [12] and the UIUC low-Reynolds-number wind-tunnel tests [13].

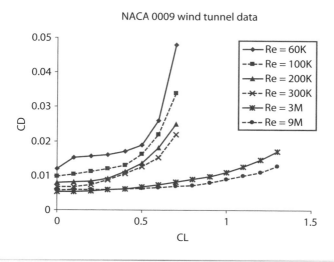

Fig. 5.9   NACA 0009 at several Reynolds numbers.

Figure 5.10 is more representative of most airfoils in the lower-Reynolds-number range. This airfoil, the Selig-Donovan 7032, is relatively predictable at Reynolds numbers above 2×10^5. The shape of the drag polar at lower Reynolds numbers differs greatly between airfoils. The data are based on the UIUC low-Reynolds-number wind-tunnel tests [13].

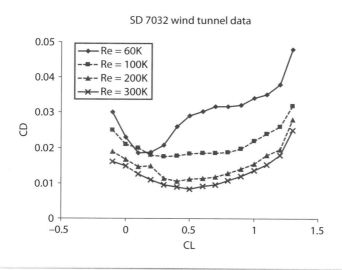

Fig. 5.10   Selig Donovan SD7032 at low Reynolds numbers.

To reduce computational memory demands and to reduce computation time, a simplified airfoil two-dimensional profile drag method can be used. Here the camber effects are taken into account by adding the lift coefficient at zero degrees angle of attack C_{l0}. The airfoil profile drag takes the form

$$C_d = \left(C_{f,\text{upper}} + C_{f,\text{lower}}\right) \cdot FF + K \cdot \left(C_l - C_{l0}\right)^2 \tag{5.35}$$

The factors K, FF, and C_{l0} can be found through regression analysis of airfoil drag polar data. The skin friction is given as the upper and lower flat-plate values, which allows different transition locations to be taken into account.

Similar methods can be employed to predict the drag of slender fuselages and bodies. For a slender fuselage the drag coefficient is

$$C_{D,\text{Fuse}} = C_f \cdot FF_{\text{Fuse}} \cdot \left(\sum Penalty\right) \cdot \frac{S_{\text{wet,Fuse}}}{S_{\text{ref}}} \tag{5.36}$$

where the pressure drag is accounted for by the form factor and the penalty factors. The fuselage form factor FF_{Fuse} is a function of the fineness ratio, which is the ratio of length to equivalent diameter. Not all fuselage sections of interest are circular in cross section, and the thickness distribution varies. The penalties' factors account for fuselage shaping that has negative impacts beyond the conditions for which the form factor was created. The sleek, rounded fuselage on a modern composite sailplane will have substantially lower drag than a square cross-section fuselage with sharp corners, for example.

Spheres are a special type of body. The drag coefficient for a sphere with respect to the unmanned aircraft reference area is found by

$$C_{D,\text{sphere}} = C_{d,\text{sphere}} \cdot \frac{S_{\text{frontal}}}{S_{\text{ref}}} \tag{5.37}$$

where the friction and pressure drag are accounted for in $C_{d,\text{sphere}}$. The sphere drag coefficient is referenced to the sphere frontal area, and so the ratio of the sphere frontal area to the wing reference area is applied to be consistent with other drag contributions. The sphere drag coefficient is 0.41 when below a Reynolds number of 4×10^5 and 0.15 when above that value. By comparison, a flat plate can be assumed to have a drag coefficient of 1.18 referenced to the plate frontal area regardless of shape. EO/IR balls used for imagery frequently have a nearly spherical shape, so that this drag contributor is important.

The aft fuselage shaping can have a large drag impact. A blunt aft fuselage with steep closure angles will have large regions of separation. This is known as boat-tail drag, and the closure angle is the boat-tail angle. Unmanned aircraft with pusher propellers often have blunt aft fuselages so that the large engine can be housed at the rear. The propeller can entrain air in such a way that separation is delayed. However, poor shaping can still lead to significant separation and large drag. Furthermore, this wake impinges upon the propeller causing propeller efficiency loss and noise. A shaft extension that separates the engine from the pusher propeller can allow a more gradual aft fuselage closure angle by placing the engine further forward, which reduces drag caused by separation.

Most UAS fuselages have large turbulent boundary-layer regions. Fuselages tend to have larger Reynolds numbers than wings because of the greater length, and so transition tends to occur at a smaller ratio of the length. Access panels, latches, screw heads, cooling holes, and other features can trip a laminar boundary layer. Also, many fuselages have sharp edges and nonsmooth cross sections, which hinder laminar flow. The early Aerosonde small UAS attempts to promote laminar flow with a nose that has a large region of favorable pressure gradient and no panel lines.

Example 5.4   EO/IR Payload Configuration

Problem:

You are asked to perform a trade study for the EO/IR ball integration. Although there are many factors that will ultimately be considered such as field of regard, the drag will be a driver. The configurations will be considered, as shown in Fig. 5.11. All drag coefficients are notional for the purposes of this exercise.

Figure 5.11a is a smooth fuselage without an EO/IR ball and has a C_D of 0.005 relative to the wing reference area, which is 0.75 m^2. The EO/IR ball has a diameter of 0.2 m.

Figure 5.11b mounts the EO/IR ball in the nose of the fuselage such that its roll stage is aligned with the fuselage centerline. Assume that a wind-tunnel test showed that the drag increased 20% over Fig. 5.11a.

Figure 5.11c is a fully exposed pan-tilt EO/IR ball located under the fuselage. Assume that the EO/IR ball has the same drag coefficient as a sphere at high Reynolds number.

Figure 5.11d is a pan-tilt EO/IR ball partially faired into the fuselage nose. Assume that the fuselage drag except for the ball is equivalent to Fig. 5.11a. Because of the fairing design, assume that the drag of the EO/IR ball is reduced by 50% relative to a fully exposed ball.

Which EO/IR ball solution should be adopted if drag is the primary consideration?

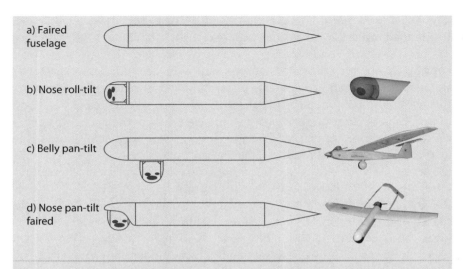

a) Faired fuselage

b) Nose roll-tilt

c) Belly pan-tilt

d) Nose pan-tilt faired

Fig. 5.11 EO/IR ball integration alternatives.

Solution:

First, the frontal area of the sphere is 0.03142 m^2.

For Fig. 5.11b, the fuselage C_D is 20% greater than that of Fig. 5.11a. Therefore, the drag coefficient is 0.006.

For Fig. 5.11c, the drag of the fuselage is equal to the faired fuselage drag plus the drag of the ball:

$$C_{D,\text{Fuse}} = C_{D,\text{Fuse,faired}} + C_{d,\text{Sphere}} \cdot \frac{S_{\text{Sphere}}}{S_{\text{Ref}}} = 0.005 + 0.15 \cdot \frac{0.03142}{0.75} = 0.0113$$

For Fig. 5.11d, the contribution of the sphere is reduced by 50%, which gives a total drag coefficient of 0.008.

The lowest drag EO/IR solution is the nose-mounted roll-tilt configuration, which is Fig. 5.11b.

5.8.2 Induced Drag

Induced drag is drag due to lift. When the lift vector of a wing panel is tilted aft relative to the flight path due to induced downwash, the aft component of the force produces drag. The induced drag coefficient C_{Di} for a wing is given by

$$C_{\text{Di}} = \frac{C_L^2}{\pi \cdot AR \cdot e} \tag{5.38}$$

Here e is the span efficiency factor. Earlier we saw how e can be estimated from the lifting-line theory. This factor can also be found by VLM,

Trefftz plane analysis, or panel methods. See Chapter 4 for a discussion of e for nonplanar wings, including biplanes.

5.8.3 Trim Drag

Trim drag is drag that is generated in order to trim the aircraft. Chapter 4 discusses the ways to provide trim for conventional, canard, tandem, and flying-wing configurations. Recall that for a conventional configuration the required horizontal tail lift coefficient for trim $C_{L,H}$ is

$$C_{L,H} = \frac{C_{m0,w} + C_{m,f} - C_{L,w} \cdot sm}{\eta_H \cdot \overline{TVC}_H} \tag{5.39}$$

The lift coefficient becomes more negative with increased wing lift coefficient and positive static margin. The wing pitching-moment coefficient is generally negative, and so this too contributes to a negative lift coefficient. The induced drag on the horizontal tail is

$$C_{Di,H} = \frac{C_{L,H}^2}{\pi \cdot AR_H \cdot e_H} \tag{5.40}$$

where e_H is the horizontal tail span efficiency. The horizontal tail profile drag can be higher at the trimmed lift coefficient as well. Similar relationships exist for other configurations as well, where the trim requirements increase the overall aircraft drag.

Note that when a VLM, Trefftz plane analysis, or CFD is applied that it is hard to discriminate the trim induced drag from the overall induced drag. All of the surfaces are analyzed simultaneously. The same is true when the profile drag for all wings and tails are found by looking up the profile drag coefficient for the lift distribution. The profile drag is calculated as a single value for the trimmed condition rather than a separable difference from a nontrimmed condition. It is still useful to consider trim drag, as the influence of trim on the overall drag can be significant regardless of drag prediction methods.

5.8.4 Interference Drag

Interference drag is drag produced by the mutual interference of two or more aerodynamic bodies. The interference drag can be a relatively minor contribution for refined designs or quite significant for a poor design. Hoerner [14] provides empirical methods, where the interference drag is broken down into several components that can be summed. A multiplication factor accounts for improvements possible with a good design. The contributing interference drag elements represent poor designs, and the interference drag factor is an improvement from this baseline. Hoerner's suggested value of 0.1 for this correction tells us that the interference drag can easily vary by an order of magnitude.

CFD was unavailable when Hoerner's methods and other empirical approaches were developed. Today CFD is used to minimize these interference effects holistically. The flow around the entire unmanned aircraft is predicted at once. With CFD, the interference drag shows up indirectly as a nonseparable part of the drag acting on the fuselage, wings, and other surfaces.

5.8.5 Wave Drag

Wave drag is drag that is caused by shocks and other compressibility phenomenon in transonic flows. In the transonic regime there are both supersonic and subsonic regions in the flow. The flow over the top or bottom of a transonic wing accelerates to supersonic speeds. The flow must slow down to subsonic speeds before reaching the trailing edge. The mechanism for slowing the flow is a shock wave, which creates a form of pressure drag and can cause separation. Wave drag is important for transonic UA such as jet-powered HALE UAS, high-speed targets, or UCAVs.

Grasmeyer et al. [15] provides a method for estimating the wave drag of wings and tails. Using an extended version of the Korn equation that includes sweep theory, the drag-divergent Mach number M_{DD} for a wing panel i can be approximated by

$$M_{DD}(i) = \frac{\kappa_a}{\left|\cos\left[\Lambda_{c/2}(i)\right]\right|} - \frac{t/c(i)}{\cos^2\left[\Lambda_{c/2}(i)\right]} - \frac{C_l(i)}{10 \cdot \left|\cos^3\left[\Lambda_{c/2}(i)\right]\right|} \quad (5.41)$$

where κ_a is the airfoil technology factor, t/c is the wing thickness to chord ratio, and $\Lambda_{c/2}$ is the wing sweep about 50% of the chord length. The factor κ_a is estimated by Grasmeyer et al. [15] as 0.87 for a NACA 6-Series section and 0.95 for a supercritical section. The critical Mach number M_{Crit} is defined by

$$M_{crit}(i) = M_{DD}(i) - \left(\frac{0.1}{80}\right)^{1/3} \quad (5.42)$$

If the flight Mach number is above the critical Mach number, then the local wave drag coefficient is

$$C_{d,wave}(i) = 20 \cdot (M - M_{crit})^3 \quad (5.43)$$

and the total wave drag coefficient is

$$C_{D,wave} = \sum C_{d,wave}(i) \cdot \frac{S_{segment}}{S_{ref}} \quad (5.44)$$

If the flight Mach number is below the critical Mach number, then the wave drag is zero.

The wing wave drag is assumed to be a more significant driver at a lower Mach number than the fuselage. In other words, the impacts of fuselage wave drag in the lower transonic region are considered to be unimportant in the determination of cruise Mach number.

Wave drag can also be estimated by transonic CFD codes. However, the simple method described here can be analyzed quickly for conceptual design purposes.

5.8.6 Cooling Drag

A key cooling mechanism for aircraft is for air to travel over hot components to facilitate heat transfer through conduction and then convection. The cooling air interaction with the object creates drag through friction and separation.

The cooling drag associated with reciprocating engine propulsion can be modeled as a power loss to the engine or as aerodynamic drag. Here we will model this as a power-dependent drag. Torenbeek [16] recommends a conceptual design method of cooling drag estimation for reciprocating engines

$$\frac{D_{\text{Cool}}}{q} = C_{\text{Cool}} \cdot \frac{P_{\text{Shaft}} \cdot T_{\text{amb}}^2}{\sigma \cdot V} \tag{5.45}$$

where the empirical constant C_{Cool} is 4.9×10^{-7} ft^2/lb-$^\circ$R^2.

For ram air ECS systems, the total drag associated with cooling is found by

$$D_{\text{Ram}} = \dot{m}_{\text{Ram}} \cdot V \cdot \left(1 - \frac{V_e}{V}\right) \tag{5.46}$$

According to Sloan [17], the ram air mass flow rate required to cool electronics can be found by

$$\dot{m}_{\text{Ram}} = 3.59 \cdot \frac{Q_{\text{cool}}}{T_{\text{comp}} - T} \text{ kg/h} \tag{5.47}$$

where the temperatures are in degrees Celsius, and the heat load is in watts. To put the ram cooling equation in convenient units, the following is attained:

$$D_{\text{Ram}} = 0.0012686 \cdot \frac{Q_{\text{cool}}}{T_{\text{comp}} - T_{\text{air}}} \cdot V \cdot \left(1 - \frac{V_e}{V}\right) \text{ lb} \tag{5.48}$$

where the drag is in pounds, the heat load is in watts, the temperatures are in degrees Fahrenheit, and the velocity is in knots. Here V_e is the exit velocity from the cooling system. A velocity recovery V_e/V value of 0.2 can be assumed if no cooling system design information is available.

A fuel-cooled ECS system, like the one used on Global Hawk, might have negligible drag. The heat is rejected through the airframe skins. The only drag effect is related to heating of the boundary layer.

5.8.7 Landing-Gear Drag

The landing-gear drag is broken into two phases of flight: cruise and landing/takeoff. The cruise condition has all of the gear in the full retraction position, if the configuration has retractable gear. The landing/takeoff condition has the gear fully extended.

The gear drag is composed of combinations of tires, struts, doors, and body cavities. The landing-gear drag coefficient can be found by adding the contributors as follows:

$$C_{D,\text{gear}} = \sum \left(C_{D,\text{tire}} + C_{D,\text{strut}} + C_{D,\text{door}} + C_{D,\text{cavity}} \right) \qquad (5.49)$$

The tire drag is estimated for the nose and main gear by

$$C_{D,\text{tire}} = C_{d,\text{tire}} \cdot N_{\text{tire/strut}} \cdot N_{\text{strut}} \cdot \frac{D_{\text{tire}} \cdot W_{\text{tire}}}{S_{\text{ref}}} \qquad (5.50)$$

The size of the nose and main gear can be different. Hoerner [14] shows a variation in tire drag coefficient between 0.12 and 0.35, depending on tire type. The drag coefficient will be reduced with the addition of fairings such as wheel pants.

The strut drag can be treated as a circular cylinder perpendicular to the freestream or as a faired strut. The circular cylinder drag coefficient varies with Reynolds number, assuming natural transition, with the critical Reynolds number centered about 3×10^6. Strut fairings resemble thick airfoils.

If retractable landing gear is used, then gear door drag is calculated for the extended gear case. The gear doors are generally aligned with the flow. A simple approximation is to calculate the equivalent flat-plate turbulent skin friction on both sides of the door to yield the drag coefficient.

For retractable landing gear, the gear bay cavity drag can be calculated for the extended gear case. The drag coefficient of the open gear bay cavity is found in Hoerner [14]. The total cavity drag coefficient can be approximated by

$$C_{D,\text{cavity}} = 0.0083 \cdot N_{\text{strut}} \cdot \frac{W_{\text{door}} \cdot L_{\text{door}}}{S_{\text{ref}}} \qquad (5.51)$$

This drag coefficient is calculated for each cavity.

The total drag for the up and down position depends on the gear configuration. If a gear is retractable, then the retracted drag coefficient is zero, and the extended drag includes the tire, strut, door, and cavity. If the gear is permanently extended, then both drag values include the strut and tire, but not the gear well cavity or the gear door. The retraction scheme can be different for the nose and main gears.

The landing-gear drag estimation is most accurate if CFD and wind-tunnel testing are used. The preceding method is a summation of drag contributors, but the overall landing-gear drag might not behave in this manner. Whole-vehicle analysis such as CFD and wind-tunnel tests capture the interactions.

5.9 Miscellaneous Drag

There are inevitably many drag contributors that are not accounted for in the preceding formulations. A user-input miscellaneous drag factor can be applied to scale the drag polar. This factor is a direct multiplier, and so it scales the minimum drag value and the drag caused by lift by the same ratio. In other words, the factor acts as a simultaneous minimum drag shift and increases the parabolic behavior of the drag polar. Alternatively, an additive miscellaneous drag coefficient can be applied. These drag sources can include antennas, panel lines, protruding mechanisms, or air gaps.

References

[1] Mueller, T. J., *Fixed and Flapping Wing Aerodynamics for Micro Air Vehicle Applications*, AIAA, Reston, VA.

[2] Schetz, J. A., *Boundary Layer Analysis*, Prentice-Hall, London, 1993.

[3] Hoak, D., Ellison, D., et al., "USAF DATCOM," Air Force Flight Dynamics Lab., Wright-Patterson AFB, OH.

[4] Roskam, J., *Airplane Design, Part VI: Preliminary Calculation of Aerodynamic, Thrust and Power Characteristics*, Roskam Aviation and Engineering Corp., Lawrence, KS, 1990.

[5] Aerospace Engineering Dept., Univ. of Illinois, Urbana, IL, http://www.ae.illinois.edu/m-selig/ads/coord_database.html.

[6] Bertin, J. J., and Smith, M. L., *Aerodynamics for Engineers*, 2nd ed., Prentice-Hall, London, 1989, pp. 247–257, 261–282.

[7] Anderson, J. D., *Fundamentals of Aerodynamics*, 2nd ed., McGraw–Hill, New York, 1991, pp. 324–350.

[8] Katz, J., and Plotkin, A., *Low Speed Aerodynamics*, 2nd ed., Cambridge Univ. Press, New York, 2001, pp. 167–183, 206–535.

[9] Grasmeyer, J., "A Discrete Vortex Method for Calculating the Minimum Induced Drag and Optimum Load Distribution for Aircraft Configurations with Noncoplanar Surfaces," Virginia Polytechnic Inst. and State Univ., VPI-AOE-242, Blacksburg, VA, Jan. 1997.

[10] NASA, "Vortex Lattice Utilization," NASA SP-405, Langley Research Center, Hanpton VA, May 17–18, 1976.

[11] Nelson, R. C., *Flight Stability and Automatic Control*, McGraw–Hill, New York, 1997, pp. 49–58.

[12] Abbot, I. H., and Von Doenhoff, A. E., *Theory of Wing Sections, Including a Summary of Airfoil Data*, Dover, New York, 1959, p. 455.

[13] Selig, M. S., Donnovan, J. F., and Fraser, D. B., *Airfoils at Low Speeds*, H. A. Stokely Soartech, Virginia Beach, VA, 1989, pp. 254, 320–324.

[14] Hoerner, S. F., *Fluid-Dynamic Drag*, S. F. Hoerner, Vancouver, WA, 1992, pp. 5-10–5-11, 8-1–8-20, 13-15–13-16.

[15] Grasmeyer, J., Naghshineh, A., Tetrault, P.-A., Grossman, B., Haftka, R. T., Kapania, R. K., Mason, W. H., and Schetz, J. A., "Multidisciplinary Design Optimization of a Strut-Braced Wing Aircraft with Tip-Mounted Engines," MAD Center Rept., MAD 98-01-01, Blacksburg, VA, Jan. 1998.

[16] Torenbeek, E., *Synthesis of Subsonic Airplane Design*, Delft Univ. Press, Delft, The Netherlands, 1984, p. 516.

[17] Sloan, J., *Design and Packaging of Electronics Equipment*, Van Nostrand Reinhold Co., New York, 1985.

Problems

5.1 Calculate the Reynolds number and Mach number for a wing section in the unmanned aircraft, which has $S_w = 8 \ \text{m}^2$, a constant wing chord length of 0.5 m, a mass of 200 kg, and a flight C_L of 0.4. Calculate these values from sea level to 30 km using the 1976 standard atmosphere.

5.2 Plot the flat-plate skin-friction coefficient as a function of transition location for a Reynolds number of 10^4, 5×10^5, 10^5, and 5×10^6. The transition location should vary from 0–100% of the flat-plate length.

5.3 Consider a wing with no sweep about the quarter-chord line. The span is 30 ft, the root chord is 3 ft, and the taper ratio is 0.5. The zero-lift angle of attack is -2 deg for all airfoils along the span. The wing has a washout angle of 3 deg at the tip (positive washout is a nose-down twist, or negative angle of incidence relative to the root), and the washout distribution is linear from the root to the tip. Use five evenly distributed panels along the semispan. What is the lift coefficient and induced drag coefficient when the wing is at 5-deg angle of attack relative to the root chord?

5.4 A wing operates with an average chord Reynolds number of 2×10^5 and an average loiter two-dimensional lift coefficient of 1.2. The airfoil is required to reach a lift coefficient of at least 1.4 without stalling. Find an airfoil that has a high-endurance parameter at the loiter condition. Recommendation: use tools and data sources provided on University of Illinois at Urbana Champagne airfoil website.

5.5 Demonstration project. Follow all applicable laws and safety precautions. Purchase several conventional configuration balsawood gliders. Make three groups of gliders. The first group has the original span. The second group has 75% of the original span. The last group has 50% of the original span. Toss the aircraft from an elevated height such that the aircraft fly at a steady speed and descent angle. (This might require practice.) An observer should time the flights. Time the durations for at least 10 flights for each group. What is the endurance difference between the groups? For relatively straight flights, calculate the flight-path angle and velocity of each group. Estimate the wing chord Reynolds numbers.

Chapter 6 Mass Properties

- Understand the mass properties methods throughout the unmanned-aircraft life cycle
- Use and develop weight estimation methods
- Employ weight management techniques

Fig. 6.1 MQ-1A landing and another preparing for takeoff. (U.S. Air Force photo by Tech. Sgt. Erik Gudmundson.)

6.1 Introduction

The mass properties discipline performs a key function within the aircraft design cycle, combining weight, center of gravity, and inertia determination with rigorous weight management and control. For conceptual and preliminary design, the foundation for industry mass properties' methods is based on historical data, primarily from manned, conventional aircraft. Mass properties analytical methods similar to those used for manned aircraft are usually applied to unmanned aircraft during the conceptual preliminary design phase through detail design.

The term *mass properties* identifies the scope of activity in that it is inclusive of the properties of mass. This includes weight, center of gravity, and moments of inertia. Every component of the UA must be accounted for and tracked by weight and location within the UA to determine center of gravity and mass moments of inertia. Accounting for mass properties in a tracking system serves many valuable purposes, including weight control and management, and can be the basis for cost analysis, bill of materials, and mass distributions. Every design decision impacts mass properties, and the role of mass properties' management must be clearly defined relative to the chief engineer and systems engineering when making configuration decisions.

Effective mass properties management and control is essential to the success of aircraft development and must receive the attention and engineering resources comparable to other major design disciplines such as aerodynamics or propulsion. However, weight control and management is a particularly challenging undertaking for unmanned aircraft. All too frequently, conceptual weight methods for unmanned aircraft do not adequately predict initial weight estimates, resulting in uncontrollable and catastrophic weight growth as the design matures. This chapter attempts to address these gaps for some common UAS design problems. The interested reader should consult SAWE's *Introduction to Weights Engineering* [1] for further information on this important design discipline.

6.2 Mass Properties Throughout the Aircraft Life Cycle

The nature of mass properties' methods changes as the design matures through conceptual, preliminary, and detailed design phases and then into flight test, full production, and flight operations. The level of detail and sophistication of the tools increases as the program progresses.

Early conceptual design mass properties efforts concentrate on estimating the UA takeoff gross weight, fuel weight, and distribution of other weights across major weights groups. The methods presented in

Chapter 3 are applicable to the initial weight estimates. These early methods help identify design space and determine concept feasibility and can provide suitable approximations that are within perhaps 20% of the true weight. However, these simple weight methods do not capture design details and requirements impacts and therefore tend to under-predict the weight.

The later stage of conceptual design employs more comprehensive weights methods reflecting the UA configuration, systems, propulsion installation, and other attributes that are more specific to the selected design approach. For example, structures' weight estimating relationships are a function of the structural geometry; materials; loads; flight regime; manufacturing concept; reliability, maintainability, and supportability (RM&S); special features; and operational capability. A top-level list of fixed weight components such as payloads and avionics suites is identified, and allocations are provided for unknown components. More detailed performance models predict the fuel burn, which is input to the weights model. Increasing the level of detail necessitates a numerical iterative solver to converge the weights. The conceptual design weights methods are almost always contained within an integrated aircraft design code that can calculate the weights for each design point. These weight methods describe the selected UA concept rather than a generic aircraft. This chapter concentrates on conceptual design methods.

In the preliminary design phase, weights estimates are based on more detailed methodologies, and new weights management processes are initiated. The weights are apportioned to lower levels and can be managed by more people. CAD modeling of the structures begins to replace the parametric geometry definition from conceptual design, requiring estimation of the individual structural parts in a bottoms-up fashion. All structural and system bottoms-up, bill of material weights must include appropriate installation and nonoptimum factors to reflect completely integrated components. The design team generates design solutions to the customer requirements, and the system weights impacts are tracked. Preliminary subsystems and avionics architecture designs include initial bill of materials. The layout and location of the aircraft components enable calculations of moments of inertia and center-of-gravity estimation. Preliminary design analysis is nearly always conducted with independent tools, and so the weights engineering function must ensure that the weights are rigorously tracked. A formal tracking system is usually initiated at this stage.

Through detailed design, fabrication, flight test, and manufacturing, mass properties details are tracked and verified through calculations and ultimately measurements. The structures are designed using high-fidelity modeling such as finite element modeling. All components are identified in detail design, followed by parts procurement as the design progresses to fabrication.

6.2.1 Conceptual Design Weights Methods

Conceptual design is broken into two phases: early conceptual design and full conceptual design. Early conceptual design methods involve simple weight estimation methods, performance estimation, and propulsion type selection, as described in Chapter 3. Here a closed-form takeoff gross weight estimation method is possible, which is a function of component fixed weights and mass fractions only. Full conceptual design utilizes more complex weight estimation methods that account for detailed system information, and a closed-form solution is not common.

The author has used early conceptual design methods dozens of times. The response time can vary from the duration of a meeting to a day. The results are remarkably close to the results of more detailed methods, despite the small investment of effort. However, the accuracy is insufficient for a professional conceptual design. These simple methods are suitable for determining the "corner of the universe" of possible designs in which the design of interest resides.

In the later stage of conceptual design, more complex weight relationships are used. This weights analysis seeks to define the takeoff gross weight and distribution of the weights across major weight groups, with more detailed breakdowns where possible. Weights can be fixed weights or calculated through functions and procedures that are dependent upon the takeoff gross weight or other aircraft characteristics. No single approach is possible for unmanned aircraft, given that these systems can vary from micro air vehicles to large HALE designs. However, weight estimation resources exist to help generate feasible conceptual weights estimates. Conceptual design weight estimation methods tend to apply to families of aircraft within a weight class and using similar design approaches.

The role of weights engineering in the conceptual design phase is often to generate weight estimation software routines that are integrated into an aircraft multidisciplinary design optimization code. This is followed by examination of the results and allocating the resulting weight components to the design disciplines or integrated product team leads. Often the weight software is based on legacy methods inherited by a new weights engineer, with cryptic or nonexistent documentation. The identity of the authors of the methods and the underlying assumptions might be unknowable. These weights algorithms can be trustingly applied to a new design problem without scrutinizing the underlying methods and their applicability, frequently leading to erroneous results. In other situations, an emerging

UAS company might need to develop weight estimation methods without the benefit or drawbacks of legacy techniques. The weights engineer will quickly find that suitable UAS conceptual weights methods do not exist.

Weight methodologies fall into the following categories: 1) statistical, 2) quasi-analytical or analytical, 3) design-based, and 4) assigned. The use of methods depends on stage of design and knowledge of the system design approach.

Statistical weights are based on data from previous component designs and are formulated such that major driving parameters can vary. Many weight estimating relationships described in this chapter are statistical methods. Statistical methods imply a technology approach that must be relevant to the problem at hand. These methods are used extensively in conceptual design and to a lesser degree in preliminary design.

Quasi-analytical and analytical methods use the physics of the design to estimate the weight. For example, a wing can be sized based on the expected aerodynamic loads, wing geometry, spar arrangement, and structural materials. The quasi-analytical methods might be of limited detail to enable integration with aircraft design software. Quasi-analytical methods are used mostly in preliminary design, though improvements in computer technology have enabled their use in conceptual design. When based on the physical phenomena of interest, such as loads and material properties, quasi-analytical methods are more suitable for extrapolation than statistical methods.

Design-based methods include actual weights as measured or estimates based on the designed part. A measured weight can be the vendor-provided component weight that is guaranteed by the contract. Designed weight can be a structural part of known geometry, thickness, and material density. For example, the weight of a component W_{Comp} is calculated from the volume V_{Comp} and density ρ_{Matl} by

$$W_{\text{Comp}} = \rho_{\text{Matl}} \cdot V_{\text{Comp}} \cdot g \qquad (6.1)$$

where g is the acceleration of gravity. Design-based methods are used mostly in the detail design phase, though measured weights of selected parts such as engines, avionics, or payloads can be used as early as conceptual design.

Allocated weights are generally those that are defined in requirements. For example, a payload weight allocation might be specified, even if no single payload suite exactly matches the allocation. Required weight allocations are generally defined in the context of a performance point, such as payload weight for a given mission profile. In other cases, the weights engineer may have no experience with a component or technology and must therefore allocate a weight that cannot be estimated using other techniques at the conceptual design level. In this way, the allocation is in effect an educated guess. It is better to provide a weight allocation for a required component rather than neglect the weight in conceptual design. Allocated weights might be a form of weight control, where allocations are used as margins.

6.2.2 Developing New Weight Estimating Relationships

In many cases suitable weight estimation procedures for unmanned aircraft simply do not exist. Parametric weight estimating relationships (WER) commonly described in aircraft design texts are applicable for unmanned aircraft classes such as general aviation, transports, and fighters. Extrapolating WERs is generally inappropriate and can lead to erroneous results because the methods were developed against a data set of aircraft that do not represent the UA of interest. Established methods that are used incorrectly are not correct. Before using a method, one should ask 1) is the WER being used for the intended purpose, 2) what is the applicable range of input parameters, and 3) what is the valid range of output weights? In many cases the established WERs are not suitably documented to rigorously address these questions, and so engineering judgment can be applied. For example, a general-aviation structure WER is probably appropriate for an all-metal UAS expected to weigh 3000 lb. The output weight estimates should be carefully scrutinized to identify noncredible outputs.

New weight techniques must be developed for most new unmanned aircraft design efforts that are not similar to manned aircraft. At the conceptual design level, parametric WERs are useful for rapid design space exploration. However, these methods must be based on statistical curve fits for data within the applicable design space, and such data are difficult to obtain. Unmanned aircraft component weights data or weight reports are rarely available outside of established companies or government labs. Further complicating matters, the sparse available data can represent vastly different technical approaches.

In the absence of rich data sources, the forms of WERs of manned aircraft can be modified based on a few data points. For example, a wing weight WER can take the form

$$W_{\text{Wing}} = F1 \times S_W^{E1} \times \left[\frac{AR}{\cos^2(\Lambda_{c/4})} \right]^{E3} \times q^{E4} \times \lambda^{E5}$$

$$\times \left[\frac{t/c_{\text{Root}}}{\cos(\Lambda_{c/4})} \right]^{E6} \times (N_Z \times W_{\text{DG}})^{E7} + C1 \qquad (6.2)$$

where *F1* is a multiplication factor, *E1* through *E7* are exponent factors, and *C1* is an additive constant. The wing parameters are as follows: *AR* is the wing aspect ratio, N_Z the ultimate load factor (multiples of acceleration of gravity, i.e., 5 g), q the dynamic pressure ($1/2^* \rho^* V^2$), S_w the wing area, t/c_{Root} the wing average thickness-to-chord ratio, W_{DG} the design gross weight, λ the wing taper ratio, and $\Lambda_{c/4}$ the wing sweep at the quarter-chord.

Here it is possible to adjust the exponents and factors to match available data. When there are insufficient data to rigorously create a curve fit, then it might be expedient to reuse the existing exponents and adjust the multiplication factor *F1* to match a known data point. Any additive constants (such as *C1*) should be eliminated if possible because these will likely yield errors if the equation is extrapolated to low weight values. If data used to generate the new curve fit do not support elimination of the additive constant, then a new value should be found. Adjusting *F1* ensures that the relationship holds only for the calibration point. The weight change with variation of other parameters might not be realistic.

Frequently, new statistical WERs must be developed for a design problem. To create these methods, the major driving parameters must be defined and captured in an equation. The WER is a curve fit to the data with variations in parameter ranges. The number of parameters should be kept to a minimum if sparse data are available. Rich data sets might enable the impact of more variables to be appropriately captured.

Let us start with an example of electric motors, for which many aircraft design texts do not have WERs. With an electric motor it can be seen that the major driving parameter is peak power within a given motor class. Also, there are significant differences between four different classes of motors: ferrite brushed motors, rare Earth magnet brushed motors, brushless inrunners, and brushless outrunners. The mass per power in kilograms per kilowatt is plotted for the four motor classes in Fig. 6.2.

Clearly the power alone is not sufficient to adequately characterize these motors. It was determined that another suitable independent variable is the maximum voltage. The weight per power of the motor is given by

$$\left(\frac{W_{Motor}}{P_{Mot,Max}}\right) = F1 \times P_{Mot,Max}^{E1} \times V_{Max}^{E2} \tag{6.3}$$

where *F1* is a constant, *E1* and *E2* are exponents, P_{Max} is the maximum output power of the motor, and V_{Max} is the maximum rated motor voltage. *F1*, *E1*, and *E2* were selected by an optimizer to minimize the average error for several motors within each class. The motor data were found in numerous model aircraft product catalogs and incorporated into

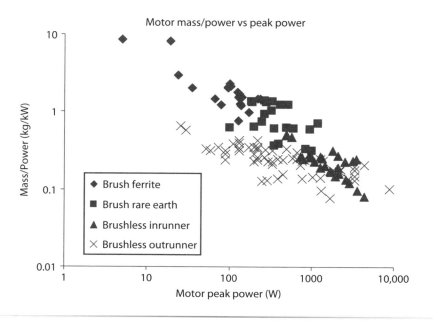

Fig. 6.2   Mass per power for four motor classes.

a database. The optimizer-selected parameters, number of motors analyzed, and average errors are presented in Table 6.1.

High-fidelity design tools can be employed to analyze the weight component of interest where limited data exist. High-fidelity methods such as CFD, finite element method (FEM) structural analysis, or electromagnetic simulations are computationally intensive and therefore ill-suited to integrated conceptual design usage. The outputs from such analyses can be packaged into surrogate methods such as data table look-up procedures or response surfaces. A series of analysis points across the driving parameters must be run offline and put into look-up routines for the conceptual design code. These offline analyses should be calibrated against known design points.

Table 6.1   Motor Weight Parameters

Motor Class	F1	E1	E2	# Motors	Avg. Error, %
Brushed ferrite	7.765	−0.632	0.596	20	21.0
Brushed rare earth	8.160	−0.961	1.166	23	22.7
Brushless inrunner	13.17	−0.610	0.067	28	25.4
Brushless outrunner	0.889	−0.288	0.1588	78	22.3

Example 6.1 Developing a New Weight Estimating Relationship

Problem:

A new weight estimating relationship is needed for three-phase brushless motor controllers. The data in Table 6.2 are available for generating the relationship. The equation takes the form

$$W_{\text{ESC}} = F_{\text{ESC}} \cdot P_{\text{Mot,Max}}^{E1} \text{ oz}$$

Table 6.2 Motor Controller Data

Power, W	Weight, oz
88.8	0.25
199.8	0.75
277.5	1
333	1
444	1.25
888	1.25
1,332	1.33
777	1.33
1,554	1.33
2,849	1.375
3,418.8	3.75
3,330	4
10,360	9.5

What are the best values of F_{ESC} and $E1$? What is the valid range of motor power?

Solution:

An optimizer is required to minimize the difference between the calculated and actual motor weight values. The objective function to be minimized is

$$\text{Objective} = \sum_{i=1}^{NESC} \left[F_{\text{ESC}} \cdot P_{\text{Mot,Max}}^{E1}(i) - W_{\text{ESC,Data}}(i) \right] \Big/ N_{\text{ESC}}$$

The two unknown parameters F_{ESC} and $E1$ are selected by the optimizer. The Microsoft Excel^{TM} Solver tool is applied. F_{ESC} is 0.0124, and $E1$ is 0.679. The average error is 31.3%.

The range of validity is motor input power of 88.8 to 10,360 W.

6.2.2.1 Weight Convergence

Iterative methods become necessary to achieve weights convergence for complex weight estimating relationships. Many components that contribute to the UA calculated takeoff gross weight are either directly or indirectly functions of the input takeoff gross weight.

$$W_{Comp} = f\left(W_{TO,Input}\right) \qquad (6.4)$$

The calculated takeoff gross weight is the summation of all of the component weights.

$$W_{TO,Calc} = \sum_{i=1}^{N_{Comp}} W_{Comp}(i) \qquad (6.5)$$

An initial takeoff gross weight guess is provided to the mass properties' code, and the difference error from the calculated weight is provided.

$$Error = \frac{W_{TO,Input} - W_{TO,Calc}}{W_{TO,Input}} \qquad (6.6)$$

A new weight guess is provided using the error information, and the difference error is provided again. This process is continued until weights convergence is achieved or the maximum number of iterations is reached. The convergence criterion is a minimum value of the takeoff gross weight error's absolute value.

Although there are many suitable convergence techniques, the secant method is a robust and rapid way to achieve weights convergence. First, two initial weight guesses are provided. These first guesses should define the upper and lower bounds of the potential takeoff gross weights. The lower bound $W_{TO,Calc}Lower$ could be the summation of all fixed weights because the UA will never weigh less than that value. The initial selected upper-bound value $W_{TO,Calc}Upper$ can be $1.5-5$ times the early conceptual design estimate from Chapter 3 methods, depending on design uncertainty. The upper-bound value might require adjustment. The upper bound should not exceed the applicable range of the weight estimating relationships. The takeoff gross weight is calculated at each of these two initial takeoff gross weight input values. Index 1 is associated with the lower-bound value, and index 2 is associated with the upper-bound value only for the initialization step that follows.

$$W_{TO,Calc}1 = f\left(W_{TO,Input}1\right) \qquad (6.7)$$

$$W_{TO,Calc}2 = f\left(W_{TO,Input}2\right) \qquad (6.8)$$

The difference functions $\Delta W_{\text{TO}}1$ and $\Delta W_{\text{TO}}2$ are defined as

$$\Delta W_{\text{TO}}1 = W_{\text{TO,Calc}}1 - W_{\text{TO,Input}}1 \qquad (6.9)$$
$$\Delta W_{\text{TO}}2 = W_{\text{TO,Calc}}2 - W_{\text{TO,Input}}2 \qquad (6.10)$$

Now a looping structure is used for N iterations. The new input takeoff gross weight is found by

$$W_{\text{TO,Input}}New = W_{\text{TO,Input}}2 - \Delta W_{\text{TO}}2 \times \frac{W_{\text{TO,Input}}2 - W_{\text{TO,Input}}1}{\Delta W_{\text{TO}}2 - \Delta W_{\text{TO}}1} \qquad (6.11)$$

Care must be taken to handle cases where the calculated new takeoff gross weight $W_{\text{TO,Input}}New$ exceeds the lower bound. In this event, one relatively robust approach is to set $W_{\text{TO,Input}}2$ to

$$W_{\text{TO,Input}}2 = 0.5 \times \left(W_{\text{TO,Input}}Upper + W_{\text{TO,Input}}Lower \right) \qquad (6.12)$$

Then $W_{\text{TO,Input}}2$ is recalculated. Now $W_{\text{TO,Input}}New$ is set to

$$W_{\text{TO,Input}}New = 0.5 \times \left(W_{\text{TO,Input}}2 + W_{\text{TO,Input}}Lower \right) \qquad (6.13)$$

This correction approach might not work for all weights formulations and design problems, and so the code developer should generate other methods of keeping the weights inputs within permissible bounds as necessary. Also, if the bounds must be adjusted too many times, this can be a sign that the weights will not converge under any circumstances.

The convergence is checked by determining if the difference between the new takeoff gross weight and $W_{\text{TO,Input}}2$ is within the tolerance. This difference is calculated as

$$\Delta W_{\text{TO,Input}}New = f\left(W_{\text{TO,Input}}New \right) - W_{\text{TO,Input}}New \qquad (6.14)$$

If the exit criterion is not yet reached, then the input variables are reassigned, and one $W_{\text{TO,Calc}}2$ is calculated again.

$$W_{\text{TO,Input}}1 = W_{\text{TO,Input}}2 \qquad (6.15)$$
$$\Delta W_{\text{TO}}1 = \Delta W_{\text{TO}}2 \qquad (6.16)$$

and

$$W_{\text{TO,Input}}2 = W_{\text{TO,Input}}New \qquad (6.17)$$
$$\Delta W_{\text{TO}}2 = f\left(W_{\text{TO,Input}}New \right) - W_{\text{TO,Input}}New \qquad (6.18)$$

The iteration loop is entered again. The iterations continue until the exit criterion is satisfied.

Optimizers can perform the weights convergence function. The optimizer is given the weights convergence maximum error as a constraint and the takeoff gross weight as a design variable. This can be done when iterative weights convergence methods are considered to be insufficiently smooth for gradient-based optimizers. The disadvantage is that the optimizer must contend with an additional constraint and design variable that can make optimization less efficient.

Example 6.2   Weight Convergence

Problem:

Use the secant method to converge the W_{TO} of an aircraft. Here we will use the linear sizing relationships used in Sec. 3.1, but the weight will be solved iteratively rather than directly. The sizing equation

$$W_{TO} = \frac{W_{PL} + W_{Avion} + W_{Other}}{1 - \left(MF_{Struct} + MF_{Subs} + MF_{Prop} + MF_{Energy}\right)}$$

is rewritten as

$$W_{TO,Calc} = W_{PL} + W_{Avion} + W_{Other}$$
$$+ \left(MF_{Struct} + MF_{Subs} + MF_{Prop} + MF_{Energy}\right) \cdot W_{TO,Input}$$

From Example 3.1, the major parameter values are as follows:

$$M_{Struct} = 0.25$$
$$M_{Subs} = 0.05$$
$$M_{Prop} = 0.05$$
$$M_{Energy} = 0.52$$
$$W_{PL} + W_{Avion} + W_{Other} = 100\,lb$$

Solution:

The parameters are put into the calculated takeoff gross weight equation so that it is only a function of input takeoff gross weight.

$$W_{TO,Calc} = 100\,lb + (0.25 + 0.05 + 0.05 + 0.52) \cdot W_{TO,Input}$$
$$= 100\,lb + 0.87 \cdot W_{TO,Input}$$

Now initial $W_{TO,Input}$ guesses are required to start the secant algorithm. We set these parameters as follows:

Summation of nonvarying weights:

$$W_{TO,Input}1 = 100\,lb$$

A safe upper bound for this problem:

$$W_{TO,Input}2 = 2000\,lb$$

The calculated takeoff gross weight is evaluated at each of the input weights.

$$W_{TO,Calc}1 = 100\,lb + 0.87 \cdot (100\,lb) = 187\,lb$$
$$W_{TO,Calc}2 = 100\,lb + 0.87 \cdot (2000\,lb) = 1840\,lb$$

The difference functions are

$$Diff1 = W_{TO,Calc}1 - W_{TO,Input}1 = 187\,lb - 100\,lb = 87\,lb$$
$$Diff2 = W_{TO,Calc}2 - W_{TO,Input}2 = 1840\,lb - 2000\,lb = -160\,lb$$

The looping algorithm can now be entered with these initial values. The new input W_{TO} is calculated.

$$W_{TO,Input}New = W_{TO,Input}2 - Diff2 \times \frac{W_{TO,Input}2 - W_{TO,Input}1}{Diff2 - Diff1}$$

$$= 2000\,lb + 160\,lb \times \frac{2000\,lb - 100\,lb}{160\,lb - 87\,lb} = 769.23\,lb$$

The difference for this new input weight is calculated as

$$\Delta W_{TO,Input}New = f\left(W_{TO,Input}New\right) - W_{TO,Input}New = 769.23\,lb - 769.23\,lb$$
$$= 0\,lb$$

Here complete convergence is attained at the first iteration. The convergence loop is not entered. This behavior is expected for linear equations. However, nonlinear equations might require several iterations.

6.2.2.2 Weight Algorithm Approaches

The most important products of an UA weights algorithm are a converged takeoff gross weight and weights of groups or components that comprise the takeoff gross weight. The methods used in the weights algorithm might depend upon more weight terms than just the takeoff gross weight, and so these must be addressed. The takeoff gross weight is dependent upon the fuel burn, and the UA fuel burn is dependent upon the takeoff gross weight.

We use the takeoff gross weight throughout this book to represent the maximum weight of the UA. Weights engineering has other ways of defining the maximum weight as well. The maximum design weight might be the product of the maximum load N_z and the maximum maneuver weight. The ramp weight is the maximum weight of the aircraft when fully loaded on the ground before fuel is burned during taxi. There are numerous other ways of defining the maximum weight.

Some weights methods require convergence of more than one weight term. For example, the calculated wing weight might be a function of the input wing weight due to the effects of inertia relief. The weight of the wing itself reduces the bending moment of the wing because the weight loads act in the opposite direction of the lift, which in turn reduces the wing bending material weight.

Although many wing weight methods do not include the wing weight term in the wing weight calculation, let us consider this scenario. The takeoff gross weight is calculated in the outer ˜convergence loop, and the wing weight is calculated in an inner convergence loop. In other words, the wing weight is converged for every takeoff gross weight input. A convergence methods such as the secant method described earlier can be used for each loop.

The fuel weight can be handled in several ways. The fuel weight or fuel mass fraction can be inputs into the weights algorithm. The values can be selected by the optimizer. The weights algorithm converges the takeoff gross weight based on these inputs, and the performance code later performs mission profile analysis based on these values. The performance code determines if there is sufficient fuel to complete the mission, and if there is, then the fuel remains at mission completion. If a fuel mass fraction is input, the fuel weight W_{Fuel} is calculated as

$$W_{\text{Fuel}} = MF_{\text{Fuel}} \cdot W_{\text{TO}} \qquad (6.19)$$

Another approach is for the performance code to be called for every takeoff gross weight iteration. The performance code outputs the required fuel at the current takeoff gross weight estimate. An advantage to this approach is that the exact fuel weight required is calculated without the need for optimizer support. A disadvantage is that only simple performance methods can be used and the computational load can be high.

Many weights methods use the empty weight rather than the takeoff gross weight. Most weight algorithm architectures allow a simple calculation of empty weight rather than a convergence loop. The empty weight W_E can be calculated as

$$W_E = W_{\text{TO}} - W_{\text{Fuel}} - W_{\text{PL}} \qquad (6.20)$$

6.2.3 Preliminary and Detail Design Mass Properties Methods

As the design progresses from the conceptual to the preliminary design phase, a rigorous weights management process becomes necessary. Modern conceptual design practices often involve an integrated multidisciplinary design environment, as described in Chapter 18. Here multiple aircraft design disciplines such as aerodynamics, performance, propulsion, geometry definition, and mass properties are linked together in a common framework. Preliminary design generally progresses to more detailed analysis methods, such as CFD and structural FEM, which are not fully integrated with one another. The author anticipates that these detailed methods will become fully integrated in the future with advances in software and computer technology. Also at preliminary design, more detailed information about the UA is known. An off-line mass properties tracking system is required starting at preliminary design and should be enhanced through the life of the program.

The mass properties tracking system is a data management tool that includes the masses and locations of all components on the UA. Usually this system takes the form of a database or a spreadsheet. Ideally the tracking system is hierarchical so that lower-level components can be tracked within groups. The locations of the center of mass of all components must be known in order to calculate the aircraft center of gravity and moments of inertia.

The groups in the hierarchy can be organized by discipline or assembly. Examples of disciplines include avionics, propulsion, subsystems, and structures. Examples of assemblies are right wing, removable payload bay, or a fuselage section aft of a manufacturing break. Ideally, the tracking system is capable of rapid cross referencing between assemblies and disciplines. The tracking system should also be capable of generating reports in standardized formats and showing changes between reports. MIL STD-1374 weight reporting standards should be used for military aircraft unless specified otherwise by a customer.

6.2.3.1 Center-of-Gravity Estimation

Center of gravity (c.g.) must be maintained within permissible bounds. The importance of c.g. is often overshadowed by weight growth concerns, but weight and c.g. are related. The c.g. will vary during flight if fuel is burned or stores dropped. The permissible center-of-gravity travel is derived from stability and control analysis and can be a function of dynamic pressure or Mach-number-related aerodynamic center shifting. The fore-aft c.g. travel boundaries can be driven by elevator function control power and static margin. The y-axis c.g. travel boundaries are driven by lateral-directional control characteristics. Unless the center of gravity of the fuel and droppable elements are located directly at the UA center of gravity, any change in their state will yield a shift in the UA center of gravity.

The c.g. is calculated using knowledge of the weights and placement of every component on the aircraft. The center of gravity of each component is located at the coordinates $[x_{CG}(i), y_{CG}(i), z_{CG}(i)]$. These coordinates are with respect to the reference datum, which is generally placed ahead of the nose and below the UA by convention. The center-of-gravity coordinates for the UA are located at coordinates (X_{CG}, Y_{CG}, Z_{CG}). The c.g. is found using the equations

$$X_{CG} = \frac{\sum W(i) \cdot x(i)}{W_{Tot}} \qquad (6.21)$$

$$Y_{CG} = \frac{\sum W(i) \cdot y(i)}{W_{Tot}} \qquad (6.22)$$

$$Z_{CG} = \frac{\sum W(i) \cdot z(i)}{W_{Tot}} \qquad (6.23)$$

where $W(i)$ is the weight of a component and W_{Tot} is the total weight of the UA at the state for which it is measured. The weights of fuel and payloads are simply weight components like any other for the purposes of c.g. estimation. The total weight can be found by

$$W_{Tot} = \sum W(i) \qquad (6.24)$$

A c.g. excursion diagram, also known as a "potato plot," is a visual tool that shows how the center of gravity varies during flight due to fuel burn or store drop events. The general form is shown in Fig. 6.3. The X axis is the center-of-gravity location as a percentage of mean aerodynamic chord, and the Y axis is either time or UA weight. The starting point is the takeoff condition, and the end point is the landing condition. The path between these lines shows the change in center-of-gravity location during flight. A discrete mass drop event results in a discontinuous line, whereas fuel burn is a continuous line with varying slope.

The location, capacity, and sequence of the fuel tanks strongly affect the shape of the c.g. travel curve. Because the UA can have a large number of

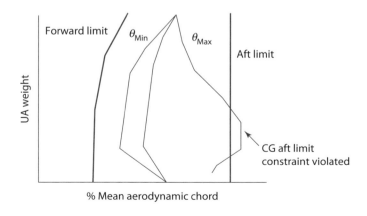

Fig. 6.3 Center-of-gravity excursion diagram.

potential mission profiles involving climb, cruise, and descent segment combinations, it might be impractical to evaluate every relevant case. One approach is to plot the c.g. travel for multiple body pitch elevation angles relative to the horizontal plane θ. Fuel and other fluids can shift with pitch angle. The behavior of the fuel system must be characterized to generate these curves. If any of the curves exceed the c.g. limits, then a redesign might be necessary. Beyond potato plots, the several stressing design cases should be identified and evaluated.

The weights engineer may support the fuel system design to assign fuel sequencing to remain within the limits. Fuel sequencing analysis should consider combinations of stores that can produce out-of-limits condition. On some aircraft, fuel might be able to be loaded in a bladder tank in the payload bay. Weight and balance considerations for this capability must be accounted for in the conceptual design phase.

Payloads such as weapons can be deployed in flight, which affects c.g. location. Any moment arm of the payload will cause the center of gravity to shift. Opening of doors can also impact the aerodynamic center, which could in turn affect the permissible c.g. bounds.

6.2.3.2 Mass Moments of Inertia Estimation

Mass moments of inertia are important for UA flight controls. Higher moments of inertia correspond with longer response times or increased control power to generate an unmanned aircraft response. Primary and cross-product moments of inertia terms are required for control law development. Each component has local moments of inertia about its own center of mass, $I_x(i)$, $I_y(i)$, and $I_z(i)$. The moments of inertia about the center of gravity and the major axes must be calculated as well: I_{x0}, I_{y0}, and I_{z0}. Also, the UA product of inertia about the y axis I_{xz0} must be calculated. The total moments of inertia are calculated as follows:

$$I_{x0} = \sum W(i) \cdot y(i)^2 + \sum W(i) \cdot z(i)^2 + \sum I_x(i) - W_{\mathrm{Tot}} \cdot \left(Y_{\mathrm{CG}}^2 + Z_{\mathrm{CG}}^2 \right)$$

$$(6.25)$$

$$I_{y0} = \sum W(i) \cdot x(i)^2 + \sum W(i) \cdot z(i)^2 + \sum I_y(i) - W_{\mathrm{Tot}} \cdot \left(X_{\mathrm{CG}}^2 + Z_{\mathrm{CG}}^2 \right)$$

$$(6.26)$$

$$I_{z0} = \sum W(i) \cdot x(i)^2 + \sum W(i) \cdot y(i)^2 + \sum I_z(i) - W_{\mathrm{Tot}} \cdot \left(X_{\mathrm{CG}}^2 + Y_{\mathrm{CG}}^2 \right)$$

$$(6.27)$$

$$I_{xz0} = \sum W(i) \cdot x(i) \cdot z(i) - W_{\mathrm{Tot}} \cdot X_{\mathrm{CG}} \cdot Z_{\mathrm{CG}}$$
$$(6.28)$$

6.3 Conceptual Design Weight Estimation Methods

UAS conceptual design weight estimation methodologies are presented in this section. These methods are appropriate for academic conceptual design studies and should not be relied upon for estimating the weight of flight hardware.

The takeoff gross weight of the aircraft is given by

$$W_{\mathrm{TO}} = W_E + W_{\mathrm{PL}} + W_{\mathrm{Fuel}} + W_{\mathrm{tfo}} + \underbrace{W_{\mathrm{Crew}}}_{=0} \qquad (6.29)$$

The zero fuel weight W_{0F} is given by

$$W_{0F} = W_E + W_{\mathrm{PL}} + W_{\mathrm{tfo}} + \underbrace{W_{\mathrm{Crew}}}_{=0} \qquad (6.30)$$

The useful load W_{Useful} is defined by

$$W_{\mathrm{Useful}} = W_{\mathrm{TO}} - W_E - W_{\mathrm{tfo}} = W_{\mathrm{PL}} + W_{\mathrm{Fuel}} + \underbrace{W_{\mathrm{Crew}}}_{=0} \qquad (6.31)$$

The empty weight of the aircraft is given by

$$W_E = W_{\mathrm{struct}} + W_{\mathrm{pwr}} + W_{\mathrm{feq}} \qquad (6.32)$$

where W_{struct} is the structures weight, W_{pwr} is the propulsion system weight, and W_{feq} is the fixed equipment weight.

6.3.1 Structural Weight Estimation

The weight of the structures group is

$$W_{\mathrm{struct}} = W_{\mathrm{Wing}} + W_{\mathrm{Fuse}} + W_{\mathrm{Nac}} + W_{\mathrm{Emp}} + W_{\mathrm{LG}} \qquad (6.33)$$

where W_{Wing} is the wing weight, W_{Fuse} is the fuselage weight, W_{Nac} is the nacelle weight, W_{Emp} is the empennage weight, and W_{LG} is the landing-gear weight.

Many structural WERs available in aircraft design texts are not applicable to most modern UAS applications. These methods are based on weights data from the 1970s and earlier manned aircraft, so that only metallic aircraft are described with occasional correction factors for composite construction. The designer often finds these resources ineffective for composite unmanned aircraft that fall outside the weight range or operating environment of reference vehicles. As discussed earlier, there is a paucity of UAS weights data available to support new structural WERs.

To remedy this difficulty, the designer can employ quasi-analytical structural sizing analyses. Wings are readily evaluated through such analyses, but fuselages are often more challenging because of the complexity of the loads, large number of hatches or other cutouts, and functionality impacts. Quasi-analytical methods are more computationally expensive than empirical methods.

Structural weights of subsonic UASs can be dominated by the minimum gauge structure. Recall that the minimum gauge is the minimum thickness of the skin material. The minimum weight of the structure if it were to experience no loads would simply be the surface area multiplied by the weight per area of the minimum gauge structure.

$$W_{Skins} = S_{Exposed} \times F_{Skins} \qquad (6.34)$$

where W_{Skins} is the skin weight, $S_{Exposed}$ is the exposed skin area, and F_{Skins} is the weight per exposed area of the minimum gauge structure.

The load-bearing structure can be conceptually segregated from the minimum gauge structural shell. For example, the fuselage can be modeled as a tube running the length of the fuselage with a diameter equal to the average structural height of the fuselage. The tube is sized to withstand the structural loads. The weight of the structural tube is added to the minimum gauge outer shell.

Modern UAS structures are usually made from composites, though some metallic structures are used as well. Wood was a common material in the early years of unmanned systems, but it is a rarity today. Small UASs weighing less than 10 lb can be constructed of resilient foam materials with some reinforcement.

The weight of a foam structure with covering can be calculated by

$$W_{Struct} = F_{Fill} \times V_{Structure} \times \rho_{Foam} + S_{Exposed} \times F_{Covering} \qquad (6.35)$$

where W_{Struct} is the weight of the foam structure of interest, F_{Fill} is the ratio of the volume filled by foam to the total volume, ρ_{Foam} is the density of the foam, and $F_{Covering}$ is the weight per area of the covering. Additional weight estimates are required for component mounting, hatches, and impact management structure. This method is often relevant for small UASs and MAVs.

6.3.1.1 Wing Weight Estimation
Parametric Wing Weight WER
Very rapid evaluation of new aircraft can utilize a wing weight mass fraction. The wing weight mass fraction of similar aircraft can be applied. Very high-altitude transonic aircraft or solar-powered aircraft tend to have very light wing loadings. A minimum wing aerial weight floor must be established based on the assumed structure type. Wing aerial weight is the structural weight for a given planform area.

Gerard [2] developed the following wing weight equation for manned sailplanes:

$$W_{\text{wing}} = 0.0038 \cdot (N_Z \cdot W_{\text{TO}})^{1.06} \cdot AR^{0.38} \cdot S_w^{0.25}$$
$$\cdot (1 + \lambda)^{0.21} \cdot (t/c)_{\text{root}}^{-0.14} \, \text{kg} \qquad (6.36)$$

where N_Z is the ultimate load factor in g, W_{TO} is the maximum takeoff gross weight in kilograms, AR is the wing aspect ratio, S_w is the wing planform area in square meters, λ is the taper ratio, and $(t/c)_{\text{root}}$ is the wing root thickness-to-chord ratio. This equation is based on 48 different composite sailplanes ranging from $250-889$ kg W_{TO}, wing spans of $13.4-26.4$ m, and aspect ratios of $17.1-38.7$. The standard deviation between the equation and the reference data is $\pm 8.4\%$. This method is suitable for any UAS bounded by these parameters, such as those in the MALE class. However, this equation only applies to wings with negligible sweep. The control surfaces such as flaps and ailerons are included in this wing weight.

Quasi-Analytic Wing Weight Methods

Until wing weights for multiple unmanned aircraft can be assembled into a single database suitable for a statistical WER development, quasi-analytical methods are most appropriate for UASs that do not fit within the manned aircraft domain. Such methods provide the greatest flexibility for new design problems. Detailed geometry, loads, and material properties can be handled directly. Macci [3] provides a semi-analytical wing weight estimation methodology that lends itself well to a software algorithm. The piecewise linear beam wing weight method is presented here based on methods of Naghshineh-Pour [4]. An overview of quasi-analytical method of wing weight estimation is presented in Chapter 7.

6.3.1.2 Empennage

The empennage weight includes the tail surfaces. The tail configurations vary greatly as shown in Chapter 4. Most existing empennage WERs address conventional horizontal and vertical tail configurations, though there is no true conventional UAS tail configuration. Also, many WERs are intended for manned aircraft and can generate infeasible answers for small UASs.

When separate horizontal and vertical tail WERs are available within the appropriate UA class but the tail is of a V-tail configuration, the following relationship can be applied:

$$W_{\text{Emp}} = W_{\text{HT}} \cdot \cos^2(\Gamma) + W_{\text{VT}} \cdot \sin^2(\Gamma) \qquad (6.37)$$

where W_{HT} is the horizontal tail weight, W_{VT} is the vertical tail weight, and Γ is the V-tail dihedral angle. The horizontal and vertical tails weights are determined from WERs. Depending on the WER input parameters, the horizontal and vertical tail equivalent geometries must be modified to offer an average tail solution.

Quite often UAS tail skins are at minimum gauge thickness. The tails also have shear members such as vertical webs or tubes. The spanwise tubes can also help take bending loads. Tails can have control surfaces, with hinge mechanisms and structural provisions for actuators. The tails also must attach to other structural elements such as fuselages or booms. A suitable WER for minimum gauge tail structures is

$$W_{\text{Emp}} = \frac{1}{6} \cdot F_{\text{Emp}} \cdot F_{\text{Cont}} \cdot S_{\text{Emp}} \cdot t_{\text{Min}} \cdot \rho_{\text{Matl}} \text{ lb} \qquad (6.38)$$

where F_{Emp} is the empennage multiplication factor, F_{Cont} is the factor for control surfaces, S_{Emp} is the empennage planform area in square feet, t_{Min} is the minimum gauge skin thickness in inches, and ρ_{Matl} is the material weight density in pounds per cubic foot. A F_{Emp} value of 1.3 and an F_{Cont} value of 1.2 can be used as initial values, but one should consider the tail configuration and construction method to select the best values. Chapter 7 provides information on structural material properties.

Another similar approach for tails is to use a constant aerial weight. This aerial weight is based on similar designs to the UA of interest. The WER is

$$W_{\text{EMP}} = WA_{\text{Emp}} \cdot S_{\text{Emp}} \text{ lb} \qquad (6.39)$$

where WA_{Emp} is the aerial weight of the tails in pounds per square feet. This aerial weight is referenced to the planform area. Palumbo [5] recommends a WA_{Emp} value of 0.5 lb/ft^2 for tactical UASs with composite tail structures, as is the case for Shadow 200. WA_{Emp} is approximately 0.8 – 1.2 lb/ft^2 for metal general-aviation aircraft and 3.5 – 8 lb/ft^2 for supersonic fighters. Such rules of thumb should be used with caution.

6.3.1.3 Fuselage Weight Estimation

Unmanned aircraft fuselages take diverse forms. Many designs have a traditional single centerline fuselage that is distinct from the wing. Others have a large central pod fuselage with separate booms that connect the tail to the rest of the aircraft. Many flying-wing designs have blended fuselages and wings, as is commonly seen in the MAV and UCAV categories. There are not sufficient data available to generate a single method that covers all potential cases.

Parametric Fuselage Weight WER

A general fuselage structural weight equation for a structure for a semimonocoque or composite shell fuselage for subsonic or transonic UAS

Table 6.3 Parameter Definitions and Values

Term	Definition	Value
F_{MG}	Main gear on the fuselage factor	1 if no main gear is on fuselage 1.07 if main gear is on fuselage
F_{NG}	Nose gear on the fuselage factor	1 if no nose gear is on fuselage 1.04 if nose gear is on fuselage
F_{Press}	Pressurized fuselage factor	1 if unpressurized 1.08 if pressurized
F_{VT}	Vertical tail on the fuselage factor	1 if vertical tail weight not included 1.1 if vertical tail weight included
F_{Matl}	Materials factor	1 if carbon fiber 2 if fiberglass 1 if metal 2.187 if wood 2 if unknown

weighing between 1 to 800,000 lb is

$$W_{Fuse} = 0.5257 \times F_{MG} \times F_{NG} \times F_{Press} \times F_{VT} \times F_{Matl} \times L_{Struct}^{0.3796}$$
$$\times (W_{Carried} \times N_Z)^{0.4863} \times V_{EqMax}^2 \text{ lb} \qquad (6.40)$$

where L_{Struct} is the structural length of the fuselage in feet, $W_{Carried}$ is the weight of the components carried within the structure in pounds, V_{EqMax} is the maximum equivalent velocity in knots, and N_Z is the load factor in g. The F parameter definitions and values are detailed in Table 6.3. This is based on a curve fit to 197 fuselages, ranging from hand-launched gliders to the largest cargo aircraft. The reference fuselages have a fineness ratio (length-to-diameter ratio) of at least 4:1. Most of the data are from manned sailplanes. The average error between predicted weight and data is 29.6%. Figure 6.4 gives insight into the correlation and data point groupings. This method can be used when better alternatives are unavailable. One caution for this equation is that metal and carbon-fiber fuselages have similar weight characteristics, which is not consistent with common expectations. This might be partially explained by the small relative sample size of carbon-fiber fuselages used in the development of the equation. This equation includes the nacelle weight for any engines contained within the fuselage.

Many UAS have twin-boom or single-boom designs that connect the tail structure to the wing or fuselage. A simple WER for the boom structures W_{Booms} is

$$W_{Booms} = 0.14 \cdot L_{Boom} \cdot W_{Cant} \text{ lb} \qquad (6.41)$$

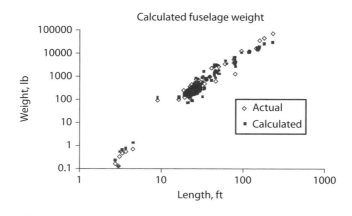

Fig. 6.4 Fuselage WER comparison with data.

where L_{Boom} is the total boom length from wing to tail attachment points in feet and W_{Cant} is the total cantilevered weight in pounds such as the tails and any systems contained within the tails. Note that this is the weight of all booms combined (relevant to one- or two-boom configurations).

Quasi-Analytic Fuselage Weight WER
As with the wing, quasi-analytical methods estimate the fuselage structural weight based on the loads. Additional empirical factors are applied to correlate the estimates with historical data. These methods account for the structural configuration, size, materials, and load conditions. Quasi-analytical fuselage weight estimation methods can provide detailed outputs that can serve as a starting point for more detailed structural sizing methods. Chapter 7 presents such a method of fuselage sizing.

6.3.1.4 Nacelles
Nacelles are structures that house the propulsion system. Nacelles can be integrated into a fuselage or pod structures protruding from a wing or fuselage, among many potential configurations. The nacelle weight group generally contains the cowling or other external aerodynamic structure and propulsion support structure. Truss structures for supporting reciprocating engines are included. The nacelles can also include elements of the air induction and exhaust system.

Nacelle WER for reciprocating engines and turboprop generally takes the form

$$W_{\text{nac}} = F_{\text{nac}} \cdot P_{\text{Max}}^{E1} \text{ lb} \tag{6.42}$$

where F_{nac} is the nacelle multiplication factor, $E1$ is an exponent, and P_{Max} is the maximum power of all engines combined in horsepower. Using methods presented in Roskam [6] when $E1$ is 1, F_{nac} ranges from 0.24 to 0.37 for reciprocating engines and is given as 0.14 for multi-engine turboprops. An F_{nac} value of 0.3 is recommended as a starting point for reciprocating engine UASs.

Jet nacelle weight can be estimated by the WER

$$W_{nac} = F_{nac} \cdot T_{Max} \, lb \tag{6.43}$$

where T_{Max} is the maximum thrust of all engines combined in pounds. Torenbeek [7] recommends a F_{nac} value of 0.055 for low-bypass-ratio turbofan engines and 0.065 for high-bypass-ratio turbofan engines on commercial transport aircraft. These values might be appropriate for HALE UASs with external engines, such as the Global Hawk. A F_{nac} value of 0.06 can be used for small turbojet and low-bypass-ratio turbofan engines with nacelles attached to the fuselage as a starting point.

6.3.1.5 Landing-Gear Weight Estimation

Conventional takeoff and landing is the dominant launch and recovery technique for UASs weighing over 500 lb, and it is frequently employed at smaller sizes. The landing-gear WERs typically used for manned aircraft can be used within their applicable range for unmanned aircraft. A more general landing-gear weight equation for all scales of unmanned aircraft is

$$W_{LG} = F_{LG} \cdot W_{TO} \, lb \tag{6.44}$$

where F_{LG} is the landing-gear mass fraction, which will vary from 0.03 – 0.06. This method obscures landing-gear configuration and dimensions. An initial F_{LG} value of 0.04 is recommended for aircraft that take off and land on paved runways. Higher values are used for carrier-based aircraft or those that operate from rough fields. Fixed landing gear is lighter than retractable gear.

General aviation, transport, and fighter/attack landing-gear WERs are recommended within the applicable range. Several methods are presented by Roskam [6], Torenbeek [7], and Raymer [8]. These WERs take into account the landing-gear length and configuration.

Landing gear for large UASs can be excessively expensive to develop, and numerous manned aircraft landing-gear options can exist. The cost is particularly high for retractable landing-gear systems. The nose gear and main gear could potentially come from different aircraft types. When landing gear is used from other aircraft, then the measured weight should be used in the weight model.

6.3.2 Propulsion System Weight Estimation

The weight of the propulsion system is

$$W_{\text{pwr}} = W_{\text{Engine}} + W_{\text{ai}} + W_{\text{Props}} + W_{\text{FuelSys}} + W_{\text{PropSys}} \qquad (6.45)$$

where W_{Engine} is the weight of the uninstalled engine, W_{ai} is the air induction weight, W_{Props} is the propeller weight, W_{FuelSys} is the fuel system weight, and W_{PropSys} is the weight of balance of the propulsion system weight. One should be careful not to double book the fuel system elsewhere.

6.3.2.1 Engine Weight

The weight of the engine comes from data of a selected engine or from weight estimating relationships for the class of engines. Weight estimating relationships are covered in this section.

Jet Engines

Jet-engine weight can be determined parametrically at the conceptual design stage if the engine has not been selected or if a new engine will be developed for the application. Parametric engine sizing is known as a "rubber engine." A word of caution is needed: few UAs can find a close engine match or can afford a new engine development. Nevertheless, the engine weight can be found by

$$W_{\text{Jet}} = T_{\text{Max}} \cdot T/W_{\text{Ref}} \qquad (6.46)$$

Some representative values of T/W_{Ref} are found in Table 6.4.

Jet engines must be installed on the airframe. Some potential installations include internal to the fuselage, outside of the fuselage in nacelles, and wing-mounted nacelles. The installation includes structural provisions, inlet ducting, and exhaust nozzles. Nacelle weight is covered under structures weights. The total installed engine weight is given by

$$W_{\text{Jet,Installed}} = F_{\text{Install}} \cdot W_{\text{Jet}} \qquad (6.47)$$

Typical installation factors F_{Install} range from 1.05 to 1.3.

Table 6.4   Thrust-to-Weight Ratio Values for Various Jet-Engine Classes

Engine Class	Thrust Range, lb	Reference Thrust/Weight
Model aircraft turbine	5–30	5
Cruise missile or target	30–1000	3.6–8.9
Business jet	800–10,000	3.4–5.3
Supersonic fighter	5000–50,000	5.5–8

Reciprocating, Rotary, and Turboprop Engines

The rubber engine sizing for a reciprocating or rotary engine is given by

$$W_{\text{Engine}} = P_{\text{Max}} \cdot P/W_{\text{Ref}} \tag{6.48}$$

The reference power-to-weight ratio P/W_{Ref} for several engine types is provided in Chapter 8. Depending on the maximum power needed or required fuel type, the engine selection for many UAS applications can be limited. The UA designer will generally need to select between a handful of engines.

The installed weight of the engine includes the engine mounting frame, mufflers, vibration isolators, and cooling system. The cowling structure is included as part of the structural weight. The installed engine weight is

$$W_{\text{Eng,Installed}} = F_{\text{Install}} \cdot W_{\text{Engine}} \tag{6.49}$$

where the installation factor F_{Install} ranges from $1.1-1.5$.

Electric Propulsion Systems

An electric propulsion system consists of a motor, gearbox, electronic speed control, and a power source. The power source can be a battery, fuel cell, or solar collection system. The power source scaling is described in Chapter 8 and is not repeated here.

$$W_{\text{EPS}} = W_{\text{Motor}} + W_{\text{Gear}} + W_{\text{ESC}} + W_{\text{PowSource}} + W_{\text{Props}} \tag{6.50}$$

The motor WER was developed earlier in this chapter. The gearbox weight is approximated by

$$W_{\text{Gear}} = F_{\text{Gear}} \cdot P_{\text{MotMax}} \tag{6.51}$$

The multiplication factor F_{Gear} depends upon the type of engine and gearbox. For small motors a reasonable value of F_{Gear} is 0.12 lb/kW. The electronic speed control WER is

$$W_{\text{ESC}} = F_{\text{ESC}} \cdot P_{\text{MotMax}} \tag{6.52}$$

where F_{ESC} is approximately 0.08 lb/kW for brushed and brushless motors.

6.3.2.2 Air Induction

The air induction system is composed of the inlets for reciprocating engines or jet engines and the associated support structure. This weight group can be calculated separately or included as part of other weight groups. Podded jet engine inlets are often included in the nacelle weight. This group can also be included within the balance of propulsion system group along with other components. Roskam [6] presents numerous manned aircraft methods applicable to large UASs. The following method can be used for UASs weighing less than 500 lb:

$$W_{\text{ai}} = F_{\text{ai}} \cdot P_{\text{Max}} \tag{6.53}$$

where F_{ai} is the air induction multiplication factor. Approximate values for F_{ai} are $0.08 - 0.24$ for reciprocating engines and $0.04 - 0.07$ for electric motors. The reader is encouraged to find more representative values for the design problem of interest.

6.3.2.3 Propellers

Propellers do not constitute a major portion of the UA weight, but the weight must be properly accounted for. Because these components are often located at the extreme fore or aft end of the aircraft, they can impact the center-of-gravity location. The General Dynamics propeller weight method described by Roskam [6] yields reasonable results even at small scales. This can be modified as

$$W_{\text{Prop}} = K_{\text{Prop}} \cdot N_{\text{Props}} \cdot N_{\text{Blades}}^{0.391} \cdot \left(\frac{D \cdot P_{\text{Max}}}{1000 \cdot N_{\text{Props}}} \right)^{0.782} \text{lb} \tag{6.54}$$

where K_{Prop} is a multiplication factor, N_{Props} is the number of propellers (assumed to be equal to the number of engines), D is the propeller diameter in feet, $P_{\text{Shaft,Tot}}$ is the total maximum shaft horsepower of all engines, and N_{Blades} is the number of blades for each propeller. Roskam recommends a K_{Prop} value of 24.0 for turboprops above 1,500 shp and 31.92 for engines below 1,500 shp. Here, a K_{Prop} value of 15 is recommended for plastic or composite propellers for engines with less than 50 shp.

6.3.2.4 Fuel System

The fuel system contains fuel tanks, pumps, fuel lines, valves, and venting, among other components for traditional fuel types. In the case of hydrogen-powered aircraft, this group includes the hydrogen storage device less the hydrogen. For battery-powered aircraft, the fuel system group can include battery installation structure but not the battery itself.

For traditional fueled systems, the WERs for the fuel system weight W_{FuelSys} generally take the following forms:

$$W_{\text{FuelSys}} = F_{\text{fs}} \cdot W_{\text{Fuel}}^{E1} \text{lb} \tag{6.55}$$

or

$$W_{\text{FuelSys}} = F_{\text{fs,Vol}} \cdot V_{\text{Fuel}}^{E2} \text{lb} \tag{6.56}$$

where F_{fs} and $F_{fs,Vol}$ are multiplication factors, V_{Fuel} is the fuel volume in gallons, and $E1$ and $E2$ are exponents. In the first equation, the following parameters can be used:

Application	F_{fs}	$E1$	Source
MALE single engine	0.692	0.67	Torenbeek [7] general-aviation methods
MALE multi-engine	1.56	0.6	Torenbeek [7] general-aviation methods
Tactical and small tactical UAS	0.05–0.1	1	Estimate

6.3.2.5 Propulsion System

The propulsion system group includes any remaining propulsion system items not covered by other weight groups. This can include engine controls, oil system, engine starters, thrust reversers, and propeller controls. Some weight estimating relationships found in other sources combine weight groups such as air induction into this broader weight group. Roskam [6] presents a number of methods for manned aircraft that can have application to MALE UAS and UCAVs. A simple method of estimating this weight group is

$$W_{PropSys} = F_{PropSys} \cdot W_{Engine} \qquad (6.57)$$

The multiplication factor $F_{PropSys}$ ranges from $0.08 - 0.35$ for several manned aircraft.

6.3.3 Fixed Equipment Weight Estimation

The fixed equipment group weight is

$$
\begin{aligned}
W_{feq} = {} & W_{Avion} + W_{FCS} + W_{Hyd} + W_{Elec} + W_{ECS} \\
& + \underbrace{W_{Oxy}}_{=0} + W_{APU} + \underbrace{W_{Furn}}_{=0} + W_{Handling} + W_{Ops} \\
& + W_{Arm} + W_{Weaps} + W_{FTI} + W_{Aux} + W_{Ballast} + W_{Paint} \qquad (6.58)
\end{aligned}
$$

where

$$
\begin{aligned}
W_{APU} &= \text{auxiliary power unit weight} \\
W_{Arm} &= \text{armament weight} \\
W_{Aux} &= \text{auxiliary gear weight} \\
W_{Avion} &= \text{avionics, instrumentation, and communication system weight} \\
W_{Ballast} &= \text{ballast weight}
\end{aligned}
$$

W_{ECS} = environmental control system weight
W_{Elec} = electrical system weight
W_{FCS} = flight control system weight
W_{FTI} = flight-test instrumentation weight
W_{Furn} = furnishings weight
$W_{Handling}$ = baggage and cargo handling equipment
W_{Hyd} = hydraulic and pneumatic system weight
W_{Ops} = operational items weight
W_{Oxy} = oxygen system weight
W_{Paint} = paint weight
W_{Weaps} = weapons provisions, launchers, and guns

Strict definitions of what items to include in each group are not well established. In reviewing historical weight statements from different airframe manufacturers, it quickly becomes apparent that different standards are adopted throughout the industry. This section attempts to define weight group definitions that are appropriate for UASs.

The basic component weights must be mounted to the airframe. The mounting structure can add 10–25% to the uninstalled component weight. The installed weight of the component $W_{Comp,Install}$ is

$$W_{Comp,Install} = f_{Install} \cdot W_{Comp} \tag{6.59}$$

where $f_{Install}$ is the weight installation factor and W_{Comp} is the uninstalled component weight. Recommended values for $f_{Install}$ range from 1.1 to 1.25. This does not include structural sizing of the airframe elements that house the components. It does include local brackets and mounting provisions. The installation weight should be included within each weight element of the fixed equipment group.

6.3.3.1 Avionics, Instrumentation and Communications System

The UAS avionics, instrumentation, and communications systems can be difficult to satisfactorily estimate without knowledge of the system architecture and selection-specific components. This is driven by the requirements and available technologies. Compared to manned aircraft, the avionics can comprise a large portion of the UA weight. The combination of avionics and communication systems can be 11–15% of the takeoff gross weight. The wide range of UAS classes means that no single avionics or communications approach is appropriate for all unmanned aircraft. Chapter 10 provides information on avionics capabilities, and Chapter 12 describes communication system performance.

The avionics, instrumentation, and communications weight can be calculated as

$$W_{\text{Avion}} = W_{\text{Avionics}} + W_{\text{Inst}} + W_{\text{Comms}} + W_{\text{Wiring}} \qquad (6.60)$$

where W_{Avionics} is the avionics weight, W_{Inst} is the instrumentation weight, W_{Comms} is the communication system weight, and W_{Wiring} is the wiring harness weight. The avionics weight, in turn, can be composed of the autopilot, UA management system, mission management system, and local processors. Rather than create a hierarchy for all avionics, instrumentation, and communication systems, some general considerations are provided. The architecture depends greatly on the UA class, mission, and design implementation.

In some cases the avionics architecture and component selection is not known at the beginning of conceptual design. The basic avionics weight equation can be used in such situations:

$$W_{\text{Avion}} = f_{\text{Avion}} \cdot W_{\text{TO}} \qquad (6.61)$$

where the multiplication factor f_{Avion} ranges from 0.06 to 0.16, and a nominal value of 0.1 is recommended. This equation is relevant because larger aircraft tend to carry heavier avionics suites. After all, it is inappropriate for a UCAV to carry MAV avionics.

Some UASs consider all or part of the communication system to be part of payload weight. There does not seem to be a consistent standard for UAS weight accounting. The author recommends any communication system that is flight critical or supports payload downlink should be considered part of the avionics weight unless specified otherwise by a customer. Any communication system that is mission-specific, such as ground communications relay or an airborne base station, can be treated as a payload.

Autopilot
Autopilot weight is not suitable for a WER. In fact, global positioning system (GPS)-based autopilots capable of full UA guidance, navigation,

Table 6.5 Notional Autopilot Weight and Functionality by UAS Class

UA Class	Functions	Typical Weight, lb
Micro UA	Autopilot, VMS, Comms, INS, ADS, PL Cont	0.05–0.5
Small UAS	Autopilot, VMS, INS, ADS, PL Cont	1
Small Tactical UAS	Autopilot, VMS, ADS, INS	2
Tactical UAS	Autopilot, VMS, INS	5–10
MALE UAS	Autopilot, INS	10–50
HALE UAS	Autopilot	50–200
UCAV	Autopilot	50–200

and control can weigh between a few grams and a few hundred pounds. Several applicable autopilots are shown in Table 6.5 Weight is a strong function of processing demands and UA scale. The technology is also advancing rapidly, ever in the direction of miniaturization.

The autopilot might be a processor only. It might also include pressure transducers and an inertial measurement unit. Beyond flight controls, an autopilot can also perform unmanned aircraft management functions, communications management, and payload control.

Air Data System

The air data system consists of air data probes, pressure transducers, and signal processors. The two primary types of probes are pitot-static to measure indicated airspeed and total pressure to measure pressure altitude. Multihole probes can measure pitot-static pressure as well as estimate angle of attack and sideslip. Pitot-static probes typically weigh between 0.06 – 1 lb. Flight-test booms that include angle of attack and sideslip vanes can weigh between 2 – 20 lb, which might be too large for small UAS applications.

GPS

GPS antenna weight varies based on performance, technology level, and militarization. Small commercial GPS antennas will typically weigh less than 0.5 lb. Selective availability antispoofing module (SAASM) military GPS antennas can weigh more than 1 lb. The GPS antenna weight might be negligible for large UAS and a major driver for MAV and small UAS classes.

Inertial Navigation System

The inertial navigation system (INS) combines an inertial measurement unit (IMU) and GPS to provide a navigation solution for the autopilot and other systems. The INS can be a stand-alone unit or integrated with the autopilot. Another application for the INSs is to provide stabilization for payloads and support sensor pointing estimation. Several INSs are shown in Table 6.6, based on marketing materials.

Processors

Processors are used for a variety of applications including flight controls, mission management, UA management system, and payload data processing. Processors can range from a single board with conformal protective coating to a series of processors contained within an EMI shielded metal box. Simple processor boards can weigh less than 0.25 lb. Processor suites suitable for SAR or SIGINT processing can weigh 25 – 180 lb and consume 0.4 – 2 kW. Computer technology is constantly improving, and so today's large systems might have negligible weight in the future. Using

Table 6.6   Example INS Weights

INS	Weight, lb	Power, W	Notes
RCCT Micro INS	0.25	3.5	
RCCT GS-111 INS	1.7	—	Based on the GS-111 autopilot
Systron Donner C-MIGITS III	2.4	18	
Kearfott KN-4073	8	35	Used on FireScout
Kearfott KN-4072	11	35	Used on Global Hawk
Northrop Grumman LN-251	12.7	30	
Northrop Grumman LN-100G	21.6	37.5	

2003 technology, UAS processor weight can be estimated as $0.5-0.7$ lb/ GFLOPS. (A GFLOPS is 10^9 floating point operations per second.)

Wiring Harnesses
Wiring harnesses include the wiring and connectors that link the systems together. Wiring harnesses can pass signals or carry power. The weight depends on wiring length, number of wires, shielding, and connector types. In the absence of mature wiring harness designs from which weights can be estimated, the following method can be used to estimate the wiring harness weight W_{Wiring}:

$$W_{\text{Wiring}} = f_{\text{Wiring}}$$
$$\cdot \left[\sum W_{\text{Avion}}(i) + \sum W_{\text{Comms}}(i) + \sum W_{\text{Payloads}}(i) \right] \quad (6.62)$$

where f_{Wiring} is the wiring harness factor and the summations indicate the uninstalled component weights of the avionics, communication system, and payloads. Appropriate values of f_{Wiring} range from $0.2-0.35$. Wiring weight is frequently underestimated by UAS developers without historical weight databases of past aircraft programs. The author has personally grossly underestimated wiring weight in conceptual design and later expended considerable effort to bring the UA W_E back down.

Line-of-Sight Communications
Numerous factors contribute to the line-of-sight communications weight. System drivers include operational environment, link range, antenna type, waveform, frequency band, encryption type, survivability features, and technology level. Vastly different line-of-sight communication systems exist for micro UA and large UASs. Chapter 12 describes line-of-sight communication system design and performance.

The weights of several line-of-sight communications systems are shown in Table 6.7. Component selection must be based on required performance, UA impacts, and cost, among other considerations.

Table 6.7 Example Line-of-Sight Communications System Airborne Weights

System	Weight, lb	Power, W	Notes
FreeWave F-Series	0.44	~30	Does not include antenna
GMS MT Series	0.48	~15	
L-3 Mini UAV Data Link	<0.5	<10	
CUBIC Mini-CDL	1.3	35	Does not include power amplifier or antenna
Enerdyne Enerlinks II DVA	1.36	<30	Does not include power amplifier or transmitter
L-3 Mini-CDL-200	<2	<30	Does not include antenna
Enerdyne Enerlinks III	3.31	<30	Does not include power amplifier or transmitter
L-3 Mini TDCL Transceiver	<19	~60	Includes directional antenna

SATCOM

SATCOM systems use pointing dish antennas or fixed antennas. SATCOM design drivers are covered in Chapter 12. High-gain dish antennas are required for high-bandwidth communications, and fixed antennas are suitable for low-bandwidth communications with low-Earth-orbit satellite constellations. The mass fraction relative to takeoff gross weight for SATCOM systems ranges from 1–10%, and so selection of hardware components is the best approach to estimating weight.

For the Global Hawk system, L-3 provides details on components contributing to the Ku-band SATCOM [9]. The Global Hawk SATCOM weights are shown in Tables 6.8 and 6.9. This represents mid-1990s technology, and so newer systems will likely have reduced weight for equivalent performance.

The total weight contributing to the Ku-SATCOM is 253 lb, and the total power is 2256 W. The L-3 Ku SATCOM system for the Predator A (MQ-1A) uses a 30-in. dish, so that the antenna weight is likely less

Table 6.8 Global Hawk Ku-band SATCOM Component Weights

Component	Weight, lb	Power, W	Notes
CAMA	85	310	Common airborne modem assembly, shared with line of sight
SATCOM RFA	26	78	
HVPS	40	1800	High-voltage power supply
HPA	56	33	High-power amplifier
SATCOM Antenna	46	35	Gimbaled 48.8-in.-diam dish, Ku-band

Table 6.9   Global Hawk UHF SATCOM Components

Component	Weight, lb	Power, W	Notes
UHF RX/TX	13	150	Transmitter/receiver
UHF PA	14	700	Power amplifier
LNA/Diplexer	3	7	Low noise amplifier
UHF SATCOM antenna	8	—	

for that application. Smaller-aperture, lower-data-rate dish antenna SATCOM systems can have airborne weights of less than 40 lb.

The Global Hawk also has a fixed-antenna UHF SATCOM link. This system reuses the CAMA just described. The weights for this system, excluding CAMA, are shown in the Table 6.9.

The total weight for the Global Hawk UHF SATCOM system is 38 lb, and the power is 857 W, excluding the CAMA. Iridium SATCOM systems can be integrated for less than 2 lb, which is suitable for small tactical UASs.

6.3.3.2 Flight Control System

The flight control system consists of the flight control actuators and mechanisms required to deflect the control surfaces. The control surfaces themselves are included in the structural weight. The flight control computer is the autopilot, which is under the avionics weight.

The actuator type can be electric or hydraulic, though the former is by far the most common in unmanned aircraft applications. The actuators are selected based on maximum torque and angular rate or linear force and response rate required. This in turn is driven by the unmanned aircraft maneuverability, flight regime, and control surface sizing. One or more actuators can be used to drive each control function. The following WER can be used for initial sizing:

$$W_{FCS} = F_{FCS} \times S_{CS} \times V_{EQ,Max}^2 \text{ lb} \tag{6.63}$$

where S_{CS} is the total UA control surface planform area in square feet and $V_{EQ,Max}$ is the maximum equivalent airspeed in KEAS. The multiplication factor F_{FCS} ranges from $0.00007-0.0002$ for electromechanic actuators on small UASs through MALE UASs. UCAVs should use manned fighter FCS weight equations if hydraulic actuation is used. These equations commonly have the forms

$$W_{FCS} = F1 \cdot (W_{TO} \cdot q_{Dive})^{E1} \text{ lb} \tag{6.64}$$

$$W_{FCS} = F1 \cdot W_{TO}^{E1} \text{ lb} \tag{6.65}$$

$$W_{FCS} = F1 \cdot (S_{CS} \cdot q_{Dive})^{E1} \text{ lb} \tag{6.66}$$

where $F1$ is a multiplication factor, $E1$ is an exponent, and q_{Dive} is the maximum dive dynamic pressure in pounds per square foot. $F1$ and $E1$ can be determined by looking at flight control system for aircraft of a similar class. The reader is encouraged to use these equation forms with the appropriate factors and exponents for the UASs class of interest.

6.3.3.3 Hydraulic and Pneumatic System

Hydraulics can be used for flight control actuators or gear retraction. Pneumatics are used for gear retraction or braking systems. The hydraulics and pneumatics are generally accounted for in the flight controls group.

6.3.3.4 Electrical System

The electrical system can consist of the generator, power rectifiers, power distribution units, converters, and power system wiring harnesses. The signal wiring harnesses can be included in this weight group in some cases. Manned aircraft electrical system weight equations commonly have the form

$$W_{\text{Elec}} = F_{\text{Elec}} \cdot \left(W_{\text{FuelSys}} + W_{\text{Avion}} \right)^{E1} \qquad (6.67)$$

where F_{Elec} is the electrical system multiplication factor and $E1$ is an exponent. Manned aircraft equations do not scale well to tactical UASs and smaller unmanned aircraft.

An alternative form is recommended for UAS:

$$W_{\text{Elec}} = F_{\text{Elec}} \cdot \left[P_{\text{PL,Max}} + W_{\text{Avion}} \cdot (P/W)_{\text{Avion}} \right]^{E1} \cdot (L_{\text{Tot}} + b)^{E2} \qquad (6.68)$$

where $P_{\text{PL,Max}}$ is the maximum payload power draw in watts, P/W_{Avion} is the avionics power-to-weight ratio in watts per pound, L_{Tot} is the overall UA length in feet, b is the wing span in feet, and $E1$ and $E2$ are exponents. Avionics power-to-weight is $10-20$ W/lb, with a typical value of 15 W/lb. The equation can be used with the notional values of F_{els}, $E1$, and $E2$:

$$W_{\text{Elec}} = 0.003 \cdot (P_{\text{PL,Max}} + 15 \cdot W_{\text{Avion}})^{0.8} \cdot (L_{\text{Tot}} + b)^{0.7} \text{ lb} \qquad (6.69)$$

6.3.3.5 Environmental Control System

The environmental control system includes the air-conditioning, pressurization, and anti-icing systems. The purpose of the air conditioning is to provide cooling airflow to the avionics, communication systems, and payloads. Pressurization systems are generally intended for the same purpose. Most UAs do not include anti-icing systems.

Detailed weights data from UAs are not available, and so WER for manned aircraft can be adapted. The General Dynamics method for fighter and attack aircraft presented in Roskam [6] can be modified as

$$W_{\text{ECS}} = f_{\text{Mod}} \cdot 202 \cdot \left(\frac{W_{\text{Avion}} + W_{\text{PL}}}{1000} \right)^{0.75} \text{lb} \qquad (6.70)$$

where f_{Mod} is a modification factor applied for the problem of interest. This factor may be unity for a UCAV, or 0.5–0.75 for tactical or MALE UASs. Another approach is to use a ratio of the empty weight.

$$W_{\text{ECS}} = f_{\text{ECS}} \cdot W_E \text{ lb} \qquad (6.71)$$

The ECS multiplication factor f_{ECS} can range from approximately 0.01 to 0.05, with a suggested value of 0.024.

6.3.3.6 Auxiliary Power Unit

Auxiliary power units (APUs) are powered generators that produce electricity and pressurized air. These devices are primarily used for ground operations such that external ground power and pressurized air sources are not required, reducing dependency on ground support equipment. APUs can be used to provide power for starting the engines.

The weight of an APU can be estimated by

$$W_{\text{APU}} = 0.338 \cdot \dot{m}_{\text{Bleed}}^{1.22} + 11.3 \cdot P_{\text{Shaft}}^{0.415} + W_{\text{Gen}} \text{ lb} \qquad (6.72)$$

where $\dot{m}_{\text{Bleed}}$ is the maximum bleed mass flow rate in pounds per minute, P_{Shaft} is the maximum output shaft horsepower, and W_{Gen} is the generator weight. This WER has an average error of 16.1% for the 35 military and civilian APUs used in its derivation.

6.3.3.7 Furnishings

An unmanned aircraft has no furnishings unless it is optionally piloted or is a partial conversion of a manned aircraft. This weight group can include instrument panels, seats for the crew and passengers, food handling provisions, emergency systems, insulation, cabin trim, sound proofing, internal lighting, and passenger windows. Dedicated UAS designs have zero furnishings weight.

6.3.3.8 Baggage and Cargo Handling Equipment

Baggage and cargo handling equipment generally do not apply to current-generation unmanned aircraft. The handling equipment is part of the ground support equipment in order to keep UA weight low. Future cargo UASs might more closely represent manned aircraft and include more cargo handling equipment. An example of a cargo handling equipment component is a removable cargo pallet. The 463L pallet measures 88 × 108 in. in length and width, respectively, and weighs 225 lb. Perhaps even more futuristic passenger-carrying unmanned aircraft will need to include baggage handling equipment.

6.3.3.9 Operational Items

Operational items are generally weights associated with passenger and crew provisions such as food, water, eating utensils, and lavatories. This weight is zero for UASs that do not carry humans.

6.3.3.10 Armament

The armament group includes protective armor and weapon support systems. Metal plating or composite armor to protect critical systems is included. Weapons bay doors and missile ejectors are included for internal weapons storage. The weight of this group depends on the weapon systems and armor approach.

6.3.3.11 Guns, Launchers, and Weapons Provisions

Guns, missile launchers, and other weapons provisions are included in the fixed system weights if permanently installed. Some examples include the following:

* Airframe-mounted guns
* Fire control radar
* Weapon system bus
* Permanently installed missile launcher

Other weapons-related systems that can be considered payloads are the following:

* Removable gun pods
* Gun ammunition
* Hardpoint-mounted missile launchers
* Missiles

Weapons can have weights impacts beyond the immediate weapon system. Some examples include backup structure, cutouts for doors, and weapon doors, all of which are included in the structures. Additionally, the weapons can require sophisticated avionics and data buses, which are at least partially accounted for under avionics weights.

6.3.3.12 Flight-Test Instrumentation

Flight-test instrumentation and the flight termination system are covered in Chapter 10. Weights of these systems are almost always limited to flight test and are not part of the operational system. Removable flight-test systems can be counted against the useful load and payload weights. Any flight-test related systems that are not removed from the UA must be included in the fixed equipment weight category. For example, a removable flight-test data recorder could be considered payload weight. A permanently installed flight-test wiring harness or strain sensors built into the skins are fixed system weights.

6.3.3.13 Auxiliary Gear

Auxiliary gear is equipment not accounted for in other fixed system weight categories. This may include maintenance tools stored on the aircraft. This weight group can be estimated as zero weight for most applications.

6.3.3.14 Ballast

Ballast is weight that serves no other purpose than to apply mass to the aircraft. The ballast might be required to balance the aircraft, particularly longitudinally to maintain the appropriate static margin. Another use of this additional weight is to match the best aerodynamic performance (i.e., L/D) with a given flight speed, as is done in sailplanes. Lastly, ballast can help offset aeroelastic phenomena such as flutter by placing masses appropriately along the wing.

In the first case where ballast is required to maintain balance, this additional weight is not a desirable outcome. The ballast might be the consequence of weight growth in another part of the aircraft or inability to place components in the desired location caused by geometric constraints. In most situations, the ballast is applied to the nose of the aircraft, as weight growth tends to make the aircraft become tail heavy over time.

6.3.3.15 Paint

Paint is a surface treatment applied to the airframe that provides coloration and often environmental protection of the structure. Many conceptual design level weight estimates ignore this weight contributor, though the weight contribution is rarely negligible. Paint usually consists of filler, primer, and multiple layers of paint, and a protective top coat. Paint can also take the form of an outer pigmented layer as part of a molded composite structure, such as gel coat. The weight of the paint $W_{\text{Paint}}(i)$ for a reference surface area $S_{\text{Paint}}(i)$ is

$$W_{\text{Paint}}(i) = F_{\text{Paint}} \cdot S_{\text{Paint}}(i) \qquad (6.73)$$

where F_{Paint} is the weight per area of the paint system. The total weight of the UA paint is the sum of all painted components. A good initial value for F_{Paint} is 0.02 lb/ft^2. Alternatively, the paint for the aircraft can be estimated by using the total exposed area as the reference area.

Another relationship is

$$W_{\text{Paint}} = MF_{\text{Paint}} \cdot W_{\text{TO}} \tag{6.74}$$

where MF_{Paint} is the paint mass fraction. Roskam [6] recommends MF_{Paint} values of 0.003 to 0.006.

Not all structures are painted. For example, metal structures might have exposed metal skins made from aluminum or other materials. Mylar-covered structures might have clear or pigmented Mylar skins. The paint weight is zero where no paint is applied.

6.3.4 Payload Weight Estimation

Payload weights can be provided as a weight allocation or be derived from performance requirements. In the case of an allocation, the payloads weight is simply tracked as a constant weight.

The definition of payloads weight varies. This weight generally includes any intelligence collection sensors such as EO/IR balls or synthetic aperture radars, for example. The payload integration provisions can be considered either airframe or payload weights, depending on the nature of the inter-face. Payload management systems and data storage devices could be considered either payload or avionics weight. Communications links that transmit payload data from the UA to the ground can be considered either payload or communications system weight. Some UASs have even claimed the flight critical avionics as payload weight, though the author discourages this practice.

The weight of an imaging (EO/IR) ball with a diameter ranging from 1.5–20.8 in. can be found by

$$W_{\text{EO/IR}} = 0.0164 \cdot f_{\text{Elec}} \cdot D_{\text{Ball}}^3 \cdot (h/D)_{\text{Ball}} \tag{6.75}$$

where $(h/D)_{\text{Ball}}$ is the ratio of the height to the diameter of the ball and D_{Ball} is the ball diameter in inches. The factor f_{Elec} is equal to 1 if all electronics are contained within the EO/IR ball, and 1.21 if a separate control box is used. The average error is 31.1% with the 50 EO/IR balls used to develop the relationship.

6.3.5 Fuel Weight Estimation

The fuel weight can be the output of a performance algorithm or provided directly as an input. If the fuel mass fraction is an input, then the fuel weight is calculated by

$$W_{\text{Fuel}} = MF_{\text{Fuel}} \cdot W_{\text{TO}} \tag{6.76}$$

Note that the fuel weight is just the weight of the fuel and does not include the fuel system weight.

6.3.6 Trapped Fuel and Oil Weight Estimation

The trapped fuel and oil weight W_{tfo} estimation can be treated as a fraction of the total fuel weight

$$W_{tfo} = F_{tfo} \cdot W_{Fuel} \qquad (6.77)$$

where F_{tfo} is the multiplication factor. The amount of trapped fuel depends on the fuel tank configuration and provisions for fuel scavenging. F_{tfo} can range from $0.01-0.05$.

6.3.7 Crew Weight Estimation

When UA fly without pilots, the crew weight is zero. Even though this is a book about unmanned aircraft, there are some circumstances for which humans will be onboard. Optionally piloted aircraft (OPA) can fly with or without pilots. In the future, UAS can carry human passengers or casualties.

6.4 Weight Management

Weight must be actively managed across the entire life cycle of the UA. Rare is the aircraft product line that loses weight over time. Weight growth is like entropy for an UA design, where weight increases over time unless order from a weights management process is applied.

The weights engineer plays a vital role in maintaining the weight and center-of-gravity limits of the UA. Successful weights engineers are empowered and persistent. Weights engineers must perform weight audits and generate reports on a frequent periodic basis.

Recall the concept of weight escalation described in Chapter 3. The *weight escalation factor* (WEF) is the ratio of change to takeoff gross weight for every pound of fixed weight added. This is also called the *growth factor*. In initial conceptual design this can be quickly calculated by

$$WEF = \frac{1}{1 - (MF_{Struct} + MF_{Subs} + MF_{Prop} + MF_{Energy})} \qquad (6.78)$$

Here all of the mass fractions linearly scale with takeoff gross weight. More complex weight estimating relationships used in the later phase of conceptual design are more difficult to use for determination of the WEF. The definition of WEF

$$WEF = \partial W_{TO} / \partial W_{Fixed} \qquad (6.79)$$

can be modified to permit numerical differentiation:

$$WEF = \Delta W_{TO} / \Delta W_{Fixed} \qquad (6.80)$$

Here small variations in fixed weights are applied. A weights engineering cliché is "weight begets weight." The WEF quantifies this phenomenon and shows how fast weight can grow and the degree of weights management challenge.

In preliminary design and especially in detail design, the UA design analysis is no longer linked in an integrated code. The time to realize the impact of a fixed weight change may be on the order of weeks instead of seconds. Take the example of an avionics component that doubles in weight. The empty weight of the UA will go up by the amount of the component weight increase. Later the structures group can resize the structure based on the new weight roll-up, and the structures weight increases to accommodate the higher loads associated with the weight increase, resulting in an even higher weight. The performance group then runs the performance analysis with the new weights and discovers that more fuel is required to perform the mission. The increased fuel weight increases the takeoff gross weight. The structure must again be resized for the new takeoff gross weight. The cycle can continue several more times until the weights converge.

Weight maturity can be described within a metric and tracked throughout the development program. The weight maturity of the system can be calculated as

$$WM = \sum_{i=0}^{i=N_{Comp}} W_{Comp,i} \cdot WM_i \Big/ W_E \qquad (6.81)$$

where WM is the weight maturity and W_{Comp} is the weight of a tracked component. All components that comprise the empty weight of the UA must be tracked at the level at which they can be practically measured. For example, an actuator, engine, or a small structural assembly can be tracked. A removable avionics box containing processors can be tracked, whereas tracking each component on the processor's board is impractical. Levels of weight maturity can be assigned based on component status. Some suggested maturity events and associated scores are shown in Table 6.10.

An important element of weight management is weight margins. A weight margin is defined as the expected weight growth between the current best estimate weight and the final anticipated weight at a defined state of the development program. The weight of interest can be the unmanned aircraft empty weight or the design gross weight. Possible end states might be the first production flight test or delivery of the product. Appropriate weight margins between a conceptual design and the final delivered product can range from 5 – 10%, though many programs experience much greater weight growth. The selected value depends on the suitability of the design tools, requirements maturity, similarity of the UA to other designs, maturity of major components, and technology risk. A relatively conventional new UAS developed by an established team can use a 5% weight margin at conceptual design.

The author recommends tracking design gross weight as the primary weight metric rather than empty weight. This permits ready inclusion of fuel weight in the design trades. Long-endurance UA have a relatively high fuel mass fraction, perhaps reaching 30 – 50% of the takeoff gross weight. Drag reduction or engine fuel consumption improvements can impact required fuel weight and can therefore impact design gross weight just as much as the elements comprising the empty weight.

Table 6.10 Weight Maturity Events and Associated Scores

Status	Description	WM Score
Allocation	An allocation is provided for a system, often without detailed analysis or substantiation. The allocation can take the form of a fixed weight or a simple mass fraction.	0
Parametric estimate	Parametric weight estimating relationships are used to estimate the weight of the component.	0.25
Detailed estimate	Analytically derived weight estimate based on detailed methods capturing the specific attributes of the component.	0.5
Vendor specification data	Weights data provided by the vendor based on their weights measurements.	0.5
Vendor contractual weight	Weight at which the vendor is contractually obligated to deliver the component.	0.75
Calculated	The weight is based on calculating the weight from a CAD model. In the case of structures, the structural geometry is defined as well as all skin thicknesses and material systems.	0.75
Measured	Component is weighed in the final configuration.	1.0

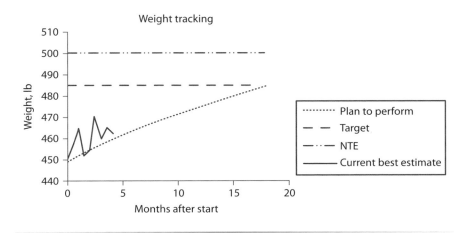

Fig. 6.5 Weight management plan tracking example.

Figure 6.5 illustrates a weight management plan. The top horizontal line is the not-to-exceed (NTE) weight, and the target weight is below. A margin of perhaps 2–4% separates these lines. The plan to perform line starts off with margin relative to the target weight at program initiation and intersects the target weight at product delivery or another program conclusion milestone. The margin reduction rate is highest at the beginning of the program, with the highest slope leading to preliminary design review (PDR) and then a reduction between PDR and crtical design review (CDR). The most fundamental information is learned about the design during the earliest phases of development. The current best estimate (CBE) line represents the current knowledge about the aircraft weight. The weight that is tracked may be empty weight or takeoff gross weight.

The author recommends that the structural sizing and performance calculations be performed using the NTE weight rather than the CBE weight, so that these analyses do not require continuous updating as the weight increases throughout the program. The difference between the CBE and NTE is the weight margin. An increase in CBE will result in a reduction in weight margin and will not require a recalculation of the flight performance and required fuel. This effectively decouples the rapidly changing weights from the long-term performance. Some development programs calculate performance at the current best estimate, which results in continuous structural weight and performance updates. While using the CBE for performance might yield a lighter aircraft at the end of the program, the development schedule can experience substantial slips and increased development cost.

A sample weight report is shown in Fig. 6.6. This report summarizes the weights groups and categories. The weight from the last report is included,

	This report	Last report	Wt change	% Change	Reason
Wing group	93.0	87.0	6.0	6.9%	New finite element model results
Empennage group	31.0	31.0	0.0	0.0%	
Fuselage group	77.5	76.0	1.5	2.0%	Front bulkhead material change
Nacelle group	15.5	11.0	4.5	40.9%	Engine mount configuration change
Landing gear group	31.0	29.0	2.0	6.9%	Larger tire size
Structure total	248.0	234.0	14.0	6.0%	
Engines	42.6	41.0	1.6	4.0%	Rear bearing change
Air induction system	4.3	3.7	0.6	15.2%	Change from composite to metal
Fuel system	23.3	25.4	−2.2	−8.5%	One fuel pump eliminated
Propulsion system	7.8	7.1	0.7	9.2%	New vibration isolation mounts
Power plant total	77.9	77.2	0.7	0.9%	
Avionics and instrumentation	50.0	47.5	2.5	5.3%	GPS antenna change to SAASM
Flight control system	38.8	43.1	−4.4	10.1%	Changed actuator vendor
Electrical system	15.5	18.1	−2.6	14.4%	Changed from NiCD to LiPo batteries
ECS	7.8	7.5	0.3	4.0%	Added surface area to cooling plate
Paint	6.2	4.1	2.1	51.2%	Customer specified paint change
Auxillary gear	1.0	0.0	New	New	Spark plug tool on board UA
Fixed equipment total	119.2	120.3	−1.0	−0.9%	
Oil and trapped fuel	2.0	2.0	0.0	0.0%	
Max fuel capacity	227.9	227.9	0.0	0.0%	
Payload at max fuel	100.0	100.0	0.0	0.0%	
Empty weight	445.1	431.5	13.6	3.2%	
Takeoff gross weight	775.0	761.4	13.6	1.8%	

Fig. 6.6   Weight report example.

as well as weight changes. The reasons for each weight change are summarized. This weight report should be updated on a frequent periodic basis and use the best current estimates. A reasonable update rate is once every two weeks, though weekly or monthly updates might be appropriate in some circumstances. The intention of the report is to provide visibility to program stakeholders such as the chief engineer or customer technical monitor.

6.5 Weight Engineering of Other System Elements

The UAS contains many elements beyond the UA, such as ground control stations, support equipment, launch and recovery equipment, and spares. The mass properties of these elements are very important for transportability and footprint.

Cargo aircraft have loading constraints. The system will generally be mounted on pallet positions within the cargo aircraft. The weight distribution of the pallet positions is critical, and each pallet position has a maximum weight.

Over road transportation takes the form of fitting on a truck bed or by being towed on a trailer. A maximum weight will be specified in each case, based on the UA type and terrain conditions. Trailers generally have a maximum tongue weight, which is the force exerted by the trailer on the trailer hitch.

The weight and physical arrangement of all components must therefore be controlled for transportability. It is possible for the transportability weight drivers to be just as significant as UA weight.

References

[1] *Introduction to Aircraft Weight Engineering*, Society of Allied Weight Engineers (SAWE), Los Angeles, 1996.
[2] Gerard, W. H., "Prediction of Sailplane Wing Weight for Preliminary Design," *Weight Engineering*, A98-37633, Spring 1998, p. 9.
[3] Macci, S. H., "Semi-Analytical Method for Predicting Wing Structural Mass", SAWE Paper No. 2282, La Mesa, CA, May 1995.
[4] Naghshineh-Pour, A. H., "Structural Optimization and Design of a Strut-Braced Wing Aircraft," Masters Thesis, Virginia Tech, Blacksburg, VA, Nov. 1998.
[5] Palumbo, D. J., "Design Evolution of the Shadow 200 Tactical Unmanned Air Vehicle," *Unmanned Systems 2000 Proceedings*, UAVSI, Orlando, FL, 2000, p. 4.
[6] Roskam, J., *Airplane Design, Part V: Component Weight Estimation*, DARcorp., Lawrence, KS, 1989.
[7] Torenbeek, E., *Synthesis of Subsonic Airplane Design*, Delft Univ. Press, Delft, The Netherlands, 1984, p. 285.
[8] Raymer, D. P., *Aircraft Design: A Conceptual Approach*, 2nd ed., AIAA, Reston, VA, 1992, pp. 395–409.
[9] "Global Hawk – Integrated Communications System," L3 Communications, Product Sheet L-3CSW 10/06, Salt Lake City, UT, 2006.

Problems

6.1   Use the secant method to converge the takeoff gross weight of an aircraft with the following properties:

  * Total nonvarying weights of 500 lb
  * Fuel mass fraction of 0.3
  * Propulsion mass fraction of 0.1
  * Structural weight relationship of $W_{\text{Struct}} = 0.2 \cdot W_{\text{TO}}^{0.8}$

6.2   Develop a WER to predict the takeoff gross weight of model sailplanes based on wing span and wing area. Use model airplane catalogs as a data source. Use at least 20 data points that cover hand-launch gliders, 2-m gliders, open-class gliders, and cross-country gliders. What are the average and maximum errors?

6.3   For the model sailplanes from problem 6.2, estimate the total weight of the radio control equipment (receiver, battery, servos, and wiring) based on recommended components. What is the average mass fraction of this equipment? Subtract the radio control equipment weight from the takeoff gross weight to yield the structures weight. Develop a WER to estimate the structures weight as a function of wing span and wing area. What are the average and maximum errors?

6.4   Repeat problem 6.2, but create separate WERs for composite and wood structures. What are the average and maximum errors?

6.5   Repeat problem 6.2, but add several manned sailplanes. Generate new WERs. What are the average and maximum errors?

6.6   Gather weight and heading accuracy data for several INS units. Use heading accuracy with GPS outage (heading drift). Look at a variety of sizes. Plot the unit weight vs GPS outage heading accuracy. (Some manufacturers include Rockwell Collins, Northrop Grumman, and Kearfott.)

6.7   Develop a WER for actuators. The WER should be a function of maximum torque and angular rate. Use hobby servos and UAS actuators. What are the average and maximum errors?

Chapter 7 / *Structures*

- Understand fundamental structural concepts
- Understand metal, composite, and other structural materials
- Learn major structural arrangements
- Use quasi-analytic structural sizing techniques

The Aurora Flight Sciences' Orion wing undergoes a limit load structural test.

7.1 Introduction

The unmanned aircraft structures provide strength and rigidity while maintaining the geometric shape. All components must attach to the structure, so that the structure holds the unmanned aircraft together. The structure is subjected to multiple load cases. The structural members perform specific roles such as reacting bending, torsion, shear, or combinations thereof. The unmanned aircraft has loads associated with launch, recovery, gusts, maneuvering, and ground handling, to name just a few examples. A wide array of materials is available for structures, including metal, wood, and composites. In this chapter, we will see how to analyze and size wing and fuselage primary structures. A brief structures review shows the building blocks for the more detailed analysis.

In conceptual design the primary information that is needed about the structures is the weight. As we saw in Chapter 6, there are few suitable parametric weight estimation methods that cover many UAS classes of interest. The structural sizing methods presented in this chapter offer an analytical approach to structural weight estimation that are based on loads, material properties, and structural geometry. This provides an intermediate step ahead of detailed finite element structural analysis.

7.2 Structural Concepts

Before providing details on aircraft structures such as wings and fuselages, let us review some basic structural concepts. This is intended to provide a background for the more detailed methods used for conceptual airframe sizing. If you are new to structures, you should consult a book on mechanics of materials or structures such as Beer and Johnston [1].

7.2.1 Structural Coordinates

It is convenient to use a structural coordinate system for beams and structural segments, shown in Fig 7.1. The y coordinate runs vertically from the origin. The x coordinate runs horizontally from the origin along the reference plane. The z coordinate runs in the general direction of the beam from the origin according to the right-hand rule. A beam cross section is parallel to the x-y plane. An origin placed at the point of a cross section that intersects the neutral axis is useful for calculating many properties. The x' and y' coordinates originate from the neutral axis and are parallel to the x and y coordinates.

7.2.2 Moments of Inertia

Moments of inertia relate how bending moments and torsion correspond to normal and shear stresses. The moments of inertia of a cross

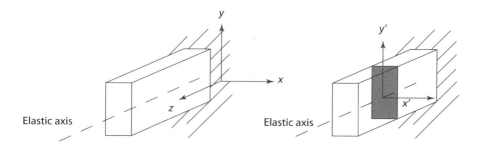

Fig. 7.1 Structures' coordinates.

section cut through a plane normal to the beam are found by

$$I_{xx} = \int y^2 \cdot dA, \qquad I_{yy} = \int x^2 \cdot dA, \qquad I_{xy} = \int x \cdot y \cdot dA \qquad (7.1)$$

The polar moment of inertia J relates how the geometry relates to the shear forces. For axisymmetric cross sections, J is found by

$$J = \int \rho^2 \cdot dA \qquad (7.2)$$

where ρ is the radial distance from a starting reference.

7.2.3 Tension and Compression

Tension is a stress that acts to stretch the column. Compression is a stress that pushes the ends of the column together. The normal stress σ under axial loading P is

$$\sigma_z = \frac{P}{A} \qquad (7.3)$$

where A is the cross-section area. Positive axial loading gives a tension stress, and negative yields a compressive stress. The compressive stress is a pressure exerted by a force on the area of the beam or column.

The normal strain ε is the deformation of the structural member per unit length. *Hooke's Law* provides a relationship for stress and strain within the material's proportional limit:

$$\sigma_z = E \cdot \varepsilon \qquad (7.4)$$

The coefficient E is called *Young's modulus* or the *modulus of elasticity*, and ε is the strain.

Buckling is frequently the limiting condition for a column in compression. The critical buckling load P_{cr} for a column pinned at both ends

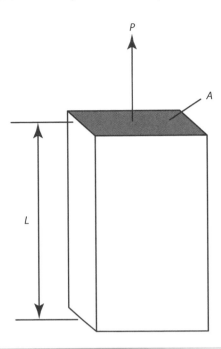

Fig. 7.2 Tension and compression geometry.

can be found from Euler's formula:

$$P_{cr} = -\frac{\pi^2 \cdot E \cdot I}{L^2} \tag{7.5}$$

where I is a moment of inertia and L is the length of the beam. The problem is illustrated in Fig. 7.2. This reveals that the critical buckling load is inversely proportional to L^2, showing that the unsupported column length must be kept short. Local stiffeners such as ribs or stringers can reduce the effective length of a portion of skin, thereby increasing the critical buckling load.

7.2.4 Shear

The average shear stress τ_{ave} across a beam with cross-section area A and with a shear force V applied normal to the beam is

$$\tau_{ave} = \frac{V}{A} \tag{7.6}$$

Hooke's Law of shearing relates the shear strain γ to shear stress by

$$\tau = G \cdot \gamma \tag{7.7}$$

where the coefficient G is the *shear modulus*.

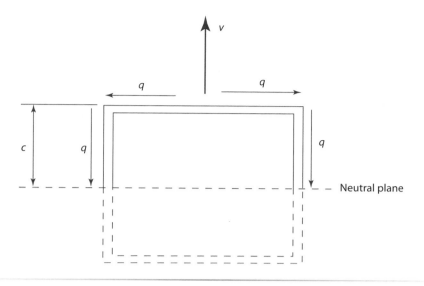

Fig. 7.3 Shear flow geometry.

The first moment with respect to the neutral axis is

$$Q = \int_{y=y_1}^{c} y \cdot dA \qquad (7.8)$$

where y_1 is the y coordinate above the neutral axis and c is the maximum height above the neutral axis. The shear flow q is

$$q = \frac{V \cdot Q}{I_{xx}} \qquad (7.9)$$

The average shear stress for a portion of the beam with width t is

$$\tau_{ave} = \frac{q}{t} = \frac{V \cdot Q}{I_{xx} \cdot t} \qquad (7.10)$$

The highest shear stress generally occurs at the neutral axis because Q is the highest at that y distance. This is the case as long as $Q(y)$ increases more rapidly as y approaches the neutral axis than $t(y)$.

A shear web of constant width b and a height of $2c$ has the following shear-stress distribution (see Fig 7.3):

$$\tau_{xy} = \frac{3}{4} \cdot \frac{V}{c \cdot b} \cdot \left(1 - \frac{y^2}{c^2}\right) \qquad (7.11)$$

7.2.5 Torsion

The shear stress due to torsion in a circular cross-section beam is

$$\tau = \frac{T \cdot \rho}{J} \tag{7.12}$$

where T is the torque, ρ is the radial distance from the center of the beam, and J is the polar moment of inertia.

The maximum shear stress τ_{max} occurs at the maximum radius, denoted by R:

$$\tau_{max} = \frac{T \cdot R}{J} \tag{7.13}$$

For a closed thin-walled section of arbitrary shape, the shear stress is

$$\tau = \frac{T}{2 \cdot t \cdot A_{Bound}} \tag{7.14}$$

where t is the wall thickness and A_{Bound} is the bounded area enclosed by the centerline of the wall thickness.

The angle of twist ϕ for a beam of length L subjected to torsion is

$$\phi = \frac{T \cdot L}{J \cdot G} \tag{7.15}$$

The polar moment of inertia for a hollow circular cylinder is

$$J = \frac{1}{2} \cdot \pi \cdot \left(R_{Out}^4 - R_{In}^4\right) = \frac{1}{2} \cdot \pi \cdot \left[R_{Out}^4 - (R_{Out} - t_{Wall})^4\right] \tag{7.16}$$

where R_{Out} is the outer diameter, R_{In} is the inner diameter, and t_{Wall} is the wall thickness.

Example 7.1   Propeller Shaft Extension

Problem:

An unmanned aircraft uses a 12-ft steel-tube shaft extension with a 6-in. outside diameter to transfer the torque from a centerline engine to a propeller located at the tail, as shown in Fig 7.4. The maximum torque occurs at 5000 rpm, where the peak power transferred is 300 hp. The factor of safety is 1.5. The ultimate shear stress is 50 ksi, G is 11 ksi, and the density is 0.28 lb/in^3.

a. What is the wall thickness if the shaft is designed for strength?
b. What is the wall thickness if the shaft is designed such that the maximum twist is 30 deg at maximum loading?
c. What are the shaft weights corresponding to each case?

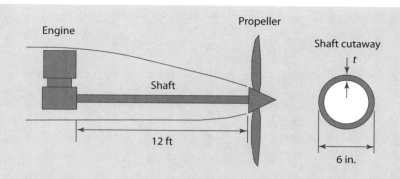

Fig. 7.4 Shaft extension.

Solution:

Torque is found by

$$T = \frac{P_{\text{Shaft}}}{\Omega}$$

where P_{Shaft} is the shaft power and Ω is the rotational rate. The desired torque units are pounds-inches, and so the power must be converted to inches-pounds per second, and the rotational rate should have units of radians per second. The torque is

$$T = \frac{P_{\text{Shaft}}}{\Omega} = \frac{300\,\text{hp}}{5000\,\text{rpm}} \cdot \left(\frac{550\,\text{ft} - \text{lb/s}}{\text{hp}} \right) \cdot \left(\frac{60 \cdot \text{rpm}}{2 \cdot \pi \cdot \text{rad/s}} \right) = 315.1\,\text{ft} - \text{lb}$$

Assuming a thin wall, the wall thickness required for strength is found by

$$t = \frac{T}{2 \cdot \tau_{\text{Ult}} \cdot A_{\text{Bound}}} \cdot FOS = \frac{2 \cdot T}{\tau_{\text{Ult}} \cdot \pi \cdot D^2} \cdot FOS$$

$$= \frac{2 \cdot 315.1\,\text{ft} - \text{lb} \cdot (12\,\text{in./ft})}{50 \times 10^3\,\text{lb/in.}^2 \cdot \pi \cdot (6\,\text{in.})^2} \cdot 1.5$$

$$= 0.00201\,\text{in.}$$

The weight of the shaft extension is

$$W_{\text{Shaft}} = t \cdot \pi \cdot D \cdot L \cdot \rho_{\text{Matl}}$$

$$= 0.00201\,\text{in.} \cdot \pi \cdot 6\,\text{in.} \cdot 12\,\text{ft} \cdot (12\,\text{in./ft}) \cdot 0.28\,\text{lb/in.}^3 = 1.525\,\text{lb}$$

The polar moment of inertia is

$$J = \frac{1}{2} \cdot \pi \cdot \left[R_{\text{Out}}^4 - (R_{\text{Out}} - t_{\text{Wall}})^4 \right]$$

$$= \frac{1}{2} \cdot \pi \cdot \left[(3\,\text{in.})^4 - (3\,\text{in.} - 0.0181\,\text{in.})^4 \right] = 0.3400\,\text{in.}^4$$

With a thin wall assumption

$$J = 2 \cdot \pi \cdot R^3 \cdot t = 2 \cdot \pi \cdot (3 \text{ in.})^3 \cdot 0.0020 \text{ in.} = 0.3403 \text{ in.}^4$$

Here it can be seen that the polar moment of inertia with the thin wall assumption is within 0.1% of the more accurate method, and so we will go with the thin wall estimate is valid.

The angular deflection is

$$\phi = \frac{T \cdot L}{J \cdot G} = \frac{(315.1 \text{ ft} - \text{lb}) \cdot (12 \text{ in.}/\text{ft}) \cdot (12 \text{ ft}) \cdot (12 \text{ in.}/\text{ft})}{(0.3403 \text{ in.}^4) \cdot (11 \times 10^3 \text{ psi})} = 145 \text{ rad}$$

Although the strength-based shaft weighs only 1.525 lb, it twists 145 radians under the full load, which is equivalent to 23.1 revolutions over its length. This shaft is more of a torsion spring than a rigid shaft. Also, the twist estimation assumes small twist angles, which is obviously violated here.

To constrain the shaft to 30 deg under load, the required polar moment of inertia is

$$J = \frac{T \cdot L}{\phi \cdot G} = \frac{(315.1 \text{ ft} - \text{lb}) \cdot (12 \text{ in.}/\text{ft}) \cdot (12 \text{ ft}) \cdot (12 \text{ in.}/\text{ft})}{(30 \text{ deg}) \cdot (\pi/180 \text{ rad}/\text{deg}) \cdot (11 \times 10^3 \text{ psi})}$$

$$= 94.54 \text{ in.}^4$$

If we use a thin wall assumption, the required skin thickness needed to provide the necessary torsional rigidity is

$$t = \frac{J}{2 \cdot \pi \cdot R^3} = \frac{94.54 \text{ in.}^4}{2 \cdot \pi \cdot (3 \text{ in.})^3} = 0.557 \text{ in.}$$

The weight corresponding to this thickness is 423.6 lb. Designing this shaft for stiffness results in a weight increase of 277 times relative to the strength-based design.

The wall thickness of 0.577 in. seems high for a thin wall assumption. If we calculate J, taking into account the inner and outer diameters, J is 71.3 in.[4] rather than 94.5 in[4]. This shows that the thin wall assumption is not valid, and the wall thickness should be found by numerically solving the following equation:

$$J_{Req} - \frac{1}{2} \cdot \pi \cdot \left[R_{Out}^4 - (R_{Out} - t_{Wall})^4 \right] = 0$$

7.2.6 Bending

The normal stress caused by bending σ_z is given by

$$\sigma_z = -\frac{M_x \cdot y}{I_{xx}} \tag{7.17}$$

where M is the bending moment, y is the vertical distance above the elastic axis, and I_{xx} is the moment of inertia about the x axis. The maximum bending stress $\sigma_{z,max}$ is

$$\sigma_{z,\,max} = -\frac{M_x \cdot c}{I_{xx}} \tag{7.18}$$

where c is the maximum y distance from the elastic axis.

The radius of curvature ρ for a beam in bending is found by

$$\rho_x = \frac{E \cdot I_{xx}}{M_x} \tag{7.19}$$

The stress for complex bending with M_x and M_y components is found by

$$\sigma_z = -\frac{M_x \cdot y}{I_{xx}} + \frac{M_y \cdot z}{I_{yy}} \tag{7.20}$$

7.2.7 Beams with Shear and Bending Loads

The maximum stress at a point on the beam cross section is

$$\sigma_{max} = \frac{\sigma_x + \sigma_y}{2} + \sqrt{\left(\frac{\sigma_x - \sigma_y}{2}\right)^2 + \tau_{xy}^2} \tag{7.21}$$

The maximum shear stress is

$$\tau_{max} = \sqrt{\left(\frac{\sigma_x - \sigma_y}{2}\right)^2 + \tau_{xy}^2} \tag{7.22}$$

7.2.8 Pressure Vessels

The volume of a circular cylinder tank with hemispherical end caps is

$$V_{Tank} = \frac{4}{3}\pi \cdot R^3 + \pi \cdot R^2 \cdot L \tag{7.23}$$

where r is the cylinder radius and L is the length of the cylinder portion.

The gauge pressure P is the difference between the inside pressure and the outside pressure. The hoop stress σ_{hoop} of the cylindrical portion of the pressure vessel is

$$\sigma_{hoop} = \frac{P \cdot R}{t} \tag{7.24}$$

The longitudinal stress σ_{long} for the cylindrical portion is

$$\sigma_{long} = \frac{P \cdot R}{2 \cdot t} \tag{7.25}$$

The maximum in-plane shear stress is

$$\tau_{\max} = \frac{P \cdot R}{4 \cdot t} \tag{7.26}$$

The principle stresses are identical for the spherical portions of the tank. These stresses are found by

$$\sigma = \frac{P \cdot R}{2 \cdot t} \tag{7.27}$$

The maximum shear stress is

$$\tau_{\max} = \frac{P \cdot R}{4 \cdot t} \tag{7.28}$$

The wall thickness on the cylindrical portion is

$$t_{w,\text{cylinder}} = \frac{P_{\text{gas}} \cdot R \cdot FOS}{\sigma_{\text{Ult}}} \tag{7.29}$$

where *FOS* is the factor of safety and σ_{Ult} is the ultimate tensile strength. Similarly, the wall thickness for the hemispherical end caps is

$$t_{w,\text{caps}} = \frac{P_{\text{gas}} \cdot R \cdot FOS}{2\sigma_{\text{Ult}}} \tag{7.30}$$

These wall thicknesses only consider the loads associated with pressurization. Flight loads and other stresses must be properly analyzed as well. The mass of the tank is then found by

$$M_{\text{Tank}} = \rho_{\text{matl}} \cdot \left\{ \left[\frac{4}{3} \pi \cdot \left(R + t_{w,\text{caps}} \right)^3 + \pi \cdot \left(R + t_{w,\text{cylinder}} \right)^2 \cdot L \right] - V_{\text{Tank}} \right\} \tag{7.31}$$

Example 7.2 Superpressure Airship Hull

Problem:

A solar-powered superpressure airship has an envelope that is fully inflated at sea level and is capable of reaching high altitude. The maximum pressure differential across the airship hull is 2 atm, which includes the factor of safety. The airship has a constant cylindrical body with hemispherical ends. The length-to-diameter ratio is 10. The envelope material has a tensile strength of 2 GPa and a density of 1400 kg/m³. The total volume is 100 m³.

$V_{Tank} = \frac{4}{3}\pi \cdot R^3 + \pi \cdot R^2 \cdot L = \frac{4}{3}\pi \cdot R^3 + \pi \cdot R^2 \cdot \left(\frac{2 \cdot R}{L_{Tot}/D}\right)$

$= \pi \cdot R^3 \cdot \left(\frac{4}{3} + \frac{2}{L_{Tot}/D}\right)$

a. What is the length and diameter of the envelope?
b. What are the envelope wall thicknesses for the cylindrical and hemispherical portions?
c. What does the envelope weigh?

Solution:

The volume is

$$V_{\text{Tank}} = \frac{4}{3}\,\pi \cdot R^3 + \pi \cdot R^2 \cdot L = \frac{4}{3}\,\pi \cdot R^3 + \pi \cdot R^2 \cdot \left(\frac{2 \cdot R}{L_{\text{Tot}}/D}\right)$$

$$= \pi \cdot R^3 \cdot \left(\frac{4}{3} + \frac{2}{L_{\text{Tot}}/D}\right)$$

Solving for radius gives

$$R = \left[\frac{V_{\text{Tank}}}{\pi \cdot \left(\dfrac{4}{3} + \dfrac{2}{L_{\text{Tot}}/D}\right)}\right]^{1/3} = \left[\frac{100\ \text{m}^3}{\pi \cdot \left(\dfrac{4}{3} + \dfrac{2}{10}\right)}\right]^{1/3} = 2.748\ \text{m}$$

The total length is

$$L_{\text{Tot}} = \left(\frac{L_{\text{Tot}}}{D}\right) \cdot 2 \cdot R = 10 \cdot 2 \cdot 2.748\ \text{m} = 54.97\ \text{m}$$

The length of the cylindrical section is

$$L = L_{\text{Tot}} - 2 \cdot R = 54.97\ \text{m} - 2 \cdot 2.748\ \text{m} = 49.47\ \text{m}$$

The envelope thickness of the cylindrical portion is

$$t_{w,\text{cylinder}} = \frac{P_{\text{gas}} \cdot R \cdot FOS}{\sigma_{\text{Ult}}}$$

$$= \frac{(2 \cdot 101325\ \text{Pa}) \cdot 2.748\ \text{m} \cdot 1.5}{2 \times 10^9\ \text{Pa}} = 0.000418\ \text{m}$$

The envelope thickness for the hemispherical caps is

$$t_{w,\text{caps}} = \frac{P_{\text{gas}} \cdot R \cdot FOS}{2\sigma_{\text{Ult}}}$$

$$= \frac{(2 \cdot 101325\ \text{Pa}) \cdot 2.748\ \text{m} \cdot 1.5}{2 \cdot 2 \times 10^9\ \text{Pa}} = 0.000835\ \text{m}$$

The mass of the tank

$$M_{\text{Tank}} = 2 \cdot \pi \cdot \rho_{\text{matl}} \cdot R \cdot \left(2 \cdot R \cdot t_{w,\text{caps}} + L \cdot t_{w,\text{cylinder}}\right)$$

$$= 2 \cdot \pi \cdot (1400\ \text{kg/m}^3) \cdot 2.748\ \text{m} \cdot (2 \cdot 2.748\ \text{m} \cdot 0.000835\ \text{n}$$
$$+ 49.47\ \text{m} \cdot 0.000418\ \text{m}) = 610\ \text{kg}$$

where ρ_{matl} is the density of the tank material. If we assume a thin-walled tank, the tank mass is

$$M_{\text{Tank}} = 2 \cdot \pi \cdot \rho_{matl} \cdot R \cdot \left(2 \cdot R \cdot t_{w,\text{caps}} + L \cdot t_{w,\text{cylinder}}\right) \qquad (7.32)$$

7.3 Materials

Airframes can use numerous types of metal, wood, and composite materials. The selection depends upon consideration of multiple factors including strength, stiffness, fracture toughness, crack propagation resistance, density, operating temperature range, impact resistance, corrosion resistance, manufacturing approach, and cost.

7.3.1 Metal

Metals are the most common structural materials for manned aircraft, but are comparatively rare in unmanned aircraft. The advantages of metals are ease of analysis, low cost, robust environmental characteristics (temperature, humidity, UV, abrasion), ease of maintenance, and good performance. Aluminum, titanium, and steel are the most commonly used metals for aircraft structures. The reasons that most UAs use composite materials instead of metals are that composites have superior strength/weight and are easier to fabricate at a small scale. Material properties for several metals are shown in Table 7.1.

Metals are homogeneous, isotropic materials, which means that the material properties are independent of orientation. This attribute greatly simplifies structural analysis compared to composites and wood.

Metal structures are often sized by the minimum gauge thickness. This is the minimum thickness that the material can be manufactured or that is acceptable for the airframe in operations. The bending loads become negligible at the wing tips, but an aluminum skin with a thickness of aluminum foil would be operationally unacceptable.

Aluminum is the most commonly used metal in aircraft structures. Although more expensive and less strong than steel, its high strength/weight and low weight/area at minimum gauge offers weight advantages.

Table 7.1 Representative Metal Material Properties

Material	σ_{TU}, ksi	τ_U, ksi	E, ksi	G, ksi	ρ, lb/in.^{3}
Aluminum	16–84	10–48	10.1–10.9	3.7–4.1	0.095–0.101
Magnesium	30–34	19–23	6.5	2.4	0.064–0.065
Steel	55–260	35–155	29–30	11	0.277–0.284
Titanium	130–170	100–105	15.5–16.5	6.2	0.160–0.174

Table 7.2 Representative Wood Material Properties

Material	σ_{TU}, ksi	τ_U, ksi	E, ksi	G, ksi	ρ, lb/in.^{3}
Spruce	8.6	1.1	1.5	0.07	0.017
Various Types	8.4–15	0.9–2.4	1.3–2.2	0.07–0.1	0.014–0.026

Aluminum generally comes in the form of an alloy, where the alloy includes zinc, magnesium, copper, and manganese metals. New lithium-aluminum alloys rival composite structures' performance.

Magnesium has similar material properties as aluminum, but it is rarely used in primary structures. This material has poor corrosion and flame resistance. One common use is aircraft wheels and brackets.

Titanium has high strength/weight, heat resistance, and corrosion resistance, making it a high-performance metal. This material is significantly more expensive than aluminum or steel, thereby limiting its use. Highly loaded structures with tight geometric constraints, such as a wing carry-through structure, often use machined titanium. Titanium is relatively brittle and difficult to machine.

Steel has high strength and low cost, but at the expense of weight. Steel has high fatigue resistance and can tolerate high temperatures. Steel tubes are often used in truss structures such as engine mounts. Steel rods can be used in landing-gear struts, especially on small UAs.

7.3.2 Wood

Wood is rarely used as a primary structural material for UAs today. Many early UASs used wood structures with fabric covering, in keeping with the most prevalent construction techniques of the day. Now, wood is most often used for secondary structures and as a core material for composite sandwich structures. Wood airframes can have greater touch labor than metal or composite structures unless advanced techniques such as laser cutting are used. Material properties for wood are shown in Table 7.2.

Some UASs weighing less than 55 lb are adaptations of model aircraft using wood structures. For example, the Hobby Lobby Telemaster trainer series have been frequently adapted to test payloads, propulsion systems, autopilots, and other systems. Although the touch labor is high for small wood UASs, the skill requirements are relatively low compared with other materials.

Wood structures are damage-tolerant and easily repaired. The damaged parts are easily identified. Frequently, the damaged component can be replaced without affecting other structural members. Based on the author's informal experiments, wood structures can be more survivable against small arms fire than composite structures. Major disadvantages include poor moisture resistance and large material property variations.

7.3.3 Composites

Composite materials offer the potential for significant weight savings over metal structures. The weight savings comes from the increased strength per weight of the materials, reduction in parts count, and elimination of many fasteners, such as rivets. An additional benefit is improved corrosion resistance. Material properties for several types of composites are shown in Table 7.3.

The weight savings experienced by using composites in practice are not as large as the material properties would suggest. Typical weight savings over metal structures are usually around 20–25%. Often, rivets or other fasteners are used to attach composite structures to other structures, which add weight compared to a simple bond. Large manufacturing variations reduce the allowable properties used to size the structure, leading to additional materials to ensure that the structure is reliable. Surface paints, UV barriers, and water barriers add to the weight as well.

Composite materials are composed of fiber and matrix materials. The fiber is the load-bearing material that drives the composite strength and stiffness. Fiber can take the form of unidirectional fibers, woven fabric, whiskers, or particulates. Unlike the matrix materials, fibers are temperature-resistant.

The strength and stiffness properties of composite materials are dependent upon the fiber orientation. That is to say, the materials' properties can be tailored to the structural loads. Most aerospace structures use multiple plies of unidirectional fiber sheets. The layers are generally arranged with fiber orientations of 0, 90, and ± 45 deg. For example, the largest load on an unswept wing is bending, followed by torsion. Here, the majority of the fibers are oriented along the spanwise direction (0 deg) to react bending, and fewer layers are oriented at ± 45 deg to react torsion.

The material properties of the composite layups are difficult to predict analytically. It is usually necessary to conduct coupon tests of a representative part. The coupons are flat rectangular sheet samples that are fabricated with the same methods as the final part. Joints that hold the structures together can be tested in a similar manner.

Table 7.3 Notional Composite Material Properties

Material	σ_{TU}, ksi	τ_U, ksi	E, ksi	G, ksi	ρ, lb/in.^{3}
Graphite-epoxy (0)	110–180	9–12	21–25	0.65	0.056
Graphite-epoxy (± 45)	16.9–23.2	43.2–65.5	2.34–2.38	5.52–6.46	0.056–0.058
Fiberglass-epoxy (E-Glass)	105	7.9	4.23	0.51	0.071
Aramid-epoxy	200	9	4.23	0.51	0.071
Boron-epoxy	195	15.3	30	0.7	0.7

The composite component allowable strength properties, known as *allowables*, are engineering limitations placed upon the structures. A layup is layers of composite material plies with a given orientation schedule, core material, and surface treatments. The allowables for each layup must be defined for a structural analysis. This is usually determined by coupon testing of a large number of samples. Based on statistical sampling of the coupons, various bases are established.

The *FOS* used in metal structures is derived from the ratio of the ultimate stress to the yield stress of common metals. The tradition *FOS* value of 1.5 corresponds to aluminum. Composites do not yield like metals; rather, they fail without warning. Therefore, the traditional *FOS* usage has little physical meaning. Composite structures are sized based on allowables.

Composites have low conductivity relative to metals, which poses challenges for electromagnetic interference (EMI), lightning protection, and electrical grounding. Metal structures use the conducting outer skins as a Faraday cage to prevent the charge from entering the interior. The charge from lightning strikes is conducted along the outside of a metal airframe. The internal electronics are protected from outside electromagnetic radiation by the metal shell. A metal structure can also act as grounding for electronics, thereby preventing ground loops that can destroy circuits. Composite structures might need conductive layers to provide these functions. Conductive meshes can be applied as a layer in the layup as one means of accomplishing this need.

The material that occupies the space between composite fibers is known as *matrix* material. The matrix holds the fibers in place and helps to distribute the load. The matrix material also plays a role in the operating temperature limits, chemical resistance (i.e., solvents), and damage tolerance, and they can serve as an environmental barrier. The matrix distributes the load from damaged fibers to surrounding fibers and can also stop crack propagation, thereby providing damage tolerance. Like pushing on a rope, fibers are unable to carry compression loads without the geometric support of the matrix. Matrix materials have much lower strength and stiffness than the fibers they support and bond together. The allowable material properties of the composite are often limited by the matrix material, particularly in delamination (separation of plies).

The most common polymer matrix materials used in aerospace applications are thermosets. These matrix materials form polymer chains in a nonreversible process when curing. The cure cycle is a schedule of temperatures and pressures, and it varies among thermoset types. A cure for prepreg thermosets typically takes 8–12 hrs at temperatures around 350°F. Epoxies used for wet layups can cure at room temperature and pressure with processing times ranging from 5 min to several hours. Many thermosets are hydroscopic, which means that they absorb water. The low toughness of many thermosets is a limiting factor in the achievable design strains.

The volume fraction, which is the ratio of the fibers to the overall composite material, is a major contributor to the material properties such as strength and density. The volume fraction is typically 60% for prepreg composites and less for wet layups.

Composites are not as damage-tolerant as metals. A dropped tool can cause delamination of the plies under the surface, which can be very difficult to detect. Nondestructive inspection (NDI) techniques, such as ultrasound, are often used to detect such damage.

Composite structures have lower density than those made of metal. Sandwich construction has a lower average density than solid layers of composites. The result is that composites can have larger effective skin thicknesses than metal structures, which detracts from usable internal volume.

Composite structures' costs can be reduced through unitized structures. Here, the part count and number of fasteners is reduced through building a large structural component as a single unit. Multiple parts can be cocured together, and stiffeners can be integral in unitized structures, for example.

Prepreg materials are thermally sensitive and have a short shelf life. The materials must be kept in refrigerator storage. This is a drawback compared to metals, which can be stored indefinitely at room temperature. Programs often face a dilemma of ordering large lots of material to prevent the risk of material availability shortages on one hand and having the material expire before the structures are ready to manufacture on the other hand.

In conceptual design, composite structures are often treated as *black aluminum*. This means that the structural sizing methods employed are intended for isotropic materials but are extended to anisotropic materials. Representative composites materials properties are simply used in place of those of metals. For less sophisticated weight estimation methods, the structure is sized for aluminum, and then a correction factor is applied for the use of composites. These approaches obscure the advantages and drawbacks of these complex materials. Furthermore, the black aluminum assumptions can be inaccurate.

The most common forms of composite fibers are graphite, aramid, boron, quartz, and fiberglass. The type of fiber used depends upon the loads, weight constraints, manufacturing approach, and cost sensitivity.

Graphite has the highest stiffness, and it is commonly used for primary structure. This material is also known as carbon fiber. Strength and stiffness have an inverse relationship, and so fibers can be tailored to give needed strength and stiffness characteristics.

Aramid has lower strength and stiffness than graphite, but it offers the potential of impact resistance and ballistic protection. Dupont's Kevlar® is a well-known example of an aramid material. This is the same material that is

Fig. 7.5 NRL SENDER has a hollow molded fiberglass sandwich structure with carbon-fiber reinforcements.

used in bullet-proof vests. Aramid can be difficult to cut, and it frays when sanded.

Boron fibers are stronger than graphite but are more expensive. This material can be applied to areas with high compression loads due to better buckling resistance. The high cost generally limits the use of boron.

Fiberglass is an early form of composite material, and the strength and stiffness are not as impressive as more modern alternatives. However, fiberglass has relatively low cost, and its dielectric properties are appropriate for radomes. Fiberglass can also be used for lightly loaded structural regions that would not benefit from more advanced materials. The NRL SENDER, shown in Fig. 7.5, was constructed primarily of fiberglass.

7.3.4 Ceramics

Ceramics have a high-temperature capability. The service temperature limit can reach 3000–4000°F. This makes ceramics useful for high-temperature engine components.

7.3.5 Rapid Prototyping Materials

Aircraft structures can be grown in a process that is called three-dimensional printing. The basic material is a plastic or resin that can be cured when two laser beams intersect over a unit volume. The part is built one layer at a time. Until recently, three-dimensional printing was suitable for rapidly generating mockups or display models. Now these materials have the necessary strength to become part of the unmanned aircraft structure. This material is generally used for areas that are relatively small, have complex geometry, and are not subjected to high loads.

7.4 Unmanned Aircraft Loads

The gust and maneuver loads are found by developing V-n diagrams. This graphical technique defines the envelope of flight equivalent airspeeds and loads that the aircraft will experience. Roskam [2] provides a methodology for developing gust V-n diagrams for Federal Aviation Regulation (FAR) Part 23 and 25 aircraft and maneuvering V-n diagrams for military aircraft. Although UASs do not strictly fall within either of these categories, either FAR Part 23 or 25 gust techniques may be appropriate when other methods are unspecified.

A major output of the V-n diagram is the maximum positive and negative load factors that are used to size the structures. These load factors can be assigned a priori based on the type of mission. There is no uniform load factor standard for UAS applications, and so it is frequently left to the company design team leadership or program office to define. Often the maximum limiting load factors are based on analogous manned aircraft types. Example analogies between manned and unmanned aircraft are shown in Table 7.4.

Using an analogous approach to manned aircraft can be overly constraining for UAS applications. For example, a manned fighter might have a positive limit load factor of 8.67, which is limited to a large extent by the physiological limitations of the human in the cockpit. An air-to-air UCAV could potentially go several times higher to gain an advantage in combat maneuvering.

Unmanned aircraft that are air-launched can undergo a pull-out maneuver. Prior to wing deployment, an unpowered platform will follow a ballistic trajectory with the nose pointed downwards as it builds up speed.

The vertical load caused by a steady-state turn depends on the bank angle:

$$n_{z,bank} = \frac{1}{\cos(\phi_{bank})} \tag{7.33}$$

The tails and supporting structure can be conservatively sized based on the largest force from a combination of angle of attack and control surface deflection.

Table 7.4 Manned Aircraft Load Factor Analogies

Manned Aircraft (MIL-A 8861)	Analogous UAS	$n_{lim,pos}$	$n_{lim,neg}$
Attack	Attack UCAV	6.00	−3
Fighter	Air-to-air UCAV	8.67	−3
Patrol, ASW, reconnaissance	ISR UAS	3.00	−1

Unmanned aircraft can be hoisted for ground operations. For example, a target drone recovered from the water can be lifted onto the ship via a crane. Usually, at least two hoist points are used—located along the fuselage on either side of the center of gravity. This has the potential to introduce bending loads on the fuselage that are opposite in direction and greater than those experienced in flight maneuvers.

Parachute deployment introduces forces and moments. The initial shock of the drogue chute or main parachute deployment can be significant. The lines are often located at a vertical distance from the center of gravity, which produces a moment.

Ground impact loads are not limited to conventional landing gear. As described in Chapter 11, many small UASs undergo belly landings where the impact shock is absorbed by the airframe.

Large UASs with high-aspect-ratio wings can be sized by a taxi bump. The wings can be laden with fuel or hardpoint-mounted payloads, which create tip-down bending loads. Upon hitting a bump, the unmanned aircraft regions near the landing gear rise while the inertia of the wings resists the upward movement.

7.5 Shell Structure Analysis

Most UAS structures of interest, such as wings and fuselages, are semi-monocoque structures. These shell structures have load bearing skins with supporting structural members. With proper problem formulation, wing and fuselage structures can be analyzed in a similar manner.

All too often a gap exists for aircraft conceptual design analysis. Parametric structural sizing methods are based on aircraft outside of the applicable range of the UAS problem, and they are therefore of dubious relevance. Conceptual design finite element models (FEM), such as NASTRAN, are used for detailed structural sizing and analysis. A FEM is simply too slow to put within an optimization loop.

Methods are needed to rapidly analyze unmanned aircraft structures based on loads, geometry, and structural materials. Enter the quasi-analytical methods. Two methods are presented. The skin-panel method is suitable for wings, which are dominated by bending loads. The boom-and-web method can be used to analyze either wings or fuselages. These methods are described such that they can be implemented in an algorithm.

7.5.1 Skin-Panel Method

The skin-panel method divides the structural segment into members that react bending moments and those that react shear. The section geometry is based upon the physical cross section. We will consider the design mode where the output is the required skin thickness for a load case.

This formulation assumes that the M_x and S_y are the dominant forces and moments. A wing has such loading. With this assumption, the bending material can be allocated to upper and lower skins that follow the outer contours. These panels are joined by a single vertical shear web that only reacts shear loads.

Step 1: Determine Panel Geometry
The coordinate system is shown in Fig. 7.6. The structure is divided into N upper panels and N lower panels as shown in Fig. 7.7. There are $N+1$ nodes each on the upper and lower surfaces. Each panel is bounded by two nodes. The node indices for panel i are i and $i+1$. The index j corresponds with the bounding cross section that is close to the origin, and the index $j+1$ corresponds to the other bounding cross section that is closer to the end of the member. The starting node coordinates corresponding to node i are $x_j(i)$, $y_j(i)$, $z_j(i)$. Similarly, the ending node coordinates are $x_{j+1}(i)$, $y_{j+1}(i)$, $z_{j+1}(i)$. The midsection is located halfway between the starting and ending sections.

The coordinates for the mid-cross-section nodes are

$$x(i) = \frac{x_{j+1}(i) + x_j(i)}{2}, \; y(i) = \frac{y_{j+1}(i) + y_j(i)}{2}, \quad \text{and } z(i) = \frac{z_{j+1}(i) + z_j(i)}{2}$$

$$(7.34)$$

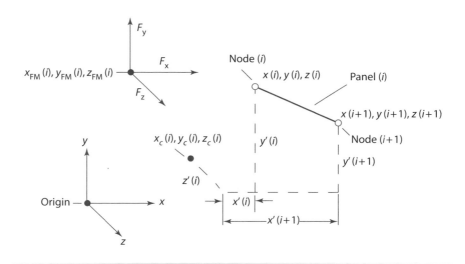

Fig. 7.6 Coordinate system.

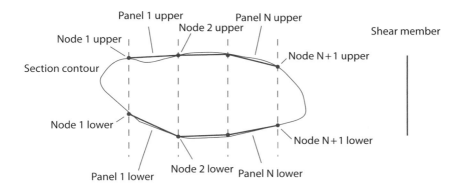

Fig. 7.7   Cross-section geometry.

The length of the web in the x-y plane b is found by

$$b(i) = \sqrt{[x(i+1) - x(i)]^2 + [y(i+1) - y(i)]^2} \qquad (7.35)$$

The coordinates of the midpoint of each panel $x_m(i)$ and $y_m(i)$ are found by

$$x_m(i) = \frac{x(i) + x(i+1)}{2}, \quad y_m(i) = \frac{y(i) + y(i+1)}{2} \qquad (7.36)$$

The section is divided into upper and lower contours. The upper contour has subscript Up, and the lower contour has subscript Low. The upper or lower surface subscripts are left off if the procedure is the same for both. The preceding procedure for calculating the segment length, node coordinates, and midpoint coordinates is performed for the upper and lower contours.

Step 2: Calculate Centroid
This method assumes that the skin thickness is constant for the upper and lower contours and that the shear material does not contribute to the centroid or moments of inertia. With these assumptions, the centroid locations x_c and y_c are found by

$$x_c = \frac{\sum_{i=1}^{N} b_{Up}(i) \cdot x_{m,Up}(i) + \sum_{i=1}^{N} b_{Low}(i) \cdot x_{m,Low}(i)}{\sum_{i=1}^{N} b_{Up}(i) + \sum_{i=1}^{N} b_{Low}(i)} \qquad (7.37)$$

$$y_c = \frac{\sum_{i=1}^{N} b_{\text{Up}}(i) \cdot y_{m,\text{Up}}(i) + \sum_{i=1}^{N} b_{\text{Low}}(i) \cdot y_{m,\text{Low}}(i)}{\sum_{i=1}^{N} b_{\text{Up}}(i) + \sum_{i=1}^{N} b_{\text{Low}}(i)} \tag{7.38}$$

From this, it can be seen that the centroid location is a function of the geometry of the upper and lower contours only.

A new coordinate system is now established to reference the node x and y coordinates to the centroid. The new coordinates x' and y' are found by

$$x'(i) = x(i) - x_c, \quad y'(i) = y(i) - y_c, \quad z'(i) = 0 \tag{7.39}$$

The coordinates of the midpoint of each panel with respect to the centroids $x'_m(i)$ and $y'_m(i)$ are found by

$$x'_m = \frac{x'(i) + x'(i+1)}{2}, \quad y'_m = \frac{y'(i) + y'(i+1)}{2} \tag{7.40}$$

Step 3: Calculate Moments of Inertia

The moments of inertia about the centroid for a panel are

$$I_{xx}(i) = \frac{b(i)^3 \cdot t(i) \cdot \sin^2(\beta)}{12} + b(i) \cdot t(i) \cdot y'_m(i)^2 \tag{7.41}$$

$$I_{yy}(i) = \frac{b(i)^3 \cdot t(i) \cdot \cos^2(\beta)}{12} + b(i) \cdot t(i) \cdot x'_m(i)^2 \tag{7.42}$$

$$I_{xy}(i) = \frac{b(i)^3 \cdot t(i) \cdot \sin(2 \cdot \beta)}{24} + b(i) \cdot t(i) \cdot x'_m(i) \cdot y'_m(i) \tag{7.43}$$

If the skin thickness is constant across all panels such that $t(i) = t$, then the moments of inertia can be normalized by t:

$$I_{xx,t}(i) = \frac{I_{xx}(i)}{t} = \frac{b(i)^3 \cdot \sin^2(\beta)}{12} + b(i) \cdot y'_m(i)^2 \tag{7.44}$$

$$I_{yy,t}(i) = \frac{I_{yy}(i)}{t} = \frac{b(i)^3 \cdot \cos^2(\beta)}{12} + b(i) \cdot x'_m(i)^2 \tag{7.45}$$

$$I_{xy,t}(i) = \frac{I_{xy}(i)}{t} = \frac{b(i)^3 \cdot \sin(2 \cdot \beta)}{24} + b(i) \cdot x'_m(i) \cdot y'_m(i) \tag{7.46}$$

The total moments of inertia normalized by skin thickness are

$$I_{xx,t} = \sum_{i=1}^{N} I_{xx,t,\text{Up}}(i) + \sum_{i=1}^{N} I_{xx,t,\text{Low}}(i) \tag{7.47}$$

$$I_{yy,t} = \sum_{i=1}^{N} I_{yy,t,\text{Up}}(i) + \sum_{i=1}^{N} I_{yy,t,\text{Low}}(i) \tag{7.48}$$

$$I_{xx,t} = \sum_{i=1}^{N} I_{xy,t,\text{Up}}(i) + \sum_{i=1}^{N} I_{xy,t,\text{Low}}(i) \tag{7.49}$$

Step 4: Find Moments About Centroid
The moments are shifted to the x',y',z' coordinate system. $M_{x'}$, $M_{y'}$, and $M_{z'}$ are

$$M_{x'} = M_x + F_y \cdot (z_c - z_{\text{FM}}) - F_z \cdot (y_c - y_{\text{FM}}) \tag{7.50}$$
$$M_{y'} = M_y - F_x \cdot (z_c - z_{\text{FM}}) + F_z \cdot (x_c - x_{\text{FM}}) \tag{7.51}$$
$$M_{z'} = M_z + F_x \cdot (y_c - y_{\text{FM}}) - F_y \cdot (x_c - x_{\text{FM}}) \tag{7.52}$$

Step 5: Calculate Axial Stress for Each Panel
The bending moment at the center of panel i is

$$\sigma_z(i) = \frac{M_{x'} \cdot \left[I_{yy} \cdot y'_m(i) - I_{xy} \cdot x'_m(i)\right]}{I_{xx} \cdot I_{yy} - I_{xy}^2} + \frac{M_{y'} \cdot \left[I_{xx} \cdot x'_m(i) - I_{xy} \cdot y'_m(i)\right]}{I_{xx} \cdot I_{yy} - I_{xy}^2}$$
$$+ \frac{F_z}{\sum_{i=1}^{N} t \cdot \left[b_{\text{Up}}(i) + b_{\text{Low}}(i)\right]} \tag{7.53}$$

This can be normalized by the constant skin thickness:

$$\sigma_{z,t}(i) = \frac{\sigma_z(i)}{t} = \frac{M_{x'} \cdot \left[I_{yy,t} \cdot y'_m(i) - I_{xy,t} \cdot x'_m(i)\right]}{I_{xx,t} \cdot I_{yy,t} - I_{xy,t}^2}$$
$$+ \frac{M_{y'} \cdot \left[I_{xx,t} \cdot x'_m(i) - I_{xy,t} \cdot y'_m(i)\right]}{I_{xx,t} \cdot I_{yy,t} - I_{xy,t}^2} + \frac{F_z}{\sum_{i=1}^{N} \left[b_{\text{Up}}(i) + b_{\text{Low}}(i)\right]} \tag{7.54}$$

Step 6: Calculate Bending Skin Thickness
The skin thickness to resist bending and axial loads is then

$$t_{\text{bend}} = \frac{Max[|\sigma_{z,t}(i)|]}{\sigma_{\text{Ult}}} \cdot FOS \tag{7.55}$$

Step 7: Calculate Shear Material Thickness

The shear member is assumed to be a single web located at the maximum thickness point. The height of the web h_{web} is found by

$$h_{web} = Max\left[y_{Up}(i) - y_{Low}(i)\right] \tag{7.56}$$

The length of the web L_{web} is found by using the same index i. Either the upper or lower surfaces can be used in this calculation.

$$L_{web} = \sqrt{\left[x_{j+1}(i) - x_j(i)\right]^2 + \left[z_{j+1}(i) - z_j(i)\right]^2} \tag{7.57}$$

The maximum shear stress occurs at the neutral axis. The neutral axis is assumed to be at the centroid. To be conservative, the shear web thickness for the entire shear web is held constant to the thickness found at the centroid. The shear flow of the web is

$$
\begin{aligned}
q = &-\left(\frac{S_x \cdot I_{xx} - S_y \cdot I_{xy}}{I_{xx} \cdot I_{yy} - I_{xy}^2}\right) \cdot \sum_{i=1}^{N} t_{bend} \cdot b_{Up}(i) \cdot x'_{m,Up}(i) \\
&-\left(\frac{S_y \cdot I_{yy} - S_x \cdot I_{xy}}{I_{xx} \cdot I_{yy} - I_{xy}^2}\right) \cdot \sum_{i=1}^{N} t_{bend} \cdot b_{Up}(j) \cdot y'_{m,Up}(i)
\end{aligned} \tag{7.58}
$$

Here, it is assumed that the upper and lower skins do not directly react the applied shear loads. The skin thickness required to handle this shear load is

$$t_{shear} = \frac{|q|}{\tau_{Ult}} \cdot FOS \tag{7.59}$$

Step 8: Calculate Bending Material and Shear Material Weight

The surface area of each panel can be approximated as a trapezoid, even though the panel is generally twisted. The z distance between the two bounding cross sections is Δz. The total weight of the bending material W_{bend} is

$$W_{bend} = \rho_{bend} \cdot t_{bend} \cdot \Delta z \cdot \left[\sum_{i=1}^{N} b_{Up}(i) + \sum_{i=1}^{N} b_{Low}(i)\right] \tag{7.60}$$

where ρ_{bend} is the bending material density.

The shear material weight W_{shear} is

$$W_{shear} = \rho_{shear} \cdot t_{shear} \cdot h_{web} \cdot L_{web} \tag{7.61}$$

7.5.2   Boom-and-Web Method

The skin-panel method makes many simplifying assumptions in order to easily calculate the required skin thickness. In the process of creating a convenient panel arrangement, the structural details are obscured. Now, we will evaluate an alternative technique that is more general and whose results more closely resemble the stresses on the true structure.

The boom-and-web method simplifies the structure in order to permit easier analysis, much like the skin-panel method. The cross section is broken into multiple panels. The skin is divided into booms that carry only axial loads and shear webs that carry only shear loads. Many structures texts, such as Niu [3] and Megson [4], provide derivations of this method. This method analyzes a defined structure, but it cannot directly find the required bending material thickness from loads.

The actual structure will have panels that carry stress directly with a thickness t_D. Here the normal stress will be carried by booms running along cross-section node lines. Between the booms, the panel area, which has thickness t, carries only shear loads. The combined thickness of the panel and booms approximates the total skin thickness of that structural segment. The idealization is illustrated in Fig 7.8.

For each structural segment, the following steps are followed:

Step 1: Determine the Geometry of a Cross Section Between the Boom Ends

There are N booms located around the perimeters of the starting and ending cross sections. Each boom should correspond with the same relative position on each cross section. Follow Step 1 of the skin-panel method to locate the x, y, z coordinates of each node. The boom locations correspond to nodes. Note that there is no distinction between upper and lower portions because the section is closed. The nodes should be defined in a clockwise fashion.

Step 2: Calculate Boom Areas from Skin Thickness

Structures can be designed such that the bending and axial loads are handled by dedicated members, and shear stress and torsion are handled

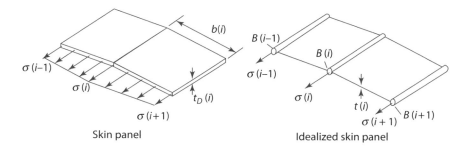

Fig. 7.8 Panel idealization.

by dedicated shear webs. Examples are a fuselage with stringers and a skin or a wing with spar caps and skins. Many composite structures have skins that handle both functions.

If the areas of the booms are already known, then this step can be skipped. If only a single load case is considered, then the boom areas can be related to the skin thicknesses and axial stresses of the adjacent panels. The relationships between the true panel thickness and the boom areas are

$$B(i) = \frac{t_D(i) \cdot b(i)}{6} \cdot \left[2 + \frac{\sigma(i+1)}{\sigma(i)}\right] + \frac{t_D(i-1) \cdot b(i-1)}{6} \cdot \left[2 + \frac{\sigma(i-1)}{\sigma(i)}\right]$$

(7.62)

If $\sigma(i)$ is nearly zero, then the ratios $\sigma(i+1)/\sigma(i)$ and $\sigma(i-1)/\sigma(i)$ will become very large. In this circumstance, boom area can take on an average value of $B(i+1)$ and $B(i-1)$.

Step 3: Calculate Centroid Location

The centroid x and y locations x_c and y_c are found by

$$x_c = \sum_{i=1}^{N_B} B(i) \cdot x(i) \bigg/ \sum_{i=1}^{N_B} B(i) \quad \text{and} \quad y_c = \sum_{i=1}^{N_B} B(i) \cdot y(i) \bigg/ \sum_{i=1}^{N_B} B(i) \quad (7.63)$$

From this, it can be seen that the centroid location is a function of the material distribution around the cross section. Here, we assume that the webs do not contribute to the centroid.

A new coordinate system is now established to reference the boom x and y coordinates to the centroid. The new coordinates x' and y' are found by

$$x'(i) = x(i) - x_c \quad \text{and} \quad y'(i) = y(i) - y_c \quad (7.64)$$

Step 4: Calculate Moment of Inertia

The moments of inertia about the centroid are

$$I_{xx} = \sum_{i=1}^{N_B} B(i) \cdot y'(i)^2 \quad (7.65)$$

$$I_{yy} = \sum_{i=1}^{N_B} B(i) \cdot x'(i)^2 \quad (7.66)$$

$$I_{xy} = \sum_{i=1}^{N_B} B(i) \cdot x'(i) \cdot y'(i) \quad (7.67)$$

Here again, it is assumed that the shear webs do not contribute to the moments of inertia.

Step 5: Calculate the Bending Stress and Axial Loads for Each Boom
Translate the bending moments to the centroid per Step 4 of the skin-panel method.

The z stress due to bending at an x' and y' location σ_z is

$$\sigma_z(i) = \frac{M_{x'} \cdot \left[I_{yy} \cdot y'(i) - I_{xy} \cdot x'(i)\right]}{I_{xx} \cdot I_{yy} - I_{xy}^2} + \frac{M_{y'} \cdot \left[I_{xx} \cdot x'(i) - I_{xy} \cdot y'(i)\right]}{I_{xx} \cdot I_{yy} - I_{xy}^2} + \frac{F_z}{\sum\limits_{i=1}^{N} B(i)}$$

(7.68)

where F_z is the force acting along the z axis.

For a horizontally or vertically symmetric fuselage ($I_{xy} = 0$), this simplifies to

$$\sigma_z(i) = \frac{M_{x'}}{I_{xx}} \cdot y'(i) + \frac{M_{y'}}{I_{yy}} \cdot x'(i) + \frac{F_z}{\sum\limits_{i=1}^{N} B(i)}$$

(7.69)

The axial load in the z direction is

$$P_z(i) = \sigma_z(i) \cdot B(i)$$

(7.70)

When panels are tapered, swept, or have dihedral, the booms can have nonzero slopes $\Delta x / \Delta z$ and $\Delta y / \Delta z$. The $P_x(i)$ and $P_y(i)$ axial-stress components are

$$P_y(i) = P_z(i) \cdot \frac{\Delta y(i)}{\Delta z} \quad \text{and} \quad P_x(i) = P_z(i) \cdot \frac{\Delta x(i)}{\Delta z(i)}$$

(7.71)

where

$$\Delta x(i) = x_2(i) - x_1(i), \ \Delta y(i) = y_2(i) - y_1(i), \quad \text{and}$$
$$\Delta z = z_2(1) - z_1(1)$$

(7.72)

The subscript 1 is the starting panel close to the root, and 2 is the end panel of the segment near the tip. From this, the total axial force of the boom is found by

$$P(i) = P_z(i) \cdot \frac{\sqrt{\Delta x(i)^2 + \Delta y(i)^2 + \Delta z(i)^2}}{\Delta z}$$

(7.73)

Step 6: Calculate the Shear Forces

The shear force acting on the panel is divided into x and y components S_x and S_y, respectively. The total shear forces are reacted by the combination of the web shear forces and the axial loads acting along the slope of the taper. This is given by

$$S_x = S_{w,x} + \sum_{i=1}^{N_B} P_x(i) \quad \text{and} \quad S_y = S_{w,y} + \sum_{i=1}^{N_B} P_y(i) \tag{7.74}$$

Solving for the web shear stress, we get

$$S_{w,x} = S_x - \sum_{i=1}^{N_B} P_x(i) \quad \text{and} \quad S_{w,y} = S_y - \sum_{i=1}^{N_B} P_y(i) \tag{7.75}$$

Step 7: Calculate the Shear Flow in Each Web

The total shear flow across panel i, $q_s(i)$, has a basic shear flow component $q_b(i)$ and an additive shear flow $q_{s,0}$:

$$q_s(i) = q_{s,0} + q_b(i) \tag{7.76}$$

The basic shear flow for a closed section with panels progressing in a clockwise manner is

$$q_b(i) = -\left(\frac{S_{w,x} \cdot I_{xx} - S_{w,y} \cdot I_{xy}}{I_{xx} \cdot I_{yy} - I_{xy}^2}\right) \cdot \sum_{j=1}^{i} B(j) \cdot x'(j)$$

$$- \left(\frac{S_{w,y} \cdot I_{yy} - S_{w,x} \cdot I_{xy}}{I_{xx} \cdot I_{yy} - I_{xy}^2}\right) \cdot \sum_{j=1}^{i} B(j) \cdot y'(j) \tag{7.77}$$

For a horizontally or vertically symmetric cross section ($I_{xy} = 0$), the basic shear flow becomes

$$q_b(i) = -\frac{S_{w,x}}{I_{yy}} \cdot \sum_{j=1}^{i} B(j) \cdot x'(j) - \frac{S_{w,y}}{I_{xx}} \cdot \sum_{j=1}^{i} B(j) \cdot y'(j) \tag{7.78}$$

When the panel geometry progresses in a counterclockwise direction, $q_b(i)$ is the negative of the preceding values.

To solve the basic shear flow around the closed cross section, an arbitrary panel must be cut such that the basic shear flow is zero. Here, the last panel is cut, and so $q_b(N) = 0$. The preceding equations for $q_b(i)$ are used for all other panels.

To solve for $q_{s,0}$, the following equation must be satisfied:

$$M_{z'} + \oint p \cdot q_b \cdot ds + 2 \cdot A \cdot q_{s,0} - \sum_{i=1}^{N} P_x(i) \cdot y' + \sum_{i=1}^{N} P_y(i) \cdot x' = 0 \tag{7.79}$$

where M_z is the moment about the z axis, p is the perpendicular distance between the member and the center of the line of action, ds is the differential closed-section perimeter length, and A is the enclosed area. This expression shows that the moment generated by the shear flow in combination with the moments generated by the booms reacts the applied moments. Solving for $q_{s,0}$ yields

$$q_{s,0} = \left. -M_{z'} - \oint p \cdot q_b \cdot ds + \sum_{i=1}^{N} P_x(i) \cdot y' - \sum_{i=1}^{N} P_y(i) \cdot x' \middle/ 2 \cdot A \right. \quad (7.80)$$

The basic shear flow integral term can be expressed as

$$\oint p \cdot q_b \cdot ds = \sum_{i=1}^{N-1} \left[S_{y,b}(i) \cdot x'(i) - S_{x,b}(i) \cdot y'(i) \right] \quad (7.81)$$

where $S_{x,b}(i)$ and $S_{y,b}(i)$ are the shear forces due to the basic shear flow in the x and y directions, respectively, acting on the web. For a clockwise panel progression, these shear-force components can be found by

$$S_{x,b}(i) = -q_b(i) \cdot [x'(i+1) - x'(i)]$$

and

$$S_{y,b}(i) = -q_b(i) \cdot [y'(i+1) - y'(i)] \quad (7.82)$$

These shear-force components are negative if the panel progression is counterclockwise.

The enclosed area is found by

$$A = \frac{|x'(N) + x'(1)|}{2} \cdot |y'(N) - y'(1)|$$
$$+ \sum_{i=1}^{N-1} \frac{|x'(i) + x'(i+1)|}{2} \cdot |y'(i) - y'(i+1)| \quad (7.83)$$

Note that this equation is applicable if the minimum and maximum points lie on the centroid. Alternatively, the area can be found by generating triangles from the centroid to the bounding boom locations for each panel.

The only undetermined term is $q_{s,0}$, which can be found by

$$q_{s,0} = \left. -M_{z'} - \oint p \cdot q_b \cdot ds + \sum_{i=1}^{N} P_x(i) \cdot y' - \sum_{i=1}^{N} P_y(i) \cdot x' \middle/ 2 \cdot A \right. \quad (7.84)$$

With all of the parameters known, $q_{s,0}$ is found. Now the shear flow for every element $q_s(i)$ can be calculated as the sum of $q_b(i)$ and $q_{s,0}$:

$$q_s(i) = q_{s,0} + q_b(i) \quad (7.85)$$

To verify that the shear flow is correct, the shear forces acting on the webs for a clockwise panel progression should equal

$$q_s(N) \cdot [x'(1) - x'(N)] + \sum_{i=1}^{N-1} q_s(i) \cdot [x'(i+1) - x'(i)] = S_{w,x} \qquad (7.86)$$

$$q_s(N) \cdot [y'(1) - y'(N)] + \sum_{i=1}^{N-1} q_s(i) \cdot [y'(i+1) - y'(i)] = S_{w,y} \qquad (7.87)$$

These right-hand-side values are negative of those just shown if the panel progression is counterclockwise.

Step 8: Calculate the Shear Stress in Each Web or Find Shear Web Thickness

The web shear stress for a known thickness t_S is

$$\tau_w(i) = \frac{q_s(i)}{t_S(i)} \qquad (7.88)$$

If this is the sizing case for the shear member, then the shear web thickness is

$$t_S(i) = \frac{|q_s(i)|}{\tau_{\max}} \cdot FOS \qquad (7.89)$$

where *FOS* is the factor of safety.

Step 9: Calculate the Actual Skin Thickness for Each Panel

The portion of the skin thickness that is caused by bending and axial stresses is from the booms, which will be denoted by t_B. The i index for the boom-panel thickness corresponds with the web index.

A simple approximation to relate t_B to the adjacent boom areas is

$$t_B(i) = \frac{B(i) + B(i+1)}{2 \cdot b(i)} \qquad (7.90)$$

The total skin thickness, if all of the shear and bending material is converted to skin, is

$$t_{\text{Tot}}(i) = t_B(i) + t_S(i) \qquad (7.91)$$

If this skin thickness is less than the minimum gauge thickness, then the skin thickness must be set to the minimum gauge value.

7.5.3 Finite Element Modeling

FEM is a detailed structural analysis method. The skin-panel method and boom-and-web method lend themselves well to conceptual and preliminary design structural sizing. However, these simple methods have coarse grids and simplifying assumptions that obscure the true behavior of the structure. Modern aerospace structures are usually sized with FEM techniques.

FEM tools break the structure into a fine mesh of structural elements. The element properties are based on the structural geometry, material properties, and structural thickness. These element properties are represented in a large matrix. Loads are applied, and a matrix solver finds the stresses and strains for each of the elements. The element mesh should be finest where the geometry or loads change most quickly. Computational loads and corresponding runtimes are high to generate a single solution, and so FEM techniques are not suitable for conceptual design with current computer technology. NASTRAN is an example FEM structural analysis tool that is in widespread use.

7.6 Wing Sizing

Unmanned aircraft conceptual design is often only concerned with the structural weight rather than details of the structure itself. Initial weight allocations are established from the conceptual unmanned aircraft sizing analysis, and then detailed FEM structural analyses are performed off-line in preliminary and detail design phases. In Chapter 6, we showed parametric weight estimating relationships based on a handful of geometric and other characteristics. In this section, you will see a semi-analytic wing sizing method that bridges the gap between parametric relationships and FEM. This semi-analytic method performs structural sizing based on the wing geometry, loads, and structural materials based on allowable stresses and deflections.

7.6.1 Wing Structural Geometry Denition

The wing geometry and loads must be defined within the wing coordinate system. The origin is at the wing root at the vertical plane of symmetry. A good x coordinate is the leading edge of the wing root. The coordinate system is specific to a wing, and so the system is redefined for each wing if there are multiple wings or other horizontal surfaces. The wing structural frame axes are parallel to those of the mass properties frame. The geometric transformation from the mass properties to the wing structural coordinate system is

$$x = x_{mp} - x_{LE,mp} \qquad y = y_{mp} - y_{LE,mp} \qquad z = z_{mp} \qquad (7.92)$$

where subscript mp denotes the mass properties frame and N is the nose of the fuselage. At a given point, the forces and moments in the fuselage frame are

$$F_x = F_{x,mp} \qquad F_y = F_{y,mp} \qquad F_z = F_{z,mp} \qquad (7.93)$$

$$M_x = M_{x,mp} \qquad M_y = M_{y,mp} \qquad M_z = M_{z,mp} \qquad (7.94)$$

The wing semispan (half of the wing) is divided into N segments along the span. The average chord $c(k)$, the panel span $b_{Seg}(k)$, and the x, y, z coordinates of the midsection quarter-chord are known for each panel.

7.6.2 Wing Loads

The primary wing loads can include the following:

- Lift loads at the maximum positive and negative lifting conditions
- Inertial forces acting on the wing due to the wing structure and components attached to the wing (This is important for the taxi bump condition and for launch loads.)
- Main landing-gear shock loads at touchdown
- Wing-mounted engine thrust and weight loads
- Hardpoint loads

Some of the loads acting upon a wing are shown in Fig. 7.9.

The lift load distribution can be calculated from a lift method such as lifting-line theory, vortex-lattice method, or other suitable lift distribution algorithm. However, it is often expedient to separate the structural sizing code from the aerodynamics code. To achieve this arrangement, a lift distribution must be approximated. If we assume an elliptic lift distribution, the lift at each panel $L(k)$ is found by

$$L(k) = L_{Max} \cdot \frac{b_{Seg}(k)}{b_{Struct}} \cdot \frac{4}{\pi} \cdot \sqrt{1 - \frac{2 \cdot z(k)}{b_{Struct}}} \qquad (7.95)$$

The maximum lift of the main wing can be estimated by

$$L_{Max} = N_{Ult} \cdot W_{TO} \qquad (7.96)$$

where N_{Ult} is the ultimate load factor. This assumes that the horizontal stabilizer downforce or canard lift is negligible. The maximum positive and negative loads are found by a velocity-load gust analysis, known as a V-n diagram.

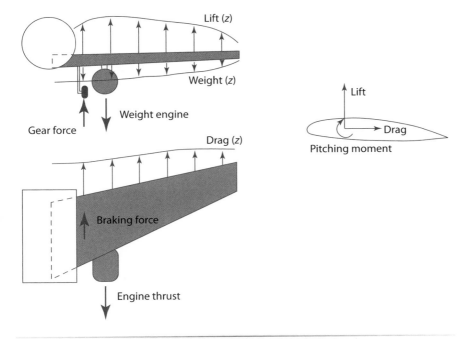

Fig. 7.9 Forces acting upon a wing.

The inertial loads on the wing act opposite of the wing positive lift force, reducing the total vertical force acting upon the wing. This is known as inertia relief. The inertial loads are adverse during taxi bump and negative lift cases.

Total forces and moments acting at a section are found through integration of the outboard panels. The forces are

$$F_{x,\text{Tot}}(k) = \frac{D(k)}{2} - \frac{W(k)}{2} \cdot N_x + \sum_{i=N}^{k+1} D(i) - W(i) \cdot N_x \qquad (7.97)$$

$$F_{y,\text{Tot}}(k) = \frac{[L(k) - W(k) \cdot N_y]}{2} + \sum_{i=N}^{k+1} [L(i) - W(i) \cdot N_y] \qquad (7.98)$$

$$F_{z,\text{Tot}}(k) = 0 \qquad (7.99)$$

The lift force is assumed to act normal to the x-z plane, which is consistent with wings with very little dihedral. Note that N_y is the acceleration acting in the y direction in the wing structures' coordinate system. This corresponds to N_z in the body frame. N_y is equal to 1 for straight and level flight. N_x is acceleration along the x axis. Positive N_x acts to slow the unmanned aircraft.

The moments about the x, y, and z axes at panel k are

$$M_x(k) = -\frac{\left[L(k) - W(k) \cdot N_y\right]}{4} \cdot b_{\text{Seg}}(k) - \sum_{i=N}^{k+1} F_y(i) \cdot \left[z(i) - z(k)\right] \qquad (7.100)$$

$$M_y(k) = \frac{\left[D(k) - W(k) \cdot N_x\right]}{4} b_{\text{Seg}}(k) + \sum_{i=N}^{k+1} F_x(i) \cdot \left[z(i) - z(k)\right] \qquad (7.101)$$

$$M_z(k) = \frac{L(k) \cdot x_{c/4}(k) - W(k) \cdot N_y \cdot x_W(k)}{2} + \frac{W(k) \cdot N_x \cdot y_W(k)}{2} - \frac{M_{\text{Pitch}}(k)}{2}$$

$$+ \sum_{i=N}^{k+1} \left[L(i) \cdot x_{c/4}(i) - W(i) \cdot N_y \cdot x_W(i) + W(i) \cdot N_x \cdot y_W(i) - M_{\text{Pitch}}(i)\right]$$

$$(7.102)$$

Here $x_W(i)$ is the x location that the weight of panel i acts upon for panel i. This can be estimated as the 50% chord point or the centroid of the panel. The pitching moment for a segment is

$$M_{\text{Pitch}}(i) = C_m(i) \cdot q \cdot c(i)^2 \cdot b_{\text{Seg}}(i) \qquad (7.103)$$

where $C_m(i)$ is the pitching-moment coefficient of panel i.

If the panel is inboard of a discrete weight $W_{\text{Comp}}(i)$ located at $z_{\text{Comp}}(i)$, $y_{\text{Comp}}(i)$, $z_{\text{Comp}}(i)$, then the following are added to the forces and moments:

$$\Delta F_{x,\text{Comp}}(k) = W_{\text{Comp}}(k) \cdot N_x \qquad (7.104)$$

$$\Delta F_{y,\text{Comp}}(k) = -W_{\text{Comp}}(i) \cdot N_y \qquad (7.105)$$

$$\Delta M_{x,\text{Comp}}(k) = -W_{\text{Comp}}(i) \cdot \left[z_{\text{Comp}}(i) - z(k)\right] \cdot N_y \qquad (7.106)$$

$$\Delta M_{y,\text{Comp}}(k) = W_{\text{Comp}}(k) \cdot N_x \cdot \left[z_{\text{Comp}}(i) - z(k)\right] \qquad (7.107)$$

$$\Delta M_{z,\text{Comp}}(k) = -W_{\text{Comp}}(i) \cdot N_y \cdot x_{\text{Comp}}(i) - W_{\text{Comp}}(i) \cdot N_x \cdot y_{\text{Comp}}(i) \quad (7.108)$$

If a panel is inboard of an engine with weight W_{Eng}, thrust T, and location z_{Eng}, y_{Eng}, z_{Eng}, then the changes to the forces and moments are

$$\Delta F_{x,\text{Eng}}(k) = -T + W_{\text{Eng}} \cdot N_x \qquad (7.109)$$

$$\Delta F_{y,\text{Eng}}(k) = -W_{\text{Eng}} \cdot N_y \qquad (7.110)$$

$$\Delta M_{x,\text{Eng}}(k) = -W_{\text{Eng}} \cdot N_y \cdot \left[z_{\text{Eng}} - z(k)\right] \qquad (7.111)$$

$$\Delta M_{y,\text{Eng}}(k) = \left(W_{\text{Eng}} \cdot N_x - T\right) \cdot \left[z_{\text{Eng}} - z(k)\right] \qquad (7.112)$$

$$\Delta M_{z,\text{Eng}}(k) = -W_{\text{Eng}} \cdot N_y \cdot x_{\text{Eng}} - W_{\text{Eng}} \cdot N_x \cdot y_{\text{Eng}} + T \cdot y_{\text{Eng}} \qquad (7.113)$$

When the landing gear impacts the ground, it has an upward impact force F_{Gear}, braking force F_{Brake}, and weight W_{Gear}. The contact point between the landing gear and the runway is x_{Gear}, y_{Gear}, z_{Gear}. We will neglect gear side forces for this discussion. Assuming that the z location of the landing-gear weight is also z_{Gear}, the changes to the forces and moments are

$$\Delta F_{x,\text{Gear}}(k) = F_{\text{Brake}} + W_{\text{Gear}} \cdot N_x \tag{7.114}$$

$$\Delta F_{y,\text{Gear}}(k) = -W_{\text{Gear}} \cdot N_y + F_{\text{Gear}} \tag{7.115}$$

$$\Delta M_{x,\text{Gear}}(k) = \left(F_{\text{Gear}} - W_{\text{Gear}} \cdot N_y\right) \cdot [z_{\text{Gear}} - z(k)] \tag{7.116}$$

$$\Delta M_{y,\text{Gear}}(k) = \left(F_{\text{Brake}} + W_{\text{Eng}} \cdot N_x\right) \cdot [z_{\text{Gear}} - z(k)] \tag{7.117}$$

$$\Delta M_{z,\text{Gear}}(k) = \left(F_{\text{Gear}} - W_{\text{Gear}} \cdot N_y\right) \cdot x_{\text{Gear}} - \left(F_{\text{Brake}} + W_{\text{Gear}} \cdot N_x\right) \cdot y_{\text{Gear}} \tag{7.118}$$

A winglet has weight W_{WL}, produces thrust T_{WL}, and generates lifting force acting parallel to the z axis L_{WL}. The winglet lifting force is negative for upward winglets. Assuming that the winglet aerodynamic center and the center of gravity are both at x_{WL}, y_{WL}, z_{WL}, and that the pitching moments are negligible, we get

$$\Delta F_{x,\text{WL}}(k) = -T_{\text{WL}} + W_{\text{WL}} \cdot N_x \tag{7.119}$$

$$\Delta F_{y,\text{WL}}(k) = -W_{\text{WL}} \cdot N_y \tag{7.120}$$

$$\Delta F_{z,\text{WL}}(k) = L_{\text{WL}} \tag{7.121}$$

$$\Delta M_{x,\text{WL}}(k) = -W_{\text{WL}} \cdot N_y \cdot [z_{\text{WL}} - z(k)] + L_{\text{WL}} \cdot y_{\text{WL}} \tag{7.122}$$

$$\Delta M_{y,\text{WL}}(k) = (W_{\text{WL}} \cdot N_x - T_{\text{WL}}) \cdot [z_{\text{WL}} - z(k)] - L_{\text{WL}} \cdot x_{\text{WL}} \tag{7.123}$$

$$\Delta M_{z,\text{WL}}(k) = -W_{\text{WL}} \cdot N_y \cdot x_{\text{WL}} + (T_{\text{WL}} - W_{\text{WL}} \cdot N_x) \cdot y_{\text{WL}} \tag{7.124}$$

The total forces and moments are calculated as

$$F_x(k) = F_x(k) + \sum_{i=N}^{k+1} \Delta F_{x,\text{Comp}}(i) + \Delta F_{x,\text{Eng}}(k) + \Delta F_{x,\text{Gear}}(k) + \Delta F_{x,\text{WL}}(k) \tag{7.125}$$

$$F_y(k) = F_y(k) + \sum_{i=N}^{k+1} \Delta F_{y,\text{Comp}}(i) + \Delta F_{y,\text{Eng}}(k) + \Delta F_{y,\text{Gear}}(k) + \Delta F_{y,\text{WL}}(k) \tag{7.126}$$

$$F_z(k) = F_z(k) + \sum_{i=N}^{k+1} \Delta F_{z,\text{Comp}}(i) + \Delta F_{z,\text{Eng}}(k) + \Delta F_{z,\text{Gear}}(k) + \Delta F_{z,\text{WL}}(k) \tag{7.127}$$

$$M_x(k) = M_x(k) + \sum_{i=N}^{k+1} \Delta M_{x,\text{Comp}}(i) + \Delta M_{x,\text{Eng}}(k) + \Delta M_{x,\text{Gear}}(k) + \Delta M_{x,\text{WL}}(k) \tag{7.128}$$

$$M_y(k) = M_y(k) + \sum_{i=N}^{k+1} \Delta M_{y,\text{Comp}}(i) + \Delta M_{y,\text{Eng}}(k) + \Delta M_{y,\text{Gear}}(k) + \Delta M_{y,\text{WL}}(k)$$

(7.129)

$$M_z(k) = M_z(k) + \sum_{i=N}^{k+1} \Delta M_{z,\text{Comp}}(i) + \Delta M_{z,\text{Eng}}(k) + \Delta M_{z,\text{Gear}}(k) + \Delta M_{z,\text{WL}}(k)$$

(7.130)

Example 7.3 Wing Loads

Problem:

We wish to calculate the wing loads for a 1000-N unmanned aircraft. The wing of interest has a single taper, a root chord of 1 m, a taper ratio of 0.5, and a span of 10 m. The wing root leading-edge location in mass properties coordinates is $x_{mp} = 5$ m, $y_{mp} = 0$ m, and $z_{mp} = 1$ m. The wing has no sweep about the quarter-chord.

In horizontal flight, the lift distribution in mass properties coordinates is

$$\frac{L(y_{mp})}{y_{mp}} = \frac{W_{TO} \cdot N_z}{b} \cdot \frac{2 \cdot c(y_{mp})}{(c_r + c_t)} \ N/m$$

Note that N_z is also in mass properties coordinates. The total wing weight is 100 N, which is distributed evenly over the span. Assume that the weights act upon the 50% chord line.

The wing lift-to-drag ratio is 40:1, and the drag is assumed to be distributed over the wing evenly. A zero-pitching moment flying-wing airfoil is used.

The wings have hardpoints, which are capable of carrying 5 kg each. The hardpoint location in mass properties' coordinates is (5 m, ± 3 m, 0 m).

The landing-gear loads all go through the fuselage.

Divide the wing into five panels per side. Calculate the forces and moments acting at the quarter-chord of the midchord of each segment. Assume that the unmanned aircraft is operating at $N_z = 1$ G.

Solution:

First, the geometry must be defined in the structures coordinates. The segment midpoint chord and midpoint centroid in structures coordinates are as follows:

Segment	Midchord, m	$x_{c/4}$, m	$y_{c/4}$, m	$z_{c/4}$, m
1	0.95	0.25	0	0.5
2	0.85	0.25	0	1.5
3	0.75	0.25	0	2.5
4	0.65	0.25	0	3.5
5	0.55	0.25	0	4.5

The lift, weight, and drag forces generated by each segment are as follows:

Segment	L(k), N	W(k), N	D(k), N
1	126.7	10	2.5
2	113.3	10	2.5
3	100.0	10	2.5
4	86.7	10	2.5
5	73.3	10	2.5

The equations for forces and moments are simplified because $N_x = 0$ and $N_y = 1$. The change to forces and moments due to the hardpoint masses (denoted by subscript hp) at the segment quarter-chord locations are given by

$$\Delta F_{y,\text{hp}}(k) = -W_{\text{hp}}$$
$$\Delta M_{x,\text{hp}}(k) = -W_{\text{hp}} \cdot \left[z_{\text{hp}} - z(k)\right] \quad \text{when } z_{\text{hp}} > z(k)$$
$$\Delta M_{z,\text{hp}}(k) = -W_{\text{hp}} \cdot x_{\text{hp}} \quad \text{when } z_{\text{hp}} > z(k)$$

The total forces and moment acting on each segment quarter-chord are

$$F_{x,\text{Tot}}(k) = \frac{D(k)}{2} + \sum_{i=N}^{k+1} D(i)$$

$$F_{y,\text{Tot}}(k) = \frac{[L(k) - W(k)]}{2} + \sum_{i=N}^{k+1} [L(i) - W(i)] + \Delta F_{y,\text{hp}}(k)$$

$$M_x(k) = -\frac{[L(k) - W(k)]}{4} \cdot b_{\text{Seg}}(k) - \sum_{i=N}^{k+1} F_y(i) \cdot [z(i) - z(k)] + \Delta M_{x,\text{hp}}(k)$$

$$M_y(k) = \frac{D(k)}{4} b_{\text{Seg}}(k) + \sum_{i=N}^{k+1} F_x(i) \cdot [z(i) - z(k)]$$

$$M_z(k) = \frac{L(k) \cdot x_{c/4}(k) - W(k) \cdot x_W(k)}{2} - \frac{M_{\text{Pitch}}(k)}{2}$$

$$+ \sum_{i=N}^{k+1} \left[L(i) \cdot x_{c/4}(i) - W(i) \cdot x_W(i) - M_{\text{Pitch}}(i)\right] + \Delta M_{z,\text{hp}}(k)$$

The values are shown in the following table:

Segment	F_x, N	F_y, N	F_z, N	M_x, N-m	M_y, N-m	M_z, N-m
1	11.25	342.63	0	−989.91	38.13	89.73
2	8.75	232.63	0	−435.62	18.13	64.48
3	6.25	135.97	0	−162.98	6.88	42.31
4	3.75	101.67	0	−50.83	1.88	23.23
5	1.25	31.67	0	−15.83	0.63	7.23

The M_x, M_y, and M_z moments must be shifted to the neutral axis for each segment for structural analysis.

7.6.3 Wing Primary Structure Sizing

Using the skin-panel method, the wing is divided into upper and lower caps and a shear web. The upper and lower skin contours are generated separately because they are discontinuous. It is reasonable to assume that the combined angle of attack and angle of incidence are negligible for defining the structures.

The wing box usually starts aft of the leading edge and ends short of the trailing edge. The upper and lower skins correspond with the chordwise locations of the wing box start and end. A single shear web located at the maximum thickness point is used to help approximate the shear material weight. The relationship between the actual wing box and the idealized wing structure for a three-panel segment is shown in Fig. 7.10.

The paneling should start from the section leading edge.

$$x(x/c) = x_{LE} + (x/c) \cdot c \tag{7.131}$$

$$y_{Up}(x/c) = z_{LE} + \frac{y_{Up}}{c}(x/c) \cdot c \tag{7.132}$$

$$y_{Low}(x/c) = z_{LE} + \frac{y_{Low}}{c}(x/c) \cdot c \tag{7.133}$$

where x_{LE} and z_{LE} are in structural coordinates.

Weights of nonstructural components must be allocated to the panels that contain them. For example, the wing can contain actuators, hardpoints, antennas, or engine nacelles. The cumulative weight of each of the components $W_{Comp}(i)$ contained within panel k is $W_{NonStruct}(k)$. These weights must be input or calculated prior to the wing weight calculation. The placement of these components on the wing is generally defined by parametric geometric relationships. If the components are not located at

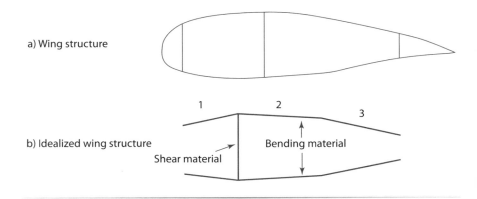

a) Wing structure

b) Idealized wing structure

Shear material

Bending material

1   2   3

Fig. 7.10   Wing paneling.

Example 7.4   Wing Panel

Problem:

We wish to calculate the required bending material thickness and shear material weight for a wing segment using the skin-panel method. The wing upper and lower coordinates for the wing box are as follows:

X, m	y_t, m	Y_b, m
0.1	0.05	−0.05
0.3	0.08	−0.08
0.5	0.07	−0.07
0.7	0.05	−0.05

The following information is known about the segment:

* Forces: $F_x = 1,000$ N, $F_y = 50,000$ N, $F_z = 0$
* Moments: $M_x = -10,000$ N-m, $M_y = 0$ N-m, $M_z = 1,000$ N-m
* Location about which the forces act: $x_{FM} = 0.25$ m, $y_{FM} = 0$ m, $z_{FM} = 0$ m
* Coordinates of the midpoint leading edge: $x_{LE} = 0$ m, $y_{LE} = 0$ m, $z_{LE} = 0$ m
* Midpoint chord is 1 m, and the segment is 1 m long (segment is not tapered).
* Assume that the airfoil angle of attack and angle of incidence are both zero.
* Material is 2024 Aluminum with a density of 0.1 lb/in.^{3}, ultimate tensile strength of 61 ksi, and ultimate shear strength of 37 ksi, and factor of safety of 1.5.

Find the following:

a. Generate wing paneling with three upper and lower panels.
b. What is the skin thickness required to counteract bending and axial loads?
c. With a single shear web at the maximum depth of the wing box, what is the shear material thickness?
d. What are the bending and shear material weights?

Solution:

a. The upper skin panels follow the upper surface. The coordinates for the upper surface panels and panel lengths are as follows (The panel coordinates for the lower surface are the same, except the *y* values are negative.):

Panel	Starting node			Ending node		
	x, m	y, m	b, m	x, m	y, m	b, m
1	0.1	0.05	0.2022	0.3	0.08	0.2022
2	0.3	0.08	0.2002	0.5	0.07	0.2002
3	0.5	0.07	0.2010	0.7	0.05	0.2010

b. The first step is to calculate the moment of inertia normalized by the skin thickness. These are

$$I_{xx,t} = 0.004501 \text{ m}^3 \qquad I_{yy,t} = 0.021149 \text{ m}^3 \qquad I_{xy,t} = 0 \text{ m}^3$$

The centroid location is

$$x_c = 0.4 \text{ m}, \quad y_c = 0 \text{ m}$$

The moments about the centroid are

$$M'_x = -10,000 \text{ N-m}, \quad M'_y = 0 \text{ N-m}, \quad M'_z = -6,479 \text{ N-m}$$

The stresses at the midpoints of the upper and lower panels are the following:

Panel	y_{mid} Upper, m	$\sigma_{z,t}$ Upper, kPa/m	y_{mid} Lower, m	$\sigma_{z,t}$ Lower, kPa/m
1	0.065	−144.4	−0.065	−144.4
2	0.075	−166.6	−0.075	−166.6
3	0.035	−77.7	−0.035	−77.7

The absolute value of the maximum stress divided by thickness is 166.6 kPa/m. The bending thickness is then found by

$$t_{bend} = \frac{Max[|\sigma_{z,t}(i)|]}{\sigma_{Ult}} \cdot FOS = \frac{166.6 \text{ kPa/m}}{420,500 \text{ kPa}} \cdot 1.5 = 0.594 \text{ mm}$$

Note that the ultimate strength was converted from kilopounds per square inch to kilopascals and the thickness is expressed in millimeters rather than meters.

c. The shear web length is equal to the segment length because the panel is not tapered. The shear web length is 1 m. The height of the shear web is equal to the maximum depth, which is 0.16 m. The shear flow equation is

$$q = -\left(\frac{S_x \cdot I_{xx} - S_y \cdot I_{xy}}{I_{xx} \cdot I_{yy} - I_{xy}^2}\right) \cdot \sum_{i=1}^{N} t_{bend} \cdot b_{Up}(i) \cdot x'_{m,Up}(i)$$

$$- \left(\frac{S_y \cdot I_{yy} - S_x \cdot I_{xy}}{I_{xx} \cdot I_{yy} - I_{xy}^2}\right) \cdot \sum_{i=1}^{N} t_{bend} \cdot b_{Up}(j) \cdot y'_{m,Up}(i)$$

The shear forces are $S_y = 50,000$ N, and $S_x = 0$ N. Because both I_{xy} and S_x are equal to zero, the shear flow equation simplifies to

$$q = -\frac{S_y}{I_{xx}} \cdot \sum_{i=1}^{N} t_{bend} \cdot b_{Up}(j) \cdot y'_{m,Up}(i)$$

Now that the bending material thickness is known, the moment of inertia is $I_{xx} = 2.675 \times 10^{-6} \text{ m}^4$. The shear flow is -389.6 kPa. The shear material thickness is

$$t_{shear} = \frac{|q|}{\tau_{Ult}} \cdot FOS = \frac{389.6 \text{ kPa/m}}{255{,}100 \text{ kPa}} \cdot 1.5 = 2.29 \text{ mm}$$

Note that the ultimate shear was converted from kilopounds per square inch to kilopascals, and the thickness was converted from meters to millimeters.

d. The bending material weight is found by

$$W_{bend} = \rho_{bend} \cdot t_{bend} \cdot \Delta z \cdot \left[\sum_{i=1}^{N} b_{Up}(i) + \sum_{i=1}^{N} b_{Low}(i) \right]$$

$$= 2768 \text{ kg/m}^3 \cdot 0.000594 \text{ m} \cdot 1 \text{ m} \cdot 1.207 \text{ m} = 1.985 \text{ kg}$$

The shear material weight is found by

$$W_{shear} = \rho_{shear} \cdot t_{shear} \cdot h_{web} \cdot L_{web}$$

$$= 2768 \text{ kg/m}^3 \cdot 0.00229 \text{ m} \cdot 0.16 \text{ m} \cdot 1 \text{ m} = 1.01 \text{ kg}$$

Note that "kg" is used to represent weight, as is commonly done. Technically kilogram is mass, and so here it represents kilogram at one times the acceleration of gravity.

the center of the panel, then the bending moment due to the component moment arm should be calculated. The sum of these bending moments for each panel is $M_{B,NonStruct}(k)$.

Wing structures can be driven by stiffness as well as strength. High-aspect-ratio wings that have sufficient strength to handle the flight loads can have excessive deflections in flight and during ground operations. Additional structural material beyond the strength-based load cases might be required to counteract these deflections, making the structure stiffness-driven.

The y deflection of a beam with $I_{xy} = 0$ can be found through integration of the equation for an elastic curve

$$y''(z) = -\frac{M_x(z)}{E \cdot I_{xx}} \tag{7.134}$$

where y'' is the second derivative of the elastic axis y location with respect to z. The slope y' is found through the integration

$$y'(z) = y'_0 - \int_0^z \frac{M_x(z)}{E \cdot I_{xx}(z)} \cdot dz \tag{7.135}$$

where y_0' is the elastic axis slope dy/dz at the wing root. The y position of the elastic axis at z is

$$y(z) = y_0 + y_0' \cdot z - \int\limits_0^z \left[\int\limits_0^z \frac{M_x(z)}{E \cdot I_{xx}(z)} \cdot dz \right] dz \qquad (7.136)$$

We are interested in calculating the deflection for a wing with a discrete number of spanwise segments. The bending moment $M_x(k)$ at the center of the segment, moment of inertia $I_{xx}(k)$, and the segment width $\Delta z(k)$ are known for each segment k. The first step in the numerical integration is to find the change in slope at segment k caused by the bending moment $\Delta y'(k)$ by integrating from the root:

$$\Delta y'(k) = -\sum_{i=1}^k \frac{M_x(i)}{E \cdot I_{xx}(i)} \cdot \Delta z(i) \qquad (7.137)$$

The change in y position of the neutral axis for a wing segment k tip $\Delta y(k)$ under loading can be approximated by numerical integration:

$$\Delta y(k) = \sum_{i=1}^k \Delta y'(i) \cdot \Delta z(i) \qquad (7.138)$$

The deflected position of each segment tip $y_{\mathrm{Tip,Defl}}(k)$ is found by adding the zero-load tip y position $y_{\mathrm{Tip}}(k)$ to the change due to loading:

$$y_{\mathrm{Tip,Defl}}(k) = y_{\mathrm{Tip}}(k) + \Delta y(k) \qquad (7.139)$$

A more general approach is to calculate the slopes for each segment without any symmetry approximations. The relationships for x'' and y'' are

$$x'' = \frac{-I_{xx} \cdot M_y + I_{xy} \cdot M_x}{E \cdot \left(I_{xx} \cdot I_{yy} - I_{xy}^2 \right)} \qquad (7.140)$$

$$y'' = \frac{-I_{yy} \cdot M_x + I_{xy} \cdot M_y}{E \cdot \left(I_{xx} \cdot I_{yy} - I_{xy}^2 \right)} \qquad (7.141)$$

These x'' and y'' relationships can be numerically integrated from the wing root to the tip to find the total deflections at the tip. The vertical tip deflection is of greater interest than the horizontal deflection. The vertical deflection is almost always larger because I_{xx} is smaller than I_{yy}, and M_x has a larger magnitude than M_y for most wings. A large vertical deflection in flight increases the effective wing dihedral, which impacts the lateral-directional handling qualities. Taxi bump loads, wing fuel loads, and negative lift loads during takeoff roll can cause wing-tip interference with the ground.

7.6.4 Wing Secondary Structure Sizing

Secondary structures are structures that do not carry the primary loads. Failure of secondary structure might not result in the loss of an unmanned aircraft. Secondary structures include the following:

- *Trailing edges*—Trailing-edge structure is usually not part of the main wing box structure. This structure can be rigidly attached to the wing box, or it can be a control surface. The trailing edge must transfer the aerodynamic loads to the wing box through attachment interfaces. The downstream end of the trailing edge comes to a sharp point, and so the structural arrangement must accommodate the upper and lower surfaces coming together.
- *Leading edges*—The leading-edge structure length is generally 5–15% of the chord length. This is not part of the primary wing box structure if leading-edge devices such as slats or slots are employed. So-called *clean leading edges* can be included in the primary wing box as the most forward cell.
- *Ribs*—Ribs are vertical members that are generally aligned normal to the span or the elastic axis. These members help prevent buckling and can distribute concentrated loads to the skins. Examples of concentrated loads that ribs can distribute include hardpoints, engine pylon mounts, and landing-gear struts.
- *Stiffeners*—These members run along the wing box much like spar caps. However, the primary function is to help prevent buckling of the upper and lower skins. Stiffeners protrude inwards from the upper and lower skins. These members are most common in metal structures. Composite structures frequently use the depth of the sandwich construction to prevent buckling.
- *Brackets*—These members help transfer concentrated loads to other structural members. Brackets can interface to servos, control surfaces, avionics components, subsystem components, or landing-gear assemblies.
- *Hatches and access panels*—To maintain the unmanned aircraft, any internal areas that require inspection or maintenance must be accessed. Reinforcements are required whenever there is a cutout in the primary structure, giving a higher weight than the uncut structure.
- *Landing-gear doors*—Landing gear that retracts into the wing must penetrate the wing outer mold line. In straight wings, this usually cuts through the main wing box. Swept wings can allow the landing gear to retract aft of the main wing box towards the wing root. In the former case, it is necessary to shift the load paths away from the lower skins and into the spars bounding the landing gear.

Table 7.5 Aeroelastic Phenomena

	Aerodynamic	Elastic	Inertial
Flutter	√	√	√
Buffeting	√	√	√
Dynamic response	√	√	√
Dynamic stability impacts	√	√	√
Control system reversal	√	√	—
Divergence	√	√	—
Control effectiveness impacts	√	√	—
Load distribution impacts	√	√	—
Static stability impacts	√	√	—

7.6.5 Aeroelasticity

High-speed aircraft and those with high-aspect-ratio wings are often sized based on aeroelastic considerations rather than static load cases. Several important aeroelastic phenomena and their dependencies on aerodynamic, elastic, and inertial forces are shown in Table 7.5. Aeroelastic effect often limits the upper bound on wing aspect ratio and the achievable aerodynamic efficiency. A rigorous treatment of aeroelasticity is beyond the scope of this book.

7.7 Fuselage Analysis and Sizing

7.7.1 Fuselage Loads

The dominant fuselage loads are produced in the following ways:

* Horizontal stabilizers and canards produce a vertical force and pitching moment.
* Vertical tails generate a side force when subjected to side force or rudder deflection. Vertical surfaces that are not symmetric about the fuselage generate torsion, along with side force, on the fuselage.
* Aerodynamic loads are generated by the pressure distribution over the fuselage. The wing induces an upwash in the forward fuselage and downwash aft of the wing. Sideslip creates side force due to the body in the front and a side force in the aft fuselage due to the fuselage and vertical stabilizer.
* The inertia of the fuselage structure and the components housed within the fuselage produce bending moments and shear forces.
* Fuselage pressurization generates tensile stress on the fuselage walls, similar to those experienced by a cylindrical pressure vessel. Most UAS

fuselages are not pressurized because there are no humans onboard. However, pressurized avionics and payloads compartments can enable blowing air for electronics cooling.

- Landing gear provides a shock on impact. The nose gear is nearly always affixed to the front of the fuselage. The main gear can be located on the fuselage or the wing. Wing-mounted main gear indirectly transfer a landing load to the fuselage via the wing/fuselage interface structure.
- The load from rail launch acceleration acts in the direction of flight and is imparted via the launcher interface. Typical forward accelerations are $5-20\ g$.
- Rocket launch loads are imparted at the rocket/fuselage interface. The direction of the loads is mostly forward, but with an upward component. When there are two rockets mounted on the fuselage sides, the resulting force usually has an inward component which acts to compress the fuselage sides.
- Depending on the unmanned aircraft size, the fuselage will be lifted by hand or with a hoist. These are ground-handling loads. When humans manipulate the unmanned aircraft, they grab convenient points on the structure. This grip generates a crush load. Hoisting involves lifting the unmanned aircraft by hoist points or in a hoist structure.
- Several key fuselage loads are shown in Fig. 7.11.

The fuselage and loads must be defined within the fuselage coordinate system. The origin is at the nose of the fuselage. Note that this coordinate system is specific to a fuselage, and it must be defined for each fuselage if multiple fuselages are used. Positive x points horizontally out the right wing, positive y is vertical, and positive z points horizontally towards the

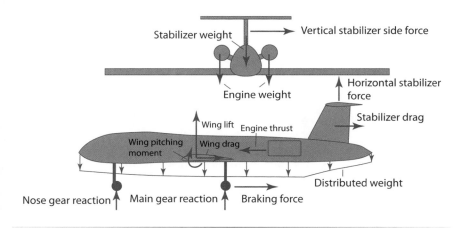

Fig. 7.11 Fuselage loads.

tail. The geometric transformation from the mass properties to the fuselage structural coordinate system is

$$x = y_{mp} \quad y = z_{mp} - z_{N,mp} \quad z = x_{mp} - x_{N,mp} \qquad (7.142)$$

where subscript mp denotes the mass properties' frame and N is the nose of the fuselage. At a given point, the forces and moments in the fuselage frame are

$$F_x = F_{y,mp} \quad F_y = F_{z,mp} \quad F_z = F_{x,mp} \qquad (7.143)$$
$$M_x = M_{y,mp} \quad M_y = M_{z,mp} \quad M_z = M_{x,mp} \qquad (7.144)$$

7.7.2 Fuselage Primary Structural Sizing

Many UAS fuselages consist of a load-bearing skin with bulkheads for reinforcement. Metal structures use stringers to help carry the bending loads along the fuselage. Composite structures use ply orientation to best carry the loads. Bulkheads provide a means of distributing point loads to the skins and stabilize the structure against buckling.

Small composite UASs might have a simple shell structure with constant skin thickness. Hatches and access panels break the shell structure and require local reinforcement. A small number of bulkheads help to distribute loads from the wing and landing impact. For these simple fuselages, weight is a function of surface area and nominal skin thickness.

More complex fuselages that are not dominated by minimum gauge thickness require loads-based structures analysis. The fuselage structure is sized from the loads, structural geometry, and materials properties. Fuselage structures can be approximated by the boom-and-web method. The fuselage is divided into multiple segments along its length, where the division points correspond with bulkhead or frame locations. Example 7.5 shows how to analyze one segment of the fuselage. This procedure can be repeated for all of the fuselage segments that comprise the fuselage structure.

Example 7.5   Boom and Web Fuselage Analysis

Problem:
Let us consider a tapered fuselage segment that has a circular cross section. The fuselage has eight booms, each with a cross section area of 10 mm². Starting from the tail end, the segment end cross section has a 1-m diam, and the beginning cross section has a radius of 1.5-m diam. The z distance between the cross sections is 2 m.

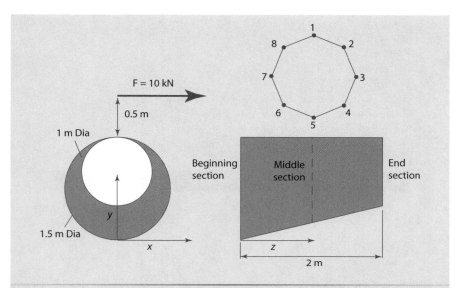

Fig. 7.12 Segment geometry, axes, and forces.

A 10-kN side force generated by the rudder deflection acts at the end cross section. The y height of the force above the top of the end cross section is 0.5 m. The problem is illustrated in Fig 7.12.

The material is 2024 Aluminum with a density of $0.1 \, lb/in.^3$, ultimate tensile strength of 61 ksi, and an ultimate shear strength of 37 ksi. The factor of safety is 1.5. The problem is illustrated in Fig 7.12.

a. Calculate centroid and moments of inertia.
b. Evaluate bending stresses.
c. Is there sufficient bending material?
d. What is the required shear skin thickness?

Solution:

The coordinates of the booms for sections 1, 2, and the middle sections are as follows:

Boom	X			Y			Z		
	1	2	Mid	1	2	Mid	1	2	Mid
1	0	0	0	1.5	1.5	1.5	0	2	1
2	0.5303	0.3536	0.4420	1.2803	1.3536	1.3169	0	2	1
3	0.75	0.5	0.625	0.75	1	0.875	0	2	1
4	0.5303	0.3536	0.4420	0.2197	0.6464	0.4331	0	2	1
5	0	0	0	0	0.5	0.25	0	2	1
6	−0.5303	−0.3536	−0.4420	0.2197	0.6464	0.4331	0	2	1
7	−0.75	−0.5	−0.625	0.75	1	0.875	0	2	1
8	−0.5303	−0.3536	−0.4420	1.2803	1.3536	1.3169	0	2	1

The centroid location of the midboom is $X_C = 0$ m, $Y_C = 0.875$ m, and $Z_C = 1$ m.

The moments of inertia are $I_{xx} = 1.563 \times 10^{-5}$ m^4, $I_{yy} = 1.563 \times 10^{-5}$ m^4, and $I_{xy} = 0$ m^4.

The area enclosed by the middle cross section is 1.105 m^2.

The moments at the centroid are $M_y = 10$ kN-m and $M_z = 8.75$ kN-m. The force acting at the centroid is $F_x = 10$ kN.

The slopes of booms and total axial forces are as follows:

Boom	dx/dz	dy/dz	dL/dz	σ_z, Pa	P_z, N
1	0	0	1	0	0
2	−0.0884	−0.0366	1.005	2.828×10^8	2828
3	−0.125	0.125	1.016	4×10^8	4000
4	−0.0884	0.2134	1.026	2.828×10^8	2828
5	0	0.25	1.031	0	0
6	0.0884	0.2134	1.026	-2.828×10^8	−2828
7	0.125	0.125	1.016	-4×10^8	−4000
8	0.0884	−0.0366	1.005	-2.828×10^8	−2828

The material properties converted to relevant units give a density of 2,768 kg/m^3, an ultimate tensile strength of 420.6 MPa, and an ultimate shear strength of 255.1 MPa. The allowable tensile strength is the ultimate tensile strength divided by the factor of safety, which yields 280.4 MPa. The majority of the boom loads are above this value, and so this indicates that there is insufficient bending material.

The total shear force that the webs must react is $S_{wx} = 12$ kN. Cutting the eighth web, the basic shear flows are as follows:

Web	q_b, N/m	F_x, N	F_y, N
1	0	0	0
2	−3,394	621	−1,500
3	−8,194	−1,500	−3,621
4	−11,588	−5,121	−2,121
5	−11,588	−5,121	2,121
6	−8,194	−1,500	3,621
7	−3,394	621	1,500
8	0	0	0

The integral of the basic shear flow is $-$ 12,803 N-m.

The shear flow at the cut web $q_{s,0}$ is 1,269 N/m.

The shear flows for each web are the summation of the q_b and $q_{s,0}$.

Web	q_s, N/m
1	1,269
2	−2,125
3	−6,925
4	−10,319
5	−10,319
6	−6,925
7	−2,125
8	1,269

The skin thicknesses due to bending and shear are as follows:

Web	Bending thickness, mm	Shear thickness, mm	Total thickness, mm
1	0.0209	0.0075	0.0284
2	0.0209	0.0125	0.0334
3	0.0209	0.0407	0.0616
4	0.0209	0.0607	0.0815
5	0.0209	0.0607	0.0815
6	0.0209	0.0407	0.0616
7	0.0209	0.0125	0.0334
8	0.0209	0.0075	0.0284

7.7.3 Fuselage Secondary Structures

The fuselage has secondary structure that can be a significant weight contributor. Some fuselage secondary structure elements include the following:

- *Brackets*—Like brackets on wings, these structures help transfer point loads to other structural members.
- *Access panels*—The hatches and small doors enable maintainers to access internal components.
- *Doors*—Doors are large hinged surfaces that open in flight. Those that apply to the fuselage can include landing-gear doors and weapons bay doors.
- *Window frames*—Internal payloads and situational awareness cameras must be able to see through the outer mold line with the necessary field of regard. This requires the addition of windows. The window material is generally nonstructural, and so all loads must be distributed around these apertures.

• *Radomes*—These dielectric surfaces are generally made of E-glass or quartz composites, which have inferior material properties relative to graphite. Radomes generally carry only the inertial load and aerodynamic loads acting upon the radome itself. The primary structure above or below the radome must carry the larger loads, though the height of the radome can substantially cut into the available structural depth.

7.8 Structures Manufacturing

UAS manufacturing methods vary substantially across unmanned aircraft classes and intended applications. Small UASs can be built using techniques from model aviation, sporting goods, or aerospace. MALE UASs often adopt construction techniques from the home-built and general aviation industries. HALE and UCAV platform construction and manufacturing use military and commercial aerospace methods. In this section, we will look at composite aerospace manufacturing methods followed by techniques that are commonly used for academic projects or simple prototypes.

7.8.1 Composites Manufacturing

Most composite UASs that enter production use molded parts. This approach provides an outer skin that houses additional internal structure. This is in contrast to prototype techniques that commonly lay up composites over a foam core, where the inside volume is consumed by the foam. The wing box contains shear webs and ribs. The fuselage structure contains bulkheads, keels, and stiffeners.

Hollow molded composite structures generally use molds created by computer numerically controlled (CNC) routers or mills. The CNC tools use digital geometry files generated from a CAD file. CNC molds are often built from high-density foams, MDF, and aluminum. Molds intended for autoclave use must be able to handle the thermal stresses.

Building the molded part takes several steps. The mold is prepared by filling any dents and then covering with a release agent to ensure that the cured part will separate from the mold. The composite plies are cut to shape using a Gerber cutter or by hand. The plies are organized according to the sequence that they are required. Next, the plies are placed on the mold in the proper location and orientation, according to the layup sequence. Lasers can project the required ply geometry onto the mold. Core material can be part of the layup sequence. Some layups require that a vacuum bag be placed over the mold to apply pressure to the layup, helping to eliminate voids and keeping the materials fixed to the mold. Most pre-preg materials must be cured at elevated temperatures, and so the mold is placed in an oven or autoclave. An autoclave controls both temperature and pressure.

Many UASs have constant chord wings. There are aerodynamic efficiency and structures weight penalties for not tapering the wings. Often the justification for the untapered wing is manufacturing cost reduction. Building a wing out of wood or metal requires many ribs, and having a single rib tool can indeed save cost for rib construction. However, the workflow of an unmanned aircraft fabrication might require that many ribs be constructed at once, which necessitates multiple rib tools. Composite structures generally have fewer ribs, if any at all. The composite wing manufacturing cost difference is negligible by tapering the wing. The wing designer is encouraged to refine the wing with taper to improve the performance. A constant chord wing can be viewed as a sign of limited design attention.

7.8.2 Student Projects and Simple Prototype Methods

Because this is an educational book, many readers are likely to work on a student project involving flight hardware. Others might wish to develop low-cost flying prototypes that will not go into production. Full manufacturing processes and tooling might be unavailable or too expensive for such endeavors. Some simplified construction techniques are described that can aid the fast-paced development. This information can provide useful ideas for prototype construction, but it is not sufficient as a sole resource.

7.8.2.1 Wood

Simple wood construction has been the primary structural approach of model aviation for years. Compared with composites or metal structures, small wood aircraft are very simple. Few tools are required, and the skills are fairly intuitive, building on basics that are taught in high school shop courses. As with any form of construction, proper safety precautions should be taken.

The wing is composed of the following structural members:

* Spars generally made of spruce
* Ribs made of balsa sheet
* Leading edge carved from a balsa plank
* Balsa sheeting applied to portions of the wing to ensure the proper shape

The fuselage is made with the following members:

* Thick balsa sheets and plank trusses used to form the fuselage
* Aircraft plywood used for fuselage bulkheads.

A thin plastic covering material is applied to the outer surface. An example aircraft built in this manner is shown in Fig. 7.13.

Fig. 7.13 Model aircraft with wood structure and plastic covering.

7.8.2.2 Composites

Simple composite structures use dry composite fabrics and epoxy. The dry fabrics are widely available in fiberglass, Aramid, or carbon-fiber materials. The epoxy system comes in two parts: resin and hardener. The epoxy components are mixed together in the correct proportions and used to wet the fabric. Lambie [5] and Noakes [6] describe some composite fabrication techniques that are suitable for student projects.

Simple composite wings can be built using a foam core. The core is cut using two templates on either end of a foam block. The lower surface is cut first with templates corresponding to the lower contour. The lower surface templates are replaced by upper surface templates, and the process is repeated. Three pieces of foam are produced from this procedure: an upper bed, a lower bed, and the wing core. Additional cutouts for spar caps or partial ribs can be made with a router.

Next, the skins are applied. Two pieces of flexible Mylar are used to provide a smooth shape to the skins. These sheets follow the upper and lower surfaces of the wing starting from the trailing edge and ending a short distance from the leading edge. These sheets are hinged at the trailing edge with tape or peel ply. The hinged pieces are placed flat in the open position. Layers of fabric are applied to the Mylar sheets and wetted with epoxy. Excess epoxy is removed with a wiper device. Foam-compatible spray adhesive is used to adhere a strip of dry fabric directly to the foam core leading edge. Strips of unidirectional carbon fiber can be applied in a similar fashion. The foam core-mounted fabric is wet with epoxy. Now, the foam core is placed in the Mylar sheets with the trailing edge at the hinge line.

A vacuum bag is used to press the composite fabric to the foam core during cure. The vacuum bag system is available from home-built aircraft suppliers or composite materials distributors. The wing materials are placed inside the vacuum bag with the clean Mylar surfaces touching the bag. The ends of the bag are sealed with putty or bag clips. The vacuum pump is turned on, and the air is removed from the bag. The bag is placed between the upper and lower foam beds to help maintain the wing shape. Heavy weights are placed on the beds to ensure that the wing does not warp.

After the wing cures, the pump is turned off, and the part is removed from the bag. The Mylar sheets are peeled away from the skins. A smooth wing with a messy perimeter is revealed. Some stray cured fibers are very sharp, and so the part must be handled with care. Most of the excess material is removed with a scissors or a cutting wheel. The remainder is removed with a grinder and then smoothed with sand paper. The trailing-edge control surfaces are cut from the wing. Often, the upper or lower skins are used as skin hinges, which can be effective for smaller unmanned aircrafts that have few flights.

Simple composite fuselages can be built in a number of ways. Examples include the following:

* Pull parts off of a male plug. A plug with the fuselage geometry is carved from wood or built from other materials. A layer or two of wet fabric is applied to the waxed plug. After this material cures, it is cut from the mold. The cut lines are then glued together. This thin shell is capable of holding the shape for the body. The shell is sanded so that it is roughened enough for additional layers to bond to it. The remaining layers are applied. After the final cure, the part is sanded and painted.
* Another method is to use a male plug to make a female mold. The female mold is made in halves. The fuselage halves are made in the female molds.
* A foam plug can be formed, and the materials applied to the outside. The foam can be removed with solvents or a rasp, leaving the composite outer shell.

References

[1] Beer, F. P., and Johnston, E. R., *Mechanics or Materials*, 2nd ed., McGraw-Hill, New York, 1992.

[2] Roskam, J., and Lan, C. E., *Airplane Aerodynamics and Performance*, DAR Corp., Lawrence, KS, 1997, pp. 593–602.

[3] Niu, M. C., *Airframe Stress Analysis and Sizing*, 2nd ed., Conmilit Press, Lrd., Hong Kong, 1997.

[4] Megson, T. H. G., *Aircraft Structures for Engineering Students*, 2nd ed., Halsted Press, New York, 1990.

[5] Lambie, J., *Composite Construction for Homebuilt Aircraft*, 2nd ed., Aviation Publishers, Hummelstown, PA, 1995.

[6] Noakes, K., *Successful Composite Techniques, A Practical Introduction to the Use of Modern Composite Materials*, 3rd ed., Osprey Publishing, Oxford, England, U.K., 1999.

Problems

7.1   A nose-gear strut takes only compressive load. It has a length of 2 ft, outside diameter of 3 in., and a wall thickness of 0.2 in. It is made of 2024 aluminum. What is the moment of inertia? What is the critical buckling load?

7.2   A shear web has a height of 0.25 m and a thickness of 0.01 m. A 100-kN shear force is applied. Plot the shear-stress distribution along the height. What is the maximum shear stress, and where does it occur?

7.3   A recovery system consists of a vertical pole and an arm with a net assembly cantilevered off to one side. The unmanned aircraft captures the net at a distance of 15 ft from the center of the pole and at a height 25 ft above the base. The unmanned aircraft weighs 400 lb and decelerates horizontally at 10 g. What torque is applied to the vertical pole from this capture? The pole's outside diameter is 2 ft, and the wall thickness is 0.5 in. The material is 2024 aluminum. What is the maximum twist of the pole along its vertical axis?

7.4   A pneumatic-powered aircraft uses pressurized air to spin a turbine. The spherical tank capacity is 10 ft^3. The maximum pressure is 5,000 psi. The material is 2024 aluminum. Use a factor of safety of 2. What is the required wall thickness? How much does the empty tank weigh? How much would it weigh if constructed of steel?

7.5   A wing has a span of 25 ft and a constant chord length of 2 ft. The wing generates 1,000 lb of lift at 1 g. Assume a linear load distribution that has zero load at the wing tip. Use five evenly spaced wing segments for each side. Calculate the shear forces and bending moments when the aircraft is loaded at 4 g. Calculate the answers at the inboard portion of the segment.

7.6   Use the wing in problem 7.5. Calculate the spar cap weight if the spar cap has a uniform thickness along the span. The spar cap is made from 2024 aluminum. Ignore inertia relief. Assume that the upper and lower spar caps are separated by 0.25 ft and have a width of 50% of the chord. Use a factor of safety of 1.5.

7.7 Use the conditions from problem 7.6. Calculate the spar cap weight if the thickness is allowed to vary for each segment.

7.8 Use the conditions from problem 7.7. Calculate the spar cap weight for each segment with inertia relief. Use panels outboard of the panel of interest for inertia relief calculations.

7.9 Use the conditions from problem 7.8. Add a payload at 50% of the span on both sides that weighs 100 lb each. Calculate the shear loads, bending moments, and spar cap weight for the 4-*g* load case. Assume that the overall lifting load is unchanged by this payload addition.

7.10 Use the skin-panel method to generate paneling for a wing section that uses the NACA 2412 airfoil. The chord is 0.5 ft. The spar caps start at 10% chord and end at 70% chord. Use three panels each upper and lower spar caps. Calculate the centroid location. Calculate the moments of inertial normalized by the skin thickness.

7.11 Using the geometry from problem 7.10, calculate the skin-panel thickness for each panel if the bending moments are $Mx' = -1,000$ ft-lb and $My' = 100$ ft-lb. Calculate the stresses at the midpoint of each panel. Assume 2024 aluminum. Use a factor of safety of 1.5.

7.12 Use the boom-and-web method to calculate the loads on a fuselage section. The tapered fuselage segment has a circular cross section and the arrangement, materials, and paneling shown in Example 7.5. The forces acting at point $x = 0$ m, $y = 1$ m, $z = 2$ m are $F_x = 5$ kN, $F_y = -5$ kN, and $F_z = 0$ kN. The moments acting at that point are $M_x = -10$ kN-m, $M_y = 5$ kN-m, and $M_z = -2$ kN-m. Calculate the bending stresses for each boom. Determine if there is sufficient material in the booms. Calculate the required shear skin thickness for each panel.

7.13 Repeat problem 7.12 with 16 booms instead of 8 booms. Each boom has 5-mm^2 cross-section area so that the total boom area remains the same. Compare the answers to see if the increased panel density makes a significant difference. Compare the stresses at every other boom, starting with the first boom. The stresses should be within 1% of the preceding example.

Chapter 8

Propulsion Systems

- Understand common propulsion types
- Explain propulsion system performance prediction
- Learn high-altitude propulsion challenges and solutions

Fig. 8.1 Hydrazine-powered Mini Sniffer III UAS. (Photo courtesy of NASA.)

8.1 Introduction

Propulsion systems are vital to the operation of UA. Drag must be overcome to maintain steady flight velocity or to accelerate, and gravity must be overcome to climb. Propulsion systems generate thrust to achieve these aims.

Airbreathing propulsion systems accelerate air through a propulsor. These propulsion systems include jets, turboprops, piston-props, ducted fans, and electric props, among others. The thrust generated by an airbreathing propulsion system is given by

$$T = \dot{m}_a(V_e - V) + (P_\infty - Pe) \cdot A_e \tag{8.1}$$

where m_a is the mass flow rate of the air through the propulsion system, V is the freestream velocity, V_e is the mass-averaged exit velocity, P_∞ is the ambient pressure, P_e is the pressure at the propulsion system exit, and A_e is the propulsion exit area.

Airbreathing and nonairbreathing propulsion types are considered in this chapter. Airbreathing propulsion mixes the oxygen in the air with fuel to support combustion. The energy from the combustion is converted to useful work via a propulsor device such as a propeller or ducted fan. Nonairbreathing engines, as the name suggests, have no dependency on air chemistry to support power generation. Rockets and battery-driven motor-propellers are two examples.

Unmanned aircraft can utilize different types of propulsion than manned aircraft. Aircraft in both domains are frequently powered by jets, turboprops, and reciprocating engines. Unmanned aircraft commonly use battery-propeller propulsion today, and occasionally fuel-cell powerplants. Solar-powered aircraft technologies are now maturing. This chapter covers these unique UA propulsion systems that are frequently omitted from aircraft design texts as well as the more conventional approaches.

8.2 Propellers

Propellers offer perhaps the most intuitive introduction to propulsion. These propulsor devices change the momentum of the air to generate thrust. Propellers are an element of a propulsion system, where shaft power must be provided by another device such as a turboshaft engine, reciprocating engine, or electric motor.

The propeller-driven aircraft performance equations shown in Chapter 3—both using reciprocating and electric motors—have a propeller efficiency term η_P. This efficiency is not a fixed term; rather, it is a function of the propeller design and operating conditions. A single average value for the flight segment of interest must be selected, but it should be based on

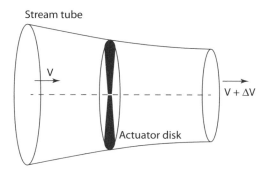

Fig. 8.2   Propeller disk actuator flow properties.

knowledge of the expected propeller capabilities. As will be shown, the efficiency for a given propeller is a function of the thrust required, altitude, and flight velocity.

The actuator disk propeller analysis method provides valuable insight into the propeller performance. In the purest form, this method shows the theoretically best efficiency that a propeller can attain if friction losses are neglected. The details of the propeller design, such as the chord and twist distribution, are neglected with this simplified approach. The propeller is treated as a disk that imparts a momentum change onto the airflow that reaches it. Therefore, the only propeller geometry parameter is the diameter. While obscuring the propeller design details appears to be very limiting on first consideration, the method is highly useful for conceptual design and much preferable to simply assuming a constant efficiency. Calibrations between practical propellers and ideal propellers further improve the utility.

For a propeller, the thrust equation can be written in terms of the propeller disk area A_p:

$$T = \frac{1}{2}\rho \cdot A_p \cdot \Delta V \cdot (2V + \Delta V) \qquad (8.2)$$

where ρ is the air density. The problem is illustrated in Fig. 8.2. The propeller area A_p is simply

$$A_p = \frac{\pi}{4} \cdot D^2 \qquad (8.3)$$

where D is the propeller diameter. The velocity difference is defined by

$$\Delta V = V_e - V \qquad (8.4)$$

V_e is the average velocity of the air acted upon by the propeller far downstream. Solving the thrust equation for the velocity difference yields

$$\Delta V = \sqrt{V^2 + \frac{2T}{\rho \cdot A_p}} - V \tag{8.5}$$

The theoretical propeller efficiency is

$$\eta_{p,ideal} = \frac{2V}{V_e - V} = \frac{1}{\Delta V/2V + 1} \tag{8.6}$$

With the preceding equations, the ideal efficiency can be found for a given propeller diameter at any altitude, airspeed (except static case), or thrust condition. This powerful result is achieved at little computational expense. However, inescapable practical inefficiencies are neglected in the actuator disk model. To remedy this deficiency, a nonideal efficiency correction can be applied. The total propeller efficiency of a nonideal propeller is

$$\eta_p = \eta_{p,ideal} \cdot \eta_{p,nonideal} \tag{8.7}$$

where the nonideal efficiency $\eta_{p,nonideal}$ ranges from approximately 85–95% based on comparisons with blade element models. A nominal value of 90% can be used.

More detailed performance models covered in Chapter 9 use the shaft power rather than the propeller efficiency. The relationship between shaft power $P_{Shaft, Prop}$ and efficiency is

$$P_{Shaft,Prop} = \frac{P_{Thrust}}{\eta_p} = \frac{T \cdot V}{\eta_p} \tag{8.8}$$

Here, the propeller efficiency is undefined at zero freestream velocity. The general form of the actuator disk model relating shaft power to thrust is

$$P_{Shaft,Prop} = \frac{T}{\eta_{p,nonideal}} \cdot \left(V + \frac{1}{2}\Delta V\right) \tag{8.9}$$

A modified version of this shaft power equation for a static thrust case ($V = 0$) is

$$P_{Shaft,Prop} = \frac{T^{3/2}}{\sqrt{2\rho \cdot A_P} \cdot \eta_{p,nonideal}} \tag{8.10}$$

The derivation of the static shaft power is left to the reader as an exercise. The nonideal efficiency is also known as the figure of merit for hovering propellers. The figure of merit typically ranges from 0.7 to 0.8 in the static thrust case [1].

The actuator disk method is suitable for conceptual design purposes only. In preliminary and detailed design phases, the author recommends the use of blade element models or computational fluid dynamics (CFD) to estimate propeller performance. Blade element models calculate the circulation distribution of the propeller using blade chord and twist distribution geometry definition. The drag-vs-lift properties of the airfoils are used directly. Some blade element models can estimate the influence of spinners and fuselages on the performance. XROTOR is a prevalent blade element analysis code.

CFD codes can be used for propeller analysis, where the flow must be modeled as a rotational field. A single blade is modeled with three-dimensional blade geometry. The transition between laminar and turbulent flow is difficult to calculate with most CFD tools, and so the transition location is usually either fixed or set to fully turbulent.

The output of the detailed propeller models is a table or plot of coefficient of thrust C_T, coefficient of power C_P, and efficiency η_P vs the advance ratio J. The advance ratio is defined as

$$J = \frac{V}{n \cdot D} \tag{8.11}$$

where n is the propeller rotational velocity in cycles per second. Note that the rotational velocity can be expressed in revolutions per minute by

$$\text{rpm} = 60 \cdot n \tag{8.12}$$

The thrust and power coefficients are defined by

$$C_T = \frac{T}{\rho \cdot n^2 \cdot D^4} \tag{8.13}$$

$$C_P = \frac{P_{\text{Shaft,Prop}}}{\rho \cdot n^3 \cdot D^5} \tag{8.14}$$

The relationship between the efficiency, thrust coefficient, power coefficient, and advance ratio is

$$\eta_P = J \cdot \frac{C_T}{C_P} \tag{8.15}$$

The propeller-driven propulsion system performance is a combination of the propeller and the power source that drives the propeller. The first step is to identify the advance ratio at which the thrust is generated. Using the definition of advance ratio, the rotational velocity can be expressed as

$$n = \frac{V}{J \cdot D} \tag{8.16}$$

Solving the thrust coefficient equation for thrust and substituting in the preceding equation gives

$$T(J) = C_T(J) \cdot \rho \cdot \left(\frac{V \cdot D}{J}\right)^2 \tag{8.17}$$

For a given propeller and flight condition, the flight velocity, density, and propeller diameter are known. The thrust is expressed as a function of the unknown advance ratio. A convergence method can be employed to identify the advance ratio at which the calculated thrust equals the required thrust.

Now, the shaft power associated with generating the required thrust is calculated. With the appropriate advance ratio substitutions and solving the power coefficient equation for shaft power, this becomes

$$P_{\text{Shaft,Prop}}(J) = C_P(J) \cdot \rho \cdot \left(\frac{V}{J}\right)^3 \cdot D^2 \tag{8.18}$$

Engine power is provided as a function of rpm. The conversion between advance ratio and rpm is

$$\text{rpm} = 60 \cdot \frac{V}{J \cdot D} \tag{8.19}$$

If the available engine power is greater or equal to the required shaft power at the selected rotational velocity (n or rpm), then the propulsion system can operate at that thrust level. The maximum thrust is the point at which required power equals the maximum shaft power available.

Engines and propellers might require gearing to match the rotational velocities for best performance. A common case is where the desired propeller operates at lower rpm than where the engine produces the best power. Engines often operate most efficiently at higher rpm values. Gear ratios of 1.5–3 are common. Gearing can occur through gears, belt drive, or fluidic clutches. Gears have efficiencies, where the lost power is rejected as heat.

The propeller tip Mach number should be kept less than one to prevent high noise levels. Lan and Roskam [2] recommends an upper tip Mach number of 0.72 for acceptable noise. Compressibility effects can also degrade propeller efficiency. The tip Mach number M_{tip} is found by

$$M_{\text{tip}} = \frac{\sqrt{(\pi \cdot D \cdot n)^2 + V^2}}{a} \tag{8.20}$$

where a is the speed of sound. The tip Mach number can also be expressed in terms of advance ratio and flight Mach number M.

$$M_{\text{tip}} = M\sqrt{\left(\frac{\pi}{J}\right)^2 + 1} \tag{8.21}$$

Propellers can be made from a variety of materials and processes. For reciprocating engines above around 20 hp, suitable propellers might be available for ultralight and general-aviation aircraft. These are generally made from wood and metal. Turboprop engines used on manned aircraft have specialized metal blades. Engines of less than 20 hp must use either model aircraft or custom propellers. Model aircraft propellers are made mostly from wood, plastics, or carbon fiber.

Rain can cause degradation on propellers without proper protection provisions. Propellers made exclusively of wood might not be capable of operations in rain. Metal plates are sometimes added to the blade leading edges to provide the necessary protection. Carbon fiber and plastic propellers are generally robust to the rain for small UAS applications.

Variable pitch propellers adjust the pitch angle of each blade near the root. Increasing the propeller pitch can increase the maximum speed achievable. The efficiency can also be improved over a broad speed range. The variable pitch mechanism adds cost, complexity, and weight over a fixed-pitch propeller.

Model aircraft propellers are identified by diameter and nominal pitch. In the United States, both characteristics are in units of inches. The first number is the diameter, and the second is the nominal pitch. For example, a 6 × 3 prop is 6 in. in diameter and 3 in. in nominal pitch. Diameter is a directly measured property that is easily understood.

Nominal pitch, on the other hand, requires some explanation. This characteristic is a measure of the blade pitch angle at 75% of the propeller radius. The pitch angle is what can be measured by placing a straight edge on the lower surface of the prop. Because propeller blade airfoil sections generally have finite thickness on the lower surface, the nominal pitch is less than the geometric pitch. The geometric pitch is the angle across the chord line connecting the leading edge and trailing edge. The following equation is used to calculate the nominal pitch from the measured lower surface angle θ_{Nom} at 75% of the blade radius.

$$\text{Pitch} = \frac{3}{4}\pi \cdot D \cdot \sin(\theta_{\text{Nom}}) \tag{8.22}$$

The geometric pitch can be calculated by adding the difference between the nominal pitch and the angle along the lower surface of the blade airfoil section. Note that no precise lower surface angle exists if the airfoil section is outwardly curved along the lower surface, as is typical for many propeller airfoil sections.

Often, professional UAS development efforts require better knowledge of the propeller performance characteristics than can be inferred from diameter and pitch designations. Detailed information about the propeller geometry is needed to predict the performance characteristics. The blade chords and geometric pitch angles should be measured across multiple stations. The propeller airfoil sections across various radial stations should be identified so that the drag can be modeled as a function of lift. This propeller outer-mold-line information can suitably define the propeller for blade element or CFD analysis.

A common error for small UAS development programs with small budgets is to select propellers based on static test results alone. Such a test typically evaluates multiple propellers on the selected engine for the UA. Maximum rpm is easily measured. Slightly more sophisticated tests will measure static thrust as well. Propellers that enable either high rpm or high static thrust values are ranked highly. The fundamental problem with this test is that static thrust performance has no meaningful relationship to cruise performance, except that rpm is reduced with larger diameter. In fact, this test will generally favor low pitch props that might be unsuitable to the flight cruise conditions.

8.3 Reciprocating Engines

Reciprocating engines are the most common form of propulsion for UAs with takeoff gross weight values between 20 and 2,500 lb. Commercial-off-the-shelf (COTS) engines are widely available between 1–200 hp. Perhaps the widespread adoption of reciprocating engines is driven by the lack of suitable jet and turboprop engines for the most prolific UAS classes.

Reciprocating engines come in many forms. The most prevalent are two-stroke and four-stroke engines. Slightly less common is the rotary engine. Two-stroke and four-stroke engines are described in this section.

Brake specific fuel consumption (BSFC) is defined as fuel flow rate per shaft power:

$$BSFC = \frac{\dot{W}_{fuel}}{P_{Shaft}} \qquad (8.23)$$

where $\dot{W}_{fuel}$ is the fuel flow rate in terms of fuel weight per time. The BSFC can be expressed in terms of thermal efficiency and the fuel heating value:

$$BSFC(\text{lb/hp} - \text{hr}) = \frac{2545}{\eta_{th} \cdot H(\text{BTU/lb})} \qquad (8.24)$$

where H is the fuel heating value in BTU/lb. Reciprocating engine thermal efficiency is typically 15%.

UAS engines come from a variety of sources ranging from model aircraft to general aviation. Although clean sheet custom UAS engine development can be undertaken, the majority of UAs use modified engines from other sources. A list of engine pedigrees is shown in Table 8.1.

Model aircraft engines mostly generate less than 12 hp, with the greatest selection around 0.5–2 hp. Two-stroke glow fuel engines are the most popular. These require a fuel-oil mixture. High power-to-weight ratio and low cost are the most important considerations. These are designed to support passive fuel systems to simplify integration for the hobbyist. Fuel consumption can be high because the typical modeler tires after flight duration of 5–20 min and the small fuel weight has negligible impact on flight performance. A single engine can be used on a handful of aircraft that have perhaps a dozen flights on each, and so maintainability is not a design driver. Although data are not readily available, a model aircraft engine will likely have mean time between overhaul values of perhaps 10–100 h for UAS operational conditions. Glow fuel has a low heating

Table 8.1 Reciprocating Engine Pedigrees

Pedigree	Power, hp	Pow/Wt, hp/lb	BSFC, lb/ hp-hr	Notes
Two-stroke model aviation	0.1–20	0.84–1.7 (glow) 1.1–1.7 (gas)	>1 (glow) 0.7–1 (gas)	Glow fuel at low power Pump gas at higher power
Four-stroke model aviation	0.5–2	0.36–1.3	>1 (glow)	Glow fuel
Hand portable lawn equipment	0.5–2	~0.5	>0.7	Low cost Pump gas
Small lawn mowers	1–10	~0.5	>0.7	Low cost Pump gas
Mopeds and scooters	5–30	—	0.5–1.5	May use fuel injection and engine controls
Unmanned target engines	10–60	~1.5	0.7–1.5	Short life, high power/ weight
Ultralights	20–120	0.55–2.1	0.6–0.8	Well-suited to UAS Two- and four-stroke Pump gas and avgas
General aviation	90–300	0.48–0.51	0.4–0.6	Mostly four-stroke, high quality, Avgas
Automotive	90–450	—	—	Requires adaptation to aviation, pump gas

value. Therefore, large fuel weight is required for long-endurance flight. The fuel energy partially drives the high-engine BSFC.

A word of caution is in order about advertised model aircraft engine specifications. Power specifications and dynamometer data from model aircraft engine manufacturers should be verified. Fuel flow data are rarely available from model aircraft engine manufacturers. The most reliable data provided are the maximum static rpm with a given prop. This is easy for manufacturers to generate, but it is also relatively unimportant for UAS analysis. Static rpm and static thrust have little bearing on flight characteristics. Independent tests are required for engineering quality data. The author assumes that advertised model aircraft engine peak power is 30% optimistic until confirmed.

Manufacturing quality is often poor for model aircraft engines relative to general-aviation engines. Manufacturing tolerances might not be tight enough to support reliability needs. The complete UA with payloads will generally cost two orders of magnitude more than the engine when model aircraft engines are used, yet the engines might be the most significant reliability driver.

Larger model aircraft engines often use pump gasoline for fuel. These have gas-oil mixtures, which is burdensome to mix under field conditions. The mixture ratio can be difficult to control precisely. Gas-fueled model aircraft engines often use spark ignition. These gas engines are similar to lawn equipment engines such as those used on leaf blowers or lawn mowers.

Model aircraft engines or other simple engines might require substantial modification and support systems for integration on UASs. Most frequently these engines do not contain provisions for generators. Two approaches are 1) put a generator in-line with the shaft and 2) place a generator parallel to the shaft via a belt drive. The loads from the generator could require improved shaft bearings or additional engine structural support. Although it is possible to add an onboard engine starter motor or use a dual-purpose starter-generator, most small and tactical UAS rely upon external starters.

Large aviation engines, such as those used on ultralights and general-aviation aircraft, are readily adaptable to UAS applications. Ultralight engines start at approximately 20 hp with mostly two-stroke cycles. Larger ultralight engines tend to be four-strokes, with the Rotax 912 and 914 series seeing widespread use on MALE UASs. General-aviation engines are less common on UASs, primarily because of the higher power that is required for many designs. However, Thielert and Austro general-aviation heavy fuel engines are earning their way onto unmanned aircraft thanks to the fuel type and low fuel consumption. General-aviation engines have mean time between overhauls ranging from approximately 1,000 to 2,000 hrs.

Fig. 8.3 Twin-cylinder two-stroke engine installation on the DRS Neptune UAS.

Engine controls can be added to simple engines in order to improve reliability and fuel consumption. The engine control unit can control throttle servo position, spark timing, cowl flap opening, or fuel injection (also called *metering*), depending on engine configuration. Measured engine properties can include shaft rotational angle, rpm, air inlet temperature, exhaust temperature, and cylinder head temperature. These sensors must be added to the engine.

The noise from a straight pipe exhaust is generally too high for ISR operations. Mufflers reduce the acoustic signature of the engine. The volume of effective mufflers can have significant impact on the UA design and should be considered at the earliest stages to ensure that the fuselage or cowling is sized appropriately. See Sec. 8.3.2 for further details of engine vibration.

Vibration is often solved with vibration isolation mounts where the engine interfaces to the aircraft, which is generally at the firewall. High engine vibration levels might require a large volume for engine movement, which complicates cooling duct design. Often, flexible ducts or non-sealed cooling flow paths are required. Large gaps have negative drag consequences.

Fuselage-mounted reciprocating engines are usually installed in a compartment that is separated from the rest of the fuselage by a firewall. An example of an aft-mounted engine is shown in Fig. 8.3. The engine is mounted to the firewall with a support structure, such as a welded tubular metal frame. An engine cowling that is mounted to the firewall or engine support structure encloses the engine bay and provides the local outer mold line aerodynamic shape. The engine cooling intake, engine exhaust, exhaust vents, and maintenance hatches pass through the cowling. The engine support systems, such as fuel pumps, ignition control, generators, and header tanks are housed within the engine bay. Some manned aviation engines are sold as *firewall forward* kits.

8.3.1 Fuel Types

Fuel quality and logistics are major considerations for UASs. Glow-fueled engines for model aircraft have a fuel type that is extremely difficult to support in the field because of a notable lack of local hobby shops in war zones. Even for pump gas, local fuel quality and chemistry can vary substantially from region to region.

For military systems, the best way around these impositions is to use heavy fuel (JP-5 or JP-8) that is already in the logistics system and has carefully controlled quality. Heavy fuel is less volatile than pump gas or glow fuel, thereby improving handling safety. This attribute is especially important on ships where an uncontrolled fire can be devastating.

The challenge for heavy fuel engines is in cracking large hydrocarbon molecules. The low volatility that makes heavy fuel safe also makes it difficult to use. A diesel cycle engine uses compression ignition at high pressure. Compression ignition requires heavy engine components to react to the sudden pressure spikes. Normally aspirated diesel engines are sensitive to altitude, humidity, and temperature variations. Turbocharged diesel engines with advanced engine controls can be designed for robust operations and high-altitude capabilities.

Spark-ignition heavy fuel engines can operate at lower pressures and have more robust operability. Precise control over spark timing and pressure ratio is required. However, advanced techniques are required to crack the large hydrocarbon molecules. The heat from an operating engine can be sufficient, but starting the engine is problematic. Many engine companies have seen that an unmodified engine can be made to run on heavy fuel if started using gasoline. Using two fuel types is a logistics disadvantage. One technique to operate a pure heavy fuel engine is to heat the fuel or parts of specialized combustion chambers. Another approach is to use injectors that spray a controlled fine fuel mist.

8.3.2 Engine Configurations

Piston engines have inherent vibration that can have negative performance consequences for payloads requiring high pointing accuracy. The nature of the vibration and methods available to mitigate vibration are dependent upon the engine configuration.

A single-cylinder-piston mass oscillates in one direction and is offset only by a rotating counterweight on the shaft. The counterweight mass can be increased, but this increases vibration orthogonal to the direction of the stroke when the piston is in midstroke.

Opposing cylinders, also known as a boxer configuration, generally have pistons that fire symmetrically, offsetting the vibration in the stroke direction. The pistons can be offset to eliminate interference with the crankshaft, which introduces a torque vibration. Also, the counterweight can oscillate orthogonally to the piston stroke.

Rotary engines have the potential for low vibration if properly balanced. Muffler integration can be a challenge for pusher configurations, as seen on the Shadow 200, which can lead to high noise levels. For the purposes of this book, general discussions of reciprocating engines apply to rotary engines as well.

8.3.3 Four-Stroke Reciprocating Engines

Most modern general aviation engines are four-stroke engines, which use the Otto cycle. Four-stroke engines require the piston to travel up and down twice per every power stroke. The process goes from induction, to compression, to power stroke, and to exhaust. The power stroke occurs every other crankshaft revolution. The cycle is shown in Fig. 8.4.

The indicated horsepower (IHP) is the piston volume – pressure integral over the entire cycle. The compression ratio is a major factor in determining the overall IHP. The spark occurs before the initiation of the power stroke because the fuel-air mixture takes a finite time to start the burn. The

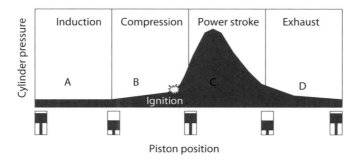

Fig. 8.4 Four-stroke engine cycle.

designer is primarily concerned with the brake horsepower (BHP), which is given by

$$BHP = \eta_m \cdot IHP \tag{8.25}$$

where η_m is the mechanical efficiency, which accounts for friction, power for pumps, generators, cooling, and other losses. The designer might not have detailed information about the indicated horsepower or the mechanical efficiency. The following equation allows the designer to use intuitive parameters to predict engine performance:

$$BHP = \frac{\eta_m \cdot BMEP \cdot N_{Cyl} \cdot c_{Cyl} \cdot rpm}{792,000} \tag{8.26}$$

where $BMEP$ is the brake mean effective pressure in psi, N_{Cyl} is the number of cylinders, and c_{Cyl} is the cylinder volumetric capacity in cubic inches. The capacity per cylinder can be found by

$$c = \frac{\pi}{4} D_{bore}^2 \cdot l_{stroke} \tag{8.27}$$

where D_{bore} is the bore diameter in inches and l_{stroke} is the stroke length in inches.

It is generally more efficient to run the engine at a high $BMEP$ and lower rpm to achieve best efficiency because the higher rpm incurs losses from friction and reduces engine life. Higher $BMEP$ engines run at higher temperatures and require cooling. Stinton [3] provides more detail on two- and four-stroke aviation engine performance.

Reciprocating engines are very sensitive to the fuel-air mixture ratio and humidity. High humidity combined with high temperatures can produce up to a 20% reduction in available power.

A normally aspirated engine relies upon ambient pressure and suction to bring air into the cylinders. This type of engine generates its maximum power at sea level, and performance degrades as altitude increases. The Gagg–Ferrar equation describes this behavior:

$$\frac{P_{Shaft}}{P_{Shaft,SL}} = 1.132 \cdot \sigma - 0.132 \tag{8.28}$$

where σ is the air density ratio relative to sea level. Engine power can be restored by means of turbocharging, where the pressure of the intake air is increased by mechanical means. This helps aircraft that fly in hot, humid conditions or at high altitudes.

The brake specific fuel consumption for a piston engine is defined as the ratio of fuel flow to the power and is usually given in the units pounds per horsepower-hour. Mises [4] found that the lubrication oil consumption ranged from 0.02–0.05 lb/hp-hr, which only applies to engines where fuel and oil are not mixed together in the fuel tank. A brake specific fuel consumption equation from Mises [4] can be modified to yield

$$BSFC = 1052 \cdot \frac{F(\sigma) \cdot \eta_v}{x \cdot BMEP} \quad (8.29)$$

where $BSFC$ is the brake specific fuel consumption in pounds per horsepower-hour, σ is the density ratio, η_v is the volumetric efficiency, and x is the air-to-fuel ratio. The density altitude function $F(\sigma)$ is estimated by

$$F(\sigma) = \frac{0.935 \cdot \sigma}{(\sigma^{1.117} - 0.065)} \quad (8.30)$$

8.3.4 Two-Stroke Reciprocating Engines

Two-stroke engines combine the intake/compression and combustion/exhaust functions of the four-stroke engines such that the entire process occurs over one crankshaft revolution rather than two. Two-stroke engines are generally lighter and less expensive than four-stroke engines, but are less efficient. For a given power level, two-stroke engines tend to be half the displacement and weight of their four-stroke counterparts. A two-stroke engine is shown in Fig. 8.3.

The brake horsepower for a two-stroke engine is

$$BHP = \frac{\eta_m \cdot BMEP \cdot N_{Cyl} \cdot c_{Cyl} \cdot \text{rpm}}{396,000} \quad (8.31)$$

Note that the constant is halved from that of the four-stroke because power is produced once per revolution on a two-stroke. The specific weight of the two-stroke engine ranges from 0.8 to 2 lb/BHP. The specific fuel consumption ranges from 0.75–2 lb/hp/hr as compared to an average of 0.5 for four-stroke engines. However, newer direct-injected two-stroke engines have BSFC values comparable to four-stroke engines. Two-stroke engine rotational rates generally vary between 5,000–9,000 rpm, which generates higher-frequency noise and requires either gearing or small-diameter propellers.

8.3.5 Turbochargers

The most common means of improving reciprocating altitude performance is by turbocharging. A turbine is driven by the hot-engine exhaust gases, which in turn powers a separate compressor that boosts the intake manifold pressure to the cylinders. Turbochargers increase the air temperature in the intake manifold through adiabatic heating, and so the engine must be designed to withstand the thermal and pressure stresses induced by the turbocharger, which ultimately results in a weight penalty. Alternatively, intercoolers with heat exchangers can cool the pressurized air. The additional complexity accompanying the turbocharger will incur cost and decrease the mean time between engine overhauls. Typically, turbocharged engines require higher octane fuels than their normally aspirated counterparts to avoid engine knock.

8.3.6 Reciprocating Engine Performance Analysis

Whether the reciprocating engine is a two-stroke, four-stroke, rotary, or more exotic type, the process for evaluating the propulsion performance is similar when it powers a propeller to generate thrust. An engine model must be established that outputs the following:

$$P_{\text{Shaft,Max}}(h, \text{rpm})$$
$$BSFC(h, \text{rpm}, \text{throttle})$$

where throttle is the ratio of P_{Shaft} over $P_{\text{Shaft,max}}$ for a given altitude. The throttle can also be called *demand*. The engine model should represent the integrated flight configuration rather than just the engine core. The methods can be based on engine empirical relationships, scaling from similar engines, detailed engine modeling techniques, or interpolation of test data.

The Gagg–Ferrar equation is an example of an empirical relationship. Here the maximum shaft power is a function of altitude alone, and no information is known about the rpm vs peak power. In this case, the actuator disk propeller model is a suitable match, as there is no rpm defined for that method either. Other empirical relationships can be applied for maximum power and BSFC.

Another approach is to scale the performance characteristics directly from data of similar engines. The scaling takes the following form:

$$P_{\text{Shaft,Max}}(h, \text{rpm}) = P_{\text{Shaft,Max,SL}} \cdot \frac{P_{\text{Shaft,Max,Ref}}(h, \text{rpm})}{P_{\text{Shaft,Max,SL,Ref}}} \tag{8.32}$$

$$BSFC(h, \text{rpm}, \text{throttle}) = BSFC_{SL} \cdot \frac{BSFC_{Ref}(h, \text{rpm}, \text{throttle})}{BSFC_{SL,Ref}} \quad (8.33)$$

where

$BSFC_{Ref}$ = brake specific fuel consumption of the reference engine
$BSFC_{SL}$ = best brake specific fuel consumption at sea level
$BSFC_{SL,Ref}$ = best brake specific fuel consumption at sea level for the reference engine
$P_{Shaft,Max}$ = maximum shaft power
$P_{Shaft,Max,Ref}$ = maximum shaft power for the reference engine
$P_{Shaft,Max,SL}$ = maximum power rating at sea level
$P_{Shaft,Max,SL,Ref}$ = maximum power rating at sea level for the reference engine

This scaling assumes that the engines have a similar rpm range. If not, then the look-up parameter rpm can be switched to the ratio of rpm to the maximum rpm.

The best engine model is based on test data from the selected engine in the configuration that will fly. Reciprocating engine tests use dynamometers and other test equipment to measure fuel consumption vs applied shaft load and throttle (or commanded rpm). The test includes instrumentation that measures rpm, fuel flow, temperatures, and perhaps other parameters. Sophisticated engine tests might put the dynamometer and engine inside an altitude chamber that can simulate atmospheric conditions throughout the operating altitude range. If an altitude chamber is not used, then the test data will require corrections for anticipated altitude behavior. The corrections can utilize empirical equations or scaling, as just described in this section.

The following process can be used to estimate the integrated engine-propeller performance:

1. From the propeller methods, calculate the shaft power $P_{Shaft,Prop}$ required by the propeller to generate the required thrust T. If a propeller map is available, then also determine the rpm associated with this condition.
2. The total shaft power required is the sum of $P_{Shaft,Prop}$ and the generator load:

$$P_{Shaft} = P_{Shaft,Prop} + \frac{P_{Gen}}{\eta_{Gen}} \quad (8.34)$$

3. Determine if the shaft power is less than the maximum shaft power. If so, then proceed to the next step. If not, then the maximum power constraint is violated.

4. Calculate the throttle for this condition:

$$\text{throttle} = \frac{P_{\text{Shaft}}}{P_{\text{Shaft,Max}}(h, \text{rpm})} \tag{8.35}$$

5. Calculate the BSFC from the engine model now that all of the parameters are known. Some engine data sets will include fuel flow rate directly instead of BSFC.
6. Calculate the fuel flow rate if required.

$$\dot{m}_{\text{Fuel}} = BSFC \cdot P_{\text{Shaft}} \tag{8.36}$$

8.4 Turbofans and Turbojets

Turbofan and turbojet engines are generically referred to as jet engines. These propulsion systems maintain thrust at high speeds and high altitudes better than propeller-driven alternatives. As a result of this behavior, the range and endurance equations for jet-powered aircraft in Chapter 3 are derived based on the assumption that thrust and thrust-specific fuel consumption are independent of airspeed. The high-speed capabilities of jet engines make them suitable for UAs flying at equivalent airspeeds greater than 200 kt and at Mach numbers greater than 0.6. Jet engines also have lower thrust lapse with altitude relative to nonturbocharged reciprocating engines driving propellers.

The simplest form of jet engine is the turbojet, shown in Fig. 8.6. The oncoming air enters the engine in an inlet, which helps diffuse (compress) the air. A compressor slows the air to near static conditions, while increasing the pressure and temperature. The compressed air enters a burner where fuel is injected and ignited. This heated air is then expanded in a turbine, which provides power to drive the compressor. This power is mechanically transferred between the turbine and compressor via a shaft. The air exits the turbine and is accelerated in a nozzle to provide high velocity flow for thrust generation.

The pressure of the air exiting the turbojet nozzle is expanded almost to ambient pressure. The pressure thrust contribution becomes negligible:

$$T = \dot{m}_a(V_e - V) + \underbrace{(P_\infty - Pe)}_{\approx 0} \cdot A_e \Rightarrow \dot{m}_a(V_e - V) = \dot{m}_a \cdot \Delta V \tag{8.37}$$

The propulsive efficiency can be modified to be a function of thrust and mass flow rate:

$$\eta_{\text{propulsive}} = \frac{2V}{V_e - V} = \frac{1}{(\Delta V/2V) + 1} = \frac{1}{(T/2V \cdot \dot{m}_a) + 1} \tag{8.38}$$

From this relationship, it can be seen that a propulsive efficiency is increased when the ratio of thrust to mass flow rate is decreased. This is accomplished when the engine generates thrust by acting on a larger capture area. Turbofans accomplish this aim by using a fan rather than just the core flow.

Turbofans are similar to turbojets, except that a second turbine is added to drive a ducted fan. The high-pressure turbine drives the compressor. The low-pressure turbine drives the fan, and it is located downstream of the high-pressure turbine. The fan turns at a much slower rate than the compressor, and so two separate shafts are used. The high-speed shaft drives the compressor, and the low-speed shaft drives the fan. In the special case of a geared turbofan, the fan has a reduction gearbox that spins the fan at a lower rate than the low-speed shaft. Figures 8.5 and 8.6 show turbofan engines.

A defining parameter for turbofan engines is the bypass ratio α, which is defined at the ratio of the mass flow through the fan $\dot{m}_f$ to the mass flow through the compressor $\dot{m}_c$:

$$\alpha = \frac{\dot{m}_f}{\dot{m}_c} \tag{8.39}$$

The bypass ratio can be used to characterize engine classes. A turbojet has a bypass ratio of 0. Low-bypass-ratio engines have bypass ratios ranging from 0.5–2. Medium-bypass–ratio engines have bypass ratios of 2–4. High-bypass-ratio engines have bypass ratios above 4.

Fig. 8.5   Cutaway model of the Pratt and Whitney PW 500 turbofan engine.

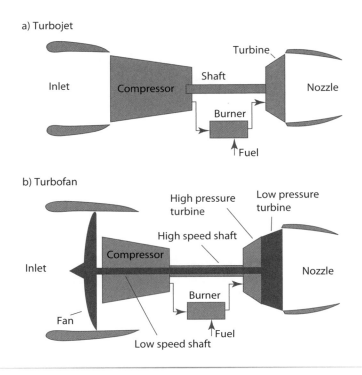

Fig. 8.6 Turbojet and turbofan major components.

A detailed description of jet-engine cycle analysis is beyond the scope of this book. Sources such as Oates [5] and Mattingly [6] provide an introduction to methods of calculating on-design and off-design performance analysis of various forms of turbojets, turbofans, and turboprops. Detailed engine component performance information used by the cycle analysis model for a fixed engine is generally proprietary to the engine manufacturer or otherwise unavailable. However, these methods are suitable for engine trade studies and performance estimation of conceptually sized engines.

Jet-engine manufacturers provide airframe manufacturers with engine decks that can be used by flight performance models. The engine maps could be an executable computer program based on calibrated cycle analysis methods or data tables based on test results. Driving parameters are flight velocity, atmospheric conditions, thrust required, mechanical power extraction, and bleed air extraction. The outputs of the engine deck can include maximum thrust available, minimum thrust, and fuel burn. Engine decks are specific to a single engine configuration. Often, the engine decks make assumptions about the airflow at the engine face and nozzle that are incompatible with the installed inlet and nozzle designs.

The propulsion performance models must provide corrections for these differences. Engine decks might be too computationally expensive to run in an integrated conceptual design tool.

A specific engine must be selected in the later stages of conceptual design rather than using a rubber engine model. Recall from Chapter 3 that a rubber engine model scales the engine properties, such as weight, based on the maximum thrust. Before entering detailed technical discussions with engine manufacturers, a refined list of candidate engines should be identified. Mature jet engines are generally selected for UAS programs because development budgets are rarely available and UA production quantities are usually insufficient to justify a commercial development by the engine manufacturer. The suitability of the available engines for the UA design must be evaluated with minimal information.

The engine manufacturer can publish information on a handful of jet-engine parameters. Publicly available information can include engine dry weight, major dimensions, sea-level static uninstalled maximum thrust, and sea-level static thrust specific fuel consumption. The weight can be directly applied within the mass properties' analysis. Dimensions of the engine will help define the UA geometry. The thrust and fuel consumption parameters must be adjusted for the UA integration, flight regime, bleed air extraction, and power extraction.

Jet-engine maximum thrust and thrust specific fuel consumption are dependent upon flight conditions, particularly altitude and Mach number. The behavior is dependent upon the engine cycle, with bypass ratio as a major driver. Mattingly [6] provides thrust lapse equations for three bypass-ratio classes. Torenbeek [7] offers the following approximation for thrust lapse during takeoff assuming no altitude variation:

$$\frac{T_{\text{Max}}}{T_{\text{Max,SLS}}} = 1 - \frac{0.45 \cdot M \cdot (1 + \alpha)}{\sqrt{(1 + 0.75 \cdot \alpha) \cdot G}} + \left(0.6 + \frac{0.11 \cdot \alpha}{G}\right) \cdot M^2 \qquad (8.40)$$

where G is a gas-generator function with a recommended range of $0.9-1.2$ for takeoff. This equation shows that the thrust lapse with Mach number increases as the bypass ratio grows.

Another approach is to generate data tables of published engine data of a similar bypass ratio. The thrust-specific fuel consumption and maximum thrust can be found in the following manner:

$$T_{\text{Max}}(h, M) = T_{\text{Max,SLS}} \frac{T_{\text{Max,Ref}}(h, M)}{T_{\text{Max,SLS,Ref}}} \qquad (8.41)$$

$$TSFC(h, M, \text{Throttle}) = TSFC_{\text{SLS}} \frac{TSFC_{\text{Ref}}(h, M, \text{Throttle})}{TSFC_{\text{SLS,Ref}}} \qquad (8.42)$$

where subscript SLS is sea-level static, Max is maximum, and Ref is the reference engine used to generate the data tables. Throttle is the ratio of thrust required to maximum thrust. This method can be expanded to use bypass ratio to interpolate between the results of multiple engine tables representing turbofan engines with several different bypass ratios.

Jet-powered aircraft can utilize an air-cycle machine environmental control system (ECS) that uses high-pressure bleed air from the jet engine. The air is usually extracted from the compressor. The air is exhausted from the aircraft at low speed and can be assumed to be negligible. For conceptual design purposes the effects on the engine can be analyzed as a reduction to the maximum thrust and an increase to the *TSFC*. The change is maximum thrust available can be modeled as

$$\Delta T_{\text{Bleed}} = -\frac{\dot{m}_{\text{Bleed}} \cdot V}{\eta_{\text{Cycle}}} \quad (8.43)$$

where $\dot{m}_{\text{Bleed}}$ is the mass flow rate of the bleed air, and η_{Cycle} is the efficiency losses on the bleed air due to the engine losses. η_{Cycle} is dependent upon the engine. The *TSFC* multiplication factor that accounts for bleed can be estimated by

$$f_{\text{Bleed}} = 1 + \frac{\Delta T_{\text{Bleed}}}{T_{\text{Req}}} \quad (8.44)$$

Power extraction can also have a significant impact on jet-engine performance. Jet engines have accessory pads that provide mechanical power for generators, starter motors, fuel pumps, and hydraulic pumps. Unmanned aircraft can have very large electrical power requirements for payloads, communications systems, and avionics. Simply not calculating the impacts of this power extraction will lead to overly optimistic performance and sizing estimates. The UA power comes from generators mounted to the engine's accessory pad. The impact to the maximum thrust available due to power extraction can be calculated as

$$\Delta T_{\text{Gen}} = -\frac{P_{\text{Gen}}}{V \cdot \eta_{\text{Gen}} \cdot \eta_{\text{Cycle}}} \quad (8.45)$$

where P_{Gen} is the generator power output and η_{Gen} is the generator efficiency. Note that η_{Cycle} might not necessarily have the same value as for bleed air, but 80% efficiency is a reasonable assumption. The *TSFC* multiplication factor due to generator power extraction can be estimated by

$$f_{\text{Gen}} = 1 + \frac{\Delta T_{\text{Gen}}}{T_{\text{Req}}} \quad (8.46)$$

The resulting equations for maximum thrust available and thrust specific fuel consumption are

$$T_{\text{Max}}(h, M, \dot{m}_{\text{Bleed}}, P_{\text{Gen}}) = T_{\text{Max}}(h, M) + \Delta T_{\text{Bleed}} + \Delta T_{\text{Gen}} \quad (8.47)$$

and

$$TSFC(h, M, \text{Throttle}, \dot{m}_{\text{Bleed}}, P_{\text{Gen}})$$
$$= TSFC(h, M, \text{Throttle}) \cdot f_{\text{Bleed}} \cdot f_{\text{Gen}} \quad (8.48)$$

Jet engines and turboprops usually operate with heavy fuel. This includes Jet A for civilian airports and JP-5 or JP-8 for military operations. Although heavy fuel compatibility is difficult to realize with reciprocating engines, it is the normal fuel type for turbines. This attribute makes jets and turboprops highly desirable for military applications.

Turbines can be very reliable and have long mean time between overhauls compared with reciprocating engines. There are relatively few moving parts to fail. The combustion is continuous rather than oscillatory, and the rpm is steady; therefore, the vibration is low relative to reciprocating engines.

With the heavy fuel compatibility, good altitude performance, ability to generate thrust at high speeds, high reliability, and long life, why are turbines not used more frequently on UASs? There are several answers to this question. In many cases turbines simply are not available with the necessary performance for smaller UASs. Reciprocating engines tend to have a lower acquisition cost and can be easier to integrate with the avionics. Also, jet-powered aircraft are not efficient for long-endurance lowspeed missions.

Candidate jet engines are generally derived from manned aircraft applications. Business jet engines range from 800–10,000 lb thrust and have medium to high bypass ratios. These engines are suitable for long-range or HALE UAS. Typical business jet-engine sea-level static *TSFC* values range from 0.5–0.8 lb/lb-hr. Commercial jet transport engines tend to be larger than what is required for many UAS applications and are therefore not used. Large high-speed UASs, such as UCAVs, can utilize military tactical aircraft engines. Often, the afterburner is not used for subsonic aircraft.

There are very few manned aircraft engines below 900 lb thrust, requiring engines built specifically for UASs. The largest markets for military small turbine engines are targets and cruise missiles, which support production runs of several hundred to a few thousand engines per UA type. The thrust class of targets (Fig. 8.7) and cruise missiles generally ranges from 30–1,000 lb. Cruise missile engines are designed to operate for a single flight, must start rapidly, and might not be required to power generators or supply bleed air. These engines tend to have a short design life and are difficult to tailor to other UAS applications such as long-duration ISR missions. Also, these engines are usually single-spool turbojets using centrifugal compressors and are therefore much less efficient than a traditional turbofan. This configuration is driven to a large extent by cost. These engines have sea-level static *TSFC* values ranging from $1-1.5$ lb/lb-hr. Targets are often recoverable and designed to fly multiple sorties, and their engines must either have suitable life or be replaceable between flights. So far, there has been little investment in efficient, long-life small turbofan engines with support for bleed air and power extraction.

At the smallest end of the jet-engine thrust spectrum are engines designed for model aircraft. For many years, hobbyists used reciprocating engine-powered ducted fans, but now, small turbojets are widely available and sufficiently affordable for this market. These engines are intended for aircraft weighing $10-55$ lb, and so the thrust class is generally less than 30 lb. These model aircraft jet engines often come with digital engine controllers. Like target and cruise missile engines, the design life is low, and the *TSFC* is high. However, these engines are suitable for very small high-speed UAS and subscale research UA.

Unmanned aircraft designed with radar-cross-section (RCS) reduction objectives appear to use jet propulsion exclusively. Examples of UCAV designs (see Fig. 8.8) include Northrop Grumman X-47A and B, Boeing X-45C and Phantom Ray, General Atomics Avenger, BAE Taranis, Dassault Filur, and Saab Sharc. Similarly, jet-powered ISR UAS with low observable design attributes include the Lockheed Martin Dark Star and Pole Cat. All of these aircraft have inlets mounted on the upper fuselage or on the upper

Fig. 8.7 BQM-74E Target uses a Williams International turbojet engine.

a)

b)

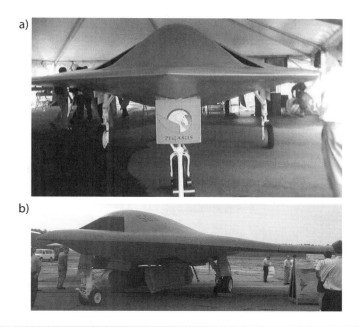

Fig. 8.8 Bifurcated upper surface inlet on the a) X-47A and single inlet on the b) X-47B.

surfaces of the wings. The engines are buried within the airframe, where the inlets can curve aggressively to shield the engine face. Engine nozzles might be partially blocked from below. Details of low-RCS installation options and other UA jet-engine installation arrangements are covered in Chapter 4.

Example 8.1 Engine Altitude Performance Estimation from Reference Engine

Problem:

You are surveying candidate engines with the objective of selecting an engine for the new UA. Although detailed information is available for a family of engines from one manufacturer, the only usable data on another promising engine come from a data sheet. These data include the uninstalled sea-level static maximum thrust and the sea-level static thrust specific fuel consumption, which are

$$T_{\text{Max,SLS}} = 5{,}000 \text{ lb}$$
$$TSFC_{\text{SLS}} = 0.6 \text{ lb/lb-hr}$$

A similar engine has the following thrust and thrust specific fuel consumption curves:

$$T_{Max} = T_{Max,SLS} \cdot \frac{\delta}{\theta}$$

$$TSFC = \theta^{0.47} \cdot (TSFC_{SLS} + 0.4 \cdot M)$$

where δ is the ratio of pressure at altitude vs sea level and θ is the ratio of temperature at altitude vs sea level. The reference engine $T_{Max,SLS}$ is 10,000 lb, and the $TSFC_{SLS}$ is 0.55 lb/lb-hr. Assuming no bleed air or power extraction, what are the uninstalled maximum thrust and thrust specific fuel consumption values at 50,000-ft MSL altitude and Mach 0.7?

Solution:

At an altitude of 50,000 ft, δ is 0.11512, and θ is 0.75187. The maximum thrust of the reference engine at the flight condition is

$$T_{Max} = T_{Max,SLS} \cdot \frac{\delta}{\theta} = 10,000 \text{ lb} \cdot \frac{0.11512}{0.75187} = 1,531.1 \text{ lb}$$

The specific fuel consumption for the reference engine is

$$TSFC = \theta^{0.47} \cdot (TSFC_{SLS} + 0.4 \cdot M)$$

$$= 0.75187^{0.47} \cdot [0.55(\text{lb/lb} - \text{hr}) + 0.4 \cdot 0.7] = 0.7259(\text{lb/lb} - \text{hr})$$

The maximum thrust of the engine at the flight condition is estimated by

$$T_{Max}(h, M) = T_{Max,SLS} \frac{T_{Max,Ref}(h, M)}{T_{Max,SLS,Ref}} = 5,000 \text{ lb} \frac{1,531.1 \text{ lb}}{10,000 \text{ lb}} = 765.6 \text{ lb}$$

The thrust specific fuel consumption of the engine at the flight condition is estimated by

$$TSFC(h, M, \text{Throttle}) = TSFC_{SLS} \frac{TSFC_{Ref}(h, M, \text{Throttle})}{TSFC_{SLS,Ref}} = 0.6 \frac{0.7259}{0.55}$$

$$= 0.7919 \text{ lb/lb} - \text{hr}$$

8.5 Turboshafts and Turboprops

Turboprop engines are quite similar to turbofans, except that the low-pressure turbine drives a propeller instead of a ducted fan. A reduction gearbox is usually required to match the speed of the propeller and low-pressure turbine. Turboshaft engines are similar to turboprops, except the shaft does not drive a propeller and is available for other uses such as powering a generator. Figure 8.9 shows the basic arrangement of a turboshaft engine.

Turboprop engines have many favorable attributes. Like jets, these engines usually operate on heavy fuels such as Jet A, JP-5, and JP-8. The

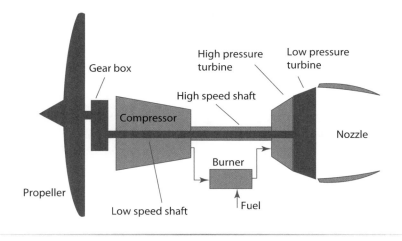

Fig. 8.9   Turboprop major components.

vibration level is low, and reliability is high relative to reciprocating engines. The performance is suitable for UAs operating between 25,000–50,000 ft, which is above the operable range of normally aspirated reciprocating engines. Quite importantly, turboprops have low weight for their power and are scalable to very large power levels. Turboprops have power-to-weight ratios of 2.1–2.9 hp/lb, and turboshafts are approximately 2.6–5.5 hp/lb. Manned aircraft turboprop engines have mean time between overhauls of approximately 3,000–4,000 hrs, which is over twice that of reciprocating engines. Turboprops are used at flight speeds less than Mach 0.6 due to propeller tip compressibility constraints.

So why are turboprops not used more frequently on UASs? Perhaps the greatest single reason is the lack of available engines in the desired power class. Another consideration is that turboprops often have higher brake specific fuel consumption compared with gasoline and heavy fuel reciprocating engines. Turboprops also have higher acquisition costs, which is detrimental when UA price is emphasized over life-cycle cost.

Most turboprop engines are intended for large general-aviation aircraft, regional commercial transports, and military transports. The smallest are built for general-aviation single- or twin-engine aircraft, and the equivalent power is greater than 450 ESHP. This engine class has much more power than is necessary for most UAS applications.

Very small turboprop engines are emerging for model aviation uses. These engines are of the same pedigree as model aircraft turbojets. These engines generally use centrifugal compressors. The engines are designed for low cost and high power-to-weight, which comes at the expense of fuel consumption, reliability, and engine life. These engines generally do not support generator power extraction or bleed air without modification.

Two notable UASs that employ turboprop engines are the General Atomics MQ-9 Predator B and the IAI Heron TP. The Predator B has a single TP-331 engine driving a pusher propeller at the aft end of the fuselage. The twin-boom Heron TP has a single 1,000 ESHP class turboprop driving a pusher propeller on the central fuselage between the tail booms.

Turboshaft engines are ideal for unmanned vertical takeoff and landing (VTOL) propulsion. The Bell Eagle Eye uses a single 641 ESHP Pratt and Whitney PW200/55 turboshaft engine to drive two tilting rotors. The Boeing A-160 Hummingbird unmanned helicopter uses a Pratt and Whitney PW207D turboshaft to power the main rotor and tail rotor.

For turboprops, most of the thrust is generated by the propeller. However, not all of the energy is absorbed by the compressor and propeller. High-speed air exits the nozzle providing a portion of the overall thrust T_{Nozzle}. The equivalent thrust power is given by

$$P_{\text{Thrust}} = (P_{\text{Shaft}} - P_{\text{Extract}}) \cdot \eta_{\text{p}} + T_{\text{Nozzle}} \cdot V \qquad (8.49)$$

where P_{Extract} is power extraction other than the propeller. The shaft power minus the extracted power is available for the propeller. Assuming that all of the extracted power goes to the generator, this relationship becomes

$$P_{\text{Thrust}} = \left(P_{\text{Shaft}} - \frac{P_{\text{Gen}}}{\eta_{\text{Gen}}} \right) \cdot \eta_{\text{p}} + T_{\text{Nozzle}} \cdot V \qquad (8.50)$$

When the engine equivalent thrust power equals the required thrust power, the required propeller thrust power $P_{\text{Prop,Req}}$ is given by

$$P_{\text{Prop,Req}} = \left(P_{\text{Shaft}} - \frac{P_{\text{Gen}}}{\eta_{\text{Gen}}} \right) \cdot \eta_{\text{p}} = P_{\text{Thrust}} - T_{\text{Nozzle}} \cdot V \qquad (8.51)$$

A *recuperator* is a device that diverts the hot exhaust gases and uses this heat to preheat the inlet air. Heat exchangers are used to transfer the heat. Heat applied from the recuperator reduces the requirement of the fuel combustion to provide the heat, thus reducing the fuel flow rate and the brake specific fuel consumption. However, the hot exhaust gas ducting and heat exchangers increase the propulsion system weight.

Turboprops, like jets, can provide bleed air to the ECS. The high-pressure air comes from the compressor. When engine decks or engine cycle models are not available, the change in maximum power available ΔP_{Bleed} can be modeled as

$$\Delta P_{\text{Bleed}} = -\frac{\dot{m}_{\text{Bleed}} \cdot V^2}{\eta_{\text{Cycle}}} \qquad (8.52)$$

The BSFC multiplication factor that accounts for bleed can be estimated by

$$f_{\text{Bleed}} = 1 - \frac{\Delta P_{\text{Bleed}}}{P_{\text{Req}}} \tag{8.53}$$

Turboprop and turboshaft performance estimates might come from engine cycle models or manufacturer-provided engine decks. The latter is preferable for performance analysis, especially when the data are based on test results.

8.6 Electric Motors

Electric motors first appeared on model aircraft and unmanned aircraft in the late 1970s. Astroflight developed the Sunrise I and II solar-powered aircraft in 1974 that used their rare Earth brushed motors. AeroVironment fielded the Pointer remotely piloted aircraft in the 1980s, which initially saw service in Operation Desert Storm in 1990. The NRL advanced electric UAS technology in the 1990s with the FLYRT, SENDER, Swallow, MITE, and Eager systems, among others. A new generation of operational small electric UASs emerged in the early 2000s in the form of AeroVironment RAVEN, Lockheed Martin Desert Hawk, and NRL Dragon Eye.

Electric motors have multiple advantages. Unlike internal combustion engines, there is very little maintenance and no consumables such as liquid fuel and lubricants. Electric motors can be stored for very long periods of time. No starters are required, and so air-launch is simplified. The performance characteristics are independent of altitude to a first order, provided that arcing does not occur. A new motor can be custom designed and built for a UAS with minimal cost relative to other propulsion types.

Electric motor systems are much simpler to design, integrate, and test than any other type of propulsor. This attribute makes electric motors a common choice for student projects and rapidly fielded UASs. In a matter of days, a single individual can select the motor system and perform initial testing. Compare this to a reciprocating engine that might require several man-years to make fully operational.

The biggest limitation for electric flight is the short endurance that can result from the power source. Namely, modern batteries have low specific energy relative to liquid hydrocarbon fuel, resulting in UAs with only 0.5–3 hrs endurance. Fuel cells are breaking this endurance barrier, as will be shown later in this chapter.

Most operational electric aircraft are small, usually weighing less than 20 lb and using motors with less than 1 kW. Reliable and supportable low-power reciprocating engines are difficult to design and manufacture. Suitable reciprocating engine choices that operate on appropriate fuel types

Fig. 8.10 Virginia Tech Forestry UAV with twin wing-mounted electric motors.

exist above this scale, yielding higher-performance UA. Small electric UASs are most often launched by hand or with the aid of elastic lines.

Electric-powered aircraft are practical at larger scales. One such example is the Virginia Tech Forestry UAV (Figs. 8.10 and 8.11), which was designed to carry a 25-lb payload for 2 hrs using lithium-ion batteries. The primary justification for using electric-battery propulsion instead of a fueled engine is that the sensitive scientific payloads could not have contamination from hydrocarbons.

Brushless motors are purely inductive. These are alternating current motors, though specialized motor controllers enable operation from direct current power sources. This configuration is known as brushless dc motor. The two main configurations of brushless motors are *inrunner* and *outrunner* motors, which are available in the hobby market at less than 10 kW.

Motor technology has made large strides over the past 30 years. During the 1970s–80s most unmanned aircraft motors were brushed rare earth or ferrite motors. Brushless inrunner motors became available in the mid-1990s, with Aveox as the market leader. These brushless motors had substantial improvement in reliability. The high operating rpm typically required gearboxes for proper matching with propellers on loitering UASs. In the late 2000s, brushless outrunner motors penetrated the market with efficient products that generated high torque at lower rpm, often eliminating the need for a gearbox.

Electric motor reliability can be high. A single shaft is suspended between two ball bearings. Brushed motors have a commutator that provides a physical electrical path to the motor core. These brushes wear down over time and must be replaced for proper performance and reliability.

Brushless motors are purely inductive, and so motor life is limited primarily by the bearings.

Electric motors can be turned on and off in flight, providing advantages for launch and recovery operations. Unmanned aircraft with hazardous propeller locations can be made airborne prior to starting the motor, which is especially helpful for hand launch and hand-release launch techniques. Air-launched aircraft can go through a sequence from the stowed to the flight configuration before starting the motor. On recovery, the motor can stop to increase the sink rate, whereas reciprocating engines have a minimum operating rpm.

The motor speed is controlled by a motor controller. In model aircraft, these devices are known as electronic speed controls (ESC). The function is to vary the voltage that is applied to the motor, thereby controlling the rotational speed. Brushed motors use transistors that switch the voltage on and off at very high rates. Three-phase brushless motor controllers vary the voltage across three power wires that interface with the motor.

Electric motors apply nearly constant torque to the propeller. Reciprocating engines, by comparison, apply torque pulses to the propeller during the power stroke. The smooth operation reduces the structural loads on the

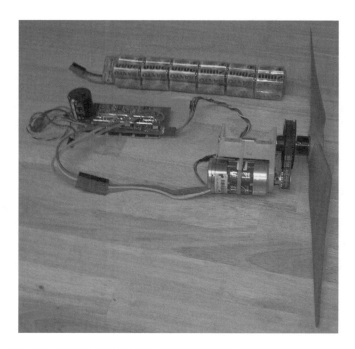

Fig. 8.11 Virginia Tech Forestry UAS electric propulsion system string consisting of a NiCd battery, electronic speed control, and geared brushless in runner motor.

propeller, which enables lower weight designs and thinner blade airfoil sections near the hub. The propellers with thin airfoil sections tend to be more efficient.

Properly balanced electric motors have low vibration levels compared to well-designed reciprocating engines. This quality can support low vibration levels at the payload for low jitter, better pointing accuracy, and reduced target location error. The cowling can be tightly integrated with the motor because little or no provision for the movement of motor vibration is needed. Also, the motor frontal area is less than a reciprocating engine producing equivalent power, permitting lower drag fairings—especially for pusher configurations.

Electric propulsion systems separate the propulsor from the power source. The propulsor is the motor-propeller combination. The power source can be a battery, solar array, turbine-driven generator, or fuel-cell stack, to name a few options. Turboprops, jet engines, and reciprocating engines with propellers combine the power source with the propulsor. The thrust-to-weight ratio of an electric motor with propeller is significantly lighter than the other propulsor-powerplant combinations. This opens up numerous configuration options for propeller location.

Electric motors have no emissions, and so the UA emissions depend upon the power source type. Batteries and solar arrays produce no emissions. Hydrogen fuel cells emit only water. Hydrocarbon-based power sources for electric motors have complex emissions that include CO_2 and water, among many other constituents. Electric UASs that are powered by batteries, solar arrays, and hydrogen fuel cells have potential for very limited environmental impact. Additionally, UAs without emissions will not contaminate sensitive air-sampling payloads for missions such as in situ chemical and biological agent detection.

The key feature of the electric motor is that the voltage is constant and the current varies. Given a steady voltage across the leads of a motor, the rotational rate will remain constant regardless of load. The higher the power load on the motor, the more current is drawn from the power source. Ultimately, the current reduces the voltage through resistance losses. This discussion assumes that the power from the motor does not exceed design limitations of a given motor. This relationship is expressed as

$$\text{rpm}_{\text{mot}} = K_\text{V} \cdot V_{\text{mot}} \tag{8.54}$$

where rpm_{mot} is the motor shaft rpm, K_V is the voltage constant, and V_{mot} is the voltage across the motor leads. Model aircraft electric motors can be more dangerous than gas engines of the same power class because the propeller does not slow down readily when a body part is hit; rather, the motor just pulls more current from the battery as it attempts to retain the rpm.

The following is a procedure for determining the current drawn from the battery of an electric UA.

1. Either the required thrust or thrust power absorbed by the propeller is known in advance. Recall that the advance ratio J for a given propeller and thrust condition is found through a table look-up procedure. The propeller rpm is found by

$$\text{rpm}_{\text{prop}} = 60 \frac{V}{J \times D} \qquad (8.55)$$

If the flight velocity and either thrust or power is known, the unknown parameter can be found by

$$P_{\text{abs,prop}} = T_{\text{prop}} \cdot V \quad \text{or} \quad T_{\text{prop}} = \frac{P_{\text{abs,prop}}}{V} \qquad (8.56)$$

2. Now, the motor shaft power and rpm for a geared system can be determined:

$$P_{\text{shaft}} = \frac{P_{\text{abs,prop}}}{\eta_{\text{gear}} \cdot \eta_p} \qquad (8.57)$$

and the shaft rpm

$$\text{rpm}_{\text{Motor}} = \text{rpm}_{\text{prop}} \cdot G \qquad (8.58)$$

where G is the gear ratio.
3. Now, the voltage and current across the motor leads can be calculated. The current across the motor leads is

$$I_{\text{Mot}} = P_{\text{Mot}} \frac{K_V}{RPM_{\text{Mot}}} \cdot I_{0L} \qquad (8.59)$$

The voltage is found by

$$V_{\text{Mot}} = \frac{\text{rpm}_{\text{Mot}}}{K_V} + I_{\text{Mot}} \cdot R_{\text{Mot}} \qquad (8.60)$$

where I_{0L} is the current required to turn the shaft from rest, and K_V is the motor voltage constant. The motor efficiency η_{Mot} is given by

$$\eta_{\text{Mot}} = \frac{P_{\text{Shaft}}}{I_{\text{Mot}} \cdot V_{\text{Mot}}} \qquad (8.61)$$

4. Determine the power required at the leads of the motor controller P_{ESC}. This power is a function of the power input to the motor and the

controller efficiency:

$$P_{ESC} = \frac{I_{Mot} \cdot V_{Mot}}{\eta_{ESC}} \tag{8.62}$$

The motor controller efficiency is a function of relative power setting. When the power setting is near 100%, the efficiency is highest. An average expected motor controller efficiency should be used when a wide power setting range is considered. Often, detailed motor controller efficiency information is unknown, and so an estimate for the class of system can be used. The efficiency for a brushed motor controller can be 85%, and brushless motor controller efficiency can be 95%.

5. The final step is to determine the effect of the power draw on the power source. The power required must be less than or equal to the maximum power capacity. For batteries, fuel cells, or hybrid electric systems, the power consumption will draw down the energy source and drive flight duration.

6. Boucher [8] can be consulted for further information on electric motor systems.

8.7 Batteries

A battery is an electrochemical device that converts stored chemical energy into electrical power. In half of the cell, positively charged ions, called cations, migrate to a cathode electrode through an electrolyte. In the other half of the cell, negatively charged ions, called anions, migrate to the anode electrode through an electrolyte, as shown in Fig. 8.13a. An electrical load can be placed between the battery leads. Batteries operate with a closed thermodynamic cycle.

Batteries can be rechargeable or single use. Nonrechargeable batteries are known as *primary*, and rechargeable batteries are called *secondary*, though this has nothing to do with redundancy. Primary cells may have superior performance characteristics and may be the best choice for single-use electric-powered aircraft. The cost of replacing batteries is generally prohibitive for multi-use UAs, and so secondary batteries are used for those applications. Secondary batteries are also required for multiday duration solar-powered aircraft that use batteries to power the motors at night.

Several battery chemistries have been used for UAS propulsion systems. The properties of several battery types are shown in Table 8.2. Nickel-cadmium (NiCd) batteries were dominant in the 1980s and 1990s. Nickel metal hydride (NiMH) made a brief appearance in the late 1990s and early 2000s. However, lithium-ion (Li-Ion) and lithium-ion-polymer (Li-Po) batteries are the main type in use today for small UAS propulsion

Table 8.2 Characteristics of Battery Chemistries

Battery Type	Theoretical Specific Energy, W-hr/kg	Practical Specific Energy, W-hr/kg	Specific Power, W/kg	Cell Voltage, V
Lead acid (Pb/acid)	170	30–50	180	1.2
Nickel cadmium (NiCd)	240	60	150	1.2
Nickel metal hydride (NiMH)	470	23–85	200–400	0.94–1.2
Lithium ion (Li-Ion)	700	100–135	250–340	3.6
Lithium polymer (Li-Po)	735	50.7–220	200–1900	3.7
Lithium sulfur (LiS)	2550	350	600–700	2.5

systems. Lithium sulfur (LiS) promises to provide improved performance once more fully matured. An early application of LiS was the rechargeable battery used on the multiday endurance QinetiQ Zephyr solar-powered UAS. It can be seen that LiS batteries have the highest theoretical and practical specific energy values, though current cells have limited charge/discharge cycles. Mueller [9] describes electric propulsion technology. Lithium-based batteries are suitable for a wide range of UAS applications, including rotorcraft (Fig. 8.12).

Technology predictions for battery performance abound. The reader may find entertainment in reading highly optimistic battery technology predictions from 10 or 20 years ago that did not materialize on the projected schedule. Despite not meeting the timelines of the futurists, batteries have made remarkable progress over the past decades. NiCd battery-powered UA from the 1980s might have achieved 20-min endurance, where today's unmanned aircraft using LiPo batteries might have practical flight durations of over 2 hrs. The designer should be wary of depending upon the next generation of batteries to enable the mission capability. Rather, it is recommended that off-the-shelf batteries be used. Despite the high operational utility provided by UASs, transportation and consumer products industries will likely drive battery technology breakthroughs.

Battery packs consist of multiple cells. These can be arranged in series, parallel, or a combination of series and parallel. The arrangement depends

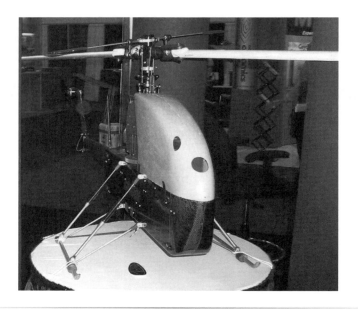

Fig. 8.12 NRL Spider battery-powered unmanned helicopter.

upon the input voltage of the motor system and maximum current required. The series battery pack voltage V_{Pack} and maximum current $I_{\text{Max,Pack}}$ are

$$V_{\text{Pack}} = N_{\text{Cells}} \cdot V_{\text{Cell}} \tag{8.63}$$

$$I_{\text{Max,Pack}} = I_{\text{Max,Cell}} \tag{8.64}$$

where N_{Cells} is the number of cells, V_{Cell} is the cell voltage, and $I_{\text{Max,Cell}}$ is the cell maximum current. The voltage and maximum current for a parallel battery pack are

$$V_{\text{Pack}} = V_{\text{Cell}} \tag{8.65}$$

$$I_{\text{Max,Pack}} = N_{\text{Cells}} \cdot I_{\text{Max,Cell}} \tag{8.66}$$

Packs that are arranged in a combination of series and parallel generally have strings of serial cells that are connected in parallel. This can be thought of as series battery packs connected in parallel. In this case, the pack voltage and maximum current are

$$V_{\text{Pack}} = N_{\text{Series}} \cdot V_{\text{Cell}} \tag{8.67}$$

$$I_{\text{Max,Pack}} = N_{\text{Strings}} \cdot I_{\text{Max,Cell}} \tag{8.68}$$

where N_{Series} is the number of cells arranged in series for each string and N_{Strings} is the number of parallel strings.

The usable energy in a battery is given by the capacity. Recall the battery energy equation from Chapter 3:

$$Energy_{Batt} = E_{spec} \frac{W_{Batt}}{g} \eta_{Batt} \cdot f_{Usable} \tag{8.69}$$

This is a parametric equation that scales the battery energy with battery weight and specific energy. This is appropriate for the early conceptual design stage. As the design matures, the battery pack is defined as well. Each cell has a capacity C_{Cell} given in ampere-hours and a cell voltage. The usable battery energy is

$$Energy_{Batt} = N_{Cells} \cdot C_{Cell} \cdot V_{Cell} \cdot \eta_{Batt} \cdot f_{Usable} \tag{8.70}$$

The battery efficiency η_{Batt} is a function of the current draw profile. The current-efficiency dependency behavior varies among the battery chemistries and the specific cell design.

In the description of the electric motor system in the preceding section, the power source was not defined. Now consider the case of a battery power source. The battery current I_{batt} is found by solving the quadratic equation

$$I_{batt} = \frac{V_{batt,spec} - \sqrt{V_{batt,spec}^2 - 4 \cdot R_{batt} \cdot P_{cont}}}{2 \cdot R_{batt}} \tag{8.71}$$

where $V_{batt,spec}$ is the zero-current voltage across the battery leads. This current draw must be less than or equal to the pack maximum current, or the pack can suffer damage. The voltage across the battery leads under load V_{batt} is

$$V_{batt} = V_{batt,spec} - I_{batt} \cdot R_{batt} \tag{8.72}$$

The useful power generated by the battery $P_{batt,use}$ is

$$P_{batt,use} = V_{batt} \cdot I_{batt} \tag{8.73}$$

and the battery power given off as heat $P_{batt,heat}$ is

$$P_{batt,heat} = I_{batt}^2 \cdot R_{batt} \tag{8.74}$$

The total power generated by the battery $P_{batt,tot}$ is

$$P_{batt,tot} = P_{batt,use} + P_{batt,heat} \tag{8.75}$$

The battery efficiency is

$$\eta_{batt} = \frac{P_{batt,use}}{P_{batt,tot}} \tag{8.76}$$

The total propulsion system efficiency is given by

$$\eta_{propulsion} = \frac{P_{abs,prop}}{P_{batt,tot}} \tag{8.77}$$

Onboard rechargeable battery systems must have charge/discharge controllers, heat exchangers, and power conditioners. This is important for solar-powered aircraft with battery energy storage. PCS [10] uses the following scaling laws for charge/discharge controllers $M_{charger}$ and heat exchangers M_{HE}:

Battery charge/discharge controller mass:

$$M_{charger} = 1.10 \cdot \frac{P_{batt}}{B_M} \text{ kg} \tag{8.78}$$

Heat exchanger mass:

$$M_{HE} = 1.10 \cdot P_{batt} \cdot \frac{(1 - \eta_{batt})}{C_M} \text{ kg} \tag{8.79}$$

The factors B_M and C_M both range between $250-350$ W/kg, depending on the technology level. The roundtrip efficiency η_{batt} is typically $60-75\%$.

8.8 Fuel Cells

Fuel cells are electrochemical devices that use chemical reactions of a fuel source and oxidizer to generate electrical power. The fuel and oxidizer are consumed in the conversion process, and the byproducts are either exhausted from the fuel cell or stored onboard the UA. The reaction between the fuel and oxidizer occurs in the presence of an electrolyte, which is not consumed. Fuel cells have an open thermodynamic cycle.

Hydrogen-powered fuel cells operate in a similar manner regardless of type, as shown in Fig. 8.13b. A catalyst converts the H_2 into a positively

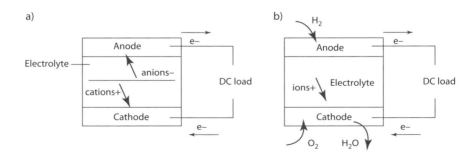

Fig. 8.13 Battery and fuel cell elements.

charged hydrogen ion and a negatively charged electron at the anode. The hydrogen ions freely pass through the electrolyte, but the electrons are blocked. This creates a voltage potential difference between the anode and cathode. The electrons are directed via porous gas electrodes embedded in the cathode and anode. The hydrogen ions combine with O_2, and the electrons returning from the electrical load to form H_2O, which can be either liquid water or water vapor. Because H_2O is the only emission, fuel cells are widely considered to be a clean power source.

Multiple individual fuel cells can be combined in a *fuel-cell stack*. Like batteries, the fuel cells can be arranged in series or parallel to produce the desired voltage or peak current. The voltage per cell is defined by the fuel-cell type, and the peak current is a function of the active surface area.

The two most common types of fuel cell are *proton exchange membrane* (PEM) and *solid-oxide fuel-cell* (SOFC) types. PEM and some SOFCs react with hydrogen (H_2) and oxygen (O_2) to generate electricity and water (H_2O). The equivalent BSFC for a PEM fuel cell can be approximately $0.15 - 0.25$ lb/hp-hr, where the reference fuel is H_2.

The temperature of the PEM fuel cell must be controlled to prevent thermal loading failure. The chemical reaction is exothermic, so that the heat generated must be rejected to keep the system within the operable limits. The electrolyte membrane must generally be kept low to prevent damage (some require temperatures less than approximately 180°F), which requires rejection of low-grade heat.

The membrane must be properly hydrated to ensure that water is created and evaporated at the same rate. If the required hydration is not maintained, the membrane will dry and crack, allowing the hydrogen and oxygen to react directly and generate destructive heating. On the other hydration extreme, excessive water will flood the electrodes and block the reaction.

PEM fuel-cell systems involve numerous support systems to ensure effective operations. Actively controlled pumps must properly hydrate the membrane. Thermal management might require heat exchangers. The incoming air might require conditioning, including dehumidification and pressurization.

Solid-oxide fuel cells (SOFCs) are high-temperature fuel cells that have the potential to run on numerous fuel types in addition to H_2. SOFCs are not as mature as PEM because of challenges with high-temperature materials. However, SOFCs offer the potential for greater efficiency.

Some key milestones in fuel-cell propulsion include the following:

- 2003, AeroVironment Hornet micro UA flies for 5 min on a fuel cell.
- 2004, AeroVironment Helios carries a fuel cell capable of generating propulsion power.
- 2005, AeroVironment successfully flies the subscale Global Observer UAS using LH2 storage.

Fig. 8.14 NRL XFC PEM fuel-cell powered UAS.

* 2009, NRL demonstrates 26-hrs endurance on the PEM fuel-cell-powered Ion Tiger research aircraft.
* 2009, NRL flies the mission-capable XFC small tactical UAS on PEM fuel cells for 6 hrs using a Protonex Technology Corporation fuel-cell stack (Fig. 8.14).

An airbreathing fuel-cell system must intake air for the oxygen required for operation. High-altitude operation might require one or more turbos to provide air at the appropriate pressure, much like a reciprocating engine. A combustor separate from the fuel cell burns H_2 gas and conditioned oxygen to drive the turbo expander, which, in turn, drives the compressor.

Fuel-cell systems can operate in a regenerative fashion. The byproduct of H_2-O_2 fuel cell operation is H_2O, which can be captured and stored. Power is provided to an electrolyzer by a separate power source, such as solar arrays, to separate this H_2O into H_2 and O_2 in a reverse osmosis process. The H_2 and O_2 are generally stored in gaseous form to avoid the losses associated with cryogenic storage. Fuel cells that perform the function of an electrolyzer when operating in reverse are known as *regenerative fuel cells* (RFC). RFCs are generally not as efficient as optimized electrolyzer systems, but they can provide weight savings and complexity reductions by combining both functions into a single element.

8.8.1 Hydrogen Storage

Hydrogen has a very high energy density relative to other fuels. Hydrogen has an energy density of 120 MJ/kg (33.3×10^3 W-hr/kg). This compares favorably with gasoline, which has 44.4 MJ/kg (12.3×10^3 W-hr/kg).

Hydrogen is notorious for being explosive. Public perceptions are driven to a large extent by the famous Hindenburg airship accident, which was primarily caused by the flammable skins rather than the H_2 lifting gas. However, many of these concerns might be unfounded. H_2 is the lightest gas and tends to diffuse quickly upon interaction with the atmosphere. Hydrogen can be stored and handled safely with proper equipment and procedures. The main methods of hydrogen storage are pressurized H_2 gas, cryogenically cooled liquid hydrogen (LH2), and chemical storage.

8.8.2 Pressurized Gaseous Hydrogen

Pressurized gaseous H_2 is stored in a pressure vessel. H_2 gas has a low density as a fuel, requiring substantial volume for long-duration UASs. The container geometry is usually a sphere or a cylinder with hemispherical end caps. The pressurized tanks must have low gas permeability to prevent loss of H_2 fuel. The mass of the H_2 gas contained within a pressure vessel is

$$M_{H_2} = \rho_{H_2} \cdot V_{Tank} = \frac{P_{H_2} \cdot V_{Tank}}{R \cdot T} \qquad (8.80)$$

where P_{H_2} is the gas pressure, V_{Tank} is the tank volume, R is the universal gas constant, and T is the gas temperature. The H_2 mass fraction MF_{H_2} is given by

$$MF_{H_2} = \frac{M_{H_2}}{M_{Tank} + M_{H_2}} \qquad (8.81)$$

where M_{Tank} is the tank mass. The tank mass depends upon the stresses associated with pressurization, scale, and configuration. Compressed gaseous hydrogen H_2 mass fractions of 6% have been realized in practice.

One method of estimating tank weight is through comparison with the characteristics of other tanks. A tank figure of merit FOM_{Tank} is defined by

$$FOM_{Tank} = \frac{P_{H_2} \cdot V_{Tank}}{M_{Tank}} \qquad (8.82)$$

This parameter has units of length.

A first principles structural sizing can provide more design detail. The volume of a circular cylinder tank with hemispherical end caps is

$$V_{Tank} = \frac{4}{3} \cdot \pi \cdot r^3 + \pi \cdot r^2 \cdot L \qquad (8.83)$$

where r is the cylinder radius and L is the length of the cylinder portion. The wall thickness on the cylindrical portion is

$$t_{w,cylinder} = \frac{P_{H_2} \cdot r \cdot FOS}{\sigma_y} \qquad (8.84)$$

where FOS is the factor of safety and σ_y is the maximum allowable stress. Similarly, the wall thickness for the hemispherical end caps is

$$t_{w,caps} = \frac{P_{H_2} \cdot r \cdot FOS}{2 \cdot \sigma_y} \qquad (8.85)$$

These wall thicknesses only consider the loads associated with pressurization. Flight loads and other stresses must be properly analyzed as well. The mass of the tank is then found by

$$M_{\text{tank}} = \rho_{\text{matl}} \cdot \left\{ \left[\frac{4}{3} \cdot \pi \cdot \left(r + t_{\text{w,caps}} \right)^3 + \pi \cdot \left(r + t_{\text{w,cylinder}} \right)^2 \cdot L \right] - V_{\text{Tank}} \right\}$$

(8.86)

where ρ_{matl} is the density of the tank material.

Example 8.2   Pressurized Tank Hydrogen Storage Mass Fraction

Problem:

A fuel-cell system requires 20.0 kg of H_2 for a long-endurance mission. The compressed hydrogen fuel tank is designed for 10,000 psi. A factor of safety of 5 is required. The H_2 gas is kept at sea-level standard temperature. Assume that the ratio of the cylindrical portion length to the radius is 2:1. Using the following materials, what are the hydrogen mass fractions?

1. ASTM-A514 steel alloy, density = 7,860 kg/m^3, yield strength = 690 MPa
2. 2014-T6 Aluminum alloy, density = 2,800 kg/m^3, yield strength = 410 MPa

Solution:

The pressure is converted from pounds per square inch to pascals:

$$P_{H_2} = 10,000 \text{ psi} \cdot \frac{1 \text{ Pa}}{1.4504 \times 10^{-4} \text{ psi}} = 6.8947 \times 10^7 \text{ Pa}$$

The volume of the tank is

$$V_{\text{Tank}} = \frac{M_{H_2} \cdot R \cdot T}{P_{H_2}} = \frac{20.0 \text{ kg} \cdot 287.05 (\text{N} - \text{m/kg} - \text{K}) \cdot 288.15 \text{K}}{6.8947 \times 10^7 \text{ Pa}}$$

$$= 2.40 \times 10^{-2} \text{ m}^3$$

Using the relationship $L = 2r$,

$$V_{\text{Tank}} = \frac{4}{3} \pi \cdot r^3 + \pi \cdot r^2 \cdot L = \frac{4}{3} \pi \cdot r^3 + \pi \cdot r^2 \cdot (2r) = \frac{10}{3} \pi \cdot r^3$$

Solving for r,

$$r = \left(\frac{3}{10 \cdot \pi} V_{\text{Tank}} \right)^{1/3} = \left(\frac{3}{10 \cdot \pi} 2.40 \times 10^{-2} \text{m}^3 \right)^{1/3} = 0.1318 \text{ m}$$

And the cylindrical segment length is 0.2637 m.

Starting with steel, the wall thickness for the cylindrical portion is

$$t_{w,cylinder} = \frac{P_{H_2} \cdot r \cdot FOS}{\sigma_y} = \frac{6.8947 \times 10^7 \, \text{Pa} \cdot 0.1318 \, \text{m} \cdot 5}{690 \times 10^6 \, \text{Pa}} = 6.587 \times 10^{-2} \, \text{m}$$

The wall thickness for the hemispherical end caps is

$$t_{w,caps} = \frac{P_{H_2} \cdot r \cdot FOS}{2\sigma_y} = \frac{6.8947 \times 10^7 \, \text{Pa} \cdot 0.294 \, \text{m} \cdot 5}{2 \cdot 690 \times 10^6 \, \text{Pa}} = 3..293 \times 10^{-3} \, \text{m}$$

The mass of the tank is

$$M_{tank} = \rho_{matl} \cdot \left\{ \left[\frac{4}{3} \pi \cdot \left(r + t_{w,caps} \right)^3 + \pi \cdot \left(r + t_{w,cylinder} \right)^2 \cdot L \right] - V_{Tank} \right\}$$

$$= 7{,}860 \, \text{kg/m}^3 \cdot \left\{ \left[\frac{4}{3} \pi \cdot (0.165 \, \text{m})^3 + \pi \cdot (0.198 \, \text{m})^2 \cdot 0.264 \, \text{m} \right] \right.$$

$$\left. - 2.4 \times 10^{-2} \, \text{m}^3 \right\} = 213 \, \text{kg}$$

The hydrogen mass fraction is then

$$MF_{H_2} = \frac{M_{H_2}}{M_{Tank} + M_{H_2}} = \frac{20.0 \, \text{kg}}{213 \, \text{kg} + 30.0 \, \text{kg}} = 8.6\%$$

Following the same procedure for the aluminum alloy gives a hydrogen mass fraction of 12%.

8.8.3 Liquid Hydrogen

Cryogenic-cooled liquid hydrogen (LH_2) is stored in an insulated vessel. Heat transfer to the LH_2 causes some of the fuel to change to the gaseous state. Spherically shaped double-walled vessels called *dewars* are effective means of storing the LH_2 at temperatures at or below 20K. The fuel cell operates from the gaseous H_2 leaving the dewar. If insufficient gaseous H_2 is available, then a heating element within the dewar boils more LH_2 to increase the flow rate. LH_2 storage in dewars offers the highest hydrogen storage mass fraction of the three alternatives described here at large scale. Colozza [11] covers hydrogen storage technologies.

The volume of the LH_2 storage is

$$V_{Tank} = \frac{M_{H_2} \cdot (1 + f_{Vol})}{\rho_{LH_2}} \tag{8.87}$$

Here, f_{Vol} is the ratio of excess volume to the total volume to allow for LH_2 boil off to H_2 gas. Colozza [11] recommends an f_{Vol} value of 7.2%. The density ρ_{LH_2} is the density of liquid hydrogen, which is 67.8 kg/m^3. If the tank is a sphere, then the radius is

$$r = \left(\frac{3/4 \cdot V_{Tank}}{\pi}\right)^{1/3} \tag{8.88}$$

If the tank is a circular cylinder, then Eq (8.83) must be solved for radius. Equations (8.84) and (8.85) provide the wall thickness. The mass of the inner tank is

$$M_{tank,inner} = \rho_{matl} \cdot \left[\frac{4}{3}\pi \cdot \left(r + t_{w,caps}\right)^3 + \pi \cdot \left(r + t_{w,cylinder}\right)^2 \cdot L\right]$$
$$- V_{Tank} \tag{8.89}$$

Insulation is required to prevent excessive boil-off rates and buildup of frost on the tank. Insulation can consist of vacuum-jacketed system, rigid closed cell foam, or other techniques. The LH$_2$ boil-off rate as a result of heat loss through the insulation $\dot{M}_{H_2}$ is

$$\dot{M}_{H2} = \frac{K \cdot A \cdot (T_s - T_{LH_2})}{t_{Ins} \cdot h_{fg}} \tag{8.90}$$

where

$K =$ thermal conductivity of the insulation
$A =$ surface area of the tank
$T_s =$ temperature of the outer surface of the insulation
$T_{LH_2} =$ temperature of the LH$_2$
$t_{Ins} =$ thickness of the insulation
$h_{fg} =$ latent heat of vaporization of LH2 (446,592 J/kg)

The weight of the insulation is equal to the product of insulation volume and density.

8.8.4 Chemical Hydrogen Storage

With chemical storage, hydrogen is combined with other elements in molecules. Usually the hydrogen is reversibly taken up by nitrides, imides, or hydrides. Metal hydrides are a common storage technique. The rate of H$_2$ release might be the most limiting factor for chemical storage. Hydride storage has H$_2$ mass fractions of approximately 1–7% relative to the hydride, where higher values correspond to nonreversible active heating. When the remainder of the tank is fully accounted for, practical hydride H$_2$ mass fractions might only be 0.36–0.68% in practice. Future developments can yield H$_2$ mass fractions of up to 5%. Depending on the metal, the hydrogen density ranges from 90–151 kg/m^3.

Another form of chemical storage is traditional fuels such as gasoline, methanol, or butane. These fuels can be efficiently stored in liquid or gaseous form. Most of these fuels require a reformer to extract the H_2 gas.

8.9 Solar Power

Solar cells use the photovoltaic effect to convert the sun's radiated power into electrical power. This power does not require onboard energy storage for peak daylight operations. However, multiday flights require that excess energy be stored to power the UA through the night. An UA that solves this energy balance can operate almost indefinitely, bounded only by reliability and component life.

8.9.1 Solar Collector Integration

Individual solar cells are combined in solar arrays. Much like a battery, the cells can be arranged in combinations of series and parallel to yield the appropriate voltage. Integrating solar arrays on an aircraft involves compromises across a number of disciplines.

A flat array panel can be mounted on the aft upper surface of the wing, perhaps between the maximum thickness point and the trailing edge. Such an approach was used on the solar-powered aircraft in Fig. 8.15. This is the region of the airfoil with the least curvature on the upper surface. An externally mounted array requires a large flat section over this region, which increases the airfoil drag and can reduce the maximum lift coefficient. Even arrays made of brittle crystalline cells can accommodate some curvature, and so very large solar-powered aircraft can have externally mounted arrays on relatively conventional airfoils. Alternatively, the array

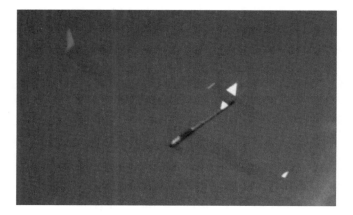

Fig. 8.15   Author's 3-m span solar-powered aircraft flying in 1995.

can be mounted inside a wing, though this requires a covering material that does not absorb light in the critical wavelength range for the solar cell. Also, any structure required to maintain the wing aerodynamic geometry, such as ribs and spars, can cast a shadow on the array.

Solar arrays cause structural challenges. Solar-powered aircraft usually have gossamer structures, and the solar arrays comprise a large portion of the wing weight. Most of the solar-array weight is aft of the wing torsional axis, which can require additional structure or active aeroelastic control along the wing to prevent flutter. The weight of the arrays can create static sag on the wings both along the span and torsionally.

Solar arrays drive wing sizing. Many solar-powered aircraft design efforts find that the wing area required to generate propulsion power is greater than the optimum wing area for minimum power flight. The wing must grow to accommodate the additional solar collection area. However, the increased area results in higher UA weight, more drag, and hence more required power. The wing sizing is often solved through optimization. Other surfaces such as horizontal tails may grow in size to increase the collection area, as shown in Fig. 8.16.

Solar arrays perform best when oriented normal to the sun vector. Aerodynamic surfaces such as wings and horizontal stabilizers perform best when nearly horizontal. The stressing conditions for solar-powered flight involve low sun angles, which coincide with early morning, late afternoon, winter, and extreme latitude flight. To collect at these low sun angles, the aerodynamic surfaces must be made more vertical to increase the normal component of the sun vector. Some approaches include the Aurora Flight Sciences Vulture Z-wing, vertically folding wing tips, and rolling vertical tails. All of these configurations increase the power required for flight, but more than compensate for this by increasing the power generated by the arrays. These configurations convert to a minimum drag state to reduce the power required to fly at night.

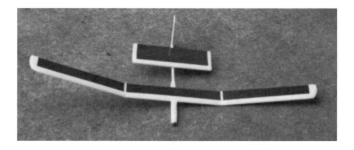

Fig. 8.16 Scale model of author's solar-powered tactical unmanned aircraft concept with an enlarged horizontal tail intended to increase the solar collection area.

8.9.2 Solar Model

Solar models calculate the solar radiation and the sun vector angles for any time of day, time of year, latitude, and altitude on Earth. Orbital mechanics predict the sun's location relative to local geospatial coordinates on Earth. Atmospheric attenuation models calculate the losses caused by atmospheric absorption.

The solar radiation in space at the radius of Earth's orbit around the sun r is

$$I = I_{mean} \frac{r_{mean}^2}{r^2} = I_{mean} \frac{[1 + \varepsilon \cdot \cos(\theta)]^2}{(1 - \varepsilon^2)^2} \tag{8.91}$$

where I_{mean} is the solar radiation at r_{mean} (the mean radius), ε is the orbit eccentricity (0.0167 for Earth), and θ is the day angle. θ is defined by

$$\theta = 2 \cdot \pi \cdot \frac{(n - 4)}{365} \tag{8.92}$$

where n is the number of days since December 31. The total solar radiation is less than that at the edge of space due to atmospheric absorption or attenuation. The equation for total solar radiation is

$$I_{tot} = C_{atten} \cdot I \cdot \cos(\psi) \tag{8.93}$$

where C_{atten} is the attenuation coefficient and ψ is the angle between the collection surface normal and the sun vector. Hall [12] provides data on the solar attenuation as a function of altitude and sun elevation angle.

The declination angle, which is the angle of tilt of the planet as viewed by the sun, is defined as

$$\delta = i_{pole} \cdot \sin\left(\frac{2 \cdot \pi \cdot day}{365}\right) \tag{8.94}$$

where i_{pole} is the tilt of the pole and *day* is the number of days since the last vernal equinox. The vernal equinox occurs on March 20, which is the 79th day of the year. On Earth, the pole tilt is 23.44 deg. The hour angle ω is

$$\omega = \pi - 2 \cdot \pi \frac{time}{24} \text{ radians} \tag{8.95}$$

where *time* is the number of hours since solar midnight. The elevation angle of the sun above the horizon *El* is

$$El = \sin^{-1}[\sin(lat) \cdot \sin(\delta) + \cos(lat) \cdot \cos(\delta) \cdot \cos(\omega)] \tag{8.96}$$

The sun azimuth angle Az is the sun's projected angle along the surface of the planet measured from North. Az is calculated by

$$Az = -2 \cdot \tan^{-1}\left[\frac{\cos(\delta) \cdot \sin(\omega)}{\cos(lat) \cdot \sin(\delta) - \sin(lat) \cdot \cos(\delta) \cdot \cos(\omega) - \cos(El)}\right]$$

(8.97)

The elevation angle of the horizon is found by

$$El_{horiz} = \sin^{-1}\left(\frac{R_E + h_{cloud}}{R_E + h}\right) - \frac{\pi}{2} \text{ radians}$$

(8.98)

where R_E is the radius of Earth (6378.1 km), h_{cloud} is the weather-dependent altitude of the cloud layer, and h is the UA altitude. The hour angle of the local sunrise at altitude is

$$\omega_{SR} = \cos^{-1}\left[\frac{\sin(El_{horiz}) - \sin(lat) \cdot \sin(\delta)}{\cos(lat) \cdot \cos(\delta)}\right]$$

(8.99)

The time of local sunrise at altitude in hours $time_{SR}$ is

$$time_{SR} = \frac{\pi - \omega_{SR}}{2 \cdot \pi} \cdot 24$$

(8.100)

The time of the local sunset at altitude is

$$time_{SS} = 24 - time_{SR}$$

(8.101)

8.9.3 Solar Power System Analysis

Multiday duration solar propulsion system analysis methods evaluate the energy balance throughout the diurnal cycle. The energy collected minus the system losses during the day must be stored to provide energy for the night flight.

The total power generated by a solar array P_{array} is found by

$$P_{array} = I_{tot} \cdot S_{array} \cdot \eta_{array}$$

(8.102)

where S_{array} is the array collection array and η_{array} is the array conversion efficiency. This assumes planar array. Losses occur in the array relative to the single cell due to wire resistance, mismatching of cells, and other factors. In the case of an array consisting of multiple cell orientations, I_{tot} is the area-averaged value. A maximum power point tracker generates the highest cell output power for a given irradiance by adjusting the current and voltage of each parallel string of cells in the array. The solar-array efficiency is defined by

$$\eta_{array} = \eta_{cell} \cdot \eta_{covering} \cdot \eta_{wiring} \cdot \eta_{tracker} \cdot \frac{N_{cells} \cdot S_{cell}}{S_{array}} \qquad (8.103)$$

where

$$\eta_{cell} = \text{solar-cell efficiency}$$
$$\eta_{covering} = \text{efficiency of the solar-cell covering material}$$
$$\eta_{wiring} = \text{efficiency of the array wiring}$$
$$\eta_{tracker} = \text{efficiency associated with the peak powerpoint tracker}$$
$$N_{cells} = \text{number of solar cells in the array}$$
$$S_{cell} = \text{top collection area of the individual solar cell}$$

Recall that I_{tot} is a function of the angle between the solar-cell normal and the sun vector ψ. The sun elevation and azimuthal angles define the sun vector in the Earth-fixed reference frame. The solar-cell normal vector is a function of the aircraft geometry, flight direction, flight location, angle of bank, and angle of attack. The total solar radiation is also a function of the time of day, time of year, latitude, and altitude. Unlike the other more conventional forms of propulsion covered in this chapter, solar-powered aircraft propulsion performance is a function of the heading.

The total system excess power available for storage P_{excess} is

$$P_{excess} = P_{array} \cdot \eta_{conv} - P_{systems} - \frac{P_{Thrust}}{\eta_{mot} \cdot \eta_{ESC} \cdot \eta_{prop}} \qquad (8.104)$$

where η_{conv} is the efficiency of the voltage conversion and power conditioning process and $P_{Systems}$ is the power sent to onboard systems such as avionics and payloads. P_{excess} is dependent upon both time and flight path.

The energy balance situation is further complicated when energy storage is considered. The preceding discussion of system power applies only to the case where the collected solar power is applied directly to propulsion and onboard systems. Except for special seasonal cases near the poles, solar power is not available throughout the entire diurnal cycle. At night, the aircraft must either glide down or use stored energy to help maintain altitude. For the latter case, batteries or fuel cells are suitable candidate solutions. Excess power produced during the day that is not needed for propulsion or onboard systems is used to charge the energy storage device. When the excess power is positive, the stored energy is

$$E_{stor} = \int P_{excess}(t) \cdot \eta_{charge} \cdot dt \qquad (8.105)$$

where η_{charge} is the efficiency of the stored energy conversion process. This integral equation should be discretized into multiple time steps across daylight time for which the excess power is positive.

The energy provided by the batteries or fuel cells at night $E_{out,night}$ is

$$E_{out,night} = \left[P_{systems} + \frac{P_{Thrust}}{\eta_{mot} \cdot \eta_{ESC} \cdot \eta_{prop}} \right] \cdot \eta_{discharge} \cdot [24 \text{ hrs} - (t_{ss} - t_{sr})]$$

(8.106)

where $\eta_{discharge}$ is the efficiency in converting the stored energy to useful energy. This equation is not an integral because power needed at night is assumed to be constant when altitude is maintained.

When the power from the sun is insufficient to maintain altitude at dawn and dusk, the energy storage device provides the power to the propulsion system and the onboard systems. Power output from the energy storage system at dawn and dusk conditions $P_{out,dd}$ is

$$P_{out,dd} = \frac{1}{\eta_{discharge}} \left(P_{array} \cdot \eta_{conv} - P_{systems} - \frac{P_{Thrust}}{\eta_{mot} \cdot \eta_{ESC} \cdot \eta_{prop}} \right)$$

(8.107)

Integrating this over the daylight hours with insufficient excess power yields the energy output from the energy storage system during dawn and dusk $E_{out,dd}$:

$$E_{out,dd} = \int P_{out,dd}(t) \cdot dt$$

(8.108)

The energy balance is satisfied when the energy stored during the day is greater than or equal to the energy output at night:

$$E_{stor} \geq E_{out,night} + E_{out,dd}$$

(8.109)

Descending at night is an alternative to using stored energy to power propulsion. The solar-powered aircraft becomes a glider when no solar power is available. The aircraft climbs during the day, providing potential energy. The potential energy is converted to kinetic energy through the descent, where the velocity is maintained to offset the drag in the gliding descent. The UA must not descend below a threshold minimum altitude for the cycle to work. This approach saves the weight of the power storage system, and it might yield a smaller UA. However, this approach might have more geographic limitations than if energy storage is applied.

Sizing of a long-endurance solar-powered aircraft requires a balance of multiple variables. The unmanned aircraft size is driven by the required solar collection area, solar-cell weight per area, solar-cell efficiency, power conditioner efficiency, energy storage specific power and specific energy, and a number of component efficiencies. These are, in turn, driven by the required altitude, flight direction, latitude, and time of year.

8.9.4 Solar-Cell Technology

Semiconducting materials in solar cells absorb photons from sunlight. The photons release electrons in the semiconductor, which produces an electrical field that can be harnessed. The semiconductor is usually made of silicon (Si) or gallium arsenide (GaAs). Characteristics of various classes of solar cells are shown in Table 8.3.

Most solar-cell semiconductors are wafers cut from factory-grown crystal ingots. These are known as crystalline cells. Silicon multicrystalline cells are low cost but have relatively low efficiency. Monocrystalline silicon cells improve the efficiency at the expense of higher cost. GaAs crystalline solar cells offer high efficiency, but they are generally more expensive and heavier for a given surface area. All crystalline solar cells are brittle and must be protected from damage. GaAs cells are the technology of choice when aircraft performance is a greater consideration than solar-cell cost.

Thin-film solar cells come in flexible sheets that can conform to the surface of a wing. These solar cells have a high specific power, but their low efficiency requires substantially greater collection area than crystalline silicone cells. Eric Raymond flew the solar-assisted Sun Seeker from California to North Carolina in 1991 by using amorphous thin-film Sanyo cells. Iowa Thin Films made a solar-powered unmanned aircraft, which successfully flew with their own amorphous thin film cells. Despite these successes, the more efficient crystalline silicon solar cells are more frequently used on solar-powered aircraft of all sizes. Geis [13] and Ralph [14] provide additional insights into solar powered aircraft design and technology.

Solar collectors can focus sunlight onto a more concentrated region. Less solar-cell coverage is then needed to produce the same amount of power, and the solar-cell weight can be reduced. If solar-cell area is not an active design constraint, and if the collector/solar-cell system specific power is greater than for solar cells alone, then solar collectors might be a feasible alternative.

Table 8.3 Properties of Solar-Cell Classes

Cell Type	Efficiency, %	Aerial Mass, kg/m^2
Multicrystalline Si	14–19	0.41–0.7
Monocrystalline Si	18–22	0.28–0.71
Monocrystalline GaAs	20–30	0.17–0.85
Thin-film Si	6–10	0.022–0.16

8.10 Hybrid Electric

Hybrid electric propulsion combines traditional powerplants with battery-electric propulsion. The motivations for hybrid electric propulsion are 1) improved reliability of a powerplant through battery backup, 2) increased electrical power available for payloads, 3) ability to operate under purely electric propulsion for low acoustic flight, and 4) improved fuel efficiency. Reliability is achieved if the system can continue operation under battery power alone in the event of a powerplant failure. For hybrid electric configurations that convert all of the powerplant power to electrical power via a generator, the battery/generator combination can flexibly provide flight power and payload power. When using only the battery-motor-prop, the powerplant no longer contributes to the acoustic signature. Finally, using the battery only for peak loads such as climb or dash enables the powerplant to be optimized for efficiency at cruise.

Several hybrid electric architectures are shown in Fig. 8.17. The first case (Fig. 8.17a) converts all of the powerplant power to electrical power via a generator. All of the propulsive power comes from a motor-propeller. A power management system controls battery charging and power draw. In the second case (Fig. 8.17b) the battery-motor is coupled to the powerplant-propeller drive shaft such that it can supplement the engine power when

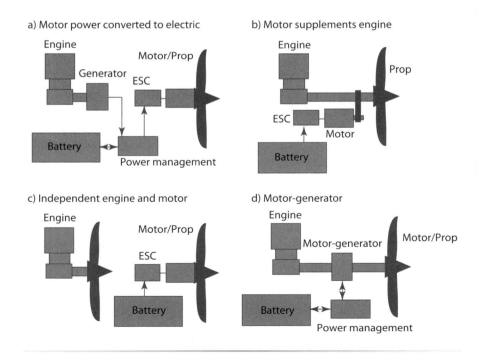

Fig. 8.17 Hybrid electric propulsion configurations.

required, such as for cruise or dash. The engine provides the propulsion power in other flight conditions. The third case (Fig. 8.17c) has independent powerplant-propeller and battery-motor-propeller systems, where the electrical propulsion system is used for peak propulsion loads or as a redundant propulsion system for increased reliability. The fourth case (Fig. 8.17d) is much like (Fig. 8.17b), except that the mechanically coupled motor also serves as a generator. As with case in Fig. 8.17a, a power management system controls battery charge and discharge functions. Bental Industries has demonstrated a hybrid electric system of this configuration.

Physically separating the powerplant and motor-propeller opens up many system options. The primary powerplant might be a reciprocating engine or a turboshaft that drives a generator, both of which are fixed weights. The fuel is generally located near the center of gravity to prevent center-of-gravity shifts as the fuel is depleted. Like pure battery-motor-propeller systems, motor-propellers have high power/weight and can therefore be located in many airframe locations.

The powerplant can be shut down temporarily for quiet electric flight and improved sensor performance. This can reduce the undetected slant range to the target, providing enhanced collection. The propeller diameter can be increased with the motor beyond what is practical with mechanical gearing from the powerplant. The larger diameter propeller might be more efficient and quieter in some flight regimes. The powerplant may only operate at a single rpm, making vibration isolation easier.

The downside is that there are efficiency losses in converting the shaft power to electricity and then back to mechanical power for the propeller. The additional generators, motors, batteries, power distribution, and power electronics add weight. Hybrid electric propulsion systems are also more complex than traditional powerplant-propeller systems.

8.11 Pulsejets

Pulse jets are a lightweight and simple form of propulsion for high subsonic speed UA. The most notorious and successful pulsejet-powered system is the German Fiesler V-1 "Buzz Bomb" cruise missile that was widely employed in World War II. The drawbacks of this propulsion system are that the fuel consumption is high, and the acoustic signature is high relative to jet engines.

With *shutter-valve* pulsejets, hinged slats, or shutters, are mounted near the lip of the engine intake. These slats are the only moving parts. Air can flow through the open slats and into the combustion chamber. Fuel is sprayed into the combustion chamber where it mixes with the air. The fuel-air mixture is ignited with a spark. The explosion forces the slats closed such that no new air can enter, and the air is expelled through the convergent nozzle. The convergent shape accelerates the exiting air velocity to nearly sonic speed. After the explosion is complete, oncoming air forces the slats open. New air partially clears the combustion chamber of remaining air and combustion products from the previous cycle. Now the cycle can be repeated. This configuration was applied to the V-1.

An alternative configuration is the *acoustic* pulse jet. Both the inlet and exhaust are rear-facing, and there are no moving parts. Lan and Roskam [2] details this pulsejet configuration. The recent Russian ENICS E85 and E95 target drones appear to use acoustic pulsejets.

Pulsejet systems have largely fallen out of favor. However, this propulsion class might still have applications to short-range battlefield reconnaissance systems, air-launched decoys, and low-cost cruise missiles. Most low-cost, high subsonic speed UA fielded today use turbojet engines.

8.12 Rockets

Rocket propulsion is rarely used on UAS application except as a launch aid or first stage in a high-speed propulsion system. The key advantages of rocket propulsion are a very high thrust-to-weight ratio relative to jets and the ability to operate at very high altitudes. The drawbacks are a short burn time and handling hazards.

Rockets are not airbreathing engines, and so the thrust equation becomes

$$T = \dot{m}_{f+O} \cdot V_e \tag{8.110}$$

where $\dot{m}_{f+O}$ is the combined fuel and oxidizer mass flow rate out the nozzle. Rockets are commonly characterized by their specific impulse, which is defined by

$$I_{sp} = \frac{T}{\dot{m}_{f+O} \cdot g} = \frac{V_e}{g} \tag{8.111}$$

The specific impulse has units of time. The thrust in terms of specific impulse is

$$T = I_{sp} \cdot \dot{m}_{f+O} \cdot g \tag{8.112}$$

The two primary classes of rockets are *liquid-fueled* and *solid* rockets. Liquid-fueled rockets have a liquid fuel such as LH2 and an oxidizer such as liquid oxygen (LOX), which are stored separately. The two components are combined in a combustion chamber and exit a convergent-divergent nozzle at high supersonic velocity. Complex fuel storage, pumps, cooled nozzle, and injectors are required for operation. Liquid-fueled rocket specific impulse values range from approximately $300-411$ s.

Solid rockets are much simpler. The solid propellant contains both the solid chemical fuel and oxidizer, but these can be stored in a stable manner until ignited. The exhaust gases exit a nozzle at high speed. The solid fuel is contained within a casing, which is generally metal for UAS applications. Simple UAS rocket launch systems can simply change out modular solid rocket boosters for every launch. Solid rocket specific impulse values range from 140–290 s, which is lower than what is achievable by liquid-fueled rockets.

Notable examples of rocket-powered UASs include the following:

* In 1936, two liquid-fueled rocket-powered unmanned aircraft unsuccessfully attempted to deliver mail across a frozen lake in New York.
* The Lockheed D-21 drone launched from a B-52 accelerated to supersonic speeds using a rocket first stage.
* The BAE Australia Nulka hovering naval missile decoy uses rocket propulsion for short-duration flights off-board the host ship.

8.13 Gliders

Gliders have no propulsion in the traditional sense. You might find it curious that aircraft without propulsion systems are included in this chapter on propulsion. Although gliders have no means of converting stored energy into thrust, these aircraft do convert potential energy into kinetic energy. In other words, gliders maintain flight velocity and support maneuvering by losing height over time. Although this seems like a major limitation compared to other forms of propulsion, there are many important applications such as decoys, cargo delivery, sensor emplacement, and warhead delivery.

Gliders are usually air-launched, though they might also be part of a multimode or multistage system. The still-air range R of a glider launched from an altitude h above the landing site is

$$R = h \cdot L/D \tag{8.113}$$

where L/D is the aircraft lift-to-drag ratio. This shows that the UA range is a function of the initial altitude and aerodynamic efficiency alone. The wing loading and aircraft weight are not factors. The aerodynamic efficiency is nearly independent of weight for similar Reynolds numbers and excluding compressibility effects.

The glider's still-air sink rate ROD is given by

$$ROD = \frac{V}{L/D} = \frac{1}{C_L^{3/2}/C_D} \sqrt{\frac{2 \cdot (W/S)}{\rho}} \tag{8.114}$$

Fig. 8.18 NASA engineer Michael Allen hand launches a motorized sailplane to demonstrate autonomous soaring. (NASA photo by Carla Thomas.)

Here, the aerodynamic efficiency is defined by the endurance parameter $C_L^{3/2}/C_D$ rather than the lift-to-drag ratio. The rate of descent is reduced with light wing loading W/S and low-altitude flight with high air density ρ. Lower descent rate results in longer flight duration.

The world does not have still air. Fortunately for gliders, some of the air motion is in the upward vertical direction or can contribute to the UA usable energy in other ways. Gliders that use this upward air motion for extended duration flight are called sailplanes, and sailplane flight is known as soaring. Columns of rising hot air called thermals are generated by uneven heating of the Earth's surface. Sailplanes that use thermals for lift are thermal soaring. When air encounters sloped terrain, it must rise along the front face of the feature, creating a vertical wind component. Sailplanes that fly in this wind field are slope soaring. Downstream of the vertical terrain feature, the air can oscillate vertically over long distances and up to stratospheric altitudes. This phenomenon is wave lift. Sailplanes can also harvest energy from flying through velocity gradients, which is called dynamic soaring.

The rate of climb (*ROC*) for a sailplane in the presence of an upward vertical wind component V_Z is

$$ROC = V_Z - \frac{V}{L/D} = V_Z - \frac{1}{C_L^{3/2}/C_D}\sqrt{\frac{2(W/S)}{\rho}} \qquad (8.115)$$

Note that rate of climb has the opposite sign of rate of descent. From this relationship, it can be seen that the vertical wind component required to sustain level flight is

$$V_Z \geq \frac{1}{C_L^{3/2}/C_D} \sqrt{\frac{2(W/S)}{\rho}} \qquad (8.116)$$

The factors that help sustained flight are light wing loading, high aerodynamic efficiency, and low-altitude operations.

Unmanned aircraft could potentially stay aloft for very long durations without any thrust-generating propulsion by only harvesting energy from upward air motion or velocity gradients. NASA (Fig. 8.18) and NRL successfully demonstrated autonomous soaring algorithms on an unmanned aircraft. The major limitations to this approach are 1) suitable conditions must be available every day, 2) gaining altitude might require exiting the target area temporarily, and 3) suitable conditions are generally not available over water.

8.14 High-Altitude Propulsion

High-altitude propulsion systems must operate in a rarefied atmosphere. Although no strict definition of high altitude exists, this regime can be thought of as altitudes from 50,000 ft to above 100,000 ft. The pressure and air density are low relative to sea level (Fig. 8.19), and little oxygen is available for combustion. Airbreathing engines experience performance degradation or must add additional systems to support high-altitude flight.

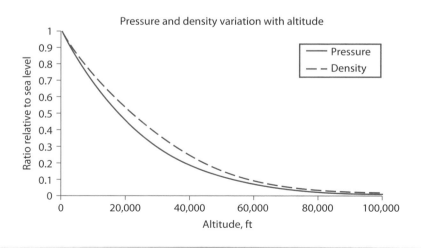

Fig. 8.19 Atmospheric properties at high altitude.

Jet engines are perhaps the most practical high-altitude propulsion system. Turbojets and low-bypass-ratio turbofans have lower thrust lapse rates than high-bypass-ratio turbofans. Jet engines retain the heat of the compressed air within the engine cycle, ultimately exiting through the nozzle such that it generates useful thrust. This compares favorably with turbocharged engines that reject much of the heat from compression as intercooler waste heat. The required jet air mass flow rate is maintained by flying fast, having adequate inlet area, and using inlet precompression. High-altitude improvements to jet engines can include wider first-stage fan blades, a robust combustor pilot flame to prevent flame-out, and higher pressure ratio compressors.

Reciprocating engines operating at high altitudes generally require multistage turbochargers, with three stages as the dominant configuration. At 65,000 ft, for example, an overall pressure ratio (OPR) of 17.8:1 is required to maintain sea-level equivalent pressure at the manifold. Each turbine stage supports a pressure ratio of around 4:1. The first compressor stage is largest so that it can have adequate capture area. Each following stage decreases in size. The compressor Reynolds numbers are low relative to sea level, degrading the efficiency. The air is cooled between compressor stages with intercoolers. Cooling fluid from the intercoolers is rejected to the airflow by radiators. These radiators can be large and therefore major drag producers for the aircraft. The turbochargers adversely affect the brake specific fuel consumption, but this is a necessary penalty for increasing the altitude operability. The turbochargers, intercoolers, and other modifications add substantial weight to the propulsion system, increasing the weight by a factor of $2-3.5$ relative to a low-altitude engine producing the same power.

Oxygen required for flight can be carried by the UA in the form of liquid oxygen (LOX). This approach eliminates the need for turbochargers and the associated high-drag heat exchangers. Carried LOX effectively eliminates altitude dependency on engine power generation, though the core engine cooling and propeller performance must be considered. The Aurora Flight Sciences Perseus A carried LOX to power a gasoline-powered reciprocating engine, which was designed to reach 80,000 ft. Perseus A flew to 50,000 ft in 1994.

Hydrazine is a monopropellant fuel that expands without combustion, enabling nonairbreathing high-altitude propulsion. The expanding gas can provide thrust directly as a rocket. Another approach is to use the hydrazine to drive a reciprocating engine. NASA successfully flew a Mini Sniffer II hydrazine-powered unmanned aircraft in 1976. The Mini Sniffer II had a reciprocating hydrazine engine.

Battery-propeller propulsion systems have no dependency on air density to generate power. On first inspection, this propulsion class would appear to be ideal for high-altitude flight, but the UA endurance is too short for much mission utility. The battery-stored energy is insufficient to enable a climb to high altitude, and so air-launch is necessary to provide flight time at altitude.

Solar-powered aircraft are well suited to high-altitude operations. These propulsion systems do not depend on oxygen for operation. The atmospheric attenuation of solar radiation decreases with altitude, so that the available power increases. However, the required power to fly increases because the flight velocity must go up in the low-density air. Solar-powered aircraft designs tend to work well at altitudes of around 65,000 ft.

8.15 Miscellaneous Propulsion Types

The most prolific and practical propulsion systems have been described up to this point. However, there is a rich diversity of propulsion options that have not been widely used. The author has designed aircraft with numerous unconventional propulsion systems. Usually propulsion techniques that apply to manned aircraft are also suited to unmanned applications. One notable exception to that rule is the human-powered aircraft propulsion system.

8.15.1 Ornithopters

Ornithopters use flapping wings to generate thrust. The flapping wings can also generate lift, especially at low speeds. The wing oscillates vertically

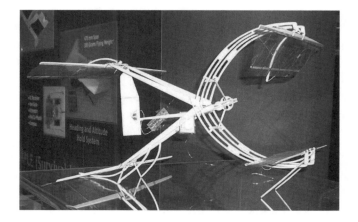

Fig. 8.20   NRL Bite Wing unmanned ornithopter.

such that the resultant combination of lift and drag forces has a net thrust component during a large portion of the cycle. Usually the wing incidence angle must change through the cycle to enable the appropriate force vector. One common way of adjusting the incidence is by wing twist, which requires a torsionally flexible wing. The wing oscillation is mechanically powered by electric motors or reciprocating engines, with mechanisms to convert the rotary motion to linear motion. Future artificial muscles that retract and extend might power ornithopters. Mueller [9] provides insight into the aerodynamics of these unmanned aircraft. Flapping wing flight can take many forms, as demonstrated in Fig. 8.20.

This propulsion technique appears to be most suited to micro UAs. The loads associated with the wing oscillations excessively penalize the structural weight, and ornithopters add considerable complexity relative to other propulsion alternatives.

8.15.2 Nuclear Propulsion

Nuclear propulsion uses the heat produced from radioactive decay in the place of combustion. A nuclear-powered jet engine or turboprop would use heat exchangers between the compressor and turbine to transfer the heat from the nuclear generator. Reciprocating engines can use heat from a nuclear source in a sterling cycle engine.

Nuclear propulsion systems have many disadvantages, as might be expected. Most obviously, the nuclear fuel is dangerous and requires special protection for humans who interact with the UA and shielding to protect onboard systems. Also, what country would want a nuclear-powered aircraft to operate in its airspace, even if permitted by international treaties? How would the population be protected in the event of a crash?

Looking at this purely from a design perspective, nuclear propulsion offers the possibility of extreme endurance. The propulsion system is likely to be much heavier than more traditional alternatives due to the radiation shielding and heat-transfer piping. However, the system weight is constant, and no traditional fuel is required.

Although no aircraft have flown with nuclear propulsion systems providing thrust, there are numerous concepts. Some notable examples include the following:

* A heavily modified B-36, called the NB-36 Crusader, flew with a nuclear heat generator from 1955–1957. The heat would have powered specialized jet engines if the program were completed.
* The Mach 3 supersonic low-altitude missile (SLAM) concept in the late 1950s was designed around a nuclear ramjet engine.
* In 2010 NRL, NASA, Aurora Flight Sciences, and Idaho National Laboratories designed a conceptual nuclear-powered unmanned flyer

that would operate within the Titan atmosphere. The powerplant is an advanced sterling cycle radioisotope generator (ASRG), which powered an electric motor. The high density of the Titan atmosphere enabled low propulsion power requirements, making the low power-to-weight ASRG practical. The flyer concept has a flight endurance of one year.

8.15.3 Flywheel Energy Storage

Flywheels are spinning masses that can transfer the rotational energy to shafts powering electric generators or propellers. The flywheel is housed within a vacuum chamber to avoid air friction losses. The energy stored in a flywheel rotor E_{Flywheel} is

$$E_{\text{Flywheel}} = \frac{1}{2} J \cdot \omega^2 \qquad (8.117)$$

where J is the moment of inertia and ω is the rotational rate.

Energy is imparted to the flywheel by applying torque from a motor. The motor can also run as a generator to extract power from the flywheel rotor. PCS [10] projects that the achievable specific energy is 130 W-hr/kg, and the specific power is 600 W/kg.

Flywheel energy storage systems are not practical for UAS applications. Flywheels spinning at high rates require precise balancing to prevent destructive vibration. The spinning mass acts as a gyroscope, which resists angular change by applying a resisting moment. Therefore, a large flywheel can impact UA flight controls. Also, the specific energy is not competitive with secondary batteries.

8.15.4 Rubber Bands

Rubber bands store energy by stretching the length of the band. When the band is stretched by winding the band, the energy is released in a rotational motion capable of turning a propeller. Flying models dating to the 19th century successfully used rubber band propulsion. The energy density of rubber bands is significantly lower than other forms of energy storage, and they are therefore not applied to practical UASs. The outdoor flight duration of typical unguided rubber band powered models is measured in seconds or minutes.

8.15.5 Microwave Power

Microwave propulsion uses directed RF waves in microwave frequencies to transmit power from the ground to the UA. A large dish or phased-array antenna on the ground transmits the microwave power to the UA, which receives it with an array of microwave-sensitive cells called a *rectenna*. This mode of operation is very analogous to solar-powered aircraft, except the energy comes from a ground-based microwave source instead of the sun. A microwave-powered unmanned airship was flown in Japan. A fixed-wing unmanned microwave powered fixed-wing UAS was flown in Canada.

Beaming high-power microwaves poses risks to air traffic and birds. The ground power transmitter tends to be large and relatively immobile. Therefore, microwave propulsion is best for static applications such as urban wide area surveillance and communications relay.

References

[1] Kohlman, D. L., *Introduction to V/STOL Airplanes*, Iowa State Univ. Press, Ames, IA, 1981, p. 22.

[2] Lan, C.-T. E., and Roskam, J., *Airplane Aerodynamics and Performance*, Design, Analysis and Research Corp., Lawrence, KS, 2003, pp. 258, 259, 291.

[3] Stinton, D., *The Design of the Aeroplane*, Blackwell Sciences, Ltd., Oxford, England, U.K., 1995.

[4] Mises, R. V., *Theory of Flight*, Dover, Toronto, 1959, p. 362.

[5] Oates, G. C., *Aerothermodynamics of Gas Turbine and Rocket Propulsion*, 3rd ed., AIAA, Reston, VA, 1997.

[6] Mattingly, J. D., Heiser, W. H., and Daley, D. H., *Aircraft Engine Design*, AIAA, Washington, D.C., 1987, p. 36.

[7] Torenbeek, E., *Synthesis of Subsonic Airplane Design*, Delft Univ. Press, Delft, The Netherlands, 1984, p. 128.

[8] Boucher, R. J., *Electric Motor Handbook*, Robert J. Boucher, Los Angeles, 1994.

[9] Mueller, T. J., Kellogg, J. C., Ifju, P. G., and Shkarayev, S. V., *Introduction to the Design of Fixed-Wing Micro Air Vehicles, Including Three Case Studies*, AIAA, Reston, VA, 2007, pp. 166–170.

[10] Power Computing Solutions, Inc. (PCS), *Electric Power System for High Altitude UAV Technology Survey*, NASA/CR-97-206337, 1997, p. 73.

[11] Colozza, A. J., "Hydrogen Storage for Aircraft Applications Overview," NASA/CR-2002-211867, Sept. 2002.

[12] Hall, D. W., and Fortenbach, C. D., et al., "A Preliminary Study of Solar Powered Aircraft and Associated Power Trains," NASA CR 3699, 1983.

[13] Geis, J., and Arnold, J. H., "Photovoltaic Electric Power Applied to Unmanned Aerial Vehicles (UAV)," *Proceedings of the 13th Space Photovoltaic Research and Technology Conference (SPRAT 13)*, 1994.

[14] Ralph, E. L., and Woike, T. W., "Solar Cell Array System Trades – Present and Future," AIAA Paper 99-1066, Jan. 1999.

Problems

8.1   Derive the static shaft power formula

$$P_{\text{Shaft}} = \frac{T^{3/2}}{\sqrt{2 \times \rho \times A_P} \times \eta_{\text{p,nonideal}}}$$

8.2   An insulated cryogenic spherical tank is used to store 20 kg of LH_2 fuel. The insulation is rigid closed-cell polyvinylchloride foam with a density of 49.8 kg/m^3 and a thermal conductivity of 0.0046 W/m-K. Assume that the aluminum has negligible insulation capabilities. The air surrounding the insulated tank is equal to the ambient condition at 10 km, which is 223.3 K. The inner tank is made of 2014-T6 aluminum alloy. The gauge pressure of the tank is 1,000 psi. Assume that the tank is completely filled. Use a factor of safety of 1.5 for the structures. The boil-off rate of the hydrogen is 5 kg/day. What is the structural thickness and weight? What is the insulation thickness and weight? What is the H_2 mass fraction of the LH_2 storage system?

8.3   A propeller has diameter of 28 in. The UA flies at an equivalent airspeed of 50 KEAS. Calculate the rpm that generates a sonic tip condition at sea level up to 100,000 ft in 10,000-ft increments.

8.4   A propeller has a 6-ft diameter. At an advance ratio of 0.6, it generates a power coefficient of 0.05. The equivalent airspeed is 60 KEAS. Plot the shaft power used by the propeller as a function of altitude from sea level up to 100,000 ft in 10,000-ft increments. If the maximum tip Mach number is 0.8, what is the maximum altitude?

8.5   A propeller has a diameter of 1.5 ft and must produce 10 lb of thrust at sea level. Use the actuator disk method to calculate the ideal propeller efficiency from airspeed of 20 to 200 kt in 20-kt increments.

8.6   A model aircraft propeller is labeled 12 × 6 in. What is the pitch, and what is the diameter? What is the nominal pitch angle at 75% radius?

8.7   A hydrogen fuel cell has a 30% thermal efficiency. What is its BSFC in pounds/horsepower-hour?

8.8   A twin-cylinder four-stroke engine has a brake mean effective pressure of 150 psi and a mechanical efficiency of 75% when operating at sea level and 6,000 rpm. The bore is 1 in., and the stroke is 1.5 in. What is the brake horsepower at this sea-level condition? Using the Gagg–Ferrar equation, what is the brake horsepower at 10,000 ft?

8.9 A jet engine with a cycle efficiency of 70% produced 40 kW of electrical power while operating at Mach 0.6 at 60,000 ft. The brushless generator is 80% efficient. The engine produces 5,000 lb of thrust with a nominal TSFC of 1 lb/lb-hr without the power extraction. Estimate the equivalent TSFC with the power extraction.

8.10 A turboprop aircraft flying at 200 kt must produce 200 lb of thrust and 5 kW of shaft power for a generator. The turboprop nozzle generates 20 lb of thrust. The propeller efficiency is 70%. How much shaft power is supplied to the propeller?

8.11 A direct-drive electric motor outputs 5 kW to the propeller at 5,000 rpm. The motor has a zero-load current of 0.4 A. The voltage constant is 70 rpm/V. The motor resistance is 0.2 Ω. The electronic speed control efficiency is 90%. What is the combined efficiency of the motor and electronic speed control?

8.12 Use the motor system from problem 8.11 A battery pack has a voltage of 72 V and a resistance of 0.1 Ω. What is the battery pack current? What is the battery pack discharge efficiency?

8.13 A NiCd battery pack has 12 cells. There are four parallel strings each having three cells in series. Each cell has a voltage of 1.2 V and a maximum current of 20 A. What is the pack voltage and maximum current?

8.14 Calculate several solar parameters for April 1 at 10:00 AM Eastern Standard Time (EST) in Washington, D.C. Calculate the day angle, solar radiation (above the atmosphere), declination angle, hour angle, elevation angle, and azimuth angle. On that day, when are sunrise and sunset?

8.15 A solid rocket motor has a specific impulse of 200 s. The average mass flow rate is 10 lb/s. What are the exit velocity and average thrust?

8.16 A glider has an endurance parameter of 20 and a wing loading of 12 lb/ft^2. What vertical wind speed is required for sustained flight at 5000 ft?

Chapter 9 Flight Performance

- Understand atmospheric models
- Calculate performance constraints
- Optimize airspeed for multiple mission segment types

Perseus-A high altitude research UAS. (Photo: NASA.)

9.1 Introduction

U nmanned aircraft must fly mission profiles that enable them to use their payloads effectively in time and space. Segments can include ingress to the mission area, loiter, and egress back to the recovery site. The altitude and airspeed of each segment are generally specified.

Unmanned aircraft performance ranges from designs that break endurance or altitude records to lackluster designs. The level of design refinement often yields aircraft that are not near the limits of what could be achieved. This might be driven by low manufacturing costs, small allocated development investment, abilities of the design team, or easily fulfilled customer requirements. In the case where performance expectations are not demanding, it is relatively difficult to not meet the requirements. The impact of configuration selection, for example, is of little importance when a gas-powered aircraft only needs to fly for 5–6 hrs.

When performance pushes the state of the art, meeting the design requirements becomes much more challenging. It is much harder to design a 24-hr endurance tactical UAS than a 6-hr equivalent. Here the design must have efficient propulsion, high aerodynamic efficiency, a high fuel mass fraction, and a flight envelope matched to the mission profile. In this chapter we will cover methods for performance estimation.

9.2 Operating Environment

The Earth's atmosphere changes with altitude, regionally around the globe, and with the weather. Despite these variations, performance is nearly always calculated against a standard atmosphere model in which the properties vary only with altitude. Several such models can be applied. The most common are the International Civil Aviation Organization (ICOA) standard atmosphere, NASA's 1976 standard atmosphere, MIL-STD-3013 and MIL-STD-310.

Atmospheric properties described in the atmosphere models include pressure, temperature, density, viscosity, and speed of sound. These properties are provided as a function of altitude. Some models use mathematical expressions to calculate these properties with simple relationships whereas others are tables of data that must be looked up and interpolated.

Here we will show a convenient method of calculating the major atmospheric properties as a function of altitude. In the troposphere, which covers the altitude range from sea level to 11 km, the temperature decreases linearly with height. In this model the lapse rate L is 6.5 K/km. The temperature T at altitude h below 11 km is given by

$$T = T_{SL} - h \cdot L \tag{9.1}$$

T_{SL} is the sea-level temperature, which is 288.15 K on a standard day. The temperature ratio θ is

$$\theta = \frac{T}{T_{\mathrm{SL}}} \tag{9.2}$$

The pressure ratio δ and density ratio σ in the troposphere are

$$\delta = \frac{P}{P_{\mathrm{SL}}} = \theta^{g/(L \cdot R)} \tag{9.3}$$

$$\sigma = \frac{\rho}{\rho_{\mathrm{SL}}} = \theta^{(g - L \cdot R)/(L \cdot R)} \tag{9.4}$$

where g is the acceleration of gravity (9.81 m/s^2) and R is the gas constant for air (287.05 N-m/kg-K). The sea-level air density ρ_{SL} is 1.225 kg/m^3, and the sea-level pressure P_{SL} is 101,325 Pa.

The stratosphere is the atmospheric region above the troposphere. It begins at 11 km and extends to 20 km. The boundary between the troposphere and the stratosphere is known as the tropopause, whose altitude h_T is 11 km. Standard atmosphere models assume that the temperature is constant throughout the stratosphere. Our model assumes that the stratospheric temperature T_S is constant at 216.65 K. The pressure in the stratosphere is

$$P = P_T \cdot \exp\left[-g \cdot (h - h_T)/(R \cdot T_S)\right] \tag{9.5}$$

The pressure at the tropopause P_T is 22,700 Pa. The density is found by

$$\rho = \frac{P}{R \cdot T} \tag{9.6}$$

The speed of sound a is found by

$$a = \sqrt{\gamma \cdot R \cdot T} \tag{9.7}$$

where γ is the ratio of specific heats, which is 1.4 for air.

The viscosity μ can be estimated using the Sutherland equation described in Chapter 5.

Example 9.1   Atmospheric Property Calculations

Problem:

Using the atmospheric relationships in this section, calculate the following properties in 1-km increments from sea level to 20-km altitude: θ, δ, σ, a, μ.

Solution:

The results are shown in Table 9.1. When making the calculations, be sure to convert any length units in kilometers to meters.

Table 9.1 Atmospheric Properties

h, km	θ	δ	σ	a, m/s	μ, kg/s-m
0	1.0000	1.0000	1.0000	340.3	1.7896E-05
1	0.9774	0.8870	0.9074	336.4	1.7580E-05
2	0.9549	0.7845	0.8216	332.5	1.7261E-05
3	0.9323	0.6918	0.7420	328.6	1.6939E-05
4	0.9098	0.6082	0.6686	324.6	1.6612E-05
5	0.8872	0.5330	0.6008	320.5	1.6282E-05
6	0.8647	0.4655	0.5384	316.4	1.5948E-05
7	0.8421	0.4051	0.4811	312.3	1.5610E-05
8	0.8195	0.3512	0.4286	308.1	1.5268E-05
9	0.7970	0.3033	0.3805	303.8	1.4922E-05
10	0.7744	0.2608	0.3367	299.5	1.4571E-05
11	0.7519	0.2232	0.2969	295.1	1.4216E-05
12	0.7519	0.1907	0.2536	295.1	1.4216E-05
13	0.7519	0.1628	0.2166	295.1	1.4216E-05
14	0.7519	0.1391	0.1850	295.1	1.4216E-05
15	0.7519	0.1188	0.1580	295.1	1.4216E-05
16	0.7519	0.1014	0.1349	295.1	1.4216E-05
17	0.7519	0.0866	0.1152	295.1	1.4216E-05
18	0.7519	0.0740	0.0984	295.1	1.4216E-05
19	0.7519	0.0632	0.0841	295.1	1.4216E-05
20	0.7519	0.0540	0.0718	295.1	1.4216E-05

Launch and recovery are often specified for an altitude and hot day conditions. "Hot day" refers to probabilistically high temperatures that decrease the density altitude. MIL-STD 310 provides atmospheric properties as a function of temperature variation.

9.3 Mission Profiles

Mission profiles are a sequential compilation of mission segments that are each intended to obtain an objective. Each segment generally burns fuel and places requirements on the airspeed. Some basic objectives include the following:

• Startup and taxi
• Takeoff
• Climb to initial cruise altitude
• Ingress
• Loiter

- Egress
- Descent to landing
- Low-speed dash
- Weapons release
- Preparation for landing
- Landing
- Reserves
- Taxi and shutdown

MIL-STD 3013 provides multiple mission profiles primarily for manned aircraft types. These can provide a solid basis for unmanned aircraft performance specifications, though modification might be required. Two relevant profiles include reconnaissance and strike. The former is relevant for surveillance

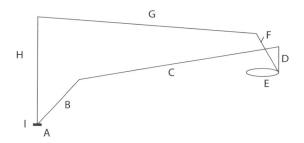

	Segment	Fuel	Time	Distance	Speed	Altitude	Thrust setting
A	Warm-up, takeoff, and accelerate to climb speed	20 min @ ground idle + 30 sec @ takeoff + fuel to accelerate from obstacle clearance to climb speed. No distance credit.					
B	Climb	——	——	——ᵃ	Min time climb scheduleᵇ	Takeoff to optimum cruise	Max continuous
C	Cruise	——	——	50 n miles	Long-range cruise	Optimum cruise	——
D	Descent	None	None	——		End cruise to mission altitude	——
E	Loiter	——	24hrs	——	Maximum endurance	Mission altitudeᶜ	——
F	Climb	——	——	——	Min time climb scheduleᵇ	Mission altitude to optimum cruise	Max continuous
G	Cruise	——	——	50 n miles	Long-range cruise	Optimum cruise	——
H	Descent	None	None	——		End cruise to landing	——
I	Reserves	——	30 min	——	Maximum endurance	Sea level	——

Notes:
ᵃCruise distances are land based and include climb distances.
ᵇClimb schedule ends at long range cruise speed/optimum cruise altitude.
ᶜMission altitude is the ISR optimum altitude (sensor performance and detectability).

Fig. 9.1   MIL-STD 3013 Airborne Warning and Control System (AWACS) profile modified for land-based subsonic ISR. Major changes denoted by underline.

and reconnaissance UASs. The latter is appropriate to UCAV. Figure 9.1 shows an ISR mission profile developed by modifying the AWACS profile.

9.4 Constraint Analysis

Unmanned aircraft must satisfy numerous flight performance constraints during the mission segments. These constraints mostly relate to the flight envelope characteristics, but can also include structures and subsystem limitations.

The propulsion system has a minimum and maximum thrust that is a function of altitude and airspeed. As seen in Chapter 8, various propulsion system types have unique characteristics. To make a performance analysis routine, generic, and adaptable, it is often desirable to have the propulsion software modules have standard interfaces for multiple propulsion types. The performance analysis generally provides an input thrust level to the propulsion module. The propulsion module should indicate if this thrust value is within the propulsion system capabilities. Other useful outputs include the fuel flow or rate of energy depletion. If the required thrust is above or below the propulsion system thrust capability band, then the engine thrust constraints are violated.

Service ceiling is defined as the altitude at which the unmanned aircraft's maximum rate of climb is 100 ft/min. The maximum rate of climb ROC_{Max} is the rate of climb that results from maximum thrust. For small angles, this is given by

$$ROC_{Max} = \frac{(T_{Max} - D)}{W} \times V \qquad (9.8)$$

The service ceiling constraint is

$$ROC_{Max} \geq 100\,\text{ft/min} \qquad (9.9)$$

The service ceiling constraint must be satisfied at some velocity at a given altitude, not at every velocity. The thrust and drag are each functions of velocity, and so the service ceiling is therefore a function of velocity. For this equation we assume that the thrust acts in the direction of flight and all propulsive drag is included in the net thrust.

The climb gradient is a measure of how steep the unmanned aircraft can climb. The flight path angle γ is the angle between the local horizon and the unmanned aircraft velocity vector. The flight path angle is a way of describing the climb gradient. It is given by

$$\gamma = \sin^{-1}\left(\frac{ROC}{V}\right) \approx \frac{ROC}{V} \qquad (9.10)$$

The climb gradient constraint is given by

$$\gamma \geq \gamma_{min} \qquad (9.11)$$

The climb gradient might be required for some stages of flight, such as after takeoff. The engine-out climb gradient is important for multi-engine aircraft. A negative engine-out climb gradient at low altitude is a dangerous unmanned aircraft design.

The unmanned aircraft might have a minimum rate of descent ROD_{Min}. This is important for descending from a high altitude and supporting a landing approach. The rate-of-descent and rate-of-climb constraints are not evaluated at the same time. The highest rate of descent ROD_{Max} occurs when the engine is at minimum thrust T_{Min}. This is given by

$$ROD_{Max} = \frac{(D - T_{Min})}{W} \times V \qquad (9.12)$$

The minimum-rate-of-descent constraint is

$$ROD_{Max} \geq ROD_{Min} \qquad (9.13)$$

High L/D platforms with light wing loading might require spoilers or other drag devices to satisfy this constraint.

The unmanned aircraft must fly faster than the minimum airspeed V_{min}. Note that the minimum airspeed might be higher than the stall airspeed because UAs typically have automatic flight control systems that impose angle-of-attack limiters or other elevator margins to avoid stall.

$$V \geq V_{min} \qquad (9.14)$$

The minimum airspeed occurs at the maximum lift coefficient $C_{L,max}$, and so it can be expressed as

$$V_{min} = \sqrt{\frac{W}{1/2 \cdot \rho \cdot C_{L,max} \cdot S_w}} \qquad (9.15)$$

Aircraft structures and control systems can also have dynamic pressure limits. Dynamic pressure is a measure of the potential loads that can be imposed on aerodynamic surfaces. The maximum equivalent airspeed $V_{eq,max}$ is the maximum sea-level airspeed. The dynamic pressure constraint is expressed by

$$V \leq \frac{V_{eq,max}}{\sqrt{\sigma}} \qquad (9.16)$$

Mach limits can be caused by aerodynamic limitations, flight control systems, air data systems, or propulsion systems. The maximum Mach M_{Max} is set at design. The constraint is expressed by

$$V < M_{max} \cdot a \qquad (9.17)$$

Recall that the speed of sound is a function of altitude in the troposphere and constant in the stratosphere.

The installed thrust available must be greater than the drag for level flight at a given airspeed. The maximum thrust level constraint is

$$D \leq T_{max} \tag{9.18}$$

Similarly, the drag must be greater than the minimum thrust to allow straight and level flight. The minimum thrust level flight constraint is

$$D \geq T_{min} \tag{9.19}$$

Finally, the altitude limits can be constrained by subsystems. For example, the fuel system might not be capable of managing low ambient pressures. The maximum altitude constraint is

$$h \leq h_{max} \tag{9.20}$$

Aeroelasticity might also drive the maximum airspeed. As we saw in Chapter 7, there are numerous aeroelastic phenomena. Two that commonly limit the upper flight speeds are flutter and aileron reversal.

The flight envelope is the velocity and altitude bands in which the unmanned aircraft can operate. The flight envelope varies with weight as fuel burns off over the mission. Interestingly, battery- and solar-powered aircraft have constant weights, so that the envelope does not vary through the mission.

Example 9.2   Flight-Envelope Constraints

Problem:

An unmanned aircraft has the following characteristics:

$$C_{L,Max} = 1.2 \qquad\qquad M_{max} = 0.7$$
$$S_w = 20 \text{ m}^2 \qquad\qquad V_{eq,max} = 150 \text{ m/s}$$
$$W = 10,000 \text{ N} \qquad\qquad h_{max} = 15 \text{ km}$$

Plot the constraint lines for the maximum altitude, maximum equivalent airspeed, maximum Mach, and minimum airspeed as a function of velocity.

Solution:

The results are plotted in Fig. 9.2. The region bounded by the four lines is the available flight envelope. Notice that on the upper speed range how the equivalent airspeed (dynamic pressure) is dominant at lower altitudes and then maximum Mach defines the upper constraint at higher altitude.

In this example all of the constraints may be put in terms of velocity directly. Other constraints such as service ceiling, maximum airspeed, flutter speed, rate of climb, and rate of descent must be found indirectly, since a closed-form solution is generally unavailable. These constraints must be evaluated at a given velocity to determine if they are violated or satisfied.

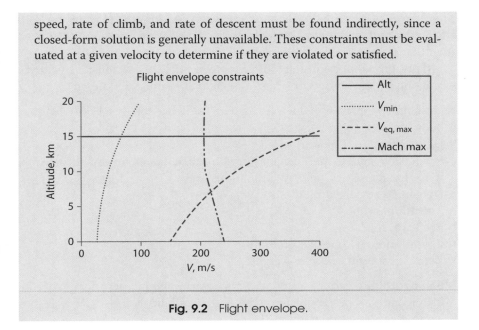

Fig. 9.2   Flight envelope.

9.5 Flight Performance Analysis

The unmanned aircraft must accomplish the required mission profiles. Often a single unmanned aircraft is required to perform more than one mission set. For example, a land-based naval UAS might be required to perform antisubmarine warfare (ASW), ISR, and strike missions—each with a different profile type. In this section we will see how to evaluate segments of mission profiles.

A performance model predicts aircraft fuel burn and optimizes flight trajectories. Simple performance methods can be combined to solve complex performance problems. Performance codes evaluate the fuel burn of the unmanned aircraft as it flies the mission profile and evaluates performance constraints.

Multiple mission segments must be evaluated. These segments can represent climb, cruise, descent, loiter, or other objectives in various sequences and combinations. It is often necessary to evaluate a mission segment by breaking it into multiple subsegments to improve the accuracy. During each segment of flight, several constraints are evaluated. These include stall, specified Mach limits, specified altitude limits, specified dynamic pressure limits, thrust for level flight, thrust for prescribed rate of climb, and engine operating limits.

A good performance code has the ability to optimize flight velocities such that the objectives are best satisfied while not violating constraints.

The figures of merit include range efficiency, endurance efficiency, climb efficiency, and maximum rate of climb. These velocities are optimized such that no constraints are violated.

Many unmanned aircraft design details must be established prior to evaluating the flight performance. The following must be known:

* Unmanned aircraft geometry such as wing area, chord lengths, body lengths
* Drag polar
* Propulsion system performance (minimum and maximum thrust, fuel flow characteristics, velocity limitations)
* Dynamic pressure limitations
* Mach limitations
* Electrical loads
* Environmental control system drag characteristics
* Zero fuel weight and fuel weight

Mission profiles can be evaluated to determine the percentage of the mission that can be completed. The first possibility lends itself well to a mission completion constraint.

The heart of the performance modeling for fueled aircraft begins with the Breguet range equation:

$$R = \frac{V \cdot L/D}{TSFC} \cdot \ell n \left(\frac{W_i}{W_f} \right) \tag{9.21}$$

This form of the Breguet equation requires that the product $V^*(L/D)/TSFC$ remains constant, which necessitates a cruise climb. Many missions are flown at constant altitude or velocity, creating inaccuracies in the Breguet range equation estimation. Climbs differing from the cruise climb can also introduce inaccuracies. The product might not remain constant if the altitude change is significant. The prediction accuracy can be improved by breaking each mission segment into subsegments and introducing corrections.

$$R = \sum R_{\text{seg}} \tag{9.22}$$

The Breguet range equation is modified to account for thrust values differing from the ideal.

$$R_{\text{seg}} = \frac{V \cdot L/D}{f_T \cdot TSFC} \cdot \ell n \left(\frac{W_i}{W_f} \right) \tag{9.23}$$

where the thrust multiplication factor f_T is found by

$$f_T = \frac{T_{\text{req}}}{D} \tag{9.24}$$

The required thrust T_{req} is found by

$$T_{req} = \left(\frac{ROC}{V} + \frac{1}{L/D} \right) \cdot W \qquad (9.25)$$

The lift to drag ratio (L/D) is replaced by the ratio of weight over drag. The required thrust becomes

$$T_{req} = \frac{ROC}{V} \cdot W + D \qquad (9.26)$$

The lift coefficient C_L is given by

$$C_L = \frac{W}{1/2 \cdot \rho \cdot V^2 \cdot S_w} \qquad (9.27)$$

The drag coefficient is found by the aerodynamics table look-up subroutine:

$$C_D = C_D(C_L, M, Re/L) \qquad (9.28)$$

The Reynolds number per unit length Re/L is found by

$$Re/L = \frac{\rho \cdot V}{\mu} \qquad (9.29)$$

We use Reynolds number per unit length rather than Reynolds number for the drag polar look-up because the Reynolds numbers vary over the unmanned aircraft. Reynolds number per unit length is a function of only altitude (ρ, μ) and flight velocity. The speed of sound, air density, and viscosity are found via standard atmosphere look-up procedures or equations.

The specific fuel consumption $TSFC$ is found via the engine deck look-up subroutine. In general,

$$TSFC = TSFC(h, M, T_{req}) \qquad (9.30)$$

The minimum and maximum thrust values at the given altitude T_{min} and T_{max}, respectively, are also output from the engine deck look-up subroutine.

The time taken to complete the subsegment range is

$$E_{seg} = \frac{R_{seg}}{V} \qquad (9.31)$$

The performance modeling must also capture the impacts of cooling drag and power extraction. The cooling drag estimation is described in Chapter 5. In general, the cooling drag is dependent upon the cooling load and the flight velocity. The relationship between drag and power, assuming thrust is equal to drag, is

$$P = D \cdot V \qquad (9.32)$$

If a generator supplies this power, then there will be power generation efficiency η_{Gen}. When this relationship is rearranged and corrected for units, power extraction equivalent drag is found by

$$D_{\text{Pow}} = 0.43687 \cdot \frac{P_{\text{Gen}}}{V \cdot \eta_{\text{Gen}}} \tag{9.33}$$

where D_{Pow} is in pounds, P_{Gen} is in watts, and V is in knots. This method is a simple approximation for the equivalent drag resulting from power extraction. The equivalent power generation drag is overcome by the engine thrust in this simple approximation.

The overall drag including all considerations becomes

$$D = \frac{C_D}{C_L} \cdot W + D_{\text{Pow}} + D_{\text{Cool}} \tag{9.34}$$

The weight calculations can occur in different orders, depending on the available inputs and required outputs. For example, the segment range and initial weight can be input and the final weight output. Or, the initial and final weights can be input and the endurance output.

The mission profiles can be calculated forward or backwards. A forward-calculated mission starts at takeoff at the takeoff weight and finishes at landing. A backwards-calculated mission is calculated in reverse order starting from landing with no fuel and then flying backwards in time to the takeoff condition. The advantage of a backwards-calculated mission is that the exact amount of fuel required to complete the mission can be found directly. A forward-calculated mission will either run out of fuel before completing the mission or extra fuel will be in the fuel tanks at landing. This chapter will cover only the forward-calculated mission analysis, though the reader is encouraged to develop his or her own backward calculation methods.

For a forward-calculated mission segment, the final weight is found by

$$W_f = \frac{W_i}{\exp\left[(f_T \cdot R_{\text{seg}} \cdot TSFC)/(V \cdot C_L/C_D)\right]} \tag{9.35}$$

The fuel burned during the segment $W_{\text{Fuel,seg}}$ is the difference between the initial and final weight.

$$W_{\text{Fuel,seg}} = W_i - W_f \tag{9.36}$$

The final weight of the segment becomes the initial weight for the next segment.

9.5.1 Mission Profile Analysis

The flight velocity for each segment will either be specified in a requirement or selected for best performance. Specified flight velocities can take the form of true airspeed, Mach number, or, indirectly, tracking of moving ground targets.

9.5.1.1 Velocity Optimization

To optimize the velocities, there are several figures of merit. These figures of merit are given such that higher values are better. For a traditional fueled aircraft, the figures of merit are as follows:

Range efficiency:

$$FOM = \frac{R_{\text{seg}}}{W_i - W_f} \tag{9.37}$$

Endurance efficiency:

$$FOM = \frac{E_{\text{seg}}}{W_i - W_f} \tag{9.38}$$

Maximum climb rate:

$$FOM = ROC_{\text{avail}} \tag{9.39}$$

Climb efficiency:

$$FOM = \frac{h_2 - h_1}{W_i - W_f} \tag{9.40}$$

For an electric-powered aircraft where the weight does not change, the figures of merit are as follows:

Range efficiency:

$$FOM = \frac{R_{\text{seg}}}{\Delta Energy} \tag{9.41}$$

Endurance efficiency:

$$FOM = \frac{E_{\text{seg}}}{\Delta Energy} \tag{9.42}$$

Maximum climb rate:

$$FOM = ROC_{\text{avail}} \tag{9.43}$$

Climb efficiency:

$$FOM = \frac{h_2 - h_1}{\Delta Energy} \tag{9.44}$$

A search algorithm can be used to find the optimal velocity for the given figure of merit in the constrained flight envelope. In a trade between speed and robustness, a bisection search algorithm is useful. Although traditional optimizers can be employed to find the best velocity, it becomes impractical to use these powerful tools to optimize velocities across a multisegment mission profile.

The first step is to determine the available flight envelope where no constraints are violated. Finding the envelope boundaries generally requires some iteration. A good initial minimum velocity estimate is the airspeed associated with $C_{L,Max}$. If a stall margin is used, then use the maximum permissible C_L (i.e., $C_L < 1.2^*C_{L,Max}$). The initial maximum velocity is the minimum of the dynamic pressure limit and the Mach limit. Depending on which constraints are violated, the minimum velocity is incrementally increased until no constraints are violated. Similarly, the maximum velocity is incrementally decreased until it falls within the feasible region.

The search begins once all of the minimum and maximum velocities fall within the feasible region. The bisection method uses five evenly distributed velocity points. Initially the first point (point 1) is set to the minimum feasible velocity, and the last point (point 5) is the maximum feasible velocity. The figure of merit (FOM) is evaluated at each of the five points. If the point with best FOM does not lie at point 1 or 5, then the point with the best figure of merit becomes the point 3. The case where the best figure of merit is at point 1 or 5 will be discussed in the next paragraph. The right and left adjacent points become points 1 and 5, respectively. Point 2 is created halfway between points 1 and 3, and point 4 is created halfway between points 3 and 5. The process continues through several iterations. The velocity associated with the best figure of merit at the end of the search is the final velocity selected.

The process is different when the best figure of merit is an end point (point 1 or 5). In the case where the best figure of merit is point 1, point 3 becomes point 5, and point 2 becomes point 3. Points 2 and 4 are generated halfway between these points as just described.

The procedure continues until the maximum number of iterations is reached.

Example 9.3   Best Range Speed

Problem:

Using the aircraft from Problem 9.2, we wish to find the best level flight range velocity. Additional parameters of interest include the following:

Installed jet engine TSFC is 0.7 hr^{-1}.
The drag polar is defined by $C_{D0} = 0.03$, $e = 0.7$, and $AR = 10$.
The segment range is 100 km.

Solution:

The figure of merit is given by

$$FOM = \frac{R_{seg}}{W_i - W_f}$$

Table 9.2 Velocity Convergence

Iteration	V, m/s					FOM				
	Pt 1	Pt 2	Pt 3	Pt 4	Pt 5	Pt 1	Pt 2	Pt 3	Pt 4	Pt 5
1	65.62	100.85	136.08	171.32	206.55	429.15	636.35	587.23	498.87	424.55
2	65.62	83.24	100.85	118.47	136.08	429.15	582.40	636.35	625.28	587.23
3	83.24	92.04	100.85	109.66	118.47	582.40	620.40	636.35	636.15	625.28
4	92.04	96.45	100.85	105.26	109.66	620.40	630.77	636.35	637.90	636.15
5	100.85	103.05	105.26	107.46	109.66	636.35	637.58	637.90	637.40	636.15
6	103.05	104.15	105.26	106.36	107.46	637.58	637.85	637.90	637.75	637.40

where

$$W_i - W_f = W_i \cdot \left[1 - \exp\left(\frac{-f_T \cdot R_{seg} \cdot TSFC}{V \cdot C_L/C_D}\right)\right]$$

The results of the velocity and *FOM* are shown in Table 9.2. Boxes are drawn around the best values.

The velocity convergence is shown in Fig. 9.3. The difference between the maximum and minimum velocities on the first iteration is 68% relative to the maximum velocity. This difference is only 4% at the sixth iteration. This method converges quickly.

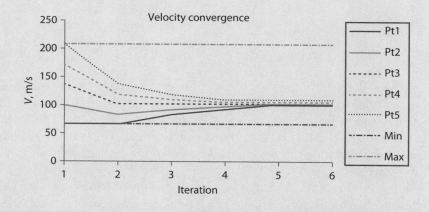

Fig. 9.3 Velocity convergence: Pt 1 is the bottom curve, Pt 2 the 4th curve from the top, Pt 3 the middle curve, Pt 4 the 2nd curve from the top, Pt 5 the top curve, Min the bottom dashed line, and Max the top dashed line.

9.5.1.2 Climb Performance

The cruise-climb altitude change is assumed equal to the altitude change required to keep the lift coefficient constant. The final subsegment altitude is found through linear interpolation or extrapolation. A good value for the reference altitude change Δh is 10,000 ft. The cruise-climb altitude change is found by

$$h_2 = h_1 + \Delta h \cdot \frac{\rho(h_1)}{\rho(h_1 + \Delta h) - \rho(h_1)} \cdot \left(\frac{W_f}{W_i} - 1\right) \tag{9.45}$$

where h_2 is the altitude at the completion of cruise climb and h_1 is the starting altitude. The rate of climb associated with the cruise climb is estimated. When a cruise-climb profile is selected, the fuel burn is first calculated without any rate of climb. The rate of climb for the cruise-climb segment is

$$ROC = \frac{h_2 - h_1}{E_{seg}} \tag{9.46}$$

The difference in fuel burn for the level flight and cruise climb conditions is found by first estimating the difference in thrust required for cruise climb.

$$\Delta T_{\text{Climb}} = \frac{ROC}{V} \cdot W \tag{9.47}$$

Next, the difference in fuel burn for the cruise climb is estimated by

$$\Delta W_{\text{Fuel,Climb}} = \Delta T_{\text{Climb}} \cdot TSFC \cdot E_{\text{Seg}} \tag{9.48}$$

This can be rearranged as

$$\Delta W_{\text{Fuel,Climb}} = \frac{h_2 - h_1}{V} \cdot W \cdot TSFC \tag{9.49}$$

Note that E_{Seg} is the endurance of the subsegment. The optimal cruise velocity is found for each new subsegment. The initial cruise altitude for the second through final subsegments is simply the altitude associated with the best cruise condition of the predecessor segment. The starting altitude is used for calculating the aerodynamic and propulsion parameters as well as the constraints.

A constant rate of climb can be specified for a mission segment. The flight velocity can be optimized for any figure of merit desired. The altitude at the conclusion of a climb segment is

$$h_2 = h_1 + ROC_{seg} \cdot E_{seg} \tag{9.50}$$

If the initial and final altitudes are specified, then the rate of climb is found by

$$ROC_{seg} = \frac{(h_2 - h_1)}{E_{seg}} \tag{9.51}$$

If the final altitude exceeds the maximum altitude, then the final altitude can be set to the maximum altitude value.

Even if a rate of climb is input, it might not be achievable. The maximum available rate of climb is

$$ROC_{\text{max,avail}} = \frac{(T_{\text{max}} - D)}{W} \cdot V \tag{9.52}$$

For descent, the minimum rate of climb (positive rate of descent) is

$$ROC_{\text{min,avail}} = \frac{(T_{\text{min}} - D)}{W} \cdot V \tag{9.53}$$

If the required rate of climb exceeds the maximum rate of climb, then the rate of climb can be set to the maximum rate-of-climb value. Similarly, if the required rate of descent exceeds the minimum rate of climb, then the rate of climb can be set to the minimum rate-of-climb value.

The mission analysis code can evaluate the climb performance without a prescribed range or endurance. The initial and final altitudes must be input, and the range and endurance will be output. The two desired climb modes are maximum rate of climb and most fuel-efficient climb. The associated velocities are found via the outer-loop figure-of-merit search procedure described earlier. For the maximum rate of climb, the maximum available thrust is used. The climb efficiency FOM is used to find the most efficient climb.

The endurance and range for a climb segment are found by

$$E_{\text{seg}} = \frac{h_2 - h_1}{ROC_{\text{seg}}} \quad \text{and} \quad R_{\text{seg}} = E_{\text{seg}} \cdot V \tag{9.54}$$

The final subsegment weight W_f is then found via the modified Breguet range equation.

9.5.1.3 Mission Profile Analysis

The mission completion criterion is a metric for determining if the unmanned aircraft with the available fuel load can complete the desired mission. The mission completion criterion evaluation method depends on if the unmanned aircraft completes the mission. If the mission is completed, this criterion is found by

$$misscomp = \frac{W_{\text{fuel}}}{W_{\text{fuelb}}} \tag{9.55}$$

where W_{fuel} is the fuel carried at takeoff and W_{fuelb} is the fuel burned during the mission.

A different procedure is used when there is insufficient fuel to complete the arbitrary mission. The mission analysis code keeps track of three different types of mission segment: range, endurance, and climb. Although all three parameters are known for each segment, only one per segment is tracked for the mission completion criteria. First, the total range, endurance, and climb are calculated:

For all range-type segments:

$$R_{tot} = \sum R_{seg} \tag{9.56}$$

For all endurance-type segments:

$$E_{tot} = \sum E_{seg} \tag{9.57}$$

For all climb-type segments:

$$\Delta h_{tot} = \sum |\Delta h_{seg}| \tag{9.58}$$

Next, the relative weighting of each metric must be assessed. The range, endurance, and climb are all put in terms of distance. Then each distance-normalized parameter is weighted. First, a denominator in terms of distance is found:

$$Denom = R_{tot} + \left(E_{tot} + \frac{\Delta h_{tot}}{ROC_{min}} \right) \cdot V_{eq,\,max} \tag{9.59}$$

ROC_{Min} is the minimum rate of climb for the service ceiling, such as 100 ft/min. Second, the relative weightings are found:

$$R_{wt} = \frac{R_{tot}}{Denom} \tag{9.60}$$

$$E_{wt} = \frac{E_{tot} \cdot V_{eq,\,max}}{Denom} \tag{9.61}$$

$$alt_{wt} = \frac{\Delta h_{tot}/ROC_{min} \cdot V_{eq,\,max}}{Denom} \tag{9.62}$$

If the unmanned aircraft depletes the fuel during any mission segment, then an error condition is triggered along with the amount of the segment that is completed. All segments that follow are not completed. There is a running summation of the completed and partially completed segments. The mission completion parameter for noncompleted arbitrary missions is

$$misscomp = R_{wt} \cdot \frac{R_{comp}}{R_{tot}} + E_{wt} \cdot \frac{E_{comp}}{E_{tot}} + alt_{wt} \cdot \frac{|\Delta h_{comp}|}{\Delta h_{tot}} \tag{9.63}$$

External payloads can add significantly to the unmanned aircraft drag. Marketing materials can specify unmanned aircraft endurance along with internal and external payload capacities. Rarely does the provided endurance correspond with external payloads. Rather, a relatively clean aircraft is assumed.

Example 9.4   Voyager's Extreme Range and Endurance Flight

Problem:

In 1988, the Rutan Voyager (Fig. 9.4) flew around the world nonstop and non-refueled, setting a world record. The following information is known about the aircraft and mission:

* The flight duration was 216 hrs (9 days, 3 min, 44 s) and covered 26,336 miles.
* The average cruise speed was 116 mph (100.7 kt), and the maximum speed was 150 mph (130.3 kt).
* The wings have a span of 108.3 ft and a wing area of 362 ft^2.
* The cruise altitude varied between 10,000 and 15,000 ft MSL, with an average altitude of 11,000 ft. Assume that the flight occurs at 11,000 ft.
* The crew had a combined weight of 303 lb, and the provisions weighed 130 lb.
* The takeoff gross weight of the UA was 9694 lb, and the empty weight was 2250 lb.
* The maximum L/D is estimated to be 27. Assume that the drag polar is

$$C_D = C_{D0} + \frac{C_L^2}{\pi \cdot AR \cdot e} \approx 0.026 + 0.012 \cdot C_L^2$$

* Assume that the engine BSFC in cruise is 0.35 lb/hp-hr and the propeller efficiency is 85%. The aircraft used two engines for takeoff, but the cruise engine had a maximum power of 110 hp.
* Assume that takeoff and initial climb to 11,000 ft consumes 2% of the aircraft takeoff weight in fuel.
* Assume that the generator extraction and cooling drag are negligible.

Calculate the total fuel burn of the aircraft.

Solution:

The fuel weight is equal to the takeoff weight minus the empty weight, crew weight, and provisions weight. This comes to 7,011 lb of fuel. The fuel burned during takeoff and the initial climb is 194 lb. Therefore, the fuel available for the mission is 6,217 lb.

Let us assume that there are four cruise segments, each with a range of 6,584 miles (5,718 n miles).

The best L/D occurs when the drag coefficient is twice C_{D0}. For this aircraft that occurs when

$$C_{D0} = \frac{C_L^2}{\pi \cdot AR \cdot e}$$

Solving our drag polar for the lift coefficient for best L/D, we get

$$C_L = \sqrt{\frac{0.026}{0.012}} = 1.47$$

Fig. 9.4 Rutan Voyager returning from its historic flight. (Photo courtesy of NASA.)

The drag coefficient is 0.0520, and the lift-to-drag ratio is 28.3. Using the definition of lift coefficient, we find that the flight velocity at this C_L is 85.7 kt at the beginning of the mission, which is far below the average airspeed of 100.7 kt (170.1 ft/s). As the aircraft burns fuel, the airspeed for best L/D will continue to be reduced. Instead we will use the average airspeed for all of the calculations.

Now we will perform the first calculation. The starting weight after the climb to altitude at the beginning of the first segment is 9,500 lb. From MIL-STD-3013 standard day atmosphere, the air density at 11,000 ft is 0.001701 sl/ft^3. The lift coefficient at the average flight speed is

$$C_L = \frac{W_i}{1/2 \cdot \rho \cdot V^2 \cdot S_w} = \frac{9,500 \text{ lb}}{1/2 \cdot (0.001701 \text{ sl/ft}^3) \cdot (170.1 \text{ ft/s})^2 \cdot (362 \text{ ft}^2)}$$

$$= 1.07$$

The drag coefficient for this lift coefficient is

$$C_D = 0.026 + 0.012 \cdot C_L^2 = 0.026 + 0.012 \cdot (1.07)^2 = 0.03966$$

The lift-to-drag ratio is

$$L/D = \frac{C_L}{C_D} = \frac{1.07}{0.03966} = 26.9$$

The conversion between *BSFC* (lbm/hp-hr) and *TSFC* (hr^{-1}) is

$$TSFC = \frac{BSFC \cdot V}{325.65 \cdot \eta_p}$$

where *V* is the airspeed in knots. The *TSFC* is then 0.1273 hr^{-1}. After we complete climb, the thrust factor f_T is 1.

The first segment calculation is

$$W_f = \frac{W_i}{\exp\left[f_T \cdot R_{seg} \cdot TSFC/(V \cdot C_L/C_D)\right]}$$

$$= \frac{9{,}500 \text{ lb}}{\exp\left[1 \cdot (5{,}718 \text{ n miles}) \cdot \left(0.1273 \text{ hr}^{-1}\right)/(100.7 \text{ kt}) \cdot (26.9)\right]}$$

$$= 7{,}261 \text{ lb}$$

The following table shows the results of the calculations:

Segment	W_i, lb	V, kt	C_L	L/D	TSFC, hr^{-1}	W_f, lb
1	9500	100.7	1.067	26.90	0.1273	7261
2	7261	100.7	0.8154	24.00	0.1273	5372
3	5372	100.7	0.6033	19.87	0.1273	3733
4	3733	100.7	0.4193	14.92	0.1273	2299

The total fuel burn is the difference between the takeoff weight and the final weight, which is 7,395 lb. The final weight is 2,299 lb.

The actual fuel mass fraction on the aircraft is 72.3%.

The jet-powered Virgin Atlantic Global Flyer beat the Voyager's speed and distance records in 2006. That aircraft had an estimated *L/D* of 37 and a fuel mass fraction of 83%.

9.5.2 Conventional Takeoff and Landing

Numerous launch and recovery techniques are covered in Chapter 11. Here the special case of runway-dependent, or conventional, launch and recovery is addressed. Although this is the norm for manned aircraft, this is only one of many options available to unmanned systems.

Various aircraft performance and design texts provide empirical methods for calculating takeoff and landing distances. These approaches are appropriate for large UASs that have characteristics similar to the applicable manned aircraft for which the equations were developed. However, unmanned aircraft frequently have attributes that make these methods questionable. For example, small UASs might have propulsion systems that do not follow the empirical relationships, and unsteady winds can strongly affect field performance.

A more general approach is to evaluate the forces acting upon the aircraft and advance the analysis through time steps. The forces can be broken into horizontal and vertical components. These forces generate accelerations that are integrated to yield ground distances. The outline of this method is presented here to highlight the approach. Other sources should be consulted for algorithm development [1].

The vertical forces acting on an unmanned aircraft that is rolling along a flat runway are

$$L + F_N + F_M = W \tag{9.64}$$

where L is the lift of the aircraft, F_N is the nose gear reaction force, and F_M is the main gear reaction force. The lift included ground effect influences. The lift force increases as the airspeed increases, which takes load off of the landing gear. Takeoff occurs once the lift is greater than the weight.

The horizontal forces acting on an unmanned aircraft rolling along a flat runway are

$$T - D - \mu_{g,N} \cdot F_N - \mu_{g,M} \cdot F_M = \frac{W}{g} \cdot a_x \tag{9.65}$$

where D is the drag, $\mu_{g,N}$ is the rolling friction for the nose gear, $\mu_{g,M}$ is the rolling friction for the main gear, and a_x is the acceleration down the runway. The rolling-ground-friction coefficients range from 0.02 to 0.3, depending on the surface type. MIL-STD 3013 provides rolling-ground-friction values for various conditions. If we assume that the rolling friction is the same for both the nose and main gear, then the horizontal forces equation simplifies to

$$T - D - \mu_g \cdot (W - L) = \frac{W}{g} \cdot a_x \tag{9.66}$$

The drag and lift are found by

$$D = 1/2 \cdot \rho \cdot V^2 \cdot S_w \cdot C_{D,g} \quad \text{and} \quad L = 1/2 \cdot \rho \cdot V^2 \cdot S_w \cdot C_{L,g} \tag{9.67}$$

where $C_{D,g}$ is the drag coefficient in ground effect and $C_{L,g}$ is the lift coefficient in ground effect.

9.5.2.1 Takeoff

Takeoff has several stages. First, the nose gear remains on the ground until the rotation speed V_R is reached. Then the aircraft rotates to a new angle of attack such that the nose wheel is off the ground and there is a higher lift coefficient. This stage continues until the liftoff speed V_{LOF} is reached where the lift is equal to the weight and the aircraft becomes airborne. Immediately after liftoff, the unmanned aircraft transitions to the initial climb-out state. Then the aircraft climbs to clear the obstacle height. The aircraft then proceeds with its climb out. The takeoff distance is the distance covered from brake release to clearing the obstacle height. The phases of takeoff are summarized in Table 9.3.

Additionally, two parameter groupings are often used: the takeoff ground roll S_G and the takeoff air distance S_A. These are given by

$$S_G = S_{NGR} + S_R \quad \text{and} \quad S_A = S_{TR} + S_{CL} \tag{9.68}$$

The total takeoff distance is

$$S_{TO} = S_G + S_A \tag{9.69}$$

The horizontal ground velocity at time step i, $V_g(i)$, is found by the numerical integration

$$V_g(i) = V_g(i-1) + a_x(i-1) \cdot \Delta t \tag{9.70}$$

The relationship between the airspeed, ground speed, and the wind speed V_w is

$$V_g = V - V_w \tag{9.71}$$

where V_w is positive as a headwind.

The distance with respect to the starting point is found by the numerical integration

$$S(i) = S(i-1) + V_g(i-1) \cdot \Delta t + 1/2 \cdot a_x(i-1) \cdot \Delta t^2 \tag{9.72}$$

For the main ground roll where the nose gear is on the ground, the acceleration a_x is

$$a_x = g \cdot \left[\frac{T}{W} - \frac{C_{D,g}}{C_{L,g}} - \mu_g \cdot \left(1 - \frac{L}{W}\right) \right] \tag{9.73}$$

Table 9.3 Takeoff Phases

Phase	Ground distance	Ending velocity	Notes
Nose wheel on ground	S_{NGR}	V_R	All gear on ground
Rotation	S_R	V_{LOF}	Nose wheel liftoff
Transition	S_{TR}	—	Unmanned aircraft airborne, but in ground effect
Climb-out to obstacle clearance	S_{CL}	V_2	Mostly out of ground effect

This can be expanded to

$$a_x = g \cdot \left[\frac{T(V, \rho)}{W} - \frac{C_{D,g}}{C_{L,g}} - \mu_g \cdot \left(1 - \frac{1/2 \cdot \rho \cdot V^2 \cdot S_w \cdot C_{L,g}}{W} \right) \right] \quad (9.74)$$

This phase begins from a dead stop and continues until V_R is reached.

Note that the density is present in the thrust and ground lift. Takeoff at higher altitudes or on hot days acts to decrease the density and thereby decreases the ground acceleration and increases the takeoff distance. Also, the thrust term is general as presented, so that it does not matter if the propulsion system is a jet, reciprocating-propeller, fuel-cell electric, or any other type with this formulation.

Once the rotation speed is attained, and the nose lifts off the ground while the main gear is still on the ground, the lift and drag coefficient have the values $C_{L,g,R}$ and $C_{D,g,R}$, respectively. The ground acceleration during rotation is

$$a_x = g \cdot \left[\frac{T(V, \rho)}{W} - \frac{C_{D,g,R}}{C_{L,g,R}} - \mu_g \cdot \left(1 - \frac{1/2 \cdot \rho \cdot V^2 \cdot S_w \cdot C_{L,g,R}}{W} \right) \right] \quad (9.75)$$

The unmanned aircraft accelerates in the rotation phase until V_{LOF} is reached and the unmanned aircraft lifts off.

In the transition phase, there is no longer any rolling resistance because the unmanned aircraft is airborne. The lift and drag coefficients change as a function of height above the ground due to flying out of ground effect. These coefficients can also vary with velocity as the angle of attack changes. The acceleration during transition becomes

$$a_x = g \cdot \left[\frac{T(V, \rho)}{W} - \frac{C_D(h, V)}{C_L(h, V)} \right] \quad (9.76)$$

During climb-out to obstacle clearance, we can assume that the unmanned aircraft is out of ground effect. The obstacle height is 50 ft for the FAR 23 and 25 manned aircraft regulations, and this obstacle height is frequently adopted by unmanned aircraft as well. The ground acceleration becomes

$$a_x = g \cdot \left[\frac{T(V, \rho)}{W} - \frac{C_D(V)}{C_L(V)} \right] \quad (9.77)$$

The field length for multi-engine aircraft is determined by the engine-out case. If the engine fails at a decision speed V_1, then the distance to clear the obstacle and the distance to stop the aircraft are equal. The total distance along the ground is called the *balanced field length*. If the engine fails before V_1 is reached, then the unmanned aircraft should brake to a stop. Otherwise, if the engine fails above V_1, then the unmanned aircraft should continue with the takeoff run. The balanced field length is generally greater than the takeoff distance.

Example 9.5 Ground Roll with Nose Gear on Ground

Problem:

An unmanned aircraft has the following characteristics at the takeoff condition:

$W = 300$ lb
$T = 100 - 0.2\,V$, where T is in pounds and V is in knots
$S_w = 15$ ft^2
$C_{L,g} = 0.5$
$C_{D,g} = 0.05$
$V_R = 40$ kt
$\mu_g = 0.03$

Assume a standard sea-level day. The headwind is 10 kt.
Calculate the ground distance S_{NGR} using a time step of 1 s.

Solution:

The density is 0.002376 sl/ft^3. The results are as follows:

Time, s	Thrust, lbf	Acceleration, ft/s^2	V_g, kt	Distance, ft
0	98.0	6.34	0.00	0.00
1	97.2	6.27	3.76	3.17
2	96.5	6.20	7.47	12.6
3	95.8	6.13	11.1	28.4
4	95.0	6.07	14.8	50.2
5	94.3	6.00	18.4	78.2
6	93.6	5.95	21.9	112
7	92.9	5.89	25.4	152
8	92.2	5.84	28.9	198
9	91.5	5.78	32.4	250

With a headwind of 10 kt, the airspeed is 10 kt faster than the ground speed. The rotation speed is reached between 8 and 9 s. If we linearly interpolate between these time steps, we find that the rotation speed is reached when $t = 8.3$ s and S_{NGR} is 209 ft.

9.5.2.2 Landing

The landing analysis is similar to that of the takeoff analysis, except the process happens in reverse. The same time-step integration can be applied for landing. More details can be found in other sources [1].

The aircraft approaches the runway at approach speed V_A. The approach speed is 1.3 times V_{Stall} for FAR 23 and 25, though some military requirements have V_A as 1.2^*V_{Stall}. Here V_{Stall} is the stall speed in the landing configuration, which can include flap deflection and extended landing gear.

After clearing the obstacle height, this again is typically specified as 50 ft, and the aircraft transitions to the touchdown velocity V_{TD}. Unmanned aircraft often do not flare like manned aircraft, as the flight control logic for flare is more difficult to implement than simply continuing the glide slope at V_A. The vertical velocity at touchdown is greater without flare, but landing gear can be designed to take the impact loads and prevent bouncing. A nonflare landing will still generally have the main gear touching the ground first. After touchdown the unmanned aircraft may undergo a rotation where the nose gear touches the runway.

Once the nose gear is on the ground, the brakes are applied for the remainder of the ground roll. The braking force is a modified rolling friction coefficient $\mu_{g,brake}$. Typical values of $\mu_{g,brake}$ range from $0.05 - 0.8$ depending on the brake system configuration and surface type. Assuming a flat runway, the acceleration during this phase is given by

$$a_x = g \cdot \left\{ \left[T(V, \rho)/W - \mu_{g,brake} \right] - \frac{1}{2} \cdot \frac{\left(C_{D,g} - \mu_{g,brake} \cdot C_{L,g} \right) \cdot \rho \cdot V^2 \cdot S_w}{W} \right.$$

$$\left. + \frac{F_N}{W} \cdot \left(\mu_{g,brake} - \mu_g \right) \right\} \tag{9.78}$$

Note that the thrust will generally be a minimum value during this stage. High-altitude UASs can have high aerodynamic efficiency, and the minimum throttle can still produce significant thrust at the landing condition. In this scenario it might be necessary to employ supplemental drag devices such as spoilers or large deflection flaps to enable a reasonably steep glide slope and landing distance.

Reference

[1] Lan, C.-T. E., and Roskam, J., *Airplane Aerodynamics and Performance*, Design, Analysis and Research Corp., Lawrence, KS, 2003, pp. 435–508.

Problems

9.1 Write a code to implement the standard atmosphere methods described in Example 9.1.

9.2   An unmanned aircraft has a drag polar defined by $C_{D0} = 0.1$, $AR = 10$, and $e = 0.7$. The weight is 500 lb, and the wing area is 30 ft^2. At 15,000 ft, its jet engine produces a maximum thrust of 200 lb, which is independent of airspeed. Using the bisection method, find the airspeed that generates the greatest rate of climb. What is the maximum rate of climb?

9.3   Use the unmanned aircraft from problem 9.2. Optimize the airspeed for maximum climb angle at 15,000 ft.

9.4   Use the unmanned aircraft from problem 9.2. Optimize the airspeed for best range performance at 15,000 ft. Assume that the TSFC is independent of airspeed.

9.5   Use the unmanned aircraft from problem 9.2. If the structure is limited to 200 KEAS, what is the maximum airspeed at 15,000 ft?

9.6   Use the unmanned aircraft from problem 9.2. The maximum lift coefficient is 1.4. Again, the structural dynamic pressure limit corresponds to 200 KEAS. Plot the minimum airspeed and dynamic pressure limits vs altitude from sea level to 40,000 ft.

9.7   An unmanned aircraft has the following characteristics: weight = 1,000 lb; wing area = 100 ft^2; sea-level thrust = 500 lb (independent of airspeed); rolling-friction coefficient of 0.02; ground-effect lift coefficient = 0.5; ground-effect drag coefficient = 0.1; and rotation speed of 60 kt. Using time steps of 1 s, calculate the distance required to achieve rotation at sea level.

Chapter 10

Avionics, Flight Software, and Subsystems

- Learn about avionics systems that are commonly used on UASs
- Learn the basic software functions
- Gain an understanding of the major subsystems

U.S. Airman with the 432nd Aircraft Maintenance Squadron works on MQ-1 Predator avionics during routine maintenance. (U.S. Air Force photo by Senior Airman Larry E. Reid, Jr.)

10.1 Introduction

A vionics and subsystems are vital to the operation of the unmanned aircraft, yet are often largely overlooked at the conceptual design phase. This is especially true of new designers and design teams. Questions about avionics architectures or cooling methods often are met with blank stares or assertions that such matters are details that can be addressed at later design phases. Conceptual designs without a basic strategy and hardware allocation will usually experience weight growth challenges and performance degradation as the design graduates to the preliminary design phase.

This chapter provides an overview of the avionics, flight software, and subsystems that are used for various classes of UASs. These topics can serve as a checklist for the design team to ensure that all systems that have size, weight, and power have been fully accounted for. The avionics section describes autopilots, inertial navigation systems, processors, and numerous other items. This is followed by flight software, which provides insight into the major software functions and architectures. Lastly, the most important subsystems are covered.

10.2 Avionics

Avionics is a juxtaposition of the words *aviation* and *electronics*. These flight electronics are considered any electrical devices used on the unmanned aircraft with the exception of the electrical power system and payloads. The electrical power system is considered a subsystem. Payloads and communications systems are treated in later chapters. Avionics components serve diverse roles such as flight control computers, inertial navigation systems, and airspace management. Some of the most important and prevalent avionics are described in this chapter.

Those who are new to UASs can take microprocessors, digital communications, and global positioning systems (GPSs) for granted. These advanced electronics are prevalent in almost all modern UASs, even at very small scales. Yet these important technologies came to maturity for widespread UAS use as late as the 1990s. The author can remember participating in NRL SENDER small UAS flight tests in 1995 that demonstrated the then-revolutionary GPS autopilots. Earlier autopilots and INS systems often used analog circuitry. Advances in avionics enabled the recent ascendency of UASs. Collinson [1] can be consulted for more in-depth treatment of avionics.

10.2.1 Avionics Design

10.2.1.1 Physical Environment

Avionics systems must be designed to be compatible with the physical environment in which they operate. Pushing these components beyond their design capabilities often leads to failures. Large UAs tend to use avionics that are purpose built for aerospace applications. However, tactical and smaller UAs are often required to have avionics that perform sophisticated roles that are simply unavailable in the aerospace avionics market. Custom avionics design or adaptation of commercial-off-the-shelf (COTS) equipment intended for other uses is common in this circumstance.

The unmanned aircraft physical environment can be demanding for electronics. The shock, vibration, and acceleration can be more stressing than what is experienced in ground-based electronics. These loads vary among unmanned aircraft and in different locations on the airframe. For example, the vibration will be higher for engine-mounted electronics than a wing tip light. Similarly, the shock loads will be greatest near the landing gear or launcher interface.

The avionics must be capable of operating within a defined temperature range. The environmental control system must be capable of maintaining avionics bay temperatures within established bounds. The avionics operate within these temperature bounds and are able to dissipate heat via the environmental control system. Boards that operate outside of their temperature limits can fail due to thermal factors. Overheating is a common problem when ground-based electronics are put in unmanned aircraft without consideration for cooling.

Enclosures are provided to protect electronics boards from the local environment surrounding the mounting location. Enclosures usually take the form of boxes. However, conformal coating of boards is also used. The enclosures provide physical mounting to the unmanned aircraft and can connect multiple boards together. Environmental threats to the electronics include electromagnetic interference, humidity, water, dust, salt, and temperature. Enclosures can have vibration isolation mounts to protect the internal components from shock and vibration.

Avionics components must adhere to geometric constraints, whether they are common standards-based or custom. Many electronics boards comply with a PC-104 form factor. Avionics boxes frequently comply with ATR (Austin Trumbull Radio) box standards. ATR box chassis come in a variety of height and width combinations, and the boxes have standard mounting and connectors that support electromagnetic interference protection and vibration isolation. ATR boxes can be housed on mounting racks on the unmanned aircraft. ATR boxes are appropriate for tactical UAS and larger unmanned aircrafts, but are too heavy for the smaller classes.

Small UASs or those that have high packaging density requirements frequently depart from aerospace standard avionics installation approaches in favor of custom or nonstandard components. Rectangular boards can be volumetrically inefficient for circular fuselage cross sections and can drive the fuselage frontal area. To solve this problem, board geometry can be tailored to the geometry. The same is true for enclosures—the box might have a nonrectangular shape to conform to the available volume. Small autopilots and other components for small UASs are simply smaller than the standards available for aerospace components.

COTS components are items that can be purchased directly with no need for further development. Merely uttering the word "COTS" can generate strong reactions from seasoned system developers that range from pleasure that the system will be affordably and quickly acquired to foreboding of reliability challenges and cost overruns. COTS implies that the system is a mature part and a nondevelopmental item (NDI). The nonrecurring engineering has already been invested, and the part should be qualified for its intended use. However, problems can arise if the COTS component is used outside of its intended design limitations or if the component is not truly fully developed. For example, a COTS image processing board used for ground applications might not be capable of operating in an airborne environment without design modifications. Military hardware that is fully developed and qualified is called military-COTS (M-COTS).

An unmanned aircraft is apportioned into bays or zones. For example, a fuselage region bounded by two bulkheads and skins can constitute a fuselage bay. Payload bays are also defined regions. The physical environment is assessed for each of these zones.

10.2.1.2 Electromagnetic Compatibility and Interference

Electromagnetic compatibility (EMC) and electromagnetic interference (EMI) are significant design drivers for robust and reliable avionics systems. EMC is the operability of the system throughout the electromagnetic spectrum in the operating environment. EMC requires that the system can function without adverse interference. Challenges to EMC are generally known as EMI. EMI is a disturbance that interrupts, obstructs, or degrades the performance of avionics systems or other electrical components. EMI can cause electronics failures or disrupt their function. EMI sources include threat jammers, unintentional jamming (i.e., friendly radars), or simply congested radio frequency (RF) backgrounds. The unmanned aircraft avionics, communication systems, and payloads are other potential sources of EMI. Another form of EMI is radio-frequency interference (RFI), which is RF interference with communications equipment.

Electromagnetic hardening is often required to ensure EMC. Shielding the avionics and wiring harnesses prevents unwanted signals from entering the system. Avionics enclosures with EMI protection are metallic with conductive seals. Shielding increases the weight substantially over non-shielded avionics and wiring harnesses. Filtering and attenuation can reject noise signals into the component. Grounding components to a common ground is frequently necessary to prevent unwanted electrical currents.

Some methods for improving EMI include the following:

* Connectors that are designed for EMI
* Sealed electronics enclosures with conductive outer cases
* Place electronics within sealed metal structure areas
* EMI shielding gaskets for connectors and enclosure seals
* Avoid ground loops (i.e., common grounding)
* Isolate electronics that might interfere with one another
* Use shielded cables in wiring harnesses (i.e., braided, foil, or metal-core conduit shielding)
* Use fiber optics rather than conductive wiring
* Use RF absorbing materials to attenuate RF interference
* Eliminate electromagnetic line of sight routes to electronics devices

Conductive outer skins can both help and hinder EMI. The skins act as a Faraday cage for outside electromagnetic energy, where the charge stays on the outer shell and does not penetrate inside. On the other hand, it can act as a resonant cavity for the internal components. Composite skins are generally less conductive than metallic structures. To help protect against outside EMI sources, composite skins are often given conductive coatings or an embedded conductive mesh.

Single event upsets (SEU) is a phenomenon where radiation causes erroneous digital signals. Digital data are a combination of 1 and 0 discrete signal values, where each represents a binary voltage state. Charged particles can change the voltage of one of these bits, which is called a *bit flip*. SEUs are well known to spacecraft designers, but are not often considered for most UAS programs. SEUs are important for high-altitude UASs where the level of exoatmospheric radiation is higher.

10.2.2 Avionics Components

Now we will explore several common avionics component types. Most of these items can be found on both manned and unmanned aircraft. However, manned aircraft components such as instrument panels or electronic pilot controls are excluded from this treatment of avionics.

10.2.2.1 Navigation Sensors

The GPS has been revolutionary in the effectiveness of UASs. Prior to the introduction of GPS, the latitude-longitude positioning had to be accomplished primarily by inertial and magnetic means. Occasionally RF navigation beacons (i.e., LORAN) or stellar navigation aided the position solution. Two early programs to use GPS include the Boeing Condor HALE UAS in 1988 and the NRL SENDER small UAS in 1995. It is remarkable that this enabling technology that is often taken for granted is so recent. Vietnam-era Teledyne Ryan Firebee UASs had flight paths that were often miles off of their flight plans due to reliance on analog-based inertial navigation.

The unmanned aircraft state can include the parameters identified in Table 10.1. An example state vector $\hat{x}$ is given by

$$\hat{x} = \begin{bmatrix} x_E & y_E & z_E & V & u & v & w & \alpha & \beta & \psi & \theta & \phi & a_x & a_y & a_z & p & q & r \end{bmatrix}^T$$

$$(10.1)$$

Table 10.1 UA State Parameters

	State variable	Notes
Position in Earth coordinates	x_E	Relative to a fixed longitude, μ
	y_E	Relative to a fixed latitude, λ
	z_E	Relative to the Earth's surface
Airspeed	V	Total airspeed
	u	Airspeed along x axis (body)
	v	Airspeed along y axis (body)
	w	Airspeed along z axis (body)
Aerodynamic angles	α, Angle of attack	
	β, Angle of sideslip	
Euler angles	Ψ, Heading angle	Relative to North, positive in West direction (counterclockwise)
	θ, Pitch angle	Relative to horizontal, positive up
	φ, Roll angle	
Accelerations in body axes	a_x	
	a_y	
	a_z	
Angular rates in body axes	p	Roll rate
	q	Pitch rate
	r	Yaw rate
Geographic position	λ, Latitude	Can be related to y_E
	μ, Longitude	Can be related to x_E

Note that the T superscript denotes that the transpose of the state matrix is shown. The state matrix has one column and many rows. This example UA state vector consists of 18 parameters, which are defined in Table 10.1. Simple navigation systems such as those used on small UASs will eliminate many of these parameters in order to reduce cost and complexity.

The velocity of the unmanned aircraft in Earth coordinate system is

$$\dot{x}_E = R_b^E V + V_w \tag{10.2}$$

where V is the velocity vector in body coordinates (u, v, w), R_b^E is the coordinate transformation from the body axis to the Earth coordinate system, and V_w is the wind velocity vector. To save space, the cosine is denoted by c, sine denoted by s, and tangent is denoted by t. The coordinate transformation is given by

$$R_b^E = \begin{bmatrix} c\theta \cdot c\psi & (s\phi \cdot s\theta \cdot c\psi - c\phi \cdot s\psi) & (c\phi \cdot s\theta \cdot c\psi + s\phi \cdot s\psi) \\ c\theta \cdot s\psi & (s\phi \cdot s\theta \cdot s\psi + c\phi \cdot s\psi) & (c\phi \cdot s\theta \cdot c\psi - s\phi \cdot c\psi) \\ -s\theta & s\phi \cdot c\theta & c\phi \cdot c\theta \end{bmatrix} \tag{10.3}$$

The expanded form of the Earth coordinate velocity is

$$\begin{bmatrix} \dot{x}_E \\ \dot{y}_E \\ \dot{z}_E \end{bmatrix} = \begin{bmatrix} c\theta \cdot c\psi & (s\phi \cdot s\theta \cdot c\psi - c\phi \cdot s\psi) & (c\phi \cdot s\theta \cdot c\psi + s\phi \cdot s\psi) \\ c\theta \cdot s\psi & (s\phi \cdot s\theta \cdot s\psi + c\phi \cdot s\psi) & (c\phi \cdot s\theta \cdot c\psi - s\phi \cdot c\psi) \\ -s\theta & s\phi \cdot c\theta & c\phi \cdot c\theta \end{bmatrix} \begin{bmatrix} u \\ v \\ w \end{bmatrix}$$
$$+ \begin{bmatrix} V_{w,x} \\ V_{w,y} \\ V_{w,z} \end{bmatrix} \tag{10.4}$$

Note that the component $\dot{x}_E$ is velocity in the east direction V_E and the component $\dot{y}_E$ is velocity in the north direction V_N. Sometimes the velocity components are not known directly from inertial measurements and must be derived from the air data system. The velocity components in body coordinates based on observed flight velocity, angle of attack, and angle of sideslip are

$$\begin{bmatrix} u \\ v \\ w \end{bmatrix} = \begin{bmatrix} V \cdot c\alpha \cdot c\beta \\ V \cdot s\beta \\ V \cdot s\alpha \cdot c\beta \end{bmatrix} \tag{10.5}$$

This can be substituted into Eq. (10.4). If the wind acts in a horizontal plane along a heading angle ψ_w and a total speed of V_w, then the wind velocity vector can be expressed as

$$Vw = \begin{bmatrix} V_w \cdot s\psi_w \\ V_w \cdot c\psi_w \\ 0 \end{bmatrix} \tag{10.6}$$

This is a valid assumption when steady winds are estimated over long time periods. Short timescale gusts have more vertical components.

The relationship between the Euler angular rates and those in the body axes are

$$\begin{bmatrix} \dot{\phi} \\ \dot{\theta} \\ \dot{\psi} \end{bmatrix} = \begin{bmatrix} 1 & s\phi \cdot t\theta & c\phi \cdot t\theta \\ 0 & c\phi & -s\phi \\ 0 & s\phi/c\theta & c\phi/c\theta \end{bmatrix} \begin{bmatrix} p \\ q \\ r \end{bmatrix} \tag{10.7}$$

If the Euler angle rates are known and the body axis rates are unknown, then this can be expressed as

$$\begin{bmatrix} p \\ q \\ r \end{bmatrix} = \begin{bmatrix} 1 & 0 & -s\theta \\ 0 & c\phi & c\theta \cdot s\phi \\ 0 & -s\phi & c\theta \cdot c\phi \end{bmatrix} \begin{bmatrix} \dot{\phi} \\ \dot{\theta} \\ \dot{\psi} \end{bmatrix} \tag{10.8}$$

The rate of change of latitude is

$$\lambda = \frac{\dot{y}_E}{R_{\text{Earth}}} \tag{10.9}$$

where R_{Earth} is the Earth's radius. The rate of change in longitude is

$$\dot{\mu} = \frac{\dot{x}_E}{R_{\text{Earth}} \cdot \cos(\lambda)} \tag{10.10}$$

The unmanned aircraft state data are estimated through inputs from many sensors. These can include inertial sensors, GPSs, magnetic compasses, and air data sensors. The various sensors can measure similar parameters but with different update rate, drift, resolution, and accuracy. For example, the GPS limits navigation position drift by providing absolute position measurements within a well-bounded error, though at a low update rate. The inertial measurement unit provides high-frequency position and rate information to the navigation algorithms, but the drift errors grow without corrections. Clearly a suitable navigation solution will incorporate the best attributes of both sensor types while overcoming their limitations.

A Kalman filter is a common solution to this challenge. It is used to optimally estimate the unmanned aircraft state based on noisy sensors with asynchronous inputs. Kalman filters weight sensor inputs based on their accuracy. Sensor errors are treated as unbiased, white, Gaussian noise.

Here we will explore a linear Kalman filter. Several matrices that are used in the Kalman filter must first be introduced.

The measurement vector $\bar{z}$ contains all of the measured parameters. If the sensors are GPS ($x_{E,GPS}$, $y_{E,GPS}$, $z_{E,GPS}$), air data system (V_{ADS}), pressure altimeter ($z_{E,Alt}$), and compass heading (ψ_{Comp}), the measurement vector becomes

$$\bar{z} = \begin{bmatrix} x_{E,GPS} & y_{E,GPS} & z_{E,GPS} & V_{ADS} & z_{E,Alt} & \psi_{Comp} \end{bmatrix}^T \tag{10.11}$$

Consider a simple state vector of

$$\hat{x} = \begin{bmatrix} x_E & y_E & z_E & V & \psi \end{bmatrix}^T \tag{10.12}$$

Note that these states and measurements are insufficient for a UA navigation solution, and these are used for illustration purposes only.

A state transition matrix Φ relates how the state vector will change over a time increment dt. The state transition matrix for this simple illustration is

$$\Phi = \begin{bmatrix} 1 & 0 & 0 & -V \cdot s\psi \cdot dt & 0 \\ 0 & 1 & 0 & V \cdot c\psi \cdot dt & 0 \\ 0 & 0 & 1 & 0 & 0 \\ 0 & 0 & 0 & 1 & 0 \\ 0 & 0 & 0 & 0 & 1 \end{bmatrix} \tag{10.13}$$

The H matrix that relates the measurement vector to the preceding state vector is

$$H = \begin{bmatrix} 1 & 0 & 0 & 0 & 0 \\ 0 & 1 & 0 & 0 & 0 \\ 0 & 0 & 1 & 0 & 0 \\ 0 & 0 & 0 & 1 & 0 \\ 0 & 0 & 1 & 0 & 0 \\ 0 & 0 & 0 & 0 & 1 \end{bmatrix} \tag{10.14}$$

The covariance matrix P quantifies the state vector estimation errors. P is calculated by the Kalman filter algorithm, as will be shown momentarily.

The measurement noise covariance matrix R accounts for the expected measurement errors. This is also known as the sequence covariance matrix. Sometimes the sensor errors are set to the standard deviation to populate this matrix. For the measurements just identified, R becomes

$$R = \begin{bmatrix} \sigma_{GPS,xy}^2 & 0 & 0 & 0 & 0 & 0 \\ 0 & \sigma_{GPS,xy}^2 & 0 & 0 & 0 & 0 \\ 0 & 0 & \sigma_{GPS,z}^2 & 0 & 0 & 0 \\ 0 & 0 & 0 & \sigma_{ADS,V}^2 & 0 & 0 \\ 0 & 0 & 0 & 0 & \sigma_{Alt,z}^2 & 0 \\ 0 & 0 & 0 & 0 & 0 & \sigma_{Comp,\psi}^2 \end{bmatrix} \tag{10.15}$$

The covariance matrix of random system disturbances Q bounds the expected errors of the unmanned aircraft. For this case it is

$$Q = \begin{bmatrix} \sigma_{x,E}^2 & 0 & 0 & 0 & 0 \\ 0 & \sigma_{y,E}^2 & 0 & 0 & 0 \\ 0 & 0 & \sigma_{z,E}^2 & 0 & 0 \\ 0 & 0 & 0 & \sigma_V^2 & 0 \\ 0 & 0 & 0 & 0 & \sigma_\psi^2 \end{bmatrix} \tag{10.16}$$

Now that the terms are defined, we can explore the linear Kalman filter algorithm. When a new measurement is made, the system goes through an update procedure. First, the Kalman gain K_k is computed.
   Update Step 1:

$$K_k = P_k^- H_k^T \left[H_k P_k^- H_k^T + R_k \right]^{-1} (\text{verify -1}) \tag{10.17}$$

The dash superscript indicates that the preceding matrix is used from before the measurement update. Next, the estimated state $\hat{x}_k$ is updated. The earlier state estimate $\hat{x}_k^-$ is updated based on the Kalman gain applied to the difference between the new measurements and the related preceding estimate.
   Update Step 2:

$$\hat{x}_k = \hat{x}_k^- \cdot + K_k \left\lfloor z_k - H_k \hat{x}_k^- \right\rfloor \tag{10.18}$$

The last step performed in processing the sensor input is to update the covariance matrix P_k.
   Update Step 3:

$$P_k = [1 - K_k H_k] P_k^- \tag{10.19}$$

The preceding steps are only performed when a measurement is made. The state estimate can be updated at a rate higher than that of new measurements. The state is projected ahead for each time step dt. This process is an extrapolation from the last measurement or last estimate.

Extrapolation Step 1:

$$\hat{x}_{k+1}^- = \Phi_k \hat{x}_k \tag{10.20}$$

Finally, the covariance matrix is updated for each time step.
Extrapolation Step 2:

$$P_{k+1}^- = \Phi_k P_k \Phi_k^T + Q_k \tag{10.21}$$

An alternative form of the covariance matrix update is

$$P_{k+1}^- = \Phi_k P_k \Phi_k^T + \Gamma_k Q_k \Gamma_k^T \tag{10.22}$$

where Γ_k is the process noise distribution matrix that maps the error inter-relationships between parameters, which will not be covered further here.

Example 10.1 One-Dimensional Kalman Filter

Problem:
Consider a rocket sled that can travel only in the x direction and cannot rotate. The unmanned aircraft state includes x_E, u, and a_x. The body points east. The following sensors are used:

Sensor	Measurement	Error	Update Rate, Hz
Accelerometer	a_x	0.1 m/s²	20
GPS	x_E	12 m	1
Velocity encoder	u	0.001 m/s	10

Generate the matrices that are used for the Kalman filter.

Solution:
The UA state vector is

$$\hat{x} = \begin{bmatrix} x_E \\ u \\ a_x \end{bmatrix}$$

We know from basic kinematics that

$$V = V_0 + a_x \cdot dt \quad \text{and} \quad x = x_0 + V_0 \cdot dt + \frac{1}{2} \cdot a_x \cdot dt^2$$

Therefore, the state transition matrix is

$$\Phi = \begin{bmatrix} 1 & dt & 1/2 \cdot dt^2 \\ 0 & 1 & dt \\ 0 & 0 & 1 \end{bmatrix}$$

Here dt should be no greater than the minimum measurement frequency. The Q matrix for this state is

$$Q = \begin{bmatrix} \sigma_{x,E}^2 & 0 & 0 \\ 0 & \sigma_u^2 & 0 \\ 0 & 0 & \sigma_{a,x}^2 \end{bmatrix}$$

The measurements are

$$\bar{z} = \begin{bmatrix} x_{GPS} \\ u_{Enc} \\ a_{x,Accel} \end{bmatrix}$$

The measurement matrix is

$$H = \begin{bmatrix} 1 & 0 & 0 \\ 0 & 1 & 0 \\ 0 & 0 & 1 \end{bmatrix}$$

The sequence covariance matrix is

$$R = \begin{bmatrix} \sigma_{GPS,x}^2 & 0 & 0 \\ 0 & \sigma_{Enc}^2 & 0 \\ 0 & 0 & \sigma_{Accel}^2 \end{bmatrix}$$

In the next extrapolation time step the state vector and covariance gains are set to the previous values ($k = k + 1$).

An inertial measurement unit (IMU) estimates the unmanned aircraft position, velocity, accelerations, angles, angular rates, and angular accelerations based purely on inertial measurements. A strap-down IMU is fixed to the unmanned aircraft. We will consider the measurements and state estimates of a strap-down IMU. Accelerometers measure linear accelerations (a_x, a_y, a_z). These can be integrated with time to yield velocity (u, v, w). The velocities and accelerations are integrated over time to give position (x_E, y_E, z_E). Gyroscopes measure the angles (θ, ψ, φ). The angular rates (p, q, r) can be found from the derivatives of the Euler angles and the Euler rates. The second derivative of the Euler angles is used to find the angular accelerations ($\dot{p}$, $\dot{q}$, $\dot{r}$).

A problem with IMUs as a navigation sensor is that the estimates drift over time. The drift can be low for advanced IMUs that use high-grade accelerometers and gyroscopes. However, miniaturized systems that are used by small UASs tend to use microelectromechanical systems (MEMS) devices that have very high drift and can provide nonsense estimates in seconds or minutes.

GPS receivers provide information about the latitude, longitude, and height above the Earth's surface. GPS triangulates synchronized encoded RF signals from multiple orbiting GPS navigation satellites. This information can be used to estimate the position in the Earth coordinate system (λ, μ, z_E). Although GPS is fundamentally a position sensor, it can also give velocity by taking the first derivative of position with respect to time.

GPS can be given much credit for the widespread use of UASs today. This amazing system bounds the position estimate within several meters instead of several kilometers for a long-endurance mission. A GPS provides position information at a relatively low update rate. GPS has a coarse/acquisition (C/A) code for civilian uses and an encrypted P code for U.S. military use. C/A code horizontal accuracy is approximately 12-m circular error probable (CEP), and P-code CEP is approximately 4.6 m [2]. GPS accuracy varies with location and a number of other conditions. The literature on the subject provides many different accuracy estimates.

An inertial navigation system (INS) combines an IMU with a GPS. The IMU provides high-frequency state estimates while the GPS provides absolute position information and bounds the drift. With these two devices an INS can estimate the unmanned aircraft's position, velocity, accelerations, angles, angular rates, and angular accelerations with bounded drift over time. The combined state vector of the INS is estimated by a Kalman filter or other algorithm.

A magnetic compass measures the angle from magnetic north, which is not exactly at the North Pole. A simple magnetic compass will measure the Euler angle ψ. Three-axis magnetic compasses measure all three Euler angles. A magnetic compass can be combined with an IMU or INS to bound drift in the Euler angles. By comparing the magnetic compass heading and airspeed to the INS speed and angles, it is possible to estimate wind speed and direction.

Early UAs and modern simple UAs may attempt to have a sensor that ensures that the unmanned aircraft is level. An N_z sensor detects the vertical acceleration. When the unmanned aircraft is level in nonaccelerated flight, the N_z sensor will measure one times the acceleration of gravity. This is necessary but not sufficient to ensure that the unmanned aircraft is level. Another approach is to use multiple electrostatic sensors that measure the electrostatic potential difference in the atmosphere. A difference between two distant sensors, say one on each wing tip, indicates that the wings are not level. This type of system was available in the 1980s on units such as the BTA autopilots. Finally, visual sensors can estimate the horizon based on contrast between the air and ground. Visual sensors can experience challenges when the unmanned aircraft is at low altitude or if there are mountains present.

Above-ground-level (AGL) sensors provide measurements of the height of the unmanned aircraft above the local terrain. The uses for AGL sensors can be ground collision avoidance, navigation by feature mapping, and support of the final stages of landing. Laser and radar altimeters are commonly used. Radar altimeters have the advantage of being able to see through clouds. Acoustic measurements can be used at close range for launch, recovery, and indoor flight. Any deviations from a vertical altimeter orientation, as is experienced during maneuvering, will create errors relative to a vertical measurement. GPS sensors provide an absolute altitude relative to a theoretical Earth, but combining GPS with terrain maps can give AGL estimates.

Visual navigation sensors are common for unmanned ground vehicles but rare for UASs. Optical flow generates a vector field of features captured by the camera. The field represents the rate of change, which is an indication of how close the object is to the camera. This can be useful for maneuvering in the vicinity of obstacles, as would be common in an urban canyon, geologic canyon, inside a building, or in a forest. Camera pixel tracking algorithms can be used to lock onto features within an image to serve as a reference point for navigation.

10.2.2.2 Landing Aids

Whether the unmanned aircraft lands on a traditional runway, captures in a net, or other recovery system is employed, landing aids are often necessary. The landing point of the unmanned aircraft must be intercepted within an acceptable tolerance. Often this accuracy is greater than what can be achieved by GPS alone. This is similar to the challenges experienced by manned aircraft that must land in zero visibility instrument flight rules (IFR) conditions.

RF-based methods generally use a narrow beam to guide the unmanned aircraft to the touchdown point. The signal may be encoded with ground-based correction information. Examples include the Sierra Nevada Tactical Automatic Landing System (TALS) and UAV Common Automatic Recovery System (UCARS).

Relative GPS positioning methods determine the relative position between two GPS receivers. Communications must occur between the two receivers by means of a data link. This is commonly known as differential GPS (DGPS). A common implementation is the real-time kinematic (RTK) method. These relative GPS approaches do not provide improvement in absolute position accuracy; rather, they improve the accuracy of the position between the two receivers. The landing platform-based receiver position offset from the touchdown point is known by the UA flight control system.

AGL sensors provide estimates of the height of the unmanned aircraft above the local surface. This differs from pressure altimeters, which provide the altitude estimates relative to mean sea level. Common AGL sensors include radar, laser, and acoustic altimeters, as described earlier.

UASs that use human pilots (AVO) in the loop for landing require landing cameras if the pilot is inside the GCS. The video from this camera must be provided to the GCS with low latency to enable pilot control response. Landing cameras can be useful even for UASs that have autonomous landing capabilities to ensure that the runway is not blocked by obstructions. For example, a camera can help the AVO identify an aircraft runway incursion or a herd of deer blocking the runway.

Machine vision systems can interpret the scene ahead of the unmanned aircraft to assist landing. For example, a landing camera might be able to distinguish major features of a runway. This information can provide guidance information to the flight control system to support the landing.

10.2.2.3 Air Data Systems

Pitot-static sensors measure the dynamic pressure by taking the difference between the total pressure and static pressure. A pitot-static probe is a cylinder that is approximately aligned with the airflow. One hole is at the forward tip, and this captures the stagnation pressure. Another hole is located on the side of the cylinder, which measures the static pressure. The pressure difference between these two holes, called ports, is equal to the dynamic pressure.

For low subsonic speeds where compressibility is negligible, the dynamic pressure q is

$$q = P_T - P_S \tag{10.23}$$

where P_T is the total pressure and P_S is the static pressure. The equivalent airspeed V_{Eq} is

$$V_{\mathrm{Eq}} = \sqrt{\frac{2 \cdot q}{\rho_{\mathrm{SL}}}} \tag{10.24}$$

The measured equivalent airspeed is known as the indicated airspeed. The true airspeed V_T is found by correcting for the density at altitude

$$V_T = \frac{V_{\mathrm{Eq}}}{\sqrt{\sigma}} \tag{10.25}$$

where σ is the ratio of density at altitude vs sea-level standard-day density. The density ratio is a function of the pressure altitude, which is provided as an output from the air data computer.

Pitot-static systems can have multiple ports. One common type is the five-hole probe. This sensor has a single stagnation pressure port in the front and four static ports on the side. In addition to airspeed, this type can provide angle of attack α and angle of sideslip β estimates by accounting for pressure differences on the four static ports. Although the angular information is helpful for the flight controls, the additional complexity can lead to lower reliability and higher maintenance burden.

Pitot-static sensors are usually flight critical, as they provide information about airspeed that cannot be obtained from an INS alone. These devices can be adversely affected by icing, water intrusion, insects, dust, and debris. Heated pitot-static systems can prevent ice accumulation, but consume substantial power and can pose a ground safety hazard (don't ever touch one). Pitot-static probes should be covered while on the ground. These covers are usually protective red bags with flags that state "remove before flight."

Vanes are devices that align themselves with the local airflow. A potentiometer measures the position of the vane. Flight-test booms often have a pitot-static system and two vanes. The alpha vane measures angle of attack, and the beta vane measures the angle of sideslip. Vanes are rarely used outside of flight test.

A barometric altimeter measures the pressure altitude. The established rate of change of pressure with respect to height is used to estimate height based on measured pressure. The pressure at the Earth's surface varies over time due to weather, and so it is important to calibrate the device prior to takeoff.

10.2.2.4 Autopilots and Flight Control Algorithms

UAS autopilots and flight control systems take many forms and have recently made large capability strides. The simplest or initial prototypes of small UAs can be little more than radio-controlled (R/C) aircraft. Modern high-performance UAs can have advanced autopilots capable of high-level autonomous flight.

Radio Control (R/C)

Traditional R/C aircraft use a direct command from a ground control box (transmitter) as a signal to directly drive the actuator. In some systems a rate gyro might lie between the unmanned aircraft's receiver and the servo to help stabilize the unmanned aircraft. Sophisticated processing can occur on the transmitter to create control surface mixing functions such as those defined in Table 10.2.

The R/C pilot essentially closes the inner flight control loops based on visual and audio feedback from direct observation of the unmanned aircraft. This is not possible for many unmanned aircraft unless external pilot handling qualities are taken into account in the design. These aircraft should be statically stable and have predictable lateral-directional behavior.

The advantage of R/C aircraft is that it saves significant development cost over autopilot integration. Highly skilled R/C pilots are plentiful,

Table 10.2 R/C Mixing Functions

Mixing name	Description
V-tail	Mixes rudder and elevator control functions into the V-tail arrangement
Elevon	Elevator and aileron function are mixed, typically on a flying-wing aircraft
Differential ailerons	Up-moving aileron deflection is greater than the down-moving aileron to reduce yaw moment due to differential drag on the wing (adverse yaw)
Elevator-flap	Elevator compensates for flap pitching moment
Flaperons	Flap and aileron functions are mixed
Crow	Outboard wing ailerons deflect up and the inboard flaps deflect down to increase drag for landing while preventing tip stalls

though their skills are generally applicable to aircraft of similar scale and speeds as model aircraft. This path can help bring an aircraft to flight test very rapidly—perhaps on the order of weeks instead of months or years. Typically this approach is applied to small UAs ($<$ 300 lb). The crash rate can be high relative to autopilot flights that are approached carefully.

RPV

Although remotely piloted vehicles (RPVs) described almost all UASs before the term unmanned aircraft vehicles (UAVs) in the 1990s, it is also identified with a level of autonomy. RPV mode is used here to characterize a type of control where the inner loops are closed by an autopilot while a ground-based pilot manages the outer loops.

Before GPS became practical in the early 1990s, it was very difficult to create a low-drift navigation solution for small UASs. Autopilots, with their limited inertial navigation capabilities, were able to close inner loops but had more difficulty with the outer loops due to drift. Ground pilots closed the outer loops by visually guiding the aircraft or by monitoring the aircraft via an onboard camera feed displayed on a ground monitor. Some modern UASs still use this method for launch and recovery.

Typical RPV inner-loop control functions include the following:

- Altitude hold (requires an altimeter)
- Heading hold (requires a magnetic compass augmented by inertial sensors)
- Wings level (requires N_z sensor, visual horizon sensor, or electrostatic potentiometer)
- Airspeed hold (requires pitot-static sensor, but may be augmented by inertial sensors)

Some modern autopilot systems use an RPV mode feature to help developers refine the inner loops prior to implementing the outer loops. The RPV mode alone is not considered to be acceptable for modern operational UAS systems.

UAS

The UAS mode closes the outer loops of the control system and permits waypoint-based navigation. This requires an acceptable unmanned aircraft position solution. The most common method of achieving the position solution is with an INS that includes a GPS. Before GPS, high-grade INS using only accelerometers and gyros provided the solution, though drift always limited the ultimate effectiveness. Stellar trackers have also been implemented, which track the unmanned aircraft's angular position relative to reference star orientation. A ground-based RF triangulation beacon network can be used, which performs in a manner similar to LORAN. The outer loop functionality includes three-dimensional waypoint navigation along a route and orbit patterns (circle, race track, figure 8, etc.).

Autonomous

The fully autonomous mode eliminates the need for operator intervention. This is a necessity for modern UASs for planned and unplanned loss of communications. Generally an autonomous flight will proceed along a preplanned route. Unmanned aircrafts must have contingency management algorithms to contend with problems with limited or no operator interaction.

Swarming

Swarming is the coordinated flight of multiple unmanned aircrafts trying to satisfy a common mission. The motivation for swarming is that multiple coordinated unmanned aircrafts might be more effective than those that are individually controlled. One example is a high – low combination where a high-altitude UAS searches a large area to find potential targets and then low-altitude UASs investigate at reduced slant range and higher resolution. Another example is a hunter – killer mission, where small, long-duration ISR UASs identify and designate targets for a faster, heavier strike UAS. A swarm of environmental UASs with methane detectors could search a large area of tundra to identify areas of the greatest permafrost melting, and then cluster the group's in situ sensors or samplers on that area. Swarming UASs can reduce the geolocation error by providing different collection geometries, which is especially useful if the error estimate ground projections are elliptical. The reasons that swarming is not employed more today are lack of UASs to cover the single unmanned aircraft mission demand and insufficient autonomy of the platforms.

Swarming unmanned aircrafts must operate routinely with high levels of autonomy. The unmanned aircrafts exchange information with one another and should be tolerant to communication outages. Airborne wireless networks (i.e., 802.11, 802.16, SECNET) allow distribution of information across the platforms. An autonomy driver is that the GCS operators cannot reasonably manage the operation of every unmanned aircraft in the swarm. Also, the swarm must be able to efficiently act upon information gathered, and human intervention might be too slow or less effective than autonomous solutions.

Control System Design

Even a basic description of control system design methodologies is beyond the scope of the book. However, some of the major design considerations are described. Different classes of UASs have different control system design drivers, as shown in the following examples:

- ISR systems require a stable camera platform that does not demand high-bandwidth turret compensation due to unmanned aircraft motion.
- Micro unmanned aircrafts need a robust gust response and the ability to navigate winds, and the UA-fixed camera should be as stable as possible. The avionics must be low cost.
- UASs that directly fly into targets such as recovery devices require very precise three-dimensional positioning along a flight path.
- Cargo UASs should have docile control laws and high reliability. The level of path tracking accuracy depends on flight modes such as normal flight and landing.
- A UCAV's low observable configurations may be unstable about one or more axes. This requires an advanced, high rate flight control system. Some designs may depart rapidly if the flight control system momentarily stops functioning.

Relaxed static margin can reduce trim drag. This is particularly pronounced for swept flying wings where the wings must provide trim as well as lift functions. However, the reduced static margin places greater demands on the flight control system. The impacts on control system weight and power consumption must be traded against the drag reduction.

The state vector used for navigation contains many more parameters than are required for flight controls. The state vector described earlier contained 18 elements. Now we will see how this can be reduced to two separate state vectors containing a total of 8 elements. These are used for the inner-loop control laws. The longitudinal and lateral-directional state vectors are separated because longitudinal and lateral-directional controls can usually be handled individually without significant coupling. The nondimensional longitudinal and lateral-directional state vectors are

$$\hat{x}_{\text{Long}} = \begin{bmatrix} \hat{V} \\ \alpha \\ \hat{q} \\ \theta \end{bmatrix} \quad \text{and} \quad \hat{x}_{\text{LD}} = \begin{bmatrix} \beta \\ \hat{p} \\ \hat{r} \\ \phi \end{bmatrix} \tag{10.26}$$

The new terms are defined as

$$\hat{p} = \frac{p \cdot b}{2 \cdot V}, \quad \hat{q} = \frac{q \cdot \bar{c}}{2 \cdot V}, \quad \hat{r} = \frac{r \cdot b}{2 \cdot V}, \quad \hat{V} = \frac{V}{V_{\text{Ref}}} \tag{10.27}$$

The control vectors are

$$\boldsymbol{u}_{\text{Long}} = \begin{bmatrix} \delta_T \\ \delta_M \end{bmatrix} \quad \text{and} \quad \boldsymbol{u}_{\text{LD}} = \begin{bmatrix} \delta_L \\ \delta_N \end{bmatrix} \tag{10.28}$$

where

δ_L = roll control (such as aileron deflection)
δ_M = pitch control (such as elevator deflection)
δ_N = yaw control (such as rudder deflection or differential spoilers)
δ_T = thrust control (such as throttle setting)

To estimate the change of state, we must analyze the linear equations

$$\Delta \dot{\hat{x}}_{\text{Long}} = \hat{A}_{\text{Long}} \Delta \hat{x}_{\text{Long}} + \hat{B}_{\text{Long}} \Delta \boldsymbol{u}_{\text{Long}} \tag{10.29}$$

$$\Delta \dot{\hat{x}}_{\text{LD}} = \hat{A}_{\text{LD}} \Delta \hat{x}_{\text{LD}} + \hat{B}_{\text{LD}} \Delta \boldsymbol{u}_{\text{LD}} \tag{10.30}$$

Here the A matrices relate how the state parameter rates are affected by the state vector. The longitudinal and lateral-directional matrices each have four rows and four columns. The longitudinal A matrix contains 20 parameters that relate to aerodynamic, geometric, propulsion, and inertial influences. The lateral-directional A matrix has 21 parameters. Parameters in the longitudinal matrices are as follows: $\{C_{D,M}, C_{D,\text{Ref}}, C_{D,\alpha}, C_{L,M}, C_{L,q}, C_{L,\text{Ref}}, C_{L,\dot{\alpha}}, C_{L,\alpha}, C_{M,\dot{\alpha}}, C_{T,M}, C_{T,V}, C_{T,\delta M}, C_{T,\delta T}, C_W, \hat{I}_{yy}, \hat{m}, M_{\text{Ref}}, \gamma_{\text{Ref}}, \varepsilon_{\text{Ref}}, \varepsilon_T\}$. The additional parameters in the lateral-directional matrices include the following: $\{A, C_{l,p}, C_{l,r}, C_{l,\beta}, C_{l,p}, C_{l,r}, C_{l,\beta}, C_{l,\delta L}, C_{l,\delta N}, C_{n,r}, C_{n,\beta}, C_{n,\delta L}, C_{n,\delta N}, C_{Y,p}, C_{Y,r}, C_{Y,\beta}, C_{Y,\delta N}, \hat{I}_D, \hat{I}_{xx}, \hat{I}_{xy}, \hat{I}_{zz}\}$

$$\hat{m} = \frac{2 \cdot m}{\rho \cdot S \cdot \bar{c}} \tag{10.31}$$

$$I_D = Ixx \cdot Izz - I_{xz}^2, \quad \hat{I}_D = \hat{I}xx \cdot \hat{I}zz - \hat{I}_{xz}^2 \tag{10.32}$$

$$\hat{I}_{xx} = \frac{8 \cdot I_{xx}}{\rho \cdot S \cdot b^3}, \quad \hat{I}_{yy} = \frac{8 \cdot I_{yy}}{\rho \cdot S \cdot \bar{c}^3}, \quad \hat{I}_{zz} = \frac{8 \cdot I_{zz}}{\rho \cdot S \cdot b^3}, \quad \hat{I}_{xz} = \frac{8 \cdot I_{xz}}{\rho \cdot S \cdot b^3} \tag{10.33}$$

$$C_W = \frac{W}{\bar{q} \cdot S} \tag{10.34}$$

The B matrices relate how the state parameter rates are affected by control inputs. The longitudinal and lateral-directional B matrices each have four rows and two columns if there are two control inputs each. In the case of a glider, the thrust control input does not exist for longitudinal control. A simple control system with only ailerons but no rudder would have no yaw control device. In each of these cases, the B matrices would have four rows and one column. Often there are multiple control effectors that offer redundancy, which requires expanded B and Δu matrices.

Control laws act in the Laplace domain rather than the time domain. The controller converts the time-domain state inputs to the Laplace domain. Based on the difference between the actual state relative to the desired state, the control laws generate control commands. These commands are acted upon by the control effectors, which generates a physical response.

10.2.2.5 UA Management Systems

An unmanned aircraft management system (VMS) performs flight critical and unmanned aircraft operation functions. This includes interfacing to command and control communication systems, subsystems, and the autopilot. Frequently high-level flight route management occurs on the VMS ahead of the autopilot. The VMS is hosted on one or more computers. The VMS and autopilot computer are sometimes combined on small systems.

10.2.2.6 Mission Management Systems

A mission management system (MMS) manages mission critical functions. These can include commanding payloads, selecting data feeds for downlink, data storage, data retrieval, data fusion, and payload downlink management. An architecture that segregates the MMS from the VMS and other flight critical systems is generally desirable for improved reliability. The MMS is hosted on one or more computers. Sometimes the VMS and MMS share a computer, though the hardware is generally separated for large UASs.

10.2.2.7 Engine Controllers

Unlike manned aircraft, UAs do not have onboard pilots that can adjust numerous engine controls by hand. Simple radio-controlled aircraft reciprocating engines or electric motor controllers only need to command throttle, so that there is little need for a more complex engine controls. A more sophisticated reciprocating engine system can have several elements adjusted such as throttle, spark advance, mixture, cowl flap settings, and variable pitch prop angle. These parameters use commanded power inputs, as well as information about rpm, crankshaft position, cylinder head temperature, altitude, and airspeed. Turbines have a similarly long list of controls and sensor data. A full authority digital engine control (FADEC) is a unit that provides for the engine monitoring, management, and control.

10.2.2.8 Computers and Processors

Processors support myriad functions. Algorithms must be executed on digital computation devices. Processors are an integral part of autopilots, INS, VMS, MMS, device controllers, and even sensors. The primary processor types are as follows:

* Microprocessors
* FPGA – field programmable gate array
* ASIC – application-specific integrated circuit
* DSP – digital signal processors

Microprocessors available to aerospace generally lag the performance available for ground-based computing. Commonly used microprocessors include 555 and PowerPC processors. Many avionics boards take a PC-104 form factor. Care must be taken to ensure that the processors are not excessively loaded.

10.2.2.9 Remote Interface Units

The computers and the devices that they control can be separated by a large distance. The number of wires required to send signals to each remote device can lead to large wiring harnesses. Additionally, an avionics box might have limited connector interfaces. These problems are solved through the use of a remote interface unit (RIU). A single connector on a box can extend a cable to a RIU located far away, which then interfaces with multiple local devices. RIUs are sometimes called remote input output (RIO) devices.

10.2.2.10 Data Bus

The data bus is the medium by which data are distributed across the avionics architecture. The data bus routes information over an electrical circuit from the data producers and consumers. The data are prioritized such that the most critical data have a high probability of getting through. Common bus types include Ethernet, MIL-STD 1760, MIL-STD 1553, and CAN.

U.S. military manned aircraft typically use a 1553 data bus for avionics and payload systems. UASs that leverage this legacy hardware might need to use this bus type. For example, a 1553 bus interfaces with weapons and electronic warfare systems.

10.2.2.11 Wiring Harnesses

Wiring harnesses are electrical wires that distribute power and electrical signals. A collection of individual wires are combined into bundles and routed through the airframe. The wires are terminated with connectors, which interface with connectors from components or other wiring harness segments. The individual wires and the bundles are usually electromagnetically shielded to prevent EMI.

10.2.2.12 Airspace Integration Systems

UASs must share airspace with manned aircraft and be tracked in peacetime and in war. During a conflict, military UASs might need to share or compete for airspace with manned aircraft, helicopters, missiles, and artillery. Several systems help support safe integration to avoid midair collisions.

The commercial UAS sector has not achieved its full potential due to unresolved airspace integration challenges. The level of coordination and oversight is overly burdensome to permit routine operations. No satisfactory approach for sense and avoidance has been widely implemented. Although the status quo is quite frustrating for entrepreneurs in the UAS community, it would only take one major incident to set the industry back years.

Exceptions include flights over international waters. UASs could be able to fly routinely over unpopulated areas once routes are established. Some countries outside the United States and Western Europe have more relaxed standards that permit greater flexibility. There is generally little distinction between small and large UASs in the developed world. Ironically, nonprofessional model aircraft enthusiasts can fly unmanned aircraft with much greater airspace access than professional UAS developers with more advanced aircraft in the same class.

Transponders transmit information about the unmanned aircraft so that other air traffic is made aware of its presence. Most transponders also receive information about other aircraft. There are a number of available transponder types. The functions include ident, pressure altitude, and position. Ident is an identifying code that can be squawked upon request from air traffic control (ATC). The code must be input at the GCS and transmitted by the transponder. Some transponders are required to report their pressure altitude, which might be different than that the true altitude estimated by the INS. ADS-B also provides the GPS position and velocity. Some common transponder types include the following:

- *Mode 3/A*—This includes ident.
- *Mode C*—This mode includes pressure altitude. Mode C and 3/A are often used together.

- *Mode S*—This transponder supports anticollision systems (ADS-B and TCAS). It protects against excessive interrogation and works with modes 3/A and C. It provides pressure altitude in 25-ft increments. It also sends a call sign and/or a permanent address code.
- *Mode 5*—This military code encrypts Mode S and ADS-B GPS position information.
- *TCAS (traffic collision avoidance system)*—This system monitors transponders of other aircraft to warn of collision hazards. It is independent of ground-based systems. The TCAS interrogates other aircraft equipped with TCAS systems and receives a response. TCAS works with Mode S transponders. The TCAS system identifies mutual avoidance maneuvers (resolution advisory) when there is a collision risk, but the pilot needs to implement the maneuver. In the case of a UAS, autonomous airborne response or a GCS-commanded response could take the necessary action.
- *ADS-B (automatic dependent surveillance-broadcast)*—This transponder provides information on ident, location (GPS), and velocity.

Lighting

Unmanned aircraft must share airspace with manned aircraft, and therefore provisions are frequently made to avoid midair collisions by illuminating the unmanned aircraft. Lighting systems used for manned aircraft can be applied to unmanned aircraft. However, the manned aircraft hardware might be too heavy or require too much power for small UASs. The two primary types of lighting are navigation and anticollision lighting. Some military UASs are required to carry IR lights that are visible by night-vision equipment but not the naked eye.

Air Traffic Detection Sensors

One criterion for unmanned aircraft integration is having an equivalent level of safety as a manned aircraft. This can be interpreted as the unmanned aircraft should be able to detect and react to other aircraft at least as good as pilots in manned aircraft. This is known as sense and avoid. Many approaches have been attempted to detect other aircraft, including optical systems and airborne radars. Optical systems can be difficult to implement because to gain sufficient resolution over the required field of view generates large quantities of data that must be processed.

Airborne radar can provide situational awareness of other aircraft and weather. The detection of other aircraft provides information about air traffic when operating in civil airspace. During wartime, airborne radar can also detect enemy aircraft. Weather radars detect precipitation. The information provided by these airborne radars is most likely sent to the ground. Airborne radars are not in common use for UASs. However, radar payloads such as synthetic aperture radars or maritime search radars are widely employed. These payload radars could have extended functionality to include air traffic detection, which would make these radars multimode.

Voice Communications

Voice relay radios rebroadcast the voice of the AVO in the GCS to air traffic control and other aircraft. The voice data usually arrive at the unmanned aircraft via the command and control link. The voice data from the GCS is parsed and sent to the voice radio where it is broadcast. Similarly, voice traffic received by this airborne radio is encoded and transmitted to the GCS. The GCS should be able to change frequency channels of the radio. Civilian voice radios operate from 30–100 MHz (vhf), and military radios operate from 250–400 MHz (uhf).

10.2.2.13 Flight-Test Equipment

Flight-test equipment includes systems that support flight-test activities. These components are not part of the baseline avionics. Flight-test equipment can include test instrumentation, flight data recorders, telemetry systems, flight termination systems, and flight-test equipment wiring harnesses. Example flight-test instrumentation includes strain gauges, thermal sensors, high-grade inertial systems, and an air data boom. Flight data recorders and telemetry systems manage the data gathered from the flight-test instrumentation. Flight-test equipment will often have a dedicated wiring harness that has limited interaction with baseline systems. Traditionally flight-test wiring is orange.

Flight termination systems (FTS) ensure safety for those on ground by preventing the unmanned aircraft from flying away. The FTS is commanded via a flight termination link. Often the lack of a flight termination link connectivity will trigger a termination command. Methods of flight termination include unmanned aircraft destruction, large control surface deflection accompanied by engine shutoff, or a recovery parachute.

10.2.3 Avionics Architectures

Flight-critical avionics architectures have 1–3 strings. Simplex is a single string, duplex has two strings, and triplex has three strings. Components that constitute a string can include the INS, air data system, autopilot, actuators set, and UA management system. If there is a failure in a flight critical component of a simplex system, then the unmanned aircraft will be lost. The duplex and triplex systems seek to improve the reliability through redundancy.

Triplex systems can gracefully handle the loss of one string because there are two backups. The three systems vote, and if one system disagrees with the other two, it is considered to be an invalid input. One of the strings is in command, and the command can be switched to another string if there is an issue. The voting scheme is no longer possible once the system is down to two strings. Another method of handling errors is to independently monitor the health of each string.

Duplex systems are less common than simplex or triplex systems because it is not possible to implement a simple voting algorithm. One simple approach to a two-string system is to enable each string to control the aircraft if the other string fails. This can be done by having two sets of control surfaces for each function (i.e., two right ailerons, two left ailerons, two rudders, two elevators), where half of the surfaces are driven by each string. If one string fails, then the other can take over seamlessly and perhaps without knowledge of the failure. It is often acceptable for the flight controls to be degraded but functional if one string fails.

Modern avionics architectures for all but the simplest UASs enforce separation of the VMS and MMS. The VMS and its flight critical systems may be flight qualified without any dependencies upon the MMS. If the MMS changes, as is the case for a new payload capability, then the VMS does not need to be requalified. This clean partitioning is sometimes blurred when the VMS and MMS share a data link, as is often the case with the common data link (CDL).

10.3 Flight Software

Many UAS development programs reach the state where a nearly completed unmanned aircraft awaits software before flight test can begin. A roll-out ceremony is held where dignitaries view an impressive unmanned aircraft obscured by machine-generated fog against the backdrop of a giant national flag. Management coaches the engineers in attendance to not mutter aloud the software schedule woes lest they ruin the carefully orchestrated event. Predictably, the customer and development team management is surprised by the vexing software issues.

The software development problems arise for many reasons. Two of the most common top-level culprits are 1) the software development scope is underestimated and 2) the software is not well understood. UASs are software-intensive systems, and so a solid understanding of this domain is essential for program success.

Much of the software that was developed by the UAS prime contractor in the past is now developed by suppliers. UAS software is becoming a commodity for flight controls, data management, and GCS software. Suppliers can support multiple programs. Using third-party software can save nonrecurring engineering cost, but the UAS prime contractor gives up some control. This supplier-provided software is sometimes called firmware.

The UAS prime contractor can focus its attention on the VMS and MMS. Large defense contractors can choose to specialize in mission system software, which runs on the MMS and the ground control station. The VMS efforts involve subsystem management and integration of various avionics components provided by multiple suppliers. The contingency management algorithms usually reside on the VMS as well.

The software that estimates the unmanned aircraft state based on sensor data generally runs on a procured device. For example, the INS contains the inertial sensors and GPS, and all of the Kalman filtering is performed on the INS. The INS provides state data to the data bus, which is routed to the data consumers. Many autopilots have built in INS as well as air data sensors and include the necessary filters to estimate the state.

Flight control software can run on an autopilot that includes the sensors or on a separate processor. The UAS prime contractor can implement the control laws themselves, though this work is increasingly performed by autopilot suppliers.

Every device must be commanded, receive data, or provide data. This data exchange occurs over a real-time operating system (RTOS). RTOS examples include the following:

* Wind River VxWorks
* Green Hills Integrity
* Green Hills Apache
* Mathworks CSLEOS
* Sysgo PikeOS
* LynxOS – SE
* LynuxWorks

Flight software should be reliable and predictable. Software that responds predictably to a set of inputs is known as *deterministic*. Aerospace software standards and certification methods help ensure solid design practices and software testing methods. The DO-178B "Software Considerations in Airborne Systems and Equipment Certification" flight software standard is one such approach. Many government contracts require that the software developers or prime contractor achieve CMMI certification.

New software development tools improve development efficiency while reducing errors. Model-based design techniques allow development in the requirements and architectural level rather than working with lines of code from the start. Autocoding generates software code based on higher-level design.

10.4 Subsystems

Subsystems are unmanned aircraft systems that are required to provide electrical power, environmental controls, and flight controls.

10.4.1 Electrical Power System

UAS payloads, communications systems, and avionics can generate large electrical power loads. The power can be supplied by engine-driven generators, batteries, fuel cells, solar cells, or other methods.

Generators act much like motors in reverse, where mechanical shaft power is converted to electrical power. The source of the mechanical input power is generally a fuel-driven propulsion system such as a jet or reciprocating engine. The power generation capacity increases with input source rpm, so that the generator performance is not constant. A generator capable of producing a prescribed power at loiter condition is larger than one sized for a peak power at higher rpm. A common mistake is to pick a generator based on the peak power described in marketing materials only to later discover that the power is substantially less at the operating conditions. Generators for UAS applications can have substantial weight, and so these systems must be defined early in the conceptual design. Not accounting for generator efficiencies properly is also problematic.

Generators can produce ac or dc current. An alternator is a generator that produces alternating current. Most aircraft systems operate from direct current, and so an alternator requires a rectifier to convert from ac to dc current.

Batteries can be used to supply all of the electrical power to the unmanned aircraft or act as a backup to the generator. Batteries have already been covered in Chapter 8 as they relate to electric propulsion systems. Despite the better energy density of NiCD, NiMH, and lithium batteries, lead acid is still frequently used for power systems. Lead acid is tolerant to large voltage spikes from a generator or alternator and can act as a filter when it floats on the power bus. Lead acid is a *deep cycle* battery.

Batteries should be charged prior to flight. For charging a battery onboard the unmanned aircraft, the chargers can be housed on the unmanned aircraft or as part of the ground support equipment. Alternatively, fully charged batteries can be installed prior to flight.

The battery can be sized based on numerous criteria. Batteries used in lieu of a generator must provide all electrical power for the duration of the most stressing flights. Backup batteries are sized to power a subset of systems in the event of a power generation failure. The backup cases can be 1) return to base from maximum radius of action, 2) controlled flight into ground, or 3) sufficient time to initiate a flight termination command.

It is desirable to shed nonessential electrical loads in an emergency. In the case of a generator failure with a battery backup power source, the flight critical systems must remain operational to ensure controlled flight. The payload could be powered down to conserve battery life.

Unmanned aircraft with electric propulsion systems often use the propulsion battery to provide power to the avionics and other systems. Small motor controllers may be equipped with battery eliminator circuitry (BEC) to supply output power at the power bus voltage.

The power system wiring distributes power from the source to the loads. This wiring should be shielded to prevent EMI impacts on other electronics. The wire gauge is driven by the peak electrical current. Large currents in the power system wiring can generate magnetic fields that interfere with magnetic compasses used for navigation.

The electrical system should support ground power for ground operations when the engine is not on. The battery may supply this power. Usually ground power is supplied so that the battery is not depleted prior to flight. The ground power cables are usually attached at a ground interface panel.

Auxiliary power units (APUs) are turbine-driven generators or fuel cells that produce electricity. These units primarily operate on the ground so that ground power is not required when the engine-driven generators are not in use. APUs are not known to be in use by UAs, which is probably due to the extra cost, weight, and system complexity. Also, there are no passengers that must be kept cool while the unmanned aircraft is on the ground using APU power.

Separation of flight-critical and non-flight-critical power busses can improve the system reliability. With proper design, a power spike from a failed payload operating on the non-flight-critical power bus cannot cause a failure of a flight-critical system. A key design case for the power system is handling the situation where the payload suite demands more power than is available to it.

The electrical power system might need to perform voltage conversions. Most systems operate on dc current, and so any ac power sources undergo an ac/dc conversion. Changing dc from one voltage to another requires a dc/dc voltage conversion. Sometimes not every electrical power consumer operates within the same voltage range. Manned aviation heritage avionics and systems typically operate on 28V dc. Model aircraft systems typically use 4.8–6 V dc. Many other electronics require 12 V dc.

A well-designed power system should have a common ground. Metal objects such as the engine or a large segment of metallic structure are suitable for grounding. If multiple grounds are used, then destructive ground loops can occur. This is not to be confused with the ground-loop

Example 10.2   Electrical Power System Sizing

Problem:

Consider a tactical UAS that flies for 12 hrs. The loads are as follows:

System	Average Power, W	Peak Power, W
Avionics	100	200
Communications system	200	300
EO/IR ball + SAR	500	700

Conduct a trade study between a battery and an engine-driven generator.

a. The generator has a weight relationship of 1 kg/kW for the generator and 0.5 kg/kW for the generator controller. The combined efficiency of the generator and generator controller is 70%. The engine brake specific fuel consumption is 0.5 kg/kW-hr.
b. The battery has a specific energy E_{spec} of 150 W-hr/kg. The battery voltage is the same as the bus voltage, and no special circuitry is required.

Solution:

The average load for all systems is 800 W, and the peak load is 1200 W.

First, let's consider the weight of the generator system. The generator and generator controller are sized for the peak loading. With a peak load of 1.2 kW, the combined weight of the generator and generator controller is 1.8 kg. The fuel consumed from generating the average power over 12 hrs is approximated by

$$W_{Fuel} = BSFC \cdot E \cdot \frac{P_{Gen}}{\eta_{Gen}} = (0.5 \text{ kg/kW} - \text{hr}) \cdot (12 \text{ hr}) \cdot \frac{0.8 \text{ kW}}{0.7} = 6.86 \text{ kg}$$

The total weight of the generator, generator controller, and fuel burned for power generation is 8.66 kg.

The battery weight for the same average power load is

$$W_{Batt} = \frac{E \cdot P_{avg}}{E_{spec}} = \frac{(12 \text{ hr}) \cdot (800 \text{ W})}{(150 \text{ W} - \text{hr/kg})} = 64 \text{ kg}$$

The generator system has a much lower weight impact than the battery power approach.

phenomena on aircraft with taildragger landing gear where the aircraft spins in yaw on the ground due to instability. Electrical ground loops occur when a voltage potential exists between two separate grounds that induces an unwanted current.

10.4.2 Environmental Control System

An environmental control system (ECS) keeps unmanned aircraft components within the required temperature limits. Humidity control is another function that some ECS solutions can provide. Cooling avionics and systems is usually a greater challenge than heating components.

Many small UASs do not have identifiable environmental control systems. Electronic components may draw little power, thus reducing the cooling requirements. Motors, power electronics, and avionics can simply be housed within an enclosed volume without cooling provisions. Unless properly designed, this approach can cause electrical components to exceed the upper temperature limits. Other small UAS might have outside air flowing directly into internal electronics bays, though this can also introduce water, sand and dust. Small UASs often have lower reliability than the larger systems, and unsuitable ECS design can be a culprit in many cases.

There are numerous ways to provide environmental controls:

* *No ECS*—No environmental control provisions are provided. This is generally a poor approach that leads to low reliability.
* *Ram air cooling*—Air enters the unmanned aircraft through an opening. The high-speed air is diffused to near the stagnation conditions, where it passes through internal heat exchangers. The air is expanded to higher flow speed and exhausted. Some ram air systems are merely holes in the airframe without isolation from internal components, which can be problematic for water and dust ingestion.
* *Conductive cooling*—Thermally conductive structures transfer heat from hot components to a cooling device such as a cooling plate or heat exchanger.
* *Fuel cooling*—The fuel is routed through hot component heat exchangers. The heated fuel can then exchange heat with the outside through heat exchangers or through the skins.
* *Air conditioner*—Engine-driven compressors drive a refrigeration cycle.
* *Sterling cycle*
* *Air cycle machine*—Turbine bleed air drives a device that provides cool airflow.
* *Spray cooling*—Fluid is sprayed on the hot component and evaporated.
* *Fans*—Internal fans blow air over hot components to remove heat. This heat is later rejected to the outside by other means.
* *Radiation*—Heat is rejected by electromagnetic emissions of high-temperature components much like a spacecraft.

The requirements placed on an ECS are less stressing if the electronics components are capable of operating over a wide temperature range, especially at the upper limit. For example, many military-grade actuators do not require active cooling for typical installations.

Electronic components generate heat in a localized fashion. This heat must be removed from the device. Blowing air over the component is most effective if cooling fins increase the surface area. Hot plates are thermally conductive fixtures that provide a thermal pathway from the component.

10.4.3 Fuel System

Unmanned aircraft that use reciprocating, jet, or turboprop engines all consume fuel, which necessitates a fuel system. A fuel system includes fuel tanks, sensors, pumps, and fuel lines. Fuel systems can be as simple as a fuel tank that feeds an engine and is pressurized by engine exhaust. Other unmanned aircraft can have multiple tanks, fuel pumps, multiple sensors, and active tank management systems.

High-fuel-mass-fraction unmanned aircraft generally require fuel systems that restrict the fuel travel in order to stay within center-of-gravity limitations. The large relative fuel weight gives the fuel burn profile the ability to change the unmanned aircraft longitudinal center of gravity. Consider a fuselage with a fuel tank in the nose and one in the tail such that the combined fuel center of gravity when both fuel tanks are full and empty corresponds with the required unmanned aircraft center of gravity. If the rear tank depletes faster than the nose tank, then the aircraft becomes nose heavy. Similarly, the aircraft becomes tail heavy if the nose tank is emptied first.

The fuel center of gravity can be managed via active pumps, tank sequencing, valves, and baffling. Ideally passive systems are used to improve reliability. A good approach is to pull fuel from a series of smaller tanks such that the tanks alternate between locations in front of and behind the required center of gravity.

Trapped fuel is fuel in the tanks that cannot be used by the propulsion system. It is difficult to avoid some amount of trapped fuel. One approach to minimizing trapped fuel is to use fuel bladders. Fuel bladders deflate as the fuel is consumed much like a children's juice drink, which can improve scavenging.

The structure can serve as the fuel tank. Integral wing tanks are known as wet wings. Care must be taken to ensure that the structural materials are compatible with the fuel. Many fuels can act as a solvent for various types of composite matrix materials. One approach is to use tank liners, which are a conformal barrier between the fuel and the structure. Liners are not structural. Coatings are layers of chemicals that are applied like paint or as a slosh treatment. Liners and coatings both provide chemical separation between the fuel and the structural materials.

Slosh loads occur when the fuel is forced against the fuel tank structure. This typically occurs when the fuel tank is partially filled by air and the unmanned aircraft experiences sudden acceleration. Launch, recovery, and gust slosh loads can be the most stressing case for fuel tank design. Foam inserts within a fuel tank can reduce the slosh loads on the tank walls, but can increase the trapped fuel weight. Porous foam allows fuel to permeate at a low rate while preventing sudden fuel motion under high loads.

A pressure differential must exist across a fuel tank to allow fuel flow. The flow at the fuel exit line can be induced by gravity acting on the fuel depth or a fuel pump. The fuel tank must be vented to allow the volume of fuel exiting the tank to be replaced by air. The tank vents should have pressure close to stagnation pressure in order to prevent a pressure drop in flight.

Fueling and defueling the aircraft can be a major driver for preflight and postflight timelines. Fuel is generally pumped from an outside tank to the unmanned aircraft fuel tanks. Each tank can have a dedicated fueling port, or the fuel system can use single-point refueling. Single-point refueling is generally preferable to reduce fueling timelines. The time to refuel the unmanned aircraft drives the refueling system flow rate. This in turn sizes the intertank fuel line diameter and tank wall structures.

Unmanned aircraft can have a maximum landing weight that is lower than the maximum takeoff weight. If the unmanned aircraft is forced to abort after takeoff, then it will not be able to land until the maximum landing weight is reached. Waiting for the fuel to be depleted can take excessive time. A fuel dump system is an alternative, where fuel pumps force fuel out of the unmanned aircraft. The hazards and environmental impacts must be weighed against the probability of UA loss when choosing to use a fuel dump system.

The operators must know the fuel state to help predict the safe flight duration. One method is to analytically predict the fuel burned from engine operations during the flight. Fuel level sensors can supplement or replace the analytical approach. Fuel level sensor types include floating and capacitive devices.

Fuel tanks can pose a risk of fire if an ignition source is introduced. Ignition sources include static electricity, wire shorts, and battle damage. Fire is possible when the tank is partially filled with fuel and air. Methods of fire prevention include self-sealing tanks, fuel bladders with no air, fire extinguishers, or inert gas systems.

10.4.4 Flight Control System

Earlier in this chapter we discussed autopilots. The remaining part of the flight control system is the actuation. Actuator types include electromechanical, hydraulic, pneumatic and electrohydraulic. Most UASs use electromechanical actuators, which are also known as servos.

Electromechanical actuators are relatively simple. These are selected for UAS systems due to low cost, ease of maintenance, and wide availability of products at the sizes of interest. In fact, tactical UASs and smaller unmanned aircraft might have no other options. Small UASs frequently use model aircraft servos, while larger UASs use aerospace quality electromechanical actuators that meet various military standards. Actuator configurations generate either linear force or rotary torque, though the rotary type is most prevalent.

Hydraulic actuators use pressurized fluid to generate a linear force on a piston. These actuators are well suited to larger UASs and high-speed UA where the control surface loads are large and there are high angular rates. Large manned aircraft with controls beyond muscle power use hydraulic actuators almost exclusively. High-speed systems such as UCAVs are likely to find off-the-shelf hydraulic actuators that have superior system performance.

Pneumatic actuators are often used for gear retraction on large UASs and for braking on nearly all wheeled UASs regardless of size. The air tanks can be pressurized prior to flight or pressurized by onboard pumps while in flight. Small UASs can leverage model aircraft pneumatic systems and have custom systems developed at low cost.

Rotary actuators have a servo arm that rotates about a shaft. A pushrod or cable system is attached to a control horn on the control surface or other hinged element. A longer control arm gives lower torque but higher rates at the control surface. Rotary actuators are sized by torque and angular speed. The torque required is the control surface hinge moment. The angular rate is driven by the control system needs. Sixty degrees per second is a common rate. The torque required of a rotary actuator T_{Act} is

$$T_{Act} = T_{Hinge} \cdot \frac{L_{Arm}}{L_{Horn}} \qquad (10.35)$$

where T_{Hinge} is the hinge moment, L_{Arm} is the servo arm length, and L_{Horn} is the control horn length. The hinge moment is found through an airfoil analysis code or wind-tunnel testing. The power produced by the actuator shaft is

$$P_{Out} = T_{Act} \cdot \Omega_{Shaft} = T_{Act} \cdot \frac{L_{Horn}}{L_{Arm}} \cdot \Omega_{Surface} = T_{Hinge} \cdot \Omega_{Surface} \qquad (10.36)$$

Here Ω_{Shaft} is the servo shaft rotational rate and $\Omega_{Surface}$ is the control surface angular rate. This shows that the ratio of the length of the control arm to the control horn represents a gearing.

Linear actuators act by changing length. Typically one end is secured to the airframe and is allowed to pivot about a point. The other end is affixed to a control horn. The linear actuator rotates as the length changes. Rather than providing torque, the linear actuator provides a linear force. The actuator linear force required is

$$F_{\text{Act}} = \frac{T_{\text{Hinge}}}{L_{\text{Horn}}} \tag{10.37}$$

Again the power produced by the linear actuator is

$$P_{\text{Out}} = T_{\text{Hinge}} \cdot \Omega_{\text{Surface}} \tag{10.38}$$

Control surface actuators are a reliability driver. To improve the actuation reliability, redundancy is often considered for a control function. One approach is to divide a control surface into two segments that are commanded by two servos in a similar manner. If one servo fails, the other can provide sufficient control. A single surface can also be driven by two actuators, where the good servo can overpower the failed servo. Another approach is for a single servo to achieve redundancy through redundant gearing.

10.4.5 Pneumatic and Hydraulic System

Pneumatic and hydraulic systems are required to support actuation functions such as brakes, retractable landing gear, opening doors, and flight control surfaces. Many modern UAS efforts seek to reduce complexity with all-electric actuation.

Jet engines, turboprops, and turboshafts drive a hydraulic compressor that pressurizes the hydraulic system. Electrically driven hydraulic pumps can also provide hydraulic pressure. The electrical pumps can be powered by a generator, battery, or even a ram air turbine for emergencies. Loss of hydraulic pressure for the primary flight control system is flight critical unless backup systems can take over.

An accumulator is a pressurized volume that provides a capacity of pressurized hydraulic fluid. The accumulator is for a hydraulic system what a battery is for an electric system. The accumulator sizing is driven by compressor outage scenarios much like the backup battery is sized by generator failure.

Hydraulic lines are the mechanism for transferring hydraulic pressure from the accumulator to the actuators. These lines must withstand the pressurization stresses. Stress concentrations exist at bends and joints. Hydraulic fluid can leak from the lines, and so care must be taken to prevent the fluid from interfering with electronics or other systems.

Pneumatic systems can be pressurized prior to flight. Pneumatic functions such as landing-gear retraction, extension, and braking have limited use in a flight operation. Alternatively, an onboard pneumatic compressor can maintain accumulator pressure. An ample tank volume that is pressurized prior to flight is much simpler than carrying a compressor onboard.

10.4.6 Anti-Ice System

Anti-ice systems can prevent ice buildup or remove ice after it forms on the unmanned aircraft. UASs are just as susceptible to icing as manned aircraft when operating under icing conditions, yet few UASs have icing provisions. Icing systems are complex, heavy, and expensive. This system detracts from flight performance and payload capacity. UASs without anti-ice systems are required to not fly in known icing conditions, though sufficiently high priority missions might justify risking the unmanned aircraft.

Ice detection systems identify the presence of ice formation. Embedded skin sensors detect a voltage change with the presence of ice. Another approach is to use a camera to view ice formation. Unmanned aircraft components that are most susceptible to icing include wings, tails, air data sensors, propellers, and the carburetor for reciprocating engines.

Glycol is an antifreeze fluid that can help prevent the formation of ice. Glycol can be secreted from holes in wings or propellers to coat the sensitive surfaces. A wing configured in this way is referred to as a *weeping wing*. The system has a glycol storage tank, fluid lines, and fluid ejectors. The tank sizing depends on the surface area to be treated and the flight duration in icing conditions.

Heaters can be applied to areas such as wings, air data sensors, or the carburetor. Resistive heating elements convert electrical power to heat. These systems can be operated to prevent ice formation or to remove ice. The electrical loads can be stressing design drivers for the generator and other electrical system components. Heated pitot-static probes are commonly used on UASs.

A deicing boot is a wing device that expands to crack the ice once it has formed. The boot is generally located near the wing leading edge and mostly on the upper surface. It can be difficult to maintain laminar flow on the wing with a deicing boot. The author is unaware of any deicing boot applications to unmanned aircraft.

10.4.7 Landing-Gear System

Landing gear is still the most dominant form of launch and recovery provision for unmanned aircraft, especially at tactical and larger UAS sizes. If wheels are used, braking and steering systems are required for all but the simplest UASs. Most UASs also require that the gear retract to minimize drag in flight if range and endurance are stressing.

The landing-gear steering is usually driven by an electromechanical actuator, though hydraulic steering actuators may be used for large UAS equipped with hydraulic systems. Most UAs have a tricycle landing-gear arrangement, so that the nose gear is steerable. Differential braking of the main gear is another viable approach to steering on the ground.

Braking systems slow the unmanned aircraft ground roll. Landing-gear brakes can be actuated with electromechanical actuators, though pneumatic systems are most prevalent. Reverse thrust can be used to slow the unmanned aircraft initially. Variable pitch propellers can adjust the blade angles so that negative thrust is generated. Jet-powered UASs could use thrust reversers, though it appears that no UAS has used this approach yet. Without reverse thrust, all of the braking energy must be absorbed by the brakes.

References

[1] Collinson, R. P. G., *Introduction to Avionics*, Chapman and Hall, London, 1998.
[2] Parkinson, B. W., and Spilker, J. J., Jr., *Global Positioning System: Theory and Applications, Vols. I-II*, Progress in Astronautics and Aeronautics, Vol. 163, AIAA, Reston, VA, 1996.

Problems

10.1   An unmanned aircraft has a bank angle of 20 deg and a pitch angle of 10 deg. The roll rate is 5 deg per second, the pitch rate is 0, and the yaw rate is 15 deg per second. Calculate the Euler angle rates.

10.2   An unmanned aircraft is flying at 300 kt in the Northwest direction at 40° North latitude. What is the rate of change in latitude and longitude?

10.3   An unmanned aircraft uses GPS, a barometric altimeter, a pitot-static airspeed sensor, rate gyros about all three axes, and a three-axis magnetic compass. Generate a measurement matrix.

10.4   For the sensors used in problem 10.3, show the largest state vector that can be observed by these sensors. Show the *H* matrix that relates the measurement vector to the state vector.

10.5   Repeat Example 10.1, except do not use velocity encoder.

10.6   A pitot-static tube reads a difference of 150 psf between the pitot and static sensors. What is the indicated airspeed?

10.7   A control surface has a hinge moment of 5 ft-lb. The actuator maximum torque is 2 ft-lb. What is the ratio of the servo arm length to the control horn length?

10.8   Using the arrangement from problem 10.7, the control surface maximum angular rate at the peak torque is 60 deg per second. What is the rotation rate of the actuator? What is the actuator shaft output power?

Chapter 11

Launch and Recovery

- Understand physics of unmanned-aircraft launch and recovery
- Analyze various launch techniques
- Explore different recovery methods

Fig. 11.1 Desert Hawk comes in for a landing. (U.S. Air Force photo by SSGT William Greer.)

11.1   Introduction

A ll unmanned aircraft must be launched. A somewhat smaller subset of these unmanned aircraft is designed to land intact. Although myriad conventional and unconventional launch and recovery techniques have been applied to manned aircraft for over a century, the range of options is greater for unmanned aviation. This expanded repertoire of techniques is largely enabled by the exclusion of pilot physical constraints and the inclusion of much smaller UA weights. The general motivations for unconventional methods are reduced system footprint, operational flexibility, performance augmentation, or a new mission enabler. Many of the techniques explored in this chapter enable runway independence and ship-based operations.

11.2   Physics of UA Launch and Recovery

11.2.1   Coordinate Systems and Conventions

The coordinate system used for launch and recovery calculations is referenced to a platform and the Earth's local surface. The platform can be either fixed on the surface of the Earth or moving. Fixed platforms can include runways or fixed points on the surface. Examples of moving platforms include a host aircraft, ship, or car.

Definitions of the launch and recovery coordinate system, velocity definition, and major angles are shown in Fig. 11.2. The right-hand coordinate system XYZ has an X-Y plane that is parallel to the Earth's surface and

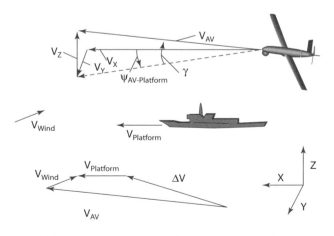

Fig. 11.2   Launch and recovery geometry and velocity definition.

passes through the platform's zero-height reference. The X axis is in the direction of platform motion projected onto the Earth's surface in the case of a moving platform. For fixed platforms, the X axis can be defined according to a relevant horizontal orientation such as runway alignment or facing a net. The Y axis is normal to the X axis in the horizontal plane. The Z axis points upwards according to the right-hand rule, and it is parallel but opposite to the gravity vector.

The origin of the coordinate system is fixed relative to the surface of the Earth, and so the X axis is parallel to the platform's heading, but the origin does not travel with the platform. If the platform changes heading, the coordinate system rotates about the Z axis such that the X axis is aligned with the platform. Only fixed heading platforms are considered here.

The platform's motion is restricted to motion along the X axis for the purposes of this discussion. By definition, the platform's side velocity component $V_{Y,\text{Plaform}}$ always has zero magnitude. Although the coordinate system does permit vertical platform motion along the Z axis, only fixed height platforms are considered here.

The UA true airspeed vector V_{AV} is the velocity of the aircraft relative to the air that it flies through. This vector is independent of wind. The UA's velocity relative to the ground in the XYZ coordinate system is the vector sum of V_{AV} and the wind vector V_{Wind}.

Two angles are defined here. The UA-platform heading angle $\psi_{AV\text{-Platform}}$ is the angle between the V_{AV} horizontal plane projection and the X axis. The UA flight-path angle γ is aircraft still-air climb angle. These parameters are defined as

$$\psi_{AV-\text{Platform}} = \tan^{-1}\left(\frac{V_{Y,AV}}{V_{X,AV}}\right) \tag{11.1}$$

and

$$\gamma = \tan^{-1}\left(\frac{V_{Z,AV}}{\sqrt{V_{X,AV}^2 + V_{X,AV}^2}}\right) \tag{11.2}$$

The velocity of the UA relative to the platform is a key parameter governing the energy of launch and recovery events. It is defined as

$$\Delta V = V_{AV} - V_{\text{Platform}} + V_{\text{Wind}} \tag{11.3}$$

It is frequently convenient to consider the horizontal and vertical components of recovery, as these velocity components are often arrested in different ways. The components are defined as

$$\Delta V_X = V_{AV} \cdot \cos(\gamma) \cdot \cos(\psi) - V_{X,\text{Platform}} + V_{X,\text{Wind}} \qquad (11.4)$$
$$\Delta V_Y = V_{AV} \cdot \cos(\gamma) \cdot \sin(\psi) - V_{Y,\text{Platform}} + V_{Y,\text{Wind}} \qquad (11.5)$$
$$\Delta V_Z = V_{AV} \cdot \sin(\gamma) - V_{Z,\text{Platform}} + V_{Z,\text{Wind}} \qquad (11.6)$$

Simplifying the vertical component by assuming horizontal wind and platform motion gives

$$\Delta V_Z = V_{AV} \times \sin(\gamma) \qquad (11.7)$$

The vertical velocity here is the UA rate of climb, which is negative for a descending approach and positive for climb. A steady rate of climb using small angle approximations is

$$\Delta V_Z = ROC \approx \frac{(T-D)}{W_{AV}} \cdot V_{AV} = \left(\frac{T}{W_{AV}} - \frac{1}{L/D} \right) \cdot V_{AV} \qquad (11.8)$$

In the case where the wind is a headwind aligned with the platform's X axis and the UA velocity is contained in the platform's X-Z plane (no Y component), ΔV_X becomes

$$\Delta V_X = V_{AV} \cdot \cos(\gamma) - V_{\text{Platform}} - V_{\text{Wind}} \qquad (11.9)$$

and

$$\Delta V_Y = 0 \qquad (11.10)$$

Further assuming small flight-path angles and using the small angle approximation, ΔV_X becomes

$$\Delta V_X \approx V_{AV} - V_{\text{Platform}} - V_{\text{Wind}} \qquad (11.11)$$

It is desirable to reduce the launch and recovery energy in order to lessen the UA loads and the footprint of any supporting apparatus. For a fixed UA weight, the only way to reduce the kinetic energy change is via ΔV magnitude reduction. Flying the UA into the wind and in the same direction of the platform travel is the best means to achieve a lower ΔV. For naval ships, the term "wind over deck" is defined as

$$V_{\text{WOD}} = V_{\text{Platform}} - V_{\text{Wind}} \qquad (11.12)$$

In the case of fixed platforms, including the Earth, the platform velocity is zero, and the only way of reducing the kinetic energy is by flying into the wind.

11.2.2 Launch

Launch involves transitioning the UA from a nonflying state to a flying state. In the case of a conventional runway launch, this can be considered *takeoff*. Nonflying does not necessarily mean that the UA is static on the ground. The nonflying vehicle could be on the deck of a moving ship or under the wing of a flying aircraft.

The flying state means that the UA is performing nominal flight operations without physical dependencies on external systems. Nominal flight can be powered or unpowered, depending on the UA type. Physical dependencies are here defined as mechanical, aerodynamic, thermo-dynamic, electrical, or other means of imparting energy to the UA. Wireless data interfaces are permissible for this definition, as many unmanned aircraft require data links with external system elements for nominal operations.

Fixed-wing aircraft require forward airspeed in order to sustain flight. The initial flight airspeed V_{Launch} must be greater than a threshold minimum airspeed V_{min} plus a safety margin to account for launch variations. The minimum airspeed for unmanned aircraft is generally higher than the stall speed V_{Stall}, which can be enforced by autopilot airspeed or angle-of-attack limiter logic.

The launch method must provide safe physical separation between the UA and the launch device and other hazards throughout the launch event. For example, the completion of a conventional wheeled takeoff results in an aircraft climbing away from the ground after clearing an obstacle height. Alternatively, pneumatic rail launchers release the aircraft in a climbing state with adequate ground clearance. Another approach with negative altitude change is successful separation from the wing of a host aircraft.

Considering launch energy gives insight into the nature of aircraft launch dynamics. The change in kinetic energy of the aircraft at the con-clusion of launch is

$$\Delta KE = \frac{W_{\text{TO}}}{2 \cdot g} \cdot |\Delta V|^2 \qquad (11.13)$$

The change in kinetic energy is referenced to a state where the UA is at rest on the launch platform.

The potential energy of the aircraft at the conclusion of launch relative to the starting condition is

$$\Delta PE = W_{\text{TO}} \cdot \Delta h \qquad (11.14)$$

The UA can use some of its own stored energy ΔE_{Stored} to assist with the launch, which generally takes the form of chemical energy from fuel or batteries. The UA will experience losses ΔE_{Losses} due to aerodynamic drag, friction, or inefficiencies in converting the stored energy to useful work. If the UA starts at rest, the change in UA energy at the conclusion of the launch is

$$
\begin{aligned}
\Delta E_{\text{AV}} &= \Delta KE + \Delta PE + \Delta E_{\text{Stored}} \\
&= \frac{W_{\text{TO}}}{2 \cdot g} \cdot |\Delta V|^2 + W_{\text{TO}} \cdot \Delta h + \Delta E_{\text{Stored}} \\
&= W_{\text{TO}} \cdot \left(\frac{|\Delta V|^2}{2 \cdot g} + \Delta h \right) + \Delta E_{\text{Stored}}
\end{aligned}
\tag{11.15}
$$

The sign of the ΔE_{Stored} term is positive because the change in the stored energy is negative when it is converted to propulsion useful work. The stored energy can take the form of fuel, propellant, or a battery, to list a few cases.

Useful work is required to change the UA's combined potential and kinetic energy. This can be accomplished via launcher useful work $\Delta E_{\text{Launcher}}$ and propulsion useful work $\Delta E_{\text{Propulsion}}$. These terms are positive when acting to increase the sum of the potential and kinetic energy terms or counter losses. The energy loss term ΔE_{Losses} is always negative and accounts for aerodynamic drag, ground friction, and inefficiencies in converting the UA's stored energy to useful work. The energy imparted to the aircraft from the launch event is

$$
\Delta E_{\text{Launch}} = \Delta E_{\text{Launcher}} + \Delta E_{\text{Propulsion}} + \Delta E_{\text{Losses}}
\tag{11.16}
$$

The following relationship is the result of setting the launch event energy equal to the change in UA energy

$$
\Delta E_{\text{Launcher}} + \Delta E_{\text{Propulsion}} + \Delta E_{\text{Losses}} = W_{\text{TO}} \cdot \left(\frac{|\Delta V|^2}{2 \cdot g} + \Delta h \right) + \Delta E_{\text{Stored}}
\tag{11.17}
$$

In the case where the aircraft does not use its own stored energy and propulsion useful work, the relationship becomes

$$
\Delta E_{\text{Launcher}} + \Delta E_{\text{Losses}} = W_{\text{TO}} \cdot \left(\frac{|\Delta V|^2}{2 \cdot g} + \Delta h \right)
\tag{11.18}
$$

This equation reveals much about the nature of launching aircraft. The launch energy imparted to the UA is proportional to the takeoff gross weight. The kinetic energy term is proportional to the relative launch velocity squared. This implies that launch speed can be a more significant launch energy driver than UA weight.

The relationship between acceleration, launch speed, and launch distance L_{Launch} has significant implications for launch system design. The kinematic equation combining these parameters is

$$L_{\text{Launch}} = \frac{|\Delta V|^2}{2 \cdot a} \tag{11.19}$$

where a is the average acceleration acting in the direction of the velocity change. This equation can be modified to express acceleration in terms of G, the multiple of the acceleration of gravity:

$$L_{\text{Launch}} = \frac{|\Delta V|^2}{2 \cdot G \cdot g} \tag{11.20}$$

This shows that the required launch length is proportional to the change in velocity and inversely proportional to the launch acceleration. It will be shown that transportable pneumatic rail launchers of reasonable length require acceleration values beyond human pilot tolerance in many applications.

Impulse-momentum is an alternative perspective that is particularly applicable to "zero-length" launch methods such as rockets. For the simplified case of straight line change in momentum, this is expressed as

$$F \cdot \Delta t = \frac{W_{\text{TO}}}{g} \cdot |\Delta V| \tag{11.21}$$

The constant launch force F is an important load case on the aircraft that will often size the structure. The right-hand side of the equation is a launch system requirement. The more general form of the impulse-momentum equation that permits variable launch force is

$$\int_0^t F(t) \cdot \mathrm{d}t = \frac{W_{\text{TO}}}{g} \cdot |\Delta V| \tag{11.22}$$

The integral form of the equation is recommended because launch force is rarely constant in practice. The peak loads are often significantly higher than the average loads.

11.2.3 Recovery

Recovery is defined as transitioning the UA from a flying state to a nonflying state. The definitions of these states are identical to those described in the preceding launch discussion. In the case of a conventional runway recovery, this phase can be considered *landing*. Alternatively, the recovery of expendable assets might result in the destruction of the UA. Regardless of recovery technique, all UAs must eventually come to rest on the ground, water, or moving platform.

UA recovery can be more challenging than launch. Launch generally requires that energy be imparted to the UA so that it can attain a threshold platform clearance, airspeed, and general heading. Recovery, on the other hand, requires that the aircraft follow a precise approach path, maintain a narrow airspeed range, and usually intersect a point on a stationary or moving recovery platform in variable wind conditions. While launch timing can wait for poor weather to pass, recovery might not enjoy the same temporal flexibility. Additionally, recovery generally occurs at slower speeds where the UA is closer to stall and more susceptible to gusts.

Recovery requires that the UA horizontal and vertical velocity components become equal to that of the recovery platform. Therefore, the energy must transition from the flying state to the at-rest state on the platform. Neglecting changes in UA stored energy and other losses, the energy that must be absorbed by the combination of platform and UA at recovery is

$$E_{\text{Recovery}} = \frac{W_{\text{Recovery}}}{2 \cdot g} \cdot |\Delta V|^2 + W_{\text{Recovery}} \cdot \Delta h \qquad (11.23)$$

where Δh is the height difference between the initiation of recovery and the at-rest state and W_{Recovery} is the UA weight at recovery. All of the recovery energy must be dissipated as heat, stored, or converted to another form of energy by a combination of the UA and the platform.

UA recovery would be simple if only the winds always blew steadily in favorable directions, and a moving recovery platform could be reliably commanded to move into the winds. Unfortunately, systems must be designed for a range of conditions, which is expressed as a recovery envelope. The recovery envelope consists of maximum wind speeds relative to the recovery platform as a function of UA heading angle. The system must be capable of operating at all points of this envelope.

The stroke length L_{Recovery} required to bring the UA to rest on the platform with a constant acceleration a acting to slow the motion is

$$L_{\text{Recovery}} = -\frac{|\Delta V|^2}{2 \cdot a} = -\frac{|\Delta V|^2}{2 \cdot G \cdot g} \qquad (11.24)$$

The time required to come to rest on the platform t is

$$t = \sqrt{-\frac{2 \cdot L_{\text{Recovery}}}{G \cdot g}} \qquad (11.25)$$

If the UA descends under the acceleration of gravity plus its initial vertical velocity relative to the platform from the initiation of recovery until the horizontal motion is arrested, the height lost during recovery relative to the platform is

$$\Delta h_{\text{Recovery}} = -\Delta Z = 1/2 \cdot g \cdot t_X^2 - \Delta V_Z \cdot t_x \qquad (11.26)$$

This height loss is an important parameter for recovery systems such as nets or suspended cables.

An impulse-momentum equation is more applicable for most vertical velocity arresting techniques. Assuming a steady initial descent until the initiation of recovery, the impulse-momentum equation is

$$\int_0^{t\,\text{Reaction}} [F_Z(t) + L_Z(t)] \cdot \partial t = -\frac{W_{\text{Recovery}}}{g} \cdot \Delta V_Z + W_{\text{Recovery}} \cdot t_{\text{Reaction}}$$

$$= W_{\text{Recovery}} \cdot \left(-\frac{\Delta V_Z}{g} + t_{\text{Reaction}} \right) \qquad (11.27)$$

If the aerodynamic lift L_Z is set to zero at initiation of recovery, the impulse-momentum equation becomes

$$\int_0^{t\,\text{Reaction}} F_Z(t) \cdot \partial t = W_{\text{Recovery}} \cdot \left(-\frac{\Delta V_{Z,\text{Impact}}}{g} + t_{\text{Reaction}} \right) \qquad (11.28)$$

Shorter recovery times and high vertical descent speeds relative to the platform generate larger upward vertical forces on the UA. Taken to an extreme, a belly landing can generate large forces on the fuselage during the time the structure compresses to absorb the impact. Conventional wheeled-landing UAs usually perform a flare maneuver where increased lift slows the vertical descent, which is followed by landing-gear shock-absorbing force at touchdown.

The recovery system reaction forces that bring the aircraft to rest initially should be quickly disengaged to prevent sending the aircraft in a rearward and upward motion. This is not a risk if the forces are generated by dissipative means such as fluid dynamic drag, heat generation, or through mechanical dampers. Methods that can store partially reversible energy, such as springs or elastic cords, must be controlled sufficiently to prevent damage to the UA.

11.3 Conventional Launch and Recovery

The vast majority of large fixed-wing unmanned aircraft types in widespread use today employ conventional launch and recovery techniques exclusively, which will also be referred to as conventional takeoff and landing (CTOL) here. For UAs weighing more than approximately 1,000 lb, the challenges of imparting launch energy and absorbing recovery energy by means other than landing gear can become impractical. UAs below this weight using landing gear frequently add alternative options such as pneumatic launchers or parachute recovery as well to support runway independent operations. There is no apparent minimum UA size for conventional launch and recovery application, as small UASs and smaller UAs have used this approach. However, other launch and recovery techniques become more prevalent at smaller scales.

Conventional launch and recovery requires a length of horizontal flat surface, called the "runway," and a further distance before the nearest obstacle of maximum specified height. The runway can take the form of a traditional paved runway, grass field, road of any quality, aircraft carrier flight deck, basketball court, or even a conference room table.

The physics of wheeled takeoff and landing is well understood for most UA applications, and well-established design methods and associated standards exist to address the design and integration. The approach for developing the landing-gear design and calculating the field performance for a tactical-to-large UASs, such as the AAI RQ-7 Shadow 200 or the Northrop Grumman RQ-4 Global Hawk, is no different than for a manned aircraft. Similarly, carrier-based unmanned aircraft such as the Naval Uninhabited Combat Aircraft Systems (N-UCAS) can be designed using the same approach as the manned counterparts. These methods are not repeated here, but can be found in texts on aircraft performance [1] and landing-gear design [2].

There are many UAS applications where traditional field performance standards are not suitable. A wheeled MAV might be capable of takeoff in 5 ft from a smooth surface, and on landing, can bounce several span lengths before settling on its back without damage. Some MAV concepts are intended to fly only indoors, and so clearing a 50-ft obstacle has no relevance when the ceiling might be only 8 ft. Gossamer solar-powered aircraft field performance is dependent upon the sun conditions. The same solar-powered aircraft will be highly sensitive to winds, making a landing circle preferable to a straight runway—in fact, the aircraft might be moving backwards relative to the ground in moderate winds. Electric aircraft might be equipped with over a dozen motors making traditional engine-out takeoff climb gradient calculations questionable. These are but a few examples.

Instead of repeating the well-established field performance methods, a more general view of wheeled takeoff and landing is provided for more direct comparison to unconventional launch and recovery techniques. For wheeled launch, the energy equation terms are now examined.

$$\Delta E_{\text{Propulsion}} + \Delta E_{\text{Losses}} = W_{\text{TO}} \cdot \left(\frac{V_{\text{Launch}}^2}{2 \cdot g} + \Delta h \right) + \Delta E_{\text{Stored}} \qquad (11.29)$$

Here all of the energy is provided by the UA's stored energy ΔE_{Stored} and converted to useful work through the propulsion system, and so no external energy sources are provided. This assumes that there are no radiated power sources such as solar. At the conclusion of the launch, the UA attains the final launch speed V_{Launch} and height above the rest condition Δh. The energy losses primarily take the form of propulsion inefficiencies from converting the stored energy to thrust, aerodynamic drag, and rolling friction.

CTOL tends to generate lower loads than alternative launch methods such as pneumatic launchers or rocket assisted takeoff. The force for launch is entirely in the form of engine thrust, which rarely exceeds the weight of the unmanned aircraft. Looking at CTOL from a stroke-acceleration perspective to gain a common comparison with other launch techniques, the stroke is

$$L_{\text{Launch}} = \frac{V_{\text{Launch}}^2}{2 \cdot a} \qquad (11.30)$$

where the acceleration a is an average and the launcher length is the line connecting the starting point to the airborne conclusion of launch. Here, the launch acceleration is generally less than one g, and the corresponding launch length is much greater than other techniques used for runway independent operations.

Conventional landing can be explained from an energy perspective much like takeoff. This energy equation is

$$\Delta E_{\text{Propulsion}} + \Delta E_{\text{Losses}} + \Delta E_{\text{Braking}} = W_{\text{TO}} \cdot \left(-\frac{V_{\text{Approach}}^2}{2 \cdot g} + \Delta h \right) + \Delta E_{\text{Stored}}$$

$$\Delta Energy_{\text{AV}} = W_{\text{TO}} \cdot \left(\frac{-V_{\text{Approach}}^2}{2 \cdot g} + \Delta h \right) + \Delta E_{\text{Braking}} - \Delta E_{\text{Stored}} - \Delta E_{\text{Losses}}$$

$$= 0 \qquad (11.31)$$

Here the kinetic energy term is negative because the aircraft is brought to rest, and the reference velocity is the approach speed relative to the platform V_{Approach}. A new term is added for the change in energy for braking

$\Delta E_{\text{Braking}}$, which is negative. The braking energy includes nonthrusting devices such as brakes or spoilers. Propulsive devices that act to slow the UA such as thrust reversers are included in $\Delta E_{\text{Propulsion}}$. The thrust will be minimized at landing, reducing the energy additions to the system from stored sources.

The deceleration can be low because of the fairly weak forces of braking and other forces acting opposite of the unmanned aircraft path. Generally the nominal average deceleration is less than 1 g. From a stroke-acceleration perspective, the stroke required for conventional landing is long relative to other recovery methods such as nets or cables. The case of arresting cables will be covered later in this chapter.

The landing-gear weight penalty for a CTOL aircraft (non-carrier-based) is typically 3–5% of the maximum takeoff or landing weight. This mass fraction can have a major impact on the sizing for a fixed set of requirements. Alternatively, if maximum performance is sought, a 3–5% increase in fuel mass fraction can have significant range and endurance benefits.

CTOL UAS sortie rates and avionics complexity are negatively impacted if the field must be shared with manned aircraft. Coordinating manned and unmanned aircraft operations in the flight pattern and on the ground require assurances that collisions will be avoided. Large MALE and HALE unmanned aircraft fly much more slowly than manned fighter and attack aircraft, requiring more time in the pattern. Future transonic, jet-powered UCAS vehicles will have similar operating conditions as many manned fighter and attack aircraft. Often this collision avoidance is implemented by separating the launch and recovery windows of these disparate systems, which disrupts the operations of both types. The inconvenience of mixed runway operations is a large motivation for runway-independent unmanned operations.

11.4 Vertical Takeoff and Landing

Vertical takeoff and landing (VTOL) aircraft use propulsive thrust to enable controlled vertical launch and recovery operations. For the systems considered here the power to drive the thrust devices is stored entirely on the UA. VTOL can enable runway-independent operations without launch and recovery equipment for many applications.

Helicopters are the most prolific and arguably most successful VTOL configuration for unmanned aircraft. Several unmanned helicopter examples are shown in Fig. 11.3. This mature technology frequently leverages manned and model helicopters although many custom unmanned helicopter designs are becoming more common. Helicopters utilize one or more lifting rotors that spin about a vertical axis. Rotors produce torque

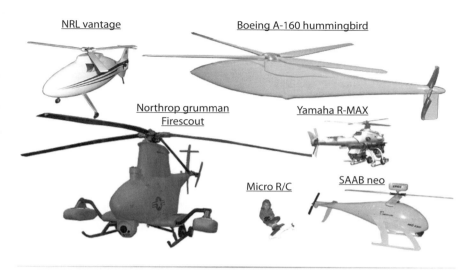

Fig. 11.3   Unmanned helicopters range from microscale to heavy lift.

that must be offset using devices such as tail rotors or additional main rotors that rotate in the opposing direction.

Beyond helicopters, there are too many alternative VTOL configurations to cover comprehensively here. A small number of VTOL UAs are shown in Fig. 11.4. These platforms use one or more vertical thrust methods. General methods include helicopter rotors, vectored jet thrust, tilt-rotors, propellers, ducted fans, dedicated lift jets, and rockets, to name a few. Propellers, rotors, lift fans, ducted fans, and other rotating propulsors can be powered by motors, gas turbines, reciprocating engines, tip jets, or other methods.

The Aurora Flight Sciences GoldenEye UAS family is a ducted fan design that can operate with optional wings (Fig. 11.5). The duct is oriented vertically for VTOL operations so that vertical thrust is provided. The body is oriented closer to horizontal for forward flight, providing a forward thrust component. Lift in forward flight is provided by duct lift, vertical thrust component, and the aerodynamic lift from the optional wing. Control vanes in the duct and throttle provide control during the VTOL mode.

The Aurora Flight Sciences Excalibur is a VTOL UAS that combines two methods of vertical lift (Fig. 11.6). The primary thruster is a pivoting jet engine located between two fuselages. The jet thrust angle is vertical for takeoff and landing and horizontal for forward flight. A single battery-powered electric lift fan is housed within the canard and another one housed in each main wing telescoping section. In forward flight the canard lift fan is covered by sliding doors, and the main wing telescoping sections are retracted. Attitude control in hover flight is controlled by the

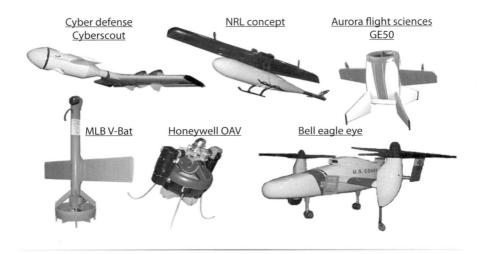

Fig. 11.4 A few of the many potential VTOL configurations.

electric lift fans. This design solution permits both VTOL and high-speed forward flight with relative simplicity.

VTOL aircraft generally have large range and endurance performance penalties due to high propulsion weight. The thrust required for VTOL can be three to five times higher than a CTOL or air-launched aircraft. The thrust-to-weight ratio of a VTOL aircraft is generally greater than 1.2, which results in a large propulsion mass fraction that detracts from the useful load.

Most VTOL aircraft must be capable of operating in a horizontal flight mode. The transition to and from vertical flight and horizontal flight is

Fig. 11.5 Aurora Flight Sciences GoldenEye 80 uses ducted fans for VTOL operations. (Photo courtesy of Aurora Flight Sciences.)

Fig. 11.6   Aurora Flight Sciences Excalibur UAS uses vectored jet thrust and electric lift fans for VTOL operations. (Photo courtesy of Aurora Flight Sciences.)

known as transition and is often the riskiest part of the flight. The transition from horizontal to vertical flight is particularly challenging. Horizontal and vertical flight often use different flight control approaches, and the flight mode transition requires a corresponding flight control transition.

Using impulse-momentum equations, the relationship between the forces and change in horizontal flight velocity at constant altitude for a transitioning UA is

$$\int \{T(t) \cdot \cos[(\theta_{\text{Thrust}}(t)] - D(t)\} \cdot dt = \frac{W}{g} \cdot \Delta V \qquad (11.32)$$

where T is the thrust, θ_{Thrust} is the thrust angle from horizontal, and D is the UA drag. The change in velocity is positive in transition to forward flight and negative for transition to hover. In this case, all of the horizontal acceleration to and from the forward-flight condition occurs above the ground.

11.5 Rail Launchers

Rail launchers provide the aircraft with a stabilizing track and launch energy to take the aircraft from rest to the flight condition. This launch technology predates conventional landing gear on flying aircraft: it was successfully employed on a Langley Aerodrome unmanned aircraft in 1896 and significantly later by the first manned aircraft in 1903. Rail launchers are common for unmanned aircraft weighing less than 500 lb. This technique enables runway-independent launch. Fig. 11.7 shows the Insitu Integrator rail launch.

The UA is secured to a shuttle that travels along the rail via guides. The UA and shuttle combined mass must be accelerated in order to reach launch speed V_{Launch}. The methods of generating this force are generally pneumatic pistons, elastic cords, rockets, flywheels, or elevated weights.

The UA is released from the shuttle assembly and clears the launcher. The initial launch velocity is sufficiently high to enable conversion of excess kinetic energy to potential energy, meaning that the UA gains some initial ground clearance altitude.

Pneumatic pistons are the most common method of providing the launch energy. Pneumatics use compressed air or another gas to power a piston. The pressurized gas is stored in an accumulator that releases into an expansion cylinder once a release valve is actuated. The expanding gas provides differential pressure across the piston face, forcing the piston to move toward the low-pressure side of the cylinder. The moving piston is mechanically linked to the shuttle directly or through a gearing system. The piston and shuttle must either be decelerated within the rail system or allowed to fly free of the launcher. The latter case can create a safety hazard.

A compressor is required to pressurize gas in the accumulator. The compressor is either fuel-powered or electric-powered. It is an additional piece of support equipment. The compressor performance is based on the accumulator pressure requirements and the minimum time to recharge.

Like other launch techniques, the rail launcher is sized by the amount of energy that must be imparted to the UA. This is a strong function of the final launch altitude and airspeed at the conclusion of launch. Ignoring UA thrust and drag, the initial release velocity relative to the platform V_{Release} is found by

$$W_{\text{TO}} \cdot \left(\frac{V_{\text{Release}}^2}{2 \cdot g} + \Delta h_{\text{Release}} \right) = W_{\text{TO}} \cdot \left(\frac{|\Delta V|^2}{2 \cdot g} + \Delta h \right) \tag{11.33}$$

where $\Delta h_{\text{Release}}$ is the height of the UA at release above the platform reference height. Simplifying by canceling the takeoff gross weight, this becomes

$$\frac{V_{\text{Release}}^2}{2 \cdot g} + \Delta h_{\text{Release}} = \frac{|\Delta V|^2}{2 \cdot g} + \Delta h \tag{11.34}$$

Solving for release velocity

$$V_{\text{Release}} = \sqrt{|\Delta V|^2 + 2 \cdot g \cdot (\Delta h - \Delta h_{\text{Release}})} \tag{11.35}$$

The acceleration length of the rail L_{Accel} is

$$L_{\text{Accel}} = \frac{V_{\text{Release}}^2}{2 \cdot a} \tag{11.36}$$

where a is the average acceleration of the combined UA and shuttle. The peak acceleration is often significantly higher than the average acceleration,

Fig. 11.7   Insitu Integrator after pneumatic rail launch. (Photo courtesy of Insitu.)

and this can size the UA structures. For launchers that jettison the shuttle from the end of the rail after UA release, the acceleration length is the same as the rail length. For launchers that contain the shuttle on the rail, the launcher length is also a function of the shuttle deceleration distance L_{Decel}:

$$L_{\text{Rail}} = L_{\text{Accel}} + L_{\text{Decel}} \tag{11.37}$$

The deceleration length is generally shorter than the acceleration length because the shuttle has much less mass without the UA and because its velocity requires less force to bring to rest. See the rail launcher geometry displayed in Fig. 11.8.

The total launcher length might exceed practical transportability and storage dimensional limits, and the rail might need to consist of multiple segments. The rail segmentation can create design challenges for pneumatic systems. Also, the shuttle must not bind across rail-segment breaks.

The launcher elevation angle θ_{Launcher} should be set to the desired flight-path angle of the UA relative to the platform. The UA will tend to follow approximately a ballistic trajectory. Ignoring UA drag and thrust, the appropriate elevation angle can be set so that horizontal velocity component of the V_{Release} is equal to the horizontal component of ΔV.

The average force acting on the combined UA and shuttle at launch F_{Shuttle} to generate the average acceleration is

$$F_{\text{Shuttle}} = (W_{\text{TO}} + W_{\text{Shuttle}}) \cdot \left[\frac{a}{g} + \sin(\theta_{\text{Launcher}}) \right] \tag{11.38}$$

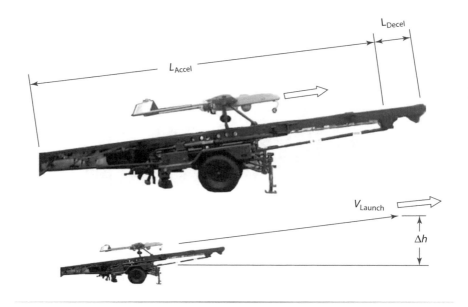

Fig. 11.8 Rail launcher geometry.

Now pneumatic pistons are considered. If the piston weight and system friction is ignored, the force acting on the piston F_{Piston} is

$$F_{\text{Piston}} = \Delta P \cdot A_{\text{Piston}} = \Delta P \cdot \frac{\pi}{4} \cdot D_{\text{Piston}}^2 \qquad (11.39)$$

where ΔP is the pressure differential, A_{Piston} is the piston face area, and D_{Piston} is the diameter of a circular piston face. The relationship between piston and shuttle forces is proportional to their relative stroke lengths

$$F_{\text{Shuttle}} = F_{\text{Piston}} \cdot \frac{L_{\text{Piston}}}{L_{\text{Accel}}} \qquad (11.40)$$

where L_{Piston} is the piston stroke length. This assumes that the shuttle deceleration is not powered by the piston. The ratio of the stroke lengths is in effect a gear ratio. The acceleration length is usually larger than the piston stroke to keep the pneumatic system size manageable.

Launchers must have a reacting mass or be fixed to the ground. Large tactical UASs often have launchers integrated into a truck or other vehicle. Stabilizing arms can provide a moment arm to prevent launcher tipping. These arms can be staked to the ground or have weights applied. Launcher weight combined with ground friction can provide sufficient anchoring effect.

Mobile rail launchers often accommodate other ground support systems such as ground power, engine start provisions, and ground data interfaces. The additional systems provide reacting mass for the launch as well as leveraging an existing trailer or truck.

11.6 Rocket Launch

Rocket boosters can be employed to launch an UA in minimal length. The rocket's high thrust-to-weight ratio and modularity can enable simple installation relative to conventional landing gear. Rocket launch is suitable when no runway is available and other techniques such as pneumatic rail launchers have excessive footprint. Also, heavier UAs and those that require high launch velocities may have no practical alternatives due to the high changes in energy. When rocket launch is combined with parachute recovery, the wings can be sized for cruise and dash performance without wing area penalties for low-speed flight. Figure 11.9 shows a target rocket launch.

Impulse-momentum provides the best view of rocket performance. The total impulse I_{Tot} is equal to the integral of the thrust over the burn time.

$$I_{Tot} = \int T(t) \cdot dt \qquad (11.41)$$

Fig. 11.9   CEi BQM-167A target rocket launch. (U.S. Air Force photo by Bruce Hoffman, CIV.)

Ignoring losses, the total impulse can change the momentum of the UA according to

$$I_{\text{Tot}} = \frac{W_{\text{TO}}}{g} \cdot \Delta V \qquad (11.42)$$

for the case where the thrust and ΔV are acting in the same direction. This is applicable for rockets accelerating an UA vertically or on a rail launcher. Rearranging, the velocity change is

$$\Delta V = \frac{g \cdot I_{\text{Tot}}}{W_{\text{TO}}} \qquad (11.43)$$

A more detailed time-step force vector analysis is generally best for calculating the actual unmanned aircraft launch performance. The forces acting on the UA are

$$\boldsymbol{F}_{\text{Net}} = \boldsymbol{T}_{\text{Rocket}} + \boldsymbol{T}_{\text{Propulsion}} + \boldsymbol{W}_{\text{Tot}} + \boldsymbol{D} + \boldsymbol{L} \qquad (11.44)$$

The UA starts at rest on the platform and, if wind relative to the platform is zero, the aerodynamic lift and drag forces are zero magnitude. The primary forces acting on the UA are the rocket thrust and the weight, though the normal propulsion system can provide thrust as well.

The rocket thrust vector is oriented upwards at angle θ_{Rocket} so that the aircraft accelerates upwards and forward simultaneously. To achieve zero-length liftoff, the vertical component of the rocket thrust must be greater than the weight of the UA and booster combined.

$$T_{\text{Rocket}} \cdot \sin(\theta_{\text{Rocket}}) \geq W_{\text{TO}} + W_{\text{Rocket}} \qquad (11.45)$$

Note that the takeoff weight does not include the weight of the booster. The rocket booster thrust line passes approximately through the aircraft center of gravity.

As the aircraft achieves some airspeed, the lift and drag forces become more significant. The initial flight-path angle can temporarily exceed the negative angle-of-attack limit, but this is a transient condition. As the rocket burn continues, the aircraft achieves sufficient energy, and the aircraft is in a permissible attitude for controllability. The booster is jettisoned after the burn is complete. See Fig. 11.10 for a diagram of the forces acting on an aircraft during a rocket-boosted launch.

Zero-length rocket launching requires a launch platform. This platform must provide adequate clearance between the rocket and the ground and be able to position the UA at the proper orientation. The geometry must prevent interference during launch. Often this launch platform serves as a cart for the UA and a maintenance stand. Figure 11.11 shows the Northrop Grumman BQM-74E stand.

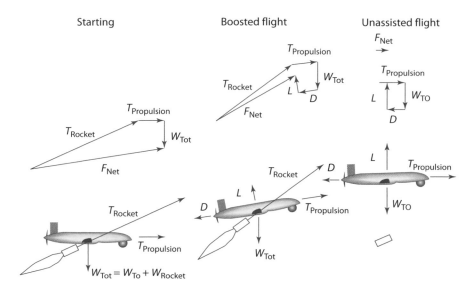

Starting   Boosted flight   Unassisted flight

Fig. 11.10 Forces acting on an aircraft during rocket-boosted launch.

1.7 Air Launch

Air launch involves releasing the UA from a host platform at altitude and with high initial airspeed relative to ground launch. The unmanned aircraft can be released from a host platform's internal bay, under wing, under the fuselage, above the fuselage, from a cargo door, or via an intermediate body, to name a few approaches. The launch method depends greatly on the mission and characteristics of both the host platform and UA.

Fig. 11.11 BQM-74E on zero-length launch stand, with approximate rocket mounting location illustrated.

Like the rocket launch technique, air launch permits the wing to be optimized for higher speed conditions when combined with parachute recovery. Air launch generally results in a higher initial altitude at the end of launch rather than ground-based methods, greatly reducing the UA's required fuel mass fraction. Furthermore, the host platform extends the reach of the unmanned aircraft, providing greater basing flexibility than ground-based systems.

Usually the UA is dropped from under the host platform. The system solution must protect the host aircraft from a collision with the dropped unmanned aircraft. To minimize this risk, the UA in the initial drop configuration should ideally be incapable of generating sufficient forces to hit the main aircraft. UA wings can be stowed or have sufficiently high wing loading to ensure an uninterrupted drop. The aircraft builds up airspeed as it descends, converting potential to kinetic energy until it attains the target launch velocity. The Naval Research Labs FINDER UAS has been successfully air-launched from the General Atomics Predator A wing hardpoint. The Teledyne Ryan (now Northrop Grumman) Firebee series of jet-powered unmanned aircraft were air-launched from under the wing of a modified C-130 for over 3,500 missions during the Vietnam War.

An UA can be launched from the top of the host platform. In this case, the UA must generate sufficient lift to clear the launch aircraft. This can be accomplished by orienting the UA such that the lift force is greater than the weight, creating an initial upward acceleration. Another measure to reduce the risk of collision is to dive and turn the host aircraft while the UA climbs and turns in the opposite direction. Such host aircraft should have short span vertical tails to provide greater clearance. Naval Research Labs used the NDM series unmanned host platform with a top-mounted release in the 1990s to test SENDER and other unmanned aircraft. Lockheed's supersonic D-21 was air-launched from the top of the modified A-12 with mixed success and later with rocket assistance from under the wing of a B-52.

UAs can be released from the aft cargo door of manned cargo aircraft. Extraction parachutes can provide a tension force to pull the UA from the cargo bay. A rail guide system can prevent interference with the host airframe, especially when the UA is close to the door dimensions. The primary advantage of this approach is that very large unmanned aircraft can be launched with relatively simple integration and potentially streamlined testing. The JPADS guided parafoil cargo delivery system operates in this manner.

Intermediary vehicles can reduce the risk to the host platform. Bodies such as chaff dispensers or sonobouy tubes have known ejection characteristics that can be leveraged to reduce the complexity of the test program.

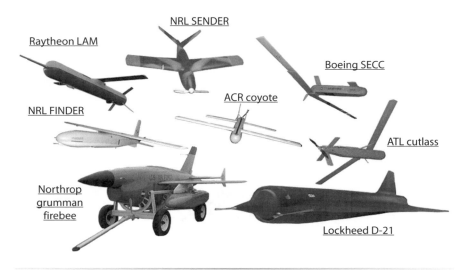

Fig. 11.12   Examples of air-launched unmanned aircraft.

The ACR Coyote, for example, fits within a sonobuoy tube to clear the host platform, followed by a reconfiguration to the flight geometry.

Balloon launch is a special case of air launch where the balloon host platform has negligible horizontal airspeed. The balloon may have some vertical airspeed, as the ascent to the release altitude might not stop at the time of UA launch. The UA accelerates downward in a nose-down attitude initially until sufficient airspeed is gained for a pull-out maneuver. The pull-out can generate high-g loads that size the wing structure. Also, the dive speed must not exceed the Mach or dynamic pressure limits of the UA. See Fig. 11.12 for examples of air-launched unmanned aircraft.

11.8 Hand Launch

Small unmanned aircraft can employ a hand-launch technique, which eliminates the need for any launch equipment to achieve runway independence. In effect, the human becomes the launch equipment. The individual launching the UA is able to provide acceptable release velocity, orientation, and angular rates to enable controlled initial flight conditions. Candidate aircraft generally have a light wing loading, light launch weight, and high static thrust-to-weight ratio. Most hand-launched unmanned aircraft weigh less than 20 lb with wing spans under 10 ft. Figure 11.13 shows several unmanned aircraft that can be hand launched.

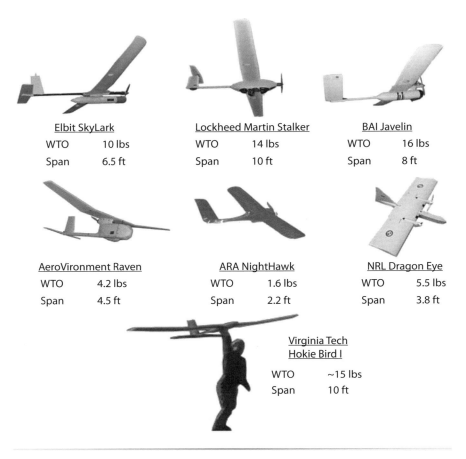

Fig. 11.13 Several unmanned aircraft capable of hand launch.

An overhead launch is the most common form (as shown in Fig. 11.14). The UA starts behind the person's back and over the head with arms mostly extended. The person may walk forward as the UA is thrown forward to build additional airspeed. Care must be taken to limit the angular rates at release. Some operators who are skilled with baseball pitching must resist the tendency to spin the aircraft.

Hand launch can drive the UA configuration. The propeller must not cause injury to the person throwing the UA, so that either a tractor configuration or a high-mounted pusher is frequently employed. Pusher propellers at the end of the fuselage are generally avoided. Also, the UA must permit the operator to grip the aircraft near the center of gravity.

Some small radio-controlled sailplanes with wing spans of 1–2 m now use a *discus launch* technique. The person grabs the sailplane by vertical

Fig. 11.14   AeroVironment Raven UAS hand launch. (U.S. Department of Defense Image; photographer not identified.)

wing-tip posts and swings it around much like a discus thrower. By the time of release, the distance between the sailplane center of gravity and the center of rotation is half a span length beyond the tip post plus the radius to the human grip. The sailplane has substantially more launch energy than with an overhead launch, with greater launch heights. A challenge for discus launch is that the sailplane has a large yaw rate at release that must be corrected. This yaw rate would pose an additional challenge for unmanned aircraft control laws.

11.9 Tensioned Line Launch

Tensioned line launch is a runway-independent method of launching UAs that minimizes transportability impacts. The system consists of a line that is fixed to a platform on one end and an UA hook on the other. The line tension can be provided via elastic materials such as bungee cord and surgical tubing, or by external means such as a winch that reels in the line.

The UA can be released from an elevated height above the reference platform or from the ground. For small unmanned aircraft such as the NRL Dragon Eye or Lockheed Martin Desert Hawk, that elevation is provided by a human. The UA can also start the launch by rolling along a flat surface level with the line until sufficient airspeed permits climb.

The forces acting on the UA during tensioned line launch are shown in Fig. 11.15. The forces acting on the aircraft in flight with a tensioned line are

$$F_{\text{Net}} = T_{\text{Line}} + T_{\text{Propulsion}} + W_{\text{TO}} + D + L \qquad (11.46)$$

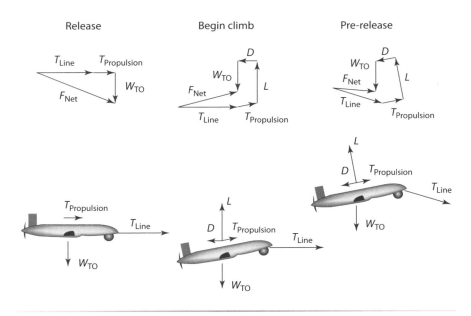

Fig. 11.15 Tensioned line launch forces.

At the initiation of launch from rest relative to the platform, the primary forces acting on the aircraft are the line tension and weight, though propulsive thrust can contribute as well. In the case of a rolling start, the takeoff weight is exactly cancelled by the landing-gear reaction forces. For release from an elevated height, the UA will descend initially unless sufficient initial upward force is provided. Figure 11.16 shows the elastic line launch of a Dragon Eye unmanned aircraft.

After the UA is in flight, the lift and drag forces become significant. The line tension vector is still relatively horizontal. The initial descent will be arrested only after the net forces acting on the UA have a positive vertical component for sufficient time. The line tension and UA minimum airspeed must be designed such that the aircraft does not impact the ground after release from an elevated height. The aircraft will continue to build up airspeed and altitude.

The force characteristics change just prior to separation from the line. The UA is in a climb with significant altitude above the line attachment to the platform. The line tension now has a downward component that acts to slow the climb rate. In the case of elastic lines, the line tension is reduced due to the line shortening. In some system configurations it is also possible that the UA has flown downrange of the line platform attachment point, which gives a tension horizontal component opposite the

Fig. 11.16   Dragon Eye UAS bungee launch. (U.S. Marine Corps photo by Lance Cpl. Brian A. Jaques.)

direction of flight. In short, the line acts to reduce the energy imparted to the UA, and it should be released.

The energy required for a tensioned line launch is

$$\Delta E_{\text{Line}} + \Delta E_{\text{Propulsion}} + \Delta E_{\text{Losses}} = W_{\text{TO}} \cdot \left(\frac{|\Delta V|^2}{2 \cdot g} + \Delta h \right) + \Delta E_{\text{Stored}}$$

(11.47)

where ΔE_{Line} is the energy provided by the tensioned line. For a line that acts only to produce useful work, the line energy contribution is

$$\Delta E_{\text{Line}} = \int_{s=0}^{\text{Final}} T_{\text{Line}}(s) \cdot \text{d}s$$

(11.48)

where s is the distance from the starting point. In the case of a spring or elastic cord, the energy stored in the line available for useful work is

$$\Delta E_{\text{Line}} = \frac{1}{2} \cdot k \cdot \Delta X^2$$

(11.49)

where k is the spring constant with units of force per distance and ΔX is the stretched distance from the nonextended state.

Safety is a major consideration for tensioned line launch systems. Elastic lines can break, sending an energetic segment in the direction of the person holding the UAs and possibly causing injury. The line tension can cause ground stakes to come loose. These stakes will accelerate towards the individual holding the UA. It is often difficult to secure a stake in soils that are too sandy or moist. A length of nonelastic line between the elastic line and the UA can offset these safety risks. A significant hazard for winches is that a person could become entangled in the lines while the winch is operating. The system must be designed for safety and properly tested.

The more energy required for launch, the greater the safety risk. Launch personnel have been known to wear bullet-proof vests to protect themselves from line breaks. If a person is holding the base end, a loss of unmanned aircraft control could send the aircraft their direction. There can be some skill to releasing the aircraft at the proper orientation, angular rates, and speed, and so launch accidents are not unheard of.

The release velocity and altitude for a given tension are controlled by the hook location on the fuselage. A hook attachment just ahead on the center of gravity on the lower fuselage will result in a high release altitude but comparatively low airspeed; this is the case with model sailplane "high start" systems. A more forward hook location will result in higher release speed and less altitude gain. The excess velocity can be used to quickly climb. In either hook configuration, it is important to ensure that the aircraft does not lose excessive altitude between the toss and initial acceleration to flight velocity. The initial sink limits the maximum wing loading for elastic line launches.

A tensioned line combined with a ground roll has advantages and disadvantages. The advantages include controllable airspeed at the time of liftoff and the ability to accommodate higher wing loading and launch weight compared to a human-elevated launch. Disadvantages are dependence upon a suitable flat surface and the complexity of UA wheel accommodation. The tensioned line can interface to a wheeled dolly to lessen the UA integration impact.

Example 11.1   Bungee Cord Sizing

Problem:

A bungee cord with a safety cable extension is used to launch a small unmanned aircraft weighing 20 N. The operator simply releases the UA from 2 m above the ground without imparting forward velocity. Soon after separation from the launch line, the aircraft must achieve at least 20-m/s airspeed and 20-m altitude, at which time the motor will engage. The bungee cord is 20 m long when not stretched, and field constraints limit the cord stretched length to 50 m. Aircraft drag losses can be neglected. What is the minimum coefficient of elasticity for the bungee cord if the UA is launched in no winds?

Solution:

This problem can be solved by looking at the tensioned cord energy equation. The energy change at the conclusion of launch is

$$\Delta E_{\text{Launch}} = W_{\text{TO}} \cdot \left(\frac{|\Delta V|^2}{2 \cdot g} + \Delta h \right) + \Delta E_{\text{Stored}}$$

The change in stored energy can be set to zero because propulsion was not used. Therefore,

$$\Delta E_{\text{Launch}} = W_{\text{TO}} \cdot \left(\frac{|\Delta V|^2}{2 \cdot g} + \Delta h \right)$$

The energy used throughout the launch is

$$\Delta E_{\text{Line}} + \Delta E_{\text{Propulsion}} + \Delta E_{\text{Losses}}$$

Propulsion was not used, and so this term is zero. The problem statement said that the drag was negligible, and so the losses term can be neglected. The line energy comes from the bungee cord. The equation for the line energy is

$$\Delta E_{\text{Line}} = \frac{1}{2} \cdot k \cdot \Delta X^2$$

The launch energy equation becomes

$$\frac{1}{2} \cdot k \cdot \Delta X^2 = W_{\text{TO}} \cdot \left(\frac{|\Delta V|^2}{2 \cdot g} + \Delta h \right)$$

All variables are known with the exception of the spring constant k. Solving yields

$$k = 2 \cdot \frac{W_{\text{TO}}}{\Delta X^2} \cdot \left(\frac{|\Delta V|^2}{2 \cdot g} + \Delta h \right)$$

The maximum stretched length is 30 m, which is found by subtracting the at-rest length from the maximum stretched length. The change in height is 18 m found by subtracting the release height from the final altitude. The spring constant is

$$k = 2 \times \frac{(20\,\text{N})}{(30\,\text{m})^2} \left[\frac{(20\,\text{m/s})^2}{2 \times (9.81\,\text{m/s}^2)} + 18\,\text{m} \right] = 1.71\,\text{N/m}$$

Fig. 11.17   AeroVironment Switchblade configuration closely resembles the gun-launched UAV (GLUAV) design.

11.10 Gun Launch

Unmanned aircraft could provide very rapid time to target if launched via a large caliber gun or as a mortar round. Missions include artillery spotting and battle damage assessment. A key advantage of such UAs is that specialized launch equipment is not required. No systems of this type have been fielded at the time of publication.

The launch loads are substantially higher than other launch techniques. CS Draper Laboratories performed initial development of the Wide Area Surveillance Projectile (WASP) unmanned that was designed to fit within a standard M-483 155-mm artillery round [3]. This unmanned aircraft required multiple wing and tail folds to fit within the shell and was designed for 16,000-g linear loads and 270-Hz rotation. These loads were distributed from the UA to a shroud and bulkheads to enable relatively conventional metal and composite airframe construction. The Aerovironment Gun Launched UAV (GLUAV) program used a folding tandem-wing configuration that closely resembles the air-launched Switchblade system shown in Fig. 11.17.

11.11 Ground-Vehicle Launch

UASs can be launched from moving ground vehicles. This eliminates the need for conventional landing gear for the purposes of takeoff. The Insitu Aerosonde was car launched for the famous transatlantic flight, enabling a lightweight airframe without gear. The ground vehicle provides the acceleration to flight speed, at which time the UA is released. Specialized racks may be placed atop the vehicle to reduce the influence of the ground-vehicle flowfield on the UA. People in the car may also physically release or throw the UA, as shown in Fig. 11.18.

Fig. 11.18 AeroVironment Puma launch from moving HMMWVV. (U.S. Army photo by Sgt. Jennifer Cohen/Released.)

11.12 Skid and Belly Recovery

Skid and belly landing is a recovery technique where the UA contacts the ground directly without the use of conventional landing gear. Either a skid device or the airframe itself is the interface to the ground, providing both landing shock absorption and friction to slow the UA to a rest.

Skids or airframe ground interface points must be designed to contend with the friction of scraping the ground. Small UASs may experience very little wear. The Desert Hawk, shown in Fig. 11.19, uses conformal skids at the ground interfaces. Larger UAs that interface with hard surfaces might require interchangeable skids that are changed after a necessary sortie interval.

Unmanned aircraft, by nature of not having a pilot, can have opposite vertical references for flight or ground operations. The payloads, for example, can face downward during flight operations, and then the UA can roll over 180 deg for landing to protect payload from damage. The IAI BirdEye 400 takes such an approach, as shown in Fig. 11.20. The winglets face down in flight and up for landing. This solution could help with

Conformal Skids

Fig. 11.19   Lockheed Martin Desert Hawk is belly landed, with conformal skids to protect the airframe from ground impact and abrasion.

internal packaging where sensors require lower hemisphere field of regard in flight. Such an arrangement is most suitable for aircraft with no camber or variable camber.

11.13 Net Recovery

Net recovery is a recovery method where the UA flies into a net, and its motion relative to the platform is arrested (see Fig. 11.21 and 11.22). This technique is commonly used for tactical UASs that are too large for simple belly landings. Unmanned aircraft larger than 1,000 lb typically avoid this technique in favor of conventional landing or parachute recovery.

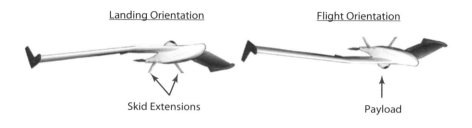

Landing Orientation   Flight Orientation

Skid Extensions   Payload

Fig. 11.20   IAI BirdEye 400 has inverted flight and landing orientations.

Fig. 11.21 Dragon Eye captured in a net. (U.S. Navy photo by Photographer's Mate 1st Class Ted Banks.)

Net recovery suffers from a reputation for damaging UAs, caused in no small part by the early U.S. Pioneer unmanned aircraft operational experience onboard battleships. The Pioneer is derived from an Israeli Army tactical UA that was originally designed for CTOL operations. The UA was acquired by the U.S. Navy and U.S. Marine Corps (USMC) and quickly adapted to support maritime net recovery. The airframe was not initially designed to withstand large net recovery loads, and the permissible relative velocity between the UA and the net was excessively limited under the circumstances. Furthermore, the early systems used remote human operators

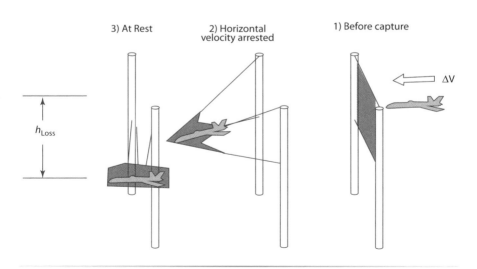

Fig. 11.22 Simple net recovery sequence.

to perform the terminal guidance into the net. The net was oriented such that the Pioneer had to fly directly towards the ship superstructure for a capture. The human remote pilots had an unreasonably high workload that frequently resulted in the UA impacting the net outside of the recovery speed envelope. Later experiments with an automatic recovery system that autonomously guided the air vehicle into the net yielded much lower damage rates.

If thrust work, losses, and changes to stored energy are neglected, the energy that the net must absorb is

$$E_{\text{Recovery}} = \frac{W_{\text{Recovery}}}{2 \cdot g} \cdot |\Delta V|^2 + W_{\text{Recovery}} \cdot h_{\text{Loss}} \tag{11.50}$$

The work to counter the UA energy comes from the tension of the four cables supporting the net.

$$E_{\text{Net}} = \sum_{i=1}^{i=4} \left[\int_{t=0}^{t=Rest} T_{\text{Cable},i}(t) \cdot \frac{\partial L_{\text{Extend},i}}{\partial t}(t) \cdot dt \right] \tag{11.51}$$

where the L_{Extend} is the linear extension of the cable supporting a corner of the net. A simple net will act as a pendulum with the suspended UA mass until the energy is dissipated through dampers or other means. The tension profile of each cable $T_{\text{Cable}}(t)$ can be defined through passive means such as springs or elastic cables, or actively controlled with a tensioner device.

More complex net systems can be used to decrease the loads on the UA or prevent the pendulum swing. If the net is able to move forward horizontally through a guide system, the UA experiences no height loss and therefore no pendulum motion. Another approach is to positively arrest the vertical motion via airbags or horizontal nets.

Nets can be combined with conventional landing to arrest the horizontal velocity after touchdown. The physics of this approach are very similar to the cable-assisted recovery on horizontal platforms described next. The primary difference is that the UA interfaces with a net via a large portion of the airframe vs a single cable with a hook.

11.14 Cable-Assisted Recovery

Unmanned aircraft can use arresting cables to arrest the horizontal velocity and, in some cases, the vertical velocity components. The arresting cables can be placed on a rigid horizontal surface or suspended by a structure. An example of the former case is the familiar carrier-based arresting system used on manned aircraft. Suspended cables are becoming increasingly popular for unmanned aircraft, but have few analogies in manned aviation. In either case, the UA becomes affixed to the cable at the initiation of recovery, and the cable provides an opposing force until the UA is at rest.

11.14.1 Arresting Cables on Rigid Horizontal Platforms

Rigid horizontal platform cable arresting systems require that the aircraft make a conventional wheeled landing and then use the cable to arrest forward motion. The ground-based cables must be low to the ground and of small enough diameter to permit the wheels to pass over them without snagging. A hook protrudes from the UA to the ground, with a geometry that will positively engage the cable while the UA moves forward. Often the hook will contact the ground to ensure that the cable is hooked. The hook extension is often angled back with a forward-facing hook. It may be retractable to reduce drag in forward flight. The geometry of the hook must protect the prop from itself and the cable. Upon engaging the UA, the cable tension slows the aircraft.

A tensioner device maintains a prescribed tension level in the cable, as shown in Fig. 11.23. This requires the paying out of cable, generally with a spool. The line tension can be controlled actively or passively. An example of an active method is a torque motor on the cable spool within the tensioner device that varies the torque level based on feedback. Passive methods include hydraulic pistons attached to the cable, elastic cables, or a torque spring on the spool. For tension control devices that store energy, such as a torsion spring or elastic line, care must be taken to prevent the aircraft from accelerating backward after the forward speed is arrested. Ideally, the tensioner acts primarily as an energy absorber, where the energy is fully dissipated after the UA is at rest.

The arresting force on the UA is the vector sum of line tensions from the lines on either side of the hook. See Fig. 11.24 for a view of an arresting hook on the AAI Shadow 200.

$$F_{\text{Cable}} = T_{\text{Cable1}} + T_{\text{Cable2}} \tag{11.52}$$

If the cable tension T_{Cable} is the same on both sides and the cable angles θ_{Cable} are symmetric, the net cable force is directed entirely in the opposite direction of the UA path.

$$F_{X,\text{Cable}} = -2 \cdot T_{\text{Cable}} \cdot \sin(\theta_{\text{Cable}}) \tag{11.53}$$

The sine proportional relationship between the cable tension and the arresting force illustrates that the cable tension does not slow the aircraft at first contact if the cable is taut between the pulleys. Using impulse-momentum relationships and assuming constant UA thrust, the cable

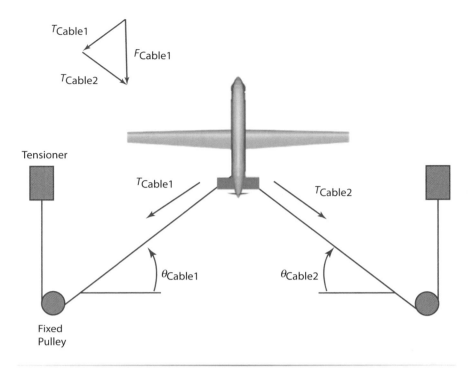

Fig. 11.23 Horizontal surface arresting cable elements and forces.

force-time profile for stopping the UA is integrated as

$$\int_0^t F_{X,\text{Cable}}(t) \cdot dt = \frac{W_{\text{TO}}}{g} \cdot \Delta V_X - T \tag{11.54}$$

where T is the UA thrust that acts along the approach path.

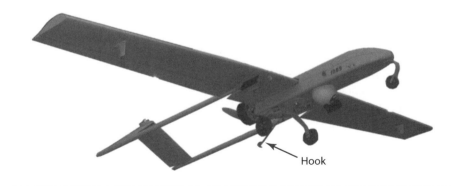

Fig. 11.24 AAI Shadow 200 variant with arresting hook.

11.14.2 Suspended Cables

Suspended cables arrest both the horizontal and vertical motion of the UA relative to the recovery platform. Recall that the arresting cables on the horizontal platform only arrest horizontal motion and act in conjunction with a conventional runway landing. Suspended cables eliminate the complexities of runway landings. The dynamics have many similarities to the net recovery technique. Suspended cable recovery differs in that the UA contacts a cable at a single hook interface. A net recovery requires that a large portion of the UA's surface area interfaces with the net.

Because arresting the horizontal motion is almost identical for suspended and nonsuspended cables, the additional consideration of arresting vertical relative motion is covered here. The suspended cable system must permit downward vertical motion of the UA after it makes contact with the cable. The UA falls under the acceleration of gravity as it loses aerodynamic lift until it finally settles. The mass and extended cable is a pendulum, and the captured aircraft will swing until the drag damps out the motion or it is stopped by external forces.

Recovery methods are possibly one of the largest opportunities for innovation in unmanned systems. Most unmanned aircraft use traditional runways, nets, parachutes, vertical landing, or belly landing recovery methods. Insitu's SkyHook^{TM} recovery system, shown in Fig. 11.25, is an exception to the traditional approaches and has been an enabler for large-scale fielding of the ScanEagle system in both land and maritime domains. The Insitu ScanEagle unmanned aircraft product line has accumulated over 500,000 flight hours, and most of this was in combat. The much larger Insitu Integrator (now the RQ-21A) was purposely designed to use this technique.

The SkyHook^{TM} is a vertically suspended cable recovery system. The vertical cable is affixed to an upper and lower boom. The primary SkyHook^{TM} implementation uses a hydraulic manlift to support the upper boom, permitting lowering for UA retrieval after landing and simplified maintenance.

The vertical rope requires an airframe-mounted hook with a horizontal extension. Quite conveniently, the UA wing can provide this functionality by incorporating a wing-tip hook. Such an approach is used with the Insitu ScanEagle and Integrator unmanned aircraft, eliminating the need for a dedicated hook extension structure. The aft-swept wings encourage the rope to slide outboard along the leading edge to wing-tip hooks. These hooks lock onto the rope after engaging.

The dynamics and loads of the combined UA and SkyHook^{TM} recovery system are highly complex. In general, the method introduces side loads on the hooked wing that differ from a net recovery. The capture forces and

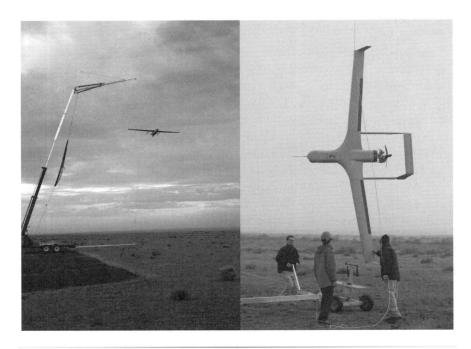

Fig. 11.25   Insitu Integrator recovering on a SkyHook™ recovery system and postcapture handling. (Photos courtesy of Insitu.)

moments acting on the aircraft are dynamic and vary direction significantly through the capture sequence, which requires significant attention when designing the UA structures and recovery system.

11.15 Parachutes

Parachutes are common recovery devices for small, tactical, MALE, and air-launched unmanned aircraft. These devices can also serve as emergency flight termination systems. Parachutes provide low descent rates and little horizontal speed relative to the recovery platform. Parachutes act either as pure drag devices or can glide, where gliding parachutes are often guided. Butler [4, 5] provides design information for parachute sizing and selection.

Nongliding parachutes reach steady-state descent speed V_T when the parachute drag is in equilibrium with the air-vehicle recovery weight, which includes the parachute weight. Assuming that the UA drag is negligible compared to that of the parachute, the relationship is expressed as

$$W_{\text{Recovery}} = D = \frac{1}{2} \cdot \rho \cdot V_T^2 \cdot C_{D,\text{Chute}} \cdot S_{\text{Chute}} \qquad (11.55)$$

where $C_{D,\text{Chute}}$ is the parachute drag coefficient with respect to the reference area S_{Chute}. The parachute drag coefficient for a round parachute ranges from

0.7 – 1.47. This reference area for a round parachute is the projection of the circle bounded by the maximum diameter D_{Chute}.

$$S_{\text{Chute}} = \frac{\pi}{4} \cdot D_{\text{Chute}}^2 \qquad (11.56)$$

The round parachute diameter required for a given terminal descent velocity and recovery weight is

$$D_{\text{Chute}} = \sqrt{\frac{8 \cdot W_{\text{Recovery}}}{\pi \cdot \rho \cdot V_T^2 \cdot C_{D,\text{Chute}}}} \qquad (11.57)$$

Gliding parachutes generally use the parafoil configuration. Parafoils are downward-arcing wings supporting a body underneath via a series of tensioned risers. Air forced in through the leading edge allows the parafoil wing to attain a chord-wise cross section that serves as an airfoil. Two control lines permit lateral-directional and airspeed control. The parafoil guidance can be performed by the main UA autopilot serving dual purposes or by a dedicated autopilot. It is possible to use the UA propulsion system to provide altitude control in support of landing.

A drogue chute is commonly used to initiate the parachute deployment. This pulls the main chute out from the aircraft and stabilizes it during inflation. The drogue and/or main parachute are often ejected clear of the aircraft via a ballistic recovery system. The initial parachute deployment loads are higher than the steady descent loads.

Deployment of the parachute imposes requirements on the UA configuration. The parachute and its lines must not interfere with the airframe or propulsion system. In the case of a pusher propeller, the parachute must clear above the prop or to the side. The lines must transition from a horizontal to vertical orientation, with the risers attaching near the aircraft center of gravity. Recessed channels along the fuselage can reduce the drag and assist with guiding the riser transition. Ducted propellers can help protect the propeller from the parachute.

The UA descending under the parachute canopy can be snagged by a helicopter or fixed-wing aircraft in what is called a midair retrieval system (MARS). The retrieving aircraft attaches to the upper portion of the parachute or between the main parachute and a higher drogue parachute. The main parachute collapses as the host aircraft bears the weight of the suspended UA. For helicopter retrievals, the UA is gently placed on the ground vertically by the host platform. Helicopters routinely captured Teledyne Ryan Firebee unmanned aircraft in this manner during the Vietnam War. Fixed-wing retrieval aircraft must secure the captured UA prior to landing.

11.16 Deep Stall

Deep stall is a steep but steady UA descent caused by fully stalled wing. This stall is generally initiated when an aft horizontal stabilizer is given a steep nose-down angle of incidence relative to the wing. The aircraft pitches up, stalling the wing. The nose would drop in a conventional stall, but the fixed stabilizer incidence prevents such a recovery. The UA descends with a high poststall wing angle of attack. The wing drag force has a large vertical component and acts much like a parachute.

This technique has been applied to free-flight model airplanes as a time-fused "dethermalizer" that prevented the models from flying away in thermals. The technique is used by Aerovironment on Pointer and other systems to enable a small recovery zone without dedicated recovery equipment. Although deep-stall recovery is mostly applied to conventional configurations, the author was able to perform this maneuver on a canard configuration with a fully pitching canard, where both the wing and canard were apparently fully stalled.

A steady descent occurs when the aerodynamic forces and weight vectors are in equilibrium, as shown in Fig. 11.26. The velocity and flight-path angle at which this occurs are a function of the aerodynamic lift and drag coefficients. The vector expression of this relationship is

$$W = L + D \qquad (11.58)$$

Alternatively this can be expressed as a scalar:

$$W = L \cdot \cos(\gamma) - D \cdot \sin(\gamma) \qquad (11.59)$$

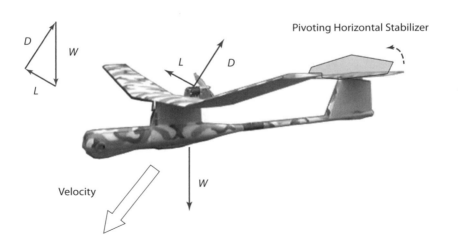

Fig. 11.26 Deep-stall recovery forces shown on Aerovironment Pointer.

Fig. 11.27   Aerovironment Raven with detached parts after recovery. (U.S. Army photo by Staff Sgt. James Selesnick/Released.)

The flight-path angle, which is negative in a glide, is

$$\gamma = \tan^{-1}\left(\frac{1}{L/D}\right) \qquad (11.60)$$

Here a low lift-to-drag ratio is desirable for minimum recovery field size.

Despite the high drag force of the stalled wing, the descent speeds can be larger than for conventional or skid landing techniques. The airframe must absorb this impact energy. Traditional shock absorbers are applicable. Aerovironment uses an innovative technique where the wing and tail detach on impact to help absorb this energy (see Fig. 11.27).

1.17   UA Impact Attenuation

Whether an UA performs CTOL, VTOL, parachute, skid/belly landing or any other technique where the UA lands on the recovery platform directly, the vertical impact must be absorbed by the UA. Alternatively, platform-based recovery devices such as nets or suspended cables perform that function.

The impulse-momentum equation for the ground impact is

$$\int_{0}^{t\,\text{Reaction}} F_Z(t) \cdot \mathrm{d}t = W_{\text{Recovery}} \cdot \left(-\frac{\Delta V_{Z,\text{Impact}}}{g} + t_{\text{Reaction}} \right) \qquad (11.61)$$

where $\Delta V_{Z,\text{Impact}}$ is the vertical velocity component at the initiation of impact. Because of the short reaction time interval t_{Reaction}, the reaction force F_Z can be quite large relative to flight loads. MAVs and small unmanned aircraft can usually be designed to handle the vertical impact loads without dedicated shock-absorption devices. The Lockheed Martin Desert Hawk UAS shown in Fig. 11.19 has a foam airframe that readily handles the ground impact. However, this becomes impractical for tactical UASs and larger UAs because of the high loads.

Shock absorbers generate a reaction force over a stroke length L_{Stroke} to more gradually react to the impact. The kinetic energy reacted by the shock absorber is a function of the vertical velocity component.

$$F_Z \cdot L_{\text{Stroke}} = \frac{W_{\text{Recovery}}}{2 \cdot g} \cdot V_{Z,\text{Impact}}^2 \qquad (11.62)$$

The shock absorber can take many forms. More sophisticated unmanned aircraft can use spring-damper systems between a skid or wheel and the airframe. Small unmanned aircraft frequently use resilient foam skids. One-time use crush zones can be employed.

Airfame-mounted airbags that deflate upon impact are frequently used in conjunction with parachute recovery systems. The airbags deploy after the UA is in the parachute descent. The airbag assembly ground reaction force and the deflation stroke length provide the impact attenuation energy. Such systems have been used on the Teledyne Ryan Compass Arrow, Teledyne Ryan AQM-34V Firebee, STN Atlas Brevel, and the Bombardier CL-289, to name a few systems.

The UA must accommodate the stowed airbag volume and weight, and the configuration must support the airbag inflation without interference. The gas to fill the airbag can be stored in pressurized vessels or generated from liquids or solids chemically. The UA integration is not trivial. For example, the airbag impact attenuation system (ABIAS) system for the Teledyne Ryan AQM-34V weighed 147.7 lb plus 8 lb of gaseous nitrogen, which is 4.6% of the maximum landing weight [6]. This mass fraction is comparable to CTOL landing gear, but only partially supports the recovery phase.

Airbag impact attenuation systems can be based on the recovery platform to achieve the same effectiveness, but without the airframe integration challenges. The problem is that the UA must hit the platform-based airbag. The airbag must either be large enough to cover the UA horizontal landing dispersion, or the UA must be guided into the airbag by other elements of the recovery system.

1.18 Nonrecoverable

Nonrecoverable unmanned aircraft are not designed to land intact. Elimination of the recovery requirement greatly simplifies the UA design and produces smaller unmanned aircraft. This is generally done for very low-cost systems or those that have mission drivers that make recovery impractical.

The simplest way to deal with recovering an UA is to use the default of destructive recovery. Almost by definition, any nondestructive recovery attempt that is unsuccessfully executed results in a destructive recovery. Operational use of cruise missiles results in target impact. Target drones, if successfully engaged with live fire, will be destroyed in flight. The Lockheed D-21 reconnaissance drone jettisoned a recoverable payload while the airframe executed a planned destruction. It is not even necessary for a nonrecoverable UA to reach the ground intact because a flight termination system can cause the airframe to disintegrate in the air.

A manufacturer's ideal scenario is a large, continuous production run of expensive nonrecoverable unmanned aircraft. Countering this interest, customer fiscal realism limits the unit cost for widely used nonrecoverable systems. A proper balance must be struck between affordability and capability. Naval Research Labs has used the term "affordably expendable" to describe small expendable unmanned aircraft such as the MITE shown in Fig. 11.28. Some concepts and applications for low-cost expendable systems include the following:

* Cruise missiles and antiradar attack
* Target drones
* Gun-launched unmanned aircraft for artillery targeting and BDA
* Rapid response air-launched ISR for ground support
* Air-launched decoys and electronic countermeasures
* Swarming micro UAs
* Unmanned aircraft that become unattended ground sensors upon ground impact

The other end of the nonrecoverable UAs spectrum covers niche, high-value missions. An early example is Lockheed D-21, where the majority of the airframe was not retrieved. More recently Defense Advanced Research Projects Agency (DARPA) explored long-range missile-launched UASs in the RapidEye program. These systems are not intended for routine use. The benefits of the time-critical intelligence, target access, or other mission effects simply outweigh the high system costs.

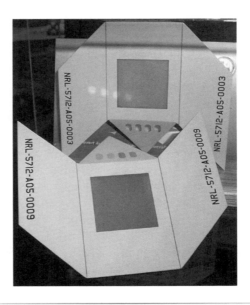

Fig. 11.28   NRL MITE is a low-cost UAS that can be affordably expendable.

11.19 Water Launch and Recovery

Water launch and recovery is very similar to CTOL, except that the UA uses water for the surface acceleration and deceleration phases instead of a solid surface. Relatively few unmanned aircraft utilize water launch and recovery despite the benefits of runway independent water-based operations. This is partially because of the small size of many unmanned aircraft as well as operational constraints.

The two primary hull configuration types are floatplanes and flying boats. Flying boats have integral hulls that comprise the main fuselage or multiple fuselages. Floatplanes are more conventional airframes that have floats added as separate bodies. Most water-launch-and-recovery-capable unmanned aircraft are flying boats.

Tactical and smaller unmanned aircraft classes might have difficulty contending with moderate to high sea states for both launch and recovery. The wave height for such aircraft is on a similar scale to the vehicle maximum dimensions. It is challenging to ensure proper orientation of the UA relative to the local water surface for both takeoff and landing.

Water recovery for ship-based fixed-wing unmanned aircraft is more common than water launch. Launch of such UAs using rail-, hand-, or even rocket-launch techniques is more convenient than water launch. A water launch requires that the UA be placed in the water and then launched, possibly requiring the ship to alter its operations. Ship recovery is generally more challenging than launch, despite mature techniques such as nets or the Insitu SkyHook^{TM}. Water landing is a viable alternative, and it does not require dedicated recovery equipment for landing. The DRS Neptune system, shown in Fig. 11.29, uses a ship-based pneumatic rail launcher and recovers via water landing for sea operations. The AeroVironment Wasp and Aqua Puma unmanned aircraft can be hand launched from a ship and water recovered.

A key challenge for unmanned aircraft that operate from water is adapting the system for water tolerance. Salt water, in particular, is highly corrosive to propulsion systems, electronics, and metal components. The system might need to separate such components from water contact, include measures for cleaning between sorties, or have interchangeable modules for frequently compromised elements. In some cases it might not be practical to turn around a unmanned aircraft after a water landing.

Retrieval of the UA after recovery imposes an operational impact on the host ship. The ship must go near the UA and stop. The ship must either directly recover the UA or send a smaller boat to aid retrieval. Either way, the ship must interrupt its ongoing operations.

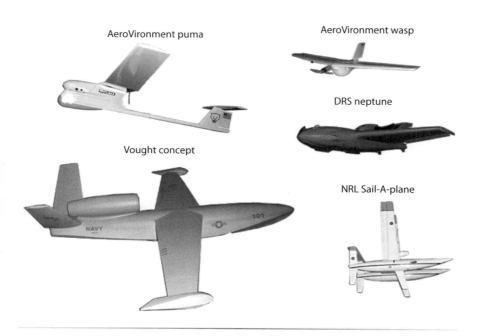

AeroVironment puma

AeroVironment wasp

DRS neptune

Vought concept

NRL Sail-A-plane

Fig. 11.29 Unmanned aircraft that support water-based operations.

A parachute recovery or deep stall combined with water surface impact provides a simple means of impact attenuation. The water penetration depth provides the attenuation stroke length. The forces countering the UA motion include buoyancy and hydrodynamic drag. Target drones such as the BQM-74 can be recovered in this manner.

Unmanned aircraft that can operate from water can also serve as unmanned surface vehicles (USVs) for the portion of the mission that is conducted on the water's surface. The USV designation is traditionally reserved for boats, but adding an air capability offers the potential for greatly expanded mission sets, extended range, and faster time to target. The NRL Sail-A-Plane, shown in Fig. 11.29, is capable of water launch and recovery, but also acts a sail-powered USV on the water's surface. The Warrior (Aero Marine) Gull is also intended to support both UAS and USV roles.

11.20   Other Launch and Recovery Techniques

An entire book could be written on all of the potential launch and recovery techniques for aircraft in general and unmanned aircraft in particular. A comprehensive coverage is simply not possible within a single chapter. A few notable other techniques are briefly listed here.

11.20.1   Air Tow

Air tow involves a powered host platform towing a trailing UA with a cable. This method is commonly used for manned sailplanes. The author is unaware of any unmanned aircraft that have used this technique. Air tow could be used to gain initial altitude or provide range extension [7].

11.20.2   Tow Line Launch

Tow line launch is similar to tension line launch described earlier. Here a person provides the line tension by running into the wind. The UA has the hook located under the UA and slightly ahead of the center of gravity to gain altitude quickly. This technique was developed for light wing loading model aircraft.

11.20.3   Moving Ground Vehicle Launch

UAs can be launched with the assistance of moving ground vehicles. The two primary techniques are rooftop mounting of the UA or towing via a cable. The former is used by the AeroSonde unmanned aircraft originally developed by Insitu, which performed the first trans-Atlantic crossing by an unmanned aircraft. An UA operating near the ground vehicle must contend with the local airflow. Vehicle-towed launch has been successfully used by manned sailplanes.

11.20.4 Partially Recoverable

The recovery energy can be substantially reduced if only a portion of the UA is recovered. A notable example is the Lockheed D-21 drone that ejected a module containing the payload and select avionics components that were recovered via a midair retrieval system. The parachute system for this module is considerably smaller that what would be required for the complete UA.

11.20.5 Dynamic Stall

Dynamic stall is where the unmanned aircraft generates temporary high lift and drag forces in the poststall regime to aid recovery. The UA geometry can be designed to generate high lift and drag coefficients at angles of attack above stall that can be used to rapidly decelerate the horizontal velocity relative to the recovery platform while maintaining lift as the speed initially decreases. Unless the propulsion system is designed to overcome the high drag forces, the UA commits to a recovery upon entering the maneuver.

11.20.6 Airship Launch and Recovery

The launch and recovery of large airships frequently requires significant ground-based launch and recovery equipment and personnel. Airships extend mooring lines that must be interfaced to the ground in order to secure the aircraft at landing. The equipment and procedures must contend with ground-based winds, including gusts.

Hybrid airships derive lift from both buoyant and aerodynamic forces. If the aerodynamic lift dominates, then the UA can perform launch and recovery operations that approximate CTOL. Hybrid airships that derive most of the lift from buoyant forces are susceptible to ground winds and must incorporate the proper ground-handling provisions.

11.20.7 Special Considerations for Ship Recovery

Maritime operations differ from ground-based operations due to the ship motion and the wind environment. Ship dynamics include vertical

Fig. 11.30 Full-scale mockup of the X-47B carrier-based unmanned aircraft.

and rolling motion caused by waves. The wind environment downwind of the ship can be turbulent because of the flow coming off of the ship structure. These factors place strong demands on the UA and its autopilot, especially for recovery operations.

Flight control systems of unmanned aircraft approaching the ship might require state knowledge of the ship position, attitude, and rates to perform a successful recovery. The ship state can be measured via an inertial navigation system (INS), and the data are transmitted to the UA via the communication system. The UA's estimate of the ship state at the time of impact might require predictive algorithms.

Carrier-based operations combine rail launchers and tensioned line recovery methods, but with additional constraints particular to aircraft carriers. For landing, the UA must be compatible with the arresting wire system, which has weight and velocity limits that drive the minimum UA size. The UA must be capable of boltering, or taking off after missing a

Fig. 11.31 Insitu ScanEagle launching and recovering on a small vessel. (Photos courtesy of Insitu.)

wire. The steep landing angle drives high descent rates with corresponding high impact energy that must be attenuated. The ship's catapult system has minimum weight aircraft limitations as well. The X-47B, shown in Fig. 11.30, is designed to operate onboard aircraft carriers. Future operational carrier-based UASs will likely use the joint precision approach and landing system (JPALS), which transmits error corrections from the ship to the UA.

The ship should be protected from impact with the UA. The Insitu ScanEagle uses a pneumatic launcher and SkyHook™ retrieval system for maritime operations (see Fig. 11.31). The recovery system is configured on the side of the ship so that the aircraft will not fly directly at the ship structure during approach. A ground operator with direct sight of the aircraft positively holds down a clear-to-land (CTL) switch to enable an automatic recovery. If the CTL switch is released, the UA will perform a wave-off maneuver where it flies away from the ship. This compares quite favorably to the Pioneer system, which had much greater recovery energy and had to fly directly towards the ship structure in order to hit the net.

References

[1] Lan, C.-T. E., and Roskam, J., *Airplane Aerodynamics and Performance*, Design, Analysis and Research Corp., Lawrence, KS, 2003, pp. 435–508.

[2] Curry, N. S., *Aircraft Landing Gear Design: Principles and Practices*, AIAA, Reston, VA, 1988.

[3] Jacobsen, S. J., and Martorana, R. T., "WASP – A Very High-G Survivable UAV," *Proceedings of Association of Unmanned Vehicle Systems International 2000 Conference*, 2000 [CD-ROM].

[4] Butler, M. C., and Montanez, R., "Design, Development and Testing of a Recovery System for the Predator™ UAV," AIAA Paper 95-1573, 1995.

[5] Butler, M. C., and Montanez, R., "How to Select and Qualify a Parachute Recovery System for Your UAV," Butler Parachutes, LLC, Online whitepaper, www.butlerparachutes.com [retrieved 25 Feb. 2008].

[6] Turner, C. T., and Girard, L. A., "Air Bag Impact Attenuation System for the AQM-34V Remote Piloted Vehicle," *Journal of Aircraft*, Vol. 19, No. 11, 1982.

[7] Murray, J. E., Bowers, A. H., Lokos, W. A., and Peters, T. L., "An Overview of an Experimental Demonstration Aerotow Program," NASA/TM-1998-206566, Sept. 1998.

Problems

11.1   A winch-launch system capable of generating constant line tension is used to launch a small unmanned aircraft weighing 20 N. The hand release height above the ground is 1.5 m, and launch personnel are required to release the UA with a minimum airspeed of 3 m/s. The minimum height above ground threshold is 0.5 m. The UA is capable of halting the initial descent and beginning the climb when 15-m/s airspeed is reached. What is the required line tension?

11.2   A horizontal rocket sled is used to launch a transonic target drone. The initial flight velocity upon separation from the sled is 500 kt. The experimental miss distance indicator payload can only withstand 3-g axial acceleration. How long must the sled rail be for the constant acceleration segment?

11.3   A quick-thinking flight-test engineer comes up with a potential remedy for recovery of an UA with nonfunctioning landing gear. He decides to purchase an inflatable playground used for children's parties. The unit is complete with a slide that can arrest both the vertical and horizontal velocity components. The UA will have a mass of 100 kg at recovery. The expected horizontal ground speed component is 30 m/s with a descent angle of 5 deg at impact. How much energy must the inflatable playground absorb? What are the likely failure mechanisms for the inflatable structure?

11.4   A 1,000-lb aircraft uses a parachute recovery system with a round chute having a drag coefficient of 1.1 chute. If the descent rate at sea level is 20 ft/s, what is the required parachute diameter?

11.5   An UA has a L/D of 10, weight of 50 lb, and an approach velocity of 40 kt. What is the thrust that will yield a descent rate of 300 ft/min?

11.6   An UA is capable of a conventional wheeled takeoff or a pneumatic launch. The launch velocity is 70 kt. The average rolling takeoff acceleration is 0.5 g, and the pneumatic rail acceleration is 10 g. The pneumatic launcher is placed horizontally. What is the difference in the effective stroke length?

11.7   A pneumatic launcher is used to launch an UA. The aircraft must clear a 50-ft obstacle, at which point it will have a steady climb with 60-kt airspeed. Assume that the propulsion and drag forces cancel. The UA is 5 ft above the ground when installed on the launcher dolly. The launcher length is 20 ft, with an acceleration stroke of 15 ft. The launcher elevation angle is

30 deg. What is the release velocity? How much energy is imparted to the UA? What is the average acceleration of the UA? What is the deceleration of the dolly?

11.8 Use the launch scenario from problem 11.7. The shuttle weighs 50 lb. The piston stroke is 5 ft, and the piston diameter is 6 in. How much force is applied to the shuttle during the acceleration phase? What is the piston force? How much pressure differential is required across the piston face?

11.9 A rocket booster is used to launch an UA. It has a total impulse of 2,000 N-s. If the required velocity change is 100 m/s, what is the maximum launch weight?

11.10 A person can hand launch a 10-lb UA to a speed of 15 kt. The wing aspect ratio is 8. If the maximum permissible lift coefficient is 1.1, what is the maximum wing loading for a sea-level launch? What is the minimum wing area and associated wing span?

11.11 A horizontal surface cable arresting system is used to recover an UA. Two tensioner devices are placed on either side of the runway, separated by 50 ft. The cable tension is constant at 1,500 lb. The UA weighs 750 lb and travels at 100 kt when it snags the cable with its tail hook. Plot the cable angle, UA velocity, and deceleration as a function of time. What is the stopping distance? What is the peak acceleration?

11.12 Using the scenario from problem 11.11, what is the required cable tension to achieve a stopping distance of 50 ft?

11.13 An UA recovers by nosing over and impacting the ground with the nose. The nose is fitted with an expendable 12-in. length crush zone that can fully compress. The aircraft builds up airspeed in the maneuver such that the impact velocity is purely vertical with a magnitude of 150 mph. Alternatively, a drogue parachute can be deployed to slow the vertical descent to 30 mph. The UA weight is 10 lb. What is the average force during impact for both cases?

11.14 Compare the wing areas required for a high-speed UAS that must be air launched or one that must be ground launched.

The UA weighs 4,000 lb. The air-launched UA can operate over a very narrow speed range and is launched at its cruise velocity of 400 kt at 30,000 ft with a cruise C_L of 0.6. The ground-launched UA has a launch velocity of 120 kt at sea level, with a maximum lift coefficient of 1.2. What are the wing areas for these two design points? What is the cruise lift coefficient for the ground-launched version if the cruise speed and altitude are the same for the two UAs?

Chapter 12

Communication Systems

- Gain an understanding of radio-frequency physics
- Learn major elements of a communications system
- Calculate link budgets
- Understand the attributes of line-of-sight and beyond-line-of-sight communications

IAI Heron TP. (Photo: U.S. Army photo by Jose Ruiz/ Released.)

12.1 Introduction

C ommunication systems provide the means to distribute data among system elements and to external entities. The majority of today's UAS use radio-frequency (RF) communications systems to transmit data wirelessly. These systems are configured for either direct line-of-sight or indirect beyond line-of-sight communications such as satellite communications (SATCOM) or airborne relay.

Because the unmanned aircraft has no pilot onboard, all actions must be autonomous or performed remotely by operators on the ground. All of the commands from ground operators described in Chapter 16 are transmitted from the ground element(s) to the unmanned aircraft. The unmanned aircraft health and status information and payload data must be transmitted from the unmanned aircraft and received by the ground for processing.

Communications engineering is a challenging discipline that requires a depth of expertise and training to perform well. This chapter provides a simplified overview of communications methods to familiarize the reader with the factors that drive communication system design.

12.2 Radio-Frequency Physics

RF communications relay information wirelessly using a portion of the electromagnetic spectrum. This is accomplished by embedding signals in electromagnetic waves. Like all electromagnetic waves, RF travels at the speed of light. The wavelength λ for a given frequency f is expressed by

$$\lambda = \frac{c}{f} \tag{12.1}$$

where c is the speed of light, which is equal to 3.00×10^8 m/s.

Figure 12.1 shows the dimensions of the full-, half-, and quarter-wavelength as a function of frequency. The half- and quarter-wavelength fractions are relevant to dipole antennas that can have physical lengths of these ratios due to ground plane interactions. We will see how different antenna configurations' performance characteristics vary with frequency in Sec. 12.5.

Communications frequencies are grouped into designated bands. The relevant bands and their characteristics are detailed in Table 12.1.

A point-to-point link that sends information in one direction only is known as a *simplex link*. A point-to-point link that communicates simultaneously in both directions is called a *full duplex* link. Both types are used in UAS applications.

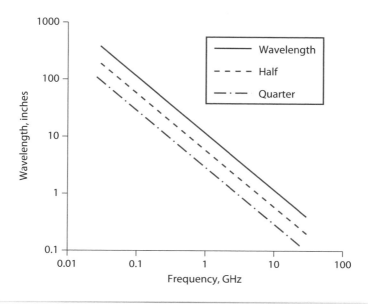

Fig. 12.1   Wavelength dimension vs frequency

Table 12.1   Frequency Bands and Associated Attributes

Band name	Frequency band	Attributes/uses
HF	3–30 MHz	• Potential long-range via ionospheric bouncing • Low data rates, suitable for analog voice
VHF	30–300 MHz	• ATC voice radios, 72-MHz radio-controlled links
UHF	0.3–1 GHz	• Vhf/uhf suitable for analog video transmission • 433.05–434.79- and 902–928-MHz (uhf) ISM bands
L	1–2 GHz	• GPS in L-band • Bandwidth-efficient digital video transmission
S	2–4 GHz	• 2.4–2.5-GHz ISM band
C	4–8 GHz	• 5.725–5.875-GHz ISM band
X	8-12.5 GHz	• Radars
Ku	12.5–18 GHz	• SATCOM • High data rate line-of-sight systems, such as CDL
K	18–26.5 GHz	• Seldom used for UAS communications
Ka	26.5–40 GHz	• High absorption • SATCOM

12.3 Elements of Communication Systems

Communication systems are composed of many elements that can be integrated in many configurations. A simplex one-way digital data link is shown in Fig. 12.2 to highlight some common elements and their functions. Such a system consists of modems, transmitters, amplifiers, and antennas.

The word *modem* is a conjunction of the words "modulate" and "demodulate." The modem modulates the input data stream onto a high-frequency carrier prior to transmission. For example, the input data stream can be digital video, and the output signal has frequency shift key (FSK) modulation, in which a data bit of 1 causes transmission of a frequency a little higher than the assigned center frequency and a 0 causes transmission of a frequency below the center frequency. Similarly, after receiving the modulated signal from the receiver the modem demodulates the data stream to a format usable by the receiving system.

The transmitter inputs the modulated signal from the modem and outputs an RF waveform. Similarly, the receiver inputs the RF waveform from the receive antenna and outputs the modulated signal.

Amplifiers increase the power of the RF signal output from the transmitter so that it can propagate over long distances. In the same way that a higher-wattage light bulb can be seen at a greater distance than a low-wattage bulb, a power amplifier that can output a higher wattage of RF power can be received at a greater distance.

The limiting factor in the receiver's ability to recover the transmitted bits is the ratio of the received signal power to the power in the noise generated by the electronics in the receiver itself. This ratio is called the

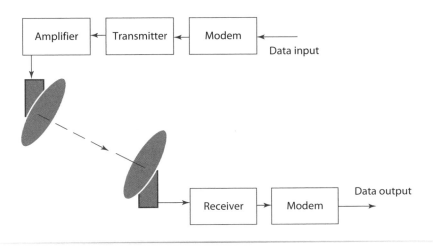

Fig. 12.2 Block diagram of communication system.

signal-to-noise ratio (*SNR*), and the noise is an inevitable result of the physics of electronic circuits, caused by the vibration of electrons in the circuits themselves. One can hear this kind of noise just by tuning an FM radio to a part of the band where there is no signal. Reliable communication can only occur if the SNR is greater than some threshold amount, which depends on the modulation being used but can be predicted theoretically. An analogy might be trying to hear a speaker in a noisy auditorium. If the speaker's voice is too soft, the noise overwhelms him, and he can't be understood. But if he increases his volume, at some point he can be understood perfectly even though the noise is still present.

Because the circuit noise is the result of an inevitable physical phenomenon, once the system designer has done all she or he can to keep the receiver noise as close to the physical limits as possible, she or he can only impact SNR through efforts to make the signal power at the receiver by as great as possible. This means that the power amplifier (PA) power should be as great as possible and that losses due to cables and antennas should be minimized. At the receiver, a special type of amplifier called a *low noise amplifier* (LNA) is used as the first step of amplification. The LNA boosts the received signal to a level that is well above the circuit noise of the following circuits while adding noise as close to the minimum physical level as possible.

The RF energy from the amplifier is radiated from the transmitter antenna, and a very small amount of the transmitted energy arrives at the receiver antenna. Antennas are the radio-frequency equivalent to optical lenses. Like a lens, an antenna can focus and intensify the signal at its source. At the other end of the link, the situation is similar to needing more magnification to view a distant light source like a remote star. As with a telescope, magnification is achieved by a great reduction in the field of view, meaning that the antenna amplifies the signal only in a very limited direction, and if it is not accurately pointed at the source of the signal, the signal will not be received. Directional antennas are usually steered via a gimbal assembly that can adjust the antenna's azimuth and elevation angles. The antenna can be housed within a radome to protect the antenna from the outside environment.

12.4 Link Budget Analysis

At the conceptual design level, link budgets are the primary communications system analysis tool used to determine whether the communication will be reliable. This analysis is generally performed by a communications engineer and often in conjunction with the communication system provider. However, small companies might not have a dedicated communications engineer, and so link analysis is frequently performed by nonspecialists. This general introduction is intended to highlight the process and describe many of the factors that drive the communication system design or selection.

12.4.1 Signal-to-Noise Ratio

When line-of-sight conditions exist, the signal strength Si at the receiver is expressed as

$$Si = P_T G_T L_T L_P G_R L_R \left(\frac{\lambda}{4\pi R}\right)^2 \qquad (12.2)$$

where
P_T = transmitted power (power)
G_T = transmitter antenna gain
L_T = signal loss through the transmitter antenna
L_P = absorptive propagation loss
G_R = receiver antenna gain
L_R = receiving signal loss from the receiver antenna through the amplifier
λ = wavelength of the carrier signal (length)
R = distance between the transmitter and receiver (length)

The noise Ni is given by

$$Ni = k \cdot T \cdot B \cdot F \qquad (12.3)$$

where
k = Boltzmann's constant (1.38054 E-23 J/K)
T = ambient absolute temperature
B = effective noise bandwidth of the receiving process
F = noise factor

F is an abstraction that allows the receiver to be modeled as a collection of ideal, noise-free components with noise added only at the very input to the receiver in a way that has the same effect as the distributed noise contributions of all of the components in the real hardware.

Combining these terms as a ratio yields the signal-to-noise ratio.

$$\frac{Si}{Ni} = \frac{P_T \cdot G_T \cdot L_T \cdot L_P \cdot G_R \cdot L_R}{k \cdot T \cdot B \cdot F} \left(\frac{\lambda}{4\pi \cdot R}\right)^2 \qquad (12.4)$$

The term

$$\left(\frac{\lambda}{4\pi R}\right)^2 \qquad (12.5)$$

is known as the free-space propagation loss. Note that the wavelength and range must have the same units of length.

The wavelength (inversely proportional to frequency) only shows up explicitly in the path propagation loss, but it indirectly affects other aspects of the communication system performance. The transmitter and receiver antenna gains are both functions of the frequency for many directional antenna types. The system losses are also strongly dependent upon the frequency.

12.4.2 Decibel Mathematics

The signal-to-noise equation will be evaluated in decibels from this point forward, and so it is worthwhile to briefly discuss decibel mathematics. The value of this method in communication engineering is that because the decibel is logarithmic, multiplication and division operations can be replaced with addition and subtraction, which allows evaluation of complicated systems without resort to a calculator. A decibel expression of quantity X, denoted by dBX, is given by

$$dBX = 10 \log_{10}(X) \tag{12.6}$$

To provide some additional insight into the definition of the $\log_{10}$ function, recall that if

$$y = \log_{10}(x) \quad \text{then } x = 10^y \tag{12.7}$$

The decibel expression of the multiplication of two terms X and Y is given by

$$10 \log_{10}(X \times Y) = 10 \log_{10}(X) + 10 \log_{10}(Y) = dBX + dBY \tag{12.8}$$

Similarly, the decibel expression of the division of these terms is

$$10 \log_{10}\left(\frac{X}{Y}\right) = 10 \log_{10}(X) - 10 \log_{10}(Y) = dBX - dBY \tag{12.9}$$

A translation of decibel and actual values is provided in Table 12.2. The range of decibels shown varies across two orders of magnitude while the referenced value spans five orders of magnitude. Addition of similar scale numbers is much simpler for engineers to perform without the aid of calculators or the oft-mentioned slide rules of engineering lore.

Notice how easy it is to approximate decibel values for other numbers, for example, see the following:

- 8 is 4 times 2, and so 8 converts to $6 + 3 = 9$ dB
- 5 is 10/2, and so 5 converts to $10 - 3 = 7$ dB

Table 12.2 Conversion of Value and Decibel of Value

Value	Decibel of value
0.001	-30
0.1	-10
1	0
2	3.0103 (approximated as 3)
4	6.0205 (approximated as 6)
10	10
100	20
1,000	30

* 9 is the square root of 81, which is roughly 10 times 8, and so 9 in dB is $(10 + 9)/2 = 9.5$ dB
* 3 is the square root of 9, and so 3 in decibels is $9.5/2 = 4.75$ dB

Remembering these manipulations allows RF engineers to perform very complex computations of system performance in their heads.

To convert from the decibel of a value back to the actual value, use

$$X = 10^{dBX/10} \tag{12.10}$$

Consider when the quantity X is a value that corresponds to power. Because power is proportional to the square of voltage, if Y is in voltage units, then to convert it to decibels requires a slightly different calculation:

$$dBX = 10 \log_{10}(Y^2) = 20 \log_{10}(Y) \tag{12.11}$$

When used to express a power level, the units "dBW" are often used. A value of 20 dBW means "20 decibels relative to 1 W," and using the preceding equation yields 100, meaning "100 times one watt" or 100 W.

For measuring the small signals present in receivers, expressing power in milliwatts is often more convenient. When expressed in decibels, power in milliwatts becomes dBm, or "decibels relative to a milliwatt." Because there are 1000 mW in 1 W,

$$dBm = 10 \log_{10}(\text{watt} \cdot 1000) = dBW + 30 \tag{12.12}$$

When a signal passes through an amplifier, the amplifier applies a gain g to the signal, meaning that if the input is Y V, the output is gY V. More typically, the gain is given in decibels $[G = 20 \log_{10}(g)]$, and the input signal is given in decibels relative to a milliwatt or decibels relative to 1 W. In this case the power level output by the amplifier is the sum of the input power in decibels relative to a milliwatt and the amplifier gain in decibels.

12.4.3 Decibel Form of Signal-to-Noise Ratio

Equation (12.4) has variables that span many orders of magnitude. Direct calculations in this manner are not intuitive. To remedy this situation, the link analysis is converted to decibel notation by taking 10 times the log of both sides. The result is a convenient addition rather than a multiplication of terms:

$$SNR(\text{dB}) = 10 \log_{10}\left(\frac{Si}{Ni}\right)$$

$$= P_T(\text{dBm}) + G_T(\text{dBi}) + L_T(\text{dB}) + L_P(\text{dB}) + G_R(\text{dBi}) + L_R(\text{dB})$$

$$+ 20 \log_{10}\left(\frac{\lambda}{4\pi R}\right) - 10 \log_{10}(1{,}000\,k \cdot T)(\text{dBm/Hz})$$

$$- 10 \log_{10}(B)(\text{dBHz}) - NF(\text{dB}) \tag{12.13}$$

Here B has units of Hertz. The noise factor term NF in decibels is $10 \log_{10}(F)$.

The decibel form of the link equation is much simpler to use than the multiplicative form. The terms themselves are smaller numbers of similar magnitude when taken as decibels. Also, the terms can be added and subtracted directly. Those performing the link analysis can discern the impacts of communications hardware without a calculator or computer. The link parameters can be effectively tracked in a simple table.

Before we show the link budget tabular method, the transmitter, receiver, and absorptive atmospheric loss terms are expanded to show more detailed contributions. Here all terms are in decibels. The transmitter loss L_T is the decibel sum of the transmitter component line losses $L_{T,\text{Line}}$, the transmitter pointing loss $L_{T,\text{Point}}$, and the transmitter radome loss $L_{T,\text{Radome}}$:

$$L_T(\text{dB}) = L_{T,\text{Line}}(\text{dB}) + L_{T,\text{Point}}(\text{dB}) + L_{T,\text{Radome}}(\text{dB}) \tag{12.14}$$

Similarly, the receiver loss L_R is the decibel sum of the receiver line losses $L_{R,\text{Line}}$, the receiver pointing loss $L_{R,\text{Point}}$, the receiver radome losses $L_{R,\text{Radome}}$, the receiver polarization loss $L_{R,\text{Polar}}$, and the spreading implementation losses $L_{R,\text{Spread}}$. Spread spectrum is not used for many common UAS applications and might therefore be neglected in such circumstances.

$$L_R(\text{dB}) = L_{R,\text{Line}}(\text{dB}) + L_{R,\text{Point}}(\text{dB}) + L_{R,\text{Radome}}(\text{dB}) + L_{R,\text{Polar}}(\text{dB})$$

$$+ L_{R,\text{Spread}}(\text{dB}) \tag{12.15}$$

Finally, the absorption losses L_P can be considered as a decibel sum of normal atmospheric absorption losses $L_{P,\text{Atm}}$ and the absorption losses due to precipitation, $L_{P,\text{Precip}}$.

$$L_P(\text{dB}) = L_{P,\text{Atm}}(\text{dB}) + L_{P,\text{Precip}}(\text{dB}) \qquad (12.16)$$

12.4.4 Link Budget Tables and Communications Properties

The link budget table (Table 12.3) shows the form of a link budget analysis. The link contributors are combined into transmitter, propagation, receiver, and noise groups. The results are summarized, and the link margin

Table 12.3 Link Budget Table

		Operation	Value
Transmitter	Tx Power, P_T	+	45 dBm
	Tx component line losses, $L_{T,\text{Line}}$	+	−1.0 dB
	Tx antenna gain, G_T	+	3.0 dBi
	Tx pointing loss, $L_{T,\text{Point}}$	+	−0.8 dB
	Tx radome loss, $L_{T,\text{Radome}}$	+	−0.5 dB
	EIRP	45.7	dBm
Propagation	Free space loss, $20 \log 10(\lambda/4\pi R)$	+	−146.4 dB
	Atmospheric absorption, $L_{P,\text{Atm}}$	+	−1.2 dB
	Precipitation absorption, $L_{P,\text{Precip}}$	+	−7.0 dB
	Total propagation loss	−154.6	dB
Receiver	Rx peak antenna gain, G_R	+	30.0 dBi
	Rx polarization loss, $L_{R,\text{Polar}}$	+	−0.5 dB
	Rx pointing loss, $L_{R,\text{Point}}$	+	0.0 dB
	Rx radome loss, $L_{R,\text{Radome}}$	+	0.0 dB
	Rx component line losses, $L_{R,\text{Line}}$	+	−1.0 dB
	Spreading implementation loss, $L_{R,\text{Spread}}$	+	−1.2 dB
	Effective carrier power	−81.6	dBm
Noise	Thermal noise density, kT	−	−174.0 dBm/Hz
	Rx noise bandwidth, BW (1 MHz)	−	60.0 dBHz
	Rx noise figure, NF	−	8.0 dB
	Effective noise power	−106.0	dBm
Summary	Available SNR	24.4	dB
	Required SNR	9.0	dB
	Net signal margin (link margin)	15.4	dB

is addressed. The terms are organized by positive and negative contributors. The order of the elements in the table is the same as the order of the system steps.

Communications engineers often use groupings of terms to describe properties of the communication system performance. These groupings can be shown in the link budget tables for convenience. The effective isotropic radiated power (*EIRP*) is the theoretical transmitter power that would be required to achieve the same field intensity as is in the highest gain direction of the antenna actually being used if it was replaced with an antenna that is an ideal isotropic radiator (a theoretical, but practically impossible antenna that radiates power uniformly in all directions), and the connection between the transmitter and antenna is lossless. The *EIRP* is the sum of the PA power, the various transmit side losses, and the antenna gain, and it can be used to calculate the electromagnetic field intensity directed at the receiver. *EIRP* is expressed by

$$EIRP(\text{dBm}) = P_T(\text{dBm}) + G_T(\text{dBi}) + L_{T,\text{Line}}(\text{dB}) + L_{T,\text{Point}}(\text{dB})$$
$$+ L_{T,\text{Radome}}(\text{dB}) \tag{12.17}$$

The total propagation loss $L_{P,\text{Tot}}$ is defined by

$$L_{P,\text{Tot}}(\text{dB}) = L_{P,\text{Atm}}(\text{dB}) + L_{P,\text{Precip}}(\text{dB}) + 20 \log_{10}\left(\frac{\lambda}{4\pi R}\right) \tag{12.18}$$

The effective carrier power P_{Carrier} at the input to the LNA is defined by

$$P_{\text{Carrier}}(\text{dBm}) = EIRP(\text{dBm}) + L_{P,\text{Tot}}(\text{dB}) + G_R(\text{dBi}) + L_R(\text{dB}) \tag{12.19}$$

The effective noise power P_{Noise} referenced to the LNA input is defined by

$$P_{\text{Noise}}(\text{dBm}) = 10 \log_{10}(1{,}000\,kT)(\text{dBm/Hz}) + 10\log_{10}(B) + NF(\text{dB}) \tag{12.20}$$

Using earlier grouping of terms, the available signal-to-noise ratio SNR_{Avail} can be expressed as the difference between the effective carrier power and the effective noise power.

$$SNR_{\text{Avail}}(\text{dB}) = P_{\text{Carrier}}(\text{dBm}) - P_{\text{Noise}}(\text{dBm}) \tag{12.21}$$

The net signal margin is most commonly known as simply the link margin. The *link margin* is the required available signal-to-noise ratio minus the required signal-to-noise ratio:

$$\textit{Link Margin}(\text{dB}) = SNR_{\text{Avail}}(\text{dB}) - SNR_{\text{Req}}(\text{dB}) \tag{12.22}$$

A 10-dB link margin is recommended. This is remarkable, as few other disciplines in the system can tolerate a 10-dB margin. A weights engineer who insists upon a 10-dB margin (10 times or 1000% margin) will likely be removed from a program, for example. The large link margin is needed primarily to combat multipath effects at the extents of the communications range.

The link budget parameters can be grouped in various other ways for pragmatic purposes. For example, communications systems vendors can provide transmitters and receivers but not the antennas. Also, the breakout of more detailed parameters for the transmitter and receiver might be unimportant to the overall system integration and might reveal competitive design details. The receiver sensitivity (or receiver threshold) is frequently the key parameter provided to specify the receiver performance and is a measure of the minimum P_{carrier} at the input of the low noise amplifier that will result in acceptable performance. If the receiver sensitivity $P_{\text{Sensitivity}}$ is known, then we can say that the link will close (that is, it will provide acceptable performance) when the received signal power is greater than the receiver sensitivity, or

$$P_{\text{Carrier}}(\text{dBm}) > P_{\text{Sensitivity}} \tag{12.23}$$

One of the most common questions asked about a UAS communication system is: What range can it achieve? The preceding equation allows us to predict this. For good performance we require

$$P_{\text{Sensitivity}} < P_{\text{Carrier}}(\text{dBm}) = EIRP(\text{dBm}) + L_{P,\text{Tot}}(\text{dB}) + G_R(\text{dBi}) + L_R(\text{dB}) \tag{12.24}$$

where

$$L_{P,\text{Tot}}(\text{dB}) = L_{P,\text{Atm}}(\text{dB}) + L_{P,\text{Precip}}(\text{dB}) + 20\log_{10}\left(\frac{\lambda}{4\pi R}\right) \tag{12.25}$$

Because wavelength is the speed of light (300,000 km/s) divided by the RF frequency, we have

$$L_{P,\text{Tot}}(\text{dB}) = L_{P,\text{Atm}}(\text{dB}) + L_{P,\text{Precip}}(\text{dB}) - 20\log_{10}(R_{\text{km}}) - 20\log_{10}(f_{\text{MHz}})$$
$$+ 20\log_{10}\left(\frac{0.3}{4\pi}\right) \tag{12.26}$$

Combining this with the inequality for $P_{\text{Sensitivity}}$ and collecting terms gives

$$20\log_{10}(R_{\text{km}}) < EIRP(\text{dBm}) - P_{\text{Sensitivity}}(\text{dm}) + L_{P,\text{Atm}}(\text{dB}) + L_{P,\text{Precip}}(\text{dB})$$
$$- 20\log_{10}(R_{\text{km}}) - 20\log_{10}(f_{\text{MHz}}) + 20\log_{10}\left(\frac{0.3}{4\pi}\right) + G_R(\text{dBi}) + L_R(\text{dB}) \tag{12.27}$$

or,

$$R_{km} < 10^{0.05 \left[\begin{array}{c} EIRP(\text{dBm}) - P_{\text{Sensitivity}}(\text{dBm}) + L_{P,\text{Atm}}(\text{dB}) + L_{P,\text{Precip}}(\text{dB}) - 20\log_{10}(f_{\text{MHz}}) + 20\log_{10}\left(\frac{0.3}{4\pi}\right) \\ + G_R(\text{dBi}) + L_R(\text{dB}) \end{array} \right]}$$

(12.28)

This relationship highlights several fundamental relationships. All else being equal, we can say the following:

* Range varies inversely with RF frequency: that is, doubling the RF frequency cuts the range in half.
* Range varies as the square root of any factor that influences the received power, which includes the output power level of the power amplifier (PA), the *EIRP*, antenna gain, and atmospheric or cable losses. A 3-dB decrease in power (a factor of 2) cuts the range by a factor of the square root of 2—or about a 30% loss in range. A 6-dB decrease in power cuts the range by a factor of 2—a 50% loss of range.
* The preceding point also says that to double the range requires finding a way to increase the power at the receiver by 6 dB.

These points highlight the importance of avoiding losses in the system wherever possible, and they enable engineers to have a powerful intuition into the behavior of the communication link.

An example link budget table is presented in Table 12.3. A collection of terms are grouped together for the transmitter, propagation, receiver, and noise. The addition or subtraction operation for each term is identified, as well as the decibel units. Finally, the major link terms are summarized. One's eye should immediately focus on the link margin.

Now the methods for determining each parameter in the link budget are provided. For more detailed analysis techniques appropriate for engineering analysis, the reader is encouraged to consult communications engineering texts.

12.4.5 Methods for Determining Parameters

12.4.5.1 Transmit Power P_T

The transmitter RF output power is typically provided by the transmitter manufacturer. Some potential limitations to the transmit power include aircraft prime power availability, personnel safety, system cost, and cooling. Depending on the band and geographic area of use, the transmit power magnitude can also be restricted by regulations. The units are decibels relative to a milliwatt or decibels relative to a watt.

Power amplifiers can exhibit significant variation in RF output power depending on the operating temperature and the RF frequency. It is important to take this variation into account in the link budget. Some power amplifiers include output leveling loops that can reduce the amount of variation considerably, but the leveling impacts the size, efficiency, and cost of the amplifier.

12.4.5.2 Component Line Losses L_T and L_R

The transmitter line loss is the total loss between the transmitter and the transmit antenna. Similarly, the receiver line loss is the total loss between the receive antenna and the receiver. Contributing factors can include the transmission line, rotary joints for steerable antennas, and filters. These losses can be -1 to -2 dB each. Effective system designs must use very low loss coaxial cables and minimize the distance between the power amplifier and the antenna.

12.4.5.3 Antenna Gain G_T and G_R

When used for transmission of an RF signal, the job of the antenna is to direct the power from the transmitter in the direction where it will do the most good. An antenna cannot do anything to increase the amount of power that comes from the transmitter. All it can do is direct more of the transmitter power in preferred directions, but increasing the power in one direction necessarily means a decrease in power in other directions.

Antenna gain is usually specified using the term "dBi," meaning decibels with respect to an isotropic antenna. An isotropic antenna is a theoretical antenna that radiates the transmitter power uniformly in all directions. As an example, an antenna with peak gain of 3 dBi will direct twice the power of an isotropic antenna in its preferred direction, but for the antenna to have gain above 0 dBi in some directions means that there must be other directions where the radiated power is less than that of an isotropic antenna. The obvious consequence of this is that when the peak gain of the antenna is very high, there is a very limited direction where this gain is delivered. Section 12.5 describes antennas in more detail. Pointing losses, described next, are losses relative to ideal pointing and are degradations to the best case gain values.

12.4.5.4 Pointing Losses $L_{T,\text{Point}}$ and $L_{R,\text{Point}}$

Antennas with gain will not be able to maintain pointing to the best gain orientation with perfect accuracy. Contributions to the error might be uncertainty in the location of the other communications element, pointing angle estimation errors, latency in pointing response, dynamics of the antenna platform, or a variety of other sources. Omnidirectional antennas have negligible pointing losses in azimuth and do not require active pointing. High gain antennas can have higher pointing losses.

12.4.5.5 Radome Losses $L_{T,Radome}$ and $L_{R,Radome}$

Antennas can require environmental protection from rain, ice, humidity, wind, and dust. Mechanically pointing antennas in particular have sensitive mechanical and electronic components that are difficult to protect from the elements. *Radomes* are environmental barriers for the antennas that are largely RF transparent, or dielectric. However, the radome is not perfectly dielectric for the communication system, and so there are losses. These losses can be approximately -0.5 to -1 dB, depending on the system design and manufacturing approach. Losses typically get worse with increasing frequency. In designs that do not use radomes, these losses are 0 dB. Another common function of a radome is to serve as an aerodynamic fairing to reduce drag.

12.4.5.6 Atmospheric Absorption $L_{P,Atm}$

Losses from atmospheric absorption of the transmitted RF energy are driven by water vapor and biatomic oxygen. The water vapor content depends on the relative humidity. The biatomic oxygen concentration is most strongly a function of altitude. The losses depend on the RF frequency, altitude of the transmitter and receiver, and range.

Figure 12.3 shows the atmospheric absorption at sea level as a function of frequency, which would be applicable to a UAS operating at low altitude. In most cases, the ground data terminal is at low altitude while the unmanned aircraft operates at a higher altitude. The line of sight slices through various altitude bands, with less absorption as the altitude increases. Detailed atmospheric absorption models are required to estimate the total absorption in this common scenario. The sea-level horizontal atmospheric absorption

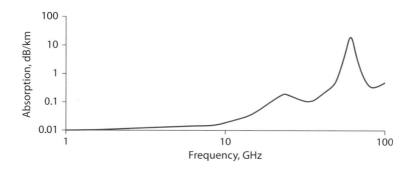

Fig. 12.3 Horizontal atmospheric absorption at sea level.

represents the worst case and can be applied for conservatism. For systems operating at frequencies less than 10 GHz with required range less than 100 km, atmospheric absorption is usually less than 0.5 dB.

12.4.5.7 Precipitation Absorption $L_{P,\text{Precip}}$

Rain can cause significant attenuation at frequencies above 5 GHz. Although the link will usually operate when no precipitation is present, the link budget analysis must be performed at the most stressing specified precipitation conditions.

Multiple rain attenuation models exist for detailed communications analysis. The Crane rain attenuation model [1] and ITU-R model are frequently applied. Professional communications analysis software often provides the user with options of selecting between the rain models.

For illustration purposes, Fig. 12.4 shows rain precipitation absorption losses as a function of frequency and rain fall rate. This is a horizontal cut through a rainstorm. Low frequencies have a clear advantage for rain losses. Links operating at S-band and at lower frequencies are much more robust than C- and Ku-band systems that are commonly used for higher data rate communications. The amount of rainfall for the maximum communications range is generally specified by the customer.

Because UASs typically operate in the lower part of the atmosphere, the distance that a line of sight signal travels through rain can be much farther than it is for satellite links, so that intuition on rain-induced loss derived from experience in the SATCOM industry can be misleading if applied to UASs. On the other hand, UASs often might not be operated in very heavy rain, and so it is important to understand the operational concept for the UAS before deciding the appropriate margin to apply for rain.

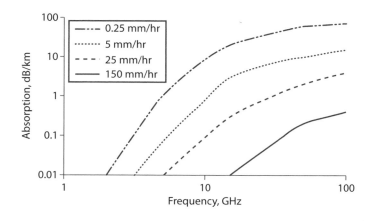

Fig. 12.4 Precipitation absorption.

12.4.5.8 Receiver Polarization Loss $L_{R,Polar}$

Electromagnetic waves have two components, an electric field component and a magnetic field component. These two wave components are always at right angles with respect to each other. Each wave component is a sinusoid whose frequency is the RF frequency of the signal.

In a linearly polarized system, the electric field orientation always stays the same. Placed in the right orientation, a linearly polarized antenna will emit a signal with the electric field oriented vertically. If the same antenna is tilted 90 deg, it will emit a signal with the electric field component oriented horizontally. If the antenna is designed with a gain pattern that results in an electric field that is oriented vertically, the antenna is described as having vertical polarization.

When a vertically polarized antenna is used at the transmitting end of the link, the maximum signal power is provided to the receiver when the receiving antenna is also vertically polarized. If either antenna is tilted, then there is a polarization loss. The reason for this loss can be understood by considering the electromagnetic field as a vector. If the vector is tilted off the vertical, it then has a vertical component and a horizontal component.

A vertically polarized receiving antenna rejects all of the energy in the horizontal component of this vector. The vertical component is decreased by the cosine of the tilt angle from what it would be without any tilt. Because the electric field is a voltage, the power loss in the vertical component in decibels is 20 times the log of the cosine of the tilt angle.

Some intuition into the behavior of polarization can be derived from realizing that visible light is a linearly polarized electromagnetic wave. Polarized sunglasses remove glare because the process of reflection of sunlight typically removes most of the vertically polarized component of the light, and polarized lenses are designed to block the horizontally polarized component.

A second type of polarization that is frequently used in antennas is circular polarization. A circularly polarized antenna actually transmits equal amplitudes of both horizontal and vertical components, but it delays one with respect to the other by a quarter of a wavelength. When this is done, the electric field component of the wave rotates through 360 deg for every cycle of the wave. An electromagnetic wave can be right- or left-hand circularly polarized (RHCP or LHCP) depending on which of the linear components was delayed by the transmitting antenna. An RHCP signal will be rejected by an LHCP receiving antenna and vice versa.

If a linearly polarized signal is received by a circularly polarized antenna, the result will be a 3-dB loss in power compared to using an antenna with matching polarization. But sometimes it can be convenient to accept this loss because the linear-to-circular configuration has the advantage that the loss is the same regardless of the tilt of the antenna. Also, in practical systems a circularly polarized antenna is not perfectly circular because the two components used to make it are not perfectly matched in amplitude. This makes the polarization somewhat elliptical, and some additional loss will result from this eccentricity.

In the case of linear polarization, the polarization loss is proportional to the square of the cosine of the misalignment angle θ_{Align}. When an aircraft is flying nose level, the misalignment angle is equal to the bank angle. Recall that the loss is a decibel value, and so $L_{R,\text{Polar}}$ is calculated by

$$L_{R,\text{Polar}} = 20\log_{10}\left[\cos(\theta_{\text{Align}})\right] \tag{12.29}$$

A 15-deg misalignment angle results in a -0.30-dB polarization loss, and 45 deg yields a -6-dB loss.

12.4.5.9 Receiver Spreading Implementation Loss $L_{R,\text{Spread}}$

Receiver spreading implementation loss is a byproduct of using a spread spectrum data link. These systems use a relatively broad frequency band to distribute the signal. The primary motivation is to reduce the probability of detection and interception of the data link. This loss is 0 dB for communications systems that do not incorporate the spread spectrum approach. Otherwise -1 to -2 dB is typical.

12.4.5.10 Thermal Noise Density kT

Thermal noise is caused by molecular vibrations and is proportional to the absolute temperature. The scaling constant is the Boltzmann's constant k, which is 1.38054 E-23 J/K. The Earth's standard average temperature is 290 K. At this temperature the following relationships emerge:

$$10\log_{10}(kT) = -204.0\,\text{dBW/Hz} \tag{12.30}$$

$$10\log_{10}(1000\,kT) = -174.0\,\text{dBm/Hz} \tag{12.31}$$

The dBm/Hz relationship is used for the link analysis covered in this chapter.

12.4.5.11 Receiver Noise Bandwidth BW

The receiver noise bandwidth BW is related to the ideal receiver bandwidth B by

$$BW(\text{dB}) = 10\log_{10}\left(\frac{B}{1\,\text{Hz}}\right) = 10\log_{10}\left(\frac{R_{\text{Data}}}{R_{\text{Data}}/B}\right) \tag{12.32}$$

where B has units of hertz^{-1}. The relationship between the bandwidth and the data rate R_{Data} depends upon the type of modulation technique applied. The ratio between the data rate and bandwidth, denoted by R_{Data}/B, is an important parameter for determining the required SNR for digital communications systems. This ratio is approximately 1 for QPSK and DQPSK. BPSK, DBPSK, and OFSK can have R_{Data}/B values of 0.5.

12.4.5.12 Receiver Noise Figure

The receiver noise figure (NF) is a measure of the receiver's noise floor and sensitivity. The value is dependent upon the receiver design implementation. Eight decibels is a representative noise figure value at X-band.

12.4.5.13 Required SNR

The required signal-to-noise ratio SNR_{Req} is driven by the type of modulation and error rates. Error rates are usually measured in terms of bit error rate (BER), or probability of bit error, for digital data links. The energy-per-bit-to-noise-power-spectral-density ratio Eb/No is an important link parameter that is related to the SNR. The required Eb/No is a function of the BER for a given modulation type. The BER is a function of the type of data that are being transmitted. BERs of interest range from 10^{-4} to 10^{-8} with an emphasis on more stringent values. The required Eb/No increases with lower BER. For the aforementioned BER range, the required Eb/No ranges from $8-12$ dB using various forms of PSK modulation and $11-16$ dB for various FSK modulation methods. Properties of various modulation techniques are covered in Sec. 12.8.

The required SNR is related to the Eb/No by the following linear relationship:

$$SNR = Eb/No \times \left(\frac{R_{Data}}{B}\right) \tag{12.33}$$

In terms of decibels, this relationship is

$$SNR(\text{dB}) = Eb/No(\text{dB}) + 10 \times \log_{10}\left(\frac{R_{Data}}{B}\right) \tag{12.34}$$

The ratio R_{Data}/B is a function of the modulation type. In cases where the ratio of data rate to bandwidth is unity, SNR_{Req} is equal to the required Eb/No. The data rates typically have units of kilobits/second or megabits/second.

Analog data links require a minimum input SNR, which is a function of the output SNR. Modulation techniques include various forms of frequency modulation (FM), pulse code modulation (PCM), and amplitude modulation (AM). Analog data of interest can include analog video (such as NTSC) and voice relay.

12.5 Antennas

RF antennas emit or receive RF energy and provide directivity. The size and configuration of the antenna are often limited by physical or operational constraints. Antenna geometry greatly affects performance and beam pattern characteristics.

A perfectly isotropic radiator emits RF energy equally in all directions spherically. Real-world applications do not utilize isotropic antennas. The RF energy is radiated more strongly in some directions and less in others. In other words, the antennas of interest have gain relative to the isotropic scenario.

An isotropic radiator illuminates 4π steradians equally and therefore has 0-dB gain. A *steradian* is a unit solid angle. The gain is given by

$$G = \frac{4\pi}{\theta_v \times \theta_h} \tag{12.35}$$

where θ_v and θ_h are the horizontal and vertical beamwidth angles in radians, respectively. When the beamwidth angles are expressed in degrees, this relationship becomes

$$G = \frac{41{,}253}{\theta_v \times \theta_h} \tag{12.36}$$

The *beamwidth* is defined as twice the angle between the maximum gain and a 3-dB drop-off in gain. Figure 12.5 shows the gain pattern of a typical antenna in two dimensions, illustrating the beamwidth. It must be

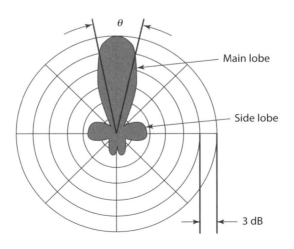

Fig. 12.5 Beamwidth definition.

remembered that the antenna pattern occupies three dimensions and is often shown in two plots: an azimuth plot (which is the gain around the horizon if the antenna is mounted normally), and an elevation plot, which is the gain from directly overhead to directly underneath along the azimuth angle that gives the maximum gain. Thus, an antenna has a beamwidth both in azimuth and in elevation. For some antennas, such as a parabolic reflector, the gain pattern for azimuth and elevation are very similar. For others, such as a quarter-wave dipole, the two patterns are dramatically different. As an example, many UASs use antennas with a pattern that is omnidirectional in azimuth (meaning that for any elevation angle it provides the same gain at any direction toward the horizon). These antennas often have an elevation beamwidth of 25 deg or so, which maximizes the gain towards the horizon where the greatest range is needed.

Another common way of defining antenna gain is in terms of the antenna effective area A_{eff}. The effective area might or might not have an intuitive relationship with the antenna's physical characteristics, depending upon the antenna configuration. However, the effective area scales with the square of linear dimensions for fixed antenna proportions. The term "effective area" is used because some types of antennas have gain that is directly proportional to their surface area, such as the parabolic reflectors used for applications like satellite TV. These antennas collect signal energy the way a pan can be used to collect rainwater. In the same way that a bigger pan collects more rain, a bigger antenna also collects more signal energy—meaning that it has more gain. The relationship between antenna gain and effective area is given by

$$G = \frac{4\pi \times A_{\text{eff}}}{\lambda^2} \qquad (12.37)$$

The relationship between the actual aperture area A_A and the effective aperture area is

$$A_{\text{eff}} = \eta_{\text{ap}} \cdot A_A \qquad (12.38)$$

where η_{ap} is the aperture efficiency. The antenna gain can now be put in terms of the actual aperture area.

$$G = \frac{4\pi \cdot \eta_{\text{ap}} \cdot A_A}{\lambda^2} \qquad (12.39)$$

From these relationships it can be shown that the wavelength plays a significant role in antenna gain. As the frequency increases, the wavelength decreases. A single antenna at 10 GHz will have 100 times the gain it has when operating at 1 GHz. The beamwidth is proportional to the square root of the gain, and so the higher frequency will have a beamwidth 0.316

that of the lower frequency. This holds for parabolic reflectors and many other antenna types. However, there are other types of tuned antennas that have gain in some narrowband of frequencies and then very little gain outside of that range.

The antenna beam pattern will have sidelobes in addition to the main lobe. These sidelobes can contribute to multipath by receiving energy reflected by terrain in a direction away from the main beam. When the antenna is used to transmit, energy is emitted in the direction of the side-lobes that can interfere with other communication systems. Sidelobes can make the system vulnerable to jamming from signal sources that are not in the direction of the mainlobe.

Nonpointing antennas can be fixed to the ground or a moving vehicle in a constant position and orientation relative to the platform. Such antennas must have a field of regard with sufficient gain to communicate line of sight with the other communications elements.

A *directional* antenna has gain that is highest in one direction. Pointing the high gain towards another antenna in the communication system must offer sufficient performance improvements over fixed antennas to justify the extra complexity of incorporating the actuation mechanisms. The direc-tional antennas must be pointed with sufficient accuracy and responsive-ness to ensure link closure. Pointing algorithms can be based on location information if this is available, but some antennas point based on received signal strength. Because beamwidth decreases with increasing gain, the higher the gain, the more strenuous the pointing demands become. Also, high-gain antennas have difficulty in initial acquisition or in reestablishing the link when using signal strength pointing techniques.

Many pointing antennas are mechanically steered using a gimbal, but another form of pointing antenna is an electronically steered phased-array antenna. Multiple antenna elements transmit or receive signals with differ-ent phase offsets. The phase offsets lead to signal cancellation in some directions and reinforcement of the signal in others. By adjusting the phases, the direction of maximum reinforcement (which is the direction of maximum gain) can be varied. One or two electronic steering axes can be accommodated by the array. The electronic steering can be sup-plemented by mechanical steering. A key advantage of phased arrays is the potential for reduced frontal area relative to dish antennas of equivalent gain. However, this may come at the expense of higher cost and higher power consumption.

Aircraft directional antennas have much in common with gimbaled pay-loads such as EO/IR balls used for surveillance. Both require adjustments in elevation and azimuth relative to the platform. The field-of-regard require-ments can be different. Line-of-sight antennas must point close to the horizon, whereas payload operations typically occur closer to nadir where the sensor is within range. Satellite antennas point upwards, with a communications field of regard almost a mirror image of the payload field of regard.

A half-wave dipole antenna has a gain of approximately 2.15 dBi at right angles to the antenna, with nulls along the length of antenna. This creates a broad "doughnut" pattern, with nulls along the antenna axis. Oriented vertically, this type of antenna provides a pattern that is omnidirectional in azimuth and is well suited for a fixed, vertically polarized airborne antenna. The gain is highest towards the horizon, which is the most stressing range case for the link budget analysis. The gain drops off as the unmanned aircraft flies over the ground data terminal, but the range is greatly reduced and less gain is required to close the link. A quarter-wave dipole is an adaptation of the half-wave dipole that requires a ground plane under its base but is better suited to mounting on the fuselage of an aircraft because it is half the height of the dipole and the feed attachment point is at the base rather than the middle of the antenna. Good antenna designers often make minor adjustments to the theoretical design to eliminate the zero-gain null that would be directly underneath the aircraft if an unmodified antenna was used. Without this modification an aircraft is unable to communicate to a ground-based receiver when it is flying directly overhead.

Parabolic reflector (dish) antennas are the most common form of directional antenna for satellite communications systems today. A parabolic reflector has the property of reflecting incoming RF energy in its beamwidth arriving at any point on the dish surface to a single focal point where a feed is mounted that collects the energy (and similarly when used for transmission, it reflects energy emitted by the feed in the direction of the antenna beam). Dish antennas have beamwidth that is a function of the wavelength and the diameter. The dish antenna gain for diameter D is

$$G = \eta_{\text{antenna}} \left(\frac{\pi \cdot D}{\lambda} \right)^2 \qquad (12.40)$$

where η_{antenna} is the antenna efficiency, which in a good design is typically 70%. The dish antenna beamwidth θ in degrees is

$$\theta = 70 \cdot \frac{\lambda}{D} \qquad (12.41)$$

Dish antennas are usually actuated in azimuth as the outer stage and elevation for the inner stage for both unmanned aircraft and ground-based antennas.

The gain of a parabolic reflector increases as the square of the frequency, unlike the gain of some other antennas like a half- or quarter-wave dipole. This leads to some important conclusions about how range varies with frequency based on the kind of antenna used:

* *Dipole or similar antenna at both ends of the link*—range decreases linearly with increasing RF frequency.
* *Dipole or similar antenna at one end of the link, parabolic reflector at the other*—range does not vary with frequency.
* *Parabolic reflector at both ends of the link*—range increases linearly with frequency.

12.6 Antenna Integration

Antenna integration must reach a suitable compromise between aerodynamic impacts, antenna coverage, and proximity with interdependent communication system components. The antenna type and required field of regard influence the integration approach. Incorporating the communication system should be a configuration driver at the earliest stages of conceptual design because later adaptation attempts for this critical system can be challenging for an ill-suited unmanned aircraft design. See Blake [2] for further reading on antennas.

Unmanned aircraft typically bristle with externally mounted antennas, creating a significant contribution to the unmanned aircraft drag. To illustrate the problem, consider an antenna consisting of a long circular cylinder of diameter d and length l. The drag coefficient of a circular cylinder is approximately 1.2 when the Reynolds number is less than 3×10^5. Here the reference area for the drag coefficient is the diameter and length, and the reference length for the Reynolds number is the diameter. The drag of this antenna D_{Antenna} is

$$D_{\text{Antenna}} = 1.2 \cdot \left(1/2 \cdot \rho \cdot V^2\right) \cdot l \cdot d \qquad (12.42)$$

Now consider the same antenna fully contained within a symmetric airfoil shape. Assuming that skin has negligible thickness, the antenna diameter is equal to the airfoil total thickness, and so the chord c_{Radome} is equal to

$$c_{\text{Radome}} = \frac{d}{t/c} \qquad (12.43)$$

where t/c is the airfoil thickness-to-chord ratio. The drag of the embedded antenna is equal to

$$D_{\text{Antenna}} = C_D(t/c, Re) \cdot \left(1/2 \cdot \rho \cdot V^2\right) \frac{l \cdot d}{t/c} \qquad (12.44)$$

Here the drag coefficient C_D of an airfoil family will be a function of the thickness-to-chord ratio and the Reynolds number. However, at Reynolds numbers between $10^4 - 10^5$ and for thickness-to-chord ratios of 15%, the airfoil drag coefficient will be approximately 0.01. It can be seen that the airfoil-shaped fairing reduces the antenna drag by an order of magnitude. The best thickness-to-chord ratio is a function of Reynolds number and fairing geometry.

Now that the benefit of aerodynamic shaping is established, other options for antenna drag reduction can be described. Some antennas are aerodynamically shaped blades, where the outside conductive surface is also the aerodynamic surface. These can be shaped as airfoils, thin ovals that approximate airfoils, or flattened diamonds. Blade antennas are simpler than antennas with fairings because the fairing and associated fittings are eliminated.

Perhaps the most synergistic approach is to place antennas inside aerodynamic surfaces that are required for other purposes, thus eliminating a dedicated drag-producing external antenna. Most omnidirectional antennas used in UAS applications have vertical polarization, which requires that the antenna be oriented vertically. Vertical aerodynamic surfaces such as vertical stabilizers and winglets are often suitable antenna housing locations, provided the airframe blockage is sufficiently low. Figure 12.6 shows winglets that house antennas on the Insitu ScanEagle and Raytheon Killer Bee unmanned aircraft. These vertical aerodynamic surfaces serve as radomes if constructed of dielectric materials. Materials such as E-Glass have less desirable structural properties than carbon fiber, and so there is often a weight penalty for load-bearing structures. Vertically polarized antennas do suffer large polarization losses if integrated into V-tails.

Low-frequency communications systems (hf, vhf, uhf) can have large antenna lengths. For example, a half-wavelength 100-MHz antenna is 57.6 in. long. This can drive the size of aerodynamic fairings such as winglets or vertical tails over more traditional design drivers such as drag

a)   b)

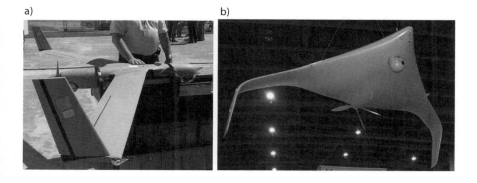

Fig. 12.6 These unmanned aircraft integrate antennas into the winglets for a low-drag installation: a) Insitu ScanEagle and b) Raytheon Killer Bee.

reduction or stability and control. A full wavelength 16-GHz Ku-band antenna, on the other hand, is less than 1 in. tall.

For sophisticated UAS development programs, integrated antenna-airframe modeling uses highly detailed finite element computational electromagnetic analysis codes. Examples of these codes include Computer Simulation Technology's MICROSTRIPES^{TM} and Ansoft's HFSS^{TM}. Like CFD, electromagnetic finite element codes discretize the three-dimensional geometry of the aircraft. The geometry can either be generated via inputs to the code's user interface or imported from CAD. Either the surface or the volume can be meshed, depending on the modeling approach. The grid size depends upon the frequency, and so high-frequency modeling of a very large aircraft can be computationally intensive. The grid generation for the aircraft can be restricted by the available processors. These powerful tools can aid in performance prediction and antenna integration design trades but are not a substitute for testing.

Many small UAS programs do not perform airframe-antenna modeling and go directly to testing. Small budgets and rapid development schedules often do not permit a time-consuming and expensive integrated antenna modeling effort. Small UASs are frequently developed by emerging companies without such modeling capabilities. Any link problems encountered during testing are fixed through antenna placement and hardware iterations. Vexing communications problems might not be satisfactorily solved without the proper analysis tools and testing.

Gain patterns for antennas integrated onto aircraft are not intuitive, especially when the wavelengths are of the same order as major airframe component dimensions. Electromagnetic waves travel along and interact with the airframe. High-frequency tends to be more optical in behavior where direct physical blockage results in reduced gain. Although analysis of low-frequency antenna integrated performance is challenging without high-fidelity tools, these antennas can often have adequate performance even in the presence of line-of-sight airframe blockage.

In-flight maneuvers impact communications. Banking maneuvers create polarization losses as described in Sec. 12.4.5.8. Another important factor is airframe interference. For example, an antenna mounted on the lower fuselage might have interference from the lower wing in the bank during portions of the turn.

Communications system testing is usually performed in a graduated series of tests. Early testing can take place using outdoor ground testing or indoor anechoic chamber testing. The former can be performed more responsively if the communications system is tested in the vicinity of the prototype build area. Also, outdoor testing is less expensive and can accommodate larger unmanned aircraft sizes than might be possible in many indoor facilities. Anechoic chambers offer a controlled environment for testing where interference from other RF sources can be largely eliminated. Finally, flight testing verifies the communications link performance in the most relevant environment.

Aircraft antenna placement depends on many factors that impact integrated antenna performance. The ultimate objective is for the antenna to provide adequate gain at all specified angular orientations of the unmanned aircraft and other communications nodes. Although clear optical line of sight between antennas will generally support good communications, the opposite is not always true. In other words, line-of-sight blockage between antennas will not necessarily dictate significantly reduced gain. The behavior depends upon frequency, antenna configuration, aircraft geometry, and aircraft orientation. No simple rules of thumb provide adequate guidance for the wide range of unmanned aircraft configurations.

Now several examples of antenna integration are provided to demonstrate options:

* Booms on twin-boom aircraft offer good blade antenna locations due to limited blockage on the sides. Variants of the IAI Heron can be seen with numerous vertical blade antennas on the booms. The Northrop Grumman FireScout also has antennas on the single boom (Fig. 12.7).
* Antennas can be physically separated from the aircraft to minimize blockage. The Northrop Grumman Hunter has an antenna separated from the upper fuselage by a faired post (Fig. 12.8).

Fig. 12.7 Northrop Grumman FireScout has a blade antenna located on the upper forward fuselage.

Fig. 12.8 Northrop Grumman Hunter has an antenna vertically offset from the fuselage.

* The lower fuselage is a suitable location for antennas that support air-to-ground communications. This position can have low blockage for high-wing aircraft. Fixed landing gear can pose a challenge.
* The upper fuselage ahead or behind the wing provides good line of sight at long range for command and control. This location might not be well suited for communications with ground elements directly below the aircraft. The FireScout has a large blade antenna on the upper forward fuselage, as shown in Fig. 12.7.

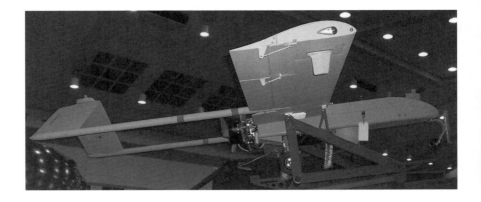

Fig. 12.9 IAI Shadow 200 has an antenna on top of the tail and under the wing.

Fig. 12.10   IAI Heron TP. (Photo: U.S. Army photo by Jose Ruiz/Released.)

* The top of vertical tails or V-tail surfaces provides good horizontal field of regard. Shadow 200 has an antenna positioned on the top of the inverted V-tail surface.
* The lower wing surface provides good field of regard both to the ground and horizontally when the aircraft is not banked. Figure 12.9 shows Shadow 200 blade antennas located on the lower wing surface.
* Antennas can be integrated into winglets. The Insitu ScanEagle and Raytheon Killer bee incorporate antennas into the winglets (Fig. 12.6).
* The IAI Heron TP can be fitted with a number of blade antennas on the tail booms and wing tips (Fig. 12.10).

12.7   Communication System Types

12.7.1   Command and Control

Unmanned aircraft command and control (C2) data links provide the operator with the ability to direct the flight and understand the unmanned aircraft state. The uplink is used for commanding the unmanned aircraft, and the downlink is used for ground receipt of health and status information.

This link is flight critical when the unmanned aircraft is incapable of landing without positive control. Securing the C2 link against jamming and unauthorized use is critical for military operations, and so antijamming and encryption capabilities are recommended. Unintentional RF interference is also a risk. A backup C2 link provides redundancy in case the primary C2 link becomes inoperable for any reason. Ideally the backup and primary C2 links use widely separated frequencies and perhaps different bands to provide frequency diversity.

C2 is generally low bandwidth, requiring only tens or low hundreds of kilobits per second ($\sim 50-200$ kbps). The low-bandwidth requirement is compatible with frequencies in the vhf and uhf spectra for line-of-sight applications. Low-bandwidth, low-Earth-orbit (LEO) satellite constellations such as Iridium might also provide suitable beyond line-of-sight C2 capabilities.

Communications hand-off between two ground control stations is common operation. One GCS might be responsible for launch and recovery operations, and the other conducts mission operations. The launch and recovery GCS has line-of-sight C2 communications with the unmanned aircraft, but the mission GCS might be either line of sight or use SATCOM. In either scenario the primary C2 must be handed from one GCS to the other. The safest approach is to have a *make-before-break* handoff, where both GCSs have positive C2 RF communications established before the first GCS relinquishes command authority. Make before break requires two C2 links. Another approach is *break before make* where the first GCS shuts down the C2 link, and then the second GCS established C2 with the unmanned aircraft. The risk of break before make is that the unmanned aircraft is not positively controlled at least momentarily.

Although the uplink and downlink sometimes use different radios, most systems use a full duplex C2 link. The C2 downlink is sometimes combined with digital payload return link. Common data link (CDL) radios are one example. Here the C2 downlink and payload data streams are multiplexed together. This approach combines mission critical and flight critical systems.

Interoperability with common ground control stations is a long-standing goal. In 2005, the Office of Secretary of Defense issued a memorandum mandating that all UASs except for small UASs would adopt the CDL standard to promote interoperability. The small UAS exemption was allowed due to their small size. Requiring all UASs to operate on the same radios and frequencies increases the problem of frequency congestion. On the other hand, disparate incompatible radios require numerous GCS types.

12.7.2 Payload Link

The payload link transmits payload data from the unmanned aircraft to a ground-receiving element. This data link is usually mission critical but not flight critical. Missions such as ISR depend upon receipt of payload data on the ground for mission success. The payload link can become flight critical if it is combined with the C2 downlink or if it is used to provide flight critical information such as a video feed from a pilot visibility camera. Digital or analog links can be employed, though digital links are needed for most types of payload data. Data rates tend to be much higher than for C2 links.

Analog links are used almost exclusively for sending near real-time analog video, usually in a NTSC format. Video quality degrades gracefully as the signal to noise is reduced below the full quality threshold. Vhf and uhf radios provide sufficient bandwidth for video transmission, and these low frequencies are robust against atmospheric absorption and rain. Many micro, small, and small tactical UASs use analog payload return links for video, though miniaturized digital data links are becoming available. The primary reasons that analog radios are used on small unmanned aircraft is low SWaP, low cost, ease of integration, and availability of unlicensed ISM bands. A major drawback for analog links is that they are not easily encrypted. Analog video standards, such as NTSC, have audio side channels that can be used to send down other payload data such as imagery metadata.

Digital payload links offer flexibility for various types of payload data. Many radios can multiplex multiple data streams together on the unmanned aircraft and then demultiplex them at the ground data terminal. Some payload links have the ability to convert analog video streams to digital video before transmission, often with the ability compress the video. Other data inputs can come from Ethernet or serial interfaces. If the data input to the radio exceeds the available data rate, the radio should have logic to handle the conflict. The unmanned aircraft software should also perform a mission management function to deconflict the data going to the radio.

Data rates for digital payload links can range from 1 Mbps for a highly compressed H.264 video stream to 50 Mbps for higher quality digital video or multiple payload streams. The low end of the data rate can be satisfied at L or S band, but higher data rates require data links operating at C, X, or Ku band. The appetite for bandwidth can be almost insatiable. Consider a hyperspectral sensor that images at hundred of bands simultaneously with dense focal plane arrays. Another example is a cluster of high-definition sensors with overlapping fields of view that is intended to provide high-resolution coverage of an entire city. Such operations could require multigigabit-per-second data rates that might not be supportable by traditional RF communications. Chapter 15 will describe methods of managing data collection rates that exceed the payload downlink capacity.

12.7.3 Air-to-Air Communications

UASs can operate cooperatively with other airborne platforms, necessitating air-to-air communications among the platforms. Having heterogeneous airborne platforms can enable many mission types, such as a hunter-killer operation where the UAS provides persistent targeting to a fast manned attack aircraft. Another common air-to-air communications application is UAS range extension via communications relay.

Combined airborne operations with other aircraft involve use of a common communications system type. Allocating frequencies for point-to-point links can be impractical and limiting, and so networked radios are desirable. Networked radios can share information with other radios on the network, provided that the nodes have the appropriate permissions. The networks should be secure and robust against jamming for military operations. Examples of operational and potential future U.S. airborne data links include Link 16, TTNT, and QNT.

Airborne communications relay can extend the effective communications range of an unmanned aircraft and support operations in mountainous terrain. The relay aircraft can be any type of airborne platform, including tethered aerostats, manned aircraft, or a UAS. To maximize range, the relay aircraft should operate at high altitude. A steerable high-gain antenna pointing at the GCS can further extend the range, but at the expense of complexity and weight. If the objective of the relay is to permit the collection unmanned aircraft to operate at low altitude in a mountainous region, then low-gain antennas can be used on both the relay and collection aircraft for communications between these platforms. The range between the airborne platforms will be on the order of $5-10$ n miles. If range extension is the objective, then the relay unmanned aircraft can use a steerable high-gain antenna that points towards the collection aircraft. The relay aircraft must know the position of the collection aircraft in order to maintain the link.

12.7.4 Voice Relay

A UAS operating in the same airspace as manned aircraft is often required to behave as if the unmanned aircraft is manned. Manned aircraft can have voice communications with the air traffic control (ATC) system, where the pilot and the control tower are able to dialog. These radio transmissions can also be heard by other aircraft, providing situational awareness.

Two methods currently exist for satisfying operator-ATC communications. First, voice communications between the unmanned aircraft operator in the ground control station and the ATC can be relayed via the unmanned aircraft. This requires a retransmission of the voice communications in both directions using the unmanned aircraft systems. The second approach is for the ground control station to have direct communications with ATC through ground control station based radios, though this has range limitations because of terrain blockage and the Earth's curvature.

ARC-210 radios are often used for military voice communications between the unmanned aircraft and ATC.

Long-range UASs controlled through SATCOM can relay ATC voice communications through the satellite link. The unmanned aircraft relays the communications via the ATC voice link. If multiple satellite hops are required, then the latency can cause effective communication difficulties.

12.7.5 Transponders

Transponders transmit data on the unmanned aircraft's position when interrogated to help avoid air-to-air collisions and support airspace management. Examples of transponders include Mode 3/A, Mode C, Mode S, and TCAS. The functionality of these transponders is covered in Chapter 10. Civilian transponders operate in vhf.

12.8 Modulation Techniques

The role of modulation is to embed a modulated signal onto a carrier signal. The carrier signal is a high-frequency sinusoidal waveform. The modulation function is performed by the modem in digital systems. The techniques have inherent properties for data rate per bandwidth and robustness to interference. Some common modulation techniques are identified in Table 12.4.

The relationship between required Eb/No and BER for a binary phase shift keying (BPSK) and quadrature phase shift keying (QPSK) modulated signals is shown in Eq. (12.45). The function $erfc(\ )$ is the complimentary error function. Note that the Eb/No is the true value and not the decibel value.

$$BER = 1/2 \cdot erfc\left(\sqrt{Eb/No}\right) \tag{12.45}$$

This can be expressed in terms of the decibel value of Eb/No.

$$BER = 1/2 \cdot erfc\left(\sqrt{10^{Eb/No(\text{dB})/10}}\right) \tag{12.46}$$

Table 12.4 Example Digital Modulation Techniques

Modulation technique	Method
Amplitude shift keying (ASK)	Change amplitude with each symbol
Frequency shift keying (FSK)	Change frequency with each symbol
Phase shift keying (PSK)	Change phase with each symbol

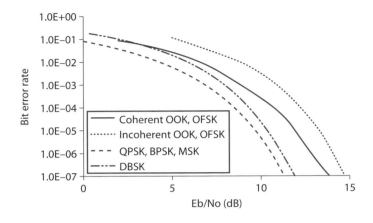

Fig. 12.11 Bit error rates for various forms of modulation.

The differential binary phase shift keying (DBPSK) modulation relationship between Eb/No and BER is given by

$$BER = 1/2 \cdot e^{-Eb/No} \qquad (12.47)$$

The BER as a function of Eb/No for several modulation techniques is presented in Fig. 12.11.

Calculating these relationships for other modulation techniques is often more complex; it is not presented here. The required Eb/No should be based on data from the communication system vendor.

12.9 Interception, Detection, and Jamming

Military communications systems must be secure and robust for effective operations. Having the RF signal detected or intercepted by an adversary is highly undesirable. Enemy detection of the signal can compromise an operation by making the adversary aware that an operation is underway. Interception of the signal can lead to an adversary's exploitation of the data, which can even include watching real-time video and countering the operation. Military data links often employ encryption, low probability of detection (LPD), and low probability of intercept (LPI) capabilities.

Furthermore, the opponent can use jamming to deny communications, necessitating jam-resistant data links. Jam-resistant approaches include spread spectrum and frequency hopping. Spread spectrum generates pseudonoise over a large frequency range to mask the signal, forcing the jammer to spread its power over the same range instead of focusing on the primary signal frequency. In frequency hopping, the primary frequency hops among frequencies in a pseudorandom manner that is difficult for a jammer to match.

12.10 RF Performance Simulation

The link budget method described in Sec. 12.4 is useful for conceptual link sizing, but only represents a single link condition. Usually the link budget is performed at what is expected to be the most stressing state. However, the link must operate under a variety of ranges, relative orientations, and atmospheric conditions. More detailed analysis should consider link performance under multiple conditions and with complete models of the integrated antennas, system dynamics, and environment.

The antenna patterns as integrated on the unmanned aircraft can be provided from antenna analysis tools and electromagnetic analysis codes described in Sec. 12.6. The data set includes gain as a function of azimuth and elevation angle in unmanned aircraft body polar coordinates.

The unmanned aircraft flight behavior is modeled over a flight trajectory. The simulation can involve one or more unmanned aircraft flight paths, including airspeed, altitude, and bank angles along a trajectory. Circling a target at maximum range is often a stressing case. The position of the ground stations and communications satellites are included in the modeling so that the antenna relative geometry can be calculated.

The attenuation models will be able to vary relevant conditions such as relative humidity and rain. These models include dependencies on geometry between link nodes. The atmospheric temperature is often modeled instead of assuming a nominal 290 K.

Detailed terrain models can be used based on a geospatial information system (GIS) data. Digital terrain and elevation data (DTED) is one prevalent format. Terrain data are important for calculating line-of-sight obstructions and multipath interference.

Every aspect of the data link that can affect link performance must be modeled. The modulation and coding behavior is included to estimate BERs. Other link characteristics can be considered such as spread spectrum and frequency hopping. Jammers can also be modeled, including the jammer type and location. Antenna pointing errors, polarization losses, line losses, and all other relevant drivers can be modeled.

The integrated simulation environment evaluates the link performance throughout the unmanned aircraft flight path. Major parameters such as link margin and other link element performance are included in software-generated reports. This detailed modeling can lead to changes in antenna location on the unmanned aircraft, amplifier selection, antenna type, or data rates.

12.11 Line-of-Sight Communications

The distance in which the ground antenna has direct communications to the aircraft with a smooth Earth's surface is governed by what is known as the *four-thirds Earth model*. The name comes from a key assumption that the RF effective radius of Earth is equal to 4/3 of the actual radius based on RF transmission characteristics. The line-of-sight range relationship comes directly from trigonometric relationships where the slant range segment is tangent to the Earth. The derivation is left to the reader as an exercise. Assuming an antenna on the surface, the line of sight slant range distance R in nautical miles for an aircraft at altitude h in feet is

$$R = 1.23 \cdot \sqrt{h} \qquad (12.48)$$

If the antenna is elevated above the ground by a height H in feet as shown in Fig. 12.12, the total line-of-sight distance becomes

$$R = 1.23 \cdot \left(\sqrt{H} + \sqrt{h} \right) \qquad (12.49)$$

At all but very close range operations, directional tracking antennas are used. As the gain increases for better link performance at long range, the pointing accuracy requirements become more demanding. A crude approach to tracking is for a human operator to physically point the antenna by hand based on knowledge of the unmanned aircraft's location relative to the ground station. This approach requires a dedicated person, and tracking performance might not be satisfactory in operational conditions.

More sophisticated systems have automatic tracking systems. A simple feedback mechanism is tracking peak RF strength, where the tracking system uses direction finding (DF) to the unmanned aircraft transmitter. More integrated tracking systems use knowledge of the unmanned aircraft position supplied by the ground station. The ground control station requires the RF link to determine the actual position of the aircraft, and so reestablishing a lost link necessitates prediction of the unmanned

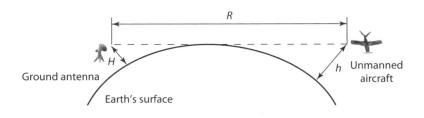

Fig. 12.12   Line-of-sight geometry.

aircraft's autonomous behavior. Tracking antennas can sever the pointing dependencies on ground control station-provided information by parsing encoded messages within the RF modulated data stream that contains unmanned aircraft position data.

12.12 Beyond Line-of-Sight Communications

Beyond line-of-sight (BLOS) communications occurs when the first transmitter and receiver are not able to communicate with direct line of sight and must use relay nodes to retransmit the signal. The lack of line-of-sight connectivity is usually due to physical blockage caused by terrain or the Earth's curvature. Excluding ionosphere bouncing at low frequencies, BLOS requires one or more relay nodes to complete the link. Each relay node receives a signal and then retransmits it to the next element in the chain. See Piscane and Moore [3] for further reading on satellite communications.

Relay nodes are typically other airborne platforms or satellites. Airborne platforms can be other UASs, airships, tethered balloons, or fixed-wing manned aircraft. An example of an airborne relay is the Northrop Grumman RQ-5A Hunter I system, which served as a relay to other Hunter platforms in the late 1990s.

Satellite Communications (*SATCOM*) is a more prevalent means of providing BLOS communications for UASs. Satellites offer the advantage of relaying information over very large distances, even globally. This has enabled the *remote split operations* model where the unmanned aircraft is launched and recovered in theatre with a LOS launch and recovery element (LRE) ground station. The mission itself is performed by a mission control element (MCE) that can be in a different continent. The Northrop Grumman RQ-4A/B Global Hawk, General Atomics MQ-1 Predator A, and General Atomics MQ-9 Reaper can operate in this fashion. The SATCOM communications architecture is shown in Fig. 12.13.

Small UASs can leverage low data rate SATCOM data links with small fixed antennas. These are suitable for C2 or health and status purposes. Nonpointing antennas greatly simplify system integration because the unmanned aircraft does not have an extra actuated system, and knowledge of the satellites is not required.

Larger UASs such as MALE, HALE, and UCAS categories incorporate higher-bandwidth SATCOM links. These are capable of not only C2, but also high-rate payload data rate return links. These communications links require high-gain antennas that can be accurately steered towards the satellite. Mechanically steerable parabolic dish antennas are the most common

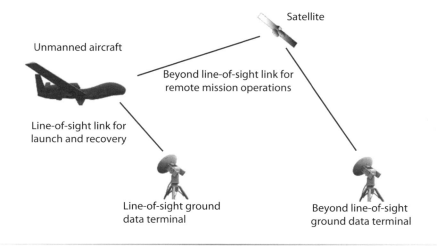

Fig. 12.13 Remote split operations communications architecture.

approach today, though electronically steered phased-array antennas are maturing. It is necessary to prevent sending strong RF to other satellites, known as adjacent satellite interference (ASI). Geosynchronous satellites can be separated by as little as 2 deg of longitude in some circumstances. To prevent ASI, the pointing accuracy and beamwidth can affect the antenna and its gimbal assembly. In fact, ensuring a narrow beamwidth to prevent ASI can be a larger driver for SATCOM dish diameter than gain requirements for closing the link budget.

Satellites capable of high-bandwidth communications usually have geosynchronous orbits. These orbits are positioned at a fixed longitude over the equator. Only limited latitude oscillations occur for properly functioning satellites. There are only a limited number of orbit slots available when sufficient longitude separation is maintained. Geosynchronous SATCOM satellites have a number of transponders, which are high-gain antennas that provide ground coverage over a designated spot on the Earth's surface. The resulting pattern is called a *spot beam*. The higher the transponder gain, the smaller the ground coverage. The transponders are usually steerable, and so the geographic coverage selection is driven by the customer demand. SATCOM satellites can be owned either by governments or by companies.

The geometry of geosynchronous orbits dictates that very high and low latitudes have limited or no coverage. These satellites orbit above the surface at an altitude of 35,800 km. The orbital radius of the satellite relative to Earth's center R_{Sat} is 42,178 km because the idealized Earth's radius R_{Earth} is 6,378 km. If the unmanned aircraft is capable of communicating with the satellite with the dish at zero elevation angle relative to local horizontal and it is near sea level, then the maximum latitudes where it can communicate with the satellite is 81.3° North or South. The elevation angle between the local horizon and the satellite ε_{Sat} is driven by atmospheric limitations, unmanned aircraft antenna integration, and unmanned aircraft maneuvering. The satellite coverage geometry is shown in Fig. 12.14. The maximum latitude Γ_{Max} where the unmanned aircraft can have continuous communications with the geosynchronous satellite located at the same longitude is

$$\Gamma_{Max} = \cos^{-1}\left[\frac{R_{Earth} \cdot \cos(\varepsilon_{Sat})}{R_{Earth} + h_{Sat}}\right] - \varepsilon_{Sat} \qquad (12.50)$$

The communication system can have a threshold elevation angle ε_{Min} below which the atmospheric and precipitation absorption losses are excessive. This sets a floor for the satellite elevation angle.

$$\varepsilon_{Sat} \geq \varepsilon_{Min} \qquad (12.51)$$

The unmanned aircraft will have a minimum elevation angle as a function of azimuth that is a result of the antenna integration. The unmanned aircraft will also perform maneuvers such as banking, climbing, and descending. The unmanned aircraft antenna elevation angle θ_{AV} is the sum of the installed antenna elevation angle $\theta_{Installed}$ and maneuvering angles $\theta_{Maneuver}$.

$$\theta_{AV}(\psi) = \theta_{Installed}(\psi) + \theta_{Maneuver}(\psi) \qquad (12.52)$$

If the worst-case value of θ_{AV} exceeds the minimum satellite elevation angle ε_{Min}, then the satellite elevation angle becomes the greatest value of θ_{AV}.

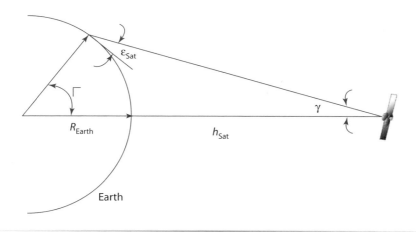

Fig. 12.14 Geosynchronous satellite geometry.

Example 12.1 Extreme Southern-Latitude SATCOM Coverage

Problem:

An unmanned aircraft using a high-bandwidth satellite link is selected to monitor the penguin population in Antarctica.

* Relative to the aircraft horizontal in its body-axis system, the elevation angle for the dish antenna is
  * 0 deg at the nose
  * 5 deg on the sides
  * Degree looking aft
* The aircraft is expected to periodically perform benign banking maneuvers of only 15 deg in level flight.
* The minimum satellite elevation angle is 6 deg.

How far south can the unmanned aircraft observe the wildlife?

Solution:

The elevation angle θ_{AV} is evaluated at several azimuth angles. At the nose this angle is

$$\theta_{AV}(0\,\text{deg}) = \theta_{\text{Installed}}(0\,\text{deg}) + \theta_{\text{Maneuver}}(0\,\text{deg}) = 0\,\text{deg} + 0\,\text{deg} = 0\,\text{deg}$$

where the maneuvering angle is zero at the nose because bank does not change the elation angle. On the side the worst-case maneuvering angle is 15 deg when the unmanned aircraft banks away from the satellite.

$$\theta_{AV}(90\,\text{deg}) = 5\,\text{deg} + 15\,\text{deg} = 20\,\text{deg}$$

And at the tail (or aft) the angle is

$$\theta_{AV}(180\,\text{deg}) = 15\,\text{deg} + 0\,\text{deg} = 15\,\text{deg}$$

The worst-case unmanned aircraft elevation angle is 20 deg, which is greater than the minimum satellite elevation angle of 5 deg. The satellite angle ε_{Sat} is therefore set to 20 deg. The elevation angle due to banking only affects the sides and not the fore and aft of the unmanned aircraft. Using the following equation,

$$\Gamma_{\text{Max}} = \cos^{-1}\left[\frac{R_{\text{Earth}} \cdot \cos(\varepsilon_{\text{Sat}})}{R_{\text{Earth}} + h_{\text{Sat}}}\right] - \varepsilon_{\text{Sat}}$$

$$\Gamma_{\text{Max}} = \cos^{-1}\left[\frac{6{,}378\,\text{km} \cdot \cos(20\,\text{deg})}{42{,}178\,\text{km}}\right] - 20\,\text{deg} = 61.8\,\text{deg}$$

The continuous coverage maximum southern latitude is 61.8°. Unfortunately, there are no penguins at this latitude.

Now intermittent communications is considered. The satellite elevation angle is driven by the minimum elevation angle of the satellite, which is 6-deg elevation. The maximum southern latitude for intermittent communications is therefore

$$\Gamma_{\text{Max}} = \cos^{-1}\left[\frac{6,378\,\text{km}\cdot\cos(6\,\text{deg})}{42,178\,\text{km}}\right] - 6\,\text{deg} = 75.4\,\text{deg}$$

Most of Antarctica's coast is north of 75° southern latitude, so that the mission is possible if temporary communications outages are permitted.

Here the maneuvering angle will generally be driven by maximum bank angle rather than climb or descent flight-path angle. If intermittent communications outages are permissible, then the minimum elevation angle selected might not be the most stressing value.

The atmospheric attenuation becomes more dominant as the low satellite elevation angle relative to the horizon means that the signal must pass through more atmosphere. High-bandwidth satellites tend to operate at Ku and Ka band where atmospheric and precipitation absorption losses are high. These losses are reduced significantly if the elevation angle is high.

Detailing the link budget requires obtaining significant proprietary technical data from the communications satellite provider. Large communications companies have established relationships and are well positioned to create the airborne link. UAS system integrators generally procure and integrate hardware from communication system companies rather than develop a custom SATCOM air data terminal.

Beyond line-of-sight communications can be a significant contributor to the system operations and maintenance costs. Cost rates depend largely on negotiations. The U.S. government will have more weight in negotiating low rates for widespread operations than a private contractor, especially for a limited-duration flight-test campaign involving a handful of aircraft. Many satellites have a fixed number of transponders that cover a fixed region on the surface. During normal times, a transponder generally points towards a highly populated region enjoying economic prosperity. In wartime, the military must pay to move the commercial satellite spot beams to the area of interest. Military-owned communications satellites are tasked by the military.

Low-bandwidth communications satellites generally have LEO. The specific satellites involved in the communications changes over time as the orbits bring them in and out of view of the unmanned aircraft. A single aircraft data terminal can have multiple satellites in view at one time with this class of system.

12.13 Frequency Management

Avoiding frequency conflicts is a major challenge for unmanned systems. The functionality of the unmanned aircraft depends on the interactions with the outside world. These necessary functions can involve RF communications with other system elements, RF communications to outside systems, or RF sensing or environmental information required for safe operations. Frequency conflicts can create loss of performance or worse—endangering human lives in other aircraft or on the ground.

Frequency usage depends on regulations and the operational environment. Frequency regulations are governed by international and regional law. The operational environment includes deconfliction with other RF emitters, avoidance of interference, and the actions of opposing forces in wartime. Constraints imposed by frequency management will ultimately preclude UASs from "darkening the skies."

Industrial, scientific, and medical (ISM) bands are unlicensed bands covered by International Telecommunications Union (ITU) ITU-R regulations. Some of these bands can be used without coordination in the United States; however, regional and international rules can vary. COTS communications products are readily available that leverage the ISM bands. Common useful frequencies for UAS applications include 433.05–434.79 MHz (uhf), 902–928 MHz (uhf), 2.4–2.5 GHz (S), and 5.725–5.875 GHz (C). Although these frequencies are easy to use because of the lack of restriction, this creates interference risks with other users on the same bands. For example, other users in S-band include cell phones, wireless LANS, and even microwave ovens. Transmit power is limited by the regulations, and so the link range can be limited.

Radio-controlled (R/C) aircraft can operate with several radio types. COTS R/C units generally use either 72 MHz with AM, FM, or PCM modulation. New systems are increasingly taking advantage of the ISM band with spread spectrum technology operating in S-band. Low-cost indoor models sometimes use IR data links. Even with the proliferation of capable low-cost autopilots, R/C is still widely used for early test flights of small UASs.

Use of licensed bands requires coordination with the frequency management organizations or the FCC. Unlicensed bands can still have limitations on radiated power. Nations have varied regulations on licensed frequencies.

Even when ISM bands and approved frequencies are selected, the frequency deconfliction problem is not over. The unmanned aircraft must contend with RF interference (RFI) from multiple RF systems on the aircraft. A fleet of common unmanned aircraft types using similar radios must not interfere with each other. Finally, the UAS must manage RF conflicts with other assets.

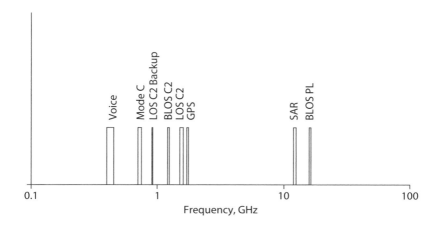

Fig. 12.15 Example frequency usage plot.

To manage RFI on the unmanned aircraft, the frequencies of all onboard RF systems must be characterized. The operating frequencies of avionics processors can also be vulnerable to RF. Often radios cause interference off of their primary frequency, known as *sideband interference*. Emitters and receivers that have frequencies close to one another should be given close scrutiny. Frequency usage plots, as shown in Fig. 12.15, are a convenient way to rapidly visualize frequencies that can cause interference. RF systems that transmit or receive RF energy can include the following:

• LOS primary C2 link
• LOS payload return link
• LOS secondary C2 link
• SATCOM C2 link
• SATCOM payload data return link
• GPS and DGPS (L1 frequency of 1575.42 MHz, and L2 at 1227.6 MHz)
• UHF or VHF voice relay
• Remote viewing terminal (RVT) link
• Transponders (mode 3A, C, S, or TCAS)
• Radar altimeter
• Communications relay payload
• Synthetic aperture radar payload
• Blue force tracking system
• Flight-test tracking and telemetry radio (flight test)
• Flight termination system radio (flight test)

If a UAS manufacturer is fortunate enough to have many UASs operating in the field, frequency congestion among those like-systems can arise.

This is sometimes referred to as the *darkening the skies* scenario. The ultimate limiting factor for the quantity of unmanned aircraft covering an area of interest might be the available channels. Communications systems can have channels that occupy a portion of the available operating frequency. The systems that operate in close proximity should not use the same channels in order to prevent interference. To complicate matters, some channels can experience interference from other channels. A channel may be known to have interference with outside sources regionally. Frequency management software can be used to assign frequencies to the various assets. This software will consider a channel or frequency compatibility matrix, permitted channel matrix, unmanned aircraft operating area, ground control station locations, and interference ranges. The channel assignments might require optimization algorithms to maximize the number of unmanned aircraft that can fly.

Many different UAS types, communications systems, and weapon systems will operate within a battlefield. The frequency management issues go well beyond channel deconfliction within a single UAS family. Those who have the full frequency picture can impose frequency constraints on a UAS operation.

References

[1] Crane, R. K., *Electromagnetic Wave Propagation Through Rain*, Wiley, New York, 1996.
[2] Blake, L. V., *Antennas*, Wiley, New York, 1966.
[3] Piscane, V. L., and Moore, R. C., *Fundamentals of Space Systems*, Oxford Univ. Press, New York, 1994.

Problems

12.1 A highly maneuverable UCAS is able to maintain high-bandwidth SATCOM with a geosynchronous satellite constellation. The UCAS is required to operate at up to 45° N and 45° S latitude. The installed antenna has a 10-deg elevation angle from horizontal in UA body axes for unobstructed antenna pointing. What is the maximum bank angle for continuous satellite communications?

12.2 What is the necessary antenna height required to maintain 100-n miles line-of-sight visibility with an unmanned aircraft with a minimum altitude of 5,000 ft above ground level?

12.3 The Earth's radius is 3,960 miles. In the "Four-thirds Earth model" the effective radius is 4/3 times the true radius. A ground-based antenna is at a negligible height above the ground. The slant range between the ground antenna and the unmanned aircraft is tangent to the Earth's surface. Derive the relationship between line-of-sight slant range in nautical miles and unmanned aircraft above ground altitude h in feet. Small terms can be neglected.

12.4 Convert 500 W to dBW.

12.5 What is the real number for 10 dB – 3 dB?

12.6 What is the precipitation loss for a 10-GHz transmitter at sea level in 25 mm/hr at 30 km communications range?

12.7 An unmanned aircraft and ground station both use vertically polarized antennas. If the unmanned aircraft is orbiting a target with a 30-deg bank angle at the same elevation as the ground station, what is the polarization loss?

12.8 A QPSK encoded data link requires a bit error rate of 10 -5. What is the required Eb/No?

12.9 A 20-in. dish antenna is used for 5-, 10-, and 15-GHz applications. Assume an antenna efficiency of 70%. What is the gain and beamwidth for each case?

Chapter 13

Physics of Remote Sensing and in Situ Measurement

- Identify useful bands of the electromagnetic spectrum
- Understand optical systems
- Learn the physics of radar and LIDAR

Fig. 13.1 NASA Global Hawk image of Tropical Storm Frank (Photo courtesy of NASA.)

13.1 Introduction

Many UAS missions require that the payload collects information about a target remotely. Such missions include reconnaissance, surveillance, and mapping, to name a few. The primary method of collecting this information is through the use of electromagnetic energy. Electromagnetic waves travel through physical space at the speed of light. In this chapter we will explore optical and radio-frequency (RF) sensors that can be used to generate imagery. This high-level coverage is intended to provide you with a basic understanding of the key principles and driving parameters. Further reading of remote-sensing texts is encouraged to gain more detailed knowledge of payload system design [1 – 3].

The UA provides a platform for the payloads to operate. This platform enables the sensor to be put in the desired collection conditions both spatially and temporally. Without payloads, the UA generally has no purpose. Even if the reader is mostly interested in UA design, an appreciation of payloads is essential to produce effective UAS solutions. A business executive once told the author in regards to a new UAS program internal proposal that "it is not about the glory of flight." The system must have utility to attract customers, and the payload systems provide that utility.

13.2 Electromagnetic Spectrum Characteristics

Unmanned aircraft use a wide range of the electromagnetic spectrum to gather intelligence. Cameras generally operate between the visible band and long-wave infrared. Radar imagers utilize uhf-band through Ku-band radio frequencies. As can be seen in Fig. 13.2, the atmosphere is opaque across much of the electromagnetic spectrum, and so sensors

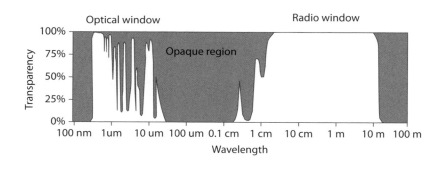

Fig. 13.2 Absorption bands.

capable of remote sensing must operate where the atmosphere is suitably transparent. This chapter covers how collection is accomplished in each of these regions.

Only a small portion of the electromagnetic spectrum is useful for remote sensing. The two primary transmission windows are the optical window ranging from $0.4 - 14$ μm and the radio window that ranges from 0.2-cm to 20-m wavelengths. Atmospheric absorption is excessive in many wavelength bands for remote-sensing purposes. The bulk of the absorption is attributed to water vapor (H_2O), which can vary substantially in the lower atmosphere depending on weather conditions. Other notable contributors include O_3, CO, CO_2, CH_4, N_2O, and O_2.

13.2.1 Visible Band (0.4–0.7 μm)

Visible band optics, as the name suggests, operates in the same spectral range as the human eye. Although monochromatic cameras exist, this band offers the ability for cameras to accurately depict the world in color. The color provides additional intelligence content by enabling color description of the targets. All other cameras described are either monochromatic or have false colors, as true colors are only defined in the visible band.

Color imagery is obtained by using multiple photodetectors that are tuned to different wavelengths within the visible spectra. These are generally red, green, and blue detectors, from which all visible colors can be derived. Color imagers are effectively multispectral sensors using only three wavelengths.

The primary limitation of color imaging systems is relatively poor night-vision performance. The target must be illuminated by visible light of some form, whether naturally by the sun or by artificial means such as a searchlight.

Color cameras have varying sensitivity to low light conditions as is experienced at dawn, dusk, or moonlight conditions. The *Lux* rating is the most commonly applied metric to define the sensitivity of the cameras to low light levels. A low Lux rating corresponds with the ability to operate with little illumination.

The electro-optic (EO) designation is frequently used interchangeably with "visible band." However, EO includes not only the visible band, but also near-infrared (NIR) and short-wave infrared (SWIR) bands.

13.2.2 Near Infrared (NIR) (0.7–1 μm)

Although the human eye cannot see in the NIR band, EO sensors often operate in this spectral band. True colors are not defined in NIR, and so these cameras are monochromatic. NIR sensors can often use the same type of glass lens and window materials as visible sensors.

13.2.3 Short-Wave Infrared (SWIR) (1-3 μm)

SWIR sensors typically operate between $1-1.7$ μm. Sensor systems in this band have good resolution for low light conditions without the need for cooled detectors. Low-cost conventional glass optics are used. SWIR sensors are not thermal imagers like MWIR and LWIR, and so they must detect reflected energy. SWIR illumination comes from hydroxyl ion emissions night glow in the atmosphere, which is $5-7$ times brighter than starlight in this band. Like NIR, SWIR is a monochromatic sensor. An advantage of SWIR over MWIR or LWIR sensors is that the optics tends to be more compact for an equivalent level of night-vision performance, assuming that the scene is properly illuminated. SWIR is not a thermal sensor for objects at normal scene temperatures, so that heat signatures against a background are not discernable.

13.2.4 Midwave Infrared (MWIR) (3-6 μm)

MWIR sensors typically operate between $3-5$ μm. These sensors detect energy that is both reflected and emitted. MWIR is particularly effective in maritime environments where the water content in the air reduces the effectiveness of LWIR. MWIR detectors can be smaller based on diffraction limitations, yielding smaller optics. Sapphire or other exotic materials are used for windows and lenses, which adds cost and manufacturing complexity to MWIR optical systems relative to EO. The MWIR detectors on the focal plane are usually cooled to low temperatures in order to achieve the necessary sensitivity.

13.2.5 Long-Wave Infrared (LWIR) (6-14 μm)

LWIR sensors typically operate between $8-14$ μ. The sensors primarily detect emitted energy from the target. The ability to detect heat signatures is particularly useful for identifying humans or running engines on unmanned aircraft. Many UAS LWIR cameras do not require cooled detector arrays, thereby reducing complexity and power consumption over MWIR alternatives. LWIR sensor must use exotic lens and window materials such as germanium and cannot image through conventional glass materials.

13.2.6 Radar Imagery Bands

Radars use the RF portion of the electromagnetic spectrum. The useful bands overlap with those used for RF communications systems because of the low atmospheric absorption. The spectral bands used by radars are described in terms of frequency instead of wavelength. Radars generally use vhf-Ka bands. The low-frequency end of the spectrum is mostly used for foliage penetration synthetic aperture radars (SARs), and X-Ka bands are used for search radars and high-resolution SARs.

13.3   Aerial Remote Sensing

Aerial remote sensing requires that the sensor field of view be placed on the target or scene of interest. Various methods of moving the sensor along the ground are driven by the remote-sensor payload configuration. Common methods of advancing the field of view are shown in Fig. 13.3.

The most intuitive type of remote sensing is the framing camera that images an entire scene at once. This is the format of familiar still imagery or full motion video. Here all of the elements of the focal plane array are active simultaneously or scanned at a very high rate such that the image appears to be taken at an instant. When the ground coverage of a framing camera advances to an adjacent area with little or no frame overlap, this is referred to as step-stare. The dwell time of framing cameras on a point on the ground is limited only by the time the scene is within the UA's field of regard rather than the forward motion of the UA.

Pushbroom collection involves a swath of resolution cells roughly orthogonal to the flight path projected on the ground that are advanced via the UA forward motion. This type of collection is common for synthetic aperture radars, multispectral sensors, hyperspectral sensors, and some imaging reconnaissance payloads. The dwell time is a function of the ground-resolution cell size and the forward motion of the UA.

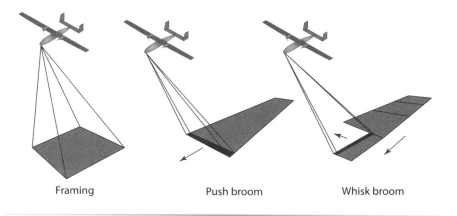

| Framing | Push broom | Whisk broom |

Fig. 13.3   Remote-sensing geometry definition.

Whiskbroom collection involves sweeping a swath of ground parallel to the flight path to form an image, and then advancing to the next image. This is sometimes used for multispectral and hyperspectral sensors. A gimbaled mirror can be used to sweep the field of view as the aircraft advances. The sensor dwell time is driven by both the forward motion of the UA and the sweep rate.

The general geometry of airborne collection is shown in Fig. 13.4. The UA is located at height h and ground range GR from the target. The look angle θ_{Look} is the angle between nadir and the line between the sensor and the target. The squint angle ψ_{Sq} is the azimuth angle between the UA's flight path and the target in the ground plane. The slant range between the UA and target SR is found by the following relationships:

$$R = \sqrt{h^2 + GR^2} \tag{13.1}$$

$$R = \frac{GR}{\sin(\theta_{Look})} \tag{13.2}$$

$$R = \frac{h}{\cos(\theta_{Look})} \tag{13.3}$$

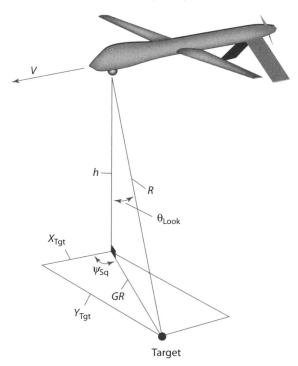

Fig. 13.4 Basic remote-sensing geometry.

The downrange distance X_{Tgt} is the distance between the UA and the target in the flight-path direction in the ground plane. The cross-range distance Y_{Tgt} is the distance between the UA and the target in the ground plane perpendicular to the flight path. Some useful ground-range relationships are

$$GR = \sqrt{X_{Tgt}^2 + Y_{Tgt}^2} \quad \text{and} \quad GR = \frac{Y_{Tgt}}{\sin\left(\psi_{Sq}\right)} \quad (13.4)$$

13.4 Optical Systems

The airborne optical system includes the detectors, optics, environmental protection (i.e., glass), actuation system, and processing optics. The UAS complete optical system involves every aspect of the system between the photons hitting the outer elements to the payload operators monitor. The interactions of all elements of the complete optical system are known as the *imagery chain*. The concept of the image chain is explained in Schott [2]. We will explore the formation of the image from the optics in this chapter. Other elements of the imagery chain are covered in Chapter 15.

13.4.1 Optical System Physics

Modern digital and analog imaging systems use a collection of individual detectors to form an image. These are arranged in a rectangular array in what is known as a focal plane array (FPA). Each detector element generates an electrical signal based on the electromagnetic energy that reaches it.

The field of view (FOV) at the focal plane array in a given direction is the angular view of the focal plane. It is defined by

$$FOV = 2 \times \tan^{-1}\left(\frac{d}{2f}\right) \quad (13.5)$$

where d is the length of the focal plane array and f is the focal length in the same length units. The horizontal and vertical FPA strips are described separately because many FPAs are rectangular in shape.

The field of regard (FOR) is the maximum angular coverage that the sensor system is capable of achieving. The FOV and FOR are identical for fixed optics with no zoom capability. The FOR for fixed optics with zoom capability is the widest FOV. For gimbaled optics, the FOR is the non-obstructed set of FOR possible within the range of motion limitations of the gimbal system.

The distance between the detectors' centers is known as the *detector pitch* or sometimes *pixel pitch*. The detector pitch P can be limited by detector materials and manufacturing technology or by optical physics limitations. The primary constraint for the latter is the diffraction limit. The light bends as it enters the lens, creating a blur spot at the detector. The energy of the point source is distributed on the focal plane with the point spread function (PSF)

$$\frac{I}{I_0} = 4 \cdot \frac{J_1^2(x)}{x^2} \tag{13.6}$$

where I/I_0 is the relative intensity relative to the center of the Airy disk of the point source and $J_1(x)$ is a first-order Bessel function of the first kind. Bessel functions are described in advanced engineering mathematics texts and are available in many engineering computation software environments. This is sometimes called the *sombrero* function. The parameter x is defined as

$$x = \frac{\pi \cdot D}{\lambda} \cdot \sin \theta \tag{13.7}$$

where θ is the angular distance from the center of the blur spot. The Airy disk PSF is plotted in Fig. 13.5.

The angle of this blur spot angle θ_{Diff} at the detector is defined as

$$\theta_{\text{Diff}} = 2.44\lambda/D \text{ radians} \tag{13.8}$$

The angle corresponds with the first x value at which the Airy disk PSF crosses zero ($x = 3.8317$). Within the circular area bounded by this angle, 84% of the total energy of the point source is captured. The projected

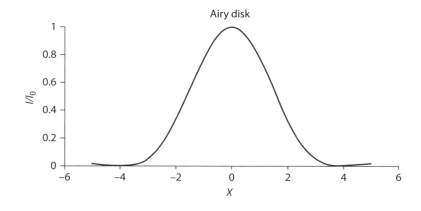

Fig. 13.5   Airy disk distribution.

diameter of this blur spot $D_{\text{Diffraction}}$ at the detector plane is

$$D_{\text{Diffraction}} = 2f \cdot \tan\left(\frac{\theta_{\text{Diff}}}{2}\right) \tag{13.9}$$

This diffraction spot is also known as the Airy disk. Combining terms yields the following relationship:

$$D_{\text{Diffraction}} = 2f \cdot \tan\left(\frac{1.22\lambda}{D}\right) \tag{13.10}$$

Generally, the detector pitch should be larger than the diffraction spot size to prevent the image blur from impacting more than one detector at a time. However, the focal plane-array detectors can be designed such that the point spread function spans two pixels. This allows object features with projected length of approximately twice the GSD to be resolved through processing.

The detector instantaneous field of view (IFOV) is the projected angle of an individual detector element. The IFOV is also known as the detector angular subtense. The IFOV α is defined as

$$\alpha = \frac{a}{f} \tag{13.11}$$

where a is the detector characteristic dimension such as maximum width or diameter. The detector size must be less than the pixel pitch because the detectors do not overlap.

Ground sample distance (GSD) is a fundamental parameter that governs achievable image quality. This parameter is a function of the camera focal plane array, optics, and collection geometry. (See Fig. 13.6 for a diagram of optics geometry.) Although other considerations are required to fully characterize the image quality, GSD can be easily calculated based upon easily understood engineering parameters. GSD is the distance between the pixels projected on the ground at slant range *SR*. Assuming that the horizontal row in the FPA is aligned with the horizon, the horizontal GSD at the center of the image GSD_H is expressed as

$$GSD_H = \frac{P}{f} \cdot R \tag{13.12}$$

where P is the detector pitch. Note that P, f, and SR have units of length. Similarly, the vertical GSD at the center of the image GSD_V is

$$GSD_V = \frac{P}{f \cdot \cos(\theta_{\text{Look}})} \cdot R \tag{13.13}$$

where θ_{Look} is the look angle.

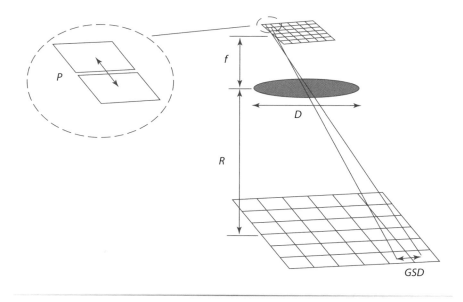

Fig. 13.6 Optics geometry.

Focal plane arrays for imagers are specified by $H_{Pix}xV_{Pix}$ format, where H and V are integers. H_{Pix} is the number of horizontal detector elements, and V_{Pix} is the number of vertical detector elements. Typical FPA arrays for full-motion video sensors are 320×240, 320×256, and 640×480. High-definition (HD) video sensors are 1080×1080 and other configurations.

Often camera data come in the form of field-of-view angles and focal-plane-array elements. The horizontal and vertical GSD equations are

$$GSD_H = 2 \cdot \tan\left(\frac{FOV_H}{2 \cdot H_{Pix}}\right) \cdot R \qquad (13.14)$$

$$GSD_V = \frac{2 \cdot \tan(FOV_V/2 \cdot V_{Pix})}{\cos(\theta_{Look})} \cdot R \qquad (13.15)$$

where FOV_H is the horizontal field of view and FOV_V is the vertical field of view.

The focal plane signal depends upon the detector quantum efficiency, detector size, integration time, and number of time-delay-integration (TDI) stages. These and other factors impact the detector signal-to-noise ratio (SNR). The detectors that comprise the FPA must be sensitive to the electromagnetic band of interest. Several common detector types are shown in Table 13.1.

Table 13.1 Common Detector Types Based on Marketing Materials Survey

Detector type	Typical pitch, μm	Typical sensitivity, μm	Notes
Charge coupled device (CCD)	—	0.55–0.8	Optical imagery
HgCDTe	—	12–18	Photovoltaic detector SWIR-MWIR capable
InGaAs (indium gallium arsenide)	25–40	0.9–1.7	SWIR
InSb (indium antimonide)	15	3.7–5	Photoconductive detector MWIR, cooled
Pt:Si	—	(1–5)	
VOx microbolometer	25–51	8–14	LWIR Generally uncooled

Refractive optics, or glass lenses, are appropriate for small cameras and applications with narrow frequency ranges. Glass lens weight can be excessive for large lens sizes because the weight grows proportionally with the cube of the diameter. Lens materials are traditionally optical glass for the optical band. At MWIR wavelengths and above, the lens materials get more exotic and expensive.

Reflective optics, or mirror lenses, do not suffer the same weight and wavelength limitations as glass lenses. Mirrors can be thin surface contours, which permits the weight to grow more closely to diameter squared, which is less sensitive to large diameter than glass lenses. These reflective devices do not require that photons pass through the material, and so the material is not sensitive to wavelength. Put another way, reflective optics can reflect all wavelengths. This is a particularly important attribute for systems with detectors in the focal plane array that are sensitive to a wide spectral range such as hyperspectral imagers.

Figure 13.7 shows some of the most common reflective optics configurations. The Herschel mount is a simple design using one parabolic reflector, but the long length makes this configuration less desirable for airborne applications. The Newtonian configuration has a simple mirror that reflects the light towards a focal plane. The folded parabolic lens reflects the light through a hole in the primary parabolic reflector with a simple mirror. The Cassegrainian lens reduces the obscuration of the simple folded parabolic lens via a folded parabolic reflector, which has a secondary benefit of shortening the lens length. Many variations on the Cassegrainian configuration that reduce aberration effects exist. The Newtonian, folded parabolic, and Cassegrainian lenses must contend with partial image obscuration.

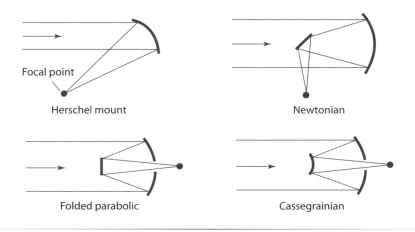

Fig. 13.7   Reflective optics configurations.

13.4.2 Scene Illumination

The detectors in the focal plane array must receive photons from the scene in order to generate the image. These photons reach the detectors via solar and thermal energy paths. Within these primary paths there are numerous ways in which photons can reach the optics.

Solar-energy paths use photons from the sun (see Fig. 13.8). We generally think of *direct reflection*, which is sunlight reflected directly off of the target along the line of sight of the sensor. Indeed, this is the dominant path during clear sky daylight conditions. However, one can observe that

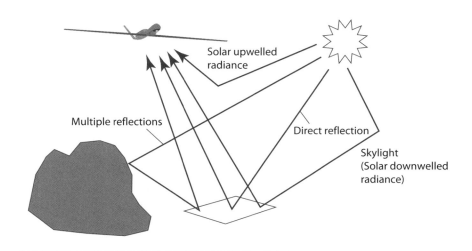

Fig. 13.8   Solar-energy paths.

sunlight indirectly illuminates scenes on overcast days, and shadowed regions are not completely dark on sunny days. This phenomenon is *skylight*, where light is scattered by the atmosphere or clouds and reaches the scene by indirect paths. Solar *downwelled radiance* is another title for skylight. This same scattering phenomenon sends photons to the detector from the background and not from target reflections, which is called *upwelled radiance*. Upwelled radiance degrades the image formed by the detector. One can observe this phenomenon when looking out a jet airliner window at cruising altitude, where the contrast of the ground scene is washed out noticeably relative to the low-altitude conditions at takeoff and landing. Upwelled radiance can dominate on hazy conditions or when thick cirrus clouds are present for high-altitude imagery. The *multiple reflections* route is solar energy that is reflected off of terrain features before reaching the target and is generally a small photon contributor.

Thermal-energy paths involve photons that originate from atmospheric emissions, scene self-emissions, or other nonsolar sources (see Fig. 13.9). All solids, liquids, and gases with temperatures above absolute zero emit energy, which is the source of photons for thermal-energy paths. With infrared imagery most people think of the self-emissions of the target, which is indeed important for LWIR, and to a lesser extent, MWIR imagery. Self-emissions can provide information about the temperature of the target as well as contrast among scene features. Self-emitted energy from the atmosphere can reflect off the target, which is called *downwelled radiance*, or go directly to the sensor, which is called *upwelled*

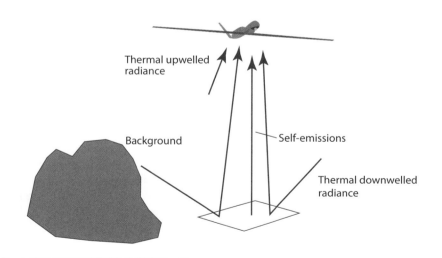

Fig. 13.9 Thermal-energy paths.

radiance. Radiance that reflects off the background or is self-emitted from the background that illuminates the target is the *background radiance.*

Solar radiation is strongest in the visible band and becomes negligible at wavelengths above $3-4$ μm. Therefore, solar illumination is used by visible, NIR, and SWIR sensors as the primary source of photons during daylight conditions. While solar radiation is active through the entire optical window (to approximately 20 μm), self-emissions become dominant around $4-5$ μm for scenes at 300 K. Sensitive MWIR detectors operating around $3-5$ μm can benefit from both thermal self-emissions and solar radiation. Solar- and thermal-energy paths are both active during daylight and transition (dawn and dusk) conditions. LWIR sensors operating around $8-12$ μm use thermal energy as the primary photon source for both day and night conditions.

The thermal emittance can be found by Planck's Law. This is expressed by the equation:

$$M = \varepsilon(\lambda) \cdot c_1 / \lambda^5 \cdot \left[\exp\left(\frac{c_2}{\lambda \cdot T}\right) - 1 \right] \qquad (13.16)$$

where

M = thermal emittance, W/cm^2-μm
$\varepsilon(\lambda)$ = emissivity of the source, which is equal to 1 for a blackbody for all wavelengths
λ = wavelength, μm
T = absolute temperature, K
c_1 = constant equal to 3.7418×10^4 w-μm^4/cm^2
c_2 = constant equal to 1.4388×10^4 μm-K

The results of Planck's Law for blackbody source temperatures of 300, 400, and 500 K are shown in Fig. 13.10. Here it can be seen that the emittance is a strong function of temperature, and the peak emittance occurs at lower frequencies with increasing temperature. The background will be near 300 K, and so thermal detector sensitivity at this temperature is important for image quality. Lava glows red, which is to say that it emits photons near 0.7 μm according to Planck's law when heated to that temperature.

Active target illumination can be used when natural illumination is insufficient for the collection conditions. Consider a color camera that operates at night. Here the background lighting condition is probably insufficient to provide meaningful imagery. A powerful flashlight on the UA pointed at the target can provide sufficient illumination to allow the camera to collect against the target scene. The problem with the visible camera and flashlight scenario is that the illumination is detectable by human observers at the target point, which can negatively impact the

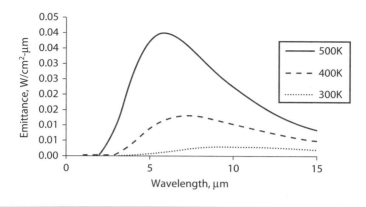

Fig. 13.10 Emittance for a blackbody radiator according to Planck's equation.

intelligence value of the imagery as a result of the alerting collection methods. The same scenario can be performed without alerting the target if nonvisible light is used.

13.4.3 Image Quality Metrics

So far we have considered the optics and detectors and used ground sample distance as a measure of resolution. Now we turn our attention to predicting the quality and utility of the imagery. Although there are numerous metrics available, we will consider the Johnson criteria and the National Imagery Interpretability Rating Scale (NIIRS).

13.4.3.1 Johnson Criteria

The three levels of target discrimination for useful imagery are detection, recognition, and identification. Detection is a reasonable probability that an imagery feature is of a general group (i.e., military vehicle or aircraft). Recognition is discrimination of a class of target (i.e., fighter vs bomber). Identification is object discrimination (i.e., Mig-29).

The Johnson criteria is a method for determining the probability of detection, recognition, and identification for imagery consumers based on the sensor's resolution. The targets are replaced by black and white line pairs, each of which constitutes a cycle. An example of the cycle equivalent of a target is shown in Fig. 13.11. These groups of cycles representing the target are known as *bar targets*, which are commonly used for optical system tests. Note that a cycle corresponds to two pixels. The

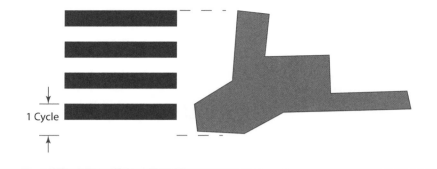

Fig. 13.11 Four-cycle representation of a target.

two-dimensional target characteristic dimension d_c is given by

$$d_c = \sqrt{W_{tgt} \cdot H_{tgt}} \tag{13.17}$$

W_{tgt} and H_{tgt} are the target width and height, respectively, as viewed by the imager.

The number of cycles is provided for 50% probability of successfully performing the detection task, which is denoted by N_{50}. The N_{50} value for detection is 0.75, recognition is 3.0, and identification is 6.0. The probability of achieving the discrimination task $P(N)$ for a given number of cycles N is

$$P(N) = \frac{(N/N_{50})^{2.7+0.7 \cdot (N/N_{50})}}{1 + (N/N_{50})^{2.7+0.7 \cdot (N/N_{50})}} \tag{13.18}$$

Results of this probability equation are plotted in Fig. 13.12.

Example 13.1   Target Identication Using Johnson's Criteria

Problem:
An optical system has an average GSD normal to a target of 2.5 in. in the collection geometry defined in a specification. The reference target is an assault rifle with a length of 40 in. and a height of 10 in. What are the probabilities of detection, recognition, and identification?

Solution:

The target characteristic dimension is found by

$$d_c = \sqrt{W_{tgt} \cdot H_{tgt}} = \sqrt{40 \, \text{in.} \cdot 10 \, \text{in.}} = 20 \, \text{in.}$$

The number of cycles across the target is the target characteristic dimension divided by twice the normal GSD.

$$N = \frac{d_c}{2 \cdot GSD_{Avg}} = \frac{20 \, \text{in.}}{2 \cdot 2.5 \, \text{in.}} = 4$$

For detection, the N_{50} value is 0.75. The probability of detection is

$$P(N) = \frac{(N/N_{50})^{2.7+0.7 \cdot (N/N_{50})}}{1 + (N/N_{50})^{2.7+0.7 \cdot (N/N_{50})}}$$

$$= \frac{(4/0.75)^{2.7+0.7 \cdot (4/0.75)}}{1 + (4/0.75)^{2.7+0.7 \cdot (4/0.75)}} = 0.9999 \, \text{(rounded down)}$$

Using a similar approach, the probability of recognition is 0.7399, and the probability of identification is 0.2169.

13.4.3.2 NIIRS Rating

The NIIRS is a series of government standard qualitative metrics that helps characterize the intelligence value of an optical system under collection conditions. Table 13.2 provides excerpts from the 1994 version of the visible NIIRS definition as presented in Leachtenauer and Driggers [1]. Separate NIIRS definitions exist for infrared, multispectral, and synthetic aperture radar, but these are not provided here.

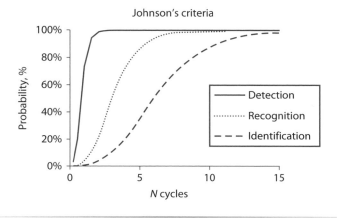

Fig. 13.12 Johnson criteria plot.

Table 13.2 Visible NIIRS Scale Excerpts (1)

NIIRS Rating	Attributes
0	• Interpretability of the image is precluded by obscuration, degradation, or very poor resolution
1	• Detect a medium-sized port facility and/or distinguish between taxiways and runways at a large airfield
2	• Detect large hangars at airfields • Detect large static radars • Detect military training areas • Detect large buildings
3	• Detect the wing configuration of all large aircraft • Identify radar and guidance areas at a SAM site by the configuration, mounds, and presence of concrete aprons • Detect a helipad by the configuration and markings • Detect the presence/absence of support vehicles at a mobile missile base • Identify a large surface ship in port by type • Detect trains or strings of standard rolling stock on railroad tracks
4	• Identify all large fighters by type • Detect the presence of large individual radar antennas • Identify, by general type, tracked vehicles, field artillery, large river crossing equipment, wheeled vehicles when in groups • Detect an open missile silo door • Determine the shape of the bow on a medium-sized submarine • Identify individual tracks, rail pairs, control towers, switching points in rail yards
5	• Distinguish between MIDAS and a CANDID by the presence of refueling equipment • Identify radar as vehicle mounted or trailer mounted • Identify, by type, deployed tactical surface-to-surface missile systems • Distinguish between SS-25 mobile missile TEL and missile support vans in a known support base when not covered by camouflage • Identify individual rail cars by type and/or locomotive type
6	• Distinguish between models of small/medium helicopters • Identify the shape of antennas on EW/GCI/ACQ radars as parabolic, parabolic with clipped corners, or rectangular • Identify the spare tire on a medium-sized truck • Distinguish between SA-6, SA-11, and SA-17 missile airframe • Identify automobiles as sedans or station wagons
7	• Identify fitments and fairings on a fighter-sized aircraft • Identify ports, ladders, vents on electronic vans • Detect the mount for antitank guided missiles • Identify individual rail ties

NIIRS Rating	Attributes
8	• Identify the rivet lines on bomber aircraft • Detect horn-shaped and W-shaped antennas mounted atop BACK TRAP and BACKNET radars • Identify a hand-held SAM • Identify joints and welds on a TEL or TELAR • Detect winch cables on deck-mounted cranes • Identify windshield wipers on a vehicle
9	• Differentiate cross-slot from single-slot heads on aircraft skin panel fasteners • Identify small, light-toned ceramic insulators that connect wires of an antenna canopy • Identify vehicle registration numbers on trucks • Identify screws and bolts on missile components • Identify braid of ropes (3–5 in. in diameter) • Detect individual spikes in railroad ties

NIIRS ratings are provided by trained imagery analysts who evaluate specific images. This rating is subjective and may vary among analysts looking at the same imagery data. By the time an imagery analyst provides such a determination, the optical system must be sufficiently mature to capture the images in collection conditions representative of operations. Qualified imagery analysts might be unavailable to assist in contractor-run design and analysis tasks. Clearly predictive analysis procedures are required to aid in the optical system design or selection.

The general image quality equation (GIQE) is a tool that links optical system parameters to an NIIRS rating. With proper parameter inputs and when used in the appropriate conditions, GIQE can have a standard error of approximately 0.4 NIIRS. Image display parameters are excluded from GIQE because it is assumed that the display monitor is properly optimized for viewing. GIQE version 4.0 is expressed as

$$NIIRS = c_0 + c_1 \cdot \log_{10}(GSD) + c_2 \cdot \log_{10}(RER) + c_3 \cdot \frac{G}{SNR} + c_4 \cdot H$$

(13.19)

where

$c_0 - c_4$ = constants defined in Table 13.3
GSD = geometric mean ground sample distance, in.
RER = geometric mean normalized relative edge response
G = postprocessing noise gain
SNR = signal-to-noise ratio of the unprocessed imagery
H = geometric mean system postprocessing edge overshoot factor

Table 13.3 GIQE 4.0 Constants Values

	c_0	c_1	c_2	c_3	c_4
$RER \geq 0.9$	10.251 (visible)	-3.32	1.559	-0.334	-0.656
$RER < 0.9$	10.751 (IR)	-3.16	2.817		

The simplest of the GIQE parameters for direct evaluation is the GSD. The other parameters require detailed knowledge of the imagery system performance, the characteristics of the scene, and the collection environment. Using example parameters provided by Thurman and Fienup [4], the relationship between NIIRS and GSD is shown in Fig. 13.13.

The RER is the slope of the normalized edge response, which is a measure of the spatial resolution. The edge response is a function of a system modulation transfer function (MTF) that accounts for many parameters in the image chain. The optics design, fabrication, alignment, and defocus are included. Other factors include atmospheric dispersion and detector design, among others. Methods of calculating RER are beyond the scope of this book.

H is the edge overshoot, which is the peak edge response generally $1-3$ pixels from the edge. This is a measure of the edge ringing resulting from image processing. Values of H used to develop GIQE range from $0.9-1.39$. The impact to the GIQE output over this range is 0.321 NIIRS, which is less powerful than GSD or RER.

The processing gain G is the increase in noise due to modulation transfer function correction or compensation (MTFC). MTFC can be used to sharpen an edge in the spatial or frequency domain. Noise increases

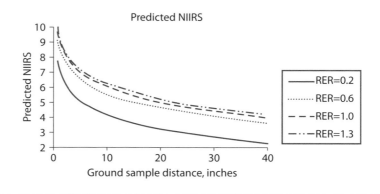

Fig. 13.13 Relationship between NIIRS, GSD, and RER using GIQE 4.0 with parameters $G/SNR = 0.2038$ and $H = 1.31$.

when sharpening is employed. Alternatively, noise can be reduced when smoothing is applied. The MTFC applied might depend on the imager characteristics, target characteristics, and the type of interpretation required. Methods for processing gain are not presented here. However, Leachtenauer and Driggers [1] note that the effect of variation of this contributor is likely to be 0.1 NIIRS or less for most applications. G is 1 when no MTFC is applied. Gain values of $1-19$ were used to develop GIQE.

SNR is the ratio of the unprocessed imagery scene differential signal level to the noise. The signal is measured in direct current, and the noise is the root mean square of the electrons due to noise. The SNR depends upon the scene irradiance and path characteristics. The values of SNR used in the development of GIQE are $2-130$.

When it is impractical to fully characterize all of the GIQE parameters but the NIIRS performance of a similar imaging system is known, changes relative to the known system can be employed. Using the GIQE, the change in NIIRS relative to a reference system (denoted by subscript 0) is

$$\Delta NIIRS = c_1 \cdot \log_{10}\left(\frac{GSD}{GSD_0}\right) + c_2 \cdot \log_{10}\left(\frac{RER}{RER_0}\right)$$
$$+ c_3 \cdot \left(\frac{G}{SNR} - \frac{G_0}{SNR_0}\right) + c_4 \cdot (H - H_0) \qquad (13.20)$$

This can be used for sensitivity analysis to determine the impacts of optical system element changes and different collection conditions.

To be relevant, the conditions under which the NIIRS applies must be specified. The target characteristics, time of collection, atmospheric conditions, and collection orientation are necessary to calculate NIIRS using GIQE parameters. For example, a small UAS with visible cameras might be capable of NIIRS 7-8 imagery during low-altitude flight operations, perhaps NIIRS 4 at 20,000 ft and NIIRS 0 at night.

Example 13.2   NIIRS Calculation

Problem:
An UA is flying at an altitude of 15,000 ft, circling around a target with a turn radius of 2,000 ft. The following properties are known about the optical payload:

- Visible band camera
- Optical zoom range of 15–0.8 deg horizontal field of view
- Focal plane array provides 640 × 480 pixels

- G/SNR is 0.3
- RER is 0.6
- H is 1.1

What is the NIIRS?

Solution:
The slant range is calculated as

$$R = \sqrt{h^2 + GR^2} = \sqrt{(10{,}000\,\text{ft})^2 + (2{,}000\,\text{ft})^2} = 10{,}198\,\text{ft}$$

The look angle is

$$\theta = \tan^{-1}\left(\frac{h}{GR}\right) = \tan^{-1}\left(\frac{2{,}000\,\text{ft}}{10{,}000\,\text{ft}}\right) = 11.3\,\text{deg}$$

At the maximum zoom condition, the horizontal GSD is

$$GSD_H = 2 \cdot \tan\left(\frac{FOV_H}{2 \cdot H_{\text{Pix}}}\right) \cdot R = 2 \cdot \tan\left(\frac{0.8\text{deg}}{2 \cdot 640}\right) \cdot 10{,}198\,\text{ft}$$

$$= 0.222\,\text{ft} = 2.67\,\text{in.}$$

Assuming that the detector spacing is the same horizontally and vertically, the ratio FOV_V/V_{Pix} equals FOV_H/H_{Pix}, which is 0.00125 deg.

$$GSD_V = \frac{2 \cdot \tan(FOV_V/2 \cdot V_{\text{Pix}})}{\cos(\theta)} \cdot R = \frac{2 \cdot \tan(0.00125\,\text{deg}/2)}{\cos(11.3\,\text{deg})} \cdot 10{,}198\,\text{ft}$$

$$= 0.227\,\text{ft} = 2.72\,\text{in.}$$

The geometric mean GSD is

$$GSD = \sqrt{GSD_H \cdot GSD_V} = \sqrt{2.67\,\text{in.} \cdot 2.72\,\text{in.}} = 2.70\,\text{in.}$$

Because RER is less than 0.9, the GIQE equation for NIIRS becomes

$$NIIRS = 10.251 - 3.16 \cdot \log_{10}(GSD)$$

$$+ 2.817 \cdot \log_{10}(RER) - 0.334 \cdot \frac{G}{SNR} - 0.656 \cdot H$$

$$= 10.251 - 3.16 \cdot \log_{10}(2.70\,\text{in.}) + 2.817 \cdot \log_{10}(0.6)$$

$$- 0.334 \cdot (0.3) - 0.656 \cdot (1.1) = 7.3$$

13.4.4 Multispectral and Hyperspectral Imagery

Multispectral imagery (MSI) and hyperspectral imagery (HSI) are classes of multiband sensors that collect imagery at multiple spectral bands. These sensors are *imaging spectrometers*, and they combine spectroscopy and remote-sensing technologies. *Spectroscopy* is the study of material spectral responses. The formal distinction between MSI and HSI classes is somewhat arbitrary, where MSI is typically less than 10 bands and HSI collects greater than 100 bands. The spectral bands are selected based on the remote-sensing objectives and target characteristics. MSI and HSI sensors are relatively new to unmanned systems and can still be considered an emerging technology at the time of this writing.

These powerful sensor classes can provide high-value intelligence not possible with visible imagery or through the use of single-band systems. Targets have a spectral signature, which is the reflectance or emissivity as a function of wavelength. We saw earlier in this chapter how scenes reflect, absorb, transmit, and self-emit electromagnetic energy. *Spectral reflectance* is the ratio of reflected energy to incident energy as a function of wavelength. The spectral reflectance characteristics varies with electromagnetic spectrum for various materials or scenes based on physical structure, chemical characteristics, temperature dynamics, atmospheric conditions, and illumination conditions. Targets of interest can be detected and identified based on matching the spectral signature, which is the general shape of the spectral curve. *Absorption bands*, which are sharp drops in reflectance, are important features in the spectral signature that act much like fingerprints. The spectral signatures of backgrounds and targets can be stored in digital libraries to aid analysis. Figure 13.14 shows the difference in spectral signature between deciduous trees and olive green paint based on data from the ASTER Spectral Library [5]. A single pixel in a HSI image will generally be a composite of multiple types of materials, making signature isolation challenging.

Much like SAR, multiband sensors might not provide intuitive products for human operators to interpret without proper training and exploitation tools. MSI and HSI data products lend themselves to automated processing

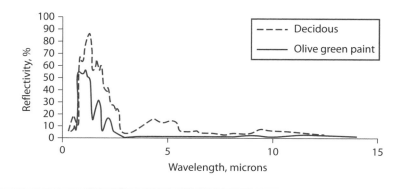

Fig. 13.14 Spectral signatures of deciduous trees and olive green paint.

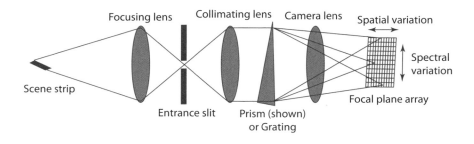

Fig. 13.15   Prism dispersive hyperspectral imager components. (Image adapted from Vagni (6).)

in order to extract the objective information such as identification of a chemical. The amount of data collected can be orders of magnitude greater than an equivalent color or monochromatic image of equivalent spatial resolution.

The basic functional elements of a dispersive hyperspectral sensor are shown in Fig. 13.15. A focusing lens projects the light onto an entrance slit, which permits a strip of light to pass through. A collimating lens straightens the light such that the rays are nearly parallel, which then proceeds to a dispersive device. The dispersive device, which is typically a prism or grating, disperses the strip of light into multiple adjacent bands. Each band is typically 10–20 nm in spectral width, and potentially as low as 1 nm. This works just like a common prism that projects the colors of the rainbow when exposed to sunlight. The dispersed light passes through a camera lens that then projects the light field onto a focal plane array. The focal plane is divided into multiple rows of detectors. Each row is sensitive to a specific wavelength band corresponding to that projected upon it. Each row of detectors views the same scene strip spatially, but at different wavelengths. Put another way, one axis of the two-dimensional focal plane array measures spatial variation whereas the other axis measures spectral variation.

Many design details and physical parameters drive the multiband sensor performance. The detector size and sensitivity must be matched to the light projected upon it. The prism must trade dispersion that improves resolving power with the associated absorption. The lens and prism materials must be compatible with the wavelengths of interest. Reflective optics can be applied to systems using gratings. Multiple grating types exist, including convex Offner spectrometers and concave Dyson spectrometers. Other systems combine gratings and prisms, such as the prism-grating-prism (PGP) configuration. Low-pass and long-pass filters can be included in the optical path to cut off light outside of the desired wavelength.

MSI and HSI data are collected in two spatial dimensions (ground coverage) and a spectral wavelength dimension. This is referred to as a *data cube*. When the temporal dimension is added, the dataset becomes four dimensional.

Because only one spatial strip can be imaged at an instant, the two spatial dimensional images must be formed by sweeping the scene strip. Image processing is often required to assemble the image from the scan patterns. Multiband sensors can use a variety of methods to sweep the sensor ground coverage (scene strip) across the ground over time. Generally pushbroom or whiskbroom techniques are employed. Whiskbroom imagery requires that a scanning mirror sweep the scene. The focal-plane-array configuration is linked to the type of sweeping employed.

13.5 Radar

Radar is an abbreviation for radio detection and ranging. Radar works by transmitting RF energy that is reflected off of the target and detected by a receiver. There are numerous forms of these active sensors.

Objects reflect or absorb RF energy to varying degrees. This is similar to how objects reflect light. Both light and radio waves are electromagnetic energy with frequencies separated by several orders of magnitude. Most optical systems require illumination of the target by a source such as the sun, or self-emissions as is the case with LWIR thermal imagers. Radar is an active sensor that requires that electromagnetic energy in the RF spectrum is transmitted at the target, reflected off of the target, and then received by the radar. One form of the radar equation is

$$P_r = \frac{P_t \cdot G_A \cdot A_{\text{eff}} \cdot \sigma}{(4 \cdot \pi)^2 \cdot R^4 \cdot L_{\text{radar}} \cdot L_{\text{atm}}} \tag{13.21}$$

where
A_{eff} = receiver antenna effective area, m^2
G_A = transmitter antenna gain
L_{atm} = two-way atmospheric loss factor due to atmospheric propagation
L_{radar} = transmission loss factor due to other sources
P_r = received signal power for a single pulse, W
P_t = transmitted signal power for a single pulse, W
R = range from the target to the antenna, m
σ = radar cross section, m^2

Note that the radar-range equation is very similar to the link equations covered in Chapter 12, which is not surprising given that both use transmit and receive antennas and send electromagnetic energy through the atmosphere. The receive power for the radar is proportional to R^4 instead of R^2 because the radar transmitted energy must go to the target and return before reaching the receiver, which is equivalent to $R^2 \times R^2$. The atmospheric loss factor L_{atm} is twice the value used for communications because the RF propagation path is twice as long.

The noise power at the receiver N_r is given by

$$N_r = k \cdot T \cdot B_N \cdot F \tag{13.22}$$

where
  k = Boltzmann's constant
  T = ambient absolute temperature
  B_N = effective noise bandwidth of the receiving process
  F = noise factor

Because this is identical to the noise power used in communication systems, details are in Chapter 12.

The signal-to-noise ratio at the receive antenna is found by the ratio of the receiver power to the receiver noise.

$$SNR_r = \frac{P_r}{N_r} = \frac{P_t \cdot G_A \cdot A_{eff} \cdot \sigma}{(4 \cdot \pi)^2 \cdot R^4 \cdot L_{radar} \cdot L_{atm} \cdot k \cdot T \cdot B_N \cdot F} \tag{13.23}$$

Recall that for many antenna types the gain is related to the aperture area and wavelength by

$$G_A = \frac{4\pi \cdot \eta_{ap} \cdot A_A}{\lambda^2} \tag{13.24}$$

Also recall that the effective aperture area is related to the actual aperture area by

$$A_A = \eta_{ap} \cdot A_{eff} \tag{13.25}$$

The signal-to-noise ratio at the receive antenna becomes

$$SNR_r = \frac{P_t \cdot \left(\eta_{ap} \cdot A_A\right)^2 \cdot \sigma}{(4 \cdot \pi) \cdot R^4 \cdot \lambda^2 \cdot L_{radar} \cdot L_{atm} \cdot k \cdot T \cdot B_N \cdot F} \tag{13.26}$$

For a given target and collection range, primary factors within the radar designer's control that impact the signal-to-noise ratio are aperture area and frequency. Shorter wavelengths, and therefore higher frequencies, improve the SNR via the $1/\lambda^2$ term. However, atmospheric losses can offset the benefit of higher frequencies. Nonetheless, most radars used by UASs operate in X, Ku, and Ka bands.

The atmospheric attenuation can be found by using the attenuation parameter α, which is measured in two-way decibel loss per kilometer. Example α data as a function of frequency altitude and rain rates can be found in Doerry [7]. The atmospheric losses are frequency dependent, with higher losses with higher frequency over the bands of interest. Rain and high humidity also adversely impact the atmospheric losses. The loss due to atmospheric propagation is

$$L_{\text{atm}} = 10^{\alpha(f) \cdot R/10} \tag{13.27}$$

The radar estimates the range to the target by

$$R = 1/2 \cdot t_{\text{RT}} \cdot c \tag{13.28}$$

where t_{RT} is the roundtrip time between transmit and receiving the RF energy and c is the speed of light.

Radars can be designed to perform many different functions that are applicable to UASs. Advanced radars can perform numerous functions in a single system, in which case they are called multimode radars. The basic functions include the following:

- *Weather radar*—Observe location and intensity of rainfall.
- *Air-to-air radar*—This type can involve search, track, and fire control functions against other aircraft.
- *Maritime search radar*—Search, track, and/or classify surface vessels.
- *Synthetic aperture radar*—Radar images targets to provide imagery intelligence.

See Stimson [8] for further reading on airborne radars.

13.6 Synthetic Aperture Radar

Like all radars, synthetic aperture radars (SAR) illuminate a target with RF energy and then process the reflected signal. Additionally, SAR uses the forward motion of the UA-mounted radar to produce the effect of a long antenna without having a physically long antenna. These imaging systems are often more complex than optical imagers due to sophisticated antenna design and signal-processing requirements. Although SAR imagery can resemble optical imagery, radar imagery phenomena can make the intelligence products both nonintuitive and provide unique information. Doerry [7] should be consulted for further reading on SAR performance and design.

A key advantage of SAR imaging sensors over optical systems is the ability to see through clouds. As we saw in Chapter 12, RF systems operating at uhf, L, X, and Ku bands have very limited degradation because of rain and humidity absorption at short range. Lower frequencies gain an advantage as range increases.

SAR sensors must provide resolution in both azimuth and range directions in order to provide imagery. The techniques for providing these two components of each resolution cell employ different approaches. The range direction is horizontal along the ground perpendicular to the flight path. The azimuth direction is horizontal along the ground parallel to the flight path. A swath of resolution cells in the range direction is swept along the flight path, or azimuth direction, to generate two-dimensional strips of imagery.

A SAR has N_r antenna elements. The pulses for each element are sent and returned serially, such that when the effective array length L is traversed all N_r elements have gone through a cycle. Only the pulse from one element is received by the array at a time. Once all of the elements have been sequenced, the range strip is formed. The image is formed by incrementing these range strips. The SAR collection geometry is shown in Figs. 13.16 and 13.17.

SARs provide range estimates for every image pixel collected without the need to geo-reference the image. These data are useful for targeting and image plotting on three-dimensional maps. The resolvable range resolution ΔR_r is given by

$$\Delta R_r = \frac{c \cdot \tau}{2} \tag{13.29}$$

where τ is the pulsewidth time of the radar. Here τ is generally less than the actual pulsewidth due to compression techniques. The resolvable range resolution can be made more generic by introducing a_{wr}, which is the actual mainlobe width with respect to the inverse of the signal bandwidth B_T. A typical value for a_{wr} is around 1.2.

$$\Delta R_r = \frac{c \cdot a_{\text{wr}}}{2 \cdot B_T} \tag{13.30}$$

From this, the required signal bandwidth $B_{T,\text{req}}$ to achieve resolution ΔR_r becomes

$$B_{T,\text{req}} = \frac{c \cdot a_{\text{wr}}}{2 \cdot \Delta R_r} \tag{13.31}$$

Alternatively, the range resolution can be defined by range bins comprising the overall range swath, which must be greater than or equal to the minimum resolvable range resolution enabled by the antenna:

$$\Delta R_r = \frac{R_{\text{Max}} - R_{\text{Min}}}{N_r} \tag{13.32}$$

where N_r is the number of range resolution cells and R_{Max} and R_{Min} are the maximum and minimum slant range distances from the UA, respectively.

The azimuth processing of a SAR is found through coherently combining of multiple pulse echoes. Arrays can be focused by applying phase correction to the returns at each element. For the pulse to be received by all antenna elements, the antenna beam pattern must be sufficiently wide for each array element to see the return of every other element. In other words, by the time that the UA traverses L, the ground patch covered by the pulse of the last array element must be visible by the first array element as shown in Figs. 13.16 and 13.18.

The azimuth resolution ΔR_a equation is

$$\Delta R_a = a_{\text{wa}} \cdot \frac{\lambda}{2 \cdot L} \cdot R \tag{13.33}$$

The mainlobe width factor a_{wa} is the mainlobe width relative to the ratio of the pulse repetition frequency to an ideal SNR gain.

The beamwidth associated with a 4-dB gain reduction $\theta_{4\text{dB}}$ is given by

$$\theta_{4\text{ dB}} = \frac{\lambda}{2 \cdot L} \text{ radians} \tag{13.34}$$

The beamwidth at 4 dB is selected to simplify the mathematics relative to another arbitrarily selected value such as 3 dB. (See Fig. 13.18 for 4-dB

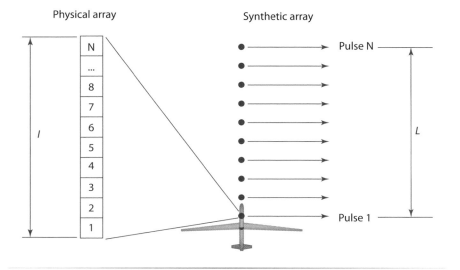

Fig. 13.16 SAR pulse sequencing.

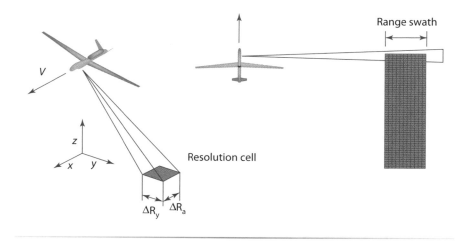

Fig. 13.17   SAR collection geometry.

beamwidth geometry definition.) By simplifying to use the 4-dB beamwidth, the azimuth resolution is simplified to

$$\Delta R_a = \theta_{4\text{ dB}} \cdot R = \frac{\lambda}{2 \cdot L} \cdot R \qquad (13.35)$$

The maximum effective array length L_{Max} is related to the length of the actual antenna l by

$$L_{\text{Max}} = \frac{\lambda}{l} \cdot R \qquad (13.36)$$

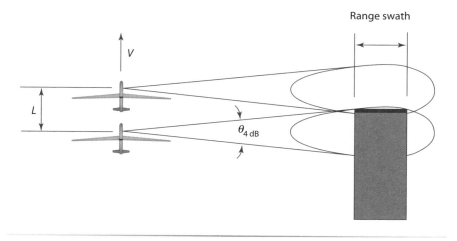

Fig. 13.18   Four-decibel beamwidth geometry definition.

Because the best resolution occurs at the maximum effective length, the best resolution becomes

$$\Delta R_{a,\min} = \frac{l}{2} \tag{13.37}$$

From this simple relationship it can be seen that resolution improves with smaller antenna lengths. Reduced antenna length is in turn made possible by higher operating frequencies. This is a remarkable difference from optical systems, which have improved resolution with increasing focal lengths.

The SAR resolution is a function of the frequency, which is constant. The range where the resolution can be achieved for defined antenna arrays and signal-processing technology is a function of transmitted power. This behavior is quite different from optical systems where resolution is linearly proportional to the slant range to first order. Depending on the system selection, the optical imaging system resolution can grow above the SAR resolution at a slant-range value. The SAR has potential to be a high-resolution long-range imagery system, whereas optical imagers have the advantage at short range.

The pulse repetition frequency f_P is generally constrained to be greater than the Doppler bandwidth of the antenna B_{Doppler}. The Doppler bandwidth is found by

$$B_{\text{Doppler}} = \frac{2}{\lambda} V_x \cdot \theta_{\text{Az}} \tag{13.38}$$

where θ_{Az} is the antenna beam azimuth angle and V_x is the flight velocity in the direction of the synthetic aperture path. Therefore, the pulse repetition frequency is

$$f_P > \frac{2}{\lambda} V_x \cdot \theta_{\text{Az}} \tag{13.39}$$

If we assume that θ_{Az} is equivalent to the 4-dB beamwidth, this relationship becomes

$$f_P > \frac{V_x}{L} \tag{13.40}$$

Strip map SAR is inherently a side-looking operation, with limited azimuth variation permitted normal to the flight path. SARs do not enable nadir imaging. These characteristics provide a bow-tie available ground coverage pattern. The squint angle is typically bounded by 45 deg to 135 deg from the nose (less than ± 45 deg from the side).

The quality of the SAR image is a function of the image signal-to-noise ratio SNR_{image}. Simply having the necessary range and azimuth resolution parameters is insufficient to generate usable imagery. The SAR image SNR is related to the receiver antenna SNR by

$$SNR_{\text{image}} = SNR_r \cdot G_r \cdot G_a \tag{13.41}$$

where G_r is the SNR gain due to range processing and G_a is SNR gain due to azimuth processing. The azimuth gain is found by

$$G_a = \frac{a_{\text{wa}} \cdot f_P \cdot \lambda \cdot R}{2 \cdot \Delta R_a \cdot V_x \cdot L_a} \tag{13.42}$$

where L_a is the azimuth processing loss relative to an ideal processing gain. The range processing gain is found by

$$G_r = \frac{T_{\text{eff}} \cdot B_N}{L_r} \tag{13.43}$$

where T_{eff} is the pulse effective duration, B_N is the input noise bandwidth, and L_r is range processing losses. By substituting Eqs. (13.26), (13.42), and (13.43) into Eq. (13.41), the image signal-to-noise equation becomes

$$SNR_{\text{image}} = \frac{P_t \cdot T_{\text{eff}} \cdot f_P \cdot \left(\eta_{\text{ap}} \cdot A_A \right)^2 \cdot \sigma \cdot a_{\text{wa}}}{(8 \cdot \pi) \cdot V_x \cdot R^3 \cdot \lambda \cdot \Delta R_a \cdot L_{\text{radar}} \cdot L_{\text{atm}} \cdot L_r \cdot L_a \cdot k \cdot T \cdot F} \tag{13.44}$$

The average transmit power P_{avg} can be expressed as

$$P_{\text{avg}} = P_t \cdot T_{\text{eff}} \cdot f_P \tag{13.45}$$

Using the average power and using the relationship $\lambda = c/f$, the image signal-to-noise relationship becomes

$$SNR_{\text{image}} = \frac{P_{\text{avg}} \cdot f \cdot \left(\eta_{\text{ap}} \cdot A_A \right)^2 \cdot \sigma \cdot a_{\text{wa}}}{(8 \cdot \pi) \cdot V_x \cdot R^3 \cdot c \cdot \Delta R_a \cdot L_{\text{radar}} \cdot L_{\text{atm}} \cdot L_r \cdot L_a \cdot k \cdot T \cdot F} \tag{13.46}$$

This powerful equation reveals many useful relationships. Image quality, as measured by SNR$_{\text{image}}$, improves with increases in average transmit power, radar frequency, aperture size, and target radar cross section. Factors that degrade image SNR are target range, flight velocity, and azimuth resolution. Momentarily we will see that when we consider the resolution cell scene RCS dependency on azimuth and range resolution, the image SNR is directly proportional to the range resolution.

The radar cross section of the scene within the resolution cell can be expressed as

$$\sigma = \sigma_0 \cdot \Delta R_a \cdot \Delta R_y \tag{13.47}$$

where ΔR_y is the range resolution on the ground plane and σ_0 is the distributed target reflectivity. The relationship between ΔR_y and ΔR_r is

$$\Delta R_y = \frac{\Delta R_r}{\cos\left(\psi_g\right)} \tag{13.48}$$

where ψ_g is the grazing angle, which is equal to 90 deg minus θ_{Look} for a flat ground plane. The grazing angle is found by

$$\psi_g = \cos^{-1}\left[\sqrt{1 - \left(\frac{h}{R}\right)^2}\right] \tag{13.49}$$

Substituting this back into the scene RCS equation gives

$$\sigma = \sigma_0 \cdot \frac{\Delta R_a \cdot \Delta R_r}{\cos\left(\psi_g\right)} \tag{13.50}$$

The distributed target reflectivity will generally be frequency dependent. The reference value $\sigma_{0,\text{Ref}}$ for a reference frequency f_{ref} can be used in the relationship

$$\sigma_0 = \sigma_{0,\text{Ref}} \cdot \left(\frac{f}{f_{\text{Ref}}}\right)^n \tag{13.51}$$

The exponent n depends on the type of target and generally varies between 0 (no frequency dependence) to 2. The target reflectivity depends upon the target geometry, material dielectric properties, and surface roughness.

Combining the previous SNR_{image} equation, the relationships for the scene RCS, and atmospheric attenuation yields

$$SNR_{\text{image}} = \frac{P_{\text{avg}} \cdot \left(\eta_{\text{ap}} \cdot A_A\right)^2 \cdot \sigma_{0,\text{Ref}} \cdot f \cdot \left(f/f_{\text{Ref}}\right)^n \cdot \Delta R_r \cdot a_{\text{wa}}}{(8 \cdot \pi) \cdot V_x \cdot \left[R^3 \cdot \cos\left(\psi_g\right) \cdot 10^{\alpha(f) \cdot R/10}\right] \cdot c \cdot L_{\text{radar}} \cdot L_r \cdot L_a \cdot k \cdot T \cdot F} \tag{13.52}$$

Several observations can be made from this relationship. The image SNR is proportional to the range resolution because smaller resolution results in a smaller power reflected from the target for given radar transmit power. More reflective background images provide higher return signals. As before, the image SNR improves with increased transmit power and aperture size, while it degrades with increased flight velocity and slant range.

Combining all of the frequency terms yields the image SNR frequency dependence

$$SNR_{\text{image}} \propto f^{n+1} \cdot 10^{-\alpha(f) \cdot R/10} \tag{13.53}$$

This is not a simple relationship between SNR and frequency, as we now have a range-dependent term. The f^{n+1} term improves image SNR with higher frequencies. However, higher frequencies have higher atmospheric absorption. The combined effect is that higher-frequency SAR systems give better image quality at short range, whereas lower-frequency systems have longer range capability.

SAR resolution is defined by the antenna geometry, pulsewidth, and radar frequency. The effective range of a SAR, defined as the range that yields

---

Example 13.3   Radar Sizing

Problem:

An X-band (9.6-GHz) SAR operates at 10,000 ft in moderate rain attenuation (0.11 dB/km). The SAR images a swath ranging from 5–5.5 km ground range directly orthogonal to the flight path. The antenna physical length is 1 m. It has a flight velocity of 100 m/s. The parameter a_{wr} is 1.2.

What is the best azimuth and range resolution that can be achieved with this system if the two resolutions are equal? What is the required radar signal bandwidth? What is the minimum pulse repetition frequency? How many array elements are required?

Solution:

The best azimuth resolution is given by

$$\Delta R_{\text{a,min}} = \frac{l}{2} = \frac{1\,\text{m}}{2} = 0.5\,\text{m}$$

The range resolution will be set to the minimum azimuth resolution. The signal bandwidth is found by

$$B_{T,\text{req}} = \frac{c \cdot a_{\text{wr}}}{2 \cdot \Delta R_r} = \frac{3 \times 10^8\,\text{m/s} \cdot 1.2}{2 \cdot 0.5\,\text{m}} = 360\,\text{MHz}$$

The number of elements required can be found by rearranging the equation

$$\Delta R_r = \frac{R_{\text{Max}} - R_{\text{Min}}}{N_r}$$

to solve for N_r, which gives

$$N_r = \frac{R_{\text{Max}} - R_{\text{Min}}}{\Delta R_r} = \frac{5{,}500\,\text{m} - 5{,}000\,\text{m}}{0.5\,\text{m}} = 1{,}000$$

suitable image SNR, is largely driven by the transmit power. Antenna cooling limits the maximum power, and therefore the achievable SAR imaging range.

Moving objects can generate a Doppler shift in their radar returns. SAR systems using a ground moving target indication (GMTI) mode can detect moving objects within the scene. The range and azimuth of the moving target are known, which can support targeting and cross-cueing to other sensor types. Although humans viewing full motion optical video can detect moving vehicles, providing the same level of target information is impractical without sophisticated vision system algorithms.

13.7 Light Detection and Ranging (LiDAR)

Light detection and ranging (LiDAR), which is also called Laser RADAR (LADAR), is an optical system, though it is covered after radar due to the similarity of operations. The physics of the operations between LiDAR and radar payload types have much in common despite the use of widely separated portions of the electromagnetic spectrum. Like radar, LiDAR emits electromagnetic pulses that are reflected off of the target and then detected by a sensor. These pulses are generated by lasers and can be spread by reflections off of a scanning mirror. The laser pulses are encoded with a signal that helps establish the range to the target for multiple resolution cells. This information can be used to generate a three-dimensional image of the scene. Passive cameras can only generate two-dimensional images in a single frame, and so LiDAR offers an extra dimension. By using the optical portion of the electromagnetic spectrum with short wavelengths, LiDAR can provide favorable contrast and resolution. LiDAR works at night, but, unlike SAR, it cannot operate through clouds.

The detectors are arranged in an array of horizontal and vertical elements. The detectors can be photodiodes that detect the reflected laser pulses from the scene. Optics such as lenses and mirrors can be employed to focus the light onto the detector array, much like cameras.

The received signal power of the LiDAR detector P_S is given by

$$P_S = P_l \cdot \frac{\rho_t \cdot A_r}{\pi \cdot R^2} \cdot \eta_O \cdot \eta_a^2 \qquad (13.54)$$

where
P_l = laser pulse power, W
ρ_t = effective Lambertian reflectivity of the target
A_r = effective collection area of the optical receiver, m^2
R = slant range, m
η_O = optical transmission efficiency of the laser system optics
η_a = atmospheric transmission efficiency

The atmospheric transmission efficiency is found by

$$\eta_a = e^{-\sigma \cdot R} \tag{13.55}$$

where the attenuation factor σ can be approximated by 0.3 km^{-1} at low altitude. Note that a conversion is necessary between kilometers and meters. Combining terms yields

$$P_S = P_l \cdot \frac{\rho_t \cdot A_r}{\pi \cdot R^2 \cdot e^{2 \cdot \sigma \cdot R}} \cdot \eta_O \tag{13.56}$$

LiDAR is flown on aircraft today for mapping and environmental monitoring. The data products include a three-dimensional map of the scene. Vegetation can be discerned from the Earth's surface when multiple images are collected. DARPA's Jigsaw LADAR program sought to exploit this capability to detect targets under a forest canopy. LiDAR can be used for environmental monitoring purposes such as evaluating coastal erosion over time or glacier dynamics. This sensor can also detect clouds, dust, and aerosols in the atmosphere.

13.8 In Situ Measurements

In situ translates from Latin as "in place," and in situ measurements are those that occur at the place of measurement. So far this chapter has described remote-sensing techniques, where the system being measured is distant from the payload. In situ measurements by UASs are measurements taken at the point at which the UA flies. UASs often compete with satellites for missions, particularly those related to Earth sciences. The ability of the unmanned aircraft to gather atmospheric data in situ can be a mission enabler.

References

[1] Leachtenauer, J. C., and Driggers, R. G., *Surveillance and Reconnaissance Imaging Systems, Modeling and Performance Prediction*, Artech House, Inc., Norwood, MA, 2001.

[2] Schott, J. R., *Remote Sensing, The Image Chain Approach*, Oxford Univ. Press, New York, 1997.

[3] Campbell, J. B., *Introduction to Remote Sensing*, Guilford Press, New York, 1987.

[4] Thurman, S. T., and Fienup, J. R., "Analysis of the General Image Quality Equation," *Visual Information Processing XVII*, Proceedings of SPIE, Vol. 6978, 69780F, 2008.

[5] Baldridge, A. M., Hook, S. J., Grove, C. I., and Rivera, G., The ASTER Spectral Library ver. 2.0, "Remote Sensing of Environment," Vol. 113, 2009, pp. 711–715.

[6] Vagni, F., "Survey of Hyperspectral and Multispectral Imaging Technologies," RTO, Technical Rept. TR-SET-065-P3, May 2007, p. 3-2.

[7] Doerry, A. W., *Performance Limits for Synthetic Aperture Radar*, 2nd ed., Sandia National Labs., Report SAND2006-0821, Albuquerque, NM, 2006.

[8] Stimson, G. W., *Introduction to Airborne Radar*, Hughes Aircraft Co., El Segundo, CA, 1983.

Problems

13.1 An MWIR camera operating at 4.5 μ has a 4-in.-diam lens and 6-in. focal length. What is the Airy disk diameter? If the square pixels are adjacent to one another, what is the minimum pixel pitch that precludes Airy disk overlap?

13.2 A payload manufacturer wishes to offer a high-definition visible imager with a very high pixel density. The detectors are sensitive to red, green, and blue light, from which all visible colors can be derived. No optical zoom is required because digital zoom will be used. The lens diameter is 6 in., and the focal length is 8 in. The focal plane measures 0.5 × 0.5 in. What is the maximum number of pixels that can be applied to the focal plane array without exceeding the diffraction limit?

13.3 A payload manufacturer wishes to offer equivalent day- and night-vision resolution by matching visible and LWIR sensors. The day sensor has a 640 × 480 focal plane array and a minimum horizontal FOV of 5 deg. The LWIR pixel pitch is 30 μ, and the ratio of D/f is 2. What is the lens diameter for the same focal-plane-array arrangement and minimum horizontal FOV?

13.4 An UA is circling a target with an orbit radius of 3,000 ft and altitude of 3,000 ft. The visible sensor has a focal plane array of 640 × 512 and a minimum horizontal field of view of 1 deg. What is the best horizontal and vertical GSD for a target in the center of the image plane?

13.5 An aircraft engine nozzle is heated to 700 K. Assume that the nozzle acts as a perfect blackbody. Plot its emittance from $1-15$ μ wavelength.

13.6 An imager has a ground sample distance of 5 in. The target has a width of 40 in. and a height of 20 in. Calculate the probability of detection, recognition, and identification using the Johnson's criteria.

13.7 An UA has a 1080 × 1080 HD visible camera. The focal length is 12 in., and the focal plane horizontal dimension is 0.25 in. The altitude is 30,000 ft above the target, and the look angle is 60 deg. RER is 1, H is 1.1, and $G/SNR = 0.2$. Calculate the geometric mean GSD and the NIIRS.

13.8 Using the information from problem 13.7, what would the NIIRS be if the focal plane array doubled the number of horizontal and vertical detectors in the same area?

13.9 Plot the SAR azimuth resolution vs antenna physical length over a range of 0.1 to 2 m.

13.10 How many SAR antenna elements are required to provide a 1-m resolution over a cross-range of 1 to 2 km?

Chapter 14

Missions and Payloads

- Understand major military, science, and civilian UAS missions and associated payloads
- Understand key drivers for the payloads

IAI Heron carries an IAI/Elta maritime radar under the fuselage. (U.S. Air Force photo by Staff Sgt. Reynaldo Ramon.)

14.1 Introduction

U ASs can perform numerous diverse missions. It is often said that unmanned aircraft are inherently well-suited to "dull, dirty, and dangerous" missions. Dull refers to long, monotonous missions that might not be eventful. Dirty is flying through environments contaminated by chemical and biological agents or radioactive environments. Dangerous applies to missions where loss of the aircraft is probable such as suppression of enemy air defenses or target drones. The missions performed by unmanned aircraft go far beyond these roles. In fact, there is substantial overlap between manned and unmanned mission capabilities. The allocation of missions between these aircraft domains depends on availability of assets, cost, and politics.

The mission is the entire reason for the existence of a UAS. Without the ability to effectively perform the mission(s) of interest, the unmanned aircraft would be nothing more than a glorified model aircraft. In fact, most successful systems use payloads that cost approximately the same amount as the rest of the unmanned aircraft.

Providers of and users of military UASs frequently speak of how these systems save lives. The scenarios range from providing critical situational awareness that identifies an ambush, identification of IEDs, or locating the enemy. Use of UASs can eliminate the need for a dangerous human-forward reconnaissance mission. Frequently the capabilities of manned and unmanned aircraft overlap, but the low-cost UAS can be fielded in larger numbers.

Use of unmanned aircraft can open up missions that would not be possible with manned aircraft due to the risk of the pilot being captured or killed. When a pilot is lost behind enemy lines, a search and rescue (SAR) team attempts a rescue mission, which puts additional lives at risk. Captured pilots can cause great political consequences. A UAS combat loss is rarely newsworthy and does not result in loss of human life.

This chapter provides an overview of the most common UAS missions. Example payloads are shown to demonstrate the physical impacts that missions have on the unmanned aircraft. Payloads are ever-changing, with a trend towards miniaturization and performance enhancements. Payload surveys conducted after this book is published will likely yield superior options.

The three main domains of UAS missions are military, scientific and research, and commercial. There may be overlap between domains, such as an oil facility airborne surveillance and military surveillance of an outpost that use similar systems and techniques. For now the missions are listed by the most prominent domain.

UASs may have the ability to accommodate multiple missions. For example, the MQ-1 Predator A and MQ-9 Reaper can perform intelligence, surveillance, and reconnaissance (ISR) and attack missions on the same sortie. Highly specialized aircraft that perform a single mission well may be a lower cost solution, but flexible platforms with multiple payload types accommodated will fill more roles. UASs with onboard mission management capabilities that allow payloads to work together in an integrated and automated fashion offer greater mission capability potential.

Myriad UAS missions have been demonstrated at least on an experimental basis. The OSD Unmanned Systems Roadmap [1] details numerous legacy and future missions. However, relatively few missions are routinely performed in operations, and most of these missions are for the military. In this chapter we will explore the broad range of potential applications, even if not performed routinely today.

14.2 Military Missions

Because the military has taken the lead in the development and operations of UASs, we will begin the overview of missions here. Military UAS missions largely overlap those performed by manned aircraft for over a century. Military interests are force protection, targeting, attack, intelligence gathering, and peacekeeping operations. One often thinks of full motion video for unmanned aircraft operations, but the potential missions are much broader.

14.2.1 Intelligence, Surveillance, and Reconnaissance

While definitions vary, *intelligence, surveillance and reconnaissance* (*ISR*) generally refers to imagery. Imagery payload products are also known as *imagery intelligence* (*IMINT*). The images can be full motion video, still images, strip maps, or multispectral layers of imagery. The imagers can be optical systems operating in a variety of bands or synthetic aperture radars. Surveillance is continuous monitoring of a target. Reconnaissance is generally a single pass collection against a target.

14.2.1.1 Surveillance

We begin the discussion of ISR with imagery balls due to the overwhelming use of this class of payload for surveillance missions. The majority of imagery balls contains both electro-optic (EO) and infrared (IR) sensors and is called an *EO/IR ball*. Other ancillary sensors such as spotting scopes, additional cameras in different bands, laser illuminators, laser markers, laser rangefinders, and laser designators can be applied as well. Alternatively, some imagery balls contain only a single sensor. Most EO/IR balls are pan-tilt configurations, where the outer stage pans in azimuth and the inner stage tilts in elevation. EO/IR payload configurations are covered in more detail in Chapter 15. The EO/IR payloads are suspended on the underside of the unmanned aircraft. A brief discussion of several common sensor types is described next. Several EO/IR balls are detailed in Table 14.1.

EO Sensor

An EO sensor is a sensor that operates in the visible or near-IR (NIR) band. If the imager is in the visible band, color video is often output. These sensors can be black and white for increased contrast and for low light imagery. EO sensors usually have zoom capabilities.

IR Sensor

The IR sensor typically operates in the short-wave IR (SWIR), midwave IR (MWIR), or long-wave IR (LWIR) band. These sensors are used for night-vision sensing or detecting thermal emissions from targets. Some IR sensors have zoom capabilities.

Spotting Scope

A spotting scope is a very narrow field-of-regard sensor used to aid with target identification. This narrow field of view is sometimes referred to as a *soda straw* view and is not appropriate for situational awareness.

Laser Illuminator

This device illuminates a target in the band of the night-vision sensor (SWIR, LWIR, or MWIR) much as a flashlight illuminates targets in the visual band. The illuminated scene can be viewed by other sensors on the EO/IR ball or by external sensors such as night-vision goggles.

Laser Markers

These markers help identify targets or points of interest to other observers. The markers use the band of the detector of interest. The functionality is

Table 14.1   EO/IR Ball Examples from Marketing Materials

EO/IR Ball	Weight, lb	Diameter, in.	Notes
Cloud Cap TASE	1.98	4.4	Single camera
Cloud Cap T2	5	~7	Dual camera
Hood Tech Atticam Multi 8000	12.2	10	Three cameras + lasers
FLIR Cobalt 190	~16.5	7.5	Multisensor + designator
L-3 MX-12D Skyball II	<55	~12	Multisensor + designator
Raytheon MTS-A (Fig. 14.1)	155	18	Multisensor + designator
Raytheon MTS-B	255	22	Multisensor + designator

Fig. 14.1 Raytheon MTS-A EO/IR payload. (U.S. Air Force photo by Staff Sgt. Cohen A. Young.)

very similar to a laser pointer, though laser markers might not necessarily use a visible band.

Laser Rangefinder
A laser rangefinder determines the range to a target. It contains both the laser and a receiver.

Laser Designator
The laser designator "paints" a target to assist in the guidance of a weapon such as a missile or laser-guided bomb. Designator lasers use encoded signals for security.

Surveillance missions most often use full-motion video. The mission payload operator (MPO) in a ground control station controls the EO/IR ball functions such as pointing, camera selection, zoom, and ancillary sensor actions. This is usually controlled through a specialized joystick. The feedback is a video stream displayed on a monitor along with key metadata. Other displays can include maps or other situational awareness and targeting aids. The end-to-end latency between the joystick inputs and the change in the video display is a critical parameter for system usability.

Very small UASs such as micro unmanned aircraft and hand-launched systems can use airframe fixed cameras or cameras that rotate about only one axis. The small unmanned aircraft scale results in high angular rates relative to larger platforms. The net effect of the UA dynamics and sensor motion constraints is jerky video outputs with limited intelligence value despite the close range to target. Special image processing software or hardware solutions can improve the operator experience.

Laser safety for rangefinders and designators is an important consideration. Designators can blind humans, so that the system must ensure a safety setting. In the United States, laser safety is regulated by the Food and Drug Administration (FDA). Between physical features such as covers or software safeguards, dangerous lasers must be prevented from operating unintentionally.

A well-known problem with EO/IR imagery is the soda straw effect, where the payload operator can see high resolution over a very narrow field of view and without broader situational awareness. If a camera lens can zoom between wide and narrow field of view, situational awareness and high resolution can be satisfied by a single camera. Otherwise, separate cameras may be required with different fields of view. These must be co-boresighted to ensure proper hand-off between imagers.

Surveillance generally involves perishable intelligence. The imagery is time sensitive, where the situation is very dynamic. Data archiving might have value after the mission, but surveillance is mostly a near real-time mission. Some examples of surveillance scenarios are as follows:

- Tracking a moving vehicle in traffic or individual in a crowd
- Determining the positions of guards prior to a compound raid
- Targeting an artillery position for an attack aircraft
- Monitoring a hostage situation

Artillery spotting is a mission related to surveillance. Here a UAS works in concert with artillery to identify targets and provide aim point corrections. The UAS can find, fix, and track the target and relay the target coordinates to the artillery unit. After the initial shots are fired, the UAS can provide miss distance information to help the artillery adjust the aim. This was a major role for the Lockheed Aquila system in the 1970s–1980s and was successfully performed by the Pioneer system in Operation Desert Storm for U.S. battleships.

The raw imagery provided by a surveillance sensor can be augmented by image processing to extract additional intelligence. Image processing can help detect and classify ground targets in clutter or camouflage based on spectral, spatial, and temporal signatures. Algorithms can compare frames to enhance the resolution. Another common technique is to lock onto features to center the image on the screen so that the target appears fixed rather than moving.

14.2.1.2 Reconnaissance and Mapping

Reconnaissance can be best defined by highlighting differentiators from surveillance. Reconnaissance uses imagery sensors to collect against more static targets or large regions of interest. Reconnaissance is strategic by nature, whereas surveillance is more tactical. The targets can be serviced periodically rather than continuously, and multiple targets can be imaged in a single sortie. Analysis of reconnaissance imagery generally takes longer to exploit and can be analyzed outside of the UAS system boundaries. The imagery products are generally still images or strip maps rather than full-motion video. Although reconnaissance imagery is often time sensitive, exploitation of the data usually does not occur in real time. Example reconnaissance missions include the following:

* Battle damage assessments
* Treaty monitoring
* Strategic target imaging
* Time-sensitive targeting

Area coverage during a sortie is more important than with surveillance, as geographic regions or large facilities can be the target. For example, Raytheon's Integrated Sensor Suite (ISS) payload used on the Global Hawk can provide several thousand square kilometers of strip map imagery or thousands of 2 km $\times$ 2 km images in a single sortie [2].

A single reconnaissance mission can image numerous targets. For example, a single Global Hawk mission in Operation Enduring Freedom imaged over 600 targets [2]. Target locations are provided in a *target deck*, and the mission planning system must generate a route that best satisfies the target requests.

Reconnaissance can use strip map and framing modes to generate the imagery. Strip map images a strip of ground instantaneously, and the forward motion of the unmanned aircraft helps form the image strip. The strip map mode is generally used for multispectral, hyperspectral, and synthetic aperture radars. Framing cameras generate traditional still images of a scene. Framing modes are used for black and white and color imagery where all of the focal-plane-array detectors can be illuminated simultaneously.

Older systems used wet film that had to be physically removed from the aircraft for development and analysis. The timeline between the image capture and the analysis was too long for time-sensitive targets, relegating these systems to strategic intelligence. Wet film offered high imagery resolution. The digital communications revolution changed everything—for the better and for the worse.

Current-generation reconnaissance payloads are digital imagers that generate still imagery or strip maps. The data collected are either stored

Table 14.2 Optical Reconnaissance Payload Examples

Reconnaissance Payload	Description
Goodrich DB-110 [3]	Dual-band visible and IR reconnaissance payload Can be pod-mounted for tactical aircraft
Goodrich SYERS-2 [3]	Multispectral system uses visible, SWIR, and MWIR bands. Supports NIIRS 7 imagery at nadir and operating altitude
Raytheon Global Hawk ISS optical payload	Visible and MWIR imagery

onboard the unmanned aircraft or transmitted to a ground element. The imagers can be single band, multispectral, or hyperspectral. Nearly all are of a roll-tilt configuration, which enables a large focal length per payload weight relative to pan-tilt sensors used for surveillance.

Example EO/IR reconnaissance payloads are provided in Table 14.2 and Fig. 14.2. Engineering information about reconnaissance payload performance is generally unavailable in open literature. Reconnaissance payloads suitable for HALE UAS may weigh approximately 500 lb.

Fig. 14.2 Lockheed Martin ALERT reconnaissance payload.

Synthetic Aperture Radars

Synthetic aperture radars (SAR) can provide operational benefits in certain circumstances relative to optical systems. These active RF payloads can penetrate clouds, fog, smoke, sandstorms, and some forms of camouflage. The collection performance does not depend upon target illumination by other sources or target self-emissions, and so SAR performance is the same for day and night operations. The radar returns from metal objects are higher than the background, which enhances detection of man-made objects. The primary SAR modes are the following:

* *Strip map*—A continuous swath of imagery is along the flight path.
* *Spotlight*—A single area on the ground is imaged multiple times to generate the equivalent of still imagery.
* *Ground moving target indicator (GMTI)*—Detect moving objects on the ground due to Doppler effects.
* *Coherent change detection (CCD)*—Detect changes between successive images.

For equivalent weight EO/IR and SAR systems, the EO/IR will generally provide superior imagery quality at close range. The SAR resolution is a function of the aperture and frequency, so that the resolution remains relatively constant until the power-limited maximum range is reached. The optical system resolution is proportional to the slant range, and so the ground spot increases in size as the distance increases. At some range the SAR will start to produce higher resolution. Therefore, the SAR can be a better stand-off imager than optical payloads. (For examples of SAR payloads, see Fig. 14.3.)

SAR systems have some drawbacks for surveillance missions. Depending on the flight time required to collect the image and the image-processing approach, SAR imagery formation can have significant image formation latency that prevents real-time imagery exploitation. The imagery can be nonintuitive compared with optical imagery, though improved resolution diminishes this limitation. Although imagery of the radar returns provides substantial information that is unavailable to optical systems, analyst or mission payload operator training is necessary to get the full benefit. The images are also black and white without intuitive color information.

SAR can often discern features that optical imagery systems cannot. For example, power lines, barbed wire fences, metal stakes, and metal lamp posts will have strong returns even at long range with a SAR because of the high radar cross section of these features. Optical imagers might not be able to detect these features due to resolution limitations. Metal vehicles can shine through forests when imaged by SAR payloads. Image fusion of SAR imagery with optical imagery can provide substantial information in one scene.

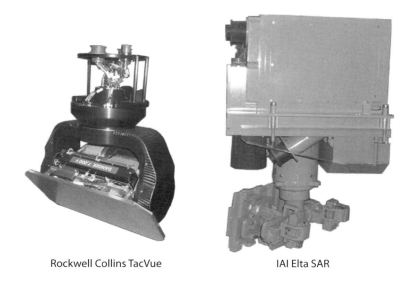

| Rockwell Collins TacVue | IAI Elta SAR |

Fig. 14.3 SAR payloads.

SARs are useful when other sensors are simply unavailable due to clouds or other obscurants. During the Balkan's campaign, Predator A and Hunter UAS were frequently unable to collect with EO/IR imagers through the persistent cloud cover.

Table 14.3 Example SAR Radars

System	Weight, lb	Max power, W	Best resolution, in.	Max range, km
EDO AN/APS-114 MTI/ SAR	69.1	407	—	20
General Atomics Lynx II	83	1000	—	50
IAI Elta EL/M-2055D	84	800	—	35
IAI Elta EL/M-2055M	176–220	2000	—	>100
imSAR NanoSAR	2	15	39	1
Northrop Grumman AN/ZPY-1 Starlite	65	750	—	—
Northrop Grumman ZPQ-1 TESAR	165	—	12	—
Raytheon ISS SAR	—	—	12	200
Rockwell Collins TacVue	<50	<300	4	25
SANDIA MiniSAR	27	—	4	—
Selex Galileo PicoSAR	22	—	<39	20
Thales I-MASTER	66	500	<39	15–20

Several SAR systems are shown in Table 14.3, which is based on marketing materials. These systems span two decades of development and show a wide difference in range and resolution performance. As with other payload classes, SAR payloads are becoming smaller over time. Modern SAR systems can weigh between 2 lb and several hundred pounds, depending upon performance requirements and design approach. Notably, the 2-lb X-band imSAR NanoSAR was successfully integrated into the Insitu ScanEagle system in 2005.

A special class of SAR that can detect targets under a forest canopy is the foliage penetration (FOPEN) radar. The foliage penetration reconnaissance, surveillance, targeting, and engagement radar (FORESTER) has been flown on the Boeing YMQ-18A (A-160 Hummingbird). FOPEN radars generally operate at lower frequencies than traditional SARs (vhf-uhf vs X-Ku band). These radars have great potential for counter-narcotic missions or jungle warfare, where more traditional ISR sensors are unable to image targets. The problem with the low frequency required to penetrate the forest canopy with low attenuation is that the long wavelengths provide poor resolution for target identification, and the false alarm rate can be high for target detection. LADAR sensors can also be used for FOPEN with higher resolution.

Hyperspectral and Multispectral Imagers

Using multiple bands to generate imagery can be effective for counter denial and deception (CDD) missions. In World War II, the United States created a fake army in England to foster the perception that the invasion force would attack at a different location on mainland Europe. The army included decoy weapons such as inflatable tanks that cost a fraction of a real tank. German reconnaissance aircraft could not discriminate between these fake tanks and the real items with the visible cameras and human eyes employed on the mission. If other bands such as VNIR, SWIR, MWIR, and LWIR had been used as well, the deception would have been easily uncovered. In other words, HSI can discriminate between decoys and targets. Use of HSI sensors can defeat many forms of camouflage by operating outside of the bands for which the camouflage was designed. For example, green paint and apparel might look entirely different in a SWIR band. NATO [4] provides additional information on the technology and uses of these sensors.

Wide-Area Airborne Surveillance (WAAS)

UASs have the ability to stay aloft over a fixed area for extended periods of time. Currently fielded MALE and HALE vehicles may have endurance of over 24 hrs per sortie with 400–2,000-lb payload capacity. Future unmanned aircraft within this class can extend the per-sortie endurance to 5–7 days, as can be seen in the Aurora Flight Sciences Orion, Boeing

Phantom Eye, and AeroVironment Global Observer fixed-wing systems. The Orion platform uses conventional heavy fuel at medium altitude, and the latter two platforms use hydrogen fuel for high-altitude operations. Various forms of unmanned airships and hybrid airships have the potential to stay aloft for multiple days as well. The persistent, unblinking eye of these extreme endurance unmanned systems is ideally suited to providing continuous coverage of a fixed area.

One such area coverage mission is called *wide-area airborne surveillance* (WAAS). For this mission a fixed area is given continuous imagery coverage. Current concepts involve multiple cameras with dense focal plane arrays and overlapping fields of view. The value of the data is that a record of past events is established, which can be analyzed to gain intelligence on how incidents unfolded. One example WAAS payload is the ITT Gorgon Stare, which is suitable for the MQ-9 Reaper. Gorgon Stare has five multimegapixel cameras.

Vast quantities of data are collected, which can exceed the bandwidth available from the RF data links. The amount of data collected is a function of the area coverage, image resolution, frame rate, and mission duration. Data management techniques such as data storage and imagery queries from airborne servers must be applied. System solutions can quickly drive data storage systems into terabytes (10^{12} bytes) or petabytes (10^{15} bytes). Chapter 15 provides analysis methodologies for the data management of such as system. Making use of such vast imagery lends itself to image-processing automation.

14.2.2 Lethal Weapons

UASs that carry weapons have much in common with guided munitions such as bombs, air-to-ground missiles, and cruise missiles. Air-launched unmanned aircraft that directly deliver warheads to targets with flight duration of tens of seconds can be considered air-to-ground missiles or bombs. An air-launched unmanned aircraft with the same end effect, but with the ability to identify the target and with flight duration of tens of minutes or hours, might be considered a loitering missile or lethal UAS. Ground-launched long-range unmanned aircraft with integrated warheads are probably designated as ground-launched cruise missiles (GLCM). What if an ISR UAS is fitted with a warhead for direct ground attack? The categorization of systems is imprecise and hotly debated.

The distinction between UASs and cruise missiles was a subject of considerable attention in the early 2000s. The Predator UAS was armed with the Hellfire missile system and made operational (see Fig. 14.4). The U.S. State Department argued that this was a cruise missile, while the U.S. Department of Defense (DOD) maintained that it was an unmanned attack aircraft. At issue were international treaties that regulated the

Fig. 14.4   Hellfire missile being loaded onto a MQ-1A Predator. (U.S. Air Force photo by Staff Sgt. Cohen A. Young.)

permissible numbers of GLCMs. In particular the Conventional Forces Europe (CFE) and Missile Technology Control Regime (MTCR) were focal areas. Is a bomb-carrying UAS simply a reusable cruise missile?

U.S. UASs that are capable of carrying weapons as well as performing ISR missions have the first letter "M" in the designation. The "M" stands for multimission. Many of these aircraft begin service as ISR unmanned aircraft, which have the first letter "R" that stands for reconnaissance. Once weapons are carried, the "M" replaces the "R." The MQ-1, -5, and -8 UASs started operations with the "RQ" designation. This designation change signifies a dramatic change in capability.

Weapons-carrying UASs can perform numerous mission types including the following:

• Suppression of enemy air defenses (SEAD)
• Destruction of enemy air defenses (DEAD)
• Deep strike
• High-value strike
• Close air support (CAS)
• Armed reconnaissance
• Air combat

An ISR UAS equipped with weapons can significantly reduce the sensor-to-shooter timeline. UASs used in the 1990's Balkan conflict were equipped with EO/IR balls that enabled payload operators to identify targets, which would then be relayed to manned strike aircraft. By the time the manned aircraft reached the target area, the targets frequently escaped. By placing weapons and target designators on the UAS, a single unmanned aircraft could complete the kill chain for time-sensitive targets. Alternatively, cooperative teams of hunter-killer UASs with enough endurance to persist over the target area could be more effective.

The kill chain can be described by find, fix, track, target, engage, assess (F^2T^2EA). These elements are described here:

- *Find*—Locate and identify the target.
- *Fix*—Establish target position.
- *Track*—Maintain the target location information as the target moves.
- *Target*—Generate target information required for the weapon system to engage the target. This could be target geolocation coordinates or laser designation.
- *Engage*—Use weapons to destroy the target.
- *Assess*—Perform battle damage assessment.

Unlike a manned aircraft, some elements of the weapon system required to complete the F^2T^2EA cycle do not reside on the unmanned aircraft with current technology. The final targeting and weapons control systems are partially implemented in the ground control station where human operators provide the final actions to employ the weapons. (See Fig. 14.5 and Table 14.4 for examples of weapons for UASs.)

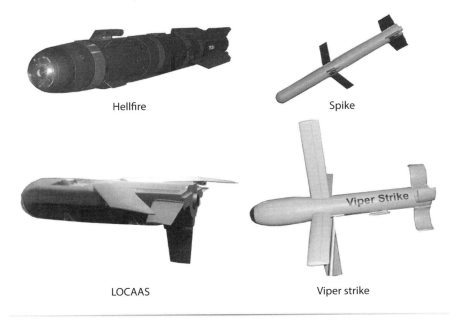

Hellfire

Spike

LOCAAS

Viper strike

Fig. 14.5 Example weapons for UASs.

Table 14.4 Sample Missiles and Bombs

Weapon	Length, in.	Diameter, in.	Weight, lb	Notes
AGM-65 Maverick	98	12	462	
AGM-114P Hellfire (LM)	64	—	98	
AGM-154 JSOW	160	—	1050	
BAT (NGC)	36	5.5	42–44	Dropped in canister, then wings deploy; Glide weapon
CLAW (Textron)	31	5.25	64	
DAGR (LM)	75	2.75	35	Guided rocket; In development
GBU-12 Paveway II	170	18	2080	
GBU-12 Paveway II	131	10.75	600	
GBU-16 Paveway II	145	14.2	1090	
GBU-31 JDAM	154	—	2036	Based on MK-84
GBU-32 JDAM	120	—	1013	Based on MK-83 bomb
GBU-38 JDAM	—	—	500	Based on MK-82 bomb
GBU-39/B SDB	70.8	—	285	Small-diameter bomb
JAGM (Raytheon)	70	7	108	In development
LOCAAS (LM)	36	—	100	Jet-powered, long range; Cancelled
Spike (Navy China Lake)	25	2.25	5.3	Air-to-ground missile
Stinger	58	2.75	23	Air-to-air missile
STS (Textron)	6	5	<10	Submunition with rotating wing

Thus far no dedicated uninhabited combat aerial vehicle (UCAV) designs have entered operational service. UCAVs differ from the previous weapon-carrying UASs in that ground attack and/or air-to-air combat are the primary missions. These platforms are more survivable, with low-observability provisions and high-speed capabilities. Perhaps UCAVs will be given AQ (attack-UAS), BQ (bomber-UAS), or FQ (fighter-UAS) designations when operational. A key mission for UCAVs is suppression of enemy air defenses (SEAD), which is a dangerous mission for manned aircraft.

Weapons employment examples are found in Table 14.5.

Guns are rarely considered for UAS armament, with greater emphasis placed on missiles and bombs. One notable exception is the QH-50 unmanned helicopter experiment, but this was never used in combat operations. Guns are not used largely because of the effectiveness of guided missiles and bombs. Gun systems would involve substantial development with questionable new utility.

Other weapon alternatives include, but are by no means limited to, the following:

- *Grenade launcher*—Grenade launchers can deliver a wide range of munitions at low cost. Effects include high explosive and fragmentation warheads. Nonlethal effects include flash-bang, smoke, tear gas, and nonlethal pellets. A common and lightweight grenade launcher is the 40-mm class.
- *Submunitions*—Submunitions (or bomblets and cluster munitions) could be of the same type used by other air-launched weapons, but directly dispersed from the UAS, which could be reusable.
- *Unguided bombs and rockets*—Although these munitions are simple to integrate, the high dispersion leads to poor target accuracy.

Antiradar UASs carry integral warheads and fly directly into enemy radars. It could be argued that these systems are themselves munitions or cruise missiles, but popular convention is to consider these systems in the UAS

Table 14.5 Weaponized UAS Examples

UAS	Weapon types
Boeing X-45A	• Bomb
Developmental Sciences SkyEye	• Rockets
General Atomics Predator A	• Hellfire missile • Stinger air-to-air missile
General Atomics Predator B	• Hellfire missile • GBU-12 bomb
Gyrodyne QH-50	• Torpedo • Machine gun
Interstate TDR-1	• Torpedo • Bombs
Northrop Grumman BQM-34	• Stubby Hobo missile • AGM-65 Maverick missile • Mk 81 and 82 bombs
Northrop Grumman Firescout (MQ-8B)	• Bat
Northrop Grumman Hunter (MQ-5A)	• Viper Strike (BAT) • BLU-108

domain. Perhaps the best known antiradar drone is the delta-wing IAI Harpy. This is canister launched via rockets, and the pusher rotary engine starts once airborne. This system carries a warhead in the nose that can destroy the radar. Direction-finding antennas guide Harpy into the target radar. Many U.S. manufacturers explored "harassment drones" in the 1970s and 1980s, which would have operated in a similar manner as Harpy. Thanks to multihour endurance, antiradar UASs can loiter until the radar is activated and then attack. Even if the radars are not destroyed, these systems discourage or deny the operations of the enemy integrated air defense system.

14.2.3 Nonlethal Weapons

Nonlethal weapons are, as the name implies, weapons that can be used without killing the target. The purpose is to modify the behavior of the target. For example, tear gas can disperse a violent crowd. Nonlethal weapons are generally used as an alternative to lethal weapons, but are still controversial. Many nonlethal payloads can inadvertently cause fatalities or serious injuries if improperly employed. This class of payload is not commonly used on UASs. Table 14.6 lists several potential nonlethal weapon types.

14.2.4 Communications Relay

Unmanned aircraft are well suited to communications relay because of the long line-of-sight range enabled by the above-ground altitude. The

Table 14.6 Nonlethal Weapons

Nonlethal weapon type	Description
Acoustic weapons	Devices that generate a painful or irritating noise
Caltrops	Devices that puncture and deflate tires; Caltrops are effective in most orientations and can be air-dropped
Electromagnetic pulse (EMP)	Electromagnetic energy that can disable electronics
Flash-bang grenades	Light flash and loud noise that alarms target
Rubber bullets or balls	Projectiles that hurt on contact but are not lethal
Smoke grenades	Smoke-generating projectiles that impair vision
Stink bombs	Projectiles that emit strong foul odor
Tear gas	Projectiles that emit airborne substance that is painful to eyes, but does not cause blindness

unmanned aircraft can relay between ground nodes that are blocked from one another by terrain features. The relay payload can also provide range extension by retransmitting the signal.

A simple ground-to-ground link is a direct relay of the signal between one ground transmitter and one or more receivers. The signal is rebroadcasted by the unmanned aircraft. Voice communications links for ground forces, such as SINCGARS, is one example.

The unmanned aircraft can serve a node in a network. One application is to serve as a cell-phone base station to act as a flying cell phone tower. The unmanned aircraft can enable a network for emergency civilian operations or military purposes. In addition to supporting the network connectivity, data from other airborne payloads can be injected into the network. Examples of RF network radios include 802.11 (Wi-Fi), 802.16 (Wi-Max), SECNET-11, and SEALANCET. Communication networks are enabling technologies for swarming UAS applications. UASs could also support communications with manned tactical aircraft and ISR aircraft through legacy links such as Link-16.

14.2.5 Electronic Warfare

Electronic warfare (EW) can be used as self-defense, as a mission enabler, or as an offensive weapon. EW roles are myriad, including electronic attack, electronic protection, and EW support. In this section we will consider electronic attack and aspects of EW support that provide mission capability. Erdemli [5] provides an extensive list of UAS EW missions and a comprehensive list of references. Example EW payloads are shown in Table 14.7.

EW is a dynamic mission area. The period between an advance in electronic warfare and an opposing electronic countermeasure can be very rapid—especially during a time of war. Threat systems such as radars and missiles adapt to advances in aircraft EW systems and vice versa.

Table 14.7 Example EW Payloads

System	Weight and power	Purpose
BAE AN/ALQ-156(V)	50 lb/425 W	Missile warning system
BAE AN/ALQ-157(M)	218 lb	Airframe-mounted IRCM
BAE AN/ALQ-214	221 lb/>1 kVA	Towed decoy system includes jammers and presents preferable target to oncoming missiles
Northrop Grumman AN/APR-39B(V)2	35 lb/200 W	Radar warning receiver

This domain if often called a "cat and mouse game," involving feverish development of electronic measures, countermeasures, and counter-countermeasures.

Electronic attack (EA) uses the electromagnetic spectrum in an offensive role. The purpose of EA is to degrade, neutralize, or destroy enemy capabilities. The name EA is sometimes used synonymously with electromagnetic pulse or directed-energy systems. Both of these can degrade or destroy enemy radars or electronics by introducing harmful electromagnetic energy. Directed energy can include lasers, radio-frequency beams, or particle beams.

Command and control warfare (C^2W) involves denying and targeting enemy communications. Disrupting an adversary's communications reduces the combat effectiveness. In Chapter 12 the subject of jam-resistant communications was described, which is simply a countermeasure to C^2W.

What is traditionally considered electronic countermeasures (ECM) is technically part of the EA category. Jamming degrades the threat system's ability to detect, track, or engage the unmanned aircraft. Jammers can raise the RF noise floor, making the aircraft signal harder to detect. Noise jammers include RF emitters and passive means such as chaff. Chaff is metallic strips that act as reflectors with the appropriate frequency response. Deception techniques seek to provide the threat system with erroneous information about the unmanned aircraft. For example, deception jammers can place a false target on a radar screen. Another form of ECM is stealth, whereby the radar is denied reception of the return pulses. Towed decoys or flares are forms of imitative deception that spoof the threats into misidentifying the decoys as the targets.

Active jammers must radiate electromagnetic power in the direction of the threat. Although small UASs with active jammers might not be able to radiate substantial power, the effectiveness of the jammer is proportional to the inverse of the square of the distance from the threat radar. Therefore, small UASs operating close to the threat can be effective.

Electronic protection (EP) is protection of friendly systems from enemy EW employment. This can also be considered electronic counter-countermeasures (ECCM). The intent is to prevent enemy detection, denial, disruption, deception, or destruction of friendly systems [5]. Examples include LPI/LPD communications systems (see Chapter 12), electromagnetically hardened avionics (see Chapter 10), and emission controls.

Electronic warfare support (ES) involves interception, identification, and locating enemy electromagnetic emissions. ES includes signals intelligence (SIGINT), though this is covered separately in Sec. 14.2.6. Here we will consider radar warning receivers that enhance survivability and support other EA capabilities.

Radar threats come in many forms. Ground-based radars are categorized as early warning (EW), height finders, target acquisition, and fire control. Fire control radars are part of a weapon system such as anti-aircraft artillery (AAA) or surface-to-air missile (SAM) systems. RWRs can also detect airborne radar from threat aircraft and their air-to-air missiles (AAM).

A radar warning receiver (RWR) detects and classifies threat radars and determines their relative position and threat status. In UAS applications, the threat information from the RWR must be either presented to operators in the GCS or acted upon autonomously by the unmanned aircraft. The UAS can ignore, avoid, or attack the threat, as examples of potential actions. RWRs must have a library of threat radar signals. RWR threat identification and localization can support targeting for weapons employment or EA.

Missile warning systems detect threats posed by missiles before or after launch. Once an approaching missile is detected, the missile warning system can activate decoys and flares to protect the unmanned aircraft. Radar-guided missiles emit encoded RF energy that can be detected by receiving antennas. Another mechanism is to detect the signatures from the missile.

Infrared countermeasures (IRCM) thwart IR seekers in SAM and AAM. A relatively simple approach is to eject heat-generating flares from the unmanned aircraft. The flares have characteristics that cause the missile seeker to follow the flare instead of the unmanned aircraft's heat signature. Airframe-mounted IR light sources can jam the IR seeker. When these light sources can be pointed at the threat missile, they are called directional infrared countermeasures (DIRCM).

Many of the ECM systems described so far are appropriate for enhancing the survivability of the UAS itself. However, UASs can provide EW support as part of a broader system of systems. In other words, UASs can improve the survivability or effectiveness of other aircraft such as manned strike aircraft. Examples of such missions include the following:

- *Stand-off jamming (SOJ)*—This is jamming outside of the engagement range of a SAM. Here a UAS jams the SAM so that another aircraft can penetrate within the threat zone. The range tends to be high, which requires high jammer radiated power and therefore a large UAS to carry and power the payload.
- *Escort jamming*—Here the EW UAS accompanies other aircraft within the threat zone. The range is shorter than SOJ, so that potentially smaller UAS can be used. The UAS should have similar flight characteristics as the aircraft it is escorting, which generally means transonic speeds. The UAS could create a chaff corridor that manned aircraft can fly within.

* *Stand-in jamming (SIJ)*—The EW UAS jams the threat radar within the threat zone. Here the UAS is at risk of being shot down by the anti-aircraft system. This role is more suited to unmanned aircraft than manned aircraft because a human pilot is not at risk. The range to the radar can be short, and so a smaller UAS can be used to generate sufficient radiated power. The SIJ UAS could be launched by the penetrating manned aircraft.

Another UAS mission is to act as a decoy. Many EW UASs served as decoys, as can be seen in Table 14.8. Decoys are UASs that are intended to appear like hostile aircraft to air defense systems. A decoy is much like an aerial target, except it is used in a combat role rather than training. Such a system flying into hostile airspace can force the enemy radars to become active or even engage the decoy. If the radar is active, it can be targeted by other systems such as manned aircraft with antiradar missiles. Also, the type and location of the radar can be identified once it is used, whereas the status might remain unknown otherwise. The radar signature of the decoy is generally tailored to closely match that of the aircraft the decoy is emulating.

14.2.6 Signals Intelligence

Signals intelligence (SIGINT) is composed of communications intelligence (COMINT) and electronics intelligence (ELINT). COMINT relates to collection of communications signals. ELINT is a collection of radar or other weapon system RF signals. SIGINT can be considered electronics

Table 14.8 UAS EW Missions

System	Mission
BAE Nulka	Nulka is a hovering rocket-powered missile decoy UAS that lures missiles away from the ship.
BAI Dragon Drone	Flew with a Rockwell Collins electronic attack payload.
Boeing ADM-20 Quail	B-52 launched jet-powered decoy provided false bomber signature.
Israeli RPV	Simulated attack profile and radar signature of manned aircraft to deceive Syrian radars in the Yom Kippur war.
NGC BQM-74 Chukar	Served as decoy to draw fire from Iraqi SAMS in Operation Desert Storm. Israeli systems drew fire from Syrian SA-6 SAMS in the Bekaa Valley, providing radar data for Israeli jammers.
NRL EAGER	This tethered electric helicopter is designed to fly for up to 1000 hrs. Tests were conducted in 1997.
NRL FLYRT	The flying radar target (FLRT) is rocket launched from a ship to distract antiship missiles. The system was demonstrated in 1993.

reconnaissance, where the intelligence gathered relates to an adversary's RF systems. Most forms of SIGINT are passive listening devices. Most SIGINT applications require comparisons of signals to libraries of emitters to ensure that the proper source is captured in an often crowded RF spectrum. Rockwell [6] provides an overview of the SIGINT market and missions. SIGINT system provider marketing materials provide system-specific capabilities.

COMINT is technical collection of communication signal by other than the intended recipient. COMINT has numerous functions including the following:

* *Spectrum survey*—Search a portion of the RF spectrum to monitor activity.
* *Specific emitter identification (SEI)*—Identify the use of a class of device or specific device.
* *Classification*—Identify activity of predetermined signals of interest.
* *Intercept*—Directly receive the signal. Ground elements that directly receive intercepted voice communications might require a linguist workstation.
* *Copy*—Record the communications signal of interest (raw or processed). The copied message can be transmitted to the ground control station for real-time exploitation or stored for later use.
* *Geolocation*—Determine the position of the emitter.

COMINT is ever-changing. Adversary communications increasingly employ low-probability-of-intercept (LPI) and low-probability-of-detection (LPD) techniques, requiring continuous collection system innovation. Along these lines, UAS systems are often required to have LPI/LPD links to prevent compromise by adversary SIGINT systems.

ELINT is technical and geolocation collection of electromagnetic signals from threat noncommunications sources such as radars. ELINT is closely related to other forms of EW such as RWRs. However, the intention is to gather intelligence about the enemy RF weapon system rather than provide self-defense services. ELINT can intercept radar signals, analyze the signals, or locate the emitters. The emitter location role is particularly useful for mobile radars that are otherwise difficult to locate, such as man portable air defense systems (MANPAD) or vehicle-mounted radars. ELINT is important for understanding the number, type, status, and location of hostile radars, helping to form an electronic order of battle (EOB). The EOB is important for future combat planning and operations. ELINT can also help identify new radar characteristics that helps guide the designs of new ECM suites.

14.2.7 Psychological Operations

Psychological operations impact the behavior of the enemy. Like non-lethal weapons, the intent is not to kill a target. Potential objectives include convincing an enemy force to surrender, preventing ambushes of friendly forces, or getting a population to depose leadership in a time of war. Psychological operations can eliminate the enemy's will to fight or help win the "hearts and minds" of civilians. Some potential UAS psychological operations methods are shown in Table 14.9.

Leaflets are retainable, can be read numerous times, are concealable, and can be distributed among people. Written content can be perceived as more credible and authoritative than other media. However, leaflets require a literate target audience, must be delivered directly to the readers, and can incriminate the holders of the leaflets, and the written message might have unintended consequences.

Airborne radio or television broadcasts do not require direct overflight of the target as with leaflet dispensing. The message can be delivered more quickly (no leaflet printing). Also, the potentially higher entertainment value may be appealing to the audience. The drawbacks are that the audience must have functioning electronic equipment (television, radio, etc.), be tuned to the correct channel or frequency, and the message must be appropriate for the audience (language, dialect, customs).

14.2.8 Aerial Targets

Aerial targets are used to simulate threat aircraft or missiles to support development of defensive systems or training. Targets can be used for air-to-air combat simulations or surface-to-air defense exercises. According to the Defense Science Board [7], the U.S. market for aerial targets is approximately $220M per year, with about 750 flights and 197 losses.

Targets take many forms, depending on the systems that they are intended to simulate. Threat systems of interest include manned aircraft (tactical aircraft, bombers, and helicopters), cruise missiles (transonic or supersonic), and even UASs. Targets support weapon system development

Table 14.9 UAS Psychological Operations Methods

Method	Description
Air-drop leaflets	Distribute leaflets containing a message.
Megaphones	Acoustically broadcast relayed voice communications.
Radio broadcasts	Broadcast messages over radio signals that can be picked up on common radios (AM or FM).

testing and training exercises. In both cases, the targets should resemble the relevant threats. Examples of air targets are shown in Fig. 14.6.

Targets are often converted manned aircraft that are beyond their service life or obsolete. When U.S. combat aircraft are retired from service, they go into long-term storage in an environment that supports preservation. A type will be converted to a drone, and the retired aircraft will be used until the availability is diminished. The type has a "Q" added to the front of the designation to indicate that it is in the unmanned configuration. For example, an F-86 becomes a QF-86. Numerous converted combat aircraft have been used, including the QF-86, QF-100, QF-4, and now the QF-16. The rationale for converting a manned aircraft to a target rather than a dedicated target drone is that many engagement phenomena of interest are dependent upon the geometric characteristics and signatures. Converted manned aircraft targets also represent the destructive effects of missile warheads on manned aircraft.

Other targets attempt to simulate the threat system characteristics affordably. This leads to smaller systems relative to the manned aircraft conversion. Most targets are either air-launched or rocket-launched from the surface. Targets that are intended to be shot down are called direct-kill targets. If the target is not destroyed in flight, it is usually recovered by parachutes. Some targets are designed to float on the water until retrieved. Targets representing tactical military aircraft have similar size and performance characteristics to cruise missiles, which draws special attention from arms control experts.

Gunnery targets tend to be much smaller and less sophisticated than missile targets. These systems are often radio-controlled aircraft with

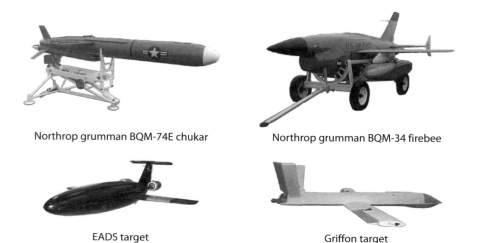

| Northrop grumman BQM-74E chukar | Northrop grumman BQM-34 firebee |

| EADS target | Griffon target |

Fig. 14.6 Targets.

upgraded guidance systems. The purpose is to imitate the visual appearance of a manned aircraft. A smaller aircraft flying close to the gun position flies slower to replicate the appearance of larger aircraft flying faster and farther away. These targets can have the appearance of enemy aircraft. Many of these airframes are made out of molded foam.

Target radar, IR, and visual signatures might need to be enhanced (increased) to represent the desired threat aircraft. Radar signature enhancement can be done through radar transponders, luneberg lenses, or corner reflectors. IR signatures can be increased through surface heating elements, flares, or other heat sources such as secondary jet engines. The visual signature can be increased through smoke generators or strobe lights.

Towed bodies can be deployed from the target to increase its survivability. A reel mechanism extends the towed body downwind of the target via a cable or tether. The minitow body used on the BQM-34 and BQM-74E targets trails the unmanned aircraft by 200 ft.

The target unmanned aircraft itself is mission equipment. Having a platform in a desired location, flying at the appropriate airspeed, and having representative signature characteristics of a threat system are major parts of the mission. Targets carry additional payloads to provide further mission utility. See Table 14.10 for some example target payloads. Potential target payloads include the following:

* Passive or active radar augmentation (i.e., corner reflectors)
* Seeker simulators
* Infrared augmentation
* Tow systems

Table 14.10   Example Target Payloads

Payload	Weight, lb	Capabilities
AN/DSQ-50A (Meggitt)	12.95	Miss-distance sensor set; estimates miss distance, time, and closing velocity; used on QF-4, BQM-34, and BQM-74E
AN/DPN-90(V)1 (Herley)	1.22	Radar transponder to augment C-band radar tracking; used on QF-4
AN/DRQ-4B		Miss distance indicator using cooperative missile telemetry and Doppler; used on QF-4
AN/APN-194	4.4	Radar altimeter, especially useful for low-flying targets; used on QF-4
AN/DSQ-57	27	Advanced radar missile scorer; used on QF-4
Mini Tow	13	Towed IR body used on BQM-34S and BQM-74E
AN/LAU-10	450	Rocket chaff dispenser; used on QF-4

- Scoring systems
- Miss distance indicator (MDI) or miss-distance sensor set (MDSS)
- IFF
- Electronic countermeasures (representing enemy aircraft EW systems)

14.2.9 Maritime Surface Warfare

UASs can perform numerous maritime surface warfare missions. Large land-based systems such as the Northrop Grumman BAMS can cover large areas through long-range, high-altitude missions. Other fixed-wing and rotorcraft UASs can operate directly from ships, providing responsive mission capabilities. Some maritime surface missions include the following:

- *Search for surface vessels*—Searches can be conducted by maritime search radars or optical payloads.
- *Provide surface traffic picture*—Provide ISR of ship traffic and relay ship AIS transponder data to the host vessel.
- *Surface vessel attack*—Use weapons such as missiles, bombs, and torpedoes to attack hostile vessels.
- *Search and rescue*—Provide optical or radar searches for people, receive emergency beacon signals, and mark location.
- *Support of amphibious operations*—Provide ISR, attack, or EW mission support.

Airborne maritime search radars detect surface vessels by radar returns in a manner similar to ship-based search radars. These radars are usually

Fig. 14.7   IAI Heron carries an IAI/Elta maritime radar under the fuselage. (U.S. Air Force photo by Staff Sgt. Reynaldo Ramon.)

Table 14.11 Maritime Search Radars

System	Weight, lb	Max power, W	Notes
Furuno 1722 Navnet	10.8[a]	2400	Flown on Aeronautics AeroSky UAV
IAI/ELTA EL/ M-2022U	251	2300	SAR strip and spot modes, ISAR, weather, and airborne target tracking
Telephonics RDR-1700B	<75	3220	SAR, ISAR, integrated AIS, weather

[a] Radome assembly only.

mounted under the unmanned aircraft such that the radar antenna can scan 360 deg in azimuth without obstruction (see Fig. 14.7 for an example). These radars are used to detect, locate, and classify surface vessels. Some maritime search radars can also generate SAR imagery, detect weather, map terrain, or other radar functions. Maritime surveillance (MS) searches for surface vessels. Maritime target acquisition (MTA) is used to acquire targeting information on a specific target. Inverse SAR (ISAR) is a mode that generates images of the ship, which can then be classified against a ship library. Example maritime search radars are shown in Table 14.11. Maritme radars operating as a SAR may have spotlight and strip imaging modes.

Large ships generally carry a transponder called an automatic identification system (AIS). These devices transmit the vessel name, position, heading, speed, and numerous other parameters. A UAS can carry an AIS receiver that collects these signals and sends them back to the host ship for exploitation. The host ship's AIS receiver is limited in range because of the proximity to the surface. The UAS provides substantial range extension by nature of the flight altitude and increased line of sight. Ship data included in AIS messages include the MMSI number (unique identifier), location (GPS), ship name, radio call sign, speed, and vessel type, among numerous other parameters.

An AIS receiver can be a stand-alone sensor or integrated with a maritime radar. Small AIS receivers can weigh less than 2 lb and consume less than 1 W. The Telephonic RDR-1700B used on the Northrop Grumman FireScout RQ-8B helicopter UAS includes an AIS mode.

The UAS can also carry other sensors such as maritime radars or surveillance sensors that may be used cooperatively with the AIS. A ship may be detected on radar but not using the AIS transponder, which can be concerning. Also, the AIS may indicate that a vessel is heading towards the battle group, and the UAS can confirm the ship's identity by imaging the vessel.

UAS LIDAR sensors can be used to support charting, mapping, and navigation in littoral and riverine environments. These data are important for surface vessels operating in these environments. In this role LIDAR is a bathymetric sensor used to explore shallow water depths. These payloads are not commonly used on UAS today.

14.2.10 Antisubmarine Warfare (ASW)

Sonobuoys are droppable sensors that float on the surface of the water primarily for ASW applications. The word *sonobuoy* is a juxtaposition of *sonar* and *buoy*. Although originally sonobuoys primarily used sonar, the sensors vary depending on how they support the mission. Sonobuoys are ejected from aircraft in circular cylinder canisters. The falling body can be slowed and stabilized by a parachute or other drag device. Upon contact with the water, the sensors and flotation devices become operational. Sonobuoys usually communicate with the ASW aircraft via RF communication links. Sonobuoys come in standard LAU-126 sizes such as A (36 in. long, 4.875 in. diameter), A/2, and G (16.5 in. long, 4.875 in. diameter). Sensors and functions include the following:

- These floating sensors can also use sonar techniques to detect, locate, and track submarines. The sonobuoys can generate the sonar field, serve as a hydrophone receiver, or both.
- Bathythermographs measure the temperature profile of the water, which can be used to help predict sonar propagation and acoustic range.
- Range-only sonobuoys detect the distance to the target by estimating the time delay of sonar pulses. SONAR receivers arranged in an array are known as air-deployed active receivers (ADAR).
- Directional command active sonobuoy systems (DICASS) detect and localize submarines to support targeting solutions.
- Search and rescue payloads can mark the location of survivors on the water surface to assist rescue.

The sonobuoys listed in Table 14.12 are designed to be ejected from the unmanned aircraft. However, the unmanned aircraft may have an integral sonobuoy such that the combined system flies to the insertion point expendably. The Sonobuoy precision aerial delivery (SPAD) system [8] is an A-size tube-launched glider with deployable wings and a G-size sonobuoy in the nose section. The glide ratio of 5.5 permits stand-off relative to a direct fly-over required for conventional sonobuoys. SPAD releases the nose payload at the insertion point, at which time the unmanned aircraft is expended.

A magnetic anomaly detector (MAD) senses variations in the magnetic field, which can be used to detect submarines below the water surface. The

Table 14.12   Example Sonobuoy Payloads

Sonobuoy	Size	Weight, lb	Sensors/Notes
AN/SSQ-36B	A	16	Bathythermograph profiles to 800 m
AN/SSQ-36B A/2	A/2	12	Bathythermograph profiles to 800 m
AN/SSQ-47	A	22	Range-only active sonobuoy
AN/SSQ-53D(2)	A	16.5	DIFAR
AN/SSQ-53F	A	19	DIFAR
AN/SSQ-62E	A	36	DICASS
AN/SSQ-77C	A	23	Vertical line array DIFAR (VLAD)
AN/SSQ-101	A	31	ADAR
AN/SSQ-125	A	36	Multistatic field source

sensor is a highly sensitive magnetometer, which is supplemented by computers for signal processing. Other uses for magnetometers on UAS include geological surveys for mining exploration, or locating unexploded ordinance, buried weapons, or vehicles under jungle canopies. Idaho National Laboratories [9] determined that a magnetometer payload system could be applied to a UAS with an integrated weight of 7.5 lb including obstacle avoidance systems.

UASs might be most effective as part of a heterogeneous system prosecuting ASW missions. Other system elements can include unmanned underwater vehicles (UUVs), unmanned surface vessels (USVs), buoys, sonobuoys, manned aircraft, manned ships, and submarines. Relative to surface-based or submersible vehicles, UASs have high speed, large field of view, large area coverage rates, and long line-of-sight range for communications. UASs could emplace UUVs, USVs, and sonobuoys for rapid response, provide communications relay and cooperatively search for submarines using airborne sensors.

UASs have the potential to support submarine attack. The use of sonobuoys and airborne sensors such as a MAD could give the UAS the ability to locate and track submarines. This information can be provided to other assets such as ASW aircraft, ships, or attack submarines to engage the enemy submarine. The UAS itself could carry torpedoes or depth charges for a direct attack role. The Gyrodyne QH-50 helicopter drone of the 1960s carried torpedoes for this purpose.

14.2.11   Chemical, Biological, Radiological, and Nuclear Detection/Sampling

Unmanned aircraft are well suited to in situ measurements of dangerous airborne plumes. The agents of interest include chemical, biological, radiological, and nuclear (CBRN) domains. This represents the "dirty" category of the "dull, dirty, and dangerous" triad of UAS specialty missions. The unmanned aircraft might be so contaminated after the mission that it must be destroyed, decontaminated, or quarantined. Using an unmanned aircraft ensures that no humans are exposed to harmful environments. Also, unmanned aircraft might be more affordably expendable than manned platforms. Some examples of UAS CBRN missions and demonstrated capabilities include the following:

* Unmanned conversions of B-17 bombers were used to fly through contaminated clouds after nuclear detonations in the 1950s to test radiation levels.
* The NRL Swallow UAS carried a fiber-optic biosensor for biological collections and identification in the late 1990s. The unmanned aircraft is electric powered and uses conventional takeoff and landing.
* The NRL Finder UAS is air-launched from a Predator UAS, which was demonstrated in 2002. After launch it is designed to fly at low altitude to collect chemical plume samples. The Finder and Predator cooperatively track the plume.
* The Defense Threat Reduction Agency (DTRA) sponsored Biological Combat Assessment System (BCAS) uses a team of ISR and biological collection Insitu ScanEagle unmanned aircraft to track, sample, and detect agents in a plume.

Multiple sensor types can aid chemical and biological detection and sampling missions. Hyperspectral and multispectral sensors can remotely sense the characteristics of chemical and biological agents in plumes or on the ground. LiDAR can provide a three-dimensional view of a plume. Samplers take in situ measurements to determine the composition of the plume or return the samples for later analysis.

14.2.12   Cargo Delivery

Tactical cargo UASs offer urgent resupply capability. Troops in contact with the enemy might need supplies to sustain combat effectiveness. Such items can include food, water, ammunition, medical supplies, emergency spare parts, batteries, and fuel.

Example Cargo UAS initiatives include the following:

* Boeing A-160 Hummingbird helicopter cargo UAS
* Lockheed/Kaman KMAX helicopter cargo UAS
* Joint precision air-drop system (JPADS) air-launched parafoil glider capable of delivering 2200 lb of payload from C-130 and other cargo aircraft
* MMIST SnowGoose ground- or air-launched powered parafoil (Fig. 14.8)

Fig. 14.8 MMIST Snow Goose powered parafoil cargo UAS on its launch vehicle.

- AeroVironment X-Glider air-launched fixed-wing cargo glider
- Textron Universal Aerial Delivery Dispenser (U-ADD, intended for UAS applications), which is an air-launched body that can house cargo
- Dynetics/ARMDEC, which was developed and demonstrated QuickMEDS, an air-launched guided medical express delivery system (MEDS) used to emplace medical supplies from a Griffon Outlaw UAS

There is a perpetual theatre and strategic airlift shortfall during times of conflict, particularly during force mobilization. UASs could supplement large air mobility assets such as C-130, C-17, and C-5. Large-capacity theatre cargo UASs, especially with STOL capabilities, would greatly offset the need for convoys. Strategic cargo UASs would supplement the C-17 and C-5 fleets for sustainment.

Unmanned cargo aircraft could provide a lower life-cycle cost. The system development cost could be lower due to lighter airframe enabled by elimination of crew provisions. The operations and maintenance costs could be lower due to manpower reduction. The human operator workload could be low because of the relatively simple mission profiles. Possibly one ground-based operator could manage multiple cargo UASs flying along established flight routes.

In-theatre cargo delivery can be both dull and dangerous. The battle of Khe San in the Vietnam War saw a desperate resupply challenge for a besieged remote fire base. Manned cargo aircraft were destroyed on the ground by mortars, forcing continued resupply to take place by inaccurate parachute drops. A commercial airfreight carrier was hit by a MANPAD missile during approach to a U.S.-controlled Iraqi airfield during Operation Iraqi Freedom. The risk to human life could be eliminated with cargo UASs flying the same missions, and the more missions could therefore be flown in dangerous combat situations.

Large cargo UASs could conceivably be unmanned conversions of established manned cargo aircraft. A single procurement would satisfy both manned and unmanned cargo aircraft, allowing substantial savings in support cost through systems commonality and through reducing the personnel costs. With the exception of the UAS-specific systems, the reliability of the aircraft would be appropriate for high value cargo. Nevertheless, it is unlikely that humans will be carried onboard such aircraft until a strong safety record is demonstrated. A cargo OPV would permit flexibility for piloted or unmanned operations. The unmanned systems could also reduce crew count by replacing the copilot or second crew shift. In this scenario the UAS avionics could control the unmanned aircraft during over-water operations while a single pilot manages takeoff and landing operations as well as emergency operations. The single pilot could rest during noncritical legs. Perhaps for the next 10–30 years a pilot would be required to fly the aircraft in crowded civilian airspace. Unmanned cargo aircraft could also fly in formation with manned cargo aircraft in an aerial convoy, which may mitigate airspace management issues and lessen communications demands.

Although adaptation of a manned cargo aircraft might provide a practical first step to cargo UASs, a new design can offer substantial advantages over traditional manned cargo aircraft. The general attributes of unmanned aircraft described in Chapter 1 apply. One important benefit for cargo UASs is the possible elimination of environmental controls (i.e., temperature and pressurization) over a large internal volume. This could permit noncircular cross sections for the cargo hold, allowing more volumetrically efficient geometries. One could imagine modular cargo modules similar to air-sea-land containers used by cargo ships. A fresh look at cargo aircraft design space enabled by unmanned technologies could lead to improvements in range, payload capacity, operational flexibility, and affordability.

Just as manned cargo aircraft can be adapted to numerous mission types, cargo UASs could do the same. Attributes that make these platforms desirable for alternative missions are the large cargo capacity, large internal volume, long range, and high power generation capability. Potential secondary missions include the following:

• Aerial tanker
• UAS launch platform

* Firefighting
* Early warning aircraft
* Humanitarian relief

14.2.13 System Emplacement

UASs can be most effective when operating as part of a broader system of systems. Unmanned aircraft may be capable of delivering other system elements to spatial and temporal conditions in a way that maximizes system performance. More specifically, UASs can emplace other system nodes where they are needed and when they are needed. Other system nodes could include unattended ground sensors (UGSs), unmanned ground vehicles (UGVs), unmanned surface vehicles (USVs), or unmanned underwater vehicles (UUVs). As a delivery means, UASs can be more affordable than manned aircraft and reduce risk to human life.

Once a heterogeneous system of systems is established, the UAS can provide intelligence elements and communications relay services. The other potential system elements—UGS, UGV, USV, and UUV classes—all operate on or under the surface. Therefore the sensor coverage and communication range is limited relative to the UAS. The UAS can link the elements together and provide communications reach back to a remote command center. The surface-based sensors can benefit from the broad area coverage capability provided by the UAS due to its altitude perch above the surface.

14.2.14 Counterimprovised Explosive Devices (C-IED), Unexploded Ordinance (UXO) Detection, and Landmine Detection

Improvised explosive devices (IEDs) are relatively unsophisticated mines or remotely detonated explosive systems. IEDs produced and emplaced by Iraqi insurgents created a large loss of life during U.S. operations in Iraq. IEDs are an effective asymmetric warfare technique, where the resources used to develop the weapons creates a large adverse impact on U.S. forces, and the cost to counter these devices is orders of magnitude greater.

Unexploded ordinance (UXO) are explosive munitions such as artillery shells, bombs, or land mines that have not detonated. The ordinance may be volatile and can explode, potentially killing or wounding civilians years after a conflict. Often the ordinance is buried underground.

The challenges of finding IEDs and UXO are very similar because both may involve ordinance buried under ground and the search area can be large. IEDs offer additional detection mechanisms due to the need to be triggered and the deliberate act of setting up the device, among other characteristics. Like finding a needle in a haystack, the problem of finding

Fig. 14.9   Honeywell Aerospace T-Hawk works in conjunction with a Foster-Miller TALON Mk II UGV to search for a simulated IED. (U.S. Navy Photo by Mass Communications Specialist 3rd Class Kenneth G. Takada.)

the devices is not as challenging as rejecting the false positives. Figure 14.9 shows heterogeneous vehicles searching for an IED. Some methods of detecting these objects include the following:

- Magnetometers can detect metal objects below the surface. Generally the magnetometer must be close to the ground.

- SAR can detect metal objects on the surface. Change detection between SAR images can help identify surface activity.
- Hypserspectral imagers can detect surface characteristics near buried devices.
- These devices can detect IED triggering mechanisms.

14.2.15 Other Military Missions

Numerous additional military missions exist for unmanned aircraft. Although an exhaustive coverage is beyond the scope of this book, some additional missions include the following:

- *Casualty evacuation (CASEVAC)*—A UAS extracts injured personnel from a combat zone to a rear area. The unmanned aircraft must include provisions for the casualty, such as a litter. Life-support infrastructure may also be added. High UAS reliability will be important for this mission, as a human life is at risk on the unmanned aircraft.
- *Combat search and rescue (CSAR)*—The UAS searches for personnel behind enemy lines, such as downed pilots. The unmanned aircraft can be equipped with emergency locator transmitter (ELT) receivers to help locate the pilot. Manned helicopters can then be called in to extract the personnel. Much like the CASEVAC mission, the UAS can be used to carry humans.
- *Helicopter support*—The UAS can be air-launched from the helicopter to provide responsive support. The unmanned aircraft can fly ahead of the helicopter to see over obstructions—particularly in an urban environment—to provide helicopter protection. In this role the unmanned aircraft can detect ambushes from RPGs, MANPADS, AAA, and small arms. The UAS can also provide ISR and targeting support.
- *Peacekeeping support*—Peacekeeping support is similar to military ISR and civilian law enforcement support in that the UAS monitors the situation on the ground. The key difference between military ISR and this mission is that war is not declared. Unlike civilian law enforcement, the scale of operations can be larger in terms of the number of people involved.
- *Border patrol*—The UAS patrols a border to identify illegal border crossings. This mission is the domain of the Department of Homeland Security in the United States. The General Atomics Predator B was employed to perform this role. High reliability is important for peacetime border patrol missions, as civilian casualties are politically unacceptable.
- *Missile defense*—UASs may detect the launch of missiles by the exhaust plume and track the trajectory of the missile flight. Launch detection is

590 *Designing Unmanned Aircraft Systems: A Comprehensive Approach*

important for theater missile defense, where UASs are likely to operate in the vicinity of short- and medium-range missiles. Hyperspectral or multispectral imagers can identify the signatures of the missile plumes. UASs can be used to track mobile transporter erector launchers (TEL) using ISR and MASINT payloads. High-altitude UASs can use high-speed missiles to hit the missiles shortly after launch, in what is called boost phase intercept (BPI). BPI is a powerful deterrent to hostile forces employing weapons of mass destruction warheads because the contents of the payload are dispersed over hostile territory.

14.3 Science and Research Missions

Scientific missions seek to gather data or perform experiments to advance the state of knowledge. Unmanned aircraft fly through the atmosphere, can perform risky operations without risk to human life, and have excellent area coverage capabilities. Additionally, custom UASs can be designed to access new flight regimes more affordably than manned aircraft, such as very high-altitude flight.

Research missions seek to use the unmanned aircraft as a means of advancing the state of the art in a technology. The technology of interest can be related to the unmanned aircraft itself, such as propulsion, aerodynamic, structures, or controls technologies.

14.3.1 Earth Science Missions

UASs have been used for various Earth science missions, but only on a small scale. Compared to the number of sorties or flight hours on military mission, Earth science utilization is almost negligible. UASs could play a more significant role in this important mission, but funding is the primary limitation. There are too few climate science missions flown by any type of aircraft—manned or unmanned—given the potential impacts of climate change and the need for data to support climate models and track changes as they occur.

UASs must often compete with manned aircraft and satellites for an Earth science mission. We begin by comparing UASs with manned aircraft. Key UAS advantages include the following:

* *Long duration*—Many UASs have single-sortie endurance of 24 hrs or more without refueling. Long-endurance manned aircraft require multiple crews and are generally large, expensive assets.
* *Dangerous operations*—Unmanned aircraft can operate in conditions that would put human life at risk in manned aircraft, enabling missions to be performed that would not be politically acceptable with humans onboard.

Examples can be flight through volcanic clouds or very low-altitude flight over methane-emitting permafrost.

* *High altitude*—Although manned aircraft such as the WB-57 and ER-2 can fly to high altitudes for science missions, UASs can stay aloft longer. UAS offer the potential of lower cost for high-altitude missions.
* *Repetition*—UASs can perform "dull" mission profiles repetitively without complaint or reduced effectiveness over time.
* *Autonomous parcel tracking*—UASs with onboard plume sensors can track plumes at low subsonic speeds.
* *High accuracy*—Autopilots might be able to provide three-dimensional unmanned aircraft positioning and timing more precisely than a human pilot.

UASs offer shorter slant-range remote sensing than satellites and can provide in situ measurements. Many of the attributes of UASs also apply to manned aircraft. Key advantages of UASs relative to satellites include the following:

* *In situ atmospheric measurements*—Satellites must remotely measure through a column of atmosphere, making measurements of a point along the column uncertain. In other words, satellites have poor vertical resolution of atmospheric properties. UASs can sample the air that they fly through and make direct measurements.
* *Responsiveness*—The flight plan of a UAS can be changed, whereas satellite Keplerian orbits are largely fixed.
* *Short slant range for remote sensing*—Flying closer to the point of observation enables smaller and often less expensive sensors for equivalent remote-sensing measurements. Additionally, the shorter slant range results in less atmospheric obscuration.
* *Payload flexibility*—UASs offer greater payload flexibility because payloads can be changed between sorties or periodically upgraded. Satellites can carry only the payload suite installed prior to launch unless serviced by manned spacecraft. UASs can carry developmental payloads that have not undergone spaceflight qualification, permitting much faster time to flight.

Other missions are as follows:

* Emplacing sonobuoys or unattended ground sensors with scientific instrumentation
* Delivering drop sondes to measure vertical profiles

Numerous UAS programs have demonstrated Earth sciences capabilities. Table 14.13 shows a few Earth sciences measurements that would be suitable for UAS missions. Despite the demonstrated utility, these missions

Table 14.13 Earth Sciences Measurements Suitable for UAS

Measurement	Applications
CO_2	Source and sink monitoring, absolute levels
Methane	Methane release from permafrost, oceans, and other sources
Temperature and humidity profiles	Local conditions for local modeling, calibration of weather models
Stratospheric ozone	In situ measurement of the ozone layer
Radiation fluxes	Determine solar radiation at altitude and geographic locations
Clouds	Composition, particles, temperatures, and processes
Glacier and ice sheets	Monitor annual changes in thickness, dynamics, surface deformation
Surface elevation	Repeat pass interferometry can detect surface changes related to geophysical events
Biomass	Determine biomass parameters such as hydration levels and vegetation density
Air pollution	Sources, evolution, and distribution of pollutants
Aerosols	Sample and identify aerosols in the atmosphere
Turbulent fluxes	Measure local turbulence levels for atmospheric model calibration

are not performed on a continuous basis. Examples of UAS demonstrated Earth sciences missions include the following:

• The AAI Aerosonde measured katabatic winds on the coast of Antarctica in 2009. Measurements included pressure, temperature, relative humidity, winds, net radiation, surface temperature, and ice thickness.
• Aerosonde unmanned aircraft flew into tropical storm Ophelia in 2005 and hurricane Noel in November 2007.
• The AAI Aerosonde gathered atmospheric and environmental data in Barrow, Alaska.
• The ACR Silver Fox flew inside the Mt. St. Helens crater to monitor the active dome in 2004. The cameras imaged small eruptions and rock falls.
• The ACR Manta provided synchronous atmospheric data in the Atmospheric Brown Cloud Scripps Maldives Campaign in 2006.
• NASA's Environmental Research and Sensor Technology (ERAST) program sought to develop high-altitude UASs and payload technology for high-altitude environmental research missions. Numerous platforms were demonstrated including the Aurora Flight Science Perseus A and Perseus B, General Atomics Altus and Altair,

Scaled Composites Raptor, and Aerovironment Pathfinder, Centurion and Helios.
* NASA operates Northrop Grumman Global Hawk, General Atomics Ikhana (Predator B variant), and tactical class UAS for a variety of Earth sciences missions.

14.3.2 Species Monitoring

UASs offer the ability to affordably cover large areas. The onboard sensors can be used to detect and classify species of interest. Animal behavior can be observed as well. Many endangered species live within dangerous environments such as the poles, and UASs operate with minimal risk to human life.

An early species monitoring mission is spotting of marine mammals. UASs can unobtrusively view marine mammals without modifying their behavior. Some commercial and military maritime activities may put marine mammals at risk, and so knowing the proximity of these creatures is important. National Oceanic and Atmospheric Administration (NOAA) flew ACR Silver Fox UAS to monitor whales in Hawaii in 2006.

Multispectral and hyperspectral sensors can prove useful in identifying specific electromagnetic signature for animals or environmental threats. Some bands are capable of water penetration. Advanced Coherent Technologies has operated multispectral sensors with water-penetrating sensors in the VNIR range to monitor whales near the surface, for example. When a large number of bands are collected, onboard processing of the collected data might be necessary to prevent excessive communications bandwidth. Automatic recognition algorithms would limit human interaction and personnel costs.

Animals with radio collars can be monitored by UASs. The above-ground altitude of the UAS can extend the communications range with the collar relative to ground-based platforms. Small fixed-wing UASs could perform this mission without disturbing the animal.

14.3.3 Flight Research UA

UASs can offer low-cost alternatives to manned flight research UA and reduce the risk to human life for unproven technologies. Unmanned aircraft can be smaller and less complex than their manned counterparts and therefore less expensive. There is no human interface, environmental systems, or ejection seats, saving perhaps 500–2,000 lb of weight per crew member. The lack of a pilot can support flight characteristics such as very high-*g* flight or high altitude without life-support systems. One could imagine that many of the manned X-planes of the 1940s – 1970s would likely have been unmanned aircraft if avionics and communications

Table 14.14 Unmanned Flight Research Aircraft

UA	Flight research objectives
X-36 (Rockwell International)	—
X-45A (Boeing)	UCAV configuration demonstrator
X-47A (Northrop Grumman)	UCAV configuration demonstrator
X-48B (Boeing)	Blended-wing body transport configuration
Falcon	Hypersonic demonstrator
HiMat	Highly maneuverable fighter design
Ikhana (General Atomics—Predator B variant)	Dense arrays of fiber-optic strain gauges for gust response monitoring
Mini Sniffer	Hydrazine monopropellant reciprocating engine testbed

technologies were adequate at the time. Furthermore, it can be argued that greater value is placed on human life today (in flight test as well as combat), making unmanned aircraft more attractive for dangerous flight tests. However, autopilots make poor heroes and inspire no one.

In many ways the research UA itself is the payload, where the flight performance or system behaviors are the purpose of the flight. Additional payloads can include test instrumentation, data recorders, telemetry systems, and flight termination systems. These systems are covered in Chapter 10 because such equipment is generally required for flight tests of most new unmanned aircraft prototypes. Flight-test equipment payloads can vary from ounces in small flight UA to hundreds of pounds in larger UA. The capabilities of the flight data systems increase with their weight. Notable example unmanned flight research aircraft are shown in Table 14.14.

14.3.4 Other Scientific and Research Missions

Additional science and research missions include the following:

- *Geoengineering*—Geoengineering is intentional human-generated climate change. The objective is to offset adverse climate change created by other sources. One key approach is to emplace aerosols in upper atmosphere. A more localized approach is to seed clouds to create rain.
- *Autonomous soaring*—Unmanned gliders exploit thermals, ridge lift, mountain wave lift, and dynamic soaring for extended-duration flight. UASs are better testbeds than manned aircraft because of the long duration and miniaturized systems.
- *Planetary flight*—Unmanned flyers can be made to operate in the atmospheres of Venus, Mars, Titan, and the gas giants.

14.4 Commercial and Civil Missions

Commercial missions are generally civilian-based endeavors that seek to support profitable operations or to support civilian government. Frequently commercial missions are funded by the companies that develop the capabilities and offered as revenue services. This potentially large market segment is currently very limited as a result of civil airspace restrictions.

14.4.1 Aerial Mapping and Surveys

UASs are suitable systems for both aerial mapping and surveys. Many manned aircraft commercial operations tend to be low cost, so that the UAS should offer improved capabilities. Manned aircraft have an additional advantage of airspace access over populated areas. UASs can offer advantages for mapping and survey operations over international waters, unpopulated regions, or in dangerous situations such as very low-altitude flight.

Hyperspectral imagers can be used to map minerals on the surface. The spectral response of the surface can be analyzed against libraries of ore characteristics. The crystalline structure and chemical composition of minerals create spectral signatures that can be identified. The spectrum of interest is typically $0.4 - 2.4$ μm. Fugro uses a modified version of the Insitu ScanEagle, called GeoRanger, to perform mineral surveys with a magnetometer.

14.4.2 Agricultural

There are numerous agricultural missions that can be performed by UASs, ranging from crop monitoring to aerial pesticide spraying. Most missions require remote-sensing payloads to provide information to farmers. Data collected from the UAS can help inform planting, watering, fertilization, and pest control decisions. Multispectral and hyperspectral sensors are particularly useful for these remote-sensing applications, where the vegetation spectral response is a function of species composition, plant stress, and canopy state. UASs could physically interact with the agricultural process, acting as an airborne tractor. Customers can include land management, natural resource, and farmer clients. UASs might be more responsive and affordable than satellites. Manned and unmanned aircraft will have much overlap for agricultural operations.

Example remote sensing missions include the following:

• Field imagery—crop disease, pests, water stress

- Water-stress detection and mapping
- Disease detection and mapping
- Crop-loss mapping for insurance assessments
- Harvest decision support
- Invasive plant assessment and monitoring
- Counternarcotics crop detection

Example physical agriculture missions include the following:

- Crop dusting
- Seeding
- Cloud seeding
- Fertilizer application
- Growth regulator application
- Defoliant application

Examples of UAS agricultural demonstrations or operations include the following:

- The RnR RPV-3 detected frost damage to crops in 2003.
- The MLB Bat sprayed pesticides in Hawaii for the U.S. Department of Agriculture (USDA).
- The Yamaha RMAX has been used to spray rice crops in Japan with considerable success.

14.4.3 Telecommunications

Much like communications satellites, UASs with communications relay payload can provide coverage over a geographic region. The payloads can be cell-phone base stations, relay nodes for point-to-point communications systems, or provide broadcast services. The advantage of UASs over satellites is that they can be launched more responsively, be more readily upgraded, and potentially be lower cost. Relative to a fixed communications tower, a UAS can provide greater coverage and can be repositioned. UASs are more responsive in times of emergency.

High-altitude, very long-endurance UASs have been considered as relay platforms in lieu of orbital UA. Candidate platforms are mostly solar powered, though beamed microwave propulsion has also been considered. The high sortie rate associated with consumable fuels adversely impacts the fleet sizing and economics. Both fixed-wing and lighter-than-air unmanned systems are frequently considered. Tethered lighter-than-air aerostats can compete with UASs for this role because aerostats can maintain a fixed position with lower cost and complexity.

14.4.4 Forest Fire Monitoring, Coordination, and Communications

The optical payloads used for ISR UAS missions are also well suited to forest firefighting. Infrared sensors—particularly thermal LWIR cameras—can be used to track fire conditions in daylight conditions and under the smoke. Identifying the fire front can help determine where manned water bombers should deliver their loads. Also, ground-based firefighters can be warned of condition changes.

UASs can establish a line-of-sight communications relay or airborne network. Forest fires often occur in sparsely populated areas or mountainous terrain that may have no established communications networks. In more populous areas the communications infrastructure can be disrupted or destroyed by the fire.

UASs could potentially serve as fire bombers in the future, which drop water or fire retardant on wildfires. Such unmanned aircraft would likely be converted manned aircraft or helicopters, though dedicated UAS types are possible. Many manned fire bombers are aircraft at the end of their service life, which places human pilots at risk, and this would be mitigated by a UAS conversion. Also UASs could potentially fly closer to the fire using the autopilot and terrain data to enable more precise placement of the fire retardant without dependency on visual cues.

Example UAS used for firefighting support include the following:

- The IAI I-View was intended to provide imagery support to firefighting operations.
- The Insitu ScanEagle supported forest firefighting operations by providing IR imagery of the flame front.
- The NASA-operated General Atomics Ikhana monitored California forest fires.

14.4.5 Law Enforcement

Civilian law enforcement uses manned helicopters and, to a lesser degree, fixed-wing aircraft for aviation applications. Helicopters can pursue cars and suspects on foot, support rescues, and search areas of interest. Fixed-wing aircraft are used for monitoring motor vehicle speed and other roles.

Many police activities occur over populated areas, where aircraft crashes could result in civilian casualties. UASs must therefore be highly reliable and unlikely to hurt people or property in the event of a crash. If these conditions are not appropriately satisfied, the UAS is suitable for operations away from populated areas and might see only limited use.

UASs must be affordable to procure and operate because many police departments are on constrained budgets. Police UAS operators and maintainers must be trained. Tactical and small tactical UASs might rival the cost of a helicopter, and so the benefit must be worth the cost.

It is conceivable that UASs may someday employ nonlethal weapons or psychological operations for civilian police purposes. Both might be effective for controlling riots without placing law enforcement lives at risk. Caltrops could be dropped ahead of fleeing vehicles to avoid a dangerous police chase. Airborne megaphones could direct people in emergency situations.

Law enforcement use of UASs can run up against regulatory and legal issues. These systems will generally operate in controlled airspace, and the current FAA regulatory environment does not support unscheduled UAS flights over populated areas. Privacy advocates naturally find police cameras in the sky problematic. Furthermore, the Fourth Amendment to the U.S. Constitution prohibits illegal search. It can be argued that UASs are not allowed to collect evidence through observation of personal property that is not obtained through a search warrant. Police forces can argue that UASs provide no more capability than is already provided by helicopters with imaging systems that are in widespread use. UAS employment of the same nonlethal weapons used by riot police would certainly be controversial. UASs are highly effective at monitoring urban areas in combat operations, but are these same techniques too invasive for civil law enforcement?

Compared with other police airborne assets including helicopters, UASs offer the ability to provide acoustically unobtrusive collection. The technology has been carefully honed through decades of military ISR use. Small UASs, in particular, can provide low-cost, close-range surveillance without being detected by the target.

UASs can provide security for events. Even at long operational slant ranges, the behavior of crowds can be monitored. Direct overflight is not necessarily required, and so the risk to people on the ground can be mitigated.

Airborne traffic monitoring is a potentially effective role for UASs. Many highways post signs announcing that traffic is monitored by aircraft. UAS could identify the speed of individual vehicles through radar or optical means, and then identify the license plate through zooming in with optical systems. Another traffic monitoring role is identifying traffic jams in urban areas or highways. The cost of UAS application should be compared with ground-based camera systems and manned helicopters.

14.4.6 Emergency Support

UASs could excel at emergency support missions. Emergency situations could result from hurricanes, volcano eruptions, floods, earthquakes, tsunamis, or major fires. Transportation infrastructure, communications networks, and emergency services may be in disarray, while many people may be at risk. Potential roles include civilian search and rescue, communications support, delivery of supplies, and audio communications.

For the search and rescue mission, UASs can provide the search function. This is much like the ISR role for military UAS, except the people of interest are civilians instead of the enemy. ISR sensors such as visible cameras, infrared cameras, and GMTI SAR could locate individuals on the ground. Once located, the people can be rescued by helicopters or other means.

UASs can establish communications networks. The types of communications relay payloads on the UAS must be compatible with the systems used by the emergency response teams on the ground. Potential forms of communications can include cell phones and push-to-talk radios.

The UASs can deliver emergency supplies to survivors, which can include food, water, medicine, shelters, life jackets, or other survival gear. Fixed-wing UASs can drop supplies with a parachute. Unmanned helicopters or other VTOL platforms can land vertically and place the supplies directly on the ground. Larger UASs are best suited for delivery of supplies due to the heavier payload weight capacity.

Much like military psychological operations, UASs can be fitted with loudspeakers to communicate with people on the ground. Here important survival instructions are communicated. Messages can include safety procedures, rendezvous directions, or requests for action. The audio message can be relayed from a ground station or prerecorded.

14.4.7 Fishing Support

One of the first missions of the ScanEagle was fish spotting for the tuna fleets, though widespread commercial employment was not realized. Tuna shoals create visible disturbances on the ocean surface that can be captured by UAS imagers. Manned helicopters provide this service today, which is a risky operation with high insurance costs.

Threats to species can be identified by UASs. For example, ghost nets are lost fishing nets that continue to catch fish for long periods of time, placing marine species populations at risk. UASs could identify and track these nets near the surface. Surface vessels could then retrieve the ghost nets.

UASs could also monitor fishing activities for the fisheries. Illegal fishing could be monitored. Once a suspect vessel is located and identified, the UAS can assist with the interception by law enforcement. ISR sensors even on small UASs can read the names of ships and monitor AIS transponders. Larger UAS can carry maritime search radars as well.

14.4.8 Other Commercial and Civil Missions

Some additional missions include the following:

* *Mail delivery and commercial freight*—Future UAS package delivery systems would closely resemble military airlift. The most likely initial roles would be routes over the ocean where the loss of the unmanned aircraft is less likely to cause loss of human life.
* *Private security*—Corporations or private citizens can someday use UASs for security or other purposes. Wealthy individuals could use UASs as part of a security strategy to prevent attacks. A facility or campus could be monitored via UASs. It is conceivable that private investigators could use UASs if the laws permitted.
* *Pipeline monitoring*—Oil and natural gas pipelines can cover hundreds or thousands of miles. Periodic surveillance from manned aircraft might be less affordable than UASs. Multispectral or hyperspectral payloads could detect leaks. Traditional ISR sensors can find downed trees or landslides. These payloads can also observe hostile forces or saboteurs intent on damaging the pipelines.
* *News*—News organizations use helicopters for coverage of rapidly changing stories such as car chases or hostage stand-offs. It is not hard to imagine that UASs could be used to provide live imagery of these events. These UAS might someday share the same airspace with police UASs.

References

[1] Office of the Secretary of Defense, *Unmanned Systems Roadmap, 2007-2032*, Washington, D.C., 2007.
[2] Chang, C. Y., and Bender, P. A., "Global Hawk Integrated Sensor Suite – Recent Upgrades and Images," AIAA Paper 2005-7006, 2005.
[3] Cox, C., Kishner, S., Whittlesey, R., and Gilligan, F., "Reconnaissance Payloads for Responsive Space," AIAA Paper RS3 2005-5004, April 2005.
[4] Vagni, F., "Survey of Hyperspectral and Multispectral Imaging Technologies," RTO, NATO, Technical Rept. TR-SET-065-P3, May 2007.
[5] Erdemli, M. G., "General Use of UAS in EW Environment – EW Concepts and Tactics for Single or Multiple UAS over the Net-Centric Battlefield," Naval Postgraduate School Thesis, Monterey, CA, Sept. 2009.
[6] Rockwell, D. L., "SIGINT: The New Electronic Warfare," *Aerospace America*, AIAA, Reston, VA, June 2004.

[7] Office of the Under Secretary of Defense for Acquisition, Technology, and Logistics, *Report of the Defense Science Board Task Force on Aerial Targets*, Washington, D.C., 2005.

[8] Gravelle, N., Schoenholtz, S., Fanucci, J., Maass, D., and Payne, J., "Design and Test of a Sonobuoy Precision Aerial Delivery System (SPAD) UAV System," *Proceedings of the 2007 AUVSI Unmanned Systems North America Conference*, Washington, D.C., Aug. 2007.

[9] Versteeg, R., McKay, M., Anderson, M. T., Johnson, R., Selfridge, B., and Bennett, J., "Feasibility Study for an Autonomous UAV-Magnetometer System," Idaho National Labs., Final Report on SERDP SEED 1509:2206, INL/EXT-07-13386, Idaho Falls, ID, Sept. 2007.

Problems

14.1 Find five EO/IR balls that weigh less than 10 lb.

14.2 Identify five jet-powered targets.

14.3 Think of three missions that can be better accomplished with heterogeneous systems than with UASs alone.

14.4 Think of three civil UAS missions not mentioned in this chapter.

Chapter 15

Mission Systems Integration

- Understand how to physically integrate payloads into the unmanned aircraft
- Create latency budgets and estimates
- Establish payload data management approaches

Fig. 15.1 MQ-9 Reaper with EO/IR ball and external weapons. (U.S. Air Force photo by Staff Sgt. Brian Ferguson.)

15.1 Introduction

I n Chapter 13 we described the physics of remote sensing, and Chapter 14 shows how various payload types can be employed for UAS missions. This chapter builds upon these topics to show payload system integration impacts on the UA and other system elements. The payload field-of-regard needs and associated flight profiles are presented for optical and radar systems. System data management as a design problem is particularly important for UASs, where the amount of data collected can exceed the downlink and management capacity. Weapon system interfaces on the UA are shown for propelled and droppable munitions or other air-launched systems. The system avionics, software, and ground element interfaces are described. The UAS only has value with proper payload integration.

15.2 Optical Payload Assembly Layout, Actuation, and Stabilization

Optical payloads are the most common type employed on UASs. Most are surveillance EO/IR balls that provide near real-time full-motion video. Some also provide targeting support, such as laser designation for guided munitions. Optical reconnaissance systems optimized for area coverage and still imagery are used less frequently, but provide high-value intelligence. Given the prevalence and importance of optical payloads, we will spend some time exploring their configurations and design attributes. The optical payload design has significant implications for the UA design, and so the system interactions are described. Leachtenauer and Driggers [1] can be consulted for further reading.

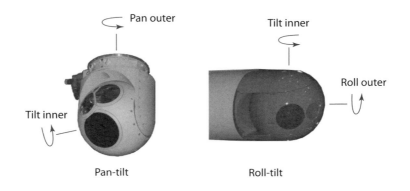

Fig. 15.2 Primary sensor actuation configurations.

Optical systems require actuation to enable pointing and mechanical stabilization. The alternative of a sensor rigidly mounted to an airframe would be fully dependent on the UA's flight dynamics to control the orientation, which would impose major limitations on the ability to collect against targets. Stabilization is increasingly important as sensor field of view decreases.

There are three major sensor actuation configurations: pan-tilt, pan-tilt-roll, and roll-tilt gimbals. Figure 15.2 shows the pan-tilt and roll-tilt configurations. The configuration selection depends on the type of imagery that is collected, weight limitations, geometric constraints, and cost. Pan is rotation in azimuth or about the vertical sensor axis. Tilt is elevation change of the sensor where 0 deg is horizontal, -90 deg is nadir, and $+90$ deg is straight up. Roll is rotation about the forward axis where 0 deg is nadir, -90 deg is left horizontal, and $+90$ deg is right horizontal. The first term in the title represents the outer axis, and the following terms are the following axes in order.

Pan-tilt sensor assemblies are the most common configuration for surveillance missions involving full-motion video. Here the outer stage is pan, and it might have continuous 360-deg rotation in azimuth. The inner stage is tilt, which provides elevation control. For ground-looking sensors mounted on the underside of the UA, the tilt stage sweeps through less than 180 deg, with less than vertical on the upper side and past nadir on the lower side. With lower-mounted pan-tilt balls, the upper elevation is blocked by the pan actuation assembly above. This configuration most closely resembles the human head in that it can look side to side and up and down, allowing the operator to apply intuition in controlling the sensor and describing the imagery.

Pan-tilt-roll sensors add a roll stage inside the tilt stage to enable the image to appear horizontal relative to the horizon even when the UA performs banking maneuvers. The primary advantage is that the high-rate sensor pointing is effectively decoupled from the UA dynamics. Roll stabilization for rejecting vibration and high-rate roll motion helps to improve user experience.

Roll-tilt sensors have roll as the outer stage and tilt as the inner stage. A key advantage to this configuration is the ability to incorporate large optics with long focal lengths relative to pan-tilt and pan-tilt-roll configurations. Roll-tilt sensors are commonly used on reconnaissance optical sensors using framing cameras or push broom techniques, synthetic aperture radars, and hyperspectral imagers. The roll stage angular travel limitations depend on the gimbal assembly. Sensors mounted on the underside of the UA can traverse approximately 180 deg in roll centered about nadir. Nose-mounted sensors can traverse 360 deg continuously. The tilt stage is generally much more limited, as these sensors tend to have a small diameter relative to the length. The tilt stage changes elevation angle of the sensor when the roll stage is pointing nadir. The tilt stage controls azimuth angle when the roll stage is at $+/-90$ deg. One difficulty with the roll-tilt sensor is that the vertical horizon direction flips when the sensor traverses nadir in roll, thus making surveillance using full-motion video less intuitive than pan-tilt sensors.

Sensors must be stabilized to maintain sensor pointing accuracy for most airborne applications, especially when the field of view is narrow. As we can see from earlier in this section, there are numerous actuation methods for sensors. The axes of control—potentially including pan, tilt, and roll—are also the stabilization axes.

The axes are controlled by actuators. Some forms include integral stepper motors, motor-belt drive, and motor-gears. There are multiple levels of stabilization ranging from low- to high-frequency responses. Low-frequency stabilization helps to keep the sensor pointed at the target during unmanned aircraft maneuvering. High-frequency stabilization is used to reject vibration and high-rate unmanned aircraft responses to turbulence.

EO/IR balls often have multiple co-boresighted sensors. These sensors can include optics of various bands (visible, SWIR, MWIR, LWIR, image intensifiers) and fields of view, laser range finders, laser designators, or laser illuminators. The packaging of the desired sensors becomes a difficult multidisciplinary challenge involving mechanical, electrical, and optics design.

An EO/IR ball is conceptually a sphere. For pan-tilt sensors, slices are taken out of the side for tilt actuation. A cylindrical upper section, the *rooftop*, controls the outer pan axis (azimuth). Sensor control and imagery-processing electronics can be housed in the rooftop. It is generally desirable to minimize the size of the rooftop to limit airframe integration impacts. The inner slice of the sphere contains the sensors and any additional mechanical stabilization stages. This slice is a convenient interface for inter-changeable sensor suites. The IAI POP payload has multiple payloads that can plug into this inner section. These slices can be line replaceable units to help supportability.

The ball must be sealed against the atmosphere to protect sensitive components. Sand, dust, and water must not penetrate the environmental housing. The payload sensors, sensor electronics, actuation, and position feedback sensors may have low tolerance to these elements. Humidity inside the ball must be maintained within limits to prevent condensation. Sand in some operational environments might be very fine, and it can get into presumably sealed spaces.

Optical windows can cold soak at altitude where the outside tempera-ture is low relative to the sea level. Water can form on the windows when the UA descends from altitude, fogging them up and impairing the functionality of the sensor. One remedy to this fogging problem is to heat the windows so that water will not condense.

The sensors are selected based on available technology and requirements. The camera core providers might be different than the lens manufacturers. Boards might need to be repackaged to fit within the geometric constraints. Often the camera component manufacturers will not sell their hardware directly, but rather go through distributors who can also serve the function of integrating other hardware. It might be necessary to have a folded optics path to fit within the form factor. In the case of current MWIR sensors, active cooling is required, and the heat must be rejected.

The length of the optics path is a major driver. Folded optics and mirror lenses help reduce length. When one considers the dimensions and weight of the core optics vs the integrated EO/IR ball, it is striking how much overhead is necessary.

When sensors are aligned, this is referred to as *co-boresighting*. Perfectly parallel sensor orientation crosses at infinity. Other applications might require that the sensors have overlapping pointing at a fixed distance. This alignment can drift over time requiring readjustment or algorithmic compensation. Co-boresighted sensors require ground testing to estimate alignment errors. This can test the target location error if the target position is precisely known. Recalibration and mechanical adjustment is often necessary to maintain tolerances. The ball experiences substantial loads, shock, vibration, and thermal variations in operations. Opening the ball might not be possible in field conditions, in which case the adjustments can take place at a depot or original equipment manufacturer. Algorithms that compensate for known alignment errors can adjust the pointing when switching the active sensor.

The EO/IR ball might have inner stabilization stages. There is angular freedom between the lens and the outer window. The window must be sized to accommodate relative movement. There is a trade between obscuration and outer ball dimensions. Shared windows reduce obscuration of the window frame, but also reduce materials choices if the spectral separation of the two sensors is too great.

Fixed domes permit relatively lightweight gimbaled sensors that do not need to withstand aerodynamic loads. The optics quality might be compromised by dome thickness variations, geometric imperfections, available material optical characteristics, and variations in sensor-dome grazing angle as a function of sensor position. Plastics such as acrylic offer a lightweight solution, but are often limited to a single band. Multiband domes might necessitate a milled and polished solid dome, which can be quite heavy and expensive relative to smaller dedicated apertures for the EO/IR ball.

Command and optics data signals must be passed between the airframe to the inner stages. Slip rings are the most traditional approach, allowing continuous pan. Early EO/IR payloads and some inexpensive miniaturized solutions use cables that must not be wound up. Cables introduce friction, unbalanced loads, and hysteresis.

Cameras, laser devices, processors, and actuators can generate heat that must be rejected. Sterling cycle coolers, conductive cooling, and fans move the heat from the sensors to the external airflow. Sensors might have limited operating times on the ground in hot conditions. This drives either rapid system check-out times or turning the sensors on in flight. Cooled sensors might need a lengthy startup time to bring the FPA within the allowable temperature range.

Small UASs might have exposed actuated sensors. These systems are driven by low cost and might have short flight durations with little weather tolerance. An example is the Lockheed Martin Desert Hawk with an exposed roll-actuated full-motion video sensor. The advantage is simplicity and no performance degradation due to a window. The drawbacks include potential low sensor life, aerodynamic loads impacts on pointing stability, and increased unmanned aircraft drag.

Roll-tilt balls can provide a very low drag installation in the nose of the unmanned aircraft. By comparison, pan-tilt payloads often protrude below the airframe OML, producing a large drag coefficient.

15.3 Sizing for Performance

Now we will consider the combined effects of optical payload sizing and UA sizing. Recall from Chapter 13 that the ground sample distance, a measure of resolution, is proportional to the slant range and inversely proportional to the focal length for a given pixel pitch:

$$GSD \propto \frac{P}{f} \cdot R \tag{15.1}$$

or

$$GSD \propto \frac{R}{f} \tag{15.2}$$

Therefore, it is necessary to either double the focal length or halve the slant range to halve the GSD.

If the camera of interest is the largest component within the payload, then the payload assembly size can be assumed proportional to the focal length. In the case of an EO/IR ball, that characteristic dimension is the ball diameter D_{Ball}.

$$D_{Ball} \propto f \tag{15.3}$$

And therefore

$$GSD \propto \frac{R}{D_{\text{Ball}}} \tag{15.4}$$

Rearranging to solve for the ball diameter yields

$$D_{\text{Ball}} \propto \frac{R}{GSD} \tag{15.5}$$

Like many electronics equipment types, the EO/IR ball weight is nearly proportional to the cube of the characteristic dimension, assuming similar proportions.

$$W_{\text{Ball}} \propto D_{\text{Ball}}^3 \propto \left(\frac{R}{GSD}\right)^3 \tag{15.6}$$

If we further assume that the payload mass fraction for a clean sheet UA design is constant, then the takeoff gross weight is proportional to the payload weight:

$$W_{\text{TO}} \propto W_{\text{Ball}} \propto \left(\frac{R}{GSD}\right)^3 \tag{15.7}$$

The power-to-weight ratio (or thrust-to-weight ratio) is nearly constant within a design approach. The engine power can be related to the other parameters by

$$P_{\text{Eng}} \propto W_{\text{TO}} \propto \left(\frac{R}{GSD}\right)^3 \tag{15.8}$$

If we neglect atmospheric sound absorption and thereby assume that the acoustic power is proportional to the propulsion power, the slant range for nondetection by ground observers is proportional to the square root of the propulsion power.

$$R \propto \sqrt{P_{\text{Eng}}} \tag{15.9}$$

Substituting yields

$$P_{\text{Eng}} \propto \left(\frac{\sqrt{P_{\text{Eng}}}}{GSD}\right)^3 \tag{15.10}$$

Solving for GSD yields

$$P_{\text{Eng}} \propto \frac{P_{\text{Eng}}^{3/2}}{GSD^3} \quad \text{and} \quad GSD^3 \propto \frac{P_{\text{Eng}}^{3/2}}{P_{\text{Eng}}} = \sqrt{P_{\text{Eng}}} \tag{15.11}$$

$$GSD \propto P_{\text{Eng}}^{1/6} \propto W_{\text{Ball}}^{1/6} \propto \sqrt{D_{\text{Ball}}} \propto \sqrt{f} \tag{15.12}$$

This relationship leads us to quite a different solution than the original relationship that we started with. When we link the UA sizing with the focal length and acoustic detection with the slant range, we see that small UAs using smaller payloads have real advantages. It is possible for a small UA to provide equivalent imagery quality as a MALE or HALE UAS, but at a fraction of the UA cost.

These are highly simplified relationships intended to explore integrated-sensor–UA impacts. Specifics of UA sizing analysis, sensor performance modeling, and acoustics detection will undoubtedly change the exponent of the GSD to focal length relationship. Nevertheless, experience has shown that small UASs have operational imagery performance that is much greater than the size alone would suggest.

This focal-length–UA sizing relationship is not unreasonable. Consider a consumer hand-held multimegapixel camera used for taking images of people. Such a camera has a small focal length in order to be hand portable. The image quality of a human face is impressive when the slant range is perhaps 10–15 ft, with perhaps thousands of pixels covering a face. Now consider the size of the telephoto lens required to generate the same image at say 20,000 ft (nearly 4 miles) slant range with equivalent quality. Such a camera system would be unwieldy and expensive. Pointing and stabilization would be challenging to maintain the same scene within the field of view. Such a camera might not be a hand-held device.

So why are large UASs used? Small payloads on small platforms might have poor stabilization due to higher angular rates, response to turbulence, and gimbal size limitations. Larger UASs will generally support longer range and endurance due to higher Reynolds-number aerodynamics and more efficient propulsion. The inertia of larger gimbal assemblies supports stabilization. Also, the UA might be required to carry multiple payloads, and many of these can weigh substantially more than the imager. Higher-altitude flight might be required for survivability or airspace integration, and therefore larger sensors are required for the long slant range. Higher-altitude flight enables broader area coverage. Large UASs can carry more sophisticated avionics, which can support better target geolocation accuracy. At long communications range, the minimum line-of-sight altitude can create a limit on the minimum slant range and drive the minimum payload size for small UASs. Finally, there might also be a perception that bigger UASs will perform better, even if the physics do not necessarily support this belief.

15.4 Field-of-Regard Requirements

Field of regard (FOR) is the full range of what can be seen by the sensor. It is the combined elevation and azimuth regions that can be viewed by

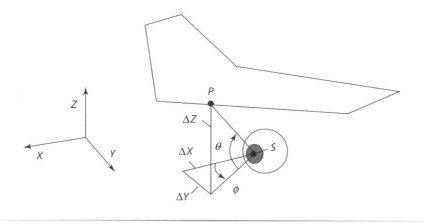

Fig. 15.3 Field-of-regard geometry.

a sensor without obstruction. The FOR depends upon sensor motion, camera fields of view, and the geometry of the UA as viewed from the sensor location. Note that FOR is the combination of all possible unobstructed fields of view for the payload. Line-of-sight field of regard is important for payload integration or high-frequency antennas.

The FOR geometry is illustrated in Fig. 15.3. The distance from sensor point S to the obstruction point P in three dimensions is given by

$$\Delta X = X_P - X_S \tag{15.13}$$
$$\Delta Y = Y_P - Y_S \tag{15.14}$$
$$\Delta Z = Z_P - Z_S \tag{15.15}$$

The azimuth angle, called the squint angle, to the obstruction ψ is

$$\psi = \tan^{-1}\left(\frac{\Delta Y}{\Delta X}\right) \tag{15.16}$$

The elevation angle to the obstruction θ is

$$\theta = \tan^{-1}\left(\frac{\Delta Z}{\sqrt{\Delta X^2 + \Delta Y^2}}\right) \tag{15.17}$$

With these relationships every known point on an UA can be evaluated to determine the obstructed and unobstructed field-of-regard regions.

Example 15.1   Field-of-Regard Mercator Plot

Problem:

Consider a simple aircraft geometry consisting of a main wing of 20-ft span and 2-ft chord, a horizontal tail with a span of 8 ft and a chord of 1 ft, a fuselage of 10 ft length and 2 ft diam, and wing two side nacelles of 3 ft length and 1 ft diam. The wings are flat plates, and the bodies are circular cylinders. The basic geometry is shown in Fig. 15.4. The coordinates for the front center of each element are shown in the following table:

Element	x, ft	y, ft	z, ft
Fuselage	4	0	0
Right nacelle	1	−4	−1
Left nacelle	1	4	−1
Wing	0	0	1
Horizontal tail	−5	0	0
Sensor	3	0	−1

Generate a Mercator plot of the sensor field of regard.

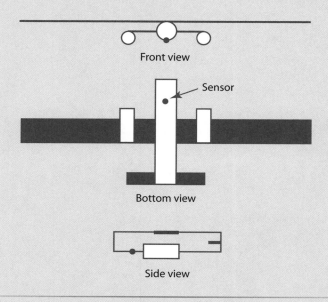

Fig. 15.4 UA geometry.

Solution:

The Mercator plot of the field of regard is shown in Fig. 15.5.

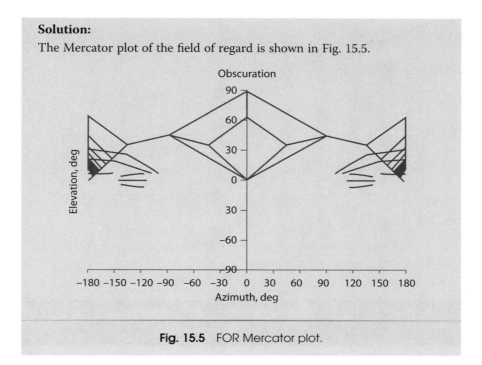

Fig. 15.5 FOR Mercator plot.

The matching of installed payload unobstructed field of regard to the mission-derived sensor collection profiles should be mathematically analyzed at multiple collection profiles. The analysis process is as follows:

1. Identify aircraft flight geometries, including collection altitude and bank angles as a function of loiter patterns.
2. Determine the range of target pointing vectors in aircraft coordinates to define field-of-regard requirements.
3. Iteratively modify UA geometry and sensor placement using Mercator plots.

Surveillance missions often involve circling around a target for extended periods of time. The collection geometry is illustrated in Fig. 15.6. The target elevation angle at +90 deg squint angle (off to the side) is given by

$$\theta = \phi_{\text{Bank}} + \theta_{\text{Look}} - 90 \deg \qquad (15.18)$$

The look angle is given by

$$\theta_{\text{Look}} = \tan^{-1}\left(\frac{h}{GR}\right) \qquad (15.19)$$

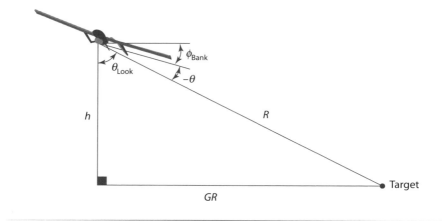

Fig. 15.6 Collection angles when circling around the target.

Example 15.2 Defining Sensor Elevation Angle for Circling the Target

Problem:

We wish to determine the required sensor elevation for a new UAS. Operational altitude of interest includes 0; 2,500; 5,000; and 10,000 ft. The turn radius will be between 500 and 5,000 ft. The flight velocity for collection will be 100 kt.

What is the sensor elevation angle for these cases?

Solution:

The bank angle as a function of turn radius at 100 kt is shown in the following figure:

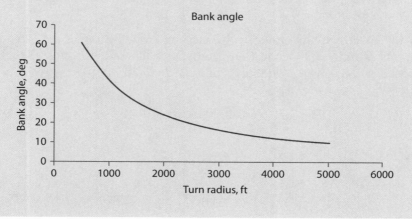

The look angle is shown in the following figure:

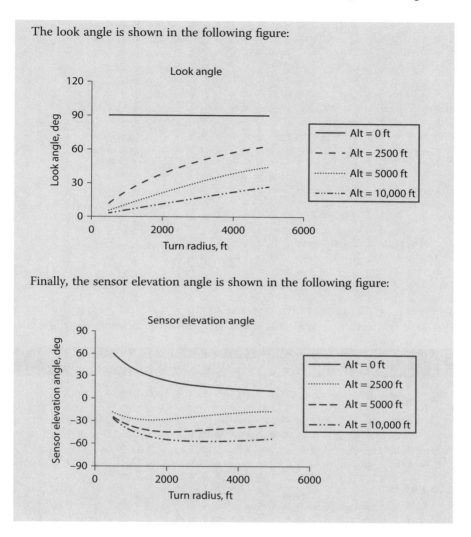

The bank angle for a level flight coordinated turn is

$$\phi_{Bank} = \tan^{-1}\left(\frac{V^2}{g \cdot GR}\right) \qquad (15.20)$$

where V is the flight velocity and g is the acceleration of gravity.

Reconnaissance collection often involves flying straight-line paths. In one reconnaissance scenario, a single target can be imaged continuously with a constant cross range, where the UA flies parallel to the target. Reconnaissance collection geometry is illustrated in Fig. 15.7.

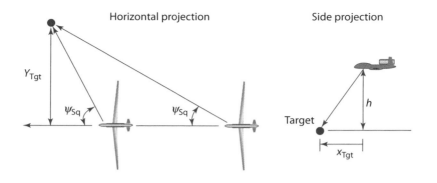

Fig. 15.7 Side-looking collection geometry.

The squint angle is given by

$$\psi_{Sq} = \tan^{-1}\left(\frac{Y_{Tgt}}{X_{Tgt}}\right) \tag{15.21}$$

where Y_{Tgt} is the cross-range distance to the target and X_{Tgt} is the target distance downrange distance that is positive ahead of the UA.

Example 15.3   Reconnaissance Mission

Problem:

The UA performs its reconnaissance mission at an altitude of 50,000 ft. We are interested in target cross-range values of 0.5, 2, 4, and 8 n miles. The target will be continuously tracked when X_{Tgt} ranges from 20 to −20 n miles.
  What is the required field of regard to perform this mission?

Solution:

The results are plotted in the following figures:

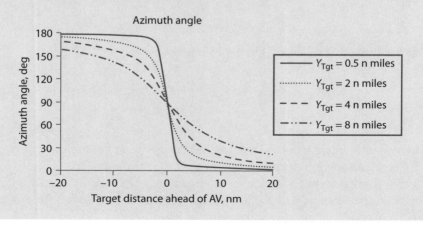

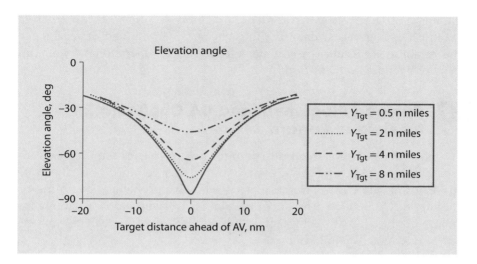

The elevation angle is

$$\theta = -\tan^{-1}\left(\frac{h}{\sqrt{Y_{Tgt}^2 + X_{Tgt}^2}}\right) \qquad (15.22)$$

Often the target is acquired in a reconnaissance mode. The UA flies towards a target area and begins the search prior to arriving. The target is acquired in the front sector, often with some cross-range component. Once the UA flight path is parallel to the target, the UA banks such that the target is at the center of the turn.

The preceding methods for calculating the field of regard consider a fixed-point sensor with infinitely narrow field of view. Real optical sensors have a nonzero field of view and a finite lens diameter. The boundaries of a sensor field of view are offset from the center of the frame. The field-of-view requirements should account for the sensor horizontal and vertical field-of-view angular width. Some obscuration can still yield useful imagery, and so identifying the percentage obscuration contours on the field-of-regard plots reveals the extent of blockage in the frame.

Although the idealized field-of-regard methods for surveillance and reconnaissance presented here provide readily quantifiable visibility requirements, more rigorous analysis can provide additional insights. One approach is to fly a simulator that includes the actuated payload against a realistic collection environment. The UA model has all of the correct flight attributes, such as flight envelope and turning radii. The sensor can represent targets through synthetic vision. The UA operator (AVO) and mission payload operator (MPO) fly the mission in the simulator. Based on the recorded simulation session's sensor pointing azimuth and elevation data, the most heavily utilized field of regard can be established. Field-of-regard data from similar operational UASs can also provide this information.

15.5 Payload Placement and UA Configuration for Field of Regard

Now that we have an appreciation for field-of-regard requirements for surveillance and reconnaissance missions, we turn to how to best integrate payloads on the UA to achieve these needs. This problem is by no means unique to UASs. The procedures are similar to laying out the canopy for pilot visibility. EO/IR ball integration is similar to placing ball gun turrets on World War II bombers.

Payload field of regard is a critical design driver, yet many UAS designs address this need late in the design process or even after the system enters service. Payload performance is the reason for the UA's existence. Airframe blockage of where the sensor needs to view degrades the system utility.

Regions where the sensor pointing is concentrated should not be obscured by the airframe. For example, a surveillance payload will spend much of its time looking on either side sector. The sensor path between these regions should also not be blocked. An example is when a pan-tilt sensor traverses from the left to the right side of the aircraft when the bank angle changes in a figure-8 orbit pattern.

The opposite is also true, that is, angular regions that the payload does not need to view can be blocked by the airframe. An inefficient requirements' developer might specify that the sensor be able to view 4π steradians (spherical), 2π steradians (hemispherical), or other very large regions. In the case of a 4π-steradian requirement, the only compliant system that the author knows of would be Wonder Woman's invisible jet. System developers will struggle with aggressive FOR requirements, often forced to develop complex systems to accommodate viewing angles that are not needed. Any viewing angles that are not used by the sensor should be available for obscuration. The obscuration can take the form of internal structure, airframe components (wings, fuselages, tails, etc.), propeller disks, external landing gear, and engine nacelles.

Here are some common forms of blockage that can adversely affect mission performance:

* Payloads located near the center of gravity on a low-wing aircraft are blocked by the lowest wing while orbiting a target. This may be particularly evident on blended-wing-body UA or flying-wing configurations with a wide root chord that fly at low altitude.

- Payloads located on the lower fuselage are blocked by engine nacelles on the wing while banking.
- Roll-tilt payloads are employed for surveillance (rather than reconnaissance), but are either incapable of pointing forward or are blocked by the forward fuselage. The target is lost if the UA maneuvers.
- Wing payload pods have fuselage obscuration on one side.
- Payloads integrated within a body have insufficient window (aperture) dimensions.

Now that we've seen some potential problems, let's explore techniques for favorable field of regard. These design options include the following:

- Use a high wing to reduce fuselage-mounted payload side sector blockage while banking.
- If a low wing is used, mount side-looking payloads (HSI, SAR, and optical reconnaissance types) ahead of or behind the wing.
- Locate surveillance payloads in the nose. This permits forward field of regard and broad side sector field of regard.
- Adequately size the windows for payloads integrated into fuselages or pods.
- If aft field of regard is required for surveillance, locate the payload below the airframe. Note that this produces large drag relative to a more integrated approach.
- Use multiple similar payloads on the UA. Although this can provide broader field of regard, the increased payload weight and cost relative to a single unit is rarely the optimum solution.

Common unmanned aircraft payloads and communications systems can be hampered when utilizing unmanned conversions of manned aircraft, making selection of a candidate challenging. The field of regard not blocked by structure is optimized for pilots, crew, and passenger visibility and is directed more upwards than downwards, which is opposite of the needs of most unmanned missions such as ISR, mapping, or ground-to-ground communications relay. Internally mounted, downward, or side-looking payload installation is typically blocked by primary fuselage structure. Satellite communications do have similar field-of-regard requirements as pilots—especially for aircraft with bubble canopies—so replacing the canopy with a radome might permit largely unobscured field of regard. Some aircraft do contain suitable payload provisions. For example, the twin-engine Diamond DA-42 MPP has a nose payload bay suitable for EO/IR balls and a belly hardpoint along the center of gravity. Single-engine propeller aircraft are dominated by tractor configurations, which partially blocks the forward field of regard and can foul sensors with oily engine exhaust. Most pusher single-engine conversions are canards, as seen on the L-3 Mobius and Proxy Aviation SkyWatch.

15.6 RF Payload Integration

In Chapter 12 we explored communication systems and how to integrate antennas on to UA. RF payloads follow most of the same approaches, particularly for communications relay or SIGINT missions. Rather than retread this now-familiar ground, this chapter covers high-power RF systems such as jammers or systems employed for directed-energy electronic attack.

High-power RF energy emitted from payloads can interfere with the primary UAS communication systems and generate EMI that can cause avionics failure or degradation. Jammers can radiate in a broad pattern if the threat radar location is unknown. This creates the need to protect the UA systems from this largely nondirectional RF source. Other jammers and directed-energy systems have narrow main lobes pointed in the direction of the target. Here RF energy can make its way to UA systems through side lobes or energy scattered by the radome or other airframe reflections.

Several methods exist to protect the UA from this potentially harmful EMI. General shielding approaches to protect avionics from EMI are covered in Chapter 10. It might not be possible to fully protect systems, and so fault-tolerant avionics might be necessary.

The EMI becomes a problem for communication systems when the payload emitter generates RF power at the communications frequencies. Even if there is frequency separation, the emitter can still produce sufficient power out of band to create interference. The problem is much worse if the communications systems and RF payloads operate at nearly the same frequency. A tightly integrated system can share the same RF systems between communications and payload operations to ensure that both are satisfied.

15.7 Airframe Mechanical Integration

All too frequently, payload accommodation is not considered in the initial design of an unmanned aircraft. Companies advertise payload volume and weight-carrying capacity, though close inspection of the aircraft reveals that internal peculiar geometry payload bays are partially populated by flight critical systems and the available field of regard is poor. Often desired window or radome locations are in the path of load-bearing structure.

Payloads can be carried internally or externally. External payload locations generally attach to hardpoints and have significant geometric

Fig. 15.8   X-45A drops a GPS-guided bomb. (NASA photo by Jim Ross.)

flexibility. However, these payloads produce drag that degrades flight performance and can introduce controllability issues that require analysis
and testing to characterize. Problems encountered by new external
payload configurations might require remediation through airframe
design changes or new flight control laws.

Hardpoints should be efficiently tied into the aircraft primary structure.
Wing hardpoints must have load paths to the wing spar(s) via rib structures
or brackets. Fuselage hardpoints can affix to existing bulkheads or keels.

Internal payloads are housed in available airframe volume or dedicated
payload bays. Internal structural mounting must be provided. Distinct,
replaceable payload bays are desirable, permitting rapid field reconfiguration. The bays should have intuitive geometry, preferably with a
constant cross section along their entire length. Rectangular cross sections
are easiest for new payload integration and can be quickly described
without confusion.

Air-launched payloads risk interference with the UA. It has been said
that "the trick is to drop the weapon without dropping the aircraft." For
dropped payloads, the risk is highest when the aerodynamic forces acting
upon the released body are sufficient to send it towards the aircraft. The
body forces are dependent upon the dynamic pressure (flight speed and

altitude), body shaping, orientation, weight, and the combined geometry of the body and host UA. This risk might be low for dense payloads dropped from slow-flying platforms. The risk is higher for release of bodies from internal weapons bays of high-speed UAs, such as UCAVs. Figure 15.8 shows a successful weapons release from a Boeing X-45A.

Self-propelled air-launched payloads pose additional challenges. Rocket-powered payloads such as missiles can generate plumes that must not adversely impact the airframe or propeller. Figure 15.9 shows the DRS Sentry HP with Spike missiles on rails located outboard of the propeller arc. Missile plumes should not degrade payloads or ablate aerodynamic surfaces. In the event of a hang fire, which means that the rocket motor fires but the body does not leave the UA, the UA stability and control will be adversely affected. A critical decision is whether to make the UA control system robust enough to handle a hang fire without losing control.

Weapons and other droppable payloads must have airframe interfaces such as racks or rails. The interface hardware should be removable from the UA if the droppable payload mission is not the primary mission, as this hardware has weight and drag impacts. The interface hardware

Fig. 15.9 DRS Sentry HP with Spike missiles.

Fig. 15.10   Hellfire missile loaded on a MQ-1L Predator. (U.S. Air Force photo taken by TSGT Scott Reed.)

provides the release mechanism, such as a solenoid or hold-back mechanism. Often power and data interfaces are required as well. If removable, the interface hardware can generally be considered payload weight. An example weapons interface is the M-299 hellfire launcher (see Fig. 15.10), which weighs 96 lb in the two-missile configuration and interfaces to the UA hardpoints by 14-in. lugs. Figure 15.11 shows the Bat glide bomb mounted on wing racks on the MQ-5A Hunter UAS.

UCAVs that have reduced radar signatures carry the weapons internally. This requires fuselage volume roughly on the center of gravity. Most UCAV designs are low-aspect-ratio delta wings or flying wings with a blended wing and fuselage. The weapons bay and jet propulsion compete for the same internal real estate. Figure 15.12 shows small-diameter bombs integrated on the Northrop Grumman X-47B mock-up.

U.S. weapons systems are usually required to comply with a MIL-STD-1760 interface and a MIL-STD-1553 data bus. These standards offer high reliability and compatibility with manned aircraft hardware. The mission management system must be designed to accommodate these standards.

UASs with weapon systems must have a fire control system. To successfully employ a weapon, the target must be found, the target must be fixed in geospatial coordinates, the targeting solution must be established with

Fig. 15.11   Northrop Grumman Bat munition mounted under the MQ-5A Hunter wing.

the weapon system, permission to fire must be granted, weapon must be launched, weapon flies to the target with tracking, the target is hit, and the warhead detonates. The UAS system—whether on the ground station, UA, or weapon itself—must perform all of these functions. Today it is necessary to have a person in the loop for many of these steps, which must be accommodated by ground station hardware and software.

Weapons integration tests start with ground tests and then progress to flight tests. Ground tests can include system integration lab (SIL) tests,

Fig. 15.12   Weapons bay door open on Northrop Grumman X-47B UCAV mock-up.

ground vibration tests (GVT), and drop tests. The flight testing can include airborne separation tests, handling qualities evaluation, guided inert launches, and lastly, guided live fire launches.

15.8 Imagery Products

For ISR UASs, the quality of the imagery products is of paramount importance. A system with the most amazing UA but poor imagery is quite useless. In this section we will consider imagery products such as full-motion video (FMV) and still imagery.

15.8.1 Imagery Chain

Photons reach the outer sensor aperture, and an image is displayed on a monitor tens or hundreds of milliseconds later. What could be simpler? The technology and processes that occur between these two states are among the most complex of the entire UAS design. Producing a successful solution to convert photons to high-quality imagery products can determine whether a system is widely fielded or just another obscure UAS.

An intuitive approach to think of the imagery steps that occur throughout the system is the *imagery chain*. Like a chain, the quality of the imagery products can only be as good as the weakest link. Schott [2] details the links of the imagery chain. The imagery chain has the following elements:

* *Input scene*—The scene is illuminated by the sun, indirect illumination, self-illumination (i.e., LWIR thermal emissions), or by the UA (i.e., laser illuminator). The scene has contrast and various spectral responses. Photons from the scene radiate in the direction of the UA. See Chapter 13 for more details on illumination.
* *Atmosphere*—The atmosphere attenuates the scene through various mechanism described in Chapter 13.
* *Optics*—The optics include any outer apertures, obscuration, and the lens system. The photons can be magnified, spectrally filtered, obscured, attenuated, and distorted through the optics.
* *Detectors*—The detectors on the focal plane array convert the photons to a usable signal. The detector sensitivity is important. The detector integration time (i.e., shutter speed) influences blur for a moving scene.
* *Image processing*—Hardware and/or software convert the focal-plane-array detector signals and form an image. The image is formed into a usable standard, which can be analog or digital. Additional processing can include compression, contrast enhancement, color adjustment, and

multiframe enhancement. Image processing can change the image format multiple times throughout the system. The processing can occur on the camera, UA, communication system components, or ground station.

* *Data transmission*—The imagery data will be transferred among multiple systems, and losses can occur at each step. Often the most critical data transmission step is from the UA to the ground station.
* *Display*—The image is displayed to the user via a user interface. This can be a computer monitor, television set, or customized display device.
* *Human eye and brain*—Finally, humans must consume the data. The eyesight, cognitive abilities, and imagery analysis training vary greatly among individuals.

Elements of the imagery chain should be balanced for an effective system. Investments should focus on the weakest links rather than the strongest links. A common fallacy is to emphasize system elements that are familiar or have created problems on past programs without taking a systems-level view of the new system. What is the point of buying a multi-million dollar EO/IR ball if a $200 ground-station monitor degrades the image? Here are some examples of a broken or degraded imagery chain:

* High-definition (HD) video is converted to analog video.
* Pixels that are collected in the focal plane array are not used.
* Digital zoom is used instead of optical zoom.
* Data are lost in multiple image format conversions.
* Imagery is compressed on the UA to the point that it is no longer useful.
* The bit error rate of the RF communication system is too high, causing lost or degraded frames.
* Visible sensors are used at night or in low light conditions.
* The target employs effective camouflage.
* The scene is obscured by haze, clouds, or fog.
* A poor performing monitor is used in the MPO workstation.
* The MPO workstation processors are overloaded.

Latency is the enemy of full-motion video. Two forms of latency are important for imagery systems. The first is the time from image capture to image display. The second is the time from the payload operator command to the display of the payload response, which is known as the *roundtrip latency*. Roundtrip latency often involves commanded change in the sensor pointing. This can be thought of as a sensor joystick command followed by the display of the sensor response. The following can add to the latency:

* *Processor cycle time*—Processors operate at a frequency usually measured in megahertz (MHz) or gigahertz (GHz). These cycle times mean that calculations do not occur instantaneously.

- *Processing time*—Imagery calculations place a large processing load on software and hardware. Processing time can be a significant contributor to latency. Image processing can occur multiple times within the imagery chain.
- *Frame rate cycle time*—Full-motion video has a frame rate of 15–30 Hz. The scene is imaged in the time between frame capture. This is a periodic latency.
- *Data bus cycle time*—A data bus operates at a frequency. Commands and feedback occur at a maximum of once per cycle. The data bus cycle time is a form of latency.
- *Communications delays*—RF signals travel at the speed of light, which can generate latency at long range. This can be pronounced for multihop satellite communications. These delays can be observed in live news coverage for events that occur on the other side of the globe.

Pointing accuracy is a measure of the estimated sensor's projected ground position to the actual ground position. The target location error (TLE) is a common metric for pointing accuracy. To estimate the ground position, it is necessary to know the geospatial position of the sensor, the orientation of the sensor, the field of view of the optics, and the terrain elevation at the target. Key sensors used to estimate the position include inertial sensors, GPS, and sensor angular position encoders. Digital terrain data can aid the estimation of the target elevation. Any differences in the time of the state estimation and the image time stamp will result in an error. The payload's ground projected location calculation can occur on the UA or in the ground station.

15.8.2 Video and Still Imagery

Imagery produced by UASs come in the forms of full-motion video and still imagery. Each form has attributes that are suitable for different mission sets. Here, we will explore the imagery products for each type.

Video is used primarily for surveillance or search missions. The observed scene changes with time, and the intelligence is time critical. While the video is often archived, these data are known as perishable intelligence because the value of the data quickly decreases as the time from collection elapses. The two major video classes are analog and digital video.

The most prevalent form of analog video is NTSC. Analog video requires analog video transmitters and analog video displays. Favorable attributes of analog video include low system latency and graceful degradation at the extents of the communication range. Analog video must be either stored on bulky video cassettes or digitized. A problem with analog video is that the full image is not collected simultaneously. Instead, every other horizontal line is collected at one interval, and then the other horizontal lines are collected in the next interval. This is known as interlacing. The problem with interlacing is that rapidly changing scenes appear as a double image when converted to a still image.

Some systems can use a combination of analog and digital video. The motivation is that some system components can generate or process only analog video. Some block cameras used in EO/IR balls output analog video. A so-called frame grabber or codec performs the conversion to digital video. The conversion can occur on the UA or on the ground.

Digital video is replacing analog video for most UAS applications. Common digital video standards include H.264, MPEG 2, MPEG 4, Motion JPEG, and JPEG 2000. H.264 is becoming a popular standard. These standards have varying degrees and methods of compression. Most use intraframe compression, where the data rate is reduced by using information about the changes between frames. The intraframe compression can induce latency of several tens of milliseconds. Recently EO/IR balls with HD video have entered the market as well.

Still imagery takes the form of continuous strip maps or rectangular images. Strip maps are generated by push-broom optical reconnaissance cameras, synthetic aperture radars, and hyperspectral sensors. Rectangular still images are generated by more conventional framing cameras. A popular still imagery standard is NITF 2.1.

15.8.3 Metadata Insertion

Metadata are information associated with imagery that provides context. Some metadata components include time of collection, location of platform, angles between the aircraft and the target, zoom angles of the optics, geo coordinates of the center of the image, and geo coordinates of the four corners of the image. Metadata associated with the image are useful for creating searchable imagery databases. The metadata can also assist in stitching the imagery to a three-dimensional software terrain map.

Various metadata standards have been used, including ESD, cursor on target (COT), key length value (KLV), and proprietary standards. The Motion Imagery Standards Board (MISB) is standardizing metadata. Table 15.1 provides excerpts from the MISB STD 0902.1 [3] minimum metadata set description.

Metadata do not necessarily need to be transmitted at the frame rate, although it is desirable to do so if communications bandwidth permits. Metadata sent at the full frame rate are known as *fast rate*. Some parameters

Table 15.1 Selected Metadata Information

Tag name	Range and units
Time stamp	Integer
Mission ID	String
Platform heading angle	0–360 deg
Platform pitch angle	+/−20 deg
Platform roll angle	+/−50 deg
Platform designation	String (i.e., Global Hawk)
Image source center	String
Sensor latitude	+/−90 deg
Sensor longitude	+/−180 deg
Sensor true altitude	−900 to 19,000 m
Sensor horizontal field of view	0–180 deg
Sensor vertical field of view	0–180 deg
Sensor relative azimuth angle	0–360 deg
Sensor relative elevation angle	+/−180 deg
Sensor relative roll angle	0–360 deg
Slant range	0–5,000,000 m
Target width	0–10,000 m
Frame center latitude	+/−90 deg
Frame center longitude	+/−180 deg
Frame center elevation	−900 to 19,000 m

change more rapidly than others and should be sent at higher update rates up to the fast rate. Such parameters include the UA orientation, sensor position/orientation, fields of regard, and target location. Parameters that can be sent at slower rates include the mission identification, platform designation, sensor identification, and security information.

The source of the metadata and the video frames can use different clocks. Ideally, the clocks are synchronized, but this is not always the case. Any systems that depend upon the time stamp will experience TLE degradation for any synchronization errors.

15.9 Software Integration

Most payloads require interfaces with both the UA and ground-station software. Payloads can be an integral part of the system, such as is common with EO/IR surveillance payloads, where the payloads are nearly always on the UA and the payload workstation is baseline in the ground control station. Most operational UASs are required to adapt to specialized payloads that might only be required for a small portion of the missions. Specialized payloads require adaptation of the system software—both on the air and ground.

The UA software interfaces generally include the payload command and control, receipt of unmanned aircraft state data, sending payload status messages, and providing the payload data output stream for storage or downlink. Sophisticated UASs have a distinct mission management system that interfaces to the payloads, data storage devices, and payload return link. Robust mission management architectures are segregated from flight-critical UA systems.

The ground payload software functionality includes the ability to command the payload, process payload state feedback, and, most importantly, process the payload's collected data to generate usable intelligence. These functions can be performed at the MPO workstation, separate mission management processors, or on other systems.

15.10 Avionics and Power Interfaces

Payload avionics interfaces usually take place across a data bus. An interface can include connectors that the payload wiring harness connects to. These connectors should be appropriately hardened for the electromagnetic and physical environment.

The data interfaces can use serial busses or networks. Common serial buses include RS-422 and the automotive heritage CAN bus. Ethernet is a popular network interface, though its use in aerospace applications can generate controversy. Military payloads, including weapons, often utilize the MIL-STD-1553 bus.

UA power is usually provided to the payloads. Payload development tests can use batteries as part of the payload, but this is rarely the case for operational systems. The UA's onboard power originates from the engine-mounted generator or, for electric or hybrid electric aircraft, the propulsion battery. The power is conditioned for the payload, which might involve an ac-dc conversion and spike reduction. Tactical and larger UAS generally provide 28-V dc power to the system. Some payloads require 12-V dc power or even ac power.

The power system must protect the UA from the payload. A common problem is that the payload draws more power than the UA can provide. A good design practice for the UA is to shut off payload power if the payload exceeds current limits. Fuses or circuit breakers can prevent damage to the UA. When practical, a separate power system for the payloads and flight critical system is desirable.

15.11 Payload Data Management

Payloads are capable of generating vast quantities of data that can outstrip the capabilities of other system elements. On the UA, the payload data must be sent to the communication downlink, compressed, stored, or deleted. Once on the ground, it may be displayed, processed, deleted, stored (temporarily or archived), retrieved from storage, or disseminated. The system must utilize payload data management techniques to ensure that the right payload data are available to the data consumers when needed.

The raw data rate associated with a digital video stream is

$$Rate_{\text{Data}} = H_{\text{Pixels}} \cdot V_{\text{Pixels}} \cdot \frac{Bits}{Pixel} \cdot Rate_{\text{Frame}} \cdot F_{\text{Compression}} \qquad (15.23)$$

H_{Pixels} and V_{Pixels} are the number of horizontal and vertical pixels, respectively. The bits/pixel, called *pixel depth*, is an average for the image and is dependent upon the image standard and if it is black/white or color. The pixel depth for color video can be $8-14$ bits/pixel. The frame rate $Rate_{\text{Frame}}$ is frequency of image generation. Recall that typical video frame rates are $15-30$ Hz. The compression factor $F_{\text{Compression}}$ depends on the image standard and the degree to which the image is compressed. Lossless video compression is the objective, but cannot be completely achieved for reasonable data rates. However, substantial video compression is possible without human operators noticing.

The data collected are often substantially greater than can be transmitted to the ground in real time. The data collected must be transmitted, stored, or discarded. Using a fluid flow analogy, the following relationship can be thought of as a continuity equation for data:

$$Data_{\text{Collected}} \leq Data_{\text{Stored}} \cdot F_{\text{Compress,Stored}} + Data_{\text{Trans}} \cdot F_{\text{Compress,Trans}}$$
$$+ Data_{\text{Discarded}} \qquad (15.24)$$

where

$$Data_{\text{Collected}} = \text{data that are output from the camera}$$
$$Data_{\text{Stored}} = \text{data that are stored on a data storage device}$$
$$Data_{\text{Trans}} = \text{data that are transmitted to the ground}$$
$$Data_{\text{Discarded}} = \text{deleted data}$$
$$F_{\text{Compress,Stored}} = \text{compression factor for stored data}$$
$$F_{\text{Compress,Trans}} = \text{compression factor for transmitted data}$$

The continuity analogy goes only so far. Unlike molecules in air, data can be copied or destroyed. That is why there is a less-than or equal-to sign in the equation.

Example 15.4 Wide-Area Surveillance

Problem:

A wide-area aerial surveillance mission (WAAS) uses 10 cameras with an overlapping field of view to provide area coverage. The following information is known about the system:

* Camera focal plane array: 1080×1080 elements
* 10 bits/pixel uncompressed
* 2:1 compression with low loss applied

    a. What frame rate is required to keep the data rate from the sensor system under 274 Mbps?

    b. If there is no communications to the ground and no data are discarded, how much data must be stored for 1 week over the target area? Assume that data are stored at the maximum sensor system data rate without further compression.

Solution:

a. The compression ratio $F_{\text{Compression}}$ is equal to $\frac{1}{2}$. The frame rate is

$$Rate_{\text{Frame}} < \frac{Rate_{\text{Data}}}{N_{\text{Cameras}} \cdot H_{\text{Pixels}} \cdot V_{\text{Pixels}} \cdot \dfrac{Bits}{Pixel} \cdot F_{\text{Compression}}}$$

$$= \frac{274 \times 10^6 \, \text{bits/s}}{10 \cdot 1080 \cdot 1080 \cdot 10 \, \text{bits/pixel} \cdot \dfrac{1}{2}} = 4.70 \, \text{Hz}$$

b. The data collected over the mission are

$$Data_{\text{Collected}} = Rate_{\text{Data}} \cdot Duration$$

$$= 274 \times 10^6 \, \text{bits/s} \cdot 1 \, \text{week} \cdot 7 \, \text{days/week} \cdot 24 \, \text{hrs/day} \cdot 3600 \, \text{s/hr}$$

$$= 1.66 \times 10^{14} \, \text{bytes}$$

The data storage requirement is given by

$$Data_{\text{Collected}} \leq Data_{\text{Stored}} \cdot F_{\text{Compress,Stored}} + Data_{\text{Trans}} \cdot F_{\text{Compress,Trans}}$$

$$+ Data_{\text{Discarded}}$$

$$= Data_{\text{Stored}} \cdot F_{\text{Compress,Stored}} + 0 + 0$$

Because there is no additional data compression prior to storage ($F_{\text{Compress,Stored}} = 1$), the data storage is equal to the data collected.

5.12 Ground Element Integration

Most modern ISR UASs have a mission payload operator (MPO) workstation. This position provides an operator with the ability to command payload functions and utilize the payload data. The MPO workstation is covered in Chapter 16. If the UAS is equipped with an EO/IR ball or perhaps a SAR, these payloads are controlled and managed from the baseline MPO workstation. There are almost limitless additional missions and associated payloads beyond the common types, and accommodating all potential types with a single system is generally not possible.

One approach to accommodating payload growth in the ground station is to have communications pass-through streams that can be used by additional workstations. The pass-through streams include payload commands and payload data return. The additional workstation must comply with an interface. Such a modular approach can partially decouple the baseline system from new missions. This can make development more rapid and potentially reduce software certification scope. The drawback is that this segregated mission system approach might be less useful than a more integrated approach.

A more modern and ambitious approach is to develop modular software that has the framework to flexibly adapt to new missions on the UA and on the ground control station. Although it is still not possible to adapt to every new payload, modular software can reduce the development time substantially for integrated payloads. Data sharing between payloads and with the UA can be accommodated, permitting sensor cross-cueing and data fusion. On the GCS, common services can permit payload data display and manipulation functions with limited new development.

Payload data that do not require near-real-time exploitation can be sent to external elements for archiving and later analysis. One such element is a tasking, processing, exploitation, and dissemination (TPED) element.

5.13 Payload Interface Control

Payload interface control documents (ICDs) are valuable tools for payload integration. Although UASs have numerous ICDs within the system boundaries, the payload ICD is documentation or design tools that mission customers will interact with the most. When properly done, these documents can provide most of the necessary information that is required to place a new payload on the aircraft. The term *document* is perhaps too limiting for this tool. The formats can include CAD models, software applications, or other forms of media. The payload ICD should contain information about mechanical, environmental, electrical, and software interfaces between the payload and the UA and other system elements.

Payloads physically mount to the UA and have geometric constraints. In general, there will be bolt patterns, vibration isolation mounts, rails, or chassis that are used to attach the payload to the airframe. The position of the mechanical fittings as well as the required tolerances should be identified in the ICD. The physical volume available to the payload should be described in drawings or potentially CAD files. The stand-off distance limits between the payloads and other UA elements should be specified to allow for vibration and deflection of the payload assembly under load.

UAs can have multiple payload bays or other mounting locations such as external hardpoints. The minimum and maximum weight capacity of each location should be identified. UAs have maximum forward and aft center-of-gravity limits, as well as lateral balance limits. The center-of-gravity limits frequently constrain the allowable payload combinations or what payloads can be expended in flight. For example, the nose is typically a poor location for weapons, since upon launching the center of gravity shifts aft. These limits can be a function of the flight weight and operating conditions. The aircraft will have a maximum operating weight that an trade payload and fuel weight. All of these considerations constitute the weight and balance elements of the ICD. The payload weight limitations by location and with fuel loading can be multidimensional and should be represented by tables or perhaps application software tools.

The payload environment should be documented. Relevant environmental considerations imposed on the payload by the UA include the following:

* Minimum and maximum temperature within an internal bay
* Method of heat rejection provided by UA (if any)
* Ambient temperature range for external payloads
* Pressure within an internal bay (for pressurized and unpressurized systems)
* Maximum accelerations in all axes
* Shock and vibration environment
* RF and EMI environment
* Payload RF emissions limitations
* Acoustics
* Magnetic environment and magnetic field limitations of payload (important if magnetic compasses are used for navigation sensors)

UAs generally provide power to the payloads, though some payload applications carry their own power supplies such as batteries. The power type and conditioning is very important. Knowledge of the power source will drive the payload power systems design. The payload power interface description should include connector types.

Payloads avionics connections usually occur on interface panels with connectors. The connector standards should be identified. If the pin-outs are custom, then the signal path for each pin should be identified.

Payloads generally require interfaces to the UA software. The communications protocol includes the expected data stream information. The UA software will provide payload commands to the payload either from the mission management system or tunneled from the ground stations. Payloads might require UA state data to avoid the weight impact of a dedicated payload INS. The payload will generate health and status information for use by other parts of the system.

15.14 Payload Modularity

Much attention has been given to the concept of payload modularity in recent years. The motivation for modularity is the difficulty experienced in adapting legacy UASs with new mission capabilities. Government program offices frequently do not have access to proprietary payload interface information to the degree necessary to integrate new payload systems without the involvement of the UAS contractors. It has been argued that standardized modularity features would enable the customer to more rapidly integrate new systems with a wider selection of contractors.

Defining modularity is difficult, but it notionally contains mechanical, electrical, and software design attributes. Payload modularity is more than a payload ICD, though the modularity features can be documented in the ICD. Modularity is a flexible approach to payload integration that drives design choices. The Insitu Integrator UAS is a pioneering system for payload modularity. It featured interchangeable nose and center-of-gravity payload bays that could rapidly adapt to new missions.

References

[1] Leachtenauer, J. C., and Driggers, R. G., *Surveillance and Reconnaissance Imaging Systems, Modeling and Performance Prediction*, Artech House, Inc., Norwood, MA, 2001.
[2] Schott, J. R., *Remote Sensing, The Image Chain Approach*, Oxford Univ. Press, New York, 1997.
[3] Motion Imagery Standards Board, "Motion Imagery Sensor Minimum Metadata Set," MISB STD 0902.1, 9 June 2010.

Problems

15.1 Use the UA geometry from Example 15.1 and the required field of regard from Example 15.3. Combine the required and available field of regard. Is this appropriate sensor integration?

15.2 An UA has a payload bay in the nose and on the fuselage near the center of gravity. The longitudinal center-of-gravity limits of the UA are 140–145 in. in the x direction from the reference datum. The maximum fuel capacity is 100 lb. The takeoff gross weight is 300 lb.

The relevant weight and x-location data are given in the following table:

Item	c.g. Location, in.	Weight, lb
UA empty	150	155
Fuel tank system centroid	140	—
Nose payload forward bulkhead	110	—
Central payload bay centroid	152	—

What is the minimum nose payload weight required to balance the aircraft for all permissible fuel and central payload weights?

15.3 Using the UA geometry from Example 15.1, generate a Mercator plot for a payload located a position $x = 6$ ft, $y = 0$ ft, and $z = 0$ ft.

15.4 Using the UA geometry from Example 15.1, generate a Mercator plot for a payload located a position $x = 0$ ft, $y = 0$ ft, and $z = 2$ ft.

15.5 An UA has a maximum sensor elevation angle of -30 deg. The airspeed is 100 kt, and the UA height above target is 7,500 ft. What is the minimum radius turn for which the target can be viewed?

15.6 A reconnaissance UA has a forward squint angle of limit 45 deg and an aft squint angle limit of 135 deg. The UA flies at an altitude of 45,000 ft. A target is observed at 10-n miles cross range. Plot the azimuth and elevation angles of the target path from first contact until it can no longer be viewed.

15.7 A long-endurance ISR UAS fleet operates for a year in the field. There are 6 orbits, which are target regions that are continuously observed. The payload is a 1080×1080 imager. The pixel depth is 12 bits/pixel. The frame rate is 30 Hz. The compression is 2:1. If all of the data are archived, how much data storage capacity is needed?

Chapter 16

Command, Control, Tasking, Processing, Exploitation, and Dissemination

- Define ground control station types
- Understand facets of mission planning
- Explore other aspects of ground control stations

Fig. 16.1   MQ-1 ground control station. (U.S. Air Force photo by Val Gempis.)

16.1 Introduction

T he first several chapters of this book cover the UA, which is what most people visualize when they think of a UAS. It is relatively easy to distinguish a Global Hawk, Predator, or a Raven from each other. Few people could as readily identify a ground control station by sight. Acquisition officials seldom if ever display desk models of ground control station shelters. Yet control stations are critical to the operation of UASs, and most operators only interface with the UA through these elements. This chapter helps to highlight the importance of this powerful system component. Further information on ground control stations can be found in Austin [1] and Fahlstrom and Gleason [2].

Currently fielded UASs require human operators in the loop. The connectivity between the operators and the UA is through RF communications links. So although *unmanned aircraft system* implies that no humans are involved, these systems might have the manpower equivalent to manned aircraft.

The operators in the control elements might not see the UA that they are controlling. The UA appears as an icon on a moving map display, the payload imagery is displayed on a monitor, and the view can be supplemented by a situational awareness camera. There is no sensation of speed or g's in banking maneuvers. The UA could be aesthetically pleasing or homely, large or small, but the icon on the map makes no distinction.

16.2 Control Element Functions and Personnel Roles

The most common form of a control element is the ground control station (GCS), which is ground based as the name indicates. Operators in the GCS directly control the UA and its payloads via RF communication systems. GCSs come in many forms and sizes depending on functionality and footprint requirements. These can range from handheld computers to entire facilities. Several forms of GCS exist, including the following:

* Launch and recovery element (LRE)
* Mission control element (MCE)
* Forward ground control station (FGCS)
* Remote viewing terminal (RVT)

The control segment may reside in an aircraft, ship, submarine, ground vehicle, or other nonstationary platform. In many scenarios, the control station cannot accurately be called a GCS when it is not fixed to the ground. The command and control functions largely overlap for stationary and moving control segments.

A tasking, processing, exploitation, and dissemination (TPED) element provides broader user interface with the UAS than a control element. TPEDs can coordinate the tasking and manage intelligence data from multiple different types of systems including manned and unmanned assets, to provide an integrated intelligence picture.

The two most common personnel roles are the UA operator (AVO) and mission payload operator (MPO). The AVO commands the UA during takeoff, flight operations, landing, and ground operations. The MPO controls the payloads and manages the data collected. Other control element roles that can be performed by dedicated personnel can include the following:

* *Mission commander*—The purpose is to ensure that the mission objectives are met.
* *Communications manager*—This person is responsible for communication system operations, which may be a significant role if complex SATCOM systems are used.
* *Intelligence specialist*—This person analyzes intelligence data beyond the functions performed by the MPO.
* *Weather observer*—This function is to predict weather for long-duration missions.

16.3 Mission Planning and Execution

The control stations must support mission planning prior to flight and then manage the mission once the UA is airborne. The planning includes flight routes and events. The user interface must be effective, or there will be high operator fatigue and UA losses.

16.3.1 Mission Planning

A complete coverage of the methods applied to GCS planning would require an entire text (one that has not yet been written). Here a few techniques are shown to demonstrate how waypoints are generated and how UA flight dynamics relate to the mission planning software.

The control segment must generate a mission plan prior to UA launch. A mission plan includes a route that consists of a series of waypoints. The waypoints are a combination of latitude, longitude, and altitude. The flight velocity and rate of climb are also defined at the waypoints. Actions such as turning on a payload, lowering landing-gear doors, or dropping a payload are also tied to waypoints.

The mission planning and control tools must generate plans that are executable by the UA. To achieve this, it is imperative to have knowledge of the UA capabilities. Some relevant UA states and modes include the following:

* Fuel load and fuel remaining
* Flight weight
* Landing gear extended or retracted
* Flap settings
* External stores (drag and weight)
* Airspace integration modes (lights, transponders)
* Generator power usage

For a given weight, altitude, airspeed, and modes, the mission planning and control tools must know the following UA capabilities:

* Maximum available rate of climb and descent
* Maximum and minimum airspeed
* Fuel burn vs airspeed
* Minimum turn radius

The mission planning and control tools must also have the necessary knowledge of the operational environment. Some potential operational environment factors can include the following:

* Point target locations or search area
* Terrain
* Political boundaries
* Three-dimensional available airspace, airspace classes, and flight corridors
* Keep-out zones
* Locations of roads and populated areas
* Threat locations and capabilities
* Weather
* Locations of friendly RF interference sources

Some aspects of the operating environment can be dynamic. Some environmental factors such as terrain, political boundaries, and population centers are static. Others such as target locations, other aircraft operations, weather, and threats can change within the duration of a sortie. The mission control function must be able to receive data updates and generate the necessary mission plan changes in while the UA is airborne. This is known as *dynamic retasking*.

The shortest distance between two points along the surface of a sphere is known as the *great circle distance*. Although the Earth is not perfectly spherical, this approximation still yields useful results. The great circle distance is important for relating flight distance between two waypoints. Mission plans are generated using waypoints, but performance calculations require knowledge of the distance flown for each segment. The spherical angular difference between two waypoints $\Delta\sigma$ is

$$\Delta\sigma = \tan^{-1}\left\{\frac{\sqrt{[\cos{(\phi_f)} \cdot \sin{(\Delta\lambda)}]^2 + [\cos{(\phi_s)} \cdot \sin{(\phi_f)} \cdot \cos{(\Delta\lambda)}]^2}}{\sin{(\phi_s)} \cdot \sin{(\phi_f)} + \cos{(\phi_s)} \cdot \cos{(\phi_f)} \cdot \cos{(\Delta\lambda)}}\right\}$$

(16.1)

where $\Delta\lambda$ is the difference in longitude, ϕ_f is the final waypoint latitude, and ϕ_s is the starting waypoint latitude. The great circle distance d_{GS} can now be found by

$$d_{GS} = R_E \cdot \Delta\sigma \tag{16.2}$$

where R_E is the Earth's radius ($\sim$6,372 km). This assumes that the altitude of the UA is negligible compared to the Earth's radius.

Coordinates are generally given in latitude and longitude, where each is expressed in the degrees, minutes, and seconds format. Latitudes are identified as north or south, denoted by N or S, respectively. Similarly, longitude is identified as east or west, denoted by E or W, respectively. For example, a latitude may be 30°20′10″S. The superscript ° is for degrees, ′ is for minutes, and ″ is for seconds. North and east are positive, and south and west are negative. Let this format be parameterized as $Deg^*Min^*Sec^*Dir$, where Deg is the degrees, Min is minutes, Sec is seconds, and Dir is the direction (north, south, east, west). Navigation requires calculations in decimal degrees. The conversion to decimal degrees Deg_{Dec} is

$$Deg_{Dec} = \text{sign}(Dir) \cdot \left(Deg + \frac{Min + Sec/60}{60}\right) \tag{16.3}$$

Example 16.1   Great Circle Distance

Problem:
A Global Hawk took off from Edwards Air Force Base and flew nonstop to the Royal Australian Air Force's Edinburgh base in Australia. The coordinates are shown in the following table:

Airfield	Latitude (Approximate)	Longitude (Approximate)
Edwards Air Force Base	34°54′20″N	117°53′01″W
RAAF Edinburgh	34°42′12″S	138°37′12″E

What was the great circle distance of this flight?

Solution:

The coordinates must first be converted to decimal degrees.

Airfield	Latitude (Approximate)	Longitude (Approximate)
Edwards Air Force Base	34.9056	−117.8836
RAAF Edinburgh	−34.7033	138.6200

Using Eq. (16.1), we get $\Delta\sigma = 120.9751$ deg.

Multiplying the angle difference in radians by the Earth's radius yields a great circle distance of 13,455 km.

Publications of the event state that the total distance flown in the historic flight was 13,219.86 km, which took 22 hrs. The difference between the calculated and publicized great circle distances can be attributable to the nonspherical shape of the Earth and airfield coordinate estimate errors in the calculations.

The mission planner and AVO interface should provide the ability to define flight patterns. The simplest are straight-line and arc segments. Orbit patterns include circular, elliptic, race-track, figure 8, and box patterns. Area coverage patterns include raster scan and zig-zag geometry. The different flight patterns are selected based on mission objectives and system constraints. Figure 16.2 shows the different flight patterns. Observation of a fixed point in low winds favors circular or race track orbits. If strong steady winds are present, the best orbit might be a race track. Complex patterns such as raster scans or figure-8s can be built up from multiple primitive straight-line and arc segments.

The orbit patterns must take into consideration the flight dynamics of the UA. A fixed-wing UA might not instantaneously turn because it requires a turn radius that is dependent upon airspeed and bank angle. Additionally the UA must ensure that the proscribed rate of climb and stall margin be maintained during the turn, and both of these constraints are functions of UA weight, mode (i.e., gear retracted or open, external stores or clean), and altitude.

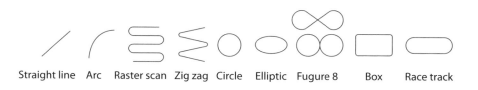

Straight line   Arc   Raster scan   Zig zag   Circle   Elliptic   Fugure 8   Box   Race track

Fig. 16.2 Flight patterns.

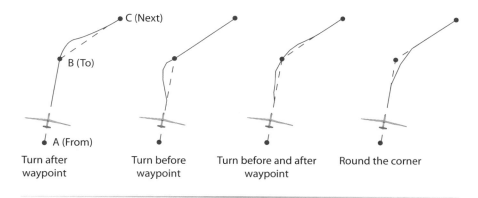

Fig. 16.3 Waypoint capture approaches.

The control segment-based mission planning process is directly tied to the specific unmanned aircraft that will fly the mission. For example, a mission plan generated for a Global Hawk HALE UAS would not be executable by a Desert Hawk SUAS. The airspeeds, turn radii, and altitudes are incompatible.

The flight plan results in a series of waypoints that are sent to the UA vehicle management system (VMS) or autopilot. Ultimately, the autopilot interprets the waypoint to fly the route and perform the actions. The mission planning system and autopilot must interpret the waypoints in the same manner.

One example of flight-plan interpretation is waypoint capture criteria. The UA will not exactly capture each waypoint. There will be some errors from position estimation and actual trajectory. The UA must fly to a three-dimensional space that is close enough to claim success at waypoint capture. The UA and control segment must have the same interpretation. The autopilot has control laws that determine how the UA behaves as it approaches or passes a waypoint when three successive waypoints are not aligned. Consider three waypoints: A is the previous waypoint (from), B is the upcoming waypoint (to), and C is following waypoint (next). If A-B and B-C are not in a straight line with each other, the UA must do one of the following, as shown in Fig. 16.3:

- After capturing point B by flying along A-B, the UA turns to intersect line B-C.
- The UA initiates a turn before point B by departing from the path A-B such that path B-C is intercepted at point B.
- The turn is initiated before point B, point B is captured, and then the UA intersects line B-C.
- The UA initiates the turn ahead of point B and takes intercepts path B-C without flying through point B, effectively rounding the corner.

Intelligence, surveillance, and reconnaissance (ISR) is widely attributed to UASs as a major mission capability. However, the mission planning and execution for surveillance and reconnaissance have important differences. Surveillance usually entails monitoring a fixed target or target area with flexible operator involvement. The mission planning for a surveillance mission can simply be to route the UA to and from the collection area where the UA will be dynamically retasked based on what is observed.

Reconnaissance involves collecting multiple targets or mapping a region of interest. The collection objectives are known a priori and can be prioritized. The mission planning system must optimize the route to collect the greatest number of highest-priority collection objectives. Efficient reconnaissance mission planning is often outside the abilities of humans, and so autorouting tools are frequently employed.

The mission planner must estimate the fuel burn or battery discharge when establishing the route. The fuel burn is a function of UA weight, altitude, airspeed, rate of climb, turn rate, onboard systems and payload power draw, and configuration. The performance methods described in Chapter 9 can be utilized in developing these fuel burn estimation algorithms. The mission planner must be alerted if the desired flight route exceeds minimum fuel remaining criteria. Similarly, the AVO tools must estimate fuel remaining based on the flight-path history and UA fuel state reporting. When the minimum fuel threshold for safe return to base is reached, which is known as *bingo fuel*, the AVO is alerted. These procedures are the same for battery-powered UA, except the state of charge is monitored instead of fuel remaining.

16.3.2 Geospatial Information Systems

Geospatial information systems (GIS) are increasingly integrated to support mission planning and control tools. A popular example is the GIS tools produced by Esri, such as ArcGIS. Some GIS services and capabilities that can be incorporated into a control segments are as follows:

- *Terrain databases*—Terrain data are elevation above the nominal Earth surface as a function of latitude and longitude. One common form is digital terrain and elevation data (DTED). Coarse DTED data are publicly available, and finer resolution is restricted. Terrain data are important for ensuring that the UA does not crash into terrain or that a minimum above ground level altitude is maintained.
- *Maps*—These include geo-registered aeronautical or road maps.

• *Ground imagery*—Geo-registered satellite imagery or aerial photography can be overlaid on the mission control displays. This provides context for flight operations. The maps and imagery can be combined.

GIS services can include almost anything that can be geo-referenced. It could include pipelines, sewer pipes, hiking trails, power lines, ocean current maps, streams, or commercial shipping lanes, as a few examples.

16.3.3 Human System Interface (HSI)

Human system interface (HSI) technology has advanced dramatically in the past two decades. Interfaces have changed from ASCII text and television monitors projecting analog video to graphics/icon driven interfaces with digital imagery displays. Mission plans can now be rapidly developed with a few mouse clicks. All of the necessary information about the operational environment is presented to the AVO and MPO without overloading the senses. The workstations are now ergonomic to reduce operator workload for improved mission success and reduced loss rates.

Many UAS losses are attributed to operator error. However, system design implementation—especially with the HSI—can directly impact the probability of occurrence of operator error. Therefore, it can be argued that many operator errors are the direct result of ineffective system design. That is not to say that operators do not cause the loss of UA due to erroneous actions. If a HSI generates excessive workload, does not effectively present critical information, does not facilitate emergency procedures, or is unintuitive, mishaps under stressful situations are a reasonable outcome.

The operators must be able to access all relevant system information rapidly when needed, but at the same time they should not be burdened by overwhelming information. One approach to addressing both considerations is to use drop-down menus. For instance, an operator might wish to inquire about the temperature within a payload bay. A menu-driven interface will allow the operator to find this information in perhaps one to three mouse clicks. Errors and warnings should be presented to the operator such that the situation can be understood quickly. The operator should not be required to monitor hundreds of parameters continuously in order to identify an issue. It is important to not clutter the user display with nominally functioning system information.

Modern mission planning systems allow the AVO to select flight-path types and orbit patterns with a high degree of automation. A flight planning software package might include a palette of orbit patterns that can be dragged over a map display with a mouse. Once created, the routes and orbit patterns should be easily modified. For example, the orientation and dimensions of a figure 8 orbit pattern can be adjusted by the mouse or a few keystrokes. The figure 8 pattern may consist of perhaps 8 – 20 waypoints, so that creating this pattern manually would produce excessive workload. The orbit should be repositioned quickly with additional mouse movements. Flight route waypoints can be placed on a map with mouse clicks and then dragged to new locations with a mouse button hold and release.

Modern mission planners are shifting from vehicle-centric mission planning to payload centric mission planning. Such tools can plan the UA flight path and sensor pointing such that the mission objectives can be best satisfied. The sensor point of interest is ensured based on the UA altitude and orientation. This often takes the form of point and click on the map.

Mission planning systems should display terrain, keep-out zones, political borders, and other relevant GIS information. Mission planning tools should prevent or alert the user to infeasible plans, such as those that will result in terrain collision.

16.3.4 Communications

Operators in the control segment must receive data from outside the system boundaries. The communications can be requests for route changes, external support of target location finding, or general situational awareness of the operational picture. External data links must support this communication.

One method is to create a text chat application that allows the operators to exchange text using the keyboard. The text data are typically exchanged over an IP protocol. Messages are terse and may consist of an abbreviated sentence or two. The risk of chat is that it might distract operators from their primary tasks.

Voice communications allow the operators to communicate without using their hands. A headset includes a microphones and an earpiece. The voice data can enter the system through a variety of means such as voice over IP (VOIP) or a dedicated data link.

A ground element can also have a traditional telephone. The phone can be connected to the outside phone network for civilian applications or a secure network for military applications. Phones can be difficult to use and can be disruptive, so that they are not suitable for frequent communications. A control element is more like an aircraft cockpit than a traditional office.

Line-of-sight ground stations may not be colocated with a military headquarters where the intelligence is consumed due to communications issues such as terrain blockage. The data must be disseminated from the ground station to the end-consumers, and tasking messages must be received.

One method of sending payload data is to broadcast the data feeds over a satellite link. Users with the appropriate access can view the information. The ground station must have access to the satellite injection points or uplink sites. Alternatively, SATCOM-equipped UASs might be able to allow end consumers to access the satellite downlink directly without going through a MCE.

Another method for coordination and dissemination communications is for the ground station to interface with networks. One example of a network is the internet. U.S. military systems can use the Secret Internet Protocol Router Network (SIPRNet).

16.3.5 Contingency Management

Combat is often described as hours of boredom punctuated by moments of terror. UAS operations are very similar. Flying predetermined waypoints for an 8–12-hr shift can be as exciting as watching paint dry. The nominal operator workload can be light for well-designed control elements. Then something goes wrong.

System failures, malfunctions, or degraded states will trigger an alarm if the system can detect the problem. If action is taken to deal with the problem, this is known as a contingency event. Contingency events must be managed by the system—often with human operators in the decision loop. Contingency event handling is dependent upon the mission objectives and operational environment. Some events that can cause alarms that require contingency events can include the following:

* Loss of communications
* Subsystem failures or malfunctioning
* Sensor measurements out of range or failed sensors
* Loss of generator—on battery power
* Low fuel (failed sensor or battle damage)
* Loss of redundancy (If a triplex flight control system loses one string, then this can trigger a contingency event.)

The following statement is perhaps more of an indication of human nature than sound system design guidance: the more sensors used to monitor the UA health, the higher the mission abort rate will be. Put another way, nonexistent sensors cannot generate alarms that require human decision. False alarms or excessively narrow nominal sensor ranges can cause an operator to generate return to base action. After all, who would want to be responsible for the loss of an expensive asset and cause harm to people or property on the ground by ignoring or overriding a system warning? Rules-based decision aids can help support the operator in taking appropriate actions.

Lost communications behavior must be planned and monitored by the ground control station. Lost communications can occur from communication system failures, environmental factors, jamming, and unintentional RF interference. In most cases, the mission will fail in a lost communications event, and often the UA will crash if the link is not reestablished. Temporary loss of link is quite common in operations, and the contingency management of this event is fundamental to safe operation.

One very common contingency action for lost communications is for the aircraft to fly to a lost communications waypoint, where it will orbit. Although a straight line path between the location that the UA is at when communications are lost and the designated point is possible, a lost communications flight path that is dependent upon phase of the mission is more desirable. The purpose of this path is to ensure safety and avoid keep-out zones. When the criteria are met, the UA flies to the initial waypoint or closest waypoint on that path, which takes the UA to the final lost communications waypoint. Numerous contingency flight paths can be defined per mission. Intermittent communications might not justify execution of the lost communications flight plan, and so a predetermined outage time can be set before executing the contingency plan. The lost communication flight path or paths must be defined at the mission planning stage.

Flight termination can be commanded by the UA as a lost communications contingency event. An UA that is not under direct control of the ground control station might congest the airspace during high-intensity operations, and it might be a liability to the broader operation. Losing communications in a military range adjoining populated areas can necessitate a termination command to ensure that the UA does not harm civilians. Chapter 10 provides a description of flight termination systems for externally commanded flight termination in a flight-test environment. Another motivation for a flight-termination contingency action is that lost communications might be an indication of other system failures that can result in the loss of aircraft. In this case it is preferable to define the specific point of impact rather than allowing it to be a random location.

The system must provide health and status information in a bandwidth-efficient manner. Depending on the system type and complexity, dozens or hundreds of UA parameters must be monitored. The air-to-ground communication system bandwidth can be constrained such that all of the monitored information cannot be downlinked to the ground at a high update rate. The UA can identify faults and report these to the control segment with a higher priority. A system tradeoff must be made to determine how much fault-management logic exists on the UA and on the control segment.

16.3.6 Payload Control

The payload control station is generally colocated with the pilot control station. The coordination between the AVO and MPO can be intense, combining data and verbal exchanges. Successfully meeting mission objectives requires skill when one operator flies the UA and the other controls the payload. This arrangement is much like driving a car with two people, where one has the steering wheel and the other has the brake and gas pedals. Some smaller UASs can combine the AVO and MPO roles for a single individual.

Tracking a moving target can be accomplished with onboard UAS algorithms or dynamic AVO interactions. The flight path depends on the sensor field of regard, wind, and the relative dynamics of the target vehicle and the UA.

Full-motion video can be used for many purposes beyond simple viewing of near-real-time scenes. Image-processing tools can use information from multiple frames to generate enhanced information. Major image-processing tasks include the following:

* *Image stabilization*—By locking onto features or a path within the scene, shaky video can be fixed about the aim point. The image borders will not correspond to the screen limits, but the central image will appear to be stable.
* *Blur correction*—Image enhancement algorithms can correct for blur.
* *Video geo-registration*—The frame center, corner points, or even every pixel ground projection location can be given a geographic coordinate. These algorithms require knowledge of the UA position, terrain data such as DTED, camera orientation (azimuth and elevation), focal-plane-array arrangement, and zoom angle.
* *Image projection on terrain*—Using geo-registration techniques, the image can be displayed with a three-dimensional appearance on a simulated ground.
* *Mosaics*—When multiple video frames are stitched together to form a larger image, a mosaic is formed. Ideally mosaics are corrected for terrain elevation, which required terrain data and video geo-registration. Mosaics can fill in the frame border voids produced by image stabilization methods. Mosaics are useful for supporting situational awareness beyond what is visible through the camera field of view.
* *High-resolution stills*—Image enhancement is achieved by comparing multiple frames to extract more information than is available

in any single frame. Impressive still images can be generated from shaky video.

* *Playback*—The video can be played backwards or forward from an earlier time. A separate screen or window for playback enables the payload operator to simultaneously monitor the real-time events on the primary video display.
* *Moving target identification and tracking*—Vehicles and other moving targets can be discriminated from the background through frame comparison. Once identified, the moving targets can be tracked.

Ground-based image systems can combine data from multiple sources to generate context for the video. Some features include the following:

* *Adding map features*—Geospatial information, such as roads, railways, rivers, pipelines, among other features, can be overlaid on the video display.
* *Marker symbols identifying targets, threats, friendly forces, or other items of interest*—MIL-STD-2525C provides standard military symbology.
* *Metadata overlays*—These data can include flight altitude, airspeed, time, camera zoom angle, slant range, and target geographic coordinates, among other information. Metadata should be synchronized with the video.

An example video exploitation system is the Pyramid Vision TerraSight^{TM}.

Most image-processing tools manipulate digital video. Disseminating video over networks also requires digital formats. Some UASs still downlink analog video to the control station, and so it is necessary to convert between analog and digital video. Frame capture tools perform this conversion, but can introduce latency.

The payload operator can identify features on the scene. For example, a target can be identified and marked on the screen. The geo-coordinates of the target can then be reported or disseminated outside of the control station. A popular format for target coordinate transfer is the cursor on target (CoT) standard.

Wide-area surveillance (WAS) is a new mission type that stresses the ability of communications links and the sensor operator. WAS uses multiple cameras with overlapping fields of view to generate imagery coverage of a large area. The payload operator display could be multiple monitors each dedicated to one camera, much like a security guard's display. A more elegant solution is to merge the imagery into a single mosaic, possibly overlaid on terrain.

Direct human monitoring of the entire WAS scene is unlikely to yield valuable intelligence. Analyzing time-triggered events is more useful. This activity can include playback of the scene or playing the scene backwards.

For example, unmanned aircraft present at the point of interest can be traced back to the origin by playing the scene in reverse. Repeated viewing of scenes across multiple days can help determine enemy behavioral patterns.

16.3.7 Security

Military UAS operations generally produce classified information that must be protected. UAS systems can also interface with networks that contain classified information. Some security provisions include the following:

* *Data encryption*—The payload data or other sensitive information can be encrypted on the UA, at the communications link, or in the ground station.
* *User authentication*—Operators log on to the system with credential authentication such as passwords.
* *Secure network interfaces*—Provisions are made to ensure that the secure networks are protected from the control segment.

16.4 Overview of Ground Element Types

Ground architectures take many forms. Small UASs can simply use a single line-of-sight ground control station to control a single UA. The architectures tend to be more complex as the UA, payload suite, and communication systems become more capable. Several ground architectures are shown in Fig. 16.4. The ground system naming convention is not standardized, though some of the more common types are described in this section. Here we will consider the following:

* Remote viewing terminal (RVT)
* Portable GCS
* Forward ground control station (FGCS)
* Line-of-sight (LOS) GCS
* Launch and recovery element (LRE)
* Mission control element (MCE)
* Tasking, processing, exploitation, and dissemination (TPED) element

The control segments do not necessarily need to be positioned on the ground. Other possibilities for control segment types include the following:

* Submarine control station
* Airborne control station (The control functions can be performed from AWACS, fighters, helicopters, or other aircraft types.)
* Ground-vehicle control station

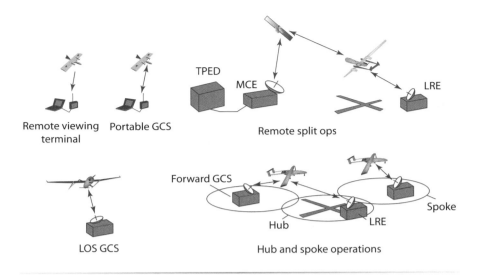

Fig. 16.4 Ground system architectures.

A key architecture driver is whether the system can operate only LOS or if it is capable of beyond-line-of-sight (BLOS) operations. Here BLOS refers to satellite communications rather than airborne relay because the latter can use LOS ground station configurations.

LOS operations generally follow one of three models:

* A LOS GCS or portable GCS communicates with one UA.
* A LOS GCS or portable GCS flies the UA, but RVTs can also receive payload data.
* The system uses a hub-and-spoke operations model. The hub consists of an LRE that is responsible for launch and recovery, as well as UA handoff between the hub and spoke. The forward GCSs fly the UA within the spoke region, but do not manage launch and recovery.

BLOS operations usually have a LRE that is in theater and an MCE that is located remotely. The LRE is responsible for launch and recovery operations and support handoff to the MCE. The MCE performs the mission operations via a satellite communications link.

16.5 Portable Ground Control Stations

Small UAS GCSs cannot be housed in shelters. The GCS consists of a computer with display and a small ground data terminal. The computers can take the form of a wearable computer, PDA, laptop computer, or several transit cases full of equipment. The most common form of computer is a ruggedized laptop. Glare can be a problem when operating outside in daylight conditions, and so glare shielding or special monitors are generally necessary for a satisfying user experience. Figure 16.5 shows an example of a portable ground control station.

These systems can be as simple as an antenna attached to a laptop computer. Other systems can be a complex rat's nest of monitors, electronics boxes, computers, antennas, joysticks, and cables. System developers may find that COTS computers and audio-visual equipment can provide many of the capabilities that are required. The problem with this approach is that setting up the PGCS can be even more complex than a home entertainment system for which the source components are intended. Frequently there are many loose cables that can be connected in incorrect ways. Such systems are frustrating to the average user or even an expert user under stress. A well-designed PGCS will integrate components such that there is only one possible way to configure the system. The effort to make this system element more usable will undoubtedly result in higher nonrecurring and recurring cost, but satisfied users can enable more longevity for the UAS product line.

Power must be provided for the GCS. Batteries are practical only for the smallest GCS. Batteries must be recharged between uses or during flight for long-duration missions. Ground power or generators are required. Batteries or power provisions can be a major driver for the PGCS footprint.

Fig. 16.5 Portable ground control station. (U.S. Army Photo by Spc. Joshua E. Powell.)

16.6 Remote Viewing Terminals

Remote viewing terminals (RVTs) provide the operator with the ability to at least view payload data (Level 2 control). RVTs can also be equipped to provide payload control (Level 3 control), which is enabled by a two-way data link. These portable systems can be the size of laptop computers or hand-held computers such as PDAs or even cell phones. Example RVTs include the L-3 ROVER (III and IV) and AAI's One System Remote Viewing Terminal (OSRVT).

16.7 Launch and Recovery Elements

Launch and recovery elements (LRE) primarily support launch and recovery operations within line-of-sight communications of the UA. These elements are also called launch and recovery systems (LRS), though LRE is the most enduring terminology. A dedicated LRE does not provide payload control functionality. However, LREs can serve as full capability line-of-sight ground stations if properly equipped.

Air traffic coordination in the vicinity of the airfield is a major LRE role. An operator in the LRE must communicate with air traffic control (ATC) to get permission to take off and land and ensure that the UA operates safely in airspace shared with manned aircraft or other UASs. Some LREs have a vhf or uhf radio that can communicate directly to the tower and other aircraft. Systems can also use the UA to relay voice communications between the LRE and ATC, such that it is functionally equivalent to having a pilot in the UA.

LREs provide control throughout the recovery. The first step is approach planning, where the UA enters the recovery site airspace and transitions to the approach path. Then there is execution of the landing approach once ATC grants permission. Finally, the UA must recover, whether on a traditional runway, special recovery system, or vertical landing. The combination of LRE and UA must have full knowledge of the runway position (latitude, longitude, elevation) at surveyed points and the runway slope.

Many systems developed as late as the 1990s use external pilots who visually guide the UA to a landing using hand controllers (as seen in Fig. 16.6). The external pilot requires significant skill and training. Such systems are falling out of favor due to improved reliability of automated landing systems.

Ground-based landing aids can include differential GPS or microwave landing systems. These systems have ground navigation devices that are at surveyed locations relative to the runway, recovery device, or landing point. Correction signals from a ground-based DGPS antenna are passed

Fig. 16.6 External pilot prepares to launch a Pioneer UAS. (U.S. Marine Corps photo by Sgt. Jennifer L. Jones.)

through the data link to the UA. A microwave landing system produces a narrow beam of microwave RF signals that enables the UA to follow the beam to the landing site. Landing aids can be integrated with the LRE or used as independent systems.

The LRE can manage ground operations such as taxi while the UA is on the ground. Operators located inside a LRE shelter rely on inputs from a situational awareness camera (if so equipped) and position estimates placed over a map display of the airfield.

LREs can support preflight and postflight activities. The distribution of functions between the LRE and ground support computers is somewhat arbitrary, as there can be much overlap. Some of the LRE preflight roles can include establishing communications links, directing startup procedures, verifying that all systems are operating nominally, and preparation for launch. Postflight roles can involve shutdown procedures and retrieval of stored data.

The LRE gives control of the UA to other control elements for mission execution. MALE and HALE systems equipped with high-bandwidth SATCOM links will transfer control to a mission control element (MCE). Another possibility is a FGCS hand-off. Whatever the control station type, it will have payload operations and UA operations functions.

Hub-and-spoke operations use a launch and recovery hub site to provide UA to service multiple spoke regions. The hub has an LRE, launch and recovery systems or runway, and maintenance facilities. The spokes have an FGCS and no launch and recovery provisions. After the UA launch, it transits to a hand-off location where the designated spoke FGCS takes control of the UA for mission operations. Once the mission is complete or the fuel state is low, the FGCS directs the UA back towards the hub where the LRE takes control and recovers the UA.

16.8 Mission Control Elements

A mission control element (MCE) provides BLOS control of the UA during mission operations other than launch and recovery. This is typically a GCS located outside of the theater of operations, where connectivity to the UA is enabled by high-bandwidth SATCOM.

The MCE is the workhorse for remote split operations (RSO). RSO minimizes the in-theater footprint by placing only the LRE and UA support in theater, while keeping the MCE operations at a distant base. This greatly simplifies the logistics burden compared to using fully staffed LOS GCS in theater.

The MCE is generally staffed by an AVO and MPO, but can also include a mission director, communications manager, or other staff. The operators rotate in shifts. If the MCE facilities are located in the home country, then the operators can go home after the shift.

Satellite communications management is a major function that is unique to this control segment type. The satellites must be scheduled to support the mission. Loss of communications or other communications events must be handled. The satellite service provider is usually outside of the UAS system boundaries.

16.9 Tasking, Processing, Exploitation, and Dissemination (TPED)

Tasking, processing, exploitation, and dissemination (TPED) elements manage tasking of ISR assets and the data that they produce. TPED can interface with myriad system types ranging from manned aircraft, UASs, or satellites. The TPED is outside of the UAS system boundaries. An

MCE can have many of the capabilities of the TPED, though the MCE is limited to the UAS that it controls. Systems that do all of the functions of TPED with the exception of tasking are called processing, exploitation, and dissemination (PED) elements. The AN/TSQ-219 Tactical Exploitation System (TES) and the Distributed Common Ground System (DCGS) are PED capabilities. Combined air operations centers (CAOCs) used in the 1999 Kosovo crisis and 2001 Operation Enduring Freedom performed many TPED functions.

The end consumers of the UAS-gathered intelligence can be senior commanders or other decision makers, targeting staff, and intelligence analysts. These parties can be contained within the TPED. If they are outside of the TPED, then the intelligence products must be disseminated to their networks or locations.

When UAS data are available only to their own GCS, this is referred to as a "stovepipe solution." A single GCS has a relatively narrow perspective of the battle space.

ISR needs often outstrip the ability of available assets to satisfy them. Efficient prioritization is important to ensure that the overconstrained requests are most fully satisfied. This tasking can come from the TPED, though other forms of tasking are common. Tasking involves providing mission objectives to the UAS. The intelligence collection requirements are based on prioritized needs and analysis of the best assets to provide the collection. The tasking involves scheduling assets, whether manned or unmanned. The detailed mission planning is left to the GCS or other control element.

Processing is manipulation of raw sensor data to help support intelligence generation. The payload products received from the UAS require additional processing to be of value. For example, multiple overlapping visual and IR camera feeds can be fused to reveal more information than if viewed individually.

Exploitation is analysis of the processed payload data to provide intelligence. This can be performed by an intelligence analyst or automated tools. Images can be annotated to identify features of interest.

Dissemination is making the exploited products available to consumers of the intelligence. One form of dissemination is archiving the intelligence products in a way that supports retrieval. Another method of dissemination is distributing products to those that need it once the information becomes available.

16.10 Hardware

Now that we've described the functionality of the various ground element types, let us look at the hardware required to achieve the required capabilities. The GCS consists of computers, user interface hardware, and shelters.

16.10.1 Computers and User Interface Hardware

The user interface hardware can vary from a wearable computer to a workstation. Laptop computers common to PGCS and RVT systems contain a keyboard, mouse function device, and a monitor. Larger systems such as STUAS, TUAS, MALE, HALE, and UCAV systems will generally adopt a workstation approach. The workstation also includes a keyboard, mouse, and at least one monitor. Workstations often have joysticks (or other form of hand controller) and headset interfaces.

The computers are generally personal computers (PCs) or Unix-based machines. Some Apple products including Macintosh computers, iPads, and iPhones are gaining popularity for small UAS applications. High-end UASs have high processing workloads, which drives the need for high-performance computing. Computers are generally stripped of all unnecessary software to minimize problems. Waiting for automatic updates to complete could result in a crash during a landing operation! The computers should reboot quickly and be robust against system failures.

UASs generate large quantities of data during a sortie. A campaign consisting of multiple flights generates much more. It is generally necessary to archive and store this information for later retrieval. Analog video, for systems that still use this format, requires cassettes for storage or must convert the video to digital format. Most other payload data are digital, which can be stored on solid state data storage devices. Fortunately, the digital storage media are continuously becoming more compact and less expensive.

The GCS will often require furniture. Very small portable GCS can simply be held by the operator while they sit on whatever surface is available, which is arguably adequate for short-duration flights. Shelter-based GCSs will have ergonomic chairs that reduce fatigue for long shifts. Other furniture can include tables for taking notes or looking at supplemental information that is not provided on the monitor.

16.10.2 Shelter Design and Facility Integration

Shelters house control elements and other facilities where humans operate. The shelter provides a physical space that is environmentally controlled.

Most shelters are transportable. The primary methods of transport are detailed in Chapter 17, though air and ground transport are quite common requirements for shelters. The transportation loads drive the shelter structural design. The geometric constraints associated with transportation can be excessively limiting on the shelter utility unless it can be reconfigured to give more volume.

Families of shelters can be procured that meet the transportability requirements. UASs are not the only applications that drive the transportable shelter market. Chandler May is one manufacturer of military shelters that are capable of air and road transport. Civilian UAS applications that only require over-land transportation commonly use recreational vehicles or fifth-wheel trailers. Shelters can be placed on the back of pick-up trucks or HMWWVVs.

Providing the appropriate environment for operator effectiveness and computers is not trivial. The temperature and humidity must be maintained within narrow ranges. Filtration systems prevent sand and dust from entering the shelter. Outside light must not interfere with viewing computer screens. Noise levels should be kept low enough to prevent operator fatigue.

The air conditioning system can be flight critical. A shelter in the desert during summer could rapidly overheat such that the human operators and GCS computers could not function. If these systems must operate in order to recover the UA, then a failure of this functionality will result in loss of the UA. Despite the significance of the air conditioner, most people do not think of this critical system when considering a UAS.

Shelters might be required to operate in a high threat environment. Electromagnetic threats can be generated by enemy weapon systems or unintentional EMI from friendly systems. In either case EMI protection can help ensure safe operation of computer and communication systems. Chemical, biological, radiological, and nuclear (CBRN) threats might demand measures for decontamination and air management. Enemy missiles can also target the RF emissions of communication systems, and so separation of the radiating antennas from the shelter can enhance personnel survivability.

Power is provided by shore power or generators. Shore power is simply electrical power provided from outside the system boundaries, such as an electrical outlet in a building. Generators are often part of the UAS. Generators produce unwanted noise and vibration that can adversely impact the environment within the shelter. Ideally, the generators can by physically separate and isolated from the shelter. Robust control segment power system design can include batteries and uninterrupted power supplies (UPS) to handle power spikes and temporary power outages.

16.11 Training

A GCS can serve the function of training systems for GCSs. Unlike manned aircraft, cockpit motion is not required to simulate an operational environment. Synthetic vision can create photo realistic environments that closely match the actual scenes seen by operators when the UA is in flight.

Aircraft emulators often simulate UA behaviors. Much like a hardware-in-the-loop simulator (HILSim), emulators use autopilots that are given simulated sensor signals generated based on the simulation environment. The UA management system and mission management system software and hardware can be incorporated as well. Using system flight hardware helps to make the simulation more realistic. Emulators can be more costly than purely software-based simulation.

Using the GCS as a training device adds realism over any solution that deviates from the operational GCS. However, operational GCS assets might be in high demand when utilization is high, as is common in wartime. A compromise is to create dedicated training simulators to aid training programs. Some simulators can be run on a laptop computer to enable a classroom training environment.

Training timelines vary significantly across UAS classes. For well-designed small systems, it is possible to train an operator on all aspects of flight operations and system maintenance in a day or week. Larger systems like HALE and MALE UASs can take months to train specialists who are qualified on more narrow roles.

16.12 Interoperability

Until quite recently nearly every UA type could talk to only a dedicated GCS type. Conversely, a GCS could only fly one UA type. The air-to-ground interfaces were closed and proprietary, and so only the prime contractor could modify the system. This arrangement is quite advantageous to contractors who wish to win sole-source contracts for system upgrades or enhancements. The government, on the other hand, would like the ability to competitively procure the UA and ground elements from different vendors in order to get the best value and foster innovation.

Control station interoperability has been a major goal for two decades. On first inspection it appears wasteful for every UAS to have a unique GCS despite large overlaps in functionality between systems. Why not have a single GCS that can fly multiple UA types? The benefits for streamlining acquisition and operations are significant, but the implementation must be effective rather than simply the least common denominator. With a published open air-to-ground interface standard, the UA and GCS developments are decoupled. Examples of interoperable ground control stations include the following:

* The common ground control station (CGCS) of the 1990s was to operate both the DarkStar (Tier III−) and Global Hawk (Tier II+) UA, though this dual capability was not realized in practice.
* The DARPA J-UCAS program intended to implement a common operating system, which included interoperable ground and airborne functionality.

The NATO STANAG 4586 standard has emerged as a viable solution to the interoperability challenge. It consists of the core UAV control system (CUCS) and a vehicle specific module (VSM). The CUCS includes processing, mission planning tools, and user interfaces. The CUCS remains substantially unchanged for flying different UA, though it can be tailored. The VSM is a module that is specific to the UA. The VSM has a standardized interface to the CUCS, and it can be implemented on the ground or on the UA. STANAG 4586 is an evolving standard with frequent revisions. There are multiple levels of interoperability defined within the standard:

* II—Receipt of payload data
* III—Payload control
* IV—UA flight-path control, exclusive of launch and recovery
* V—Launch and recovery control

The first widely fielded STANAG 4586 compliant ground station is the Army One System that is built by AAI. The AAI Shadow system is the first system to use the Army One System. Marketing materials imply that the ground station is compatible with the AAI Shadow 200, AAI Aerosonde, Pioneer, General Atomics Warrior, Northrop Grumman Hunter, Bell Eagle Eye, and Northrop Grumman FireScout.

STANAG 4586 allows private messages between UA-specific user interface software and the UA. In essence, these messages bypass the messages contained within the standard interfaces. This allows unique UA functionality to be enabled. The downside is that only 4586 ground stations with that unique software can fly the UA. It could be argued that excessive use of private messages goes against the spirit of interoperability.

STANAG 4586 may not be the end of the UAS interoperability journey. The CUCS is viewed as monolithic by some, where the interface is fixed and not easy to modify. To enable rapid development and competition, the user interface and analysis tools can be further broken into subcomponents. Naturally, contractors who do not have existing CUCS contracts are in favor of making the user interface more modular.

With currently available and developmental interoperable ground control stations, the entire UAS system can now be procured from multiple contractors. Future contract opportunities can include only UA and payload scope, along with a requirement for interoperability with an existing control segment. UAS providers might focus only on control segments with the hope of interfacing to multiple different UA programs.

There has been a recent interest in accommodating heterogeneous systems. In the case of purely UA-based solutions, the platforms might have different sensors and flight characteristics. An example is a hunter–killer system where an ISR aircraft searches and identifies the target and an attack aircraft engages the target with weapons. Another form of heterogeneous UA combinations is a high-low combination consisting of a HALE UAS that can search large areas efficiently and a smaller low-altitude UAS that can investigate targets at reduced slant range.

Expanding heterogeneous systems to other unmanned vehicle domains, the other systems can include unmanned ground vehicles (UGV), unmanned surface vehicles (USV), autonomous underwater vehicles (AUV), or unattended ground sensors (UGS). Although UASs are moving towards the STANAG 4586 standard, other systems classes now favor the JAUS standard for command and control.

References

[1] Austin, R., *Unmanned Aircraft Systems: UAVS Design, Development, and Deployment*, AIAA, Reston, VA, 2010, pp. 183–195.
[2] Fahlstrom, P. G., and Gleason, T. J., *Introduction to UAV Systems*, 2nd ed., Fahlstrom and Gleason, Columbia, MD, 1998, pp. III-1–III-30.

Problems

16.1 Convert the coordinates of Los Angeles International Airport (LAX) to decimal degrees.

16.2 Convert $-45.1022°$ latitude to the degrees, minutes, seconds format.

16.3 Calculate the great circle distance between Washington Dulles and London Heathrow airports.

16.4 What ground control stations are most appropriate for flights lasting 5 min, 1 hr, and 24 hrs?

Chapter 17

Reliability, Maintainability, Supportability, and Transportability

- Explore the reliability and availability of UASs
- Understand the deployment of UASs
- Define supportability for UASs

Sgt. John Signorelli attaches a
Dash-95 air cart to a RQ-4
Global Hawk at Beale Air
Force Base. (U.S. Air Force
photo by Staff Sgt. Bennie
J. Davis, III.)

17.1 Introduction

E ffective UAS design must consider much more than system performance. The operational effectiveness is also measured by
reliability, maintainability, supportability, and transportability,
which are often collectively called the "ilities." The U.S. DOD also characterizes the military impacts of a new system in terms of doctrine, organization, training, maintenance, logistics, personnel, and facilities
(DOTMLPF). This is a modern acronym that describes the life-cycle
costs and broader impacts of a new materiel solution. In this chapter we
will cover many of these important system operational considerations.

17.2 Reliability

Current-generation UASs are generally not as reliable as manned aircraft.
System reliability is a result of specific design and programmatic decisions.
A major reason for this condition is a perception that UASs should cost
less than manned aircraft, and the corresponding development and acquisition budgets are therefore smaller. The necessary design effort, levels of
redundancy, acquisition of high-reliability components, and testing are
usually far less than would be acceptable for manned aircraft. However,
there is no fundamental reason why UASs could not achieve the same or
better reliability as their manned counterparts.

Perhaps the appropriate balance between affordability and loss rates is
now established for ISR UA operating in combat. Increasing reliability
decreases loss rates, but increases the development and acquisition costs.
However, the current levels of reliability limit the potential roles and operational environments for UASs. For example, future human-carrying UASs
used for casualty evacuation (CASEVAC) or troop transports should have
equivalent levels of safety as other manned aircraft. UASs are unlikely to
darken the skies in civil airspace until the rate of crashes and uncontrolled
flight is reduced to acceptable levels for public safety. Reliability is a matter
of public confidence. UASs carrying weapons pose greater risk than ISR
UASs, requiring corresponding improvements in reliability.

Military UAS specifications will generally provide reliability and availability requirements. Contractors proposing new UAS designs will usually
be unable to prove these parameters based on demonstrated results.
Instead it is necessary to provide evidence that the design supports the
required reliability and availability attributes. It is often unspoken that
the desired capabilities will not be achieved until the system is highly
mature—perhaps with hundreds of thousands of cumulative flight hours.
However, government specifications are frequently unclear about when
the reliability will be achieved, with the implication that first delivered

systems will meet the requirements. Yet a contractor offering that provides the necessary level of component life testing and developmental flight testing to achieve the specified reliability is probably unaffordable to the government. Without adequate reliability design attention during system development, the first fielded systems will have poor initial reliability. The new system operators have the burden of finding design flaws while also conducting missions.

The reliability of a UAS can be estimated through analytical means. Every component or system that the UAS depends upon can be assigned a reliability value. Examples include flight critical avionics, data links, propulsion systems, the AVO workstation, and even the AVO. The system reliabilities are combined in a reliability model. See Fig. 17.1 for a graph of the UAS reliability trends.

The reliability of components arranged in series is

$$R = \prod_{i=1}^{N} R_i \tag{17.1}$$

The total reliability of components arranged in parallel is

$$R = 1 - \prod_{i=1}^{N} (1 - R_i) \tag{17.2}$$

The reliability is related to the incident rate. This can be expressed by

$$R = \exp\left[-\lambda \cdot t\right] \tag{17.3}$$

$$R = \exp\left[-\lambda_1 \cdot t\right] \cdot \exp\left[-\lambda_2 \cdot t\right] \cdots \cdots \exp\left[-\lambda_N \cdot t\right] \tag{17.4}$$

$$R = \exp\left[-(\lambda_1 + \lambda_2 + \cdots + \lambda_N) \cdot t\right] \tag{17.5}$$

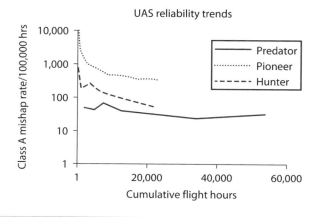

Fig. 17.1   UAS reliability improves with time.

where λ is the incident rate and t is the time interval. The incident rate is the inverse of the mean time between failures.

$$\lambda = \frac{1}{MTBF} \tag{17.6}$$

The *MTBF* can be found from the reliability over a time interval

$$MTBF = -\frac{t}{\ell n(R)} \tag{17.7}$$

Example 17.1   Simplex and Triplex Avionics

Problem:
A trade study seeks to quantify the reliability of a triplex low-cost INS architecture vs a high-end single-string INS. The low-cost INS has an *MTBF* of 200 hrs, and the high-end INS has a *MTBF* of 10,000 hrs. Which is more reliable for a 24-hr mission?

Solution:
The reliability of the simplex system is

$$R = \exp\left[-\lambda \cdot t\right] = \exp\left[-\frac{t}{MTBF}\right] = \exp\left[-\frac{24\text{ hrs}}{10,000\text{ hrs}}\right] = 99.76\%$$

The reliability of each string of the triplex system is

$$R_i = \exp\left[-\frac{t}{MTBF}\right] = \exp\left[\frac{-24\text{ hrs}}{500\text{ hrs}}\right] = 95.31\%$$

The overall reliability of the triplex system is

$$R = 1 - \prod_{i=1}^{N}(1 - R_i) = 1 - (1 - 0.9531)^3 = 99.98\%$$

The triplex system with low-cost INS is more reliable.

The mishap rate *MR* is

$$MR = \frac{N_{\text{ClassA}}}{FH_{\text{Tot}}} \cdot 100,000\text{ hrs} \tag{17.8}$$

where N_{ClassA} is the number of class A mishaps and FH_{Tot} is the cumulative fleet flight hours. A Class A mishap is defined as significant damage or loss of the UA.

The mean time between failures $MTBF$ is

$$MTBF = \frac{FH}{N_{Ab,Mx} + N_{Cx}} \tag{17.9}$$

where FH is the flight hours flown, $N_{Ab,Mx}$ is the number of aborts due to maintenance issues, and N_{Cx} is the number of cancellations. It is also given by

$$MTBF = \frac{1}{\lambda} \tag{17.10}$$

The mission reliability $R_{Mission}$ is

$$R_{Mission} = 1 - \frac{N_{Ab,Mx}}{N_{Sorties}} \tag{17.11}$$

where $N_{Sorties}$ is the number of sorties launched.

During design, the reliability can be estimated through various design techniques, such as fault tree analysis and failure modes effects criticality analysis (FMECA). Component reliability analysis should be based on data sources for the exact component or similar components. Based on the results of these analyses, specific components can be changed or enhanced, and the system architectures can be modified. The U.S. military published MIL HDBK-338B for fault tree analysis methods, and MIL-HDBK-217F provides a procedure for reliability prediction of electronic equipment.

Reliability data can be collected during system development and testing. A fleet leader is an initial system that accumulates operating hours in order to help identify reliability issues. Accelerated life testing puts components through numerous simulated cycles to assess reliability, identify problems, and assess wear.

A first step in improving reliability in operations is to understand reliability drivers. This requires collection and interpretation of reliability data. Component failures or UA losses should be tracked and investigated. Such information can be logged into databases that can be queried for analysis and reporting.

The next step in the path to higher reliability is to improve the reliability of components and systems. The sources of reliability drivers can be quickly highlighted through Pareto analysis. Here, the component or systems that impact the reliability are ranked from greatest to least contribution. This is a bar chart where the components are listed on the horizontal axis, and their reliability metric (mean time between failures or reliability rate) is the vertical axis. Greatest attention is placed on the components with the worst reliability. A Pareto chart of the system reliability drivers can facilitate architecture trades. Even with significant uncertainty in component reliabilities, a Pareto analysis can demonstrate the major drivers and quantify differences between architectures.

Postincident investigations can be a useful tool for finding the causes of reliability problems. A *root cause analysis* seeks to find the source of the failures. The culprits might be design flaws, poor quality control, ineffective operating procedures, vendor part changes, or a host of other issues.

Fixing reliability issues is expensive. Then again, the impacts of lost UA or cancelled contracts can be even greater. Enhancing system reliability is an unavoidable investment for enduring programs. Program budgets should include effort to fix reliability problems in development and operations.

Components or systems have reliability measured in several ways. These include the following:

- *Mean time between failures (MTBF)*—This is a measure of the time between any failures, including degraded functionality. For example, the *MTBF* of an engine might represent a broken part that still permits the UA to make a safe landing.
- *Mean time between critical failures (MTBCF)*—This is a measure of the rate of complete failure of a component or system.
- *Mean time between overhauls (MTBO)*—This is the mean time between major maintenance actions of the component or system. For example, an engine overhaul involves removal from the UA and often rebuilding of the engine.

Reliability models require component reliability data. However, reliability data for UAS components are notoriously difficult to obtain, especially for smaller systems without certified civilian or military manned counterparts. Large UASs such as HALE vehicles or UCAVs can utilize manned aircraft aerospace grade components.

The best approach to improving the reliability with a paucity of data is to design an architecture that is tolerant of low-reliability components. Often the reliability does not drive design choices during development, but is calculated afterwards for an established design after the state of maturity is high. If the reliability is driven by other system considerations rather than being a driver, the component reliability requirements can be much higher than what the state of the art or budgets will allow. Unfortunately, system reliability falls short for many UAS programs.

UASs have Class A mishap rates on the order of $10^1 - 10^3$ losses per 100,000 flight hours. By comparison, general-aviation and tactical military aircraft are typically 10^0, regional/commuter airlines are 10^{-1}, and large airlines are 10^{-2} losses per 100,000 flight hours. The comparatively high UAS loss rates are caused by a number of factors just discussed, but the motivating force is affordability. There is no fundamental reason why

UASs cannot achieve airline reliability performance, but such systems would take too long to develop and be very expensive relative to today's norms.

17.2.1 Power and Propulsion

The 2003 OSD reliability study [1] found that propulsion and power cause 32–37% of UAS losses, which is the single biggest driver (Fig. 17.2 shows the results from 2002). Most UASs use nonaviation reciprocating engines or adaptations of ultralight engines that frequently experience reliability challenges. The picture improves dramatically when manned aircraft turboprops and jets are used. Short-lived target drone and cruise missile jet engines are generally not suitable for long-endurance UAS applications. Table 17.1 shows a comparison of some of these systems.

Small UAS engines can be derivatives of hobby engines, where reliability is not a driver compared to high power-to-weight and low cost. The duty cycle of a hobby engine might be 10–100 flights with duration of about 5–10 min each. The engines are often maintained after a few short flights. Quality control is often poor relative to aerospace standards in order to achieve a low-cost target.

17.2.2 Avionics and Flight Controls

Avionics and flight controls constitute the second largest reliability driver for UAS [1]. Several components that make up the flight control

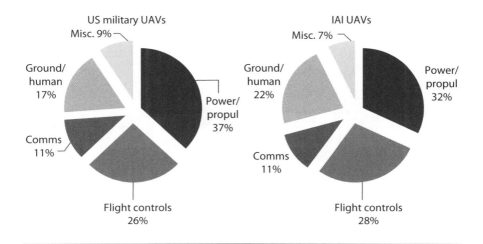

Fig. 17.2 UAS reliability by system in 2002 (1).

Table 17.1 Notional Propulsion System Reliability

Component	MTBF, hrs	MTBO, hrs	Notes
Electric propulsion motor and controller	5–30,000	5–1,000	Brushless motors and low power loading increases life
Small reciprocating engines	1–2,000	5–250	Nonaerospace engines, utlralights
General-aviation reciprocating engines	1,000–10,000	1,000–2,000	Certified engines
Jets—aerospace	100,000–150,000	5,000	Business jets and larger engines
Jets—cruise missiles, targets	——	0.5–20	Also includes model aircraft engines

system include the autopilot flight control logic, processors, actuators, inertial sensors, and air data sensors. Components generally come from aerospace suppliers, low-cost UAS systems providers, and hobby equipment. Redundancy is commonly employed with flight control systems to increase reliability. See Table 17.2 for some components that make up the flight control system.

The reliability of autopilots can vary by orders of magnitude across UAS applications. Autopilots may be very inexpensive or of aerospace quality. The autopilot may experience failures due to low-reliability hardware, flawed algorithms, or operations outside of the intended flight regime. Today, aerospace quality autopilots can be found that weigh less than 5 lb, but the costs of quality control can make these units substantially more expensive than other alternatives.

Reliability data for hobby servos generally do not exist. These data must be obtained through accelerated life tests or observing trends in the field.

Table 17.2 Notional Avionics and Flight Controls Reliability

Component	MTBF, hrs	Notes
Low-cost autopilots	1–5,000	For small UAS applications
Flight-critical avionics—aerospace	5,000–20,000	Includes autopilots and VMS
INS—aerospace	10,000–20,000	
Non-flight-critical avionics—aerospace	>300	Includes payload electronics
Model aviation servos	2–1,000	
Servos—aerospace	1,000–5,000	5,000 hrs common for UAS servos

The remedy to poor hobby servo reliability is often switching to a different model or performing stressing acceptance tests. Hobby servos generally have plastic cases that are unsuitable for EMI in military operational environments. It is advisable to have actuator redundancy when using low-reliability actuators for all but the smallest UAs.

Another approach is to use aerospace quality electromechanical actuators for UASs. These are available in sizes suitable for tactical or larger UA. Smaller systems such as small UASs might not have many suitable choices. Large, high-speed UASs such as UCAVs can use hydraulic actuation.

Redundant architectures are often necessary to achieve appropriate reliability. Single string systems are the norm for tactical and smaller UASs, where the cost, weight, and complexity of redundancy are usually unjustified. Larger, more expensive UASs employ duplex and triplex systems.

Products of nonaerospace industries, such as the automotive industry, are often used with the hope of improving system reliability at low cost. The life span of an automobile may be 5000 hrs, which is equivalent to 200,000 miles if the average speed is 40 mph. Automotive parts may or may not be suitable for a UAS application, but the automotive reliability can serve as an intuition-building benchmark.

Future UASs will likely employ advanced reliability management systems. Fault detection, isolation, and accommodation (FDIA) methods identify and respond to in-flight failures. Health and utilization monitoring systems (HUMS) track the duty cycles and health of systems.

17.2.3 Communication Systems

The OSD reliability study [1] found that communications systems accounted for 11% of UAS failures. Loss of communications does not necessarily result in the loss of an UA, especially if the link can be reacquired. The communications hardware can fail, which is often attributed to antenna mechanisms and the power amplifier. Inadequate environmental controls can increase the failure rate significantly over a properly designed system. Other sources of failure can be environmental such as RFI from radars, other communication systems, or jamming.

17.2.4 Ground Control and Human Interfaces

Human error, which can account for 17–22% of UAS system failures [1], is caused by improper action or inaction by human operators. Although the range of skills and capabilities of humans varies greatly, it is unlikely that human attributes are going to fundamentally change in the foreseeable future. Human errors can be thought of as system design failures because

Table 17.3   Notional Ground Control and Human Interface Reliability

Component	MTBF, hrs	MTBO, hrs	Notes
GCS workstations	1,000–10,000	—	
Human operators	~650,000	8–12	Requires removal from system after shift and frequent breaks

the system interacts with humans in such a way that can result in failures. For example, requiring a highly skilled external pilot to land the UA via direct control and visual cues will lead to a higher loss rate than a properly designed auto-land capability, yet the former would be considered human error. AVO training quality should have a large impact on human error reduction.

Some human error can be explained by lack of situational awareness in the GCS. Unusual vibration, noise, smells, and other sensory inputs that would be available to the pilot of a manned aircraft are not available on the ground. The UA's situational awareness camera, if so equipped, may have a narrow field of view compared to what a pilot in the cockpit would observe. Notional reliability of GCS and the operators within are shown in Table 17.3.

17.2.5 Airframe

Operational UAS airframes should be expected to have very high reliability relative to other contributors when properly designed and operated under design conditions. Advanced prototypes that operate in challenging domains or that use new structures technologies will have substantially lower structural reliability. It is not unusual for a student-built airframe typical of student competitions to break apart in flight, which serves as a great learning experience. Fortunately, most professionally

Table 17.4   Notional Airframe Reliability

Component	MTBF, hrs	MTBO, hrs	Notes
Hobbyist or student project	1–200	—	Strongly dependent on experience
Prototype structure—professional	1–1,000	—	Depends on technology
Small UAS structure	1–5,000	2–1,000	From MAVs through tactical UAS
Large UAS structure	50,000–200,000	500–5,000	MALE and larger UAS

developed UASs fare better. Table 17.4 shows notional airframe reliablity for several types of structures.

17.3 Availability

Availability is often specified for UASs. This can take the form of inherent availability, achieved availability, or operational availability. A high availability translates to being able to operate the system when it is needed.

The inherent availability is the probability that the system when operated under defined conditions and in an ideal support environment will operate satisfactorily at any point in time as required. This does not include preventive or scheduled maintenance, logistics delays, and other factors. The inherent availability A_i is defined by

$$A_i = \frac{MTBF}{MTBF + \overline{M}_{ct}} \qquad (17.12)$$

where $\overline{M}_{ct}$ is the corrective maintenance time, which is similar to mean time to repair.

The achieved availability is similar to inherent availability but includes preventive maintenance. The achieved availability A_a is defined by

$$A_a = \frac{MTBM}{MTBM + \overline{M}} \qquad (17.13)$$

where $MTBM$ is the mean time between maintenance and $\overline{M}$ is the mean active maintenance time.

Finally, the operational availability is the probability that a system operating under defined conditions in an operational environment will operate satisfactorily when called upon. Operational availability A_O is defined by

$$A_O = \frac{MTBM}{MTBM + MDT} \qquad (17.14)$$

where MDT is the maintenance down time, which includes active maintenance time (M), logistics delays, and administrative delays.

Curiously, increasing reliability can reduce the availability. For example, if a triplex system must have all three strings functional, then there is a higher probability that one string will be nonfunctional than a single-string system. The UA will likely remain on the ground if one of the strings malfunctions prior to launch.

Another technique for assessing the availability A is

$$A = \frac{FH}{FH_{Sched}} \qquad (17.15)$$

where FH is the flight hours flown and FH_{Sched} is the scheduled flight hours.

Example 17.2   Reciprocating Engine vs Turboprop

Problem:

A system developer is conducting a propulsion trade study between an advanced heavy fuel reciprocating engine and a turboprop. Through a detailed analysis that isolates the propulsion system, the following parameters are found:

Propulsion Type	MTBM, hrs	MTBF	$\overline{M}_{ct}$, hrs	$\overline{M}$, hrs	MDT, hrs
Reciprocating	100	100	2	3	10
Turboprop	5000	5000	1	3	50

What are the inherent availability, achieved availability, and operational availability for these two propulsion types?

Solution:

Propulsion Type	A_i, %	A_a, %	A_o, %
Reciprocating	98.04	97.09	90.91
Turboprop	99.98	97.94	99.01

Almost any component can impact system availability. A dead laptop battery, broken fuel pump, jammed latch, missing wrench, fried antenna steering motor, rat-gnawed ground cable, depleted spare actuator supply, sporadic RFI, sick AVO, contractual crew rest rules not satisfied, or almost innumerable other seemingly minor problems can force mission cancellation. When one considers the exhaustive list of activities and systems that must work properly to complete a scheduled mission, it is amazing that mission availability can be as high as is common. Yet many UASs have been able to achieve high availability. For example, AAI's RQ-7 Shadow 200 achieved greater than 96% availability by 2007.

17.4 Maintainability

Maintenance is the upkeep of the system. This includes repairs and scheduled maintenance. The degree to which the UAS enables maintenance is known as *maintainability*. Maintenance activities can occur at the site of operations, depot, and the original equipment manufacturer (OEM).

The system can be maintained by contractors, government personnel, or a combination of the two. Systems procured by methods other than a program of record will generally emphasize contractor maintenance. The same is true for services contracts. The government customer will invest in training a cadre of government maintainers for a program of record.

One maintainability metric is the mean time to repair (*MTTR*), which is defined as

$$MTTR = \frac{T_{\text{Rep,Tot}}}{N_{\text{Rep}}} \tag{17.16}$$

where $T_{\text{Rep,Tot}}$ is the sum of the repair time and N_{Rep} is the number of repair activities.

Another common maintenance metric is maintenance man-hours per flight hour (MMH/FH). The MMH/FH is much lower for small simple systems than large complex systems. A small UAS may have a MMH/FH of less than 1, whereas large VTOL UAS, HALE UAS, or UCAV can have values closer to manned aircraft. The RQ-2 Pioneer has a MMH/FH of 5.87. Aging aircraft require more maintenance due to corrosion and other factors. New manned military fixed-wing aircraft or helicopters can have a MMH/FH of 1-17, and an aging fleet can have a MMH/FH of 10-27.

Maintenance can be performed at multiple locations and by different organizations. Organization level (O-level) maintenance is performed at the site of operations. Intermediate level (I-level) is performed outside of the operational unit. Depot-level maintenance involves heavy maintenance away from the forward operations. Finally, the OEM performs specialized maintenance, repair, and overhaul activities at industrial facilities. It is often practical for UASs to have a two-level maintenance concept where there is only O-level and OEM or depot maintenance. This eliminates the need to set up in-theater I-level facilities and staffing.

Ideally avionics and other line replaceable units (LRUs) that require frequent removal should be installed one deep. One deep means that no other components must be removed to provide access. This reduces the time to perform maintenance and lessens the risk of incorrect connector attachment.

The UA should allow for easy access for ground inspections and maintenance. Hatches, doors, and removable fairings provide such access. The wheel wells for retractable gear are another convenient entry point. Fuel bays might be more difficult to inspect due to sealing, but any fuel system component that can fail (such as fuel level sensors, pumps, and valves) must be accessible.

Minor structural repairs should be supportable in the field. Small UAS composite structures can become cracked, delaminated, or crushed. This damage can be repaired with tools and materials provided in a field

Fig. 17.3 A U.S. Airman with the 432nd Aircraft Maintenance Squadron works on MQ-1 Predator avionics during routine maintenance. (U.S. Air Force photo by Senior Airman Larry E. Reid, Jr.)

composite repair kit. This kit may include composite cloth, shears, epoxy, sand paper, and gloves. Minor repairs can keep an UA flight-worthy. Some training is beneficial, though improvisations might be sufficient in combat conditions.

Most maintenance activities should occur inside an environmentally controlled space. Maintenance facilities such as hangars or portable maintenance shelters enable maintainers to perform their duties within a comfortable temperature range, with adequate lighting, and away from precipitation. Outdoor maintenance activities in rain, below freezing, in the hot desert, or on a ship deck in high sea states should be avoided. Facilities may be unavailable for small UAS operations that are carried in backpacks, but are more practical for tactical and larger UASs.

The maintenance personnel require technical documentation to support the in-field maintenance activities. The exception is initial deployments that are managed by the system prime contractor, where the development engineering and technician team will provide on-site support. Documentation usually comes in the form of hard-copy maintenance manuals. This can also be software based and accessible on mobile computing devices such as laptops and tablets. Figure 17.3 shows an airman performing routine maintenance.

Maintenance does not just apply to the UA. Air conditioners, generators, ground support equipment, launch and recovery equipment, and many other parts of the system must all be maintained. Spares must be provided for these maintenance activities. Table 17.5 gives some examples of routine maintenance actions.

17.5 Supportability

Supportability covers activities required to sustain operations. This includes support equipment, spares management, and other items. Support activities can be performed by the customer (i.e., military or civilian government organization) or the contractor. Customer support is often called integrated logistics support (ILS), and contractor-based support is called contractor logistics support (CLS).

Ground support equipment (GSE) is equipment that is not housed within the GCS or UA, but is required for operations. GSE is generally a vehicle, mechanical system, thermodynamic system, or computer. Some examples include the following:

* Battery-charging equipment for charging the UA or external computer batteries is used.
* Fueling equipment is needed for fueling the UA, generators, and motorized support equipment. Fueling equipment must have grounding provisions to prevent fires.
* Preflight computers (sometimes called UA test controllers) are used to facilitate the UA startup external to the GCS. Figure 17.4 shows a Global Hawk vehicle test controller used for preflight.

Table 17.5 Example Maintenance Actions for Field and Depot/OEM

Maintenance Action	Field	Depot/OEM
Engine	Oil change, spark plugs, parts replacement, engine replacement	Engine overhaul
Avionics	LRU swap out, LRU functional check, system check-out	Board level repair
Airframe	Inspection, cosmetic repair, replacement	Major airframe repair
GCS	CPU or monitor swap out	Rebuild components
Communications	Diagnostics, component replacement	Major repairs

Fig. 17.4 Sgt. John Signorelli attaches a Dash-95 air cart to a RQ-4 Global Hawk at Beale Air Force Base. (U.S. Air Force photo by Staff Sgt. Bennie J. Davis, III.)

- Ground cooling or heater carts or equipment environmentally control avionics and engines while the UA is on the ground.
- Engine start equipment helps start the engines in order to save weight of an onboard starter. Jet engines and turboprops often use compressed air to spin the turbine. Reciprocating engines need external starting motors to spin the propeller.
- Aircraft towing equipment can move the UA on the ground when it is not able to self-taxi.
- UA without landing gear can be moved on dollies, as is common for targets or large runway independent systems. The dollies can also serve the function of a field assembly station or maintenance stand.
- Mechanical manipulation equipment help ground personnel move and position heavy airframe components. This is necessary for lifting any components that are beyond the capabilities of personnel.
- Storage cases store the UA or other system components when they are not in use. These usually serve the function of transportation cases as well. Cases help prevent damage and exposure to the elements.

It is a good practice, and often a requirement, that all UA components are capable of a two-man lift. This usually translates to 150 lb per component without mechanical assistance. Carts, cranes, winches, jacks, hoists and other mechanical devices must manipulate the larger components. These devices must be part of the system. Humans have a large variation in lifting abilities, as can be seen in specifications for say a 5th-percentile female vs a 95th-percentile male.

In the case of a pneumatic-launched and net-recovered tactical UAS, the aircraft must be positioned on the launcher prior to launch. After recovery in the net, the aircraft must be removed from the net and transported to a maintenance/storage location. Ground carts can be built with the provisions to support these actions.

UA often require assembly and disassembly in the field. This is required for configuring the UA after transportation or storage. Large airframe segments might need to be replaced or removed for major maintenance or repair activities. Figure 17.5 shows two mechanical devices that lift and aid wing positioning for a MQ-9 Reaper wing installation.

Fig. 17.5 U.S. airmen and civilian employees, with the 27th Special Operations Aircraft Maintenance Squadron, rotate and attach the left wing of the MQ-9 Reaper unmanned aerial vehicle (UAV) at Cannon Air Force Base. (U.S. Air Force photo by Airman 1st Class Maynelinne De La Cruz.)

Fig. 17.6 An airman prepares the RQ-4 for launch using the UA test controller while reviewing technical orders. (U.S. Air Force photo by Staff Sgt. Bennie J. Davis, III.)

The preflight computers aid UA preflight operations and engine start. The functional differences between these support computers and the GCS is somewhat arbitrary. The support computers can be directly linked to the UA via a cable without requiring RF communications. The AVO in the GCS may be busy developing flight plans, so that ground support personnel can offload much of the preflight work through these devices. Additionally, the ground support personnel have their eyes directly on the UA and can perceive issues better than the AVO inside a shelter. Preflight computers can command engine start and must be positioned to ensure personnel safety. Laptops usually interface to panels that are covered by hatches that are sealed before launch. The Global Hawk preflight computer, called a UA test controller, is shown in Fig. 17.6.

Spare parts, known as *spares*, must be available to support maintenance actions and to ensure system availability. The spares may consist of small hardware items or very large systems. Minor spares can include filters, spark plugs, or brake pads. More significant spares may include engines and payloads. Any component or system that fails or must be replaced will necessitate installation of a spare part to bring the system back to flight readiness. The components may have a maximum life limit and must be replaced after that level of usage is reached. Other components must be overhauled. To determine scheduled maintenance or replacement time, it is necessary to maintain records to track the history of the components. The required quantities of spare parts are determined through projections of the component replacement rates. Figure 17.7 shows an engine replacement on a Predator A, where the replacement engine is a spare.

Initial spares should be included with the initial deployment. The spares list and quantities should be based on usage projections for a specified time period. After some period into the operations, the logistics support will be able to keep the spares supplies replenished. Failure to provide adequate spares will result in low mission availability due to nonflight status of UA or other system elements.

Interchangeable parts simplify logistics. An interchangeable part can be used directly without any modification. By comparison, a replaceable part requires adjustments in the field to facilitate a fit. A highly modular aircraft with interchangeable parts might introduce difficulties with tracking a single aircraft by tail number. Modularity benefits for life and availability are exemplified by the old story about an axe that has been in the family for hundreds of years: it has only had five blades and ten handles.

Support software can be used to track component usage and failure rates. This helps support ordering replacement parts to ensure high

Fig. 17.7 MQ-1 Predator undergoes an engine change at Creech Air Force Base. (U.S. Air Force photo by Lance Cheung.)

availability. Such software is only as good as the data it operates on. Support information such as part replacement and spares quantities must be entered by support personnel.

UA should be adequately rugged for support operations. Structures might be required to either withstand a tool drop or be repairable after a tool is dropped on it. Aircraft skins should not be permanently damaged from human touch. These considerations often drive the minimum gauge thickness of the aircraft. Gossamer structures or lightweight high-performance UA will have fragile structures that must be handled with care. Some back-packable airframes may be stepped on or have heavy components placed upon them.

17.6 Footprint

The ground footprint can be amazingly large for even the simplest of unmanned systems. What immediately comes to mind are the UA and GCS shelters. With more consideration one can think of launch and recovery equipment and some ground support equipment. These items are only the tip of the iceberg.

The basic operational system elements that contribute to the footprint can include the following:

* UA
* Ground control stations
* Launch equipment (i.e., pneumatic rail launcher)
* Retrieval equipment (i.e., nets, microwave landing system)
* Generators
* External air conditioner for shelters
* Air compressors
* Ground data terminals

Support equipment and operational items include the following:

* Maintenance shelters
* Fuel supply
* Fueling system (storage tanks, hoses, pumps, grounding straps, fueling vehicle, cart)
* Batteries, battery chargers
* Tools (hand tools and power tools, spare tools) and tool storage
* System test equipment
* Ground maintenance computer system
* UA start computers
* Carts, dollies, and ladders

* Wheel chocks and tow bars
* Generic diagnostic equipment (ohmmeter, spectrum analyzer)
* Optical targets (for payload calibration and alignment)
* Ground cooling equipment for UA
* Portable weather station
* Ground-based air traffic radar
* Hoses and cables
* Consumable hardware (nuts, bolts, fasteners)
* Repair kits (composite materials, epoxy, vacuum bags, vacuum pumps, shears)
* Floodlights and flashlights
* Prime mover for the UA with towbar for large UAS
* Ground vehicles to set up antennas on hills (might need to be off-road vehicles)
* Cranes, jacks, hoists (engine removal), or other mechanical manipulating equipment
* Equipment for pounding stakes into the ground
* Fire extinguishers and safety equipment
* Spares, spares cases
* Operation-specific communications equipment
* Local voice communications (hand-held devices)
* Protective equipment such as gloves and respirators for dealing with dangerous chemicals, composite repairs, or replacing fluids in systems
* Engine start equipment
* Battery disposal provisions
* Outside lighting—floodlights and flashlights
* Manuals and checklists

Transportability provisions include the following:

* UA storage cases
* Other hardware storage cases
* System-specific trailers

Personnel provisions include the following:

* First aid kits
* Cleaning equipment (grease removal, glass cleaner, towels, brushes)
* Office supplies, forms
* Personal communications (satellite phones, internet)
* Administrative support computers (timecards, reporting)
* Potable water
* Bathroom facilities

- Sleeping facilities (tents, barracks)
- Food (mess hall, MREs, coffee machines)
- Firearms, ammunition, and storage
- Photography equipment

Not all of these items will be used in every system. The footprint for a MAV can be contained within a backpack, whereas a HALE UAS can take up an entire hangar and substantial tarmac space. Many of the required items may be available at the operations location and are therefore not considered to be part of the UAS footprint. An assessment of the UAS footprint must define what is contained within the system boundaries.

17.7 Logistics and Transportability

UASs must be transported to the point of operations and then sustained. Once in theater, the operations must be supported with spare parts, fuel, and other consumables. These activities are known as logistics.

Systems are often transported on aircraft for long distance deployments. Common U.S. military transports include the C-130, C-17, and C-5. MIL-HDBK-1791 provides guidance on interfacing with military cargo aircraft. Commercial airlift options include commercial aircraft cargo holds, freighter conversions of airliners, and commercial variants of military cargo aircraft such as Lockheed L-100 or Ukrainian-built Antonov aircraft. Small UASs might be required to travel like luggage on regular commercial flights or on general-aviation aircraft. MIL-STD-1366E provides valuable data on transportation via highways, rail, water, fixed-wing aircraft, and rotary-wing aircraft.

Military cargo aircraft use pallets, which are unitized rectangular plates that can be loaded individually. Pallets generally slide on the cargo aircraft floor, guided by a grooved rail system. The prolific 463L pallet can be used on the C-130, C-17, and other cargo aircraft. In a C-130 loading configuration, it measures 108 in. wide and 88 in. long. The 463L pallet has become a standard unit of transportability within the U.S. military. Logisticians think in terms of numbers of pallets or pallet positions. Where possible, it is desirable to fit system components on a single pallet rather than requiring two contiguous pallets. This allows the hardware to be loaded on the pallets rather than be manually placed on a pallet that is already on the aircraft.

Transportation containers must protect their contents during transport. The shock, acceleration loads, vibration, and thermal environment must be well understood for each of the transportation modes. Foam padding and shock isolation mounts can help isolate the components. The packaging volume must account for the stroke of the internal component as it

Fig. 17.8 Airmen with the 432nd Aircraft Maintenance Squadron at Creech Air Force Base prepare a RQ-1 Predator for deployment in support of Operation Unified Response. (U.S. Air Force photo by Staff Sgt. Alice Moore.)

moves in the box. The box frequently outweighs the contents housed within. Figure 17.8 shows a predator in its transportation container being loaded onto a flat bed truck.

Loading and unloading a cargo aircraft will generally require powered mechanical systems. Well-equipped bases will generally have all of the required support equipment for cargo aircraft ground operations, but this might not be the case for austere locations. A K-loader is an elevated platform that can lift and roll 463L pallets straight into the back of the cargo aircraft without a tilt angle. These machines avoid the awkward ramp angles. C-130s may have a winch that pulls the cargo up the ramp. If this is insufficient, the UAS might need to provide its own cargo loading provisions such as a prime mover or forklift.

UASs can be air-dropped from cargo aircraft. The 463L pallets can slide out the back of the aircraft pulled by an extraction parachute. Once clear of the host aircraft, an unguided round parachute or a guided parafoil deploys to bring the pallet safely to the ground. The impact loads can be critical structural sizing conditions for the system components.

In-theater air transport frequently relies on helicopters. The U.S. military helicopter inventory includes CH-46, CH-47, CH-53, CH-60, and the V-22 tiltrotor. Helicopters can carry the UAS internally or as an external sling load. The internal helicopter load-out is similar to that of a cargo aircraft. A sling load is accomplished by holding the system in a cargo net that is tethered to the helicopter. Internal loads are generally preferable to sling loads.

The most practical transportation for some UA is to self-deploy with a ferry flight. Very large UASs might be too big to fit within a cargo aircraft, and they might also have very long-range capabilities. The Global Hawk is an example of a UAS capable of self-deployment. Optionally piloted vehicles (OPVs) can have a manned ferry flight to the operating site. OPVs with pilots can make multiple stops if necessary and face fewer airspace restrictions than unmanned aircraft. Self-deploying UA must be followed by other system elements through other transportation means unless this hardware is carried as cargo aboard the ferrying aircraft.

UA can be transported externally on host aircraft. The Boeing Phantom Eye was placed atop a Boeing 747 for transport from the plant to the flight-test location. This approach is very similar to the method employed for the space shuttle. Air-launched UAS types could be transported on the wing pylons of the host platform.

Ship, rail, and ground transportation modes are often required. The TEU ISO container families, also known as air-sea-land containers, provide a flexible means of addressing all three modes of transport. These rugged, stackable containers can be placed on container ships, trains, and trucks. Air-sea-land containers are manipulated by cranes.

Other over-road transport methods might be required for military and civilian systems. U.S. military ground transports include HMWVV, high mobility trailer (HMT), and military flatbed trucks. Civilian ground transport can utilize trucks, sport utility vehicles, recreational vehicles, and specialized trailers. Elements of the system such as the GCS shelter, maintenance shelters, spares storage, and launch and recovery equipment can be integrated with the ground vehicles. Over-road transportation imposes limitations on width, height, and axle weight.

The spares logistics can be satisfied through a dedicated logistics system or through commercial services. The U.S. military Transportation Command (TRANSCOM) is an example of a dedicated logistics organization. Other operations that leverage a contractor logistics support system or civil operations can be resupplied through commercial shipping companies such as FedEx or UPS.

The type of fuel used by the UA can have large implications for logistics. Military UASs that use heavy fuel can integrate well with the fuel types used by other military systems. Ground forces and Navy ships use JP-5, and Air Force aircraft generally use JP-8. Aerospace grade turboprops and jets generally run on JP-5 and JP-8, as well as Jet-A commercial jet fuel. Few reciprocating engines use heavy fuel, though more are becoming available in recent years. Depending on the heritage, other reciprocating engines might use AVGAS, pump gas (also called mogas), glow fuel (model aviation fuel), or a gas-oil mixture. These types of fuel are not supported by the military and must be provided through special logistic means.

17.8 Organization, Training, and Personnel

In this section we will address three items from DOTMLPF: organization, training, and personnel. Whether the system is military or civilian, the organization, training, and personnel considerations must be addressed. Creating a cadre of trained UAS operators with the necessary support is not a casual endeavor.

The UAS operational organization can take many forms that depend upon system maturity and ownership. System development contractors take the lead in flight test and in operations for systems that are rushed to the field. Initial activities will have a strong developmental engineering and technician presence. As contractor flight operations become more routine, it is common for the development team to dwindle as a dedicated contractor flight operations team takes over. In-theater contractor personnel are called field service representatives (FSRs). The government-owned contractor-operated (GOCO) and contractor-owned contractor-operated (COCO) service models use FSRs to conduct operations. The Boeing-Insitu ScanEagle operations used services models for multiple deployments with numerous customers.

Program-of-record (POR) systems are owned and operated by the military or civilian government agencies. Although contractor operations support may play a role, the intent is for the government personnel to operate the system. These personnel must be trained and qualified for the duties. The tasks should be simplified so that new operators can quickly become proficient. The operators come from established career tracks such as pilots or aircraft maintenance specialists. Often the number of government personnel needed to support operations is greater than for contractor support. Two reasons are that contractors may work longer hours and may be qualified in more than one function. For example, a Predator operation requires 30 active duty personnel or 9 contractors [2].

The forward-deployed personnel or those operating at MCEs must be supported, whether contractor or government employees. The infrastructure apportioned to UAS operations must be supported by logistics, security, and administrative functions. Ideally, the system operators should not be encumbered by duties other than system operations and basic administrative functions such as timesheets.

Training is required for system operators and maintainers once the system becomes operational. Small systems can have a single training program that covers all aspects of system operations including AVO, MPO, and maintenance activities. Simple system training programs could take place in as little time as a day or week. Large UAS systems have more functionally specialized training that can take several weeks to complete. Training programs for AVOs and MPOs are usually simulation-based, and the maintainer training uses system hardware. A course curriculum includes reading material, exercises, and tests. Trained instructors are often pulled from the operator community. The operator training includes not only basic system operation, but also tactics. The training program should culminate in training flights. Highly demanding tasks such as an external pilot necessitate a high degree of screening because few have the hand-eye coordination and intuition required to safely land an UA based on visual and audio cues.

17.9 Facilities

UAS operations require the use of facilities. The facilities can be part of the system or provided by external entities. Many facilities may be outside of the system boundaries, such as crew rest shelters, food, medical, and recreational facilities. The GCS can be a dedicated shelter or integrated into a ship, aircraft, or building. The base can provide ground support equipment, fuel, and power. All of these considerations affect operational costs, footprint, and system transportability.

Personnel accommodations can vary from a foxhole to a hotel room. Small UASs that are man-portable are organic to the unit, and the operators are simply soldiers. These operations might be over-the-wire where the soldiers leave the forward operating base on a patrol. The system can also operate behind-the-wire, where austere accommodations such as a tent and mess facility are available. Larger systems that operate from main operating bases are more secure, and the semipermanent presence may allow for better accommodations such as barracks and cafeterias, along with hospitals and recreational facilities. The MCE used in remote split operations is generally located in the home country or an allied nation. The MCE is on a base and can be integrated inside a building or in a transportable shelter. In either case, the operators can peacefully go home after conducting combat operations on the other side of the world.

Fig. 17.9 U.S. airmen with the 432nd Aircraft Maintenance Squadron stand near an MQ-9 Reaper UA on the day of the first U.S. Air Force noncontractor launch of an MQ-9. (U.S. Air Force photo by Lawrence Crespo.)

The base can provide additional UAS support to enable operations. Fuel used by the UA, ground support equipment, and generators is often provided by the base or ship. The base can supply electrical power, which eliminates the need for generator usage. Potable water is usually outside of the UAS boundaries. Base-provided ground support equipment such as a ground tug, engine start carts, de-icing equipment, and fuel trucks can greatly reduce the transportability footprint.

Hangars or other UA shelters are used for system maintenance and storage. These facilities could be a tent or trailer for small UASs or a full-size aviation hangar facility for larger UASs. A large buildup of forces in a new campaign may create a demand for hangar space that far exceeds the available facilities. This was the case during OEF and OIF. Portable hangars can be shipped via cargo aircraft, ships, or other transportation means. An example of a portable hangar for MQ-9 Reaper operations is shown in Fig. 17.9.

Military hangars at established bases are often sized to accommodate large aircraft (B-52, B-2, P-3, C-130, etc.). Very large aircraft such as solar-powered aircraft or the Boeing Condor will exceed the dimensions of widely available facilities. When the government considers the life-cycle costs of a new UAS, the military construction (MILCON) should be given consideration.

Ships are the facilities for ship-based UAS operations. The launch and recovery equipment, support facilities, and control station may be dispersed across the vessel. The ship may supply electrical power and fuel. Enclosed spaces may be provided for the control station and maintenance activities. The littoral combat ship (LCS) provides modular container spaces that house these elements for UAS operations inside the ship. Otherwise, it is necessary to provide topside shelters that are capable of withstanding the maritime environment. Crew living quarters and mess facilities are likely to be provided for the UAS crew.

17.10 System Responsiveness

The UAS will usually have responsiveness requirements. Responsiveness metrics can include the following:

* Time to achieve system readiness from arrival at site
* Time to flight from stored state
* Time to flight from a high readiness level (i.e., warm start)
* Turnaround time
* Time to target from receipt of tasking

A site will have a finite crew, and each crew member will have a set of system skills. System assets required to achieve flight can include ground stations, ground support equipment, launchers, and consumables. The UA will be repositioned and possibly reconfigured through the launch process.

A timeline analysis is one tool used to help choreograph the complex tasks and assets. It is common to track individuals and the UA. The sequence and duration of specific tasks performed by individuals are identified. This analysis can take the form of Gantt charts that are used for program schedules, except the tasks last for minutes rather than days. The timeline analysis reveals the sequence of system preparation tasks, identifies who will perform the work, and defines system asset states.

The timeline analysis can also apply to an entire mission, which may consist of more than one UA. This is sometimes called a day-in-the-life (DITL) analysis. The DITL may span the time from system activation (i.e., receipt of tasking) to the completion of the mission. The end state might include shutdown of all UA, dissemination of payload information, and postmission reporting.

Example 17.3 Cold-Start Launch Timeline Analysis

Problem:

Two individuals are available to prepare a UAS for flight operations from a cold start upon receiving mission tasking. One is an AVO, and the other is a system maintainer. The UA is electric powered and weighs 30 lb, and so it can be carried by a single person. A pneumatic rail launcher is required to launch the UA. Cold start is defined as the UA is in the box, and all systems are shut off.

a. Generate a timeline analysis for the system launch.
b. How long does it take from receipt of tasking to launch readiness?
c. Would any cross-training between the AVO and maintainer reduce the timeline?

Tasks include, in no particular order, the following:

Task	Minimum duration, min	Required Role	Predecessor Tasks
Pressurize launcher	10	Maintainer	Power-up generator, 1 min to initiate
Generate mission plan	5	AVO	Boot up computer
Assemble UA	2	Maintainer	——
Final system checkout	2	Both simultaneously	Pressurize launcher, peak charge UA battery, place UA on launcher, UA self-diagnostics
Boot up computer	3	——	Power-up generator
Boot up UA	2	——	——
UA diagnostics	2	Maintainer	——
Establish RF communications	1	Both simultaneously	Boot up UA, boot up computer
Power-up generator	2	Maintainer	
Peak charge UA battery	5	——	Power-up generator, 1-min initiation, can't be moved while charging
Place UA on launcher	3	Maintainer	Assemble UA

Solution:

The following figure shows one answer to the problem, though it might not be unique. The tasks are divided by the AVO and the maintainer. Here it can be seen that the procedures take a total of 15 min. The maintainer has more tasks than the AVO. However, it is not apparent that cross training will reduce the time.

References

[1] Office of the Secretary of Defense, *Unmanned Aerial Vehicle Reliability Study*, Feb. 2003.
[2] Drew, J. G., Shaver, R., Lynch, K. F., Amouzegar, M. A., and Snyder, D., *Unmanned Aerial Vehicle End-to-End Support Considerations*, RAND Corp., Arlington, VA, 2005, pp. 44, 45

Problems

17.1   The *MTBF* values for several flight critical components on the UA are as follows: communications = 15,000 hrs; engine = 5,000 hrs; and avionics = 20,000 hrs. Based only on these numbers, what is the expected loss rate?

17.2   Consider a single high-reliability actuator driving a single elevator and two less reliable actuators driving a split elevator surface. If one side of the split elevator fails, the UA has degraded operations, but it will not cause a loss. The high-reliability actuator has a *MTBF* of 20,000 hrs, and the less reliable actuators have a *MTBF* of 5,000 hrs. Which approach gives the best reliability for over 3,000 hours?

17.3 A system developer is conducting a payload trade study between two airborne radars. Through a detailed analysis that isolates the radar payloads, the following parameters are found:

Radar Model	MTBM, hrs	MTBF	$\overline{M}_{ct}$, hrs	$\overline{M}$, hrs	MDT, hrs
Radar 1	20	5000	2	1	5
Radar 2	5000	2000	15	10	50

What are the inherent availability, achieved availability, and operational availability for these two radars?

17.4 A wing has a transportable length of 80 ft. What types of cargo aircraft can be used to transport this structure?

17.5 Compare the cargo floor area and available cargo volume of the C-130, C-17, and C-5.

17.6 Use the timeline analysis in Example 17.3. Now, the pressurize launcher task takes only 1 min. The final checkout requires only the AVO. The UA can be moved while being peak charged. Redo the timeline analysis to minimize the time to system launch from receipt of tasking.

Chapter 18

System Synthesis and Mission Effectiveness

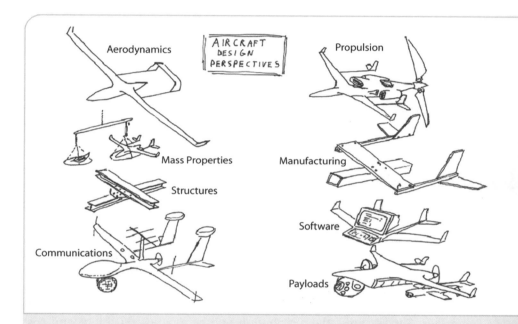

Fig. 18.1 Various discipline perspectives of the Insitu Integrator UAS, drawn by the author at the beginning of the program.

- Define system-level performance metrics
- Balance system elements and fleet sizing
- Understand elements of operational analysis
- Define the role of system engineering

18.1 Introduction

In this chapter we will see how the aircraft and other system disciplines come together through multidisciplinary design. Unlike a manned aircraft, unmanned aircraft sizing is a dependent upon system-level trades that involve ground elements. The ground infrastructure also depends on the UA size, onboard systems, and mission capabilities.

The UA is but one element of which the unmanned aircraft *system* is composed. System architectural decisions and component selection can have varying degrees of impact on the UA, ground elements, operational utility, and even systems outside of the UAS boundaries. In this chapter we consider the overall system synthesis.

We will explore attributes of balanced systems and provide methods of quantifying system performance metrics. Elements of operations analysis and optimization are presented. UA sizing methodologies that leverage discipline methods presented earlier in the book are described at the end of the chapter.

18.2 Balancing the System

A system must be well-balanced in order to achieve success. But what is meant by well-balanced? This means that the system elements have the right proportions, allocations of attributes, and can satisfy the operational need. The system architecture must be appropriate for the mission, and the mission performance should be matched to the need. The manpower requirements and footprint should be in proportion to the utility. The cost should be in keeping with the customer affordability and value expectations. Furthermore, the distribution of functionality, reliability, and cost throughout the system should be appropriately apportioned. The conservatism or aggressiveness of technologies employed should support the program goals. All of these criteria are difficult to quantify or define more precisely, which often leaves customers using adjectives to describe the systems.

Perhaps a balanced system can be better defined by examples of unbalanced system characteristics. Figure 18.1 shows what an UA might look like if various design disciplines got the upper hand in system design trades. This is inspired by the classic K. D. Woods image. Some examples include the following:

- A very expensive payload is carried by a very inexpensive UA.
- A very expensive UA carries inexpensive and low-capability payloads.
- A high-tech UA is controlled by a low-tech ground station.
- A strategic UA is controlled for long periods of time by a laptop.

- The mission systems onboard the UA are unable to send mission data to the ground.
- A tactical UAS requires 10 C-130s to deploy.
- A surveillance UAS makes too much noise at the collection altitude.
- A UCAV is not sufficiently survivable to reach the target.

18.3 System Architecture Selection

A fundamental output of system concept definition is the system architecture. System architecture is the composition and organization of system elements and communication networks. It can also entail the allocation of tasks among hardware and software in various system locations. Several system architecture questions include the following:

- What types of ground stations are required (LRE, MCE, FGCS, RVT, TPED, etc.)? What are the required quantities? Who are the users? How will the ground stations communicate with the UA and with one another?
- Will the control stations be integrated with other systems (i.e., a ship, helicopter, aircraft, tank)? What are the physical, electrical, software, and data interfaces?
- What are the types and quantities of operators in the ground stations? What automation features are required for a manageable workload?
- Will a single operator control only one or many UA?
- Is the UA capable of SATCOM or only line-of-sight communications?
- Which SATCOM service is selected? Does it require pointing and scheduling? What parts of the system manage this external interface?
- If equipped with SATCOM, is the UA controlled beyond line of sight, or is the SATCOM link only for data dissemination?
- What forms of external communications are required (ATC, air operations center, etc.)?
- Who are the consumers of the collected intelligence? How will data be provided to them?
- What are the payload product formats, and how will it be produced (on the UA or in a ground station)?
- How will payload pointing and stabilization be accomplished? Will there be postcollection enhancements?
- How will the payload data be managed throughout the system? Is onboard processing and storage necessary? Is the amount of collected data greater than the communications bandwidth?
- What is the required reliability and availability? How will this be achieved (i.e., all simplex vs redundancy in the flight control system and multiple data links)?

- What support equipment is necessary, and how does this fit within the logistics and footprint requirements?

The system architecture must be decided at the beginning of a program. It is not possible to design the UA until the architecture is established because the payloads, communications system, and avionics are largely driven by the architecture selection. The UA sizing affects the launch and recovery equipment and the support concept.

System architecture selection should be a multidisciplinary system engineering process, though often the major architecture decisions are made by a single individual at the beginning of the program. The disciplines that must be represented, at a minimum, are communication systems, payloads (mission systems), aircraft design, and RM&S (reliability, maintainability, and supportability). However, at the beginning days of a program it is often impractical to bring together such a team. The architecture selection often rests on the shoulders of an individual who has a working knowledge of all of the disciplines and is capable of finding missing information. The architecture selection process also involves finding representative hardware because SWAP, cost, availability, and maturity should guide the decision process.

18.4 System Performance Metrics

System performance metrics are useful for quantifying or describing system performance. These metrics are often called key performance parameters (KPPs) or top-level requirements (TLRs). A system requirements document or other form of specification can include hundreds or thousands of requirements that must be met for compliance. KPPs are more top level and distill what is most important. Therefore, these metrics are few in number and intuitive. Example KPPs can include the following:

- Total endurance (persistence)
- Maximum radius
- Ferry range
- Time to target
- Mission availability
- Payload capacity
- Sensor performance [i.e., NIIRS at collection conditions, target location error, weapon circular error probable (CEP)]
- Sensor effects (i.e., distinguish between a broom and a rifle)
- Undetectable slant range (visual and acoustic)
- Survivability metrics (radar and IR detection ranges, target access)
- Area coverage rates
- Radius of action or communications range

- Turnaround time
- Number of personnel required for operations
- Interoperability with existing systems

18.5 Operations Analysis Modeling

Operations analysis modeling can consist of isolated analysis of a system element or a simulation of the system in a geospatial simulation. Some examples of operations analysis tasks include the following:

- Target coverage
- Survivability analysis
- War games
- Payload effectiveness in operational environment
- Communications analysis
- Flight performance requirements analysis
- Fleet sizing
- System comparisons
- Basing studies
- Logistics analysis
- Manpower evaluations.

18.5.1 Fleet Sizing for Target Coverage

The number of UA required for an operation is a function of the endurance and ingress/egress speed. The UA cost is largely driven by these parameters. The personnel, GCS, and support equipment are dependent on the number of aircraft and the sortie rate. For example, the LRE must be active for any launch and recovery actions.

Frequently, long-endurance UASs undergo basing studies. This involves calculation of the time on station (TOS) for various combat radii. The combat radius is the distance from the launch and recovery location to the target. The number of UA N_{AV} is calculated by

$$N_{AV} = \frac{24 \text{ hrs}}{TOS \cdot A_0} \tag{18.1}$$

Here, A_0 is the operational availability (see Chapter 17). The *TOS* depends on the flight profile. Ingress and egress typically occur at a higher speed than loiter so that fuel burn can be minimized for the distance covered. Therefore, the fuel burn rate is higher for ingress and egress than for loiter. By comparison, the loiter segment airspeed is set such that the minimum fuel per time is consumed. Let us consider the special case in which the loiter, ingress, and egress all occur at the same airspeed V. The time on station is calculated by

$$TOS = E_{\text{Tot}} - 2 \cdot \frac{R}{V} \tag{18.2}$$

where R is the mission radius and E_{Tot} is the total endurance.

Example 18.1 Fleet Sizing for Continuous Coverage of Target

Problem:

A customer is generating requirements for its next-generation long-range reconnaissance UAS. The flight speed during ingress and egress is 300 kt. Fuel is burned at the same rate for transit and loiter phases. The total UA endurance is 40 hrs. The expected availability is 90%. What is the required number of UA to perform continuous target coverage as a function of mission radius?

Solution:

Radius, n miles	TOS, hrs	Num AV
0	40.00	1
500	36.67	1
1000	33.33	1
1500	30.00	1
2000	26.67	1
2500	23.33	2
3000	20.00	2
3500	16.67	2
4000	13.33	2
4500	10.00	3
5000	6.67	4
5500	3.33	8

18.5.2 Area Coverage

Area coverage is an important criterion for many missions such as maritime surface surveillance, reconnaissance or large regions, and environmental surveys. While flying high, fast, and with a wide field of view improves the area coverage rate, the payload characteristics often put limits on the maximum rate. For example, the sensor resolution requirement can set the maximum altitude that the UA can fly. Communication systems might have bandwidth limitations that restrict the maximum rate of payload data that can be transmitted to the data consumers in real time.

The ground swath D covered by a sensor is

$$D = h \cdot \left[\tan\left(\theta_{\text{Look}} + \frac{1}{2} \cdot FOV\right) - \tan\left(\theta_{\text{Look}} - \frac{1}{2} \cdot FOV\right) \right] \qquad (18.3)$$

where θ_{Look} is the look angle and FOV is the field of view. Details on these angles can be found in Chapter 13. The ground coverage rate A_{Rate} by a sensor at that look angle and collection angle is

$$A_{\text{Rate}} = D \cdot V \qquad (18.4)$$

The ground area covered A is

$$A = A_{\text{Rate}} \cdot T_{\text{Collect}} \qquad (18.5)$$

where T_{Collect} is the collection time.

Example 18.2 Maritime Area Coverage Rates

Problem:

A Navy program office is conducting a trade study of multiple tactical UASs vs a smaller number of large, high-altitude UASs. The aircraft have the following relevant characteristics:

	Altitude, ft	θ_{Look}, deg	FOV, deg	V, kt	T, hrs
Tactical UAS	10,000	45	5	70	24
HALE UAS	50,000	45	5	250	24

How many tactical UASs are required to provide the same area coverage as the HALE UAS?

Solution:

The following are calculated for the two aircraft:

Aircraft	Swath D, ft	A_{Rate}, n miles^2/hr	Area covered, n miles^2
Tactical UAS	1749.77	20.15	483.49
HALE UAS	8748.87	359.74	8633.75

The number of tactical UASs required is the ratio of its area coverage to that or the HALE UAS. This ratio is 17.86. If we consider that we will need integer aircraft, we will round up to 18 tactical UASs to perform the same area coverage as the HALE UAS.

18.5.3 Payload Utility

The payload utility can be measured in terms of payload capacity or payload effectiveness. Although these parameters are important, payload utility must be considered in combination with UA performance. After all, an UA that can carry 1000 lb for 10 min is less useful than one that can carry 10 lb for 10 hrs. Payload utility is often described by curves showing the endurance as a function of payload weight (or vice versa).

Ideally, a UAS will be specified by payload effectiveness rather than just weight. For example, specifying the NIIRS rating will allow the UA to trade optics design, imagery processing, UA size, and acoustic performance. Specifying the payload weight can yield an integrated system performance that does not yield the desired effect. One possible drawback of not specifying payload capacity directly is that multiple payloads can be used, and it is difficult to ensure that there will be sufficient allowances for all payload types. Also, there can be unforeseen future payloads that might be precluded if the payload capacity is too low.

18.6 Survivability

The primary methods of detecting UASs are by radar, acoustic, infrared, and visual means. Although it would be desirable for an UA to be undetectable under all circumstances, this is not usually practical or even possible due to physics, technology limitations, and cost.

Survivability is the ability of the UA to survive in a combat environment. This includes avoiding enemy fire and surviving weapons that engage the UA.

The kill chain is a model for considering the multiple steps that are required for a shoot-down. First, the UA must be detected by the air defense system. Once detected, the UA must be targeted by a weapon system. Next, the weapon system engages the UA. Finally, the weapon system must destroy the UA. If any one of these links in the kill chain is broken, then the UA will survive. Ball [1] and Jenn [2] provide more detailed treatment of survivability.

18.6.1 Radar

The radar-range equation is

$$R_0 = \left(\frac{P \cdot A^2 \cdot \sigma}{4 \cdot \pi \cdot \lambda^2 \cdot N} \right)^{1/4} \qquad (18.6)$$

where

R_0 = range where $S/N = 1$, m
P = radar power, W
A = radar antenna area, m^2
σ = target radar cross section, m^2
λ = radar wavelength, m
N = system noise
S = signal, W

The only term within the control of the UA design is the radar cross section (RCS). The radar range is a function of the RCS to the 1/4th power. To cut the range by half, the RCS must be decreased by a factor of 16. Methods of reducing the UA RCS through configuration design are discussed in Chapter 4. Low RCS can also be achieved through special materials that help absorb the radar's RF energy. A detailed treatment of RCS reduction is beyond the scope of this book.

The radar signature is generally discussed in terms of decibels of square meters, or dBsm. Much like the link budget analysis described in Chapter 12, the radar equation is usually handled as additions of decibel terms to help with intuition of terms that vary by several orders of magnitude.

There are numerous types of threat radars. Early warning (EW) radars detect aircraft at long range, but with limited directional resolution. EW radars typically operate in the vhf-uhf bands due to low atmospheric absorption. Height finders, as the name implies, determine the height of the oncoming aircraft. Tracking radars track the aircraft flight path and are generally queued from the EW radars. Finally, target fire control radars help target the surface-to-air missiles (SAMs) or antiaircraft artillery. Ground-based radars can also provide guidance to SAMS. As the radars proceed down the kill chain, the frequency generally increases. Many nations have an integrated air defense system (IADS), where the radars can share information about threat aircraft.

18.6.2   Acoustic

The acoustic signature is very important for small, tactical, and MALE military UASs that perform surveillance missions. Put simply, the targets will change their behavior if they know that they are being observed, and that is rarely beneficial to the UAS operation. UASs with high acoustic signatures must either increase the slant range to an undetectable distance with reduced payload performance or fly close to the target that is aware of its presence. Lore has it that enemy forces dubbed one operational UAS the "Lawnmower of Death" because of its large acoustic signature. A low acoustic signature is also important for civil and scientific UAS operations where community noise must be considered. Small UASs with poor sound attenuation have a distinctive high-frequency sound that, although not necessarily louder than general-aviation aircraft or lawn equipment, can be annoying to those on the ground. Details of small unmanned aircraft acoustic characteristics can be found in the AMA handbook on the subject [3].

In other circumstances a large acoustic signature can be beneficial. A loud UAS can deter or suppress adverse action. A known UAS presence is a show of force that can make an enemy reconsider their chances of success. It can also provide an opportunity for the enemy to surrender. In Operation Desert Storm Iraqi soldiers surrendered to a Pioneer UAS that was providing artillery spotting support for a battleship, which marks the first instance of humans surrendering to a robot. Civil and scientific UAS noise might not be a major consideration if the UA operates over the ocean or over unpopulated areas.

The acoustic-range equation is

$$R = \sqrt{\frac{3 \cdot G \cdot P_0 \cdot e^{-\alpha \cdot R}}{4 \cdot \pi \cdot I_T}} \qquad (18.7)$$

where

G = acoustic antenna gain
P_0 = radiated acoustical power in narrowband, W
α = atmospheric attenuation at the acoustic frequency, 1/km
I_T = hearing sound intensity threshold, W/cm^2
R = range, km

The detection mechanism is generally the human ear, though acoustic detection systems are also possible. This is not a closed-form equation for range, and so it must be solved for by finding the range that makes the right-hand and left-hand sides of the equation equal. The secant method used for converging the weights in Chapter 6 is one such method.

The detection range is a function of frequency. The product of P_0 and G is a function of UA orientation (azimuth and elevation) relative to the observer. If we neglect the attenuation terms, it can be seen that the acoustic detection range is proportional to the square root of the radiated acoustical power. Adding in the attenuation makes the detection range drop off faster.

The radiated acoustic power is related to the engine power and muffler configurations for reciprocating engines. Decreasing the engine power is perhaps the easiest way of reducing the acoustic signature. This can be accomplished through better aerodynamic efficiency (high $C_L^{3/2}/C_D$) and a lighter wing loading (low W/S). The engine noise can be attenuated by intake silencers and mufflers. It is simply easier to reduce the radiated acoustic power of a 5-hp engine than a 50-hp engine.

Propellers are another major source of noise. Propeller noise depends upon revolution rate (rpm), disk loading, number of blades, and blade shaping (chord, twist, and sweep distribution). The noise is much lower if the tip Mach number is subsonic. Tractor propellers are quieter than pusher propellers because they operate in undisturbed air, though it is possible for the prop wash to generate airframe noise. Pusher propellers operate in the wake of the airframe, resulting in higher noise due to wake interaction with the rotating blades. Pusher prop noise can be reduced with careful shaping of the nacelles and any wings or tails ahead of the prop.

The wavelength λ of a sound wave is

$$\lambda = \frac{a}{f} \tag{18.8}$$

where a is the speed of sound and f is the frequency. The wavelength of the sound relative to the airframe features can affect the performance of blocking features. It is easier to physically block high-frequency noise than low-frequency noise.

Jet engines have relatively high noise levels. Most jets have a broad frequency white noise caused by the interactions of the exhaust with the airflow, among other sources. Small turbines make both a high-frequency whistling sound and broad frequency noise. The high-frequency noise attenuates most rapidly with distance such that a low-frequency rumbling is heard at longer range. Jet noise is directional.

18.6.3 Infrared

The infrared range equation is

$$R = \sqrt{\frac{\tau_O \cdot \tau_A \cdot J_0 \cdot \pi \cdot D^2 \cdot G \cdot D^*}{4 \cdot S/N \cdot \sqrt{\theta/IFOV \cdot A_d}}} \tag{18.9}$$

where

τ_O = transmission of the IR optics
τ_A = transmission of the atmosphere
J_0 = radiant intensity of target above the background, W/sr
D = IR aperture diameter, m
G = processing gain
S = signal power, W
D^* = S/N per unit S for detector, $(Hz^*m^2)^{1/2}/W\alpha$
N = system noise, W
θ = scan rate, rad/s
$IFOV$ = instantaneous field of view of one detector, rad
A_d = detector sensitive area, m^2

Because humans have no ability to see in the infrared domain, we only have IR detection systems to worry about. For a given detection system, the primary UAS design feature that can be controlled is the radiant intensity J_0. The radiant intensity is directional, and so the UA can be tailored such that a low heat signature is presented to the detection system. This can be accomplished through heat source blockage and flight profiles relative to the threat. Burying the jet engines in the fuselage or other parts of the airframe is one approach. Tail surfaces can provide some blockage as well, as can be seen on the A-10 manned attack aircraft.

The best way of minimizing the IR signature is to reduce the intensity of the heat source. The biggest heat source is the engine, whether it is a turbine or reciprocating engine. Like the acoustic signature, the heat source is a function of the power or thrust required. Having high aerodynamic efficiency, low wing loading (low speed), and small payload weight reduce the engine size.

18.6.4 Visual

The visual-range equation is

$$R = \frac{1}{2 \cdot \beta} \cdot \sqrt{\left[\ell n\left(\frac{C_\theta}{C_0}\right)\right]^2 + 4 \cdot \alpha \cdot \beta \cdot a \cdot L \cdot B} - \frac{\ell n(C_\theta/C_0)}{2 \cdot \beta} \qquad (18.10)$$

where

β = atmospheric attenuation, 1/km
C_θ = required visual contrast
α = required visual contrast curve fit parameter
C_0 = target inherent contrast
A = conversion between radians and minutes of arc
L = target dimension, m
B = visual acuity gain due to binoculars
R = range, km

The visual detection mechanism is generally the human eye, which may be aided by binoculars. The observer is usually on the ground. An airborne observer may also be of interest in the case of a fighter pilot that wishes to shoot down the UA. Use of binoculars generally occurs when the observer has been queued by some other means, such as hearing the UA.

The parameters in the control of the UA designer are the target dimension and the contrast. The target effective size is a function of the wing span, overall length, and total area. It goes without saying that a smaller UA will have a smaller target dimension. The visual contrast can be improved with appropriate camouflage and other techniques.

18.7   Systems Engineering

No treatment of UAS development is complete without explaining the role of systems engineering. Systems engineering is a formal engineering discipline that considers system elements as a part of the whole system. INCOSE [4] defines systems engineering as "an interdisciplinary approach and means to enable the realization of successful systems."

If one reads classic aircraft design texts, there is much discussion of system synthesis and the importance of including all of the systems in the design. Yet there is rarely mention of systems engineering. Classical aircraft design is filled with systems engineering, though it is not given that label. It may be said that all good aircraft designers practice systems engineering, but systems engineers are not necessarily aircraft designers. Aircraft design is a highly specialized field that requires a depth of knowledge across multiple disciplines. The aircraft design techniques for managing interdisciplinary interactions are a form of systems engineering.

Systems engineering is also a process-oriented discipline in its own right that transcends aircraft design. Systems engineering practices apply equally to UASs, spacecraft, or automobiles. Key systems engineering functions include the following:

* Decomposing high-level requirements into more detailed requirements
* Allocating requirements to design teams
* Managing design margins
* Developing requirement verification and test plans
* Supporting complex trade studies across multiple systems
* Analyzing risk
* Defining and managing interfaces
* Establishing and tracking system performance metrics
* Ensuring that the requirements are satisfied
* Maintaining proper configuration management

This book is about UAS design rather than just aircraft design. Therefore, it must be said that systems engineering is a critical discipline for the successful design of a balanced UAS that meets the requirements. The entire system must be designed with a requirements focus, where cross-segment compromises and implementation trades are important.

18.8   Optimization

System design is rarely a closed-form problem with a single answer. When a feasible solution space exists, there is a continuum of viable design points or multiple islands of designs that can satisfy the constraints. Search algorithms such as optimizers are generally required to find the best designs within a constrained design space.

Most design problems have an objective function, constraints, and design variables. This is true whether we are designing UA, payloads, or integrated systems.

The objective function is a value that represents what we wish to improve. Depending on the formulation, the optimizer seeks to minimize or maximize the objective function. An example of an objective function is system cost, where the optimizer seeks to minimize cost while also achieving mission requirements.

Constraints are the boundaries of the problem. These represent limitations that, when crossed, yield an unacceptable design. Constraints can be design rules or requirements.

Design variables are parameters that the optimizer controls. These are unknowns that are allowed to vary. Design variables have upper and lower bounds, and the optimizer selects values within these bounds.

Optimum solutions often reside on one or more constraint lines. Unless margins are proscribed in the UA sizing routine, a highly constrained design might have small margins. Such designs may encounter problems after the conceptual design phase. In other words, optimized solutions can have limited design flexibility in practice when any changes result in violated constraints. A nonoptimum solution that is not at the constraint boundaries can be easier to modify through preliminary and detail design.

UA or system optimization is usually a conceptual design activity. An optimizer may make several thousand or even millions of function calls to find the optimum solution. Even with remarkable advances in computer technology, the analytical methods employed for optimization are still relatively simple to minimize computational burden. For example, the structures sizing methods may use empirical weights equations or semi-analytical techniques rather than finite element methods (FEM). More rigorous design techniques are required before parts can be built. Switching methods nearly always results in changes in results and sensitivities, and so refining the optimum to the nearest 1% rarely represents a true design improvement.

The design problem might as well be a black box to the optimizer. The optimizer merely changes design variable numbers and observes the objective function outputs and constraint satisfaction. Based on this information, it decides how to modify the design variables. Optimizers have no inherent appreciation for the design problem or intuition. Engineers who set up optimization problems are frequently frustrated when the optimizer can't "see" the solution even though it is obvious. This frustration is merely a symptom of a problem formulation that is not well suited to the optimizer that is used. Optimizers tend to exploit design assumptions and modeling errors, and so the optimizer-selected solution should be evaluated with care.

There are numerous optimizers that can be applied to UAS problems. Here we will explore gradient-based methods and genetic algorithms.

Additionally, we will consider the design of experiments design space exploration technique.

18.8.1 Gradient Optimizers

Gradient-based methods rely on derivatives of the objective function with respect to the design variables, which are known as *gradients*. These methods are also called calculus-based methods. Gradient methods are more informally described as *hill climbers* because the path of the steepest favorable gradient is selected. Gradient-based optimizers require the existence of derivatives of the objective function with respect to design variables.

The path of the steepest gradients is followed incrementally until either an optimum solution is reached or constraints prevent further improvement. Using the hill climbing analogy, the path up the hill may be blocked by fences that represent constraints. Upon reaching a constraint line, the optimizer will select a new path that follows the constraint lines until the best solution is found or additional constraints block the path. The hill climbing analogy is a three-dimensional scenario, but aircraft optimization problems frequently have 10–20 design variables and about the same number of constraints.

The optimal solution by the gradient-based optimizer might be the local optima, not necessarily the global optima. Gradient methods start from a point in the design space, and the end result is typically dependent on the starting conditions. Within the vicinity of local optima, gradient methods can be very efficient at finding the refined local optima. If we consider the hill climbing analogy again, we may start to climb to the summit of one hill but the next hill over is twice as high. A simple gradient-based optimizer is unable to see the next hill over.

Example 18.3   Wing Sizing

Problem:

We wish to size the wing for a jet-powered MALE UAS.

Excluding the wing, all other UA empty weight contributors combined are 800 lb. This includes the fuselage, tail, and systems weights. The payload weight is 250 lb. Assume that the wing weight is 4 lb/ft^2, regardless of aspect ratio or size.

The combined drag of the fuselage and tails is 0.01 with a reference area of 12 ft^2. For the purposes of this example, we are neglecting trim drag and tail sizing dependencies on the wing.

For simplicity, we will use a constant chord wing (taper ratio of 1). The wing C_{D0} is 0.01, and the span efficiency factor e is 0.8.

Assume that the jet-engine thrust specific fuel consumption at the cruise condition (TSFC) is 0.8 lb/lb-hr.

The required landing speed is 60 KEAS or less. The UA must be able to land at this speed while at the takeoff gross weight. The maximum lift coefficient for the minimum landing speed is 1.8.

The sizing flight profile includes a climb to altitude, loiter segment of 24 hrs, and descent to landing. Assume that the climb segment consumes 3% of the TOGW in fuel burn, and the final descent consumes negligible fuel. Loiter occurs at 30,000 ft ($\rho = 0.0008903$ sl/ft^3).

The wing must be transported in a box with interior usable length of 25 ft and width available for the chord of 5 ft. The wing comes apart into two sections such that the assembled span is equal to twice the disassembled span.

Find the aspect ratio and wing area that minimize the takeoff gross weight. The weights must converge to within 0.5%.

Solution:

We will use the Microsoft Excel^{TM} Solver tool for this problem. Excel is a widely used spreadsheet application that is often used by engineers for quick problem solving. Spreadsheets have limitations for complex design problems, where software languages such as C, C++, C#, Visual Basic, Java, and older languages such as Fortran become more manageable. Most engineering students and professionals use Excel, and so we will employ its built-in optimizer to quickly find a solution.

Solver is an add-in within Excel, and so it must be enabled prior to its first use. To find specific instructions for installation for your version of the software, search the help feature within Excel. In Excel 2007, Solver is installed by clicking the **Microsoft Office Button**, then clicking **Excel Options**, followed by **Add-Ins available**. In the **Manage** box, select **Excel Add-ins** and click **Go**. Select the **Solver Add-in** checkbox, and click **OK**.

Once installed in Excel 2007, Solver can be found under the **Data** tab, within the **Analysis** group. In older versions of Excel, it is found under the **Tools** menu.

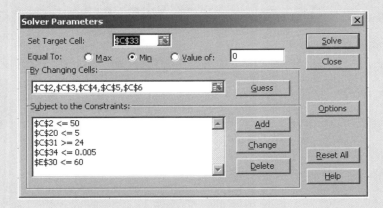

Click the Options button. Use all of the default settings, except change the maximum iterations to 400 and the tolerance to 2%.

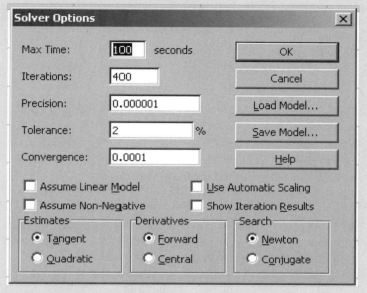

We have the following constraints:

$V_{Land} \leq 60\,KEAS$

$b \leq 50\,ft$ (twice the box length)

$c \leq 5\,ft$

$E \geq 24\,hrs$

$$\frac{|W_{TO,Calc} - W_{TO}|}{W_{TO,Calc}} \leq 0.5\% \qquad \text{(weight convergence)}$$

$C_{L,Loiter} \leq 1.2$

All of these constraints should be added to the Solver optimizer under the area entitled "Subject to the Constraints."

The design variables are b, AR, $C_{L,Loiter}$, W_{TO}, and MF_{Fuel}. The cells for these design variables should be added to the area entitled "By Changing Cells." You will need to arbitrarily select initial values to start the optimization. The starting conditions should be reasonable for the optimization to succeed.

The objective function is $W_{TO,Calc}$, which we seek to minimize. The cell containing $W_{TO,Calc}$ should be included in the box "Set Target Cell." The radio button "Min" should be checked because we wish to minimize the takeoff gross weight.

The weight formulation is

$$W_{TO,Calc} = W_{Fuse,Tail} + W_{Wing} + W_{PL} + MF_{Fuel,Tot} \cdot W_{TO}$$

The wing weight is calculated by

$$W_{\text{Wing}} = 4 \cdot S_w$$

where the wing weight is in pounds and the wing area is in square feet.

The drag coefficient is

$$C_D = C_{D,\text{Fuse,Tail}} \cdot \frac{S_{\text{Fuse,Tail}}}{S_w} + C_{D0,\text{Wing}} + \frac{C_L^2}{\pi \cdot AR \cdot e}$$

The lift to drag ratio L/D is

$$L/D = \frac{C_L}{C_D}$$

Recall from Chapter 3 that the endurance of a jet-powered aircraft is

$$E = \frac{L/D}{TSFC} \cdot \ln\left(\frac{1}{1 - MF_{\text{Fuel}}}\right)$$

The MF_{Fuel} for the endurance segment is the total fuel mass fraction minus the fuel mass fraction required for climb.

The landing equivalent airspeed $V_{\text{Eq,Land}}$ is found by

$$V_{\text{Eq,Land}} = \sqrt{\frac{2 \cdot W_{\text{Land}}}{\rho_{\text{SL}} \cdot S_w \cdot C_{L,\text{Land}}}}$$

The landing weight W_{Land} is equal to W_{TO} in this example.

A nonunique optimizer solution is shown in the following figure:

	A	B	C	D
1				
2		b	50	ft
3		AR	16.09785	
4		CL loiter	0.836644	
5		MFFuel	0.504949	
6		WTO	3410.088	lbs
7				

The landing speed is 60 KEAS, which indicates that the wing area is limited by the landing speed. The span is 50 ft, showing that the box constraint is limiting the span. The calculated chord is 3.1 ft, which is less than the box constraint. The calculated endurance is exactly 24 hrs. The weights difference between the input W_{TO} and $W_{\text{TO,Calc}}$ is 0.5%.

Your results will likely vary from these. The gradient-based optimizer solution depends upon the starting conditions. If the optimizer does not yield satisfactory results, you can adjust the starting conditions.

You may encounter a problem with the weights convergence. In this problem the burden of weights convergence is placed on the optimizer. It must work hard to ensure this constraint is satisfied that the fuel mass fraction might be higher than is required for the 24-hrs endurance. Later in this chapter we will see a design procedure that solves this problem.

18.8.2 Genetic Algorithm

Genetic algorithms (GAs) are an increasingly popular form of optimizer. GAs have attributes that are quite different from gradient-based optimizers. They are capable of finding global optima within challenging design spaces. However, this comes at some expense of efficiency in finding the refined optima. Goldberg [5] is an accessible resource for learning about GAs.

A GA is used to flexibly handle a wide range of design problems under circumstances where gradient-based methods might have limitations. The utility and limitations of GAs can be elucidated best by comparison with the competing gradient methods. Genetic algorithms handle the design problem in a significantly different manner. GAs do not require the existence of gradients. A nonsmooth design space with discontinuities is acceptable. GAs are adept at finding the global optima because the search is based on a population of design points, not a single point. However, refinement of the global optima is inefficient relative to calculus methods within the vicinity of the optima.

GAs do not operate on design variables directly. Rather, the design variables are coded in binary strings of zeros and ones. These strings are combined in larger strings that include all of the design variable strings together. The GA operates on these strings, and then decodes the strings into the parameters prior to evaluating the objective function. The objective function is called the fitness function in GA parlance, to complete the biological analogy of survival of the fittest. The string operators are reproduction, crossover, and mutation, which are analogous to genetics. Each string represents a member of the population. There are multiple individuals in the population for each generation.

GAs are not limited to real numbers like gradient optimizers. Other number types such as integers and Boolean (true or false) are also allowed. Integer design variables might be items such as number of engines, propulsion type (i.e., $1 =$ jet, $2 =$ reciprocating engine), configuration selection (i.e., $1 =$ flying wing, $2 =$ conventional, $3 =$ canard), number of aircraft in the fleet, or selection of a part number. This permits the optimizer the flexibility of selecting the system architecture and system characteristics. Boolean parameters are useful for selecting between two choices like manned vs unmanned. With this design variable flexibility, GAs are capable of solving problems that are not possible with gradient optimizers.

GAs do not handle constraints directly. Violated constraints must be converted to penalty functions that act to degrade the fitness function. In other words, the penalty caused by a violated constraint will yield a less fit solution. By comparison, many calculus methods may handle constraints directly.

The GA starts by generating a population with N_{Pop} members. Each individual has a genotype, which is a collection of genes. Each gene corresponds to a design variable. The gene is represented by a string of 1s and 0s. Each of these binary values is called an allele. There are N_{Allele} per gene.

Now let us see how to convert from a gene to a real number. Consider a gene that has four alleles. The gene's allele string *Allele*() is given by 1101, as an example. We give each of the alleles an index i, which starts at 1 on the right and finishes at 4 on the left. The design variable has a minimum value val_{Min} and a maximum value val_{Max}. The value *val* is found by

$$val = val_{\text{Min}} + \frac{\sum_{i=1}^{N_{\text{Allele}}} Allele(i) \cdot 2^{i-1}}{2^{N_{\text{Allele}}} - 1} \cdot (val_{\text{Max}} - val_{\text{Min}}) \qquad (18.11)$$

Let us say that the variable has a minimum value of 2 and a maximum value of 6. The preceding allele string has the value

$$val = 2 + \frac{(1 \cdot 2^0 + 0 \cdot 2^1 + 1 \cdot 2^2 + 1 \cdot 2^3)}{2^4 - 1} \cdot (6 - 2) = 5.467 \qquad (18.12)$$

The collection of design variables with physical meaning is known as the phenotype. Once all of the genes in the genotype are converted to the phenotype, the fitness of the system can be evaluated.

The three main operations are reproduction, crossover, and mutation. All three operations are performed against the genotype.

Reproduction is bringing forward an individual to the next generation. The probability that the individual will reproduce is based on its fitness. This is much like a roulette wheel, where the fittest individuals take up a larger angle of the wheel than unfit individuals.

Crossover is where two individuals exchange portions of their genotype. The position along the string of alleles is randomly selected. The two individuals selected for crossover are also randomly selected. The probability of crossover is defined by the user. Not all individuals will experience a crossover.

Mutation is the switching of a single allele value within a genotype. A mutation will cause a 1 to switch to a 0 or vice versa. The value of mutation is that it helps the GA to keep from having an excessively small gene pool variation. The probability of mutation is defined by the user.

Example 18.4   Simple Genetic Algorithm

Problem:

Consider a simple parabolic drag polar.

$$C_D = C_{D0} + \frac{C_L^2}{\pi \cdot AR \cdot e}$$

Here C_{D0} is 0.02, AR is 10, and e is 0.7. Use a genetic algorithm to find the lift coefficient for best L/D. Evaluate lift coefficients from 0 to 1.5. The number of alleles for the C_L gene is 5. The population size is 4.

Solution:

For this simple problem we know from Chapter 3 that the best L/D occurs when

$$C_L = \sqrt{C_{D0} \cdot \pi \cdot AR \cdot e}$$

In this problem the best C_L is 0.663, and the maximum L/D is 16.579. We will keep this optimum in mind when we see how the GA performs against this problem.

The GA output for the first and second generations are shown next:

1. The first generation is selected based on random generation of alleles. The genotype is converted to the phenotype (C_L values), and then the fitness (L/D) is evaluated for each individual. The average L/D is 13.598, and the best L/D is 15.733.
2. Based on a roulette wheel selection, individuals 1 and 2 each reproduce one copy, individual 3 produces no copies, and individual 4 produces two copies. These copies start the second generation. The best L/D remains the same because no new genes are present. The average L/D increases due to elimination of the least fit individual.
3. Individuals 1 and 3 are selected for crossover at allele position 2 (includes alleles 1 and 2). The individuals that have the crossover and the allele position are both selected by random numbers. The individuals that exchange genes are placed next to one another for clarity.
4. The crossover is completed, and the fitness is reevaluated. The crossover yields two new individuals. One is less fit, and the other is the most fit of all individuals. The best L/D is now 16.383.
5. Allele 5 of individual 4 experiences a mutation, where it changes from a value of 1 to a value of 0. This changes the C_L from 1.355 to 0.581. This yields a L/D of 16.43, which is the most fit individual. Note that this one mutation is the most effective of all that could have occurred in this generation.

1) First Generation

Individual	Alleles					CL	L/D	# selected
	5	4	3	2	1			
1	1	0	0	1	1	0.919355	15.73315	1
2	1	1	0	1	1	1.306452	13.38389	1
3	0	0	1	1	0	0.290323	12.18165	0
4	1	1	1	0	0	1.354839	13.0941	2

Avg L/D 13.5982
Best L/D 15.73315

2) Reproduce for Second Generation

Individual	Alleles					CL	L/D	
	5	4	3	2	1			
1	1	0	0	1	1	0.919355	15.73315	Old 1
2	1	1	0	1	1	1.306452	13.38389	Old 2
3	1	1	1	0	0	1.354839	13.0941	Old 4
4	1	1	1	0	0	1.354839	13.0941	Old 4

Avg L/D 13.82631
Best L/D 15.73315

3) Regroup for Crossover

Individual	Alleles					CL	L/D	
	5	4	3	2	1			
1	1	0	0	1	1	0.919355	15.73315	Individual 1&3
3	1	1	1	0	0	1.354839	13.0941	Crossover at allele 2
2	1	1	0	1	1	1.306452	13.38389	
4	1	1	1	0	0	1.354839	13.0941	

Avg L/D 13.82631
Best L/D 15.73315

4) Perform Crossover

Individual	Alleles					CL	L/D
	5	4	3	2	1		
1	1	0	0	0	0	0.774194	16.38321
2	1	1	1	1	1	1.5	12.26353
3	1	1	0	1	1	1.306452	13.38389
4	1	1	1	0	0	1.354839	13.0941

Avg L/D 13.78118
Best L/D 16.38321

5) Mutation

Individual	Alleles					CL	L/D	
	5	4	3	2	1			
1	1	0	0	0	0	0.774194	16.38321	
2	1	1	1	1	1	1.5	12.26353	
3	1	1	0	1	1	1.306452	13.38389	
4	0	1	1	0	0	0.580645	16.43439	Allele 5 mutated

Avg L/D 14.61625
Best L/D 16.43439

18.8.3 Design of Experiments

A design of experiments (DOE) is a method of exploring the design space, but it is not an optimizer. A DOE will use one or more design variables. Each design variable will be swept through a range of values. The spacing between design variable values can be even, though more sophisticated approaches are often employed to ensure efficient design space exploration. Beyond finding the best design point, a DOE can examine trends (and verify the trends make physical sense), show interactions between variables, and illustrate the design space.

Example 18.5   One-Dimensional DOE

Problem:

Using the drag polar in problem 18.4, perform a DOE to find the L/D behavior with C_L. We will impose a constraint of $C_{L,max} = 1.5$. The L/D cannot be evaluated above $C_{L,max}$. Use 11 evenly spaced C_L values from 0.0 to 2.0.

Solution:

The C_L values, L/D, and constraint satisfaction are shown in the following table:

C_L	L/D	$C_{L,max}$ Constraint
0.0	0.00	Satisfied
0.2	9.17	Satisfied
0.4	14.67	Satisfied
0.6	16.50	Satisfied
0.8	16.29	Satisfied
1.0	15.27	Satisfied
1.2	14.04	Satisfied
1.4	12.83	Satisfied
1.6	—	Violated
1.8	—	Violated
2.0	—	Violated

From the values shown, the best L/D is 16.50, which occurs at a C_L of 0.6.

18.9 Design Environments

The design environment can change as the design progresses. Alternatively, a single design environment can be used throughout the system development.

Traditionally, a conceptual design tool is used to size the UA. All methods used by the sizing tools are tightly integrated, but may be of low fidelity. The conceptual design tools may use optimizers or have the ability to sweep major design variables. The outputs of the tool are the UA gross geometric definition, mass properties' allocations, engine sizing, aerodynamic characteristics, and performance analysis.

Once the conceptual design tool is complete, then each major discipline performs more detailed analysis in their specialized tools. For example, the structures group uses FEM, the aerodynamics group uses a multiple tools including CFD, the propulsion ground relies on engine cycle analysis tools and test data. The subsystems groups size the electrical power system, actuators, and environmental control system based on the architecture and specialized methods. The geometry is defined in a computer-aided design (CAD) environment where the UA loft occurs and the parts are generated. These methods have long computation times that are ill suited to optimization tools.

These more detailed methods traditionally are not integrated. For example, the impact of an environmental control system design change is not automatically reflected in the weights, propulsion power and bleed air extraction, or performance. Systems engineering, mass properties, and the technical leadership (chief engineer and integrated product team leadership) maintain configuration control and track changes to the system across the disciplines.

The impacts of no direct design tool and design data integration are poor design impact visibility and long design cycles. The future of design is more integrated design tools that reside in a common design environment. Such tools share common geometry definition, mass properties definition, and support discipline-specific design methods. The design environments can support conceptual sizing and optimization. The initial definition and design allocations can then be used for the higher-fidelity methods employed later. Such design environments require tools with appropriate interfaces.

The UAS design includes payloads, communications systems, ground stations, launch and recovery equipment, and support equipment. These systems must also undergo conceptual, preliminary, and detail design phases. The design and analysis methods will progress from simple to high-fidelity tools in a manner similar to the UA.

18.9.1 UA Sizing

There is not a single UA sizing methodology that can handle all cases. The aircraft design methods vary substantially depending on UA class, level of resources applied, and company capabilities. The author has found that nearly every design problem is different, and this affects the problem formulation and methods that are selected. The constraints and design assumptions change the order of analysis and design tools.

For example, a jet engine can be defined, or it can be scaled to the required thrust level. Scaling an engine is called *rubber-engine* sizing because the engine changes in dimensions and weight based on the required thrust. Rubber-engine sizing is equivalent to building the perfect engine for the design rather than simply adopting available engines. The jet-engine nacelle dimensions are dependent upon the jet engine dimensions. The nacelle size affects the nacelle parasite drag. This drag influences the total thrust required from the engines to overcome the drag to achieve field performance, service ceiling, or top speed. Therefore, there is interdependency between the engine size requirements and the geometry.

In other design problems, an engine can be selected prior to UA sizing. This defined engine can be used to define the nacelle geometry before the aerodynamic analysis is performed. The order of the geometry, aerodynamics, and propulsion analysis are different between the rubber-engine and fixed-engine cases.

The following design information is usually available before optimization begins. This information can be derived from requirements or the result of system design trade studies:

Geometry	• General configuration • Geometry rules (tail volume coefficient) • Packaging constraints (transportation modes, storage box dimensions)
Mass properties	• Avionics and communication systems size, weight and power (SWAP) • Payload SWAP
Systems	• Subsystem configurations (electrical power system, environmental control system, hydraulics and pneumatics)
Propulsion	• Propulsion type • Specific engine (if known)
Performance	• Mission profile • Field performance requirements
Aerodynamics	• Airfoil families • High lift device arrangement • Laminar flow rules (i.e., all turbulent assumption or free transition)
Launch and recovery	• Launch and recovery techniques • Required launch and recovery performance
Structures	• Materials type and general structural arrangement • Load cases of interest
Mission effectiveness	• Sensor field-of-regard requirements • Survivability features and requirements

The design variables should be few in number and have high impact. For example, span and aspect ratio will have a greater design impact than span location of a taper change. The following parameters are often used as design variable, though generally a subset is applied.

Geometry	• Span, wing area, aspect ratio, or root chord (any two) • Taper distribution • Semispan locations of taper changes • Sweep angles (leading edge, quarter-chord, trailing edge, or other x/c) • Fuselage length • Fuselage shaping (fineness ratio, distribution of upper surface, lower surface, sides, and cross-section shaping parameters)
Mass properties	• Fuel weight or fuel mass fraction
Propulsion	• Thrust or power (sea level uninstalled) • Thrust-to-weight or power-to-weight ratios • Propeller diameter

The following constraints can be used in the UA design:

Performance constraints	• Stall speed or minimum airspeed • Climb gradients or climb angles (engine out) • Rate of climb • Dynamic pressure limits • Mach limits • Balanced field length • Landing field length • Service ceiling • Mission profile satisfaction (climb, range, endurance segments)
Weight constraints	• Maximum weight (landing weight, takeoff gross weight) • Weight convergence
Geometric constraints	• Maximum dimension (i.e., MAV must have less than 6-in. maximum dimension) • Storage/transportation box constraints • Folding geometry for airborne deployment or storage • Field-of-regard requirements for payloads and communications systems—line-of-sight blockage • Landing-gear geometric constraints (tip-back)
Structures	• Tip deflections • Aeroelastic constraints (flutter, divergence, aileron reversal) • Special load case satisfaction (i.e., tool dropped on skin without damage)
Systems	• Fuel volume required must not exceed the available fuel volume. • Avionics and payload power requirements must not exceed available power. • Avionics and payload must be cooled.

Propulsion constraints	• Thrust required must be less than thrust available at every flight condition. • Propeller rpm constraints • Motor or engine torque or power limits
Stability and control	• C.G. boundaries • Elevator for trim at maximum C_L • Engine out lateral-directional trim

The objective function should be selected to emphasize the most important system characteristics or performance. Finding a single quantifiable parameter that captures the essence of system desirability is not easy. Sometimes a compound objective function is required where several attributes are additively applied. Each attribute is scaled and weighted to ensure the appropriate balance. Minimizing the takeoff gross weight is perhaps the most common objective function because it drives down life-cycle cost, size, and fuel consumption in a balanced fashion. Potential objective functions might include the following:

* Minimum weight
* Minimum size
* Minimum fuel burn
* Minimum cost (acquisition, operations, or life cycle)
* Mission effectiveness (area coverage, time on station, target access)
* Compound objective functions

Although not every problem can be approached in the same way, a reasonable analysis methodology will now be presented. This methodology is intended for use with an optimizer that selects design variable values and checks constraints. The order of analysis is illustrated in Fig. 18.2.

The first step is to calculate the UA geometry because nearly all methods depend on the geometry definition. Many relevant geometric relationships are presented in Chapter 4. Usually the geometry is highly parameterized such that relatively few geometry terms are sufficient to define the complete configuration. Some of the geometry inputs are fixed, and others are design variables selected by the optimizer.

The optimizer will select two major defining wing parameters (span, aspect ratio, wing area, root chord) that can be used to calculate the other two. This assumes that the taper distribution is known, though the wing taper ratio (and taper distribution) can also be selected by the optimizer. Subsonic conventional configuration designs will generally have zero sweep. Transonic and supersonic designs may have the wing sweep selected as a design variable. The airfoil or airfoil family can be known a

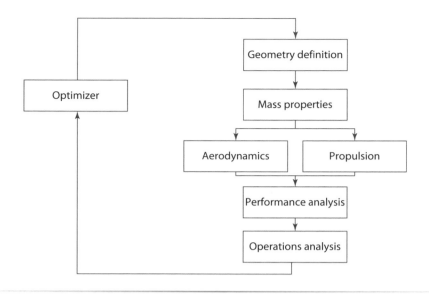

Fig. 18.2 UA sizing approach.

priori. The optimizer may also select the wing thickness-to-chord ratio (t/c) or t/c distribution along the span as design variables.

The tails can be sized based on tail volume coefficient methods. The tail moment arm is known from the wing and tail positions on the fuselage. The horizontal and vertical tail areas, depending on tail configuration, are calculated based on the wing area, span, and mean aerodynamic chord. The nondimensional tail geometry parameters (aspect ratio and taper ratio) define the tail span and chords.

The fuselage geometry can be treated simply or with detailed definition during sizing. A simple definition can treat the fuselage as a fixed length and with a length-to-diameter ratio. More complex fuselage geometry treatment defines the upper, lower, and side contours as well as the cross sections. Advanced methods can modify the fuselage contours based on fuselage contents and functionality. The wing and tail locations on the fuselage are defined. Geometry of nacelles and other bodies are treated in a similar manner as the fuselage.

The volumes of the wings, tails, and bodies are calculated. The volume and the allocation of volume are important to ensure that the payload bays, fuel, avionics, and other components fit within the available space.

If a conventional landing gear is used, then this geometry must be defined as well. The landing-gear position and strut lengths are calculated, and constraints are evaluated. The space required for retractable landing gear must also be determined, if so equipped.

Next the mass properties are calculated. Many of the methods described in Chapter 6 can be employed. The mass properties' methods should include an internal weight convergence procedure such as the secant method. Many optimizers have difficulty finding an optimum solution when burdened by weights convergence.

Either the fuel weight or the fuel mass fraction is input as design variables selected by the optimizer. Electric-powered aircraft may use battery weight or battery mass fraction. This might seem curious because the fuel or battery size required is a function of the performance analysis, which has not yet been performed. This is not a problem because if the energy source is insufficient for the mission then the mission will not be completed. The lack of mission completion will show up as a violated constraint. If the energy source is greater than what is required, then the UA will be heavier than necessary and likely will not be selected as the optimum.

The structures are sized in the mass properties' routines. Parametric weight estimating relationships or semi-analytic methods such as those described in Chapter 7 can be applied. The loads are dependent upon aerodynamics, which have not been performed. The structures' sizing can assume a lift distribution and fuselage aerodynamic forces or call aerodynamics routines for the purposes of finding these loads. The structural constraints such as deflections or aeroelasticity are evaluated.

The mass properties' routines may serve as a configuration management tool, where the SWAP of all components and systems are tracked. The required volume and power can be found through this analysis. If the weights and locations of the entire UA are known, then it is possible to estimate the center of gravity and moments of inertia as well.

The major output of the propulsion analysis is an engine deck that can be used by the performance analysis. If the engine geometry is not defined by the time that the geometry analysis is performed, then any propulsion-related geometry definition such as nacelle sizing should also be performed at this time. As described in Chapter 8, the engine decks provide fuel consumption as a function of altitude, airspeed, and thrust required for installed propulsion systems. The decks also provide minimum and maximum thrust limits as a function of altitude and airspeed. The impacts of bleed air and power extraction can also be incorporated into the engine deck. Generally, this deck takes the form of a multidimensional table.

The aerodynamics code outputs drag polars that are used by the performance analysis. The drag polars are tables of drag coefficients vs lift coefficients for several Reynolds number per unit length values. Chapter 5 covers drag polars and aerodynamic analysis methods in more detail. A drag polar is required for every aerodynamic configuration. Configurations might be a clean aircraft, flaps and gear down, and external stores, for example.

The stability and control (S&C) characteristics can be evaluated during sizing. The tail geometry can be sized from control requirements rather than tail volume coefficients, or the defined tail geometry can be checked against critical conditions. Some critical conditions can include engine-out lateral-directional control for twin-engine aircraft and elevator control power.

The performance analysis can finally be performed now that the geometry, takeoff gross weight, fuel weight, engine deck, and drag polars are defined. The performance analysis incrementally steps through the mission profile. Along the way, the airspeed can be optimized to best satisfy the segment objectives. The flight performance constraints are evaluated at each segment. If the mission is not completed, the point at which the fuel is depleted is estimated.

Finally, any additional analyses such as operations analysis are performed after the UA is sized. This analysis may include fleet sizing, cost analysis, target coverage, and survivability. The results can be incorporated into the objective function.

Other system elements such as the support equipment and UA launch and recovery equipment can be included in the sizing analysis. The launcher and retriever may have weight or dimensional constraints. The UA wing loading, propulsion characteristics, and maximum loads can impact this equipment. The same is true of support equipment required to provide ground handling and facilitate maintenance.

References

[1] Ball, R. E., *The Fundamentals of Aircraft Combat Survivability Analysis and Design*, AIAA, Reston, VA, 1985.
[2] Jenn, D. C., *Radar and Laser Cross Section Engineering*, AIAA, Reston, VA, 1995.
[3] *Sound and Model Aeronautics, A Handbook for Model Clubs*, The Academy of Model Aeronautics, Reston, VA, 1988.
[4] Systems Engineering Handbook, ver. 2a, INCOSE, 2004.
[5] Goldberg, D. E., *Genetic Algorithms in Search, Optimization, and Machine Learning*, Addison Wesley Longman, Reading, MA, 1989.

Problems

18.1   An UA has an ingress/egress flight speed of 80 kt. Fuel is burned at the same rate for transit and loiter phases. The total UA endurance is 15 hrs. The expected availability is 95%. What is the required number of UA to perform continuous target coverage at a radius of 100 n miles?

18.2   Using problem 18.2, vary the ingress/egress flight speed from 50–200 kt in 50-kt increments. Assume that the total endurance remains unchanged. Calculate the required number of UA.

18.3   An UA has an equivalent airspeed of 50 KEAS. The UA has a look angle of 45 deg and an vertical field of view of 10 deg. Show the area coverage rate as a function of altitude from 5,000- to 70,000-ft in increments of 5,000 ft.

18.4   Show the required reduction in RCS required to reduce the detection range to 3/4, 1/2, and 1/4.

18.5   Use the EO/IR ball weight equation from Chapter 6, assuming that all electronics are contained within the ball. Assume that the ball diameter is equal to 1.5 times the focal length. The EO/IR ball height-to-diameter ratio is 1.4. The minimum EO/IR ball diameter is 5 in. The ratio of focal length to lens diameter is 2. The focal plane is 525×525, with a focal plane width of 0.1 in. The UA images in the nadir direction. Assume that the UA has a detection range given by $R_{Detect} = 2,000 \cdot \sqrt{P_{Eng}}$ ft, where the engine power P_{Eng} is in horsepower. The UA has 20 lb of nonvarying weights plus the weight of the EO/IR ball. The weight escalation factor is 4. The maximum power-to-weight ratio for the UA is 0.1 hp/lb, and the UA cruises at 50% power setting. The ground sample distance must be 4 in. or less. Use the Excel Solver tool to optimize the collection altitude to yield the lightest UA that is acoustically undetectable. The minimum altitude is 2,000 ft, and the maximum altitude is 30,000 ft. Compare this to the discussion in Sec. 15.3.

18.6   Solve problem 18.5 using a genetic algorithm.

18.7   Solve problem 18.5 using a design of experiments with 11 altitude points.

Chapter 19

Cost Analysis

- Understand phases of the UAS life cycle
- Explore cost-estimating methodologies
- Understand funding opportunities for UAS systems

Small UAS, VTOL UAs, optionally piloted aircraft, and fixed-wing ultra-long-endurance UAs have widely varying costs. (Photo courtesy of Aurora Flight Sciences.)

19.1 Introduction

The cost of a UAS depends greatly upon the system class and acquisition context. A small UAS could be wholly developed under a $100K SBIR such that initial flights are achieved. A hobbyist could get the UA flying autonomously for hundreds of dollars. Alternatively, a system in the same weight class would cost tens or hundreds of millions of dollars to develop as an operational program of record. This chapter provides elements of cost-estimation methodology with some historical UAS cost data. Although a complete detailed life-cycle cost-estimating methodology is beyond the scope of this chapter, several resources are identified for those who endeavor to undertake the challenge. Factors that drive UAS cost relative to manned aircraft are presented.

The purpose of cost modeling is to estimate the cost of a UAS program or sale price of a UAS. The cost estimation can occur at conceptual design, during a proposal, or for competitive analysis. The tools employed and cost fidelity depend upon the objective.

19.2 Cost Modeling

Conceptual design typically uses top-down cost analysis methods. Cost numbers are based on system parameters known within the conceptual design framework, such as UA structures' weight, engine thrust, or sortie rate. The cost information is an output of the design analysis. The cost estimates can also be incorporated into an objective function to support optimization. A simple cost optimization seeks to minimize life-cycle cost while not violating any constraints.

The RAND Corporation has developed numerous top-down aircraft cost estimation models for military aircraft. The DACPA model, which has numerous revisions, estimates labor hours and materials costs for fighters, transports, bombers, and other military aircraft. Roskam [1] provides similar manned aircraft cost-estimating methods. These models are based upon data from numerous manned aircraft acquisition programs. Unfortunately, there is not a similar model for UASs.

A single cost model that is applicable to all UAS classes can be expected to have accuracy limitations. Simple renderings of Global Hawk, Predator, and Shadow 200 are shown to the same scale in Fig. 19.1. Small UASs and MAVs are dwarfed by the Shadow 200, and so the image does not capture the full range of UAS classes. The UA span several orders of magnitude in size and weight. The ground stations, launch and recovery equipment, and communications architectures are also quite different across all UAS applications.

Cost estimation using empirical equations, called cost-estimating relationships (CERs), are much like weight estimation using weight-estimating

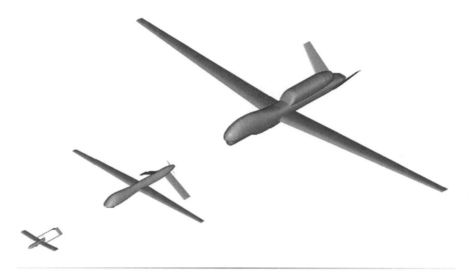

Fig. 19.1 UAS system size comparison (left to right: Shadow 200, Predator A, Global Hawk).

relationships (WER). The equations must be used within their applicable range, or the results will have questionable validity. Using CERs for large manned aircraft of the 1970s and earlier might not adequately predict the cost of a modern composite small UAS.

When developing CERs, it is appropriate to exclude failed programs that resulted in cancellation. After all, no new program should plan to fail. There are few UAS programs that were delivered on time and on budget while meeting all requirements. However, there are numerous UAS programs that have delivered useful products. Excluding notable failures improves the cost trends for programs that are expected to have some success.

A proposal or program planning effort requires a more detailed bottoms-up approach to cost modeling. This typically uses an integrated master schedule (IMS) tool that resource-loads a detailed schedule. Costs of individual hardware items are included and are generally based on vendor quotes. A single task will have a defined duration and resources assigned to it. The resources include labor, which is defined by labor categories and loading. Other resources may include purchased parts, consumable materials, contract services, and travel. Though labor intensive, a detailed bottoms-up cost based on experience and knowledge of the new program is preferable to top-down methods.

Established companies have databases of past program financial information that is applied to estimate new program costs. Other times, most of the corporate knowledge of UAS cost modeling resides within individual expertise of the employees. The latter scenario is far less desirable because individuals may leave companies and the knowledge is hard to replicate.

That is why most mature organizations seek to create proprietary cost-estimating tools based on past program data. Such tools can generate cost estimates with a solid foundation so that new programs are less likely to experience cost overruns caused by underpredicted program costs. These tools can form the basis for cost justifications in proposals.

Several sources of top-level cost information are presented in this chapter, though more detailed cost information is more difficult to come by. UAS costs have not been studied to the same extent as manned aircraft, though sufficient information exists to establish bounds of the cost range.

19.3 Life-Cycle Cost Approach

The phases of an idealized UAS program are development, manufacturing, operations, and disposal. These contributions of the life-cycle cost are shown in Fig. 19.2. The program phases occur in the order presented, though there may be considerable overlap. The relative magnitudes of the cost per phase shown in the figure are notional, though this is representative of typical aircraft program costs per phase. The research, development, test and evaluation (RDT&E) phase includes the engineering, prototyping, and testing. The manufacturing phase covers the production of the aircraft. The operations phase covers the time where the UAS is in operational service, which includes support. Finally, the systems are disposed of after the service life is complete. Frequently the disposal cost is neglected in cost modeling.

A UAS cost model can estimate the life-cycle cost of UAS programs. The life cycle consists of 1) RDT&E, 2) manufacturing, 3) operations, and 4) disposal phases. Each phase of the life cycle is modeled individually, with relevant phenomena captured. The cost model tool can be run as a stand-alone application or as an integrated analysis method of a custom UAS design code. When cost models are used in a multidisciplinary design optimization environment, the life-cycle cost or acquisition cost is often the objective function.

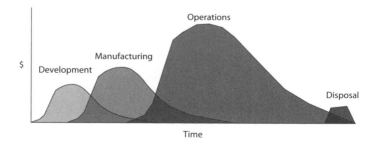

Fig. 19.2   UAS life-cycle cost phases.

The cost is the cost of contractor services and products to the government, including contractor profit. The contractor portion of the life-cycle cost is

$$C_{LC} = C_{RDTE} + C_{Manf} + C_{Ops} + C_{Disp} \qquad (19.1)$$

The life-cycle government services and products is

$$GFX_{LC} = GFX_{RDTE} + GFX_{Manf} + GFX_{Ops} + GFX_{Disp} \qquad (19.2)$$

A government program office will often utilize the services of technical advisors. These consultants are known as systems engineering and technical analysis (SETA) contractors. They represent the interests of the sponsor and may provide technical expertise that the sponsor does not possess. It is not uncommon for SETAs to outnumber government employees at design reviews. The cost for SETA support comes out of the government budget and therefore is not a cost for the prime contractor.

Government-furnished equipment (GFE) is hardware items that are provided by the government. The costs of these items fall on the program office rather than the contractor. Locating, procuring, and transporting the GFE can consume significant effort and cost. A program that involves substantial GFE procurement may require dedicated personnel in the program office.

The total program life-cycle price CT_{LC} is

$$CT_{LC} = C_{LC} + GFX_{LC} \qquad (19.3)$$

The cost methods presented later cover the contractor costs and not government costs.

19.4 Program Influences on Cost

Before providing details of the individual cost contributors, we will evaluate several program influences on the UAS life-cycle cost. Without proper context it is not possible to properly estimate cost. The program cost can vary by several orders of magnitude for a given UA size and weight depending upon the program assumptions.

UA costs are not exclusively functions of weight. Small UAS costs are almost independent of airframe size and weight. The software and avionics are more significant drivers than airframe weight at small scales. For example, a small autopilot could cost more than the airframe for man-portable UASs. The same autopilot could be used in a much larger UAS, with lower overall cost influence. The autopilot cost will generally follow an autopilot weight trend. However, autopilot functionality, reliability, ruggedness, and manufacturability can make an autopilot of fixed weight vary in cost by an order of magnitude.

Bottoms-up cost estimates are required for development contracts. Every bottoms-up cost estimate starts with a program definition. The following steps are used to initiate the effort:

1. The system requirements, concept of operations (CONOPs), or research objectives are defined.
2. The payload type, weight, power, and configuration are identified, which also drives the UA design.
3. The communications requirements are defined.
4. The development timeline is determined. Without external schedule drivers or funding profile constraints, the project will assume a natural schedule.
5. The number of UA and systems are determined.

Once the preceding technical requirements, programmatic considerations, and other drivers are identified, a baseline system is defined. This is essentially a conceptual design effort prior to release of the proposal. The system architecture is designed, and major aircraft design characteristics are generated.

The technologies utilized drive the costs. The technology readiness level (TRL) of major components must be considered, as lower maturity translates to higher risk. Areas that will have high cost sensitivity to TRL include the following:

1. Airframe materials
2. Propulsion type
3. Avionics
4. Mission payload
5. Launch and recovery system

The development costs are composed of the following:

1. Design
2. Procurement
3. Materials
4. Tooling
5. Parts fabrication
6. Assembly
7. Development tests
8. Integration
9. Documentation and drawings

Once the program is adequately defined, the bottoms-up cost estimation can proceed. This is performed through obtaining inputs from experts various areas of the program.

Initial bottoms-up cost estimates are nearly always wrong. These estimates are generally based on best-case scenarios with insufficient margin to account for uncertainties. Consider estimating the cost of building a wall in a house using the bottoms-up methodology. One might count the square feet of drywall, number of studs required, and the quantity of nails. This won't adequately cover the costs. There will be credit card bills, car trips to the hardware store, bad nails, bent nails, warped studs, studs cut incorrectly, and the builder may hit his/her hand with the hammer. People get sick, mistakes are made, accidents occur, and there are defective materials. Realism factors are required to account for the unknown. Sometimes a factor of $2-3$ is appropriate to account for the unforeseen cost drivers.

Application of a realism factor accounts for all uncertainties contained within the program, except for factors outside the control of the program. The realism factor does not reflect inflation, vendor price increases, requirements changes, or other external factors. A management reserve factor is recommended to account for external considerations. The management reserve is applied to materials and labor. It is common practice for management reserve to be removed during contract negotiations. When unforeseen (or unforeseeable) problems arise during program execution, the lack of management reserve results in a cost overrun.

In developing a bottoms-up cost estimate, the program manager will ask engineers and other personnel for estimates of the labor required to complete a proposed task. The labor hour estimates will vary significantly between estimators. Whatever person-hour estimate is provided is multiplied by a factor to account for unexpected events, sick leave, key personnel availability, and a host of other considerations. Some successful program managers have applied a total factor of 3.14 (an easy to remember factor of π) to raw labor estimates, which does not include the management reserve.

It is difficult to automate the original labor hour estimates provided by the estimators. There are many considerations that the individual estimators must consider. Factors such as personnel experience, workload, organization structure, complexity of the task, and historical experience with similar efforts are used to find the labor estimate. An effort may take significantly different labor for two different organizations. Creating the logic and assembling the data to produce the initial labor estimate is a key challenge for developing a detailed bottoms-up UAS cost model.

Personnel experience is very significant to the program outcome. Intelligence and experience are not the same. Work experience requires both time and exposure to a sufficient number of programs. An intelligent but inexperienced engineer might look at a failed component and require days to develop a methodology to create a new part. An experienced engineer may consider the same problem and suggest a specific material and thickness as a quick and conservative replacement. The experienced engineer is many times more efficient in such circumstances.

Corporate and technology experience is important. Much of the corporate experience is captured in individuals who worked multiple programs. It takes an individual approximately 10 years of continuous similar-system experience to achieve proficiency. This experience does not necessarily need to be at the same organization, but it must be with similar technology. Often experienced personnel will provide higher labor and schedule estimates than those with less experience, which can lead to a higher bid. Although the higher estimate likely reflects reality, a less experienced company may provide a lower bid due to unsubstantiated optimism.

The program cost is highly dependent upon the funding profile. Ideally, the cost will follow a bell-curve shape. The peak activity is generally at first flight. Programs rarely are permitted to follow an unconstrained development schedule. Typically, a sponsor will want the minimum schedule, minimum cost, capped cost, or fixed funding profile. The three primary aspects of program management are cost, schedule, and technical performance, but at least one of the three must be flexible during program execution.

A flat funding profile will extend the program schedule and increase the overall program cost. In the beginning, it will be difficult to spend the funds, though some efforts will get ahead of schedule. In the middle, tasks will get behind and materials will not get ordered due to lack of funds. Finally, ample funds in the last phase will permit some schedule recovery. The total funding for a flat funding profile will be higher than for a bell-shaped profile.

Small UAS developments typically take 2–5 years to go from requirements to operational or technical evaluations. Accelerated programs are generally very stressful, and use much overtime. There is a premium paid for rapid development. Longer programs can be inefficient, and personnel continuity can be lacking. Optimal, minimum cost developments will generally take 3–4 years. The optimal development timeframe is shorter for smaller UAS programs.

The first year of development is characterized by creating plans, defining the work, ramping up for execution, and then designing the system. The second year or two activities include parts fabrication, prototype assembly, and development tests, which together consume the greatest portion of the development funds. Most unforeseen problems are encountered in the second phase. The final year is occupied with flight tests. Transition to operations occurs after these activities.

Budgets represent a snapshot of a program. Budgets must be reevaluated through time as events occur. Funds must be reallocated across the program to best achieve the budget and program objectives. Budgets and resource allocations are rarely static as the program is actively managed.

Once a program is underway, the budget may be cut. The scope of the program must be adjusted accordingly with the new budget. A system that undergoes large budget cuts will likely not attain the same level of operational maturity or will sacrifice capability. Budget cuts can also delay the development, and the program can risk obsolescence in this rapidly changing field.

Many UAS programs have been advanced concept technology demonstrators (ACTD) or more recently joint concept technology demonstrations (JCTD), which have different cost behavior than traditional acquisitions. Examples of ACTD/JCTD programs include Predator, DarkStar, and Global Hawk. In an ACTD/JCTD, the government provides the contractor with a fixed budget and asks the contractor to deliver the best capability within the funding. The requirements are flexible, and the contractor has much latitude in program execution. The final product of an ACTD/JCTD is driven by available funding and technology, whereas a traditional procurement has numerous fixed requirements.

Requirements-based acquisitions have traditionally had a lower success rate than ACTD/JCTDs for UAS systems. Traditionally, the requirements tend to be lengthy and may be independent of practical implementation means within available technology. The government requirements' negotiation process with numerous stakeholders often leads to adopting many requirements that could be easily satisfied in isolation, but can be overconstraining to system developers when combined.

Two common contract types are firm fixed price (FFP) and cost plus (CP). With an FFP contract, the contractor guarantees delivery of a product of constant requirements at a fixed price. Often FFP contracts have negative outcomes when the contractor fails to deliver due to development complications. The two most common CP contract types are cost plus award fee (CPAF) and cost plus fixed fee (CPFF). These contract types typically generate more desirable results because the risk is shared by industry and government. With CPAF, the award fee is based on contractor performance.

The government has created programs that require contractor cost sharing. The NASA Environmental Research and Sensor Technology (ERAST) program of the 1990s included cost-sharing arrangements. A competition for prize money is another approach, where only the best contractors receive compensation for their efforts. In either case the contractor must be able to provide much of their own funding. Taken as an isolated contract, this is a losing proposition for a company that wishes to stay in business. However, there may be broader strategic value in advancing the UAS technology.

A contractor program management organization (PMO) oversees the program execution. This is mirrored by the government program office. The contractor PMO interfaces with the customer and provides program leadership to the system development. The PMO usually consists of a program manager, contracts, finance, schedulers, security, and administrative support. Some of these functions may be indirect labor. On a simple program many of these roles can be performed by the program manager. Large programs may have a large PMO with multiple levels of management. Layered program management often takes the form of integrated product teams (IPTs).

Table 19.1 UAS Development Costs (2)

System	Estimated development cost, $M
Aquila	$868
Hunter	$189.2
Outrider	$268.5
Predator	>$209.9
Global Hawk	$370.3 (UA) + $272.6 (ground station)
DarkStar	$326.9 (UA)

19.5 UAS Cost Data

UAS cost data can be obtained from a variety of sources. Government sources include Congressional Budget Office (CBO), General Accounting Office (GAO), Office of the Inspector General (OIG), R-2 Budget Justifications, Office of the Secretary of Defense (OSD) studies, and Department of Defense (DOD) reports. RAND Corporation has also evaluated UAS programs and associated costs. Aerospace publications and company press releases are other valuable resources.

Cost data for UAS development are limited. The GAO [2, 3] provided information of system development costs shown in Table 19.1. Several cautions are in order. The Aquila and Outrider programs were cancelled. The Hunter program was cancelled in 1996, but continued to operational status afterwards. Hunter heavily leveraged existing Israeli UAS designs. The Predator development was incomplete at the time of the GAO report publication. Some of the systems identified, including Predator, had contractor-funded development that might not be included in the numbers. The Global Hawk and DarkStar shared a common ground station at the time of the GAO report.

Table 19.2 FY05-07 UAS RDT&E and Procurement Costs for Selected Systems (4)

System	RDT&E (FY05–07) $M	Procurement (FY05–07) $M	Qty of UA
Shadow 200	$54.4	$502.6	33
Fire Scout	$258.4	$37.6	4
Predator	$207.7	$798.2	47
Global Hawk	$958	$1223.2	15
J-UCAS	$1100.2	—	—

These numbers can easily be taken out of context. The RDT&E continues for most programs after the initial fielding. This comes in the form of incremental system enhancements and major block upgrades. A snapshot of the RDT&E and procurement for program of record systems based on DOD [4] data for FY05-07 is shown in Table 19.2. Note that the procurement costs can include more than just UA, but only UA quantities are provided.

Inflation must be taken into account when considering older cost data. The cost escalation factor (CEF) is the ratio of current-year costs to then-year costs. The Consumer Price Index (CPI) is a convenient metric for estimating the rate of inflation. It is not a perfect metric because materials and labor do not necessarily follow the same trend. The CPI is published by the Bureau of Labor Statistics, and it goes back to 1913. The CEF is found by

$$CEF(TY) = \frac{CPI(Now)}{CPI(TY)} \qquad (19.4)$$

where *TY* is then-year and *Now* is the year of interest.

Example 19.1   Cost Escalation Factor

Problem:
The following UA costs and system costs come from OSD [5].

UA	Aircraft cost (then-year), $M	System cost (then-year), $M	Reference year
Dragon Eye	$0.0285	$0.1303	2004
RQ-7A Shadow	$0.39	$12.7	2004
RQ-2B Pioneer	$0.65	$17.2	2004
RQ-8B FireScout	$4.1	$21.9	2004
RQ-5A Hunter	$1.2	$26.5	2004
MQ-1B Predator	$2.7	$24.7	2004
MQ-9A Reaper	$5.2	$45.1	2004
RQ-4 (Block 10) Global Hawk	$19.0	$57.7	2004
RQ-4 (Block 20) Global Hawk	$26.5	$62.2	2004

Using the consumer price index, estimate the prices in 2010 dollars.

Solution:
The annual average CPI was 188.9 in 2004 and 218.1 in 2010. Therefore, the CEF is 1.154.

UA	Aircraft cost (2010), $M	System cost (2010), $M	CEF
Dragon Eye	0.0329	0.150	1.154
RQ-7A Shadow	0.450	14.7	1.154
RQ-2B Pioneer	0.750	19.8	1.154
RQ-8B FireScout	4.73	25.3	1.154
RQ-5A Hunter	1.38	30.6	1.154
MQ-1B Predator	3.12	28.5	1.154
MQ-9A Reaper	6.00	52.0	1.154
RQ-4 (Block 10) Global Hawk	21.9	66.6	1.154
RQ-4 (Block 20) Global Hawk	30.6	71.8	1.154

19.6 Preacquisition Costs

The government and industry expend substantial effort on a new UAS program prior to the start of a contract. The government must develop requirements and prepare for the acquisition. Contractors spend money developing their product offerings, performing proposal preparation ahead of the request for proposal (RFP), and conducting business development.

The government's preacquisition costs are related to the following items:

* Studies, which can include operations analysis evaluations and analysis of alternatives (AoA)
* Industry surveys, which is an assessment of industry capabilities generated through research, requests for information, and contractor site visits
* Requirements development
* Preparation of the RFP
* Establishing the program office that must be staffed with government and SETA contractors
* Potentially organizing and running a fly-off
* Conduct source selection to award contract
* Respond to potential protests

Industry preacquisition costs are related to the following:

* Internal research and development (IRAD) and other noncontract product development costs
* Marketing and business development to inform the government of the product capabilities and benefits relative to the competition
* Potentially participating in a fly-off
* Bid and proposal (B&P) costs associated with writing the proposal
* Legal costs for potential protests

The B&P costs for a program of record acquisition can cost several million dollars for even smaller UAS programs. A RFP might require hundreds or even thousands of pages for the proposal response, including detailed cost data. This necessitates a large proposal development team that can include outside consultants. Each competitor must make a large B&P investment, but only one will win. The B&P costs get rolled into the overhead rates of the contractors, and so the government ultimately pays for this expense through future contracts with the companies that continue to perform work for the government.

19.7 Research, Development, Test, and Evaluation Cost

The RDT&E phase consists of all materials and services necessary to design, integrate, and test the developmental UAS. UAS program RDT&E phases either seek to mature the UAS system such that it is ready for production or to demonstrate a capability. The RDT&E phase is the precursor to the system manufacturing. The cost of the RDT&E phase is

$$C_{\text{RDTE}} = C_{\text{Des,RDTE}} + C_{\text{DevTest,RDTE}} + C_{\text{HW,FT,RDTE}} + C_{\text{FTO,RDTE}}$$
$$+ C_{\text{Fin,RDTE}} + C_{\text{Fac}} + C_{\text{Fee,RDTE}} \tag{19.5}$$

The subscripts are Des for design, DevTest for development tests, HW,FT for flight-test hardware, FTO for flight-test operations, Fin for finance, Fac for facilities, and Fee for fee. All elements of the system must be included in the system RDT&E costs. These include the UA, payloads, ground control stations, aircraft launch and recovery equipment (ALRE), and support equipment. These elements come together during system tests and flight tests.

RDT&E Design
The system design cost is found by

$$C_{\text{Des,RDTE}} = C_{\text{AV,Des,RDTE}} + C_{\text{PL,Des,RDTE}} + C_{\text{SW,Des,RDTE}} + C_{\text{GS,Des,RDTE}}$$
$$+ C_{\text{ALRE,Des,RDTE}} + C_{\text{Supt,Des,RDTE}} \tag{19.6}$$

The UA design engineering consists primarily of engineering labor. This cost is

$$C_{AV,Des,RDTE} = Hr_{AV,Engr,RDTE} \cdot R_{Engr} \tag{19.7}$$

where $Hr_{AV,Engr,RDTE}$ is the number of hours required to design the UA and its systems during RDTE and R_{Engr} is the burdened engineering hourly rate (less fee). The DACPA and Roskam [1] methods estimate the labor hours based on W_{Ampr}, dynamic pressure, number of RDT&E airframes, and a difficulty factor related to the aircraft complexity. There is very limited publicly available data for unmanned aircraft design hours.

Throughout this chapter we will see labor costs, which take the form

$$C = Hr \cdot R \tag{19.8}$$

where C is the cost, Hr is the man-hours worked, and R is the hourly rate. The labor rate is the burdened labor rate without fee. Recall that the fee is calculated separately. This non-fee-burdened labor rate is

$$R = R_{Basic} \cdot (1 + Fringe) \cdot (1 + Overhead) \cdot (1 + G \& A) \tag{19.9}$$

where

$$R_{Basic} = \text{basic rate}$$
$$Fringe = \text{personnel-related costs such as vacation and medical insurance contributions}$$
$$Overhead = \text{facilities and administrative related costs}$$
$$G \& A = \text{general and accounting}$$

The cost of purchased parts, materials, and subcontract costs without the fee is

$$Cost_{Procured} = Cost_{Basic} \cdot (1 + G \& A) \tag{19.10}$$

The fringe, overhead, and G&A costs vary considerably across companies, making a single cost model based only on system characteristics difficult. In other words, the same UAS will cost different amounts if developed or built at different companies.

This RAND and Roskam [1] methods are limited by their generality, where little consideration is given to the details of the program. For example, a complex antenna performance requirement may profoundly affect the UA design effort, but the weight-based method would not distinguish this from simpler design problems. Only applying a difficulty factor may change the design hours.

Given the paucity of UAS design labor data, perhaps future studies could collect a detailed buildup of engineering hours across disciplines to improve the accuracy of future cost models. For conceptual and preliminary design, the engineering hours can be estimated from metrics of past programs if available. For the airframe detailed design, the engineering hours can be estimated partially by the product of number of drawings produced and the number of hours per drawing. The number of drawings often has a relationship to UA weight and complexity. Companies and other development organizations often track hours per drawing metrics. Producing labor-hour estimating relationships is possible, provided that sufficient program data are available.

To give insight into the difference in scope among UA programs, we will consider the autopilot development for different system classes. The data presented in Table 19.3 are based on the author's judgment. This assumes that the company's engineering team is working with an autopilot vendor that implements the control laws. Even though the autopilot will perform many of the same functions for all system classes (waypoint tracking, orbit patterns, launch and recovery, etc.), the effort will vary substantially. The level of rigor in the design, amount of testing, control law tuning, and functionality improve with UA size. This is an example of proportionality, where the subsystem development effort is scaled to the overall program scope.

Ideally commercial off-the-shelf (COTS) components do not require additional NRE because the development should be complete. However, it is often necessary to modify COTS components to work within the UAS environment. Sourcing the best COTS components takes labor to define requirements, meet with vendors, analyze options, conduct trade studies, and procure the items. Candidates for COTS include engines, avionics, communications equipment, payloads, subsystems, actuators, ALRE, shelters, and computers. Developing a new engine could cost as much as the remainder of the UAS program, and so COTS engines are generally selected for new UASs. Government off-the-shelf (GOTS) are government-owned equipment that can be procured or provided without modification.

The ground-station CER relates the development cost to the empty weight of the aircraft, based on the average of data from the Global Hawk and DarkStar CGS development. The cost per UA empty weight of the total ground station development is assumed to the $20,000/lb in FY04, where the ground-station hardware is the total less the ground-station

Table 19.3 Approximations for Autopilot Development Effort

UAS class	External contracts, $K	Internal engineering, hrs
Student project	0.5–5	40–500
Small UAS	5–300	40–1,000
Tactical UAS	500–2,000	1,000–4,000
Large UAS	1,000–10,000	1,000–20,000

software design costs. This assumes that 20% of the ground-station development effort for the referenced program is dedicated to software development. The CGS average development cost per empty weight would be $25,000/lb (FY04) if software is included. This relationship includes GCS developmental testing.

$$C_{\text{GCS,Dev,RDTE}} = \$20,000 \cdot W_E \cdot CEF(2004) \qquad (19.11)$$

The applicability of this method to small UAS is unknown, though it likely underestimates the cost for smaller UAS ground-station development.

The UA software design is

$$C_{\text{SW,Des,RDTE}} = C_{\text{AV,SW,Des}} + C_{\text{GS,SW,Des}} \qquad (19.12)$$

The software costs are found using the software estimation methods described later.

Little data are available for the development cost of ALRE and support equipment, and so no CERs are presented. There are many possible system implementations for ALRE, each with different costs. For example, a conventional runway launch and recovery will have different development costs than if a pneumatic launcher and net is employed. A bottoms-up cost estimate is recommended for ALRE and support equipment.

RDT & E Development Tests

The system development testing can include, but is not limited to, wind-tunnel testing, systems testing, structural testing, propulsion testing, radio frequency (RF), electromagnetic interference (EMI), mechanical interface/interference, simulations, and software testing. This is found by

$$C_{\text{DevTest,RDTE}} = C_{\text{AV,DevTest,RDTE}} + C_{\text{ALRE,DevTest,RDTE}} + C_{\text{AV,SW,IT}}$$
$$+ C_{\text{GS,SW,IT}} \qquad (19.13)$$

The UA development test costs $C_{\text{AV,DevTest,RDTE}}$ can use the DACPA or Roskam methods corrected for cost escalation. Little information is available on UAS development testing. The DACPA and Roskam [1] relationships were developed for aircraft with relatively little software, so that the software integration and test should be added as a separate cost. Data are lacking on ALRE development tests, and so a bottoms-up method is recommended.

Development testing can be a large cost driver with significant variation between programs. RF testing is used as an example here. RF testing is necessary for evaluating the antennas and avionics performance and interactions. This testing activity can occur in a RF test chamber or on an open range. The tests can be as little as two people over a one-week period for

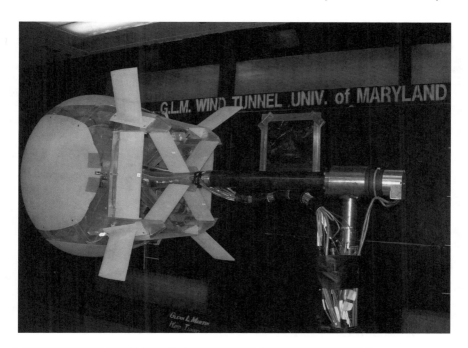

Fig.19.3 Aurora Flight Sciences GoldenEye low-speed wind-tunnel test. (Photo courtesy of Aurora Flight Sciences.)

small UASs. Alternatively, for complex RF issues, the testing can take considerably longer and use sophisticated facilities and become a major development cost.

Wind-tunnel testing costs are prohibitive for many small UAS programs. Using computational fluid dynamics (CFD) or accelerated flight tests are often more economical. A low-speed wind-tunnel facility capable of testing a small UAS at flight Reynolds number costs approximately $50–200 K per week. This cost covers wind-tunnel facilities and support, not engineering time. A high-speed wind-tunnel campaign can cost around $1–2 M. Approximately 40 hrs of wind-on time is typical to characterize a small UAS in the wind tunnel. A method of estimating the wind-tunnel test costs is

$$C_{\mathrm{WT}} = T_{\mathrm{WT}} \cdot (R_{\mathrm{WT}} + N_{\mathrm{Pers}} \cdot R_{\mathrm{Pers}}) + C_{\mathrm{Travel}} + C_{\mathrm{Model}} \qquad (19.14)$$

where T_{WT} is the tunnel entry time, R_{WT} is the cost rate for occupying the tunnel, N_{Pers} is the number of contractor personnel deployed for the test, C_{Travel} is the travel costs, and C_{Model} is the cost of the wind-tunnel model. A large program can have multiple wind-tunnel entries. (See Fig. 19.3 for a photograph of a wind tunnel test.)

Other developmental tests can include the following:

- Mechanical interface or interference tests are highly dependent upon the number of moving parts and the mechanical complexity. Fixed-wing UA have relatively little mechanical testing relative to helicopters.
- The electrical system tests consist of powering all systems and testing operational range.
- Hardware-in-the-loop (HIL) simulations
- Iron bird tests are used. Electrical and mechanical components are placed on a full-scale fixture that represents the UA. Often an actual airframe will take the place of the fixture.
- Ground vibration testing (GVT)
- Wing load tests are used. Wings are tested to limit load, ultimate load or to destruction.
- Drop tests (landing gear)
- Brake testing
- Jet-engine tests, static and in propulsion wind tunnel
- Engine dynamometer testing, altitude chamber testing for reciprocating engines
- Flying propulsion testbed
- Surrogate vehicle avionics testing (alternative flight vehicle)
- Fuel system testing
- Anechoic chamber testing (RF communications, EMI, EMC)
- Outdoor communications range testing

RDT & E Flight-Test Hardware

The cost of the hardware provided for flight test is

$$C_{HW,FT,RDTE} = C_{AV,Sys,RDTE} + C_{Manf,RDTE} + C_{Matl,Manf,RDTE}$$
$$+ C_{Tool,RDTE} + C_{QC,RDTE} + C_{GS,HW,RDTE} \qquad (19.15)$$

The aircraft systems cost for the flight-test hardware is found by

$$C_{Sys,RDTE} = C_{Eng,RDTE} + C_{Avion,RDTE} + C_{Subs,RDTE} + C_{Payl,RDTE} \qquad (19.16)$$

The engine cost can be found through parametric methods [1] for large engines or through direct engine cost inputs.

The avionics, subsystems, and payload costs are found using custom methods. Drezner and Leonard [6] detail the Global Hawk and DarkStar procurement costs by airframe, avionics, propulsion, and other costs. These costs can be divided by the associated category weight estimations to develop cost per weight values.

The RDT&E manufacturing labor costs are found by

$$C_{Manf,RDTE} = Hr_{Manf,RDTE} \cdot R_{Manf} \qquad (19.17)$$

where $Hr_{\text{Manf,RDTE}}$ is the RDT & E manufacturing hours. Large UASs can use the DACPA or Roskam [1] methods.

Model aircraft and home-built aircraft can provide some insight in the absence of historical data of UAS labor hours. Model aircraft contain very little quality control relative to a UAS because this is only for a hobby application. UASs are typically more rugged than model aircraft, which drives up the cost. A plot of labor hours vs empty weight for model aircraft and home-built aircraft is shown in Fig. 19.4. These hours appear to be approximately one order of magnitude less than what is required to fabricate an operational UAS airframe of a similar weight class. A relationship for estimating the labor hours for simple airframes is

$$Hr_{\text{Manf,RDTE}} = 10.4 \cdot W_E^{0.605} \cdot f_{\text{Matl}} \qquad (19.18)$$

where f_{Matl} is 1 for metal, 1.98 for composites, 1.88 for wood, and 1.11 for metal tube and fabric construction. This relationship assumes that simple UAS airframes are equivalent to model aircraft or homebuilt aircraft of equivalent weight. This does not include labor hours for UAS-specific systems, avionics, communications, and payloads.

More processes are required for UASs over what is encountered in model aircraft or home-built aircraft. For example, the MALE UAS could potentially be replaced with a home-built airframe. The home-built equivalent is not designed against a UAS specification. The traceability of the manufacturing process would drive up the home-built cost. The traceability might cover when a particular kit was produced, location, workers who fabricated the aircraft, and materials batches. These records would be maintained for later retrieval. A rough estimate for the difference in airframe

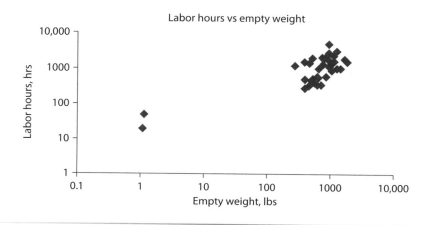

Fig. 19.4 Labor hours for model aircraft and home-built aircraft (7).

costs at a given weight for UASs and home-builts is a factor of 2–3 for prototype UASs and 4–5 for production-ready UASs.

The labor hours for the materials handling at the kit manufacturer are not included in the advertised labor estimate. Home-built aircraft have more quality control than a model aircraft because a Federal Aviation Administration (FAA) representative must inspect the aircraft before flight.

UASs must contend with a harsh electromagnetic environment relative to model aircraft and home-built aircraft. UAS airframes are usually tightly packed with avionics, communications equipment, power systems, and payloads. Substantial design and testing is required to contend with EMI and radio-frequency interference (RFI). The electromagnetic design and test challenge is greater for UASs capable of ship-borne operations.

The packaging density of UA components is a cost driver. The EMI and RFI issues become more complex with high-density packaging. Additionally, the systems integration is complicated when systems are not easily accessible. Integration is difficult when systems and wiring harnesses must be removed to access components during integration.

The RDT&E manufacturing materials costs can be found through CERs. The Roskam [1] CER has the form

$$C_{\text{Matl,Manf,RDTE}} = f\left(W_{\text{Ampr}}, V_{\text{eq,Max}}, F_{\text{Comp}}, N_{\text{AV,RDTE}}\right) \cdot CEF \qquad (19.19)$$

This manned aircraft method is most applicable to UASs within the weight and performance capabilities of manned aircraft. No UAS materials cost data for the RDT&E phase are publicly available, and so method outputs should be carefully scrutinized.

Another method takes airframe materials data from model aircraft and home-built aircraft to devise a custom CER. The custom method requires significant adjustment factors to come within the order or magnitude

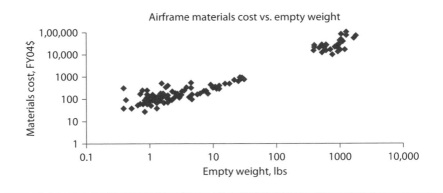

Fig. 19.5   Materials costs for model aircraft and home-built aircraft (7).

required for UAS programs. The airframe materials costs of model and home-built aircraft vs empty weight is shown in Fig. 19.5.

The cost of a model aircraft kit may be only 50% materials. The remaining costs are taken up by packaging, labor, marketing, and other factors. Model aircraft kits are generally incomplete, so that this might offset some of the difference. A correlation for these 2004 data is

$$C_{\text{Manf,Matl,RDTE}} = 16.6 \cdot W_E \cdot f_{\text{Matl}} \cdot CEF \qquad (19.20)$$

where f_{Matl} is 1 for metal, 2.23 for composites, 2.34 for wood, and 1.89 for metal tube and fabric construction.

Materials for a UAS program can be considered at either a top level or a detailed level. At a top level, the materials can be assumed to be a fixed ratio of the manufacturing cost or as dollars per pound. When considered in detail, the materials types and quantities are identified, and cost estimates are applied. Scrap rates, handling, and other considerations are evaluated.

Materials procurement costs must be considered in addition to the materials costs. Procurement consists of identifying suppliers, negotiations, placing orders, and accepting delivery. The labor will be greater for larger, more complex UAS developments. The procurement costs are considered as part of overhead in the methods applied here, which is accounted for in the burdened manufacturing rates.

The RDT&E tooling labor cost is found by

$$C_{\text{Tool,RDTE}} = Hr_{\text{Tool}} \cdot R_{\text{Tool}} \qquad (19.21)$$

where Hr_{Tool} is the tooling labor hours. This can be estimated using the DACPA or Roskam [1] methods for large UAs. The labor-hour data for model aircraft and home-built aircraft presented earlier include tooling hours.

Roskam [1] recommends that the quality control hours are approximately 13% of the manufacturing labor for manned aircraft.

The ground-station hardware costs for RDT&E include the costs of all ground-station types used for flight-test activities. Often a flight-test engineering center is set up for engineers to monitor telemetry data. These costs are dependent upon the system configuration and flight-test objectives.

Ground-station costs can vary significantly, depending on the application, technology, requirements, and cost expectations. Ground stations can be as simple as a laptop and a transceiver. On the other end of the spectrum, ground stations can be complex architectures consisting of multiple nodes distributed around the world. Although the functionality of the simplest and most complex ground stations has substantial overlap, the hardware and software implementations are quite different.

There is a loose correlation between ground-station costs and the UA costs. It is possible that a multimillion-dollar Global Hawk AV could be controlled from a $2000 laptop ground station, or that a multithousand-dollar Dragon Eye could be controlled from a multimillion dollar Global Hawk ground station. However, there are issues with affordability and risk. It would be irresponsible to risk a strategic Global Hawk with an inexpensive tactical ground station. Similarly, it would be viewed as unaffordable to have an expensive Global Hawk ground station controlling a very inexpensive Dragon Eye. Ground stations for operational UASs can consume approximately 20–50% of the UAS acquisition cost.

RDT & E Flight-Test Operations

The scope of the flight-test program is a function of the system class, program type, customer expectations, and flight-test range. The flight-test program may begin with ground-handling operations such as low-speed and high-speed taxi tests for wheeled aircraft. Next, there are the airborne flight tests. This begins with initial flights that are generally of short duration. Envelope expansion flights test various regions of the flight envelope. Later in the program, mission systems such as payloads are tested. So far these are all developmental tests. Some programs continue on to operational demonstrations such as military utility assessments.

The RDT & E flight-test operations cost is a buildup of labor, facilities, and consumables.

$$C_{\text{FTO,RDTE}} = C_{\text{Pers,FT,RDTE}} + C_{\text{Loc,FT,RDTE}} + C_{\text{Chase,FT,RDTE}}$$
$$+ C_{\text{BLOS,FT,RDTE}} + C_{\text{POL,FT,RDTE}} \tag{19.22}$$

The cost of labor for the flight-test operations is

$$C_{\text{Pers,FT,RDTE}} = F_{\text{Flt,FT,RDTE}} \cdot Hr_{\text{FT,RDTE}} \cdot N_{\text{Pers,FT,RDTE}}$$
$$\cdot (1 - F_{\text{GFX,FT,RDTE}}) \cdot R_{\text{FT,RDTE}} \tag{19.23}$$

where the flight time factor $F_{\text{Flt,FT,RDTE}}$ is the ratio of total active time to flight time. The labor rate during flight test $R_{\text{FT,RDTE}}$ is a weighted average of various labor category rates. The government-furnished factor $F_{\text{GFX,FT,RDTE}}$ is the ratio of government-furnished supplies, labor, equipment, or services.

Much of the flight-test costs are consumed by labor. Often a minimum of five people are required to support a small UAS flight-test activity. These may include an avionics engineer, a propulsion engineer, a test pilot, a backup test pilot, and a test director. The number and skills of the flight-test personnel will vary from one program to another. Large UAS systems have a large team that is deployed to the flight-test location.

The range support personnel costs are generally small for small UAS flight-test programs. Approximately two range support personnel will work part time in support of the flight-test activity. The range safety personnel will be on hand around the time of flight, which is only a small percentage of the time the flight-test team is at the flight-test location. Larger UAS programs will have a team of range support personnel dedicated to the flight-test campaign.

A large flight-test expense comes from obtaining a flight clearance. The UAS system must be approved for flight by an organization composed of expert engineers covering multiple disciplines. This might take the form of a government flight competency board. The approval organization typically requires extensive project documentation and multiple briefings. The number of development engineers required to support this activity is approximately equal to the number of approval organization members. During this time, the time of the approval organization members and the time for the engineers supporting the effort must be paid for. It is possible to spend more money on passing the approval gates for small UAS than on design engineering. The same board process is generally required regardless of UA size.

The flight-test range location cost is

$$C_{\text{Loc,FT,RDTE}} = F_{\text{Flt,FT,RDTE}} \cdot Hr_{\text{FT,RDTE}} \cdot \left(1 - F_{\text{GFX,Loc,FT,RDTE}}\right) \cdot R_{\text{Loc,FT,RDTE}}$$

$$(19.24)$$

where $R_{\text{Loc,FT,RDTE}}$ is the rate in dollars per hour for flight-test range usage. There are numerous flight-test ranges that support UAS flight-test operations. Several examples are shown in Tables 19.4 and 19.5. The flight-test culture and ease of airspace access varies greatly among locations.

Table 19.4 U.S. Government Flight-Test Ranges and UAS Airspace

Test range	Location	Notes
Boardman Bombing Range	Oregon	
China Lake	California	Navy missile test range
Creech AFB	Nevada	
Dugway	Utah	
Edwards AFB	California	Dry lakebed
Eglin AFB	Florida	
NASA Dryden	California	Co-located with Edwards AFB
NAS Patuxent River	Maryland	Webster auxiliary field used for UAS
White Sands Missile Range	California	

Table 19.5 Private and Civil Test Ranges and UAS Airspace

Test range	Location	Notes
Avon Park	Florida	
New Mexico State University	New Mexico	
University of North Dakota	North Dakota	Uses Grand Forks AFB airspace

Some flight-test operations might require a chase aircraft. The chase aircraft cost is

$$C_{\text{Chase,FT,RDTE}} = Hr_{\text{Chase,FT,RDTE}} \cdot (1 - F_{\text{GFX,Chase,FT,RDTE}}) \cdot R_{\text{Chase,FT,RDTE}}$$

(19.25)

Small UAS programs generally do not utilize a chase aircraft because the UA is hard to spot and poses a collision hazard if contact is lost. The rate for the chase aircraft $R_{\text{Chase,FT,RDTE}}$ includes the use of the aircraft and the crew onboard the aircraft. This rate might be around $1000 per hour for a general-aviation aircraft with two crew members.

The beyond-line-of-sight (BLOS) communications costs for RDT&E flight tests is found using methods described in the operations section.

The petroleum, oil, and lubricants (POL) cost for the RDT&E flight test is

$$C_{\text{POL,FT,RDTE}} = Hr_{\text{FT,RDTE}} \cdot R_{\text{POL}}$$

(19.26)

where the hourly rate of POL can be estimated by

$$R_{\text{POL}} = \dot{m}_{\text{Fuel}} \cdot \frac{P_{\text{Fuel}}}{\rho_{\text{Fuel}}}$$

(19.27)

where ρ_{Fuel} is the fuel density in pounds per gallon, P_{Fuel} is the price per gallon, $\dot{m}_{\text{Fuel}}$ is the fuel flow rate in pounds per hour.

A flight-test campaign for system development requires numerous flights. Optimistically, two flights per week can be expected early in a test program. The flight tempo usually increases later in the test program.

The program cost is generally less expensive when more than one UA is used for the flight-test effort. Although there are costs associated with building the additional UA, there are savings in other areas. First, the learning curve reduces the UA costs. Second, greater risks can be taken in the flight-test program, and so less testing is required prior to first flight. Probability of program success also increases with multiple prototypes. Prototypes have a tendency to crash. If the one and only prototype crashes, the program office might elect to cancel the program rather than spending months waiting for a new UA.

RDT & E Financing
The cost of financing the RDT&E phase is

$$C_{\text{Fin,RDTE}} = C_{\text{RDTE}} \cdot Fin \tag{19.28}$$

The finance rate *Fin* is typically 5–10%. This is sometimes called the cost of money.

RDT & E Facilities
The cost of new facilities development for the RDT&E phase is

$$C_{\text{Fac,RDTE}} = C_{\text{RDTE}} \cdot F_{\text{Fac}} \tag{19.29}$$

where the facilities factor F_{Fac} can range from 0–20%, depending on company growth [1].

RDT & E Profit
The fee that the contractor gets is a function of the contract performance. The fee is

$$Fee = Fee_{\text{BFee}} + F_{\text{PerfCon}} \cdot Fee_{\text{AFee}} \tag{19.30}$$

where the base fee and award fee are part of the contract. The contract performance factor F_{PerfCon} is based on contract performance. This assumes a cost plus award fee contract type. By setting the award fee to zero, the contract type is effectively cost plus fixed fee. The overall cost of the fee is

$$C_{\text{Fee,RDTE}} = C_{\text{RDTE}} \cdot Fee \tag{19.31}$$

Note that the combination of the cost without fee and the fee is called the price. Using the cost terminology to describe fee is used here for convenience.

System documentation can include design reports and operations manuals. The latter requires professional technical publications' writers and engineering support. The cumulative unburdened cost of the publications is

$$C_{\text{Docs}} = N_{\text{Pages}} \cdot R_{\text{Pages}} \cdot R_{\text{Writers}} \tag{19.32}$$

where N_{Pages} is the document page count, R_{Pages} is the pages/hour rate, and R_{Writers} is the writer labor rates. A notional page rate is 0.1–0.25 pages/hour. Prototype systems may have no operations manuals, whereas program-of-record systems may have hundreds or thousands of pages of documentation.

Software
The Constructive Cost Model (COCOMO) algorithm [8, 9] can be applied to estimate the software development costs associated with UAS programs.

This model estimates the total effort and schedule for software development based on a number of factors. This algorithm can be applied to both the UA and ground-station software.

The labor effort is a function of both the software size and multiple project attributes. The initial labor effort is estimated as a function of delivered lines of code. A series of correction factors are applied to the initial labor effort estimate to come up with the corrected labor effort. These correction factors cover 14 project attributes including required software reliability, product complexity, analyst capability, programming language experience, use of modern programming practices, and required development schedule. The labor distribution is spread over product design, detailed design, code and unit test, and integration and test program phases. Jalote [9] recommends various factor values for multiple project sizes.

The software schedule can also be found using the COCOMO algorithm, which uses a Rayleigh distribution of the labor effort [9]. The distribution of the software development schedule over the product design, programming (which includes the detailed design and code and unit test categories for the effort), and integration can be found via values recommended by Jalote [9] for multiple project sizes.

A much less sophisticated method for estimating software labor hours Hr_{SW} is

$$Hr_{SW} = F_{SW} \cdot SLOC_{New} \qquad (19.33)$$

where $SLOC_{New}$ is the new source lines of code (SLOC) and F_{SW} is a productivity metric in hours per SLOC. The productivity includes all phases of software development from requirements generation, coding, testing, and integration. Based on the author's experience, it is practical to achieve F_{SW} values of $0.5 - 1$ hr/SLOC.

The software size (i.e., SLOC) is a key input for the software estimating algorithms. Unfortunately, no UAS software size metrics are publicly available. The software development has great potential to drive the schedule and cost of UAS programs, so that any software metrics would be very beneficial for UAS cost estimation. Future research should analyze UAS software development on past programs.

UAS software size and cost is very difficult to estimate. The software requirements, amount of reuse, programmer capability, level of software testing, documentation, and external flight-test organization approvals are major drivers. Software size data are typically kept within each UAS program and not published. The software is estimated anew for each program based on software requirements.

There is generally some relationship between autopilot weight and software size, though it has not been quantified. For example, a simple miniature autopilot probably won't run a 500,000-SLOC operational flight program.

19.8 Production Cost

UA cost is often described in terms of the unit flyaway price (UFP). The UFP is the price to the customer of everything that leaves the ground with the exception of fuel. This includes the airframe, avionics, communication systems, and payload suite. As a rule of thumb, the airframe and its major subsystems may constitute approximately 25% of the unit flyaway price. The avionics, including the communication system, is another 25%. The payload takes up the remaining 50%.

Recall that the UA empty weight includes the airframe, subsystems, propulsion, and communications. It encompasses all but the payload and fuel weight. The OSD UAV roadmap estimates that the cost per pound of empty weight is $1,500/lb in $FY04. Using this model, the UA cost without payloads is

$$C_{AV} = \$1,500 \cdot W_E \cdot CEF \tag{19.34}$$

Technomics [10] developed more detailed methods, which are presented here in a modified format to be consistent with other estimating relationships. The original source should be consulted for assumptions, data sources, and statistical correlations. The cost of the first UA in 2003 dollars is given by

$$C_{AV1} = \$118,750 \cdot (E_{Max} \cdot W_{PL})^{0.587} \cdot \exp[-0.010 \cdot (FF_{Year} - 1900)]$$
$$\cdot f_{Prod} \cdot CEF \tag{19.35}$$

where FF_{Year} is the year of first flight and E_{Max} is the maximum endurance in hours. Here f_{Prod} is 0.3981 if it is a production unit and 1 if it is a development or demonstration unit. Note that the UA cost estimate is based on the payload capacity and endurance rather than the more traditional empty weight. Technomics [10] offers an alternative method based on the maximum takeoff gross weight W_{TO}:

$$C_{AV1} = \$12,550 \cdot W_{TO}^{0.749} \cdot f_{Prod} \cdot CEF \tag{19.36}$$

Here, f_{Pod} is 1.379 if it is a production unit and 1 is a development or demonstration unit.

The payload cost is about as much as aircraft for appropriate system balance. High-cost payloads make UASs nonexpendable. EO/IR balls tend to be one of the most expensive elements on the UA. Typical sensors cost between $8,000–15,000/lb in 2010 dollars. The sensors are expensive, and a new integrated payload involves substantial development. The UAV Roadmap [5] estimates that payloads of all types cost $8,000/lb in

$FY04. This can be expressed as

$$C_{\text{PL}} = \$8000 \cdot W_{\text{PL}} \cdot CEF \tag{19.37}$$

Technomics [10] provides a cost-estimating algorithm for UAS EO/IR payload cost $C_{\text{EO/IR}}$.

$$C_{\text{EO/IR}} = \$290 \cdot 10^9 \cdot \alpha_{\text{avg}}^{-0.830} \cdot \exp[-0.169 \cdot (FU_{\text{Year}} - 1900)] \cdot f_{\text{Tracking}} \cdot CEF \tag{19.38}$$

where α is the average resolution of the optical and infrared sensor in microradians and FU_{Year} is the year in which the first unit is used on the intended application. The tracking system factor f_{Tracking} is 1 if there is no tracking system and 6.23 if there is a tracking system. Here a tracking system might include a moving target indication system or laser rangefinder or designator.

Technomics [10] also provides a cost estimate for ground systems and equipment. This covers all ground equipment including the GCS, ground-based communications equipment, aircraft launch and recovery equipment (ALRE), vehicles, shelters, and support equipment. The cost of the ground segment C_{Ground} in 2003 dollars is

$$C_{\text{Ground}} = \$433{,}400 \cdot R_{\text{Comms}}^{0.507} \cdot f_{\text{Mobile}} \cdot f_{\text{Hand}} \cdot CEF \tag{19.39}$$

where R_{Comms} is the typical range between the GCS and UA in nautical miles. The mobility factor f_{Mobile} is 1.49 if the system is tactically mobile and 1 if not. The factor for hand-carried systems f_{Hand} is 0.706 if the system can be carried by hand and 1 if not.

Manufacturing Overview
Now let us consider more detailed manufacturing cost estimation. The manufacturing cost is

$$C_{\text{Manf}} = C_{\text{AV,Des,Manf}} + C_{\text{HW,Manf}} + C_{\text{FTO,Manf}} + C_{\text{Fin,Manf}} \tag{19.40}$$

Manufacturing Design
The hours of AV design are found using the same methods as for RDTE, except the number of aircraft in the program is applied rather than the number of flight-test UA. The cost of the UA design is then found by

$$C_{\text{AV,Des,Manf}} = HR_{\text{AV,Engr,Manf}} \cdot R_{\text{Engr}} - C_{\text{AV,Des,RDTE}} \tag{19.41}$$

Manufacturing Hardware

The cost of the hardware for the manufacturing is

$$C_{HW,Manf} = C_{AV,Manf} + C_{GS,Manf} + C_{ALRE,Manf} + C_{Supt,Manf} \qquad (19.42)$$

The cost of the UA during manufacturing is

$$C_{AV,Manf} = C_{AV,Sys,Manf} + C_{Manf,Manf} + C_{Matl,Manf} + C_{Tool,Manf}$$
$$+ C_{QC,Manf} \qquad (19.43)$$

The UA systems cost for the manufacturing phase is found using the same methods described for the RDT&E phase prototypes, with the quantities equal to the number of production standard aircraft. The manufacturing hours are found using the same methods as described for RDT&E, with the number of aircraft adjusted to that of the program. The aircraft manufacturing cost during the manufacturing phase is

$$C_{Manf,Manf} = HR_{Manf,Manf} \cdot R_{Manf} - C_{Manf,RDTE} \qquad (19.44)$$

The materials cost for manufacturing phase is found using the same methods for RDT&E, using the total program quantities of UA and subtracting the cost of the RDT&E phase materials. Similarly, the tooling costs are found using the same methods as for the RDT&E phase, but using the total program UA quantities and subtracting the RDT&E contribution. The quality control hours use the same methods as the RDT&E phase, but use the manufacturing cost during the manufacturing phase as the reference.

The ground-station hardware cost during manufacturing is

$$C_{GS,Manf} = \sum CP_{Type} \cdot N_{Type/Sys} \cdot N_{Sys} \qquad (19.45)$$

where CP_{Type} is the cost per ground station type, $N_{Type/Sys}$ is the number of that type per system, and N_{Sys} is the number of systems procured. The MCE costs are not necessarily tied directly to the number of systems because a single MCE could potentially control the aircraft from more than one system. MCEs may be fractionally allocated to systems.

Manufacturing Flight-Test Operations

The flight-test operations cost during manufacturing follows the references to the cost of the RDT&E flight-test costs, except the emphasis is on acceptance flights of production systems vs testing flight qualities of the design. Depending on the UA type, the acceptance flight duration could be less than 1 hr or perhaps up to 100 hrs. The overall flight-test rate per flight hour from the RDT&E phase is multiplied by the number of acceptance test flight hours per aircraft and the number of production standard aircraft to determine the manufacturing flight-test operations costs.

Manufacturing Finance

The finance costs during manufacturing may use the same finance rate as during RDT&E.

19.9 Operations Cost

Operations Overview

The total cost of operations is

$$
\begin{aligned}
C_{\text{Ops}} = {} & C_{\text{POL}} + C_{\text{Pers,Dir}} + C_{\text{Pers,Ind}} + C_{\text{ConMatl}} + C_{\text{Misc}} + C_{\text{BLOS}} \\
& + C_{\text{Deploy}} + C_{\text{Spares}} + C_{\text{Depot}} + C_{\text{Replace}} + C_{\text{Fee}}
\end{aligned}
\tag{19.46}
$$

Operations Fuel, Oil, and Lubricants

The POL costs constitute a large portion of the operations costs. POL costs are found by multiplying the fuel cost per hour by the total flight time:

$$
C_{\text{POL}} = FH_{\text{Tot}} \cdot R_{\text{POL}} \cdot C_{\text{Vol}}
\tag{19.47}
$$

where FH_{Tot} is the total flight hours, R_{Fuel} is the average gallons/hr of fuel burned, and C_{Vol} is the cost/gallon of fuel at the operations site. The cost/gallon of fuel can fluctuate dramatically in short time periods. Fuel is generally much more expensive in forward-deployed locations that near peacetime population centers. The total flight hours include training flights as well as operational flights.

Operations Direct Labor

The direct labor for the operations is found by

$$
C_{\text{Pers,Dir}} = C_{\text{Maint,Pers,Dir}} + C_{\text{Operators,Dir}}
\tag{19.48}
$$

In a UAS deployment, the costs are driven by the number and type of personnel rather than required hours. Say that a maintainer is only required for one hour per day. That individual must deploy for the duration of the deployment even if all but one hour of the shift is spent playing solitaire. In fact, two people will likely be required to fill that role because the maintainer will not work seven days a week. The number of a personnel class N_{Pers} can be estimated by

$$
N_{\text{Pers}} = Hr_{\text{Load}} \Big/ T_{\text{Shift}} \cdot \left(1 - \frac{D_{\text{Leave}} + D_{\text{Transit}}}{D_{\text{Deploy}}} - \frac{7 - D_{\text{On}}}{7} \right)
\tag{19.49}
$$

where
$$Hr_{\text{Load}} = \text{daily hours required for the personnel type}$$
$$T_{\text{Shift}} = \text{hours per shift (generally } 8-12 \text{ hrs/shift)}$$

D_{Leave} = number of days leave during the deployment (sick leave or rest and recuperation)

D_{Transit} = average transit time to the field in days

D_{Deploy} = duration of the deployment in days

D_{On} = number of days per week that the person will work

This requirement is rounded up to the next integer value because fractional people are not possible. Often cross-training personnel can cut down on lightly loaded individuals. This reduces the quantity of deployed personnel.

Example 19.2 Mission Payload Operator (MPO) Role Staffing

Problem:

A UAS deployment will require an MPO to manage the payloads. The operation will cycle UA over the target area such that one is on station at all times. There is target coverage of 24 hours per day, 7 days a week. Only one MPO is required at any time. The labor rules are that the MPO can be on duty for 8 hours a day, 5 days a week. The MPO gets 15 days of leave during the deployment, including sick leave. The deployment personnel transit time is 3 days in each direction from the continental United States. The deployment lasts 1 year.

How many MPOs are required for this deployment?

Solution:

Solving the equation gives

$$N_{\text{Pers}} = Hr_{\text{Load}} \Big/ T_{\text{Shift}} \cdot \left(1 - \frac{D_{\text{Leave}} + D_{\text{Transit}}}{D_{\text{Deploy}}} - \frac{7 - D_{\text{On}}}{7}\right)$$

$$= 24 \text{ hrs} \Big/ 8 \text{ hrs} \cdot \left(1 - \frac{15 \text{ days} + 6 \text{ days}}{365 \text{ days}} - \frac{7 - 5}{7}\right) = 4.56 \text{ people}$$

Rounding up to avoid fractional people yields 5 MPOs.

The daily hourly loading for maintainers is estimated by

$$Hr_{\text{Load,Maint}} = FH_{\text{Day}} \cdot MMH/FH \tag{19.50}$$

where FH_{Day} is the daily average flight hours and MMH/FH is the maintenance man-hours per flight hour.

The aircraft maintenance personnel direct labor cost is

$$C_{\text{Maint,Per,Dir}} = N_{\text{Maint}} \cdot D_{\text{Deploy}} \cdot R_{\text{Maint}} \tag{19.51}$$

where N_{Maint} is the number of maintainers and R_{Maint} is the maintenance personnel rate in dollars per day.

The cost of the operators depends upon the system architecture and staffing philosophy. As we saw in Chapter 16, the ground control station could consist of a back-packable laptop or multiple ground stations in a remote split ops arrangement. The personnel roles can include an UA operator, a mission payload operator, mission director, intelligence analysts, or other categories. The rates for each of these categories vary. The general equation for operator cost is

$$C_{\text{Operators,Dir}} = N_{\text{Operators}} \cdot D_{\text{Deploy}} \cdot R_{\text{Operators}} \qquad (19.52)$$

Here again, the labor rate for operators is in dollars per day.

The labor rates depend upon the personnel qualifications and the contract type. GSA has published approved rates for 2001–2006. These are shown in Table 19.6. These rates apply to contractor support of operations, as is common in services-based models. Military-operated systems will use uniformed personnel.

Engineering support to an operation can be considered direct cost. If a system malfunctions in the field and the correction is beyond the manuals and the deployed personnel training, then technical support is necessary. Usually this support comes from the prime contractor. The engineers are usually located at the manufacturer's facilities, though they can be temporarily deployed to fix difficult problems. Usually a level of engineering

Table 19.6   UAS Support Labor Rates by Category, 2001–2006 (11)

Labor category	GSA rate, $/hr
Operations director	126.47
Operations engineer	95.70
Mission manager	119.85
UA operator/pilot	89.90
UA maint. tech.	59.91
Sensor operator	77.90
Program manager	154.50
Project leader	125.86
Technical specialist III	128.52
Technical specialist II	95.72
Technical specialist I	73.33
Technical editor	90.39
Technical media specialist II	82.05
Technical media specialist I	49.68

support is provided in an operations and maintenance contract. A level of customer engineering support is possible for contractor-owned systems, such as military program of record UASs.

Operations Indirect Labor

The operations indirect labor includes administrative and programmatic activities. This can be estimated by

$$C_{\text{Pers,Ind}} = F_{\text{Pers,Ind}} \cdot C_{\text{Ops}} \tag{19.53}$$

where Roskam [1] suggests an indirect labor factor of 20% for manned aircraft.

Operations Consumable Materials

The consumable materials cost during operations is estimated by

$$C_{\text{ConMatl}} = Hr_{\text{Flt,Tot}} \cdot MMH/FH \cdot R_{\text{ConMatl}} \tag{19.54}$$

The consumable materials rate R_{ConMatl} depends on the type of aircraft. Note that R_{ConMatl} has units of dollars per maintenance man-hour. This material is what is required to support the maintenance.

Example items include engine filters, spark plugs, replacement parts, and composite repair materials.

Operations Miscellaneous

The operations phase miscellaneous cost is estimated by

$$C_{\text{Misc}} = F_{\text{Misc}} \cdot C_{\text{Ops}} \tag{19.55}$$

where F_{Misc} is a miscellaneous factor.

Operations BLOS Communications

The BLOS satellite communications has the potential to become a significant cost driver. A cost model can estimate the BLOS costs based on hourly rates or cost of data throughput. The hourly rate-based BLOS costs are estimated by

$$C_{\text{BLOS}} = R_{\text{BLOS}} \cdot T \tag{19.56}$$

where T is the time for which the service is used. The data throughput-based BLOS costs are estimated by

$$C_{\text{BLOS}} = R_{\text{BLOS,Mb}} \cdot DR_{\text{BLOS}} \cdot T \tag{19.57}$$

where the $R_{\text{BLOS,Mb}}$ is the dollars per megabit of data transmitted. The time T differs between developmental flight testing and operations. For test flights, this time is

$$T = Hr_{FT,RDTE} \cdot F_{T,BLOS} \tag{19.58}$$

where $F_{T,BLOS}$ is the fraction of the total flight time used for satellite communications. Similarly, for operations

$$T = Hr_{Flt,Ops} \cdot F_{T,BLOS} \tag{19.59}$$

The rates for satellite communications depend upon the provider and the contract negotiations. Relatively low-bandwidth systems with tens or hundreds of kilobits per second may cost approximately \$60–500 per hour, depending on the service plan. Rates for high-bandwidth links are difficult to find. However, the rates can be expected to be higher than for low-bandwidth links. The negotiations for high-bandwidth satellite services for UAS operations are generally negotiated by the military.

Operations Deployment

The deployment costs include the cost of the cargo aircraft, system pack-out labor, site surveys, personnel travel, infrastructure upgrades, and system setup labor:

$$C_{Deploy} = C_{Cargo,Flt} + C_{Setup} + C_{Packout} \tag{19.60}$$

The cargo aircraft cost is found by

$$C_{Cargo,Flt} = R_{Cargo} \cdot T_{Cargo,Flt} \cdot N_{Cargo,Flt,Sys} \cdot N_{Sys,Deploy} \tag{19.61}$$

where

$$T_{Cargo} = N_{Hops} \cdot \left(\frac{Range_{Cargo}}{V_{Cargo}} + T_{L\&R,Cargo} \right) \tag{19.62}$$

and

$$N_{Hops} = Truncate \left(\frac{Dist_{Deploy}}{Range_{Cargo}} \right) + 1 \tag{19.63}$$

The number of cargo aircraft by type required for deployment for each of the systems is shown in Table 19.7.

Table 19.7 UAS Cargo Aircraft Requirements

System	C-130	C-17	C-5
Global Hawk [12]	—	1	1
Predator [12]	5	1	1
Shadow 200	3	1	1

The pack-out labor cost is found by

$$C_{\text{Packout}} = R_{\text{Pers}} \cdot N_{\text{Pers,Packout}} \cdot T_{\text{Packout}} \qquad (19.64)$$

Similarly, the setup time is found by

$$C_{\text{Setup}} = R_{\text{Pers}} \cdot N_{\text{Pers,Setup}} \cdot T_{\text{Setup}} \qquad (19.65)$$

Operations Spares
The operations phase spares cost is estimated by

$$C_{\text{Spares}} = F_{\text{Spares}} \cdot C_{\text{Ops}} \qquad (19.66)$$

where Roskam [1] suggests a spares factor of 12% for manned aircraft.

Operations Depot Maintenance
The operations phase depot maintenance cost is estimated by

$$C_{\text{Depot}} = F_{\text{Depot}} \cdot C_{\text{Ops}} \qquad (19.67)$$

where Roskam [1] suggests a depot maintenance factor of 13% for manned aircraft.

Operations UA Replacement
The aircraft replacement costs are found by estimating the number of aircraft that are lost during training and operations flights. The replacement cost is

$$C_{\text{Replace}} = N_{\text{AV,Loss}} \cdot \text{CP}_{\text{AV,Manf}} \qquad (19.68)$$

where $\text{CP}_{\text{AV,Manf}}$ is the cost per UA plus payloads. The number of UA lost is found by

$$N_{\text{AV,Loss}} = \left(N_{\text{AV,LossYr,Trn}} + N_{\text{AV,LossYr,Ops}} \right) \cdot N_{\text{Yr,Ops}} \qquad (19.69)$$

The number of aircraft lost in training flights per year is

$$N_{\text{AV,LossYr,Trn}} = \frac{LR_{\text{Accident}}}{100,000} \cdot Hr_{\text{Yr,Trn,Flt}} \qquad (19.70)$$

where LR_{Accident} is the loss rate due to accidents. Similarly, the number of aircraft lost in operational flights in wartime per year is

$$N_{\text{AV,LossYr,Ops}} = \frac{(LR_{\text{Accident}} + LR_{\text{Kill}})}{100,000} \cdot Hr_{\text{Yr,Ops,Flt}} \qquad (19.71)$$

where LR_{Kill} is the combat loss rate due to hostile actions such as enemy fighters, antiaircraft systems, or even friendly fire. Chapter 17 provides information on UAS loss rates that can be used as reference.

Other Operations Costs

The operations costs listed may not be comprehensive. Other potential costs include the following:

* Training programs
* Military construction
* Facilities leases or construction
* Food, potable water, and personnel accommodation
* Security services
* IT support

19.10 Commercial Development

Many of the widely fielded UAS systems were initially developed with company funds rather than government contracts. When a company self-funds a new product, this is known as *commercial development*. This model is quite similar to that used by commercial aviation. The company must assess market demand and hopefully secure launch customers that will buy the product once operational. The commercial development model has almost become the norm for UASs when the government requires a technology readiness level of 6–7 to qualify for a bid. The government does not offer to support the development of multiple UASs, but expects the companies to reach this level of maturity through their own means.

19.11 Services Contracts

Services contracts are becoming a common means of fielding UASs. The customer pays by the flight hour or similar metric, but does not acquire the system. UAS operations are performed by the contractor, who provides all required personnel such as the AVO, MPO, and maintainers. The contractor must appropriately include the cost of the system, spares, labor, and consumables into the negotiated rate per flight hour. The burden of ensuring that the system is reliable and supportable falls upon the service provider and not the customer, which greatly reduces customer risk. The level of investment is much lower for the customer when compared to a program of record development. The customer is also likely to have access to the latest technologies and capabilities.

An example of a cost difference for a services contract vs a military-operated system is hazard insurance. Contractor operators generally require hazard insurance when they are put in hazardous situations such as extreme environments or war zones. Insurance may be a net advantage for UAS operations relative to manned aircraft. Humans on the aircraft are at greater risk than humans on the ground. High insurance rates for manned aircraft may help justify the use of UASs for services contracts.

19.12 UAS Company Startups and Investments

Investors and entrepreneurs who are reading this chapter are likely motivated to create new products that will penetrate the multibillion-dollar UAS market. There are notable examples of small UAS companies that became major forces in the industry such as AAI, AeroVironment, and Insitu. However, if company founders and investors fully appreciated the risks and their chances of success, there probably would not be many new UAS companies.

The defense budgets for UASs have grown annually for several years. A startup company can make a compelling argument that their product will capture a market segment within 5–10 years. This is possible, as can be seen by notable success stories. The author was fortunate enough to be at Insitu during an exciting growth period that saw the ISR services market filled with ScanEagles and the Integrator came into existence.

But for every success, there are many more failures. Every year at UAS trade shows a handful of new companies will debut. Although it is exciting to see new companies and their products, if you think carefully you will notice that about the same number of companies is missing. Other small companies move their display models to the booth of a larger company and change the logos painted on the tails as the result of acquisitions.

Investors provide venture capital to a company in return for a portion of the company ownership in the form of stock shares. The return on the investment will come in the form of stock dividends or stock sales. Generally, outside investors are interested in selling the stock after they have grown in value in what is known as an *equity event*. This usually takes the form of a company sale or initial public offering (IPO). Most UAS companies are bought out by larger companies, though AeroVironment (stock ticker AVAV) successfully underwent an IPO in 2007.

Large defense contractors are aggressive about acquiring small UAS companies. A defense contractor can achieve only so much prestige without a UAS capability. It is often faster and easier to acquire a company rather than grow a UAS capability organically. However, it is difficult to make a profit unless the acquired capabilities and systems are appropriate for the market. The parent company may be able to add depth of program management, systems engineering, mission systems, and key technologies. The corporate culture of the larger company may be an uncomfortable fit for the more entrepreneurial UAS company, which can stifle development of new products. The greatest value of the UAS company is usually its people. When properly managed, large companies can create an impressive UAS portfolio through acquisitions of smaller companies.

References

[1] Roskam, J., *Airplane Design, Part VIII: Airplane Cost Estimation: Design, Development, Manufacturing, and Operating*, Roskam Aviation and Engineering Corp., Ottawa, KS, 1990.

[2] *Unmanned Aerial Vehicles: DOD's Acquisition Efforts*, General Accounting Office (GAO), GAO/T-NSIAD-97-138, Washington, D.C., April 1997.

[3] *Unmanned Aerial Vehicles: DOD's Demonstration Approach Has Improved Project Outcomes*, General Accounting Office (GAO), GAO/NSIAD-99-33, Washington, D.C., 1999.

[4] Program Acquisition Costs by Weapon System, Department of Defense Budget for Fiscal Year 2007, Dept. of Defense, Washington, D.C., Feb. 2006.

[5] *Unmanned Aircraft Systems Roadmap, 2005-2030*, Office of the Secretary of Defense (OSD), Washington, D.C., 2005.

[6] Drezner, J. A., and Leonard, R. S., *Global Hawk and Darkstar*, RAND Corp., Santa Monica, CA, Vol. 1, 2002, pp. 36, 37.

[7] Armstrong, K., *Choosing Your Homebuilt, The One You'll Finish and Fly*, Butterfield Press, Templeton, CA, 1993.

[8] Boehm, B., *Software Engineering Economics*, Prentice – Hall, Englewood Cliffs, NJ, 1981.

[9] Jalote, P., *An Integrated Approach to Software Engineering*, Springer-Verlag, New York, 1991, pp. 84–126.

[10] Cherwonik, J., "Unmanned Aerial Vehicle System Acquisition Cost Estimating Methodology," Technomics, 37th DOD Cost Analysis Symposium, Feb. 2004, http://209.48.244.135/DODCAS%20Archives/37th%20DODCAS%20(2004)/Intermediate/UAVArmyforweb.pdf.

[11] Authorized Federal Supply Schedule Pricelist for Unmanned Aerial Vehicle Systems and Related Services, Contract Number: GS-24F-0028L, 2001 – 2006, GSA Federal Supply Service, Adroit Systems, Inc., Alexandria, VA.

[12] Jones, C.A., *Unmanned Aerial Vehicles (UAVs), An Assessment of Historical Operations and Future Possibilities*, Air Command and Staff College, AU/ACSC/0230D/97-03, Maxwell AFB, AL, 1997, pp. 31, 39, 42.

Problems

19.1 Use the consumer price index to create a cost escalation factor function. What is the cost escalation from 1988 to 2007?

19.2 Using data from Table 19.1, generate a CER for system development cost as a function of UA takeoff gross weight.

19.3 Using the data from Table 19.2, estimate the average procurement cost per UA. Note that the procurement costs include more than just the UA. Make an assumption about the fraction of the system cost for ground and support elements.

19.4 Use a spreadsheet to develop a resource-loaded schedule for a wind-tunnel test. The UA is a fixed-wing small UAS. The purpose is to estimate drag. Consider the number of people who are required, their labor rates, how the rates are burdened, wind-tunnel model fabrication, and facility costs.

19.5 An UA has an empty weight of 5,000 lb and a payload weight of 1,000 lb. What is the expected unit flyaway price in production in 2004 dollars?

19.6 A UAS deployment will require maintenance personnel to support the UA. The operation will cycle UA over the target area such that one is on station at all times. There is 24 hrs per day of target coverage, 7 days a week. The average maintenance man-hours per flight hour are 4 hrs/hr. The labor rules are that the maintainer can be on duty for 12 hrs a day, 4 days a week. The maintainer gets 20 days of leave during the deployment, including sick leave. The deployment personnel transit time is 4 days in each direction from the continental United States. The deployment lasts 2 years. How many maintainers are required for this deployment? The contractor-employed maintainer burdened rate is $70/hr, assuming a 1,800-hr work year. What is the total cost of this labor category for the operation?

19.7 Demonstration project. Follow all applicable laws and safety precautions. (This might be best performed as a team project.) Build a radio-controlled model aircraft from a kit. Document the hours spent building the structure, propulsion system integration, control system integration, finishing, simulation training, and for first flight activities. As a bonus, build a second aircraft kit of the same type. Compare the hours for the second model.

Chapter 20

Product Definition and Requirements Development

- Learn the methods for creating competitive products
- Understand technology impacts on achievable requirements
- Learn how to establish balance in system capabilities and requirements

A Boeing operator removes a ScanEagle from the recovery system (USMC photo by Sgt. Guadalupe M. Deanda, III).

20.1   Introduction

S ystem development involves activities from both the customer and the system provider. The system capabilities may be in direct response to customer requirements or defined by a company that anticipates customer needs. The product design of the system invariably is controlled by the developing company or organization. A system is only successful if its capabilities and the customer interests overlap simultaneously, at the right price, and with acceptable risk.

20.2   Market Surveys and Competitive Analysis

Whether you are a customer defining a competitive acquisition or a system developer trying to create a product offering, market surveys and competitive analyses are essential for understanding the competitive landscape. Of the approximately 1300+ historical UAS efforts, only a small number are currently active programs, and a subset of these are relevant to any particular acquisition. New contenders can arise suddenly, and long-established entities exit the market. Tracking the competitive environment is a dynamic endeavor.

Several key capability attributes can be used to help narrow the search. Not all of these factors are relevant for all UAS initiatives, but Table 20.1 is a suitable starting point.

When considering the potential competitors, the current capabilities of these systems as well as potential capability growth should be evaluated. For example, a UAS that is within 25% of an endurance requirement can be upgraded to include additional fuel capacity, wing extensions, or aerodynamic refinement. Propulsion fuel types can be modified from gasoline to heavy fuel through an engine upgrade program.

The market research data can come from a variety of sources. These may include the following:

* *Direct interaction with the UAS provider*—government inquiries of industry, or industry–industry collaborations that are usually covered under proprietary information agreements
* *Company press releases and marketing material*–This is information that the company provides for public consumption. Press releases tend to increase leading up to a competition as company-funded capability demonstrations attempt to entice the customers.
* *Media*—Periodicals such as newspapers, aviation magazines, websites, and television news can provide information about systems of interest.
* *Conference papers*—Technical presentations at professional conferences may include a surprising level of technical detail.

Table 20.1 Discriminating UAS Attributes

Attribute	Discussion
Maximum endurance	Endurance is generally specified in requirements. UA with 1-, 5-, or 12-hrs maximum endurance are not within the competitive range for a 24-hrs endurance requirement. Surveys of electric propulsion will seldom show mission-capable UA with more than 3-hrs endurance.
Maximum payload capacity	Requirements will state the payload capacity or integrated payload performance. Systems capable of accommodating suitable payloads should be included in the survey.
Launch and recovery method	Launch and recovery can be thought of as conventional takeoff and landing, runway independent, hand launched, air launched, and VTOL. Some requirements such as ship compatibility can include both VTOL and other runway independent techniques.
Transportability and footprint	Transportability includes modes of transportation such as ground, cargo aircraft, or man-portable techniques.
Propulsion class	Propulsion types can be characterized by the fuel type. Fuels can include heavy fuel or other liquid hydrocarbons such as pump gas and hydrogen. Other propulsion types include solar power and batteries. Many requirements explicitly state the fuel type or energy source so that the systems fit within the established logistics operations.
Low observables	What degree of radar, acoustic, visual, and IR signatures do the systems have? The largest differentiator is often radar observability.
Maximum altitude	This is the service ceiling that the UA can reach. Systems are grouped into low altitude ($< 10,000$ ft), medium altitude ($15,000-30,000$ ft), and high altitude ($> 50,000$ ft).
Maximum airspeed	The maximum dash speed or maximum cruise Mach number.
Communications	The discriminator is line of sight only and satellite communications (SATCOM). Most UASs are capable of low-bandwidth SATCOM, and so the discrimination threshold is high-bandwidth SATCOM.
Maturity	Is the system a paper design, a prototype, or operational?

- *Books*—Sources such as *Jane's* [1] provide surveys of numerous types, though the two-year publication cycle often gives dated information.
- *UAS market research organizations*—Forecast International and other companies sell detailed UAS market analysis that includes forecasts and system profiles.
- *Government-provided UAS resources*—The OSD Unmanned Systems Roadmap [2], Defense Sciences Board studies, and other government studies provide a wealth of data. Freedom of Information Act (FOIA) requests can provide useful contract information.

UAS marketing materials are notorious for providing confusing data that can create difficulty in ascertaining actual system capabilities. For example, the maximum payload and maximum endurance values are typically provided. However, these two parameters are not achieved simultaneously. Marketing materials are not suitable data sources for engineering analysis, but this literature might be all that is available.

A plot of competitor's payload weight vs endurance is particularly useful for ISR systems competitive analysis. This can be crudely estimated with a very small amount of information and enhanced as more information becomes available. Some of the methods described in Chapter 3 are suitable. The required parameters are average fuel burn rate, maximum takeoff gross weight, and empty weight. Of these, the fuel burn rate is the most difficult to estimate if it is not provided. If the maximum fuel capacity is given, then the average fuel burn rate is approximated by

$$\dot{W}_f \approx \frac{W_{\text{fuel,max}}}{E_{\text{max}}} \tag{20.1}$$

If the maximum fuel weight is unknown, it can be assumed that the fuel-tank capacity is sized such that it can carry the entire useful load, and the entire payload or fuel is located near the required center of gravity, and then the maximum fuel weight is approximated by

$$W_{\text{fuel,max}} \approx W_{\text{TO}} - W_E \quad \text{or} \quad W_{\text{fuel,max}} \approx W_{\text{PL,max}} \tag{20.2}$$

If the UA has a nose-mounted payload that is required to maintain center-of-gravity limits, but the tank is sized such that a full tank plus nose payload equals the useful load, then the maximum fuel weight can be approximated by

$$W_{\text{fuel,max}} \approx W_{\text{TO}} - W_E - W_{\text{PL,nose}} \quad \text{or}$$
$$W_{\text{fuel,max}} \approx W_{\text{PL,max}} - W_{\text{PL,nose}} \tag{20.3}$$

Often sufficient information is available about the UA to estimate the fuel burn based on aerodynamic and propulsion parameters. The fuel burn rate for a jet is

$$\dot{W}_f \approx \frac{W_{\text{avg}}}{L/D} \cdot TSFC \tag{20.4}$$

where W_{avg} is the average weight of the mission, L/D is the lift to drag ratio at the loiter condition, and $TSFC$ is the thrust specific fuel consumption at loiter. Similarly, the fuel burn rate for a propeller aircraft is estimated by

$$\dot{W}_f \approx \frac{W_{\text{avg}}}{\eta_p \cdot L/D} \cdot V \cdot BSFC \tag{20.5}$$

where V is the loiter velocity, η_p is the propeller efficiency, and $BSFC$ is the brake specific fuel consumption at the loiter condition.

Now the relationship between payload weight and endurance can be generated. The payload weight as a function of fuel weight is

$$W_{\mathrm{PL}} = W_{\mathrm{TO}} - W_E - W_f \tag{20.6}$$

The endurance is

$$E = f \cdot \frac{W_f}{\dot{W}_f} \tag{20.7}$$

where f is a factor that accounts for higher fuel burn rate activities other than loiter and where f is greater or equal to 1.

20.3 Customer Requirements Generation

Customer requirements generation is an art that balances user mission needs, historical requirements precedents, system affordability, fielding timelines, industry capabilities, and technology assumptions. There are usually numerous stakeholders that may have disparate interests. The seeds of many cancelled programs were sown in generation of unbalanced requirements. Creating effective requirements is a challenging endeavor that requires diplomacy (political savvy), systems engineering skill, and engineering knowledge of the applicable UAS technologies.

Early requirements analysis is a best practice to determine the feasibility and suitability of the requirements. These studies can be performed by a customer engineering team or by the industry teams that are likely to offer system solutions. Requirements should adapt to reflect the outcomes of such studies. Those performing the analysis should be knowledgeable about the available technologies to yield meaningful results. Ideally, there must be an understood implementation path for every requirement.

Good requirements reflect a useful capability that meets customer needs and the ability of industry to develop the system. The requirements content will largely determine the number of bidders. Lowering the bar or making the requirements more generic will increase the number of competitors but might yield less useful capabilities for the customer. Making requirements stressing, narrowly defined, or reflecting specialized advanced technologies will reduce the number of viable competitors or could lead to a sole-source contract. When a very narrowly defined requirement that matches only one system is released from the customer, potential competitors might believe the competition is "wired" for one team and often choose to not bid.

An example of a flexible UAS requirement is to specify the duration of target coverage without specifying the endurance of the individual UA. This gives the system provider the flexibility of offering a single large UA with sufficient endurance to cover the target or having multiple UA with shorter endurance provide the coverage in multiple sorties. The risk that the customer faces is that the offerings might have less capability than is desired. In this example, the end customers may strongly prefer a single UA that has 26-hrs endurance to provide a 24-hrs combat air patrol rather than 12 UA with 5-hrs endurance to provide the same coverage (assumes 1-hr ingress and egress segments). If the RFP is written such that the 5-hr endurance UA best meets the source selection criteria, the customer will be the proud owner of an UA that spends 40% of its operational flying time in ingress or egress, whereas the operators perform frequent launch and recovery cycles.

A common mistake is to create infeasible requirements or requirements that require an excessive technology push to reflect stated or perceived user needs. Such requirements can cause a program failure due to cost overruns, schedule slips, or unsatisfied requirements. The requirements developer and end user should consider the possible outcome of having no new capability due to a cancellation vs modifying the requirements to be more achievable. The Lockheed Aquila program is an example of requirements that did not match available technology. The program ended with large cost growth and schedule delays. It is simply inappropriate to emphasize user needs without a viable implementation path; this risks producing nothing when something useful could have been provided instead.

Other programs can't get started because the requirements process does not converge. Some requirements efforts take many years and can be terminated without a program start. It is often said in aerospace that "better is the enemy of good enough." A 90% agreed-upon requirement set that goes to acquisition with expediency is preferable to a 100% requirement that never goes to procurement.

A single acquisition can serve the needs of multiple customers. Joint acquisitions address the needs of more than one U.S. DOD service. Combining systems into a single common acquisition has potential to save money. At the same time the cost and complexity of the system is greater than if a single customer is addressed.

A notable example of the risks associated with joint procurements is Alliant Techsystems Outrider program. This program began as an Army tactical battlefield system, but later became a joint program with the U.S. Navy. The Navy imposed more stringent EMI requirements on the UA to enable operations in the vicinity of shipboard radars. The original lightweight composite fuselage was then changed to a heavier aluminum structure to provide the necessary shielding. This weight growth necessitated a larger engine. The nonvirtuous weight spiral resulted in failure to meet key requirements, cost growth, and then program cancellation. Later, the Army developed the AAI Shadow 200 system for its use exclusively. The U.S. Marine Corps later procured the Shadow system as well.

Every requirement has a cost-bearing impact on the system. The requirements developer should consider the affordability implications of every word. Much of the system life-cycle cost is locked in during requirements generation.

Requirements tradition, legacy, and precedent play a role in requirements development. Requirements organizations accustomed to development of manned aircraft may simply apply requirements language to the UA without much consideration of the system attributes. The requirements document may state that the system must comply with a host of external documents such as MIL-STD, MIL-HDBK, JSSEG, FAR, and other documentation without clarification or caveats. These documents can be hundreds of pages each and might be inapplicable to the unmanned system. Requirements developers should take on the responsibility of parsing the large requirements documents to only state the relevant set. For example, program office and contractor teams should not waste time and budget evaluating ejection seat or airborne pilot interface requirements for an unmanned system. A sense of proportionality should be established to reflect the nature of the system. A UCAV or strategic HALE UAS might have requirements that more closely approximate those of a manned aircraft than would be the case of a hand-launched UAS. The outcome of excessive requirements is excessive system unit cost, excessive system engineering resources, and schedule delays due to requirements modifications when common sense is finally established. Voluminous requirements will also discourage small, innovative companies from bidding due to the large investment required to write a competitive proposal.

The number of contributors to the requirements has a direct effect on the scale and scope of the requirements. A small, skilled, and empowered requirements group will tend to generate more streamlined requirements than a building full of engineers. The streamlined requirement set will likely yield a lower cost system with more chance of program success. On the other hand, a requirements team that is insufficiently staffed or that does not possess the necessary expertise will likely generate poor requirements. Clearly, there is a balance in the requirements team size.

The strategy for the system technology level must be understood and balanced against the budget and development timeline. A rapid acquisition or quick reaction capability will likely incorporate off-the-shelf technologies. Long-term development efforts with large budgets may consider significant technology investments. Research organizations such as DARPA might be required to pursue only high-risk, high-payoff technologies. The purpose of demonstrator aircraft such as X-planes is to prove enabling technologies for future systems. Highly challenging requirements can discourage potential competitors or possibly lead to unrealistic technical optimism in the proposals. This optimism can lead to poor program outcomes when realism inevitably sets in.

All requirements must be verified. The major forms of requirements verification are testing, inspection, and analysis. All forms of verification consume resources, but testing is often the most expensive and time consuming. Any requirements that are not verified during development must be rectified.

Government programs for operational UASs usually come in two forms: programs of record (POR) and services. Of the thousands of UAS initiatives, only a small handful ever achieve program of record status. The U.S. programs of record since 1960 are shown in Table 20.2.

The customer can also use a services model for deploying UAS. The contractor operates the system but does not deliver it to the customer. A benefit to the customers is that they do not need to train their personnel or affect their organization in other ways. The latest systems can be used without the time or expense of a POR acquisition. A customer can procure services under operations and maintenance (O&M) budgets, which may be more available than acquisition budgets. The customer can give the contractor system performance requirements that must be satisfied. The most successful ISR services effort to date has been the Insitu ScanEagle system, which deployed with several customers.

20.4 Developer Product Definition

Successful product definition is a challenging endeavor. The probability of penetrating the market and achieving widespread fielding is low indeed. A company-developed UAS must oust an established system, win a competitive acquisition, win a services contract, or obtain a sole-source contract. None of these paths are easy. But once a system is fielded and accepted by the operator community, many years of production are likely to follow.

First, the company must understand its potential customers. The best way to convince a customer to procure the systems or services is to offer a capability that is not available elsewhere. Successful companies anticipate customer needs ahead of the customer. Merely responding to a customer RFP as the first contact means that the company is behind the competition. The company should communicate the system capabilities to future customers early in the process so that the final requirements reflect those

Table 20.2 U.S. Program of Record UAS Systems Since 1960

System	Designation	Service dates	Customers	Notes
Gyrodyne QH-50	QH-50	1960s	USN	
Teledyne Ryan Firebee	A/BQM-34	1960–	USAF	Target, decoy, and reconnaissance
Northrop Grumman Chukar	BQM-74E	1990s–	USAF	Target and decoy
AeroVironment Pointer	FQM-151	1988–2000s	USA, USMC	Surveillance
General Atomics Predator A	MQ-1A	1995–	USAF	Accelerated fielding
Pioneer UAV, Inc. Pioneer	RQ-2	1986–late 2000s	USN, USMC	Land- and ship-based operations
Lockheed/Boeing DarkStar	RQ-3	——	USAF	Cancelled before operational
Northrop Grumman Global Hawk	RQ-4A/B	2001–	USAF	Accelerated fielding in 2001
Northrop Grumman Hunter	RQ-5A	1999–2000s	USA	Program cancelled, but LRIP systems saw service
Alliant Techsystems Outrider	RQ-6	——	USA	Cancelled before entering service
AAI Shadow 200	RQ-7A/B	2002–	USA, USMC	
Northrop Grumman Firescout	RQ-8	2010–	USN	
General Atomics Reaper	MQ-9	2001–	USAF	
MMIST Snowgoose	CQ-10	2000s	——	
AeroVironment Raven	RQ-11	2002–	USA	
General Atomics Gray Eagle	MQ-12A	——	USA	
AeroVironment DragonEye	RQ-14	——	USMC	Originally developed by NRL
DRS Neptune	RQ-15	——	——	Water landing
Honeywell T-Hawk	RQ-16	——	USA	FCS Class I
MTC SpyHawk	MQ-17	——	USMC	Cancelled Tier II demonstrator
Boeing Hummingbird	MQ-18	——	——	Unmanned helicopter
AAI Aerosonde	MQ-19	——	——	
Insitu Integrator	RQ-21A	Future	——	STUAS winner

capabilities. This improves the probability of winning the contract in a competitive source selection.

The current capabilities and future directions of the competition should be understood. Many companies will simultaneously forecast the market and position themselves competitively within a market segment. It is not enough to have a good system relative to the need; it is necessary to have the best system relative to the competition.

Remember that your competitors will improve their existing products, and some will develop new systems to meet emerging opportunities. It is unlikely that you are the only one to notice the capability need of interest to the customer. However, your company may be the only group to act upon it.

A winning system should have key discriminators relative to other systems. These discriminators can take the form of a design philosophy, performance metrics, enabling capabilities, or system maturity.

If the product provides a unique capability that cannot be easily countered by others, there is a possibility of a sole-source contract. Although rare, such opportunities are enabled by wise technology investments, design point selection, and market vision. The best way for a contractor to win a competition is to not have a competition occur at all. Government competition advocates discourage sole-source contracts because, by definition, these contract awards discourage competition. However, the possibility of such opportunities encourages industry to invest without initial government funding to address government needs.

Unlike a manned military aircraft acquisition such as a fighter, tanker, or trainer, it is possible for small companies to successfully develop UAS products that achieve widespread military sales. Such an endeavor still requires capital, but two entrepreneurs starting a company in a garage with some angel investors can beat aerospace giants. The agility and efficiency of small companies can be a major advantage. There are several major success stories such as AAI, BAI, Aerovironment, and Insitu, but most small companies are not so fortunate.

An effective way of penetrating the market is to pursue an unfulfilled capability. These open market segments should provide real utility, as not all system design points are useful. In some cases payload, propulsion, or avionics technologies may make sudden strides that enable a new UA class that was not possible before. A good example is the Insitu ScanEagle system, which fulfilled a market niche for small ship-based unmanned operations made possible through miniaturized subsystems and the enabling SkyHook^{TM} retrieval system. Highly useful and unique capabilities can quickly shape the market.

A company may wish to penetrate a crowded market, but has no suitable legacy product that can be leveraged. If the market is already populated with flying systems, the lack of system maturity is a major weakness. It is unlikely that a product will win a competition if the system utility is on par with the competitors, but with lower maturity. The new system capabilities or affordability must be much better than that of the competition in order to overcome the higher development risk perception.

The new product should leverage strengths and be within the company's ability to develop to the point that it will be picked up by a customer. Companies will usually have key patents, trade secrets, analysis capabilities, or experience that can be leveraged. Additionally, the financial and personnel resources must be made available to perform the development.

There is a downside to leveraging company intellectual property if it does not generate a good design. Often a company will be wed to a particular unconventional UA configuration. A quick survey of widely fielded systems will reveal that nearly all successful products are relatively conventional and unremarkable in appearance. Although a company may have expertise in a novel configuration, it is worth considering other design alternatives to generate a competitive solution. Unusual designs can create the perception of risk to those responsible for acquisition decisions.

Ideally, UA should not appear too common either. The vast majority of tactical systems use a conventional twin-boom pusher configuration, as described in Chapter 4. Yet another UA of this layout will be difficult to distinguish, especially if the performance is not particularly noteworthy. Often clever fuselage functional design, tail aesthetics, and overall stylistic details can make a product stand out.

Established companies may have legacy systems that can be leveraged in product offerings. This provides both opportunities and risks. Existing systems may form the basis for a modified system that can be demonstrated rapidly through reuse of key elements. Showing a direct lineage to a successful product is a powerful way to demonstrate product maturity. However, the legacy system might impose system limitations that would not exist for a new design. These limitations create competitive disadvantages.

Adapting an old system might require more investment than starting over if the modifications are extensive. Say a company wishes to modify a 6-hr endurance platform with conventional landing gear to a 24-hr endurance mission with runway independence. The provisions for runway independence (i.e., pneumatic launcher and net) may necessitate configuration and structural changes. The longer endurance will demand enhancements in aerodynamics, fuel mass fraction, and/or propulsion efficiency. Aerodynamic enhancements could take the form of a wing-span increase for higher aspect ratio or general parasite drag reduction. The fuel mass fraction can be increased by adding fuel volume and finding weight reduction opportunities. Added fuel weight might require additional structural modifications. Propulsion efficiency improvements can be achieved through changing the engine (i.e., four-stroke instead of two-stroke engines, or switching to heavy fuel from gas), changing the propeller, reducing the cooling drag, or adding advanced engine controls. All of this activity might be more work than simply starting over with a clean sheet design.

Another advantage of leveraging fielded products is that the brand might be sufficient to win contracts, even if the system has been extensively modified or even completely redesigned. The product name and general appearance of an established design can be carried over to a clean sheet design. Manned aircraft analogies include the Boeing F/A-18 E/F vs the F/A-18 C/D, or the Lockheed Martin U-2 S vs the U-2C. In these examples the aircraft look the same and have the same name, but the newer designs are quite different in terms of weight class, size, engines, avionics, and performance capabilities.

A product family is frequently a successful strategy. The family may use a similar configuration but at different scales, payload capacity, and flight performance capabilities. Common avionics, software, ground elements, and communications architectures can spread much of the system development cost across multiple UA. One of the products in the family is likely to match an emerging program requirement. Smaller UA members of the family can be developed first to reduce initial development costs. The General Atomics family includes the Gnat, Predator A, Reaper (originally called Predator B), Gray Eagle, and Avenger (originally called Predator C). BAI created the Viking family of UAS that includes the 100, 300, and 400. The Viking 400 won a large SOCOM contract.

Program-of-record systems tend to achieve a configuration stasis and do not adapt quickly to new technologies. Oftentimes these systems are in operations for many years after becoming obsolete. Despite the disparity in capabilities between new systems and legacy systems, it is difficult to supplant the established designs. One reason is that program-of-record systems represent an investment in doctrine, organization, training, maintenance, logistics, personnel, and facilities (DOTMLPF). Changing the paradigm requires a sizable investment by the government, and therefore the justification must be compelling. Large improvements in capability are necessary.

Obsolete or poor performing program-of-record systems can be replaced with either services contracts or new program of record systems. Services contracts are generally more accessible to new systems.

Technologies tend to follow an S curve. Initial investment brings about modest new capabilities. Then the rate of capability enhancement increases with time. Finally, the rate of improvement levels off, and further improvement comes only with great effort.

One could argue that manned aircraft are nearing the final stages of the S curve. The rate of airframe capability improvement is slowing, and each improvement increment demands larger investments and more time. Recent advancements include composites, jet-engine technologies, digital avionics, low observables technologies, and improved design analysis techniques such as CFD and FEM structural analysis. Many of these technologies are applied to current-generation manned aircraft. At the top of the S curve, the timeline driving the need for replacement manned aircraft will be longer. For example, variants of the C-130 and B-52 are still in service today despite being originally developed in the 1950s.

UASs are arguably at the middle of the S curve. Manned aircraft technologies have not fully migrated to unmanned aircraft. UAS technologies are still rapidly improving. These include payload, communications, processing, small propulsion systems, launch and recovery systems, and autonomy software. Improved technologies can create opportunities for new UAS designs. Companies that recognize and pursue these opportunities can be quite successful.

The difference between great and average UASs is often nuanced. Some remarkable systems appear to be inspired rather than designed. A pure systems engineering approach might not yield an exceptional system. Designs are often envisioned by a single designer, and then a larger team works to refine the concept.

The path to a widely fielded system is rarely direct. A system might be gradually modified over time to meet a variety of missions or research objectives. Funding may be sporadic and lead the project in unusual directions. A system might begin under investor funding, internal research and development, small business innovative research (SBIR), corporate partner sponsorship, or even as a student research project. The system can be used for payload testing or aerospace sciences research. Operations might alternate between military and civilian applications. An example is the Insitu ScanEagle system began as a commercial UAS intended to spot tuna for fishing vessels, but today it is widely fielded by the military.

All aspects of the system must be considered in product definition, not just the UA. Payload selection and performance can play just as significant a role as any UA attribute. The system software, communications, ground elements, and supportability provisions all play critical parts in product design.

The product leadership and design team should know the customer. Every customer is guided by biases, experience, perceptions, and personalities. The customer attributes may affect preferred technical approaches or even vendor selection. It is unwise to unnecessarily act counter to the customer preferences. This can be done without adverse consequences if the customer can be convinced of your approach.

Suppliers produce system components that are used by the UAS prime contractor. A supplier benefits from knowledge of the prime contractor systems as well as new government programs. Suppliers should anticipate emerging government programs and provide systems that will be used by the system integrators. UAS suppliers can provide avionics, communication systems, payloads, actuators, subsystems, software, shelters, support equipment, or virtually any component that may be purchased by others. Suppliers can also provide software tools that are used by development teams.

The objective of a supplier is to develop products that will be purchased by UAS system providers. A successful supplier will have their product as part of the winning UAS offering. The best way to achieve this objective is to have the products on as many UAS contenders as possible. The prime contractor, on the other hand, will likely wish to lock up the product if it yields a competitive advantage.

Consider a UAS actuator supplier. This company should have a deep understanding about UAS requirements that flow down to the actuators. These can include the EMI environment and interface standards. By interacting with system developers, they will understand the needs for servo bandwidth, reliability, torque curves, and operational environment. This valuable information can influence new actuator product development.

20.5 Government Acquisition

Government acquisition processes are continuously evolving, quite complex, and variable from one government organization to another. The methods and terminology also vary substantially among various nations. Only a brief description of U.S. DOD procurement is attempted here.

When one thinks of procurement, the traditional program-of-record competition followed by a system development and fielding usually come to mind. The government begins with studies, requirements development, stakeholder negotiations, and budgetary planning. The program is communicated to industry initially through requests for information (RFI) and industry meetings. The government issues a RFP to industry, and the industry bidders respond with a proposal. The government performs a source selection and selects a winner. The winner is announced, and if the other bidders do not successfully protest the outcome, the contract is finalized. The system is developed with the oversight of the Government Program Office and is fielded after testing is complete. This major simplification of the process masks dozens of program milestones and events.

Very few UASs directly go through this traditional procurement process. Many systems mature under other development and acquisition routes before transitioning to operational service. Some systems were developed as technology demonstrators but proved so useful that the prototype aircraft were rushed into operational service. Others were largely developed using company funds and then fielded directly under sole-source contracts due to their unique capabilities. One increasingly popular model is for the government to procure services from contractors rather than procuring the systems directly. Unmanned systems are advancing so rapidly that traditional procurement cycles are often out of step. Also, urgent wartime needs can drive emerging systems into the field in an expedited manner.

The old advanced concept technology demonstrator (ACTD) and current joint capability technology demonstrator (JCTD) processes seek to very quickly develop concepts for system demonstrations. The objective is to advance promising technologies and systems but without the burden of creating a full program-of-record acquisition. The contractor is given only top-level performance objectives rather than detailed requirements. The RQ-4A Global Hawk was originally developed under an ACTD and was sent to the field for Operation Enduring Freedom. Systems developed under the ACTD and current JCTD process are representative of operational systems.

Research demonstrators are intended to prove technologies but are not generally suitable for operations. Examples of research demonstrators are the X-planes, of which the X-36, X-45A, X-47A, and X-50 were flying unmanned systems. Not all research vehicles attain the X-plane designation. The Northrop Grumman X-47B changed into a Navy demonstrator for carrier based operations, and so these demonstrators have precedent for extended use.

Contractor-owned, contractor-operated (COCO) services contracts pay contractors for ISR services rather than for procurement of the systems. The contractor generally provides all system elements, operators, and spares to conduct the operations. The government pays the contractors for hours flown, often with incentives for high availability and mission reliability. This model requires that the contractor designs and manufactures the system, with the costs built into to the hourly rate for operations. Because the government does not pay for procurement, this can fit under more readily available operations and maintenance budgets. A variation on COCO is government-owned, contractor-operated (GOCO) where the systems are procured by the government and operated by the contractor.

Another path to fielding a system is to use company funds to mature a system that the government later procures. General Atomics has famously followed this model with the Predator A and Reaper systems and is now attempting to do the same with the Avenger. A highly unique and capable system that is urgently needed in wartime can be procured via a sole-source contract. In effect, the company circumvents the formal government requirements generation function by creating what it believes the government needs. It must be the right system at the right time, and

Table 20.3 TRL Definitions

TRL level	Definition
1	Basic principles observed and reported
2	Technology concept and/or application formulated
3	Analytical and experimental critical function and/or characteristic proof of concept
4	Component and/or breadboard validation in laboratory environment
5	Component and/or brassboard validation in relevant environment
6	System/subsystem model or prototype demonstration in a relevant environment
7	System prototype demonstration in an operational environment
8	Actual system completed and "flight qualified" through test and demonstration
9	Actual system flight proven through successful mission operations

so this approach can be risky. This method is capital intensive, and not many organizations have both the expertise and necessary financial resources. The developing company generally owns the data rights to the system, often leading to other business opportunities for system support and upgrades. This arrangement is beneficial to the contractor but can create challenges with the government when they later wish to modify the system.

Some recent government procurements have emphasized system maturity. The required system TRL may be 6 or 7. (Table 20.3 gives definitions for the different TRL levels.) This maturity requirement dictates that the system be developed through company funds or previous contracts. This places a great financial burden on competitors to mature their systems in addition to the costs of writing the proposal. The unsuccessful bidders might not be able to recoup this large investment.

Funding early system development would enable the government to ensure that the design meets the operational needs and improves supportability outcomes. Another common consequence of high TRL requirements is that U.S. prime contractors often use foreign systems with sufficient maturity. Depending on desired system capabilities, customers may have the luxury of selecting from many mature systems.

References

[1] Munson, K., *Jane's Unmanned Aerial Vehicles and Targets*, Jane's Information Group Limited, England, U.K., 1999.
[2] *Unmanned Systems Roadmap, 2007-2032*, Office of the Secretary of Defense, Washington, D.C., 2007.

Problems

20.1 Compare the Boeing F-18 E/F to the F-18 C/D, and the Lockheed Martin U-2C to the U-2S. How similar are these designs?

20.2 Compare the Northrop Grumman Global Hawk Block 10, Block 40, and BAMS. How similar are these designs?

20.3 Compare the Northrop Grumman X-47A to the Northrop Grumman X-47B. How similar are these designs?

20.4 Compare the Boeing X-45A to the Northrop Grumman X-45C. How similar are these designs?

20.5 Using open-source company marketing materials, compare the endurance vs payload capacities of the following small tactical UAS systems: AAI Aerosonde Mk 4.7, AeroMech Fury, Insitu ScanEagle, Insitu Integrator, Northrop Grumman Bat, and Raytheon Killer Bee.

20.6 Compare the endurance vs payload capabilities of the Lockheed Martin/Kaman K-Max, Boeing A-160, and Northrop Grumman Fire-X.

20.7 Compare the capabilities of the General Atomics Predator A, Gray Eagle, and Reaper with other MALE systems. Other systems include the L-3 Mobius, BAE Herti, BAE Mantis, IAI Heron, IAI Heron TP, Elbit Hermes 900, and Elbit Hermes 1500.

20.8 Think of a market niche that is not filled by other UASs.

20.9 Bonus problem: Develop a highly capable UAS that revolutionizes the industry.

INDEX

Note: Page numbers with f represent figures. Page numbers with t represent tables.

SUPPORTING MATERIALS

Many of the topics introduced in this book are discussed in more detail in other AIAA publications. For a complete listing of titles in the AIAA Education Series, as well as other AIAA publications, please visit www.aiaa.org.

AIAA is committed to devoting resources to the education of both practicing and future aerospace professionals. In 1996, the AIAA Foundation was founded. Its programs enhance scientific literacy and advance the arts and sciences of aerospace. For more information, please visit www.aiaafoundation.org.